Activate solutions to problems online at www.loghia.com

Solutions to unsolved problems of this book (Summary problems) are available online, free of charge, for 12 months by registering at the Loghia Publishing website and entering the ***Activation Code*** reported below.

Follow these instructions:

1. Go to www.loghia.com and click on "MyLoghia" on the top menu.
2. Access your account or create a new one.
3. Click on "Activate digital contents", enter the "Activation Code" reported below and click "Activate now" (this step will be necessary only the first time).
4. To access the solutions, click on "Your Digital Contents" in your personal MyLoghia page. Select the book. Then select the Chapter and the Problem you want to read the solutions for.

You can access the solutions from all your devices (computer, laptop, tablet, mobile). Be aware that access to solutions will be available for 12 months from the activation date.

Scratch off the protective coating below to reveal your ***Activation Code***

ATTENTION

This book cannot be sold or returned or changed if the *Activation Code* is visible

Guide to NMR Spectral Interpretation

A Problem-Based Approach to Determining the Structures of Small Organic Molecules

Antonio Randazzo
University of Naples Federico II

ISBN: 978-88-95122-40-3

Printed by Print Progress Srl - via A. Sogliano, 34 - 80141 Napoli (Italy)

First published in 2018

Editing assistance: Melissa Stauffer, PhD
Cover design: Gennaro Maddaloni

www.loghia.com

This book is dedicated to my wife Claudia, and to my daughters Giulia and Francesca

CONTENTS

PREFACE

The most difficult part of studying a scientific subject is acquiring the ability to put the theory into practice. This is especially true when dealing with NMR spectra. This book is designed to facilitate this process by providing undergraduate and graduate students with methods, strategies, and explanations to unravel the intriguing combinations of chemical shift, integration, and multiplicity of NMR signals. In particular, it is written to guide students to learn how to determine the chemical structures of small organic molecules from their 1D and 2D NMR spectra through a problem-based learning approach. This book is not intended to replace, but to complement a standard NMR theory textbook, leaving out topics that are more suitable for advanced courses. Readers are assumed to already have a basic knowledge of organic chemistry.

The book is organized in a way that basic ^{1}H- and ^{13}C-NMR concepts are introduced and immediately applied in a number of problems, which are solved and discussed in a step-by-step fashion. It contains almost exclusively real NMR data and it describes how to interpret the chemical shift, intensity and splitting pattern of the proton and carbon NMR signals (Chapters 1-5), paying attention to the effects of magnetically non-equivalent nuclei (Chapter 4). The role of the solvent is also explained (Chapter 6), and a description of the interpretation of the most common two-dimensional NMR experiments is reported in Chapter 7. Chapter 8 is dedicated to strategies of structural elucidation, while Chapter 9 contains exclusively summary problems with increasing levels of difficulty.

Solutions to summary problems are provided online to prevent students from instantly getting to the answers before they have worked hard to solve the problems.

Unfortunately, the interpretation of printed NMR spectra has some limitations, because, for example, it does not permit zooming into the spectral regions where the NMR signals are particularly crowded. So, in order to give the feeling of an "electronic" exploration of the NMR spectra, the book is accompanied by additional unprocessed NMR data sets (online content), which students can download and study by themselves. The spectra can be processed by a number of NMR software programs, some of which are freely licensed to NMR students worldwide (*i.e.*, iNMR, Bruker, Jeol, etc.).

Teachers will have electronic access to all of the figures in this book, in order to easily prepare their lessons. They will also be provided with a number of additional summary problems, not present in the book, to engage students in practical exercises during the course, or to be used during final examinations.[1]

Students and teachers are very welcome to point out any error or inaccuracy they might notice in this book. This would be of help in enhancing, edition by edition, its quality and usefulness.

Antonio Randazzo

[1] In order to receive access to this additional content, teachers should contact the Publisher at the email address info@loghia.com, providing information about their course and the Activation Code reported on the first page of the book.

Acknowledgments

The author is grateful to Renzo Bazzo, Nunzia Iaccarino, Mariateresa Giustiniano, Martino Forino, Bruno Pagano, and Jussara Amato for reading the book and for useful discussions. Diego Brancaccio is acknowledged for the acquisition of some of the NMR spectra. The Biological Magnetic Resonance Data Bank (BMRB) is also acknowledged for permitting the use of the spectra of metabolomic standard compounds contained in the database (Eldon L. Ulrich; Hideo Akutsu; Jurgen F. Doreleijers; Yoko Harano; Yannis E. Ioannidis; Jundong Lin; Miron Livny; Steve Mading; Dimitri Maziuk; Zachary Miller; Eiichi Nakatani; Christopher F. Schulte; David E. Tolmie; R. Kent Wenger; Hongyang Yao; John L. Markley; Nucleic Acids Research 36, D402-D408, 2008. doi: 10.1093/nar/gkm957).

CHAPTER 1

Basic concepts of NMR spectroscopy

1.1 Introduction

Nuclear magnetic resonance (NMR) spectroscopy has become the dominant method for structural analysis of organic compounds. It relies on a very complex theoretical basis that requires a strong knowledge of mathematics and physics; therefore, it is not a surprise that there are plenty of excellent books on NMR theory. However, there are very few practical guides to the interpretation of NMR spectra. Thus, this book has been designed to lead readers through practical strategies of the interpretation of 1D and 2D NMR spectra of small organic molecules in solution. It is to serve this purpose that in this chapter only a brief outline of NMR theory is reported, limited to the concepts that are relevant to the explanations provided further along in the book.

1.2 Magnetically active nuclei and energy levels

NMR is a spectroscopic technique that relies on the magnetic properties of the atomic nucleus. Nuclei have positive charges and many of them behave as if they were spinning. Anything that is charged and moves generates a magnetic dipole and produces a magnetic field, so nuclei behave like tiny bar magnets positioned along the spin-rotation axis. Nuclear magnetic dipoles are characterized by spin quantum numbers (I) that determine some spectroscopic features of the nuclei. There are three classes of nuclei:

1. Nuclei with an *even* number of neutrons and an *even* number of protons. These nuclei have no spin (I = 0), so they do not act as magnets and do not interact with the applied magnetic field. Therefore they are inactive from an NMR standpoint. For example, ^{12}C and ^{16}O belong to this class.
2. Nuclei with an *odd* number of neutrons and an *odd* number of protons. These nuclei have an integer spin (*i.e.*, I = 1, 2, 3), so they are magnetically active. The most prominent examples of nuclei of this group are ^{2}H and ^{14}N, both having I = 1.
3. Nuclei having an *odd* mass number. These nuclei have a half-integer spin (*i.e.,* I = 1/2, 3/2, 5/2), so they are magnetically active. The most important nuclei of this class are ^{1}H, ^{13}C, ^{15}N, ^{19}F, and ^{31}P, all having I = 1/2. Spectra of nuclei with I = 1/2 can be easily interpreted.

Since some nuclei act as bar magnets, they can be oriented by an external magnetic field. In NMR spectroscopy, the magnetic field is applied by superconducting solenoids, cooled by liquid helium. The type of orientation of nuclei in a magnetic field is defined by their magnetic quantum number (m), which takes integer values from −I to +I. For example, the nucleus of the deuterium atom (deuteron), having I = 1, is characterized by three nuclear orientations with respect to the applied magnetic field (m = -1, 0, +1), the nucleus of ^{23}Na (I = 3/2) is characterized by four orientations (m = -3/2, -1/2, +1/2, +3/2), and so on. This book is focused on proton (^{1}H) and carbon (^{13}C) NMR, both having I = 1/2. Therefore, both nuclei have two nuclear orientations with respect to the applied magnetic field (m = +1/2 and m = −1/2). When m = +1/2, the nuclear magnetic dipole assumes a parallel orientation (lower energy state) with respect to the field, while when m = -1/2 it assumes an antiparallel orientation (higher energy state). The energy difference (ΔE) between the two orientations is related to the strength of the applied magnetic field (B_0) and to the gyromagnetic ratio (γ) of the nuclei:

$$\Delta E = \gamma h \pi B_0 / 2\pi \qquad \textbf{(Eq. 1.1)}$$

where:
ΔE = energy difference
γ = gyromagnetic ratio
h = Planck's constant
B_0 = magnetic field

Thus, the stronger the magnetic field, the larger the difference in energy between the two nuclear states. A graphical representation of this concept is reported in Figure 1.1.

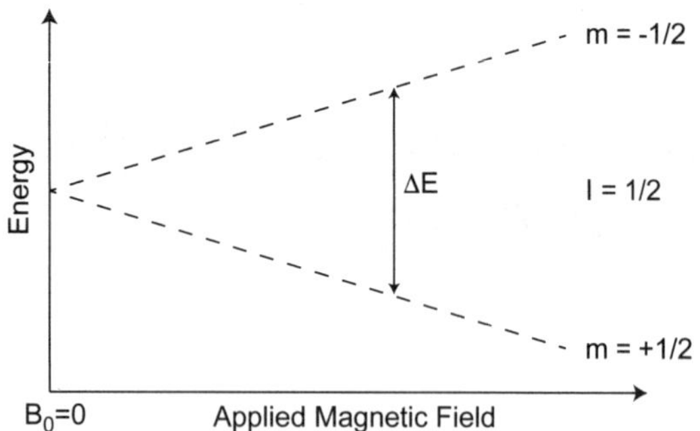

Figure 1.1 Dependence of nuclear energy levels on the strength of the applied magnetic field.

In addition, the larger the gyromagnetic ratio, the larger the difference in energy. For example, protons have a gyromagnetic ratio four times greater than the nuclei of carbon atoms, and therefore the difference in energy between the two nuclear states is four times larger for protons than for carbon nuclei.

1.3 Boltzmann distribution and NMR sensitivity

NMR experiments are performed on a sample containing a huge number of molecules, only a fraction of which contributes to the signal being measured. For example, let's consider a sample of chloroform ($CHCl_3$); when subjected to the applied magnetic field, all the protons in the sample (there are as many protons as chloroform molecules) orient their magnetic dipoles along the magnetic field. Some of them will be oriented parallel and others antiparallel to the applied magnetic field. The distribution of the protons in the two nuclear energy states is described by the Boltzmann equation (Eq. 1.2):

$$\frac{N_\beta}{N_\alpha} = e^{\frac{\Delta E}{kT}} \qquad \textbf{(Eq. 1.2)}$$

where N_β and N_α represent the population of nuclei in upper and lower energy states, respectively, k is the Boltzmann constant, T is the absolute temperature (°K), and ΔE is the difference of energy between the two orientations.

This equation indicates that the population of the nuclei is almost equally distributed between the two nuclear states at room temperature, with only a very small excess of nuclei in the lower energy state. Unfortunately, only this little difference in population between the two energy levels produces the NMR signal (see next section); therefore, NMR is a spectroscopic technique with low sensitivity. However, by using stronger magnetic fields, it is possible to increase the value of ΔE (see Eq. 1.1) and, consequently, the difference in populations (see Eq. 1.2), increasing significantly the sensitivity of this spectroscopy.

1.4 The NMR signal

It is possible for a proton to absorb a photon of electromagnetic energy and to be promoted from the lower to the higher energy state. This absorption of energy is called *resonance*. Therefore, it is common to say that the nuclei *resonate* when they are excited.

The energy of the electromagnetic wave must exactly match the difference in energy between the two states. The energy of an electromagnetic wave is described by the Planck equation:

$$E = h\nu_0 \qquad \textbf{(Eq. 1.3)}$$

where h is Planck's constant and ν_0 is the resonance frequency of the wave. Therefore, the electromagnetic wave must have a frequency equalling $\Delta E = \gamma h \pi B_0 / 2\pi = h\nu_0$. In NMR spectroscopy the difference of energy between the two nuclear states is small, and so the energy of a radio frequency (RF) wave is sufficient to promote the excitation of the nuclei. Please note that, since the energy difference of the two nuclear states of protons is about four times larger than that of carbon nuclei, the frequency used to excite protons must be four times larger. In spite of the fact that NMR spectrometers can generally excite two or more type of nuclei, they are classified according to the average frequency at which they excite ^1H (*i.e.*, 300 MHz, 600 MHz, 800 MHz spectrometers). Obviously, for example, a spectrometer operating at 800 MHz possesses a much stronger magnetic field than one operating at 300 MHz.

In modern NMR spectroscopy, an intense short RF pulse (of about 10 μs) is used to excite all the nuclei (for example, protons) of the molecule simultaneously. Once excited, each nucleus returns to thermal equilibrium, releasing the RF that it has absorbed. The emitted RF decays in a short period of time (typically from 100 ms to 5 s) and produces a signal called *free-induction decay* (FID) (Fig. 1.2, top). This is called the *time-domain* NMR spectrum because the signal is a function of time. This spectrum is very difficult to interpret, so it is converted into a *frequency-domain* spectrum (Fig. 1.2, bottom) by a mathematical operation called *Fourier Transformation* (FT), which is performed by software programs. The frequency-domain spectrum contains as many peaks (also called signals) as the number of different frequencies the sample has absorbed. For example, in the case of chloroform, because there is only one type of proton (see next section), only one RF is absorbed and the frequency-domain spectrum contains a single signal.

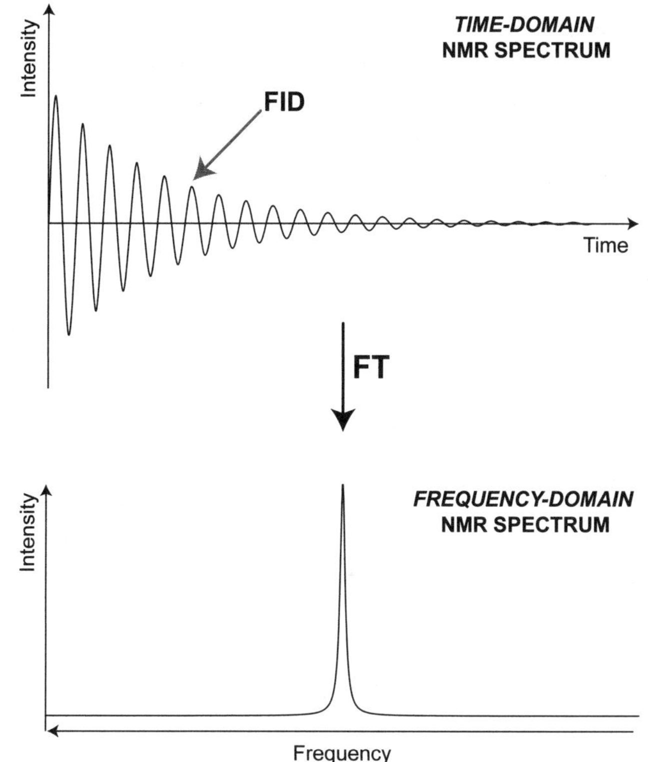

Figure 1.2 Time- and frequency-domain NMR spectra.

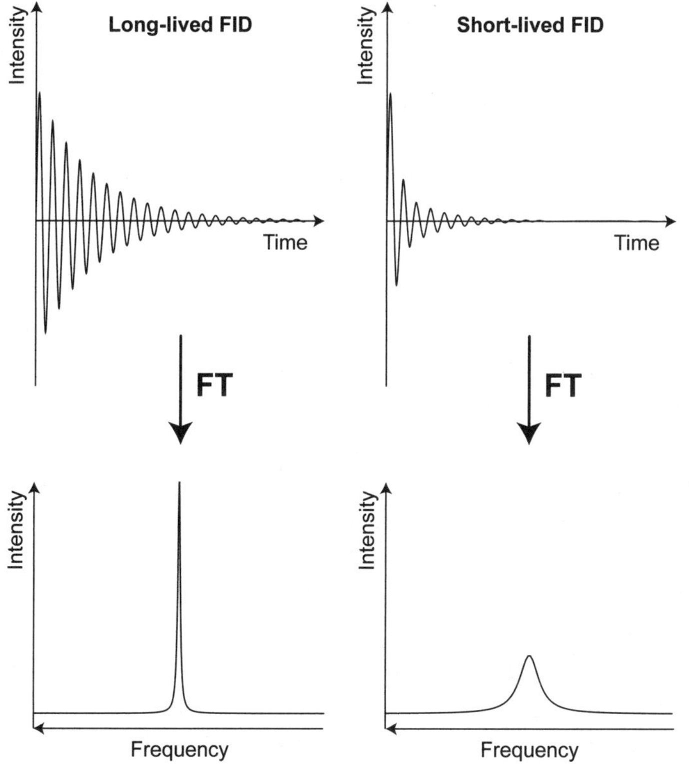

Figure 1.3 The broadness of the NMR signal depends also on the length of the FID.

As mentioned above, the lifetime of the FID can vary from 100 ms to 5 s, depending on the relaxation mechanism of the nuclei. Generally, a short-lived FID produces a broad signal in the NMR spectrum (Fig. 1.3).

As discussed in the previous section, NMR is intrinsically a low-sensitivity spectroscopy. This is due to the small population difference between the two energy states that, after excitation, emits a weak RF. Sometimes this RF can be of the same magnitude of background noise, which is produced by the electronics of the NMR instrument and by the environment. The resulting frequency-domain spectrum is characterized by a scribbled baseline like that shown in the top of Figure 1.4. This spectrum is characterized by a small *signal-to-noise ratio*, which is measured by dividing the intensity of the signal by the average intensity of the noise in the spectrum. The higher the signal-to-noise ratio, the better the NMR spectrum will appear.

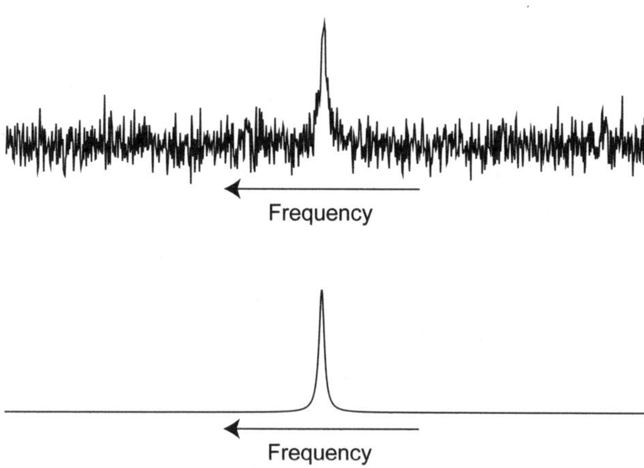

Figure 1.4 Comparison between signals with low (TOP) and high (BOTTOM) signal-to-noise ratio.

An elegant method to improve the signal-to-noise ratio involves the acquisition of many identical spectra, and then summing them together. So, for example, if you sum one hundred spectra, the signal in the spectrum increases in direct proportion to the number of spectra summed, that is, the signal will be one hundred times more intense. However, since noise is random, it increases with the square root of the number of spectra recorded. Hence, the noise will increase by only a factor of ten. Note that, in order to double the signal-to-noise ratio of a spectrum, you have to sum four spectra.

1.5 Chemical shift

Nuclei are not isolated systems, but they are placed inside a cloud of electrons. Particularly, hydrogens have s-electrons (in a spherical orbital) surrounding the nucleus. These electrons, when subjected to an external magnetic field, start to circulate, producing a local magnetic field (B_i), which opposes the applied one (B_0) at the nucleus (Fig. 1.5).

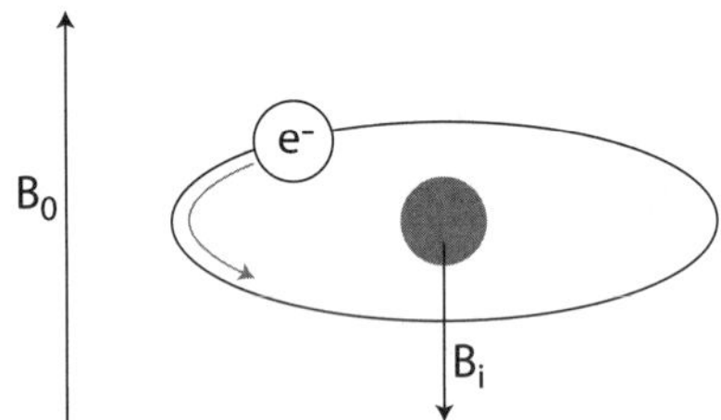

Figure 1.5 B_0: applied magnetic field. B_i: magnetic field induced by an electron surrounding the nucleus (gray circle).

This means that the magnetic field really experienced by the nucleus is smaller than that applied, and the true field can be described as $B = B_0 - B_i$. Since the nucleus experiences a smaller field, its two energy levels turn out to be closer, and the frequency necessary for its excitation is smaller as well. This phenomenon is called *diamagnetic shift* or *up-field shift.*

Electrons in p-orbitals have no spherical symmetry and produce an increased magnetic field at the nuclei; thus, they resonate at a higher frequency (*paramagnetic shift* or *low-field shift*). In ^{1}H-NMR there are no p-orbitals, so that only a small range of nuclear shielding is observed and, therefore, a limited range of absorbance frequencies is registered.

Interestingly, the chemical environment of an atom can influence its electron density through the polar effects of substituents. Electron-donating groups, for example, lead to increased *shielding,* while electron-withdrawing substituents lead to *deshielding* of the nucleus. This means that, depending on the extent of shielding, nuclei will experience different local magnetic fields and will resonate at slightly different frequencies. This is one of the main reasons why you can distinguish the different hydrogens of a molecule.

Unlike infrared and UV-visible spectroscopy, where absorption peaks are uniquely located at a particular frequency (or wavelength), the position of different NMR resonance signals is dependent on the external magnetic field strength. Since two magnets will not have exactly the same field, a given proton may resonate at slightly different resonance frequencies in the two different spectrometers. One method of solving this problem is to measure the frequency of resonance with respect to the frequency of a reference compound (standard) added to the sample. Such a standard should be chemically unreactive and easily removed from the sample. Moreover, it should give a single sharp NMR signal that does not interfere with the resonances normally observed. The frequencies for ^{1}H are usually referenced against tetramethylsilane (TMS) or 4,4-dimethyl-4-silapentane-1-sulfonic acid (DSS), the latter having much higher water solubility (Fig. 1.6).

TMS DSS

Figure 1.6 Chemical structures of TMS and DSS.

The variation of NMR frequencies from the standard is called *chemical shift.* The chemical shift (δ) is calculated from the following equation:

$$\delta = \frac{\nu_{sample} - \nu_{reference}}{\nu_{reference}} \qquad \textbf{(Eq. 1.4)}$$

where ν_{sample} is the absolute resonance frequency of the sample and $\nu_{reference}$ is the absolute resonance frequency of the standard reference compound, measured in the same applied magnetic field (B_0). Since the result of the difference in the numerator is expressed in hertz (Hz), and the denominator in megahertz (MHz), it is preferable to express δ in parts per million (ppm). Therefore, a nucleus that absorbs at 600 Hz higher than the reference compound (resonating at 300 MHz) has a chemical shift of 2 ppm:

$$\delta = \frac{600 \text{ Hz}}{300 \times 10^6 \text{ Hz}} = 2 \times 10^6 = 2 \text{ ppm}$$

According to the above definition of δ, the reference standards will resonate at 0 ppm.

From the analysis of the chemical shift equation (Eq. 1.4), it is apparent that the resolution of NMR spectra will increase with the strength of the applied magnetic field. In fact, for a spectrometer operating at 300 MHz, 1 ppm corresponds to 300 Hz; for a spectrometer operating at 500 MHz, 1 ppm corresponds to 500 Hz, and so on.

For the majority of organic compounds, the ^{1}H chemical shift scale ranges from 0 to 12 ppm. By convention, the scale of chemical shift has increasing values from right to left (Fig. 1.7).

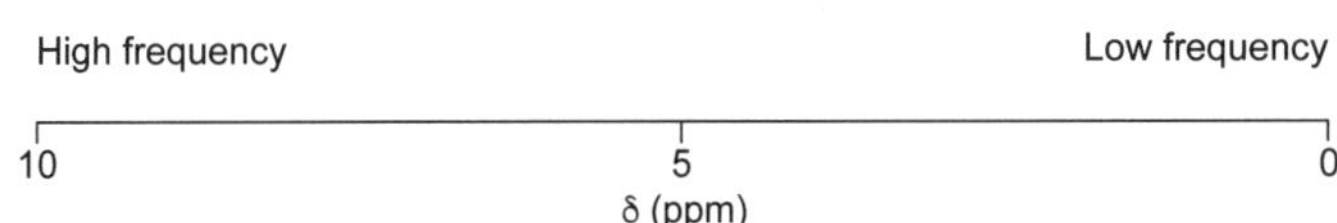

Figure 1.7 Proton NMR chemical shift scale.

Please note that the NMR chemical shift depends on the relative orientation of the molecule with respect to the axis of the applied magnetic field. When NMR is performed in solution, where the molecules can assume all possible orientations, the chemical shift is fully averaged.

1.6 Effect of the neighboring atoms

As mentioned above, an important factor influencing the chemical shift is the electron density surrounding the nucleus. In this regard, the electronegativity of neighboring groups plays an important role. A nucleus in the vicinity of an electronegative atom has a reduced electron density and therefore it will be deshielded. For example, in methyl halides (CH_3X), the chemical shift of the methyl protons (that resonate as one signal, see next chapter) is higher with fluorine (4.27 ppm) than with iodine (2.16 ppm), reflecting the greater electronegativity of the substituent. Table 1.1 shows a very good correlation between the chemical shift of the methyl (CH_3) and the electronegativity of the substituent (X).

Table 1.1 Relationship between chemical shift and electronegativity of halogen substituents in methyl halides.

X	CH_3 (ppm)	Electronegativity (Pauling units)
F	4.27	3.98
Cl	3.06	3.16
Br	2.69	2.96
I	2.16	2.66

Please also note that the further away the electronegative atom from the hydrogens in consideration, the lower its effect. For example, consider the chemical shift of ethyl and propyl halides (Table 1.2). In the ethyl halide, the methyl resonates at a smaller chemical shift than the methylene. In the propyl halides, the chemical shifts of the methylenes and methyl decrease progressively as they get farther away from the electronegative group.

Table 1.2 Relationship between chemical shift and the distance of the electronegative substituent in ethyl halides (CH_3-CH_2-X) and propyl halides (CH_3-CH_2-CH_2-X) (values are in ppm).

	CH_3-CH_2-X (a) (b)		CH_3-CH_2-CH_2-X (a) (b) (c)		
X	CH_3 (a)	CH_2 (b)	CH_3 (a)	CH_2 (b)	CH_2 (c)
F	1.35	4.55	0.97	1.68	4.30
Cl	1.33	3.47	1.06	1.81	3.47
Br	1.66	3.37	1.06	1.89	3.35
I	1.88	3.16	1.03	1.88	3.16

(values are in ppm)

1.7 Effect of the bonding electrons

The bonding electrons can also lead to shielding and deshielding effects when subjected to an external magnetic field (*magnetic anisotropic effect*). This is the case with the π-electrons that tend to circulate at a right angle to the applied field. This circulation is favored in determined orientations of the molecule with respect to the magnetic field. For example, in an aromatic ring, the ring current through the hyperconjugated system is favored when the aromatic moiety is oriented perpendicular to the magnetic field. The circulation of the electrons induces a local magnetic field that opposes the applied one (B_0) inside the ring (shielding zone), while increasing the field outside (deshielding zone). The shielding areas can be imagined as cones aligned with the external field. The remaining spaces are characterized by deshielding effects (Fig. 1.8).

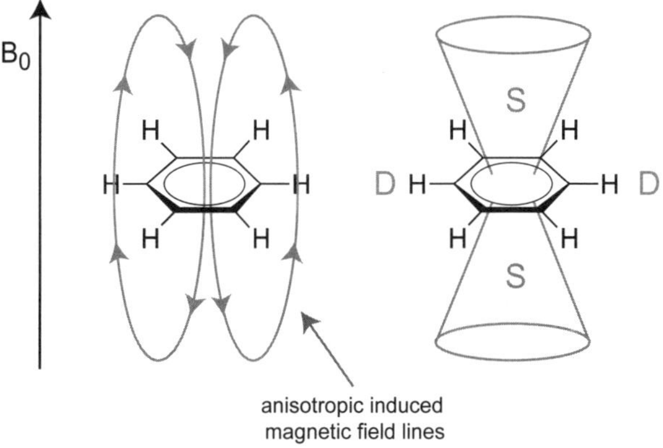

Figure 1.8 Anisotropic effect of aromatic rings. S and D indicate shielding and deshielding zones, respectively.

This phenomenon is particularly evident in [18]annulene, where the inner protons are strongly shielded and therefore resonate at -1.8 ppm, whereas the outer protons are strongly deshielded, resonating at 8.9 ppm (Fig. 1.9).

-1.8 ppm

8.9 ppm

[18]annulene

Figure 1.9 Anisotropic effect in [18]annulene.

The ring current effects explain why the aromatic protons resonate at higher frequencies (7-9 ppm) than one would expect for hydrogens bound to sp^2 carbons, whose electronegativity is not enough to justify such a deshielding.

The case of alkene groups is similar to that of aromatic ones. The circulation of π-electrons is favored when the alkene group is oriented perpendicular to the applied field, generating a local magnetic field that is parallel to the external field at the alkene protons (Fig. 1.10). Therefore, the actual magnetic field experienced by the hydrogens is larger than that applied and the proton signals are shifted to higher frequencies (4.5-6.5 ppm). Because the electrons in these systems are less mobile than the aromatic ones, the deshielding effect is more limited.

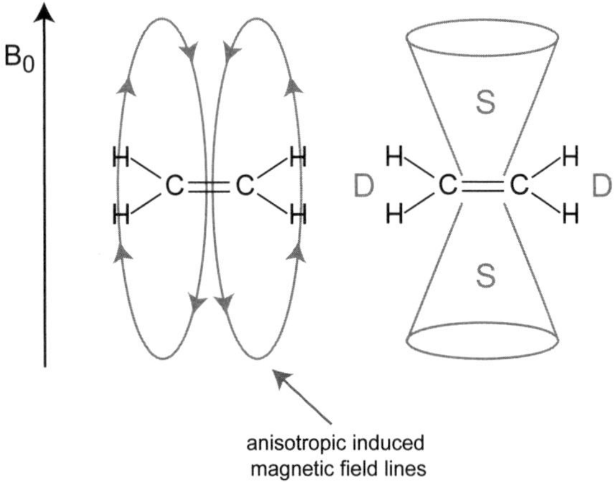

Figure 1.10 Anisotropic effect of alkenes. S and D indicate shielding and deshielding zones, respectively.

On the contrary, the π-electrons in alkynes are more free to circulate around the symmetry axis of the triple bond. So, when the alkyne is parallel to the applied field (Fig. 1.11), the circulation of π-electrons generates a local magnetic field which opposes the one applied at the acetylenic protons. Hence, they are located in the shielding zone and therefore are characterized by lower chemical shift values (2–3 ppm) than expected from the electronegativity of the sp carbons.

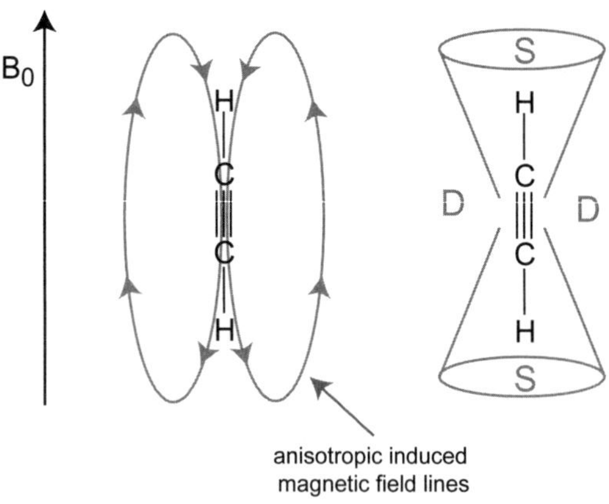

Figure 1.11 Anisotropic effect of alkynes when parallel to the magnetic field. S and D indicate shielding and deshielding zones, respectively.

Please note that each molecule, when oriented in different ways with respect to the magnetic field, may produce different effects on the considered protons. For example, when the alkyne is oriented perpendicular to the magnetic field, the acetylenic protons are deshielded as in alkenes (Fig. 1.12). However, this deshielding component is smaller when compared to the shielding obtained in the parallel orientation, since the electrons are less free to move. Hence, overall, the anisotropic effect of the molecule when oriented parallel to the field will dominate.

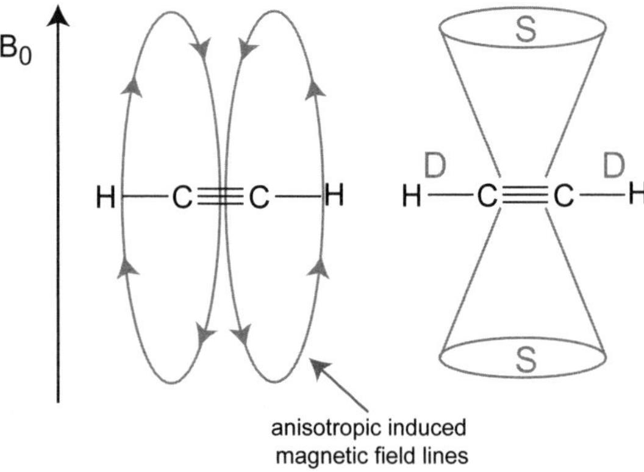

Figure 1.12 Anisotropic effect of alkynes when perpendicular to the magnetic field. S and D indicate shielding and deshielding zones, respectively.

In order to have an idea of the approximate ^1H chemical shift ranges for some functional groups commonly found in organic molecules, please look at the scheme shown in the inside front cover flap.

CHAPTER 2

Equivalence of nuclei and signal intensity in ^{1}H-NMR spectroscopy

2.1 Chemical shift equivalence of nuclei

Chemical shift equivalent nuclei are sets of nuclei all having exactly the same chemical shift values.

In order to determine whether or not two nuclei are chemical shift equivalent you must:

1. Analyze the symmetry of the molecule;
2. Determine their stereochemical relationship (topicity);
3. Consider time-averaging processes.

2.1.1 Symmetry of the molecule

If there is an element of symmetry in the molecule, then the symmetric atoms are *chemically equivalent* and therefore they will be also chemical shift equivalent.

For example, *cis*-1,2-dichloroethene and 1,1-dichloroethene possess a plane of symmetry (Fig. 2.1). As a result, the two hydrogens in each molecule are equivalent to each other and resonate at the same chemical shift. Analogously, *trans*-1,2-dichloroethene possesses an axis of symmetry (Fig. 2.1) and therefore, also in this case, the two hydrogens resonate at the same chemical shift. In contrast, *cis*- and *trans*-1-bromo-2-chloroethene do not possess any symmetry element and the NMR spectra will contain two distinct proton signals.

Nuclei which do not possess a symmetry relationship can be chemical shift equivalent only by chance.

2.1.2 Topicity

Topicity describes the stereochemical relationship of two or more atoms in a molecule. Depending on their relationship, such atoms can be homotopic, enantiotopic, diastereotopic, or heterotopic.

Homotopic and enantiotopic hydrogens resonate at the same frequency, while diastereotopic and heterotopic hydrogens show signals at different ppm values. This difference is due to the fact that, in contrast with homotopic and enantiotopic hydrogens, diastereotopic and heterotopic ones experience different chemical environments.

In order to determine the stereochemical relationship of two hydrogens, it is possible to mentally replace each atom with a "dummy" atom (X) and then compare the two structures. If they are identical, the hydrogens are homotopic; if they

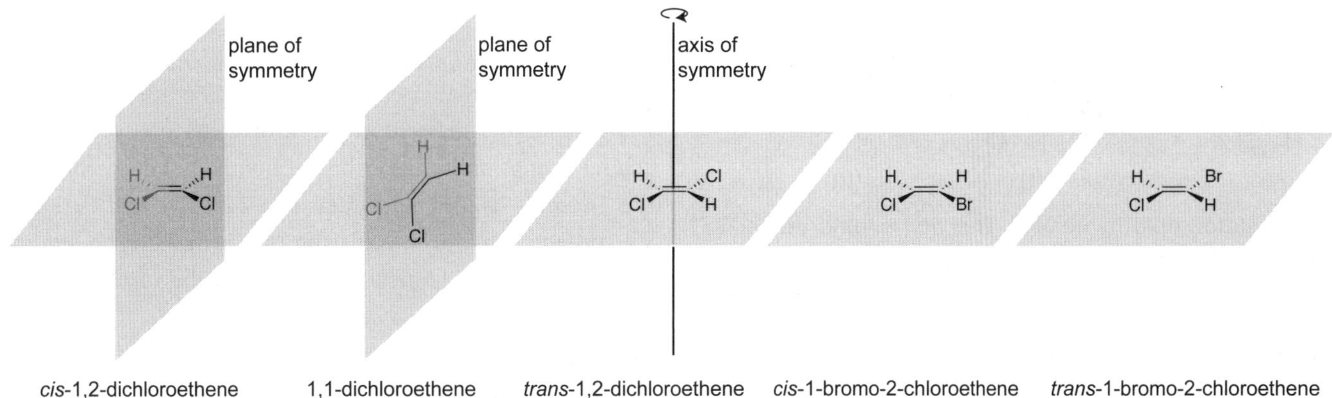

Fig. 2.1 Examples of symmetry and asymmetry in small organic molecules.

are enantiomers, the hydrogens are enantiotopic; if they are diastereomers, the hydrogens are diastereotopic; if they are structural isomers, the hydrogens are heterotopic.

Homotopic hydrogens:

Homotopic hydrogens are chemically equivalent and are typical of symmetric molecules. For example, the hydrogens of the methylene of propane are equivalent because their substitution leads to identical molecules (Fig. 2.2). Homotopic hydrogens are chemical shift equivalent.

Figure 2.2 Homotopic hydrogens.

Enantiotopic hydrogens:

Enantiotopic hydrogens are generally found in methylene groups (CH_2) bonded to two different substituents. This is the case, for example, with bromoethane (Fig. 2.3). The replacement of one hydrogen of the methylene by a dummy atom (X) forms one enantiomer. The replacement of the other hydrogen forms the opposite enantiomer. Enantiotopic hydrogens are chemical shift equivalent.

Figure 2.3 Enantiotopic hydrogens.

Diastereotopic hydrogens:

Generally, diastereotopic hydrogens can be found in CH_2 groups in chiral molecules. For example, 1-bromo-2-chloropropane possesses an asymmetric center (Fig. 2.4).

Figure 2.4 Diastereotopic hydrogens. The asymmetric center is indicated by an asterisk.

As you can see from the substitution test, the two hydrogens of the methylene are diastereotopic. As mentioned above, diastereotopic hydrogens are not chemical shift equivalent since they experience two different chemical environments generated by the presence of the asymmetric center. However, it should be noted that the greater the distance of diastereotopic hydrogens from the asymmetric center, the smaller their differences in chemical shift values. In some cases, they may resonate at the same chemical shift, even if they remain diastereotopic. For example, consider methionine and glutamic acid (Fig. 2.5).

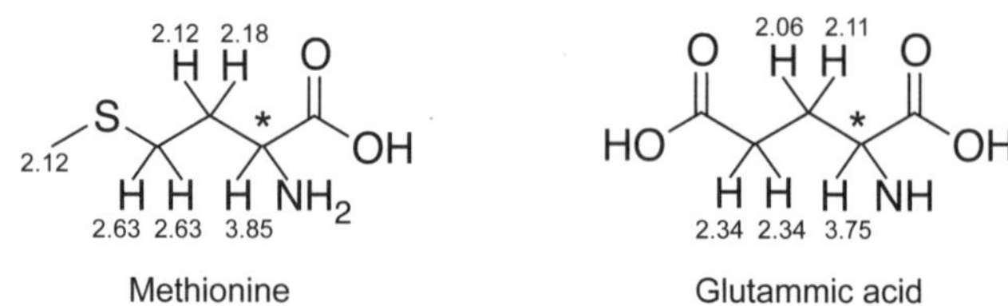

Figure 2.5 Proton chemical shifts (ppm) of methionine and glutamic acid. The asymmetric centers are indicated by asterisks.

They both possess an asymmetric center (Cα), and therefore the hydrogens of the two methylenes present in both molecules are diastereotopic. However, as you can see from the chemical shift values reported in the figure, while the two β-hydrogens resonate at two different chemical shift values, the diastereotopic γ-hydrogens are actually chemical shift equivalent in both molecules.

Not all CH_2 groups in chiral molecules are diastereotopic. In general, CH_2 groups are diastereotopic when they are part of a chiral molecule unless the CH_2 group is on a symmetry rotation axis. This is the case, for example, with (1*R*,3*R*)-1,3-dimethylcyclopentane (Fig. 2.6), where H_a and H_b are homotopic.

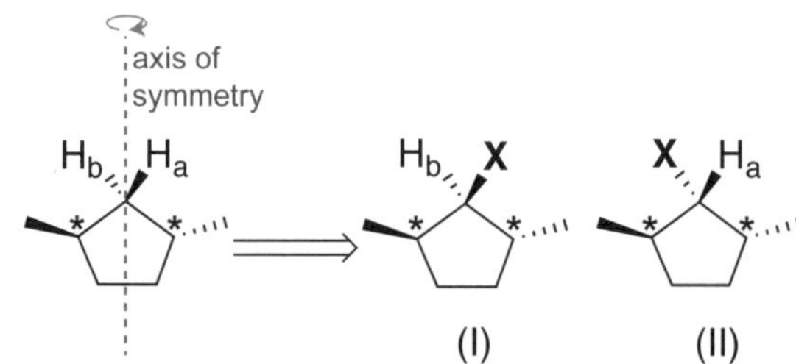

Figure 2.6 Non-diastereotopic hydrogens in a chiral molecule. The asymmetric centers are indicated by asterisks.

Interestingly, diastereotopic hydrogens are present also in achiral compounds like 3-bromopentane. The carbon bearing the bromine is not asymmetric because it has two identical substituents. However, replacing one hydrogen renders the two substituents non-identical (Fig. 2.7) and the resulting molecules possess two asymmetric centers.

Figure 2.7 Diastereotopic hydrogens in an achiral molecule.

Heterotopic hydrogens:

If in a molecule the replacement of two hydrogens with "dummy" atoms (X) leads to two structural isomers, those hydrogens are heterotopic. Heterotopic hydrogens are not chemical shift equivalent.

2.1.3 Time-averaging processes

So far we have thought of the molecules as possessing rigid structures. However, this is not true; in fact, most of the molecules analyzed so far can change their conformation because they are flexible. Any change of conformation changes the chemical environment of the nuclei, and so also their frequency of resonance. In general, each conformation will be observed by NMR as long as its lifetime is long enough to be sampled. The speed of sampling of an NMR spectrometer (also termed the *NMR time-scale*) is related to the resonance frequencies of the nuclei. In proton NMR spectroscopy, the acquisition of the signal takes about three seconds. Therefore, each conformation that exists for at least three seconds will be sampled. However, if a molecule changes conformation faster than every three seconds, an averaged chemical environment will be sampled. This is what happens to the hydrogens of a methyl, which, rotating very fast, average their chemical environment and produce a single signal in the proton NMR spectrum. Analogously, the diastereotopic hydrogens of a methylene free to rotate may be chemical shift equivalent (for further explanations see Section 6.5 in Chapter 6).

Please note that, while methyls are always free to rotate, methylenes could be locked in a rigid structure like that of a cyclic molecule, thus giving rise to two different proton signals.

IMPORTANT: *Chemically equivalent* nuclei behave the same as one another chemically, but they may not have the same NMR properties, that is, they may not be *magnetically equivalent.* Two protons are magnetically equivalent when they are chemically equivalent and are coupled to other nuclei in the molecule by the same *coupling constants.* Coupling constants will be discussed in Chapter 3, and the concept of magnetic equivalence will be described further in Chapter 4.

2.2 The NMR signal

Each NMR signal is characterized by three features: line-shape, line-width, and intensity. The line-shape is determined by the process of free induction decay (FID), which, being approximately exponential, generates Lorentzian shapes. The line-width is instead very variable and is related to the mechanism of relaxation of the nuclei and to the dynamic of the molecule. The intensity, in contrast, is the *integral* (area under the curve) of that signal. The intensity of the signal is proportional to the number of hydrogens generating the signal.

If you consider a molecule containing two sets of equivalent hydrogens like chloroacetone, it will display two signals in the proton spectrum (Fig. 2.8). The first signal is generated by the three protons of the methyl, and the second one by the two hydrogens of the methylene. As noted in the previous Section 2.1.3, both set of hydrogens are chemical shift equivalent because they average their chemical environment (note that the hydrogens of the methylene resonate at a higher value of ppm than the methyl because of the deshielding effect of the chlorine).

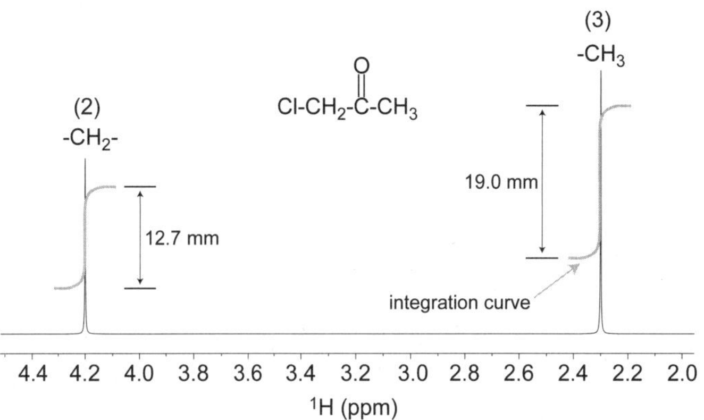

Figure 2.8 Simulated ^{1}H-NMR spectrum of chloroacetone. The stepped curves used for the integration of the signals are shown in gray.

Since the intensity is related to the number of hydrogens that generate the peak, the signal of the methylene and the methyl will have different integrals. Many NMR software programs can integrate the area underneath the signal. The integration is often displayed as a stepped curve, where the step-height is proportional to the area. You can measure the step-heights by a ruler and calculate the relative signal intensities setting the smallest step-height to unity and convert the other values proportionally. In other words, you can divide all the step-heights by the smallest value measured. In the example shown in Figure 2.8, you have to divide 19 mm by 12.7 mm. The result is about 1.5, meaning that the relative signal intensities is 1.5/1. Since we cannot have a fraction of an atom, we can suppose that the true numbers of hydrogens producing the signals are probably 2 and 3 (or 4 and 6, etc.). For chloroacetone the actual values are, of course, 2 and 3. The integration can also be indicated as a number close to the signal. In this book, the relative integrations of the proton signals will be reported as numbers in parentheses on the top of each signal.

Clearly, since the intensity of the NMR signals is proportional to the number of resonating hydrogens, the intensity is also proportional to the concentration of the sample. For example, if you consider the spectrum of a mixture containing equal molar amounts of chloroform and bromoform (Fig. 2.9), the signal intensities of the two molecules will be the same. On the other hand, if the mixture contains double the concentration of bromoform compared to that of chloroform, the signal of the bromoform will be twice as intense as that of chloroform (Fig. 2.9).

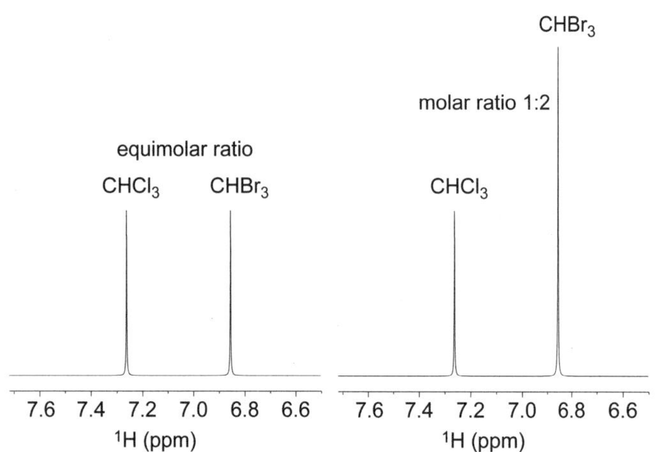

Figure 2.9 Simulated ^1H-NMR spectra of an equimolar solution (left panel) and 1:2 molar solution (right panel) of chloroform and bromoform.

This allows us to determine the relative concentrations of two or more compounds in a sample.

Problem 2.1

How many proton signals would you expect for the following molecules? What will be their relative intensities?

CH_4	CH_3CH_3	$CH_3CH_2CH_3$	$CH_3CH_2CH_2CH_3$
(I)	(II)	(III)	(IV)

Solution

(I) The methane has four hydrogens experiencing the same chemical environment. This is clear considering its three-dimensional structure (a pure tetrahedron), where many elements of symmetry are present. Therefore, a single signal is expected in the proton spectrum.

(II) Ethane can be considered as formed by two methyls (-CH_3). These rotate very fast with respect to each other, thus averaging their chemical environment. Also for this reason, this molecule can be considered symmetric (where a plane of symmetry bisects the C-C bond), so all hydrogens of the molecule are chemically equivalent and resonate as a single signal.

(III) Also propane can be considered symmetric, where the plane of symmetry passes through the carbon of the methylene (CH_2). Thus, the two methyls are equivalent to each other and generate a single signal integrating for six protons. The hydrogens of the methylene are homotopic and generate another signal integrating for two protons. Hence, the spectrum of the propane will contain two signals with a relative intensity ratio of 6:2 (or better 3:1).

(IV) Butane is also a symmetric molecule. The plane of symmetry is located between the two central methylenes. Thus, the two methylenes will generate a signal integrating for 4 protons and the two methyls will generate another signal integrating for six protons. The two signals will have a 4:6 (or 2:3) intensity ratio.

Please note that in this example and in the following ones many protons may "couple" to each other. For didactic reasons this is not commented on here because it is beyond the scope of this chapter. The "coupling" between nuclei will be treated in the next chapter.

Problem 2.2

How many signals would you expect in the ^1H-NMR spectrum of the following molecules? What will be their relative intensities?

Cl Cl (I) Cl (II) Cl Cl (III)

Solution

(I) 1,3-Dichloropropane is symmetric and the plane of symmetry passes through the central methylene.

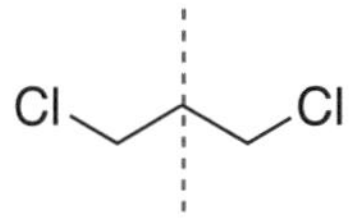

Cl Cl

Therefore, two signals are expected, having relative intensities 4:2 (or 2:1).

(II) 1-Chloro-3-methylbutane possesses an isopropyl moiety [-CH(CH$_3$)$_2$]. In this molecule, the isopropyl is able to rotate freely and average its chemical environment. Therefore, the two methyls will generate only one signal (integrating for six protons). The methine (bearing the two methyls) will generate another signal and the remaining two methylenes will produce two more signals. Therefore, the spectrum will contain four signals of relative intensities 6:1:2:2.

(III) This molecule shows two signals in the proton spectrum, having a relative intensity ratio of 3:1 (the first for the methyl and the second for the methine).

Problem 2.3

How many proton signals would you expect for the following molecules? What will be their relative intensities?

(I) 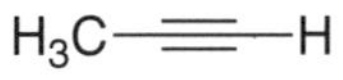 (II) 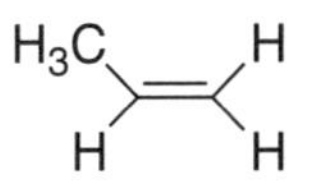 (III) 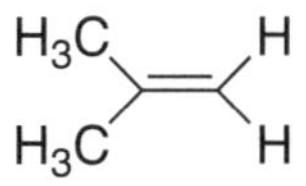 (IV)

Solution

Molecules I and IV are symmetric.

(I) 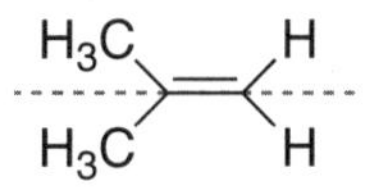 (IV)

Therefore they provide one and two signals, respectively, in the ^{1}H-NMR spectrum. In the case of IV the relative intensity ratio of the signals will be 6:2 (or 3:1). On the other hand, II and III are not symmetric and will provide two and four signals, respectively. Remember that the methyl generates only one signal. Thus, the relative intensity ratios will be 3:1 and 3:1:1:1, respectively.

Problem 2.4

For each of the following molecules, indicate (if present) the element(s) of symmetry and the equivalent hydrogens. Label equivalent hydrogens with a letter.

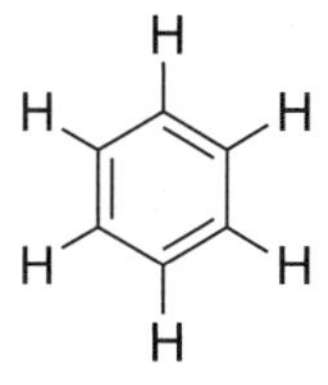

(I) 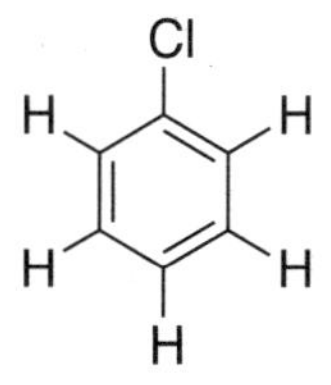 (II) 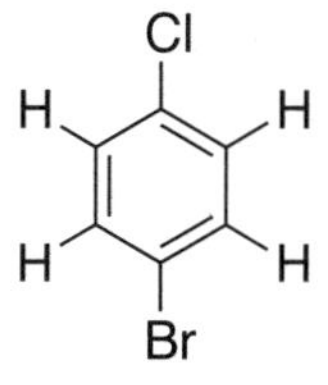 (III)

Solution

All three molecules possess at least one element of symmetry.

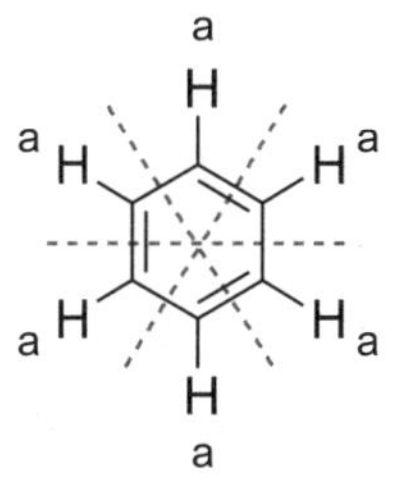

(I) 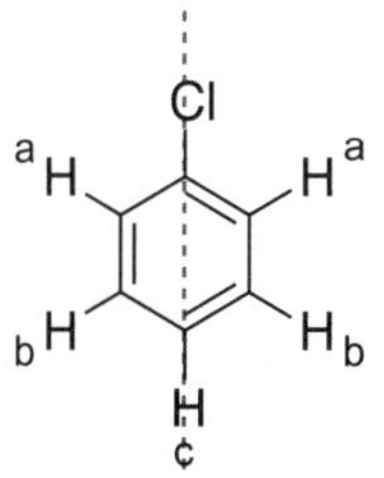 (II) 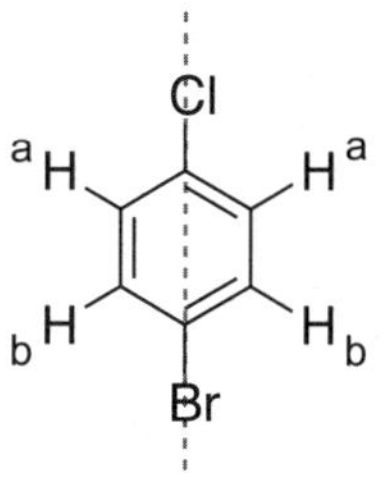 (III)

Benzene (I) is a highly symmetric molecule, and therefore all its hydrogens are equivalent. As for chlorobenzene (II), it is characterized by a single plane of symmetry, and therefore three types of equivalent hydrogens (labeled “a”, “b”, and “c”) are present in the molecule. A similar case is that of 1-bromo-4-chlorobenzene (III), which, also possessing a plane of symmetry, has two types of hydrogens (labeled “a” and “b”).

Problem 2.5

Using the substitution test to determine the topicity, decide whether the hydrogens “a” and “b” in each of the molecules reported below are homotopic, enantiotopic, or diastereotopic.

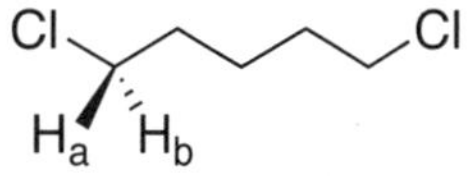

(I) 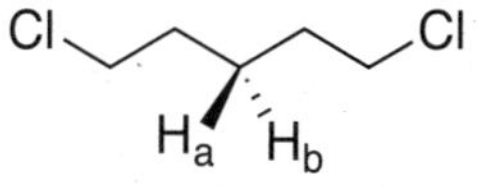 (II)

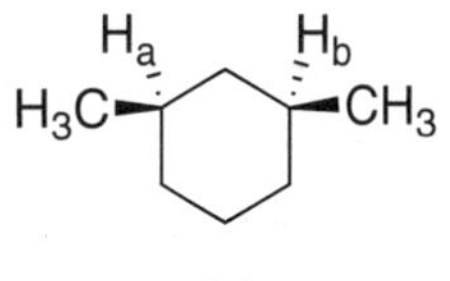

(III) 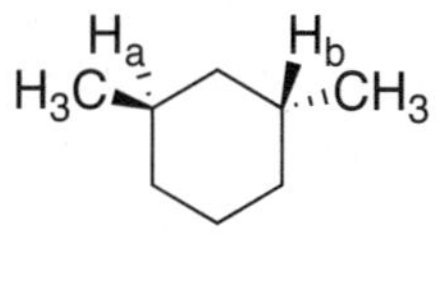 (IV)

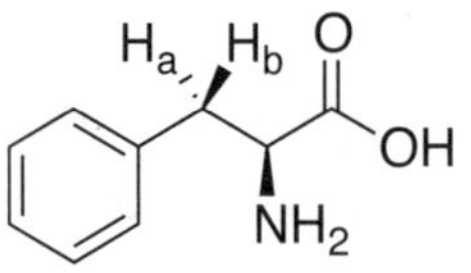

(V)

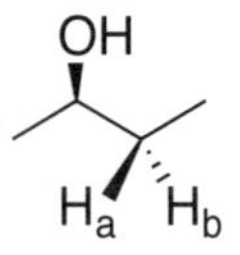

(VI)

Solution

As mentioned above, in order to determine the topicity of two hydrogens, it is possible to replace each hydrogen with a "dummy" atom (X) and then compare the two obtained structures. If they are identical, the protons are homotopic; if they are enantiomers, the protons are enantiotopic; if they are diastereomers, then the protons are diastereotopic. Remember that homotopic and enantiotopic hydrogens are chemical shift equivalent, while diastereotopic ones resonate at different frequencies.

(I)

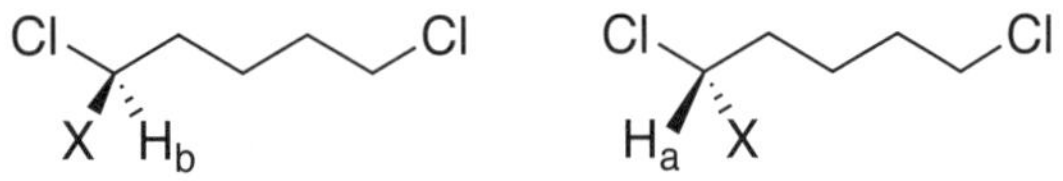

The two resulting molecules are enantiomers, so the two hydrogens are enantiotopic.

(II)

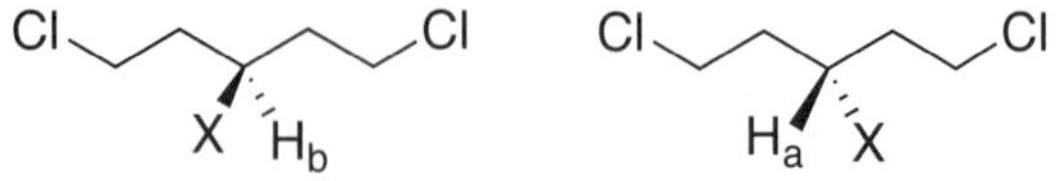

In this case the substitution provides two identical molecules, so the two hydrogens are homotopic.

(III)

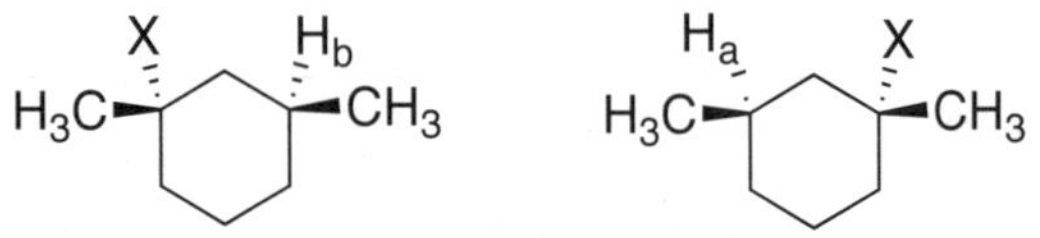

In this case the substitution provides two enantiomers, so the two hydrogens are enantiotopic.

(IV)

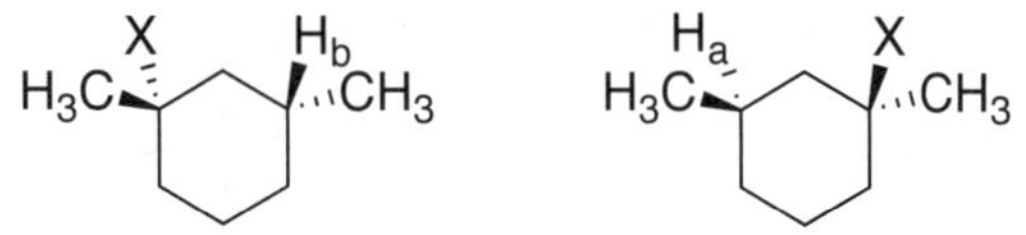

Here the substitution provides two identical molecules, so the two hydrogens are homotopic. It is easy to see this if you imagine flipping over a vertical axis one of the two molecules and superimposing it on the other. Remember that when you flip the molecule, the substituents that used to point out of the paper plane, then will point in, and vice versa.

(V)

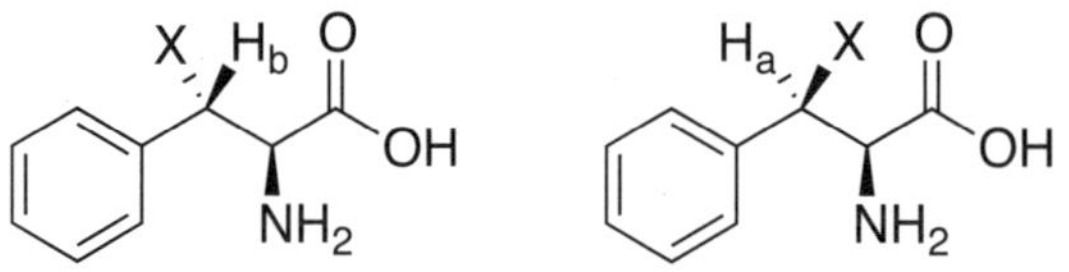

In this case, the two molecules turn out to be diasteromers, thus, the two hydrogens are diasteromeric.

(VI)

For the same reasons reported for (V), the two hydrogens are diasteromeric.

Problem 2.6

Do the following molecules provide the same ^{1}H-NMR spectrum?

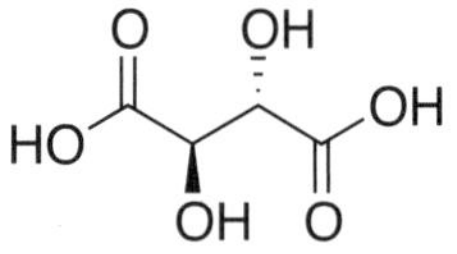

(I) 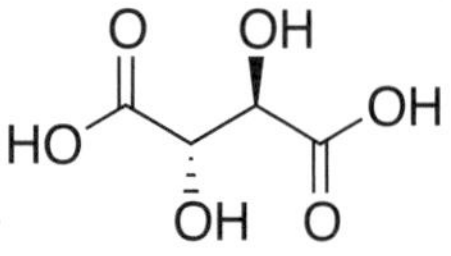 (II)

Solution

The S,R-tartaric acid (I) and R,S-tartaric acid (II) are the same molecule (*meso*). It is easy to see this if you redraw the molecules in the following way:

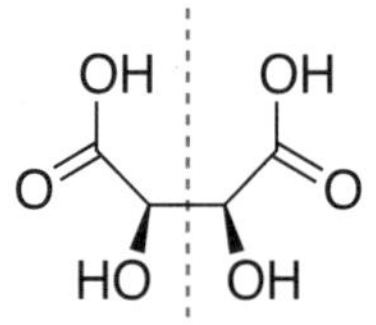

(I) 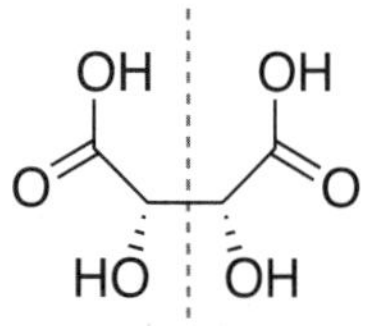 (II)

Hence, of course, they have the same spectra.

Problem 2.7

The chemical shifts of nitrobenzene and phenol are reported in gray next to each hydrogen:

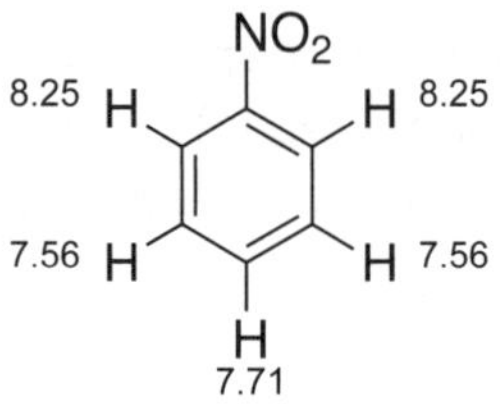

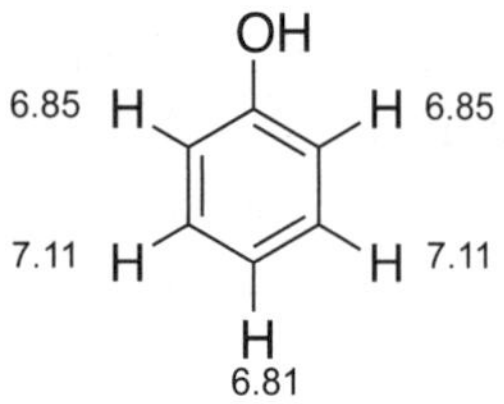

Nitrobenzene (NO_2): 8.25, 8.25 (ortho); 7.56, 7.56 (meta); 7.71 (para)

Phenol (OH): 6.85, 6.85 (ortho); 7.11, 7.11 (meta); 6.81 (para)

Can you give an explanation of those different values?

Solution

You should remember from organic chemistry that the nitro group (-NO_2) decreases the electron density at the *ortho* and *para* positions of the phenyl group through a resonance withdrawing effect.

NO_2 δ+ δ+ δ+   OH δ- δ- δ-

Therefore, these positions are less shielded and the protons resonate at higher values of ppm. On the contrary, the hydroxy group (-OH) is a strongly activating, *ortho/para* directing substituent. Therefore, the resonance allows a particularly high electron density to be positioned at the *ortho* and *para* positions, that therefore will be more shielded and the protons will resonate at lower values of ppm.

Summary Problems

Solutions to these problems are available online. Find your Activation Code and instructions on the first page of this book.

Problem 2.8

How many different types of hydrogens are there in the following compounds?

$CH_3CH_2CH_2Br$ (I) CH_3-$\underset{}{C}HCH_2Br$ with CH_3 on C2 (II) CH_3-CCH_2Br with two CH_3 on C2 (III)

Problem 2.9

Indicate which molecule(s) display(s) only one signal in the ^{1}H-NMR spectrum.

(I) Cl–C_6H_4–Cl (para)   (II) H_3C–C(=O)–CH_2Cl (III) CH_3-C(CH_3)$_2$-CH_3

Problem 2.10

Consider the structures reported in Problem 2.9. Which molecule will be characterized by an NMR spectrum having:

1) signal(s) at around 7 ppm;
2) signal (s) at around 1 ppm;
3) signal(s) between 1.5 and 4 ppm.

Problem 2.11

For the molecules reported below indicate:

1) how many signals do you expect in the proton NMR spectrum;
2) the relative signal intensities;
3) if it possible to distinguish the molecules by NMR.

$CH_3\overset{O}{\overset{\|}{C}}OCH_2CH_3$ (I) $CH_3CH_2\overset{O}{\overset{\|}{C}}OCH_3$ (II)

Problem 2.12

Draw the structures of organic molecules:

1) with 2 homotopic hydrogens;
2) with 2 enantiotopic hydrogens;
3) with 2 diastereotopic hydrogens.

Problem 2.13

Which of the following molecules is characterized by a proton spectrum having signals at 9.79, 2.46, and 1.11 ppm?

$CH_3CH_2\overset{O}{\overset{\|}{C}}H$ (I) $CH_3CH_2OCH_3$ (II)

Problem 2.14

Two molecules characterized by the same molecular formula (C_3H_6O) are propionaldehyde and acetone. Can they be distinguished by ^{1}H-NMR ? Explain why.

Problem 2.15

How many different signals do you expect for the following compounds?

(I) (II) (III)

Problem 2.16

For each of the molecules reported below, indicate in which of the following chemical shift ranges you expect them to resonate (please note that all three molecules provide a single NMR signal).

1) 3-4 ppm
2) 4.5-5.5 ppm
3) 7-8 ppm

(I) (II) (III)

Problem 2.17

Each of the following compounds exhibits a single ^{1}H-NMR signal. Approximately at which ppm would you expect each compound to resonate?

(I)

CH_3CH_3

(II)

(III)

$N(CH_3)_3$

(IV)

Problem 2.18

The aromatic protons of tyrosine resonate at 6.75 and 7.02 ppm. Can you assign these values to the pertinent hydrogens? Explain.

Problem 2.19

How many different types of hydrogens are there in the following compounds? Label equivalent hydrogens with a letter.

(I) (II) (III)

Problem 2.20

How many different types of hydrogens are there in the following compounds? Label equivalent hydrogens with a letter.

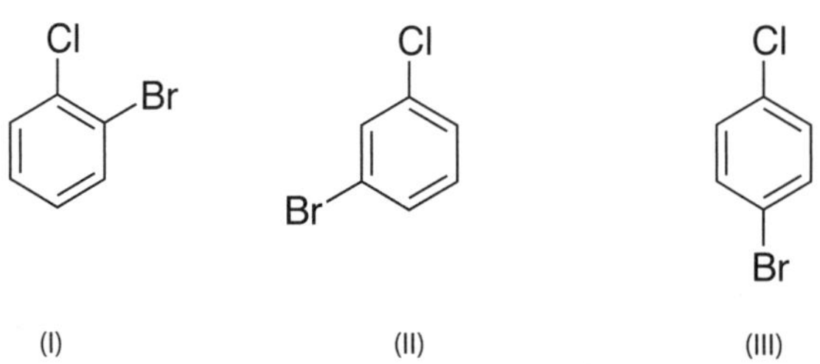

Problem 2.21

How many different signals do you expect for the following compounds?

(I) (II) (III)

CHAPTER 3

Spin-spin coupling in ^{1}H-NMR

3.1 Signal multiplicity

As mentioned in the previous chapters, electrons affect the magnetic environment of protons. However, the nuclei of nearby atoms (having themselves magnetic moments) do the same.

Consider for example an NMR tube containing a solution of a molecule of generic formula:

```
    Ha  Hb
     |   |
  X-C---C-Y
     |   |
     X   Y
```

where X and Y are generic substituents.

Taking into account all the molecules contained in the tube, for the Boltzmann distribution, in about 50% of those molecules the H_a protons will be accompanied by H_b protons having their magnetic moments aligning (so strengthening) B_0, while the other 50% of the molecules will have the magnetic moments of H_b protons opposing (and so weakening) B_0 (see the top of Fig. 3.1). Therefore, H_a protons in half the molecules experience an increased magnetic field and therefore their signal will be slightly shifted to a higher ppm value. On the contrary, in the other half, the signal of H_a protons will be slightly shifted to lower ppm values. Thus, the signal of the H_a protons is split into two sub-peaks with approximately equal areas (relative intensity ratio 1:1) (Fig. 3.1, bottom). A signal with this shape is termed a *doublet* (d), while an unsplit signal is termed a *singlet* (s). The resonance frequency of a doublet is measured at the very center between the two sub-peaks (that is, where the uncoupled signal would have resonated).

This kind of interaction between two nuclei occurs through chemical bonds, and can typically be observed up to 3/4 bonds away. This phenomenon is called *scalar coupling*.

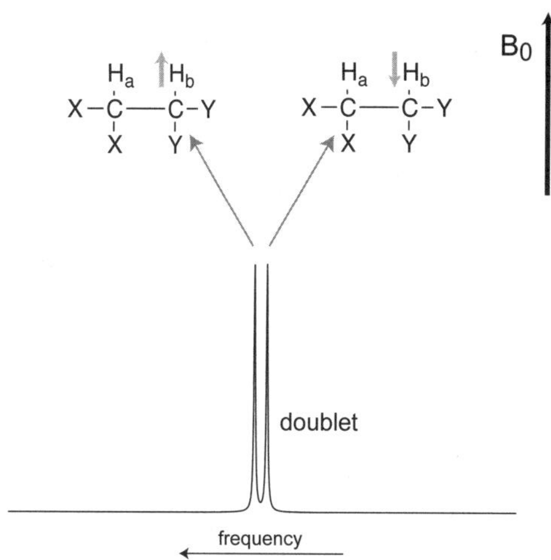

Figure 3.1 NMR signal of H_a. The scalar coupling to H_b forms a doublet. The relative intensity of the sub-peaks in the doublet is 1:1.

The distance (in Hz) between the sub-peaks is termed the *coupling constant*. By convention, coupling constants are indicated as $^{x}J_{ab}$, where x is a number indicating the number of intervening bonds between the coupled nuclei (*a* and *b*). Coupling constants are independent of the magnetic field and depend on the nature and the number of intervening bonds between the coupled nuclei, on conformation of the molecule, and on the nature of the substituents.

A given hydrogen (H_a) can also be scalar coupled to two other nuclei (b_1 and b_2) (Fig. 3.2). In this case, the local magnetic field experienced by H_a will be affected by the orientation of the nuclear magnetic moments of both H_{b1} and H_{b2}. In particular, four arrangements of equal probability of H_{b1} and H_{b2} are possible (Fig. 3.2):

1) Both magnetic moments of H_{b1} and H_{b2} strengthening the magnetic field. In this case H_a will resonate at a higher frequency.

2) The magnetic moment of H_{b1} strengthening and that

of H_{b2} weakening the magnetic field. In this case, the two orientations are canceled with each other and the local magnetic field experienced by H_a is virtually unaffected, and no variation of chemical shift will be observed.

3) Similar to the previous arrangement, the magnetic moment of H_{b1} weakening and that of H_{b2} strengthening the magnetic field. Also in this case, the orientations are canceled with each other, resulting in no variation of the chemical shift of H_a.

4) Both magnetic moments of H_{b1} and H_{b2} weakening the magnetic field. In this case, H_a will resonate at a lower frequency.

So, statistically, 25% of the molecules will have each of the above described arrangements, and each quartile generates a sub-peak. Hence, since two arrangements are equivalent (2 and 3), the probability of observing an H_a having no variation of the chemical shift is double that of having a lower or a higher chemical shift. Therefore, the signal of H_a will be split into three equally spaced sub-peaks having relative intensities of 1:2:1. A signal so shaped is called a *triplet* (t) (Fig. 3.2). The resonance frequency of a triplet corresponds to the central sub-peak.

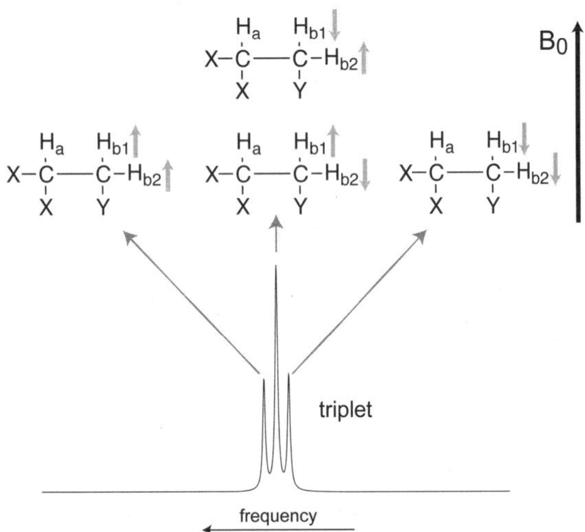

Figure 3.2 Scalar coupling of H_a to H_{b1} and H_{b2} forming a triplet. The relative intensity of the sub-peaks in a triplet is 1:2:1.

The same rationale can be used to explain the origin of a *quartet* (q) (or *quadruplet*), which is generated when a given nucleus (H_a) is coupled to three nuclei (H_{b1}, H_{b2}, and H_{b3}) (Fig. 3.3). In this case, H_a senses the eight different arrangements of the magnetic moments of the three H_b protons. However, altogether, they can generate only four different local magnetic fields, because some of the arrangements are equivalent (the two central groups in Fig. 3.3). Therefore, the relative intensity of the sub-peaks in a quartet is 1:3:3:1. Please note that no single arrangement gives a complete cancellation of the magnetic moments, and therefore the resonance frequency of the quartet, as in the case of the doublet, is measured at the very center between the two central sub-peaks.

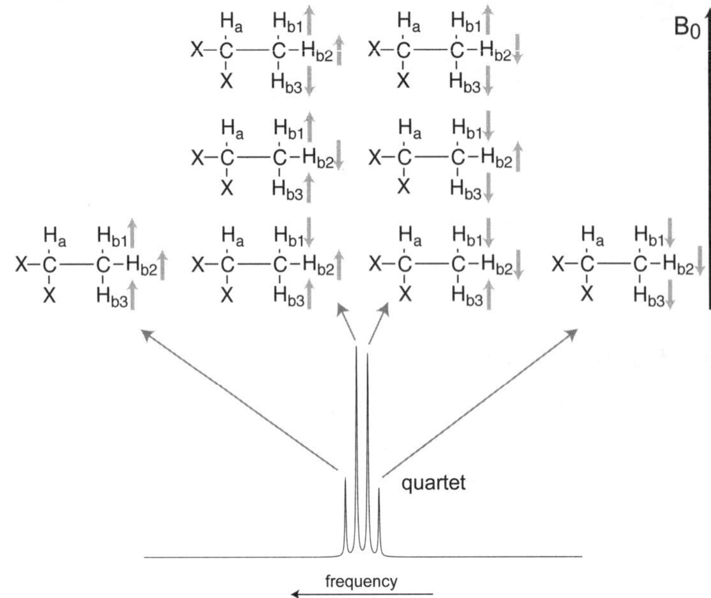

Figure 3.3 Scalar coupling of H_a to H_{b1}, H_{b2}, and H_{b3} forming a quartet. The relative intensity of the sub-peaks in a quartet is 1:3:3:1.

An empirical rule can be retrieved from the above observations: *the NMR signal for a nucleus having n neighbors is split into n+1 sub-peaks*. This rule is known as "n+1 rule". The relative intensities of the sub-peaks may be retrieved from Pascal's triangle (Fig. 3.4).

N. of coupled atoms (I = 1/2)	Relative intensities of sub-peaks	Name of multiplet	Shape
0	1	*singlet*	
1	1 1	*doublet*	
2	1 2 1	*triplet*	
3	1 3 3 1	*quartet*	
4	1 4 6 4 1	*quintet*	
5	1 5 10 10 5 1	*sextet*	
...	... ...	...	...

Figure 3.4 Relative intensities of the sub-peaks in some NMR signals according the Pascal's triangle.

Actually, this rule applies only to nuclei having the nuclear magnetic spin quantum number I = 1/2 (like ^{1}H, ^{13}C, ^{15}N, etc.) and coupled to other nuclei with the same coupling constant.

A more general rule to predict the multiplicity of a signal is the following: *n protons will split the NMR signal into 2nI + 1 sub-peaks*, where *I* is the spin quantum number (also in this case the coupling constants to the other nuclei must be the same). Therefore, for example, if a hydrogen is coupled to two other hydrogens (that have I = 1/2):

$$2nI + 1 = 2 \cdot 2 \cdot 1/2 + 1 = 3$$

The results of the equation is 3, meaning that the multiplicity of the signal is "triplet". If the result was 2, then the multiplicity of the signal would be "doublet", and so on.

In summary, the NMR signals are characterized by a fine structure termed *multiplicity* (or splitting), which is defined by the number of sub-peaks and by their relative intensities. The distance between the sub-peaks is constant (the coupling constant). A signal characterized by a fine structure can be also named with the generic term *multiplet.*

In all these examples, only H_a protons have been taken into consideration. What happens to the H_b protons?

If H_a protons are coupled to H_b protons, H_b protons are also coupled to H_a.

Thus, in the case of the generic molecule:

Ha Hb
X–C—C–Y
X Y

H_b resonates as a doublet as well, being coupled only with H_a (Fig. 3.5). The coupling constant of this doublet is the same as the doublet generated by H_a.

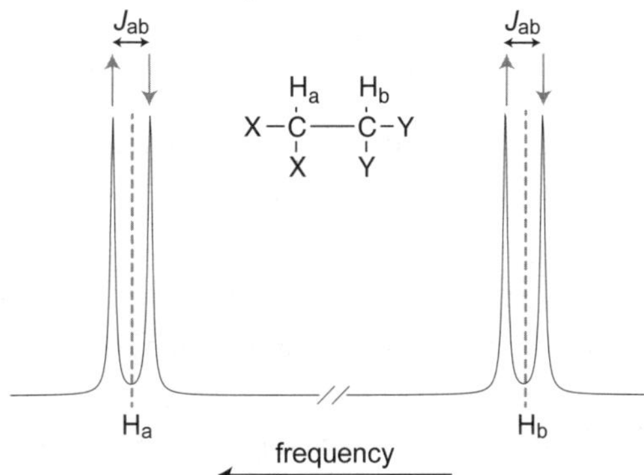

Figure 3.5 Simulated spectrum having two doublets of equal intensity. The relative positions of the two signals are arbitrary and would in practice depend on the electronegativity of X and Y.

In the case of

Ha Hb1
X–C—C–Hb2
X Y

both the scalar coupling of the two H_b protons to H_a and the scalar coupling between H_{b1} and H_{b2} should be taken into account (Fig. 3.6). These latter hydrogens are enantiotopic, hence they are chemical shift equivalent. However, it is very important to note that **the coupling between equivalent protons (although present) is not observable**. Therefore, H_{b1} and H_{b2} appear coupled only to H_a, generating a doublet with double intensity with respect to the triplet that arises only from H_a. Obviously, the coupling constants of the doublet and the triplet are the same.

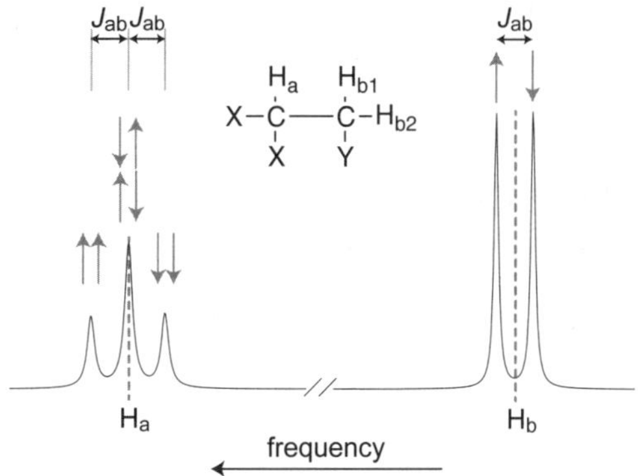

Figure 3.6 Simulated spectrum having a doublet and a triplet of relative intensities 2:1. The relative positions of the two signals are arbitrary.

Finally, in the case of the generic molecule:

Ha Hb1
X–C—C–Hb2
X Hb3

the three H_b hydrogens are homotopic, and thus they resonate at the same chemical shift, so they do not show coupling to each other. Hence, they appear coupled only to H_a, generating a doublet with three times the intensity of the quartet signal from H_a. Also in this case the coupling constants of the doublet and the quartet are the same (Fig. 3.7).

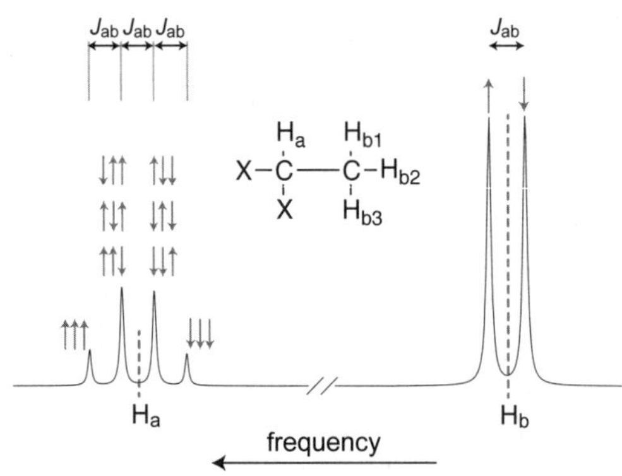

Figure 3.7 Simulated spectrum having a doublet and a quartet of relative intensities 3:1. The relative positions of the two signals are arbitrary.

As mentioned above, coupling constants depend on the nature and the number of intervening bonds between the coupled nuclei, on the conformation of the molecule, and on the nature of the substituents. Generally, the higher the number of intervening bonds, the lower the coupling constant. In saturated molecules, for example, the geminal coupling constant ($^2J_{ab}$) is typically 10-15 Hz. On the other hand, vicinal coupling constants ($^3J_{ab}$) range from 0 to 14 Hz (Fig. 3.8), depending on the dihedral angle (θ) .

$^2J_{ab}$ = 10-15 Hz

$^3J_{ab}$ = 0-14 Hz

Figure 3.8 Coupling constants of geminal (top) and vicinal (bottom) protons.

The Karplus equation (Eq. 3.1) describes the correlation between the coupling constant (*J*) and the dihedral angle (θ):

$$J(\theta) = A\cos^2\theta + B\cos\theta + C \qquad \text{(Eq. 3.1)}$$

where *A*, *B*, and *C* are empirically derived parameters whose values depend on the atoms and substituents involved. Being based on the trigonometric function cosine, the magnitude of the coupling is generally larger at angles of 0 and 180°, and smaller when the torsion angle is close to 90° and 270° (Figs. 3.9 and 3.10).

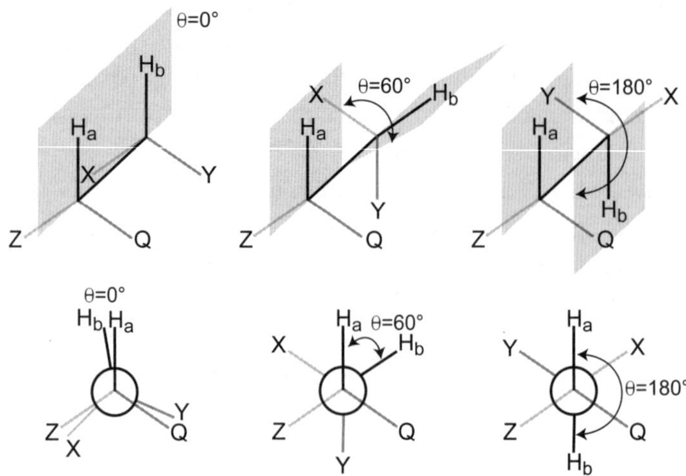

Figure 3.9 TOP: Dihedral angles seen as intersecting planes (in gray); BOTTOM: corresponding Newman projections.

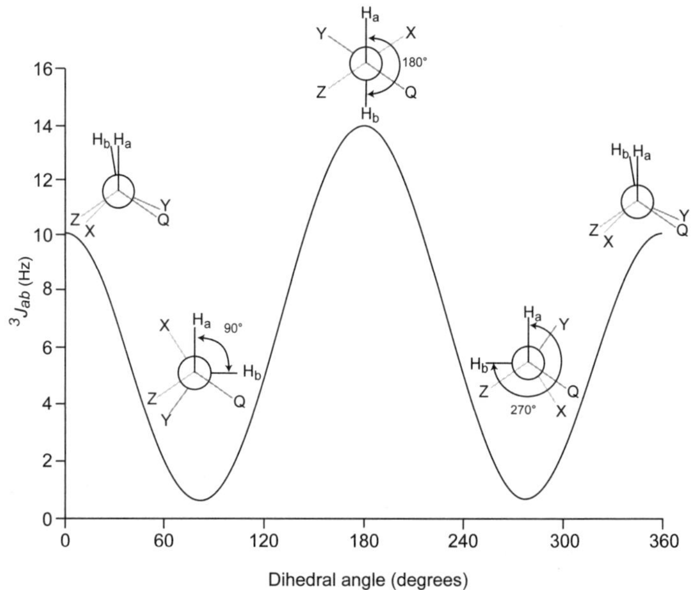

Figure 3.10 Graphical representation of the Karplus equation (Eq. 3.1).

What happens to flexible molecules characterized by free rotation around the C-C single bond?

By free rotation, two vicinal hydrogens can assume all possible dihedral angles. If the rotation is faster than the NMR time-scale (see Chapter 6), an average coupling constant of about 7 Hz is measured.

Some problems already discussed in terms of hydrogen equivalence in the previous chapter will be revisited below in terms of signal multiplicity.

Problem 3.1

Predict the signal multiplicity for the protons of the following molecules:

CH_4	CH_3CH_3	$CH_3CH_2CH_3$
(I)	(II)	(III)

Solution

As already explained in Problem 2.1, molecules I and II are characterized by a single NMR signal, that, therefore, must be a singlet. On the other hand, molecule III is characterized by two signals. The hydrogens of the methylene are homotopic, and therefore they resonate at the same chemical shift. Nevertheless, they couple to the six equivalent hydrogens of the two methyls, generating a septet (n+1 rule). The two methyls are equivalent and, due to the coupling to the two hydrogens of the methylene, they resonate as single triplet.

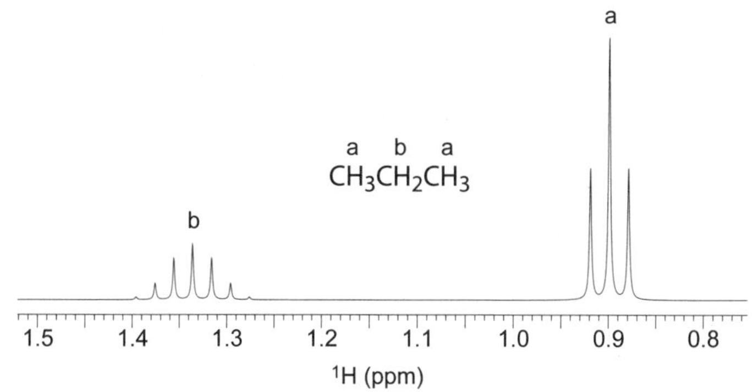

Please note that, generally, methyls resonate at chemical shift values lower than methylenes, that, in turn, resonate at lower values than methines (this is true only in alkanes). As already discussed in the previous chapter, the relative signal intensity will be 2(septet):6(triplet).

Problem 3.2

Predict the multiplicity of the proton signals of the following molecules:

Cl Cl (I)

Cl Cl (II)

Solution

(I) Following the same rationale used in the last problem (3.1), it is clear that the 1,3-dichloropropane generates a triplet (more deshielded because of the attached chlorines) and a quintet, with relative intensities 4:2 (or 2:1), respectively.

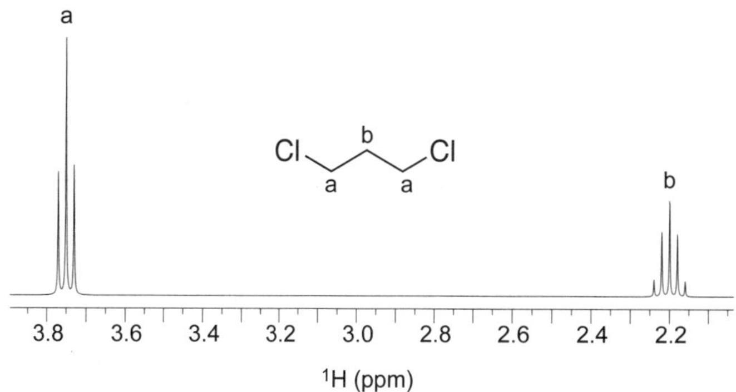

(II) The case of 1,1-dichloroethane is very simple. Two signals are expected: one from the methine resonating as quartet (a) (more deshielded), and one from the methyl (b), resonating as doublet, with relative intensities 1:3, respectively.

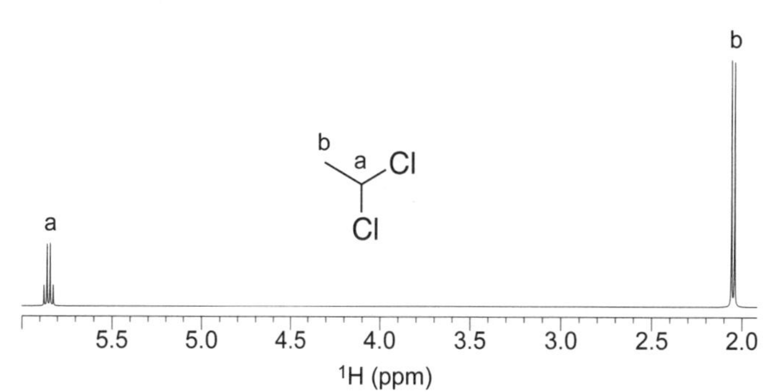

Problem 3.3

Predict the splitting patterns for *S*,*R*-tartaric acid (please do not consider the hydrogens bonded to the oxygen atoms).

O OH
HO OH
OH O

Solution

As already explained in Problem 2.6, this molecule is symmetric, and therefore the hydrogens (a) and (b) are chemical shift equivalent. Hence, the spectrum contains only a singlet.

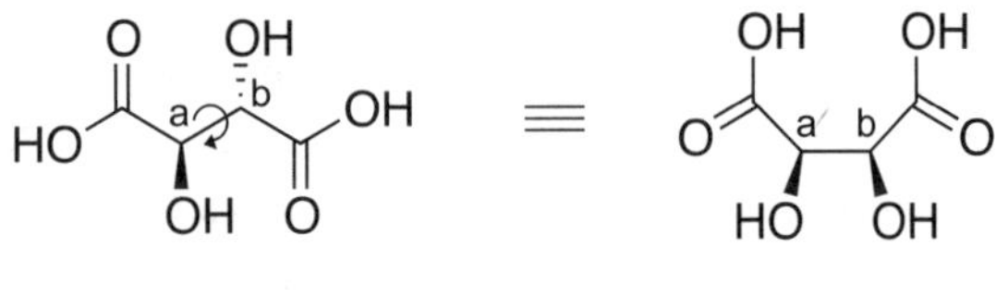

Problem 3.4

Propose a plausible structure for the compound having molecular formula C_7H_7Br, whose proton NMR spectrum is reported below (numbers on the signals indicate the relative signal intensities).

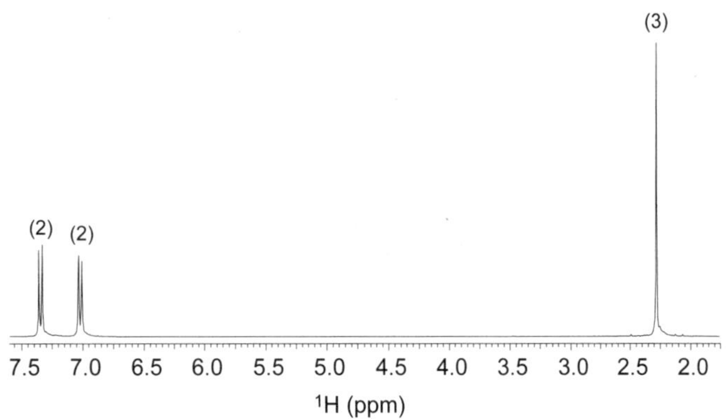

Solution

The signals around 7-8 ppm clearly suggest the presence of an aromatic moiety in the molecule. Unfortunately, it is not possible, at this stage, to determine with certainty which kind of aromatic system it is. Different scenarios can be supposed; you can start from the most probable one and try to speculate on that. If, going ahead, you realize that your hypothesis is wrong, you must begin again with another reasonable aromatic system and try again recursively until you identify the correct compound. In this case, you should first analyze the molecular formula of the compound. Neither nitrogen, nor oxygen, nor sulfur atoms are present in the molecule. Therefore, it is not possible to hypothesize the presence of a heteronuclear aromatic system. Thus, it is reasonable to suppose that the analyzed molecule is a benzene derivative. If this is true, the presence of only two doublets (of equal intensities) strongly suggests that it must be *para*-heterodisubstituted:

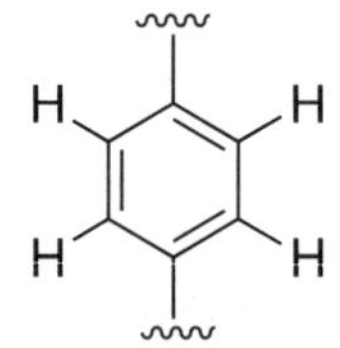

This moiety counts for C_6H_4, which means that the one carbon, one bromine, and three hydrogen atoms remain to be positioned. Keeping in mind that you should also identify the two substituents of the aromatic ring, you could suppose that -Br is one of those, and a methyl (-CH_3) is the other one.

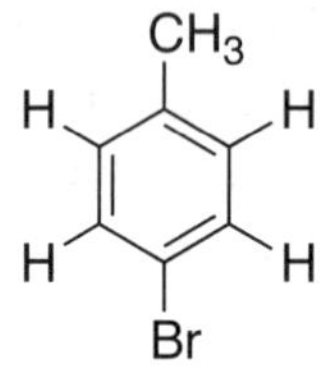

The presence of a singlet resonating at around 2.3 ppm is in agreement with this hypothesis (i) for its chemical shift (a methyl bonded to an sp^2 carbon generally resonates around 2 ppm), (ii) for its multiplicity (singlet) and (iii) for the relative intensity ratio (3:2) with respect to the other two aromatic signals present in the spectrum. Therefore, taking into account the molecular formula and the appearance of the NMR spectrum, you can conclude that the molecule is 4-bromotoluene.

Problem 3.5

Propose a plausible structure for the compound having molecular formula C_3H_7Br, whose proton NMR spectrum is reported below (numbers on the signals indicate the relative signal intensities).

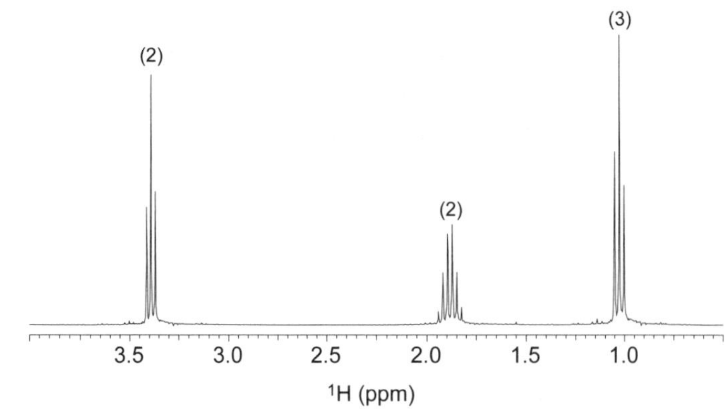

Solution

The signals are confined between 1 and 3.5 ppm, so you can exclude the presence of an aromatic system (this is much more convincing also considering the number of carbon atoms present in the molecule!). In total, the spectrum contains only three signals, namely two triplets and one sextet, of relative intensities of 3:2:2. Considering that the building blocks of organic molecules are methines (CH), methylenes (CH_2) and methyls (CH_3), you can imagine that the signals integrating for two hydrogens could be generated by methylenes, and the signal integrating for three hydrogens can be generated by a methyl. This is also in agreement with the number of carbons and hydrogens present in the molecular formula. In order to determine how to link these moieties together, you can analyze the chemical shift and the multiplicity of the signals. The chemical shift of the signal at 3.38 ppm can be justified only assuming that this methylene is bonded to the bromine. Then, the fact that it is a triplet suggests that it is coupled to two other hydrogens, which probably belong to another methylene. This methylene cannot generate other than the sextet at 1.87 ppm (the only remaining signal integrating for two protons). It remains to complete the structure of the molecule by linking the remaining methyl to the moiety identified so far to obtain CH_3-CH_2-CH_2-Br. To definitively confirm that the molecule is 1-bromopropane, it is very important to verify the consistency of all the information available. Particularly, the multiplicity of the sextet is consistent with the fact that the central methylene is coupled to other 5 protons (two from the other methylene and three from the methyl). Also, the multiplicity of the triplet is in agreement with the proposed structure, being the methyl coupled only to the central methylene. Finally, the chemical shift values of all three signals are consistent with the chemical environments of each proton.

Problem 3.6

A compound of molecular formula $C_4H_{10}O$ gives a proton NMR spectrum as shown below. Determine its structure (numbers on the signals indicate the relative signal intensities).

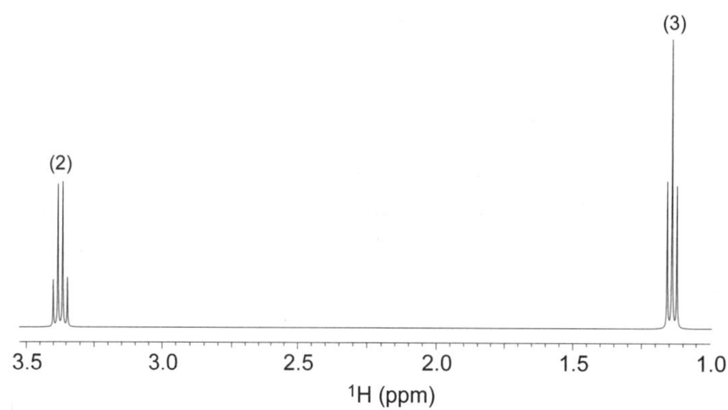

Solution

This spectrum seems very simple; however, it hides a trap. In fact, in the molecular formula there are 10 hydrogens to be assigned and there are only two signals of relative intensities of 2:3 in the spectrum. This means that the molecule is symmetric. Since only one oxygen atom is present in the molecule, it must lie on the element of symmetry of the molecule, namely, it should be the junction of the two symmetric parts of the molecule. Therefore, the spectrum could arise from the molecule C_2H_5-O-C_2H_5. Looking at the formula written in this way, the problem seems much easier. The signal at 3.35 ppm could arise from the methylenes (see the integration) that, as suggested by their chemical shift, are bonded to the oxygen atom (-CH_2-O-CH_2-). The multiplicity of this signal (quartet) clearly suggests that the methylenes are also bonded to methyls (CH_3-CH_2-O-CH_2-CH_3), which resonate instead as a triplet. Everything is consistent, and therefore the molecule is certainly diethyl ether.

Problem 3.7

Propose a plausible structure for the compound having molecular formula C_9H_{12}, whose proton NMR spectrum is reported below (numbers on the signals indicate the relative signal intensities).

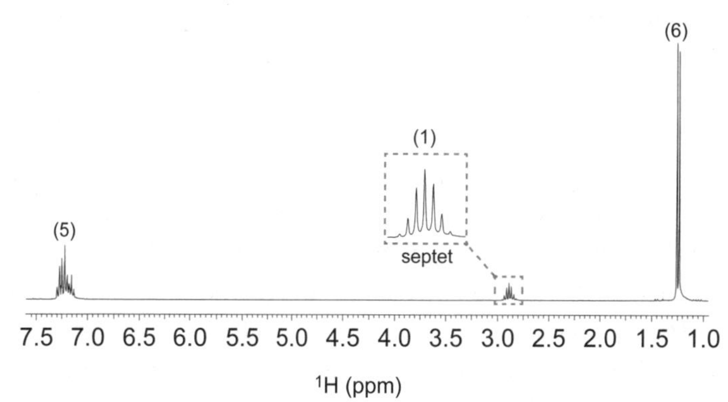

Solution

This problem is apparently more complex than the previous ones. As indicated by the integral values, there is more than one signal between 7.0 and 7.5 ppm. These signals are severely overlapped and it is not possible to define their multiplicity. However, their chemical shifts suggest that there is an aromatic moiety in the molecule. This could be consistent with the presence of a phenyl moiety that contains five hydrogens.

H H
H H
H

This moiety counts for C_6H_5, so you should try to determine the remaining part of the molecule (C_3H_7). This refers to the two remaining signals in the spectrum: a septet (integrating for one proton) and a doublet (integrating for six protons). These last signals can be justified assuming the presence of an isopropyl, which is made by two equivalent methyls (the doublet) bonded to a methine, which, in turn, generates the septet.

H_3C H CH_3
C

Hence, the compound is isopropylbenzene.

H_3C H CH_3
C
H H
H H
H

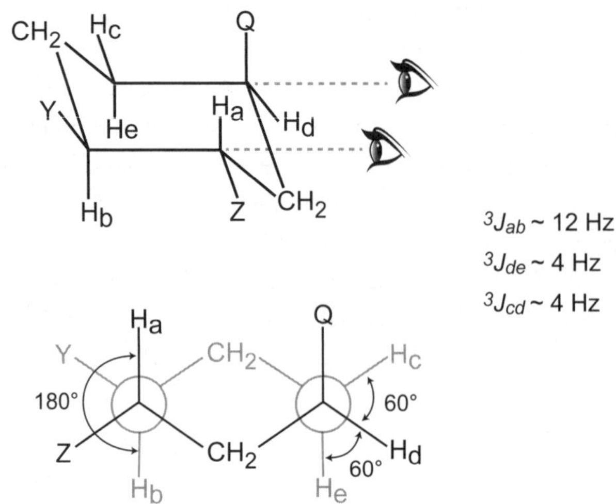

Figure 3.11 Top: A substituted cyclohexane (Q, Z, and Y are generic substituents). Bottom: the same molecule in Newman projection.

What is the multiplicity of a signal of a proton that couples to two different protons with two different coupling constants?

In order to properly answer this question, it is very useful to use a *splitting diagram* (also called *tree diagram*), that rationally describes the origin of multiplets. In the splitting diagram, the couplings are applied sequentially (it is not important which coupling you describe first). Each coupling splits the lines in two (according to the coupling constant), halving their intensity. For example, a splitting diagram that describes a doublet is reported in Figure 3.12.

The original signal resonating at ν_1 is symmetrically split in two. So, for example, if the nuclei are coupled by a constant of 10 Hz, each sub-peak of the doublet is moved respectively 5 Hz to the left and 5 Hz to the right with respect to ν_1.

The splitting diagram for the formation of a triplet is reported in Figure 3.13. In this case the hydrogen *a* is coupled by the same coupling constants to two hydrogens (H_b and H_c). In the figure, the splitting due to nucleus *b* is drawn first (you could also first describe the one due to nucleus *c*, it does not matter). Then, each of the obtained lines is further split symmetrically in two for the coupling to the nucleus *c*. Because $J_{ab} = J_{ac}$, the two central lines will be overlapped and will generate a sub-peak with double intensity with respect to the other lines. Therefore, the triplet (having relative sub-peak intensities 1:2:1) will be obtained.

3.2 Coupling constants in rigid systems and splitting diagrams

In conformationally rigid systems, the relative positions of hydrogens do not change and the coupling constants are not averaged anymore. This is the case with substituted cyclohexanes, for example, where the hydrogens have quite fixed relative orientations and provide characteristic coupling patterns. In fact, the 1,2 *trans*-diaxial protons (for example H_a and H_b in Fig. 3.11) form a dihedral angle of about 180° and, according to the Karplus equation (Eq. 3.1), they exhibit a large coupling constant (10-14 Hz). On the other hand, the equatorial/equatorial and axial/equatorial hydrogens (H_c/H_d and H_d/H_e, respectively, in Fig. 3.11) form dihedral angles of about 60°, giving rise to a J in the range of 3-5 Hz.

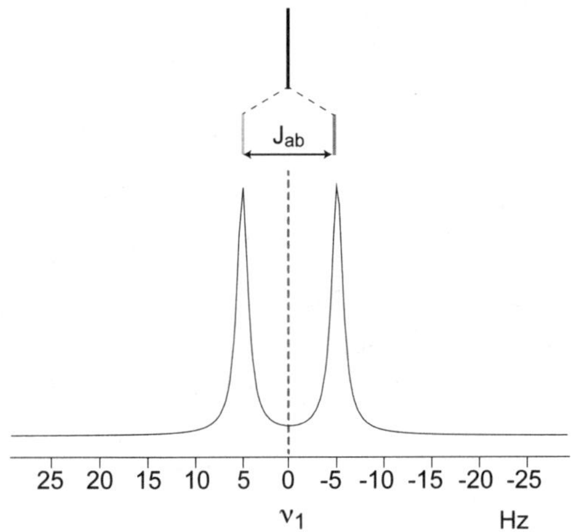

Figure 3.12 Splitting diagram for the formation of a doublet (nucleus "a" coupled to nucleus "b").

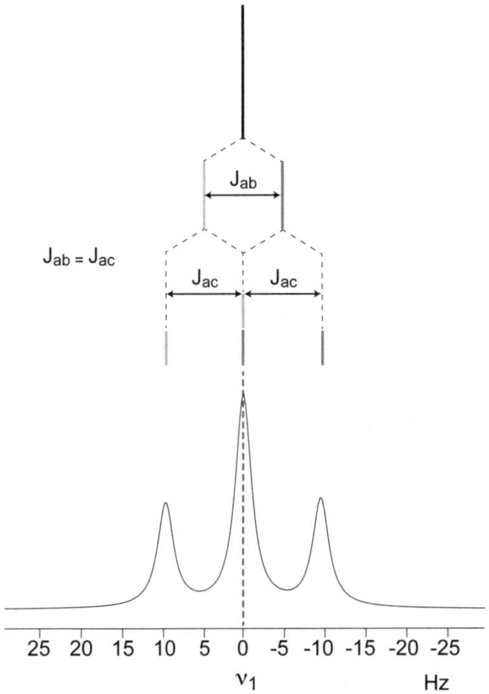

Figure 3.13 Splitting diagram for the formation of a triplet.

Now that the splitting diagram has been explained, it is possible to easily discuss the case of a given hydrogen *a*, having different coupling constants to *b* and *c* (Fig. 3.14). Assuming that $J_{ab} > J_{ac}$, the result is a four-line multiplet called a *doublet of doublets* (dd). Please note that this multiplet is very different from the quartet, since the relative intensities of the sub-peaks are 1:1:1:1 (and not 1:3:3:1), and the distances between the sub-peaks are not equal (in the quartet all the sub-peaks are equidistant) (Fig. 3.15).

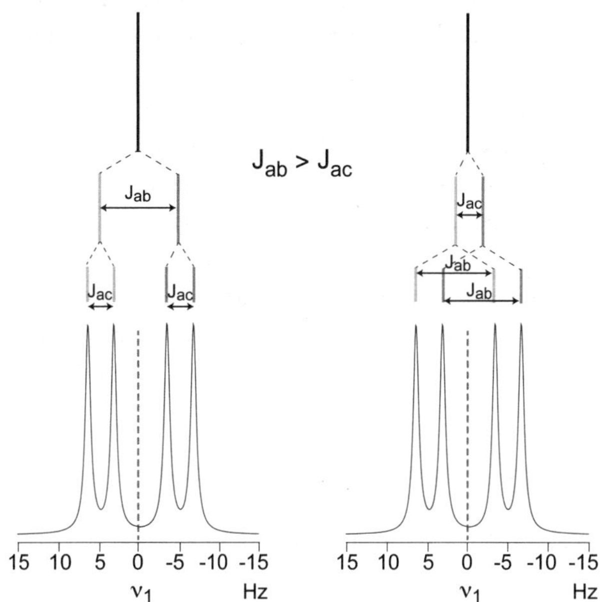

Figure 3.14 Splitting diagrams for the formation of a doublet of doublets. In the left and right panels the order of the splitting is different, but the resulting multiplet is the same.

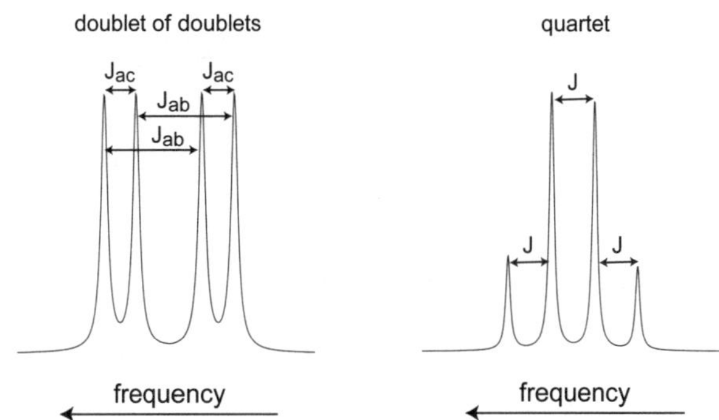

Figure 3.15 Differences between a doublet of doublets and a quartet. The doublet of doublets is characterized by a relative intensity of the sub-peaks of 1:1:1:1 and by two different coupling constants. The quartet is instead characterized by a single coupling constant and by a relative intensity of sub-peaks of 1:3:3:1.

REMEMBER: *The generic "n+1 rule" (to predict the multiplicity of the signals) and Pascal's triangle (to predict the relative intensities of the sub-peaks) can be used only when a given nucleus is coupled to other nuclei with the same coupling constants. If the coupling constants are different, only the splitting diagram can be used to predict the correct shape of the multiplet.*

Hence, while the coupling constant in all multiplets described so far (*i.e.*, doublet, triplet, quartet, quintet, etc.) can be measured between any two adjacent sub-peaks (see for example the right panel of Fig. 3.15), in multiplets formed by different coupling constants they must be measured taking into account the rationale of the splitting diagram (Fig. 3.15, left panel). So, in the case of a doublet of doublets, the large constant will be measured between the first and the third (or between the second and the fourth) sub-peaks, while the small one is measured between the first and the second (or the third and the fourth) sub-peaks.

The possible combinations of multiplets having different coupling constants are almost uncountable. By way of example, Figure 3.16 illustrates two similar but substantially different coupling patterns: a *triplet of doublets* (td) and a *doublet of triplets* (dt).

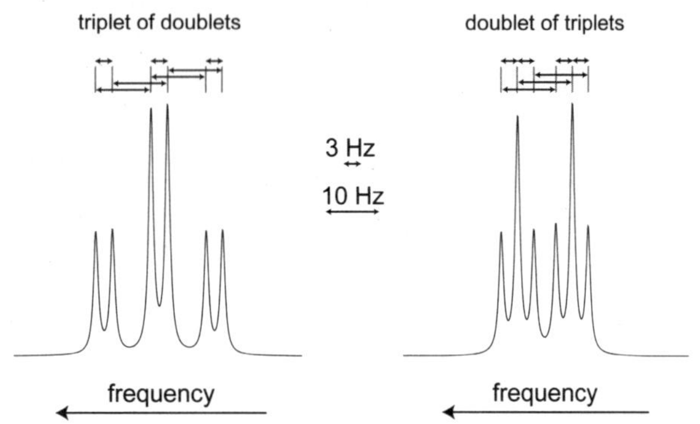

Figure 3.16 Examples of a triplet of doublets and a doublet of triplets. Short and long double-headed arrows indicate how to measure the couplings of 3 and 10 Hz, respectively.

The difference is that the triplet of doublets is substantially a triplet (having a large coupling constant, for example 10 Hz), where each sub-peak is further split in two by a small coupling (for example 3 Hz). The doublet of triplets, instead, is substantially a doublet (having a large coupling constant, for example 10 Hz), whose sub-peaks are split into triplets having a small coupling constant (for example 3 Hz).

Note that the names of the multiplets are assigned considering first the larger coupling constant. Thus, in the above examples, when the larger coupling constant forms the triplet, the name of the multiplet is "triplet of ...". On the other hand, when the larger coupling constant forms the doublet, the name of the multiplet is "doublet of ...".

The possibility of measuring coupling constants can allow the determination of the relative stereochemistry of two or more asymmetric centers in a rigid molecule. For examples, consider α- and β-D-glucose pentaacetate (Fig. 3.17).

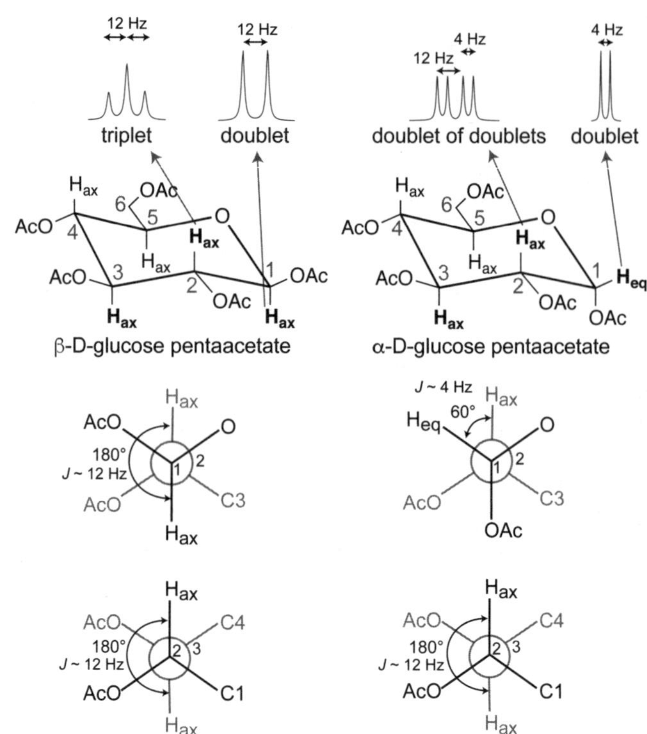

Figure 3.17 Different multiplicity observed for H-1 and H-2 in α- and β-D-glucose pentaacetate.

In β-D-glucose pentaacetate, all the acetoxyl groups linked to the six-membered ring are equatorial, and thus all hydrogens are axial. On the other hand, the α-D-glucose pentaacetate is characterized by all axial hydrogens, except for the anomeric one, which is equatorial. These two molecules have the same molecular formula and therefore they cannot be distinguished by mass spectrometry. However, they can be easily distinguished by NMR, simply by analyzing the multiplicity of the hydrogens in positions 1 and 2. The anomeric proton in the β-D-glucose pentaacetate is scalar coupled only to the adjacent H_{ax}. Both hydrogens are *trans*-diaxial forming a dihedral angle of about 180°. This means that H-1 generate a doublet in the NMR spectrum with a coupling constant of about 12 Hz (Fig. 3.17). Analogously, the hydrogens in positions 2 and 3 are *trans*-diaxial, so that H-2 is coupled by the same coupling constant to two hydrogens, and therefore it will form a triplet. On the contrary, H-1 in α-D-glucose pentaacetate is equatorial and forms a dihedral angle of about 60° with H-2. Thus, H-1 will form a narrower doublet (J ~4 Hz) than that observed for the β-D-glucose pentaacetate (J ~12 Hz). Furthermore, in this case, H-2 generates a multiplet having two different coupling constants, small (~4 Hz) to H-1 and large (~12 Hz) to H-3, and therefore it will generate a doublet of doublets. Therefore the two molecules are easily distinguishable by NMR.

Different from saturated systems, unsaturated ones have characteristic coupling constant patterns that are not regulated by the Karplus equation. In particular, geminal coupling constants are very small ($^2J_{gem}$ = 0-3 Hz), while vicinal ones are: $^3J_{trans}$ = 12-19 Hz and $^3J_{cis}$ = 6-11 Hz. These values are sufficiently different to be used to properly assign the resonances and/or to determine how a double bond is substituted.

Let's consider for example the spectrum of vinylbromide (Fig. 3.18).

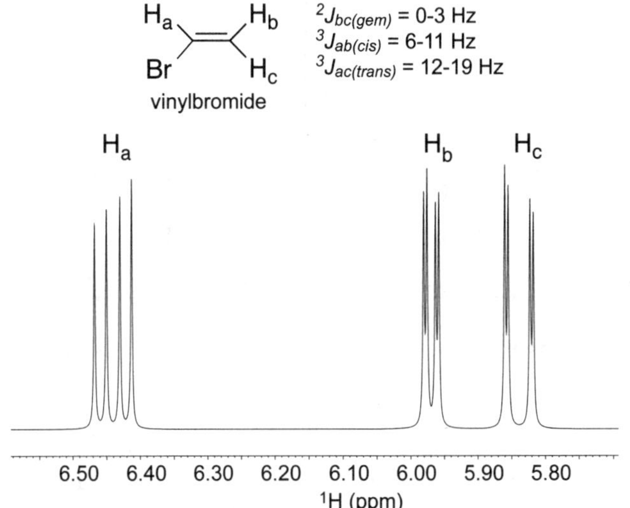

Figure 3.18 Simulated spectrum of vinylbromide. The intensities of the three doublets of doublets are affected by the "roof effect" that will be explained in Chapter 4.

The three hydrogens are coupled to each other by different coupling constants, and thus they generate three doublets of doublets (dd). Please note that the relative intensities of the sub-peaks of the doublets of doublets are not exactly in the ratio 1:1:1:1. This is due to a so called "roof effect" that will be explained in the next chapter; therefore, at this stage, disregard this distortion. The signal at 6.44 ppm is characterized by coupling constants of 7 and 15 Hz, the signal at 5.97 ppm by constants of 2 and 7 Hz, and the signal at 5.84 ppm by constants of 2 and 15 Hz. These data are sufficient to properly assign each signal to the pertinent hydrogen. In fact, the signal at 6.44 ppm cannot be assigned to H_b or H_c, since, both being geminals, they must have at least one small coupling constant. Therefore, this signal can be assigned only to H_a. The signal at 5.84 ppm can be assigned to H_c; in fact, such a large constant can be justified only by assuming a *trans* relationship with H_a. H_b is then assigned considering the *cis* relationship with H_a or it can simply be assigned by exclusion.

Interestingly, in cycloalkenes smaller than cyclohexene, the vicinal coupling constants show substantially reduced values (Fig. 3.19). This characteristic can be used as diagnostic feature to determine the structure of an unknown molecule.

Figure 3.19 Vicinal coupling constants in cycloalkenes of different size.

So far, only the couplings between geminal and vicinal nuclei have been discussed. Actually, in electron-rich systems it is possible to observe also the so-called *long-range* couplings. For example, ^4J couplings are visible in allylic, propargylic and allenic systems (Fig. 3.20). Even longer range couplings (^5J) are observable in homoallylic, homopropargylic and homoallenic systems (Fig. 3.20). ^5J and ^4J couplings can also be observed in aromatic rings (Fig. 3.21).

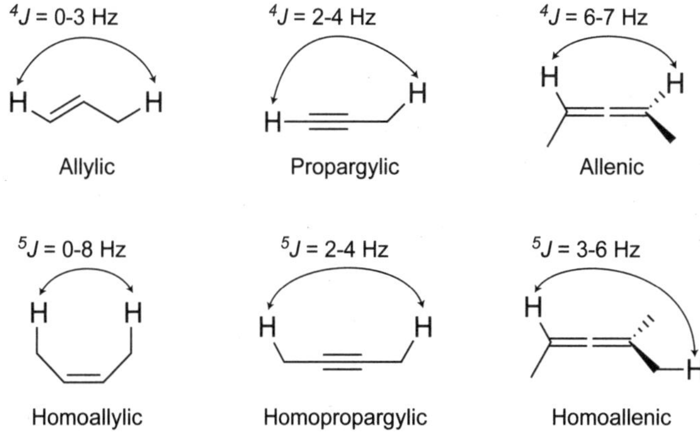

Figure 3.20 Long-range couplings in unsaturated molecules.

It should be noted that the very small coupling constants are very often difficult to detect and measure.

Proton-proton couplings over more than three bonds across saturated carbons (sp^3) are usually too small (< 1 Hz) to be detected. However, there is an exception in cyclic compounds. When an H-C-C-C-H chain adopts a so-called W geometric alignment, it is possible to observe a small ^4J coupling, also termed "W-coupling" (Fig. 3.22).

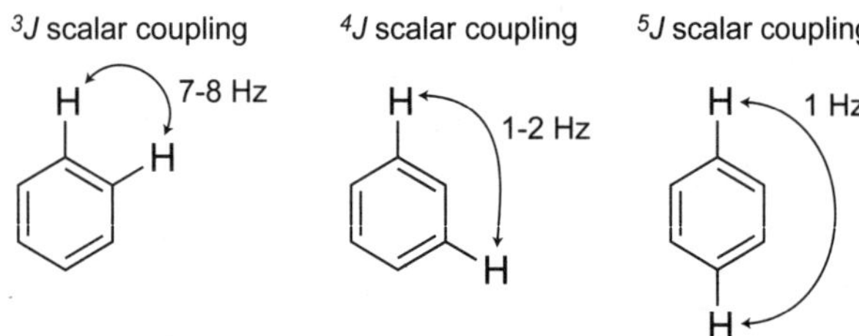

Figure 3.21 Coupling constants in aromatic rings.

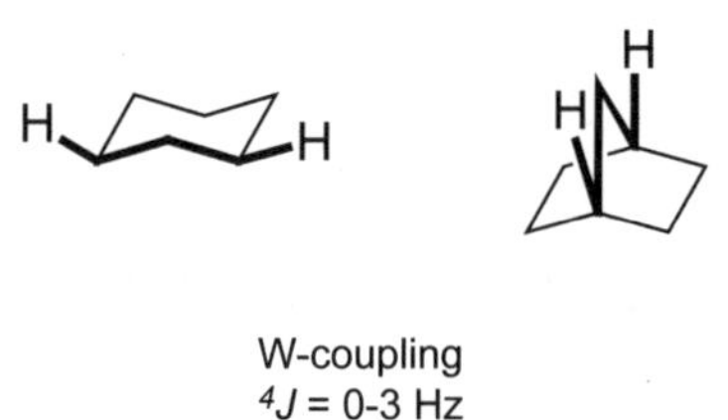

Figure 3.22 Long-range W-coupling in rigid systems.

Problem 3.8

Draw the splitting diagram for a quartet (q) with a coupling constant of 8 Hz, and for a doublet of a doublet of doublets (ddd) with coupling constants of 14 Hz, 8 Hz, 2 Hz.

Solution

To solve this problem, it is useful to use a sheet of squared paper, where, for example, each box represents one Hertz.

The quartet (q) arises from coupling with equal coupling constants to three protons.

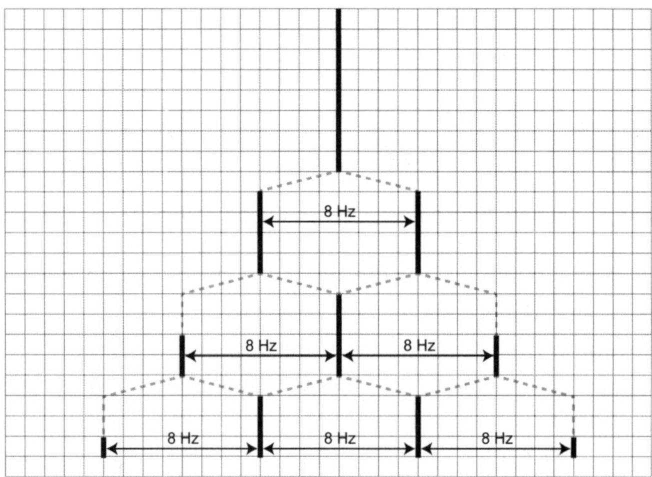

Please note that the intensity of each sub-peak is progressively divided by two. Thus, the intensities of the central final sub-peaks (three boxes high) are the result of the sum of the two halves of the previous sub-peaks.

A doublet of a doublet of doublets (ddd) also derives from the coupling to three protons; however, in this case, with different coupling constants. The result is a pattern of eight lines, all having equal intensities. Generally, it is convenient to start the splitting with the largest coupling constant.

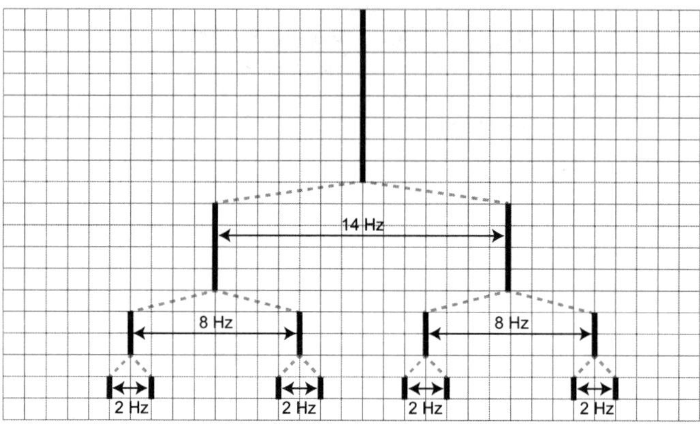

The smallest *J* value of a ddd is always the distance between the first and second line (or the last and next to last line). The middle *J* value is always the distance between the first and third line (or between the last and third from last line). The largest *J* value may be either the distance between the first and fifth line or, alternatively, it can be measured as the distance between the first and last lines minus the sum of the smallest and middle J values.

Problem 3.9

Predict the multiplicity of hydrogens *a*, *b* and *c*, of the following molecule (X, Y, and Z are generic substituents).

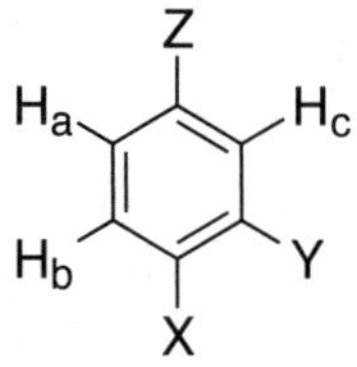

H_a, H_b, H_c, X, Y, Z

when:

1) the ortho coupling constants are about 8 Hz, and meta and para couplings are not observable;

2) the ortho coupling constants are about 8 Hz, and meta and para couplings are observable (2 Hz).

Solution

1) H_a and H_b are coupled to each other and both resonate as doublets. On the contrary, H_c does not couple to any proton and resonates as singlet.
2) In this case, H_a and H_b are still coupled to each other, but they are also coupled to H_c (by a smaller coupling constant). Therefore, their multiplicities will be doublets of doublets, and H_c will resonate as a triplet (with a 2 Hz coupling constant), because it is coupled to two protons by the same coupling constant.

Summary Problems

Solutions to these problems are available online. Find your Activation Code and instructions on the first page of this book.

Problem 3.10

In the Problem 2.11, you have already seen how it is possible to distinguish the following molecules by NMR.

$CH_3COCH_2CH_3$ (C=O) (I) $CH_3CH_2COCH_3$ (C=O) (II)

Can you now predict the multiplicity of the signals?

Problem 3.11

Can you predict the ^{1}H-NMR spectrum of the following molecules?

CH_3CH_2CHO (I) $CH_3CH_2OCH_3$ (II)

Pay particular attention to the description of the multiplicity of the signals.

Problem 3.12

What will be the signal multiplicity of the following molecules? Can you also predict the relative signal intensities?

CH_3-$CH(CH_3)CH_2Br$ (I) CH_3-$C(CH_3)_2CH_2Br$ (II)

Problem 3.13

The following compounds are characterized by the same molecular formula. Can they be distinguished by NMR? Explain why.

Br, Br, Br (I) Br, Br, Br (II) Br, Br, Br (III) Br, Br, Br (IV)

Problem 3.14

Predict how many signals you would expect to see in the proton NMR spectrum of the following molecule.

H_3C, CH_3, H_3C, H, H, H

Please describe carefully their expected chemical shifts (range), intensities, and multiplicities, describing the magnitude of the coupling constants involved.

Problem 3.15

Which of the following molecules is consistent with the NMR spectrum reported below?

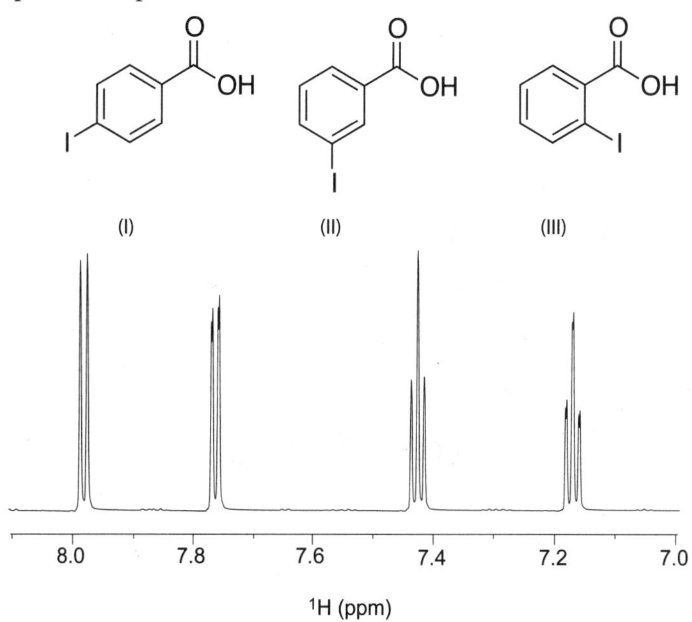

Problem 3.16

Describe how to use proton NMR spectra to distinguish between the following compounds.

Problem 3.17

Are you able to determine which molecule has generated the proton NMR spectrum reported below? Please explain.

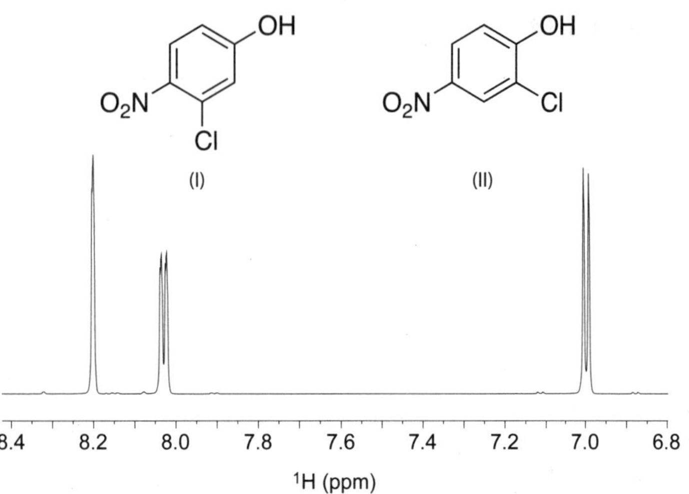

Problem 3.18

Which of the following molecules is consistent with the NMR spectrum reported below?

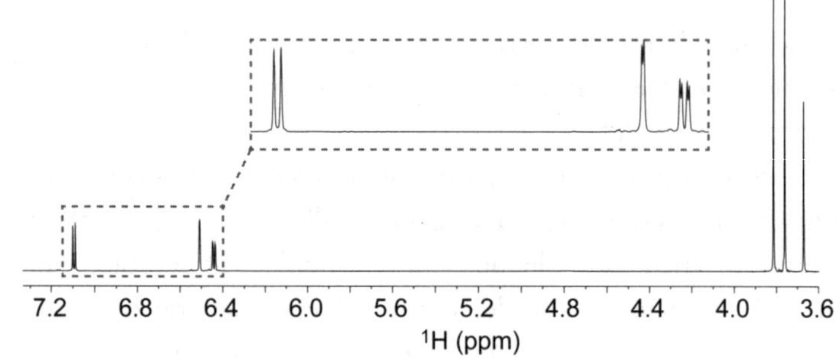

Problem 3.19

Determine the structure of the compound having molecular formula $C_8H_{18}O$ that provides the following NMR spectrum (numbers on the signals indicate the relative signal intensities).

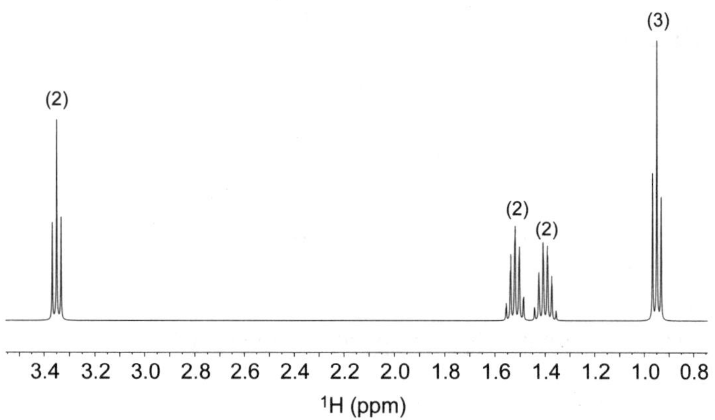

Problem 3.20

A compound of molecular formula $C_8H_8O_2$ gives a proton NMR spectrum as shown below. Assign each peak in the spectrum (numbers on the signals indicate the relative signal intensities).

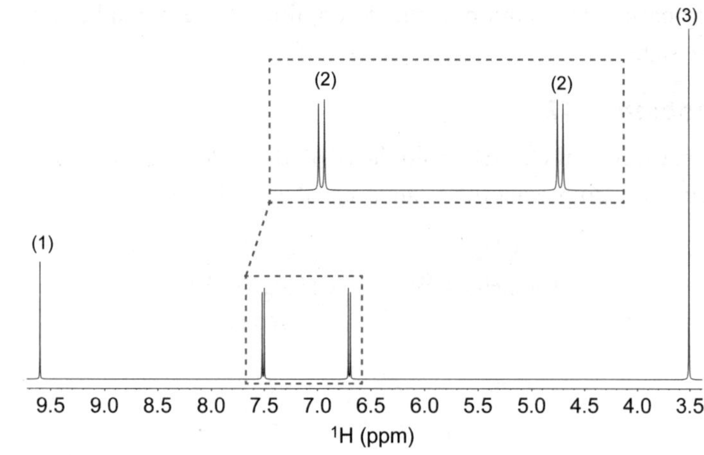

CHAPTER 4

Pople notation and classification of ^{1}H-NMR spectra

4.1 Definition of spin system

A group of protons that are coupled to each other forms a so-called *spin system*. For example, the five protons of chloroethane (CH_3-CH_2-Cl) form a spin system. Please note that it is not necessary for all nuclei within a spin system to be coupled to all the other nuclei. For example, in 1-chloropropane (CH_3-CH_2-CH_2-Cl), the methyl protons are not directly coupled to the protons of the methylene bearing the chlorine. However, the methyl is coupled to the central methylene, which, in turn, is coupled to the methylene bearing the chlorine. So, altogether these seven protons form a spin system. Organic molecules can also have more than one spin system. For example, 3-hexanone (Fig. 4.1) has two distinct spin systems, represented by the ethyl and propyl groups. In fact, the nuclei belonging to one system are not coupled to the nuclei of the other one. Ethyl acetate also has two spin systems (Fig. 4.1). The first consists of the five hydrogens of the ethyl group, and the second of the three hydrogens of the methyl. Although the latter group resonates as singlet, it is still considered a spin system, because the three hydrogens are actually coupled to each other, even if their coupling is not observable.

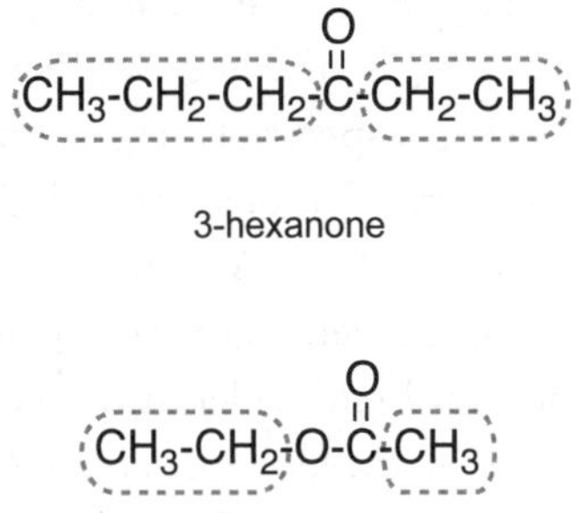

Figure 4.1 Spin systems (in dashed gray lines) of 3-hexanone and ethylacetate.

Problem 4.1

Identify the spin systems of the following molecules.

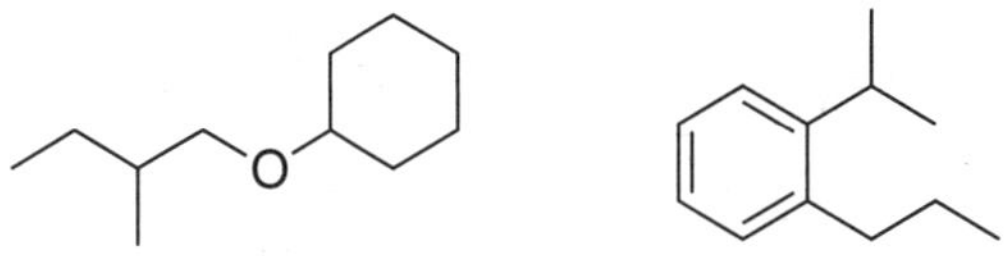

Solution

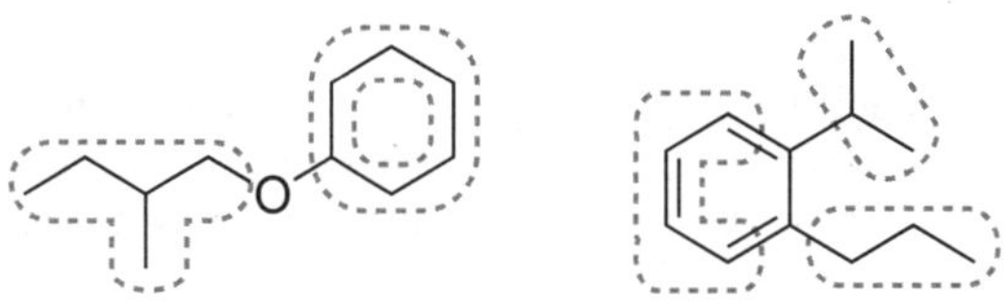

The spin systems are surrounded by gray dashed lines.

4.2 Pople notation for coupled nuclei

In order to properly describe a spin system, it is possible to use the so-called Pople notation (from the Nobel laureate John Pople), which assigns a capital letter of the Roman alphabet to each nucleus of the spin system. Generally, the letters used tend to be limited to A, B, C, M, N, X, and Y. The letters are assigned according to the following rationale: if the resonance frequency difference ($\Delta\nu$, expressed in Hz) between a pair of coupled nuclei is of the same order of magnitude as their coupling constant (J) (*i.e.*, the value of the ratio $\Delta\nu/J \leq 10$), adjacent letters of the alphabet are used (for example AB); on the other hand, if the frequency difference ($\Delta\nu$) is much greater than their coupling constant (that is, if the value of the ratio $\Delta\nu/J > 10$), distant letters in the alphabet (for example AX) are assigned to the nuclei. In order to assign the correct Pople notation to a spin system, it is necessary to know the operating magnetic field. In fact, the resonance frequency difference ($\Delta\nu$) of two signals is strictly related to the applied field (see Chapter 1), while the coupling constants are independent of it. For example, let's imagine we have two protons coupled by a coupling constant (J) of 5 Hz and resonating at 2.00 and 2.10 ppm, respectively. They are 0.1 ppm apart. Such a distance corresponds to 30 Hz ($\Delta\nu$) using a spectrometer operating at 300 MHz (see Chapter 1). In this case, the ratio $\Delta\nu/J$ is equal to 6, so the two protons must be assigned adjacent letters in the alphabet (AB). On the other hand, if the proton spectrum of the same compound is acquired using a spectrometer operating at 900 MHz, the difference between the two signals is 90 Hz and the ratio $\Delta\nu/J$ is equal to 18, so the nuclei are assigned distant letters in the alphabet (AX). For simplicity, Pople notation reported in most of the examples below refers to molecules observed at 300 MHz.

Examples of AB and AX spin systems are reported in Figure 4.2.

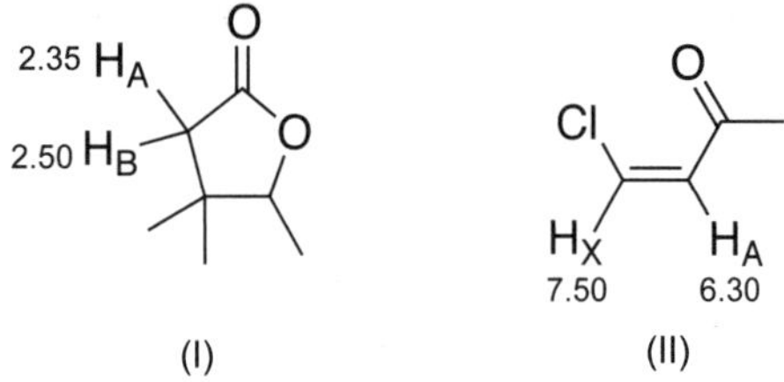

Figure 4.2 Examples of AB (I) and AX (II) spin systems. Pople notation is indicated by uppercase subscript letters added to each hydrogen. Chemical shifts are reported next to the hydrogens.

In the case of molecule I, the hydrogens H_A and H_B resonate at 2.35 and 2.50 ppm, so they are 0.15 ppm apart. If the spectrum is acquired at 300 MHz, this distance corresponds to 45 Hz. Since H_A and H_B are geminal, their coupling constant is about 14 Hz, so the ratio $\Delta\nu/J$ is about 3.2. Therefore, the correct Pople notation to describe this spin system involves the use of close letters of the alphabet (AB). In the case of molecule II, the two hydrogens experience two different chemical environments, resonating at 6.30 (H_A) and 7.50 (H_X) ppm. If the spectrum of molecule II is acquired at 300 MHz, they are 360 Hz apart. Since the two hydrogens have a *cis* relationship, they are coupled by constants of about 9 Hz, so their $\Delta\nu/J$ value is about 40, and hence distant letters in the alphabet must be used (AX) to describe the spin system. Please note that, generally, the first letters in the alphabet are assigned to nuclei generating signals at lower chemical shifts.

Pople notation can be also used to describe systems containing more than two nuclei. A couple of examples are described below. Specifically, the notation ABX describes two nuclei (H_A and H_B) whose relative $\Delta\nu/J$ ratios are smaller than 10, and a third nucleus (H_X) whose $\Delta\nu/J$ ratios with respect to H_A and H_B are larger than 10. This spin system is typical of many amino acids, like, for example, phenylalanine (compound I in Fig. 4.3). In this molecule the two methylene hydrogens on the β-carbon are diastereotopic and resonate at similar chemical shifts (3.11 and 3.27 ppm). They are 0.16 ppm apart, that corresponds to 48 Hz ($\Delta\nu$) at 300 MHz. Being geminal, they are coupled by a large coupling constant (~14 Hz), and thus the ratio $\Delta\nu/J$ is ~3.4. Hence, the two hydrogens must be assigned adjacent letters in the alphabet (AB). In contrast, the amino acidic hydrogen (Hα), resonating at 3.98 ppm, is 0.71 and 0.87 ppm apart from the other two signals. These differences correspond to 213 and 261 Hz (at 300 MHz), respectively. Since the Hα has vicinal coupling relationships with both β-hydrogens (please at this stage ignore the hydrogens bonded to heteroatoms), its coupling constants won't be larger than 14 Hz. Hence the ratio $\Delta\nu/J$ is larger than 15 and 19, respectively, and the letter X, which is distant from the letters A and B of the β-hydrogens, must be used. In summary, this is an ABX spin system.

Figure 4.3 Examples of ABX and AMX spin systems.

Another Pople notation used to describe three hydrogens in a spin system is, for example, AMX. This notation can be assigned to the olefinic hydrogens of molecule II in Figure 4.3. In fact, the hydrogens of the methylene resonate at 5.15 (H_A) and 5.65 (H_M) ppm, respectively, whereas the methine resonates at 6.65 ppm (H_X). The first two hydrogens are coupled by a very small coupling constant (3 Hz; see Chapter 3) and they are 150 Hz apart (at 300 MHz), therefore their $\Delta\nu/J$ value is about 50. The NMR signal of H_X is 450 and 300 Hz distant from the signals of H_A and H_M, respectively. H_X is coupled to H_A and H_M by coupling constants of about 9 and 15 Hz (see Chapter 3), respectively, so the values of the $\Delta\nu/J$ ratio are about 50 and 20, respectively. Hence, distant letters of the alphabet must be used (AMX) to describe the three nuclei in the system. Please note that the central letters of the alphabet are generally used only in the presence of more than two nuclei in the spin system and that there is no relationship between the distance of the letters in the alphabet and the value of $\Delta\nu/J$.

A set of equivalent nuclei is represented by the symbolism like A_n, or X_n, where "n" indicates the number of equivalent nuclei in the set. For example, A_2X indicates a spin system containing two equivalent nuclei (A) and a third non-equivalent nucleus (X). Some examples of spin systems involving sets of equivalent nuclei are shown in Figure 4.4.

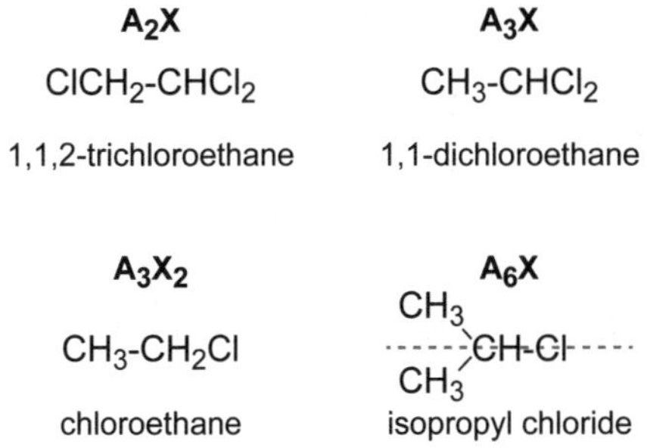

Figure 4.4 Examples of spin systems involving equivalent nuclei. The dashed line in isopropyl chloride indicates the element of symmetry of the molecule.

Problem 4.2

Use Pople notation to define the aromatic spin systems of the following molecules at 300 MHz. The chemical shifts (δ) are reported next to each proton in the figure.

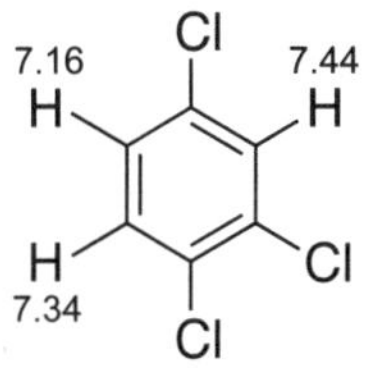

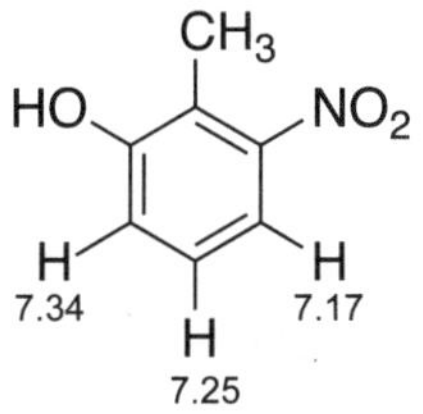

Solution

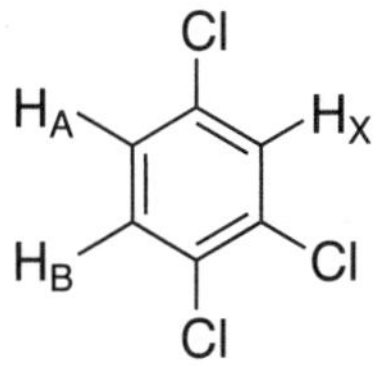

(I) The proton at 7.44 ppm is coupled only to the proton resonating at 7.16 ppm. Since the coupling involves four bonds (J_{meta}), the constant is small (~2 Hz). The operating field is 300 MHz, so these signals are 84 Hz apart (Δν). Therefore, the ratio $\Delta\nu/J$ is ~42, and distant letters in the alphabet must be used to describe this pair of nuclei. On the other hand, the protons at 7.16 and 7.34 ppm are 54 Hz apart and are coupled by a coupling constant of ~8 Hz (J_{ortho}). So, the value of the ratio $\Delta\nu/J$ is about 6.7 and close letters in the alphabet are used. Hence, this is an ABX spin system.

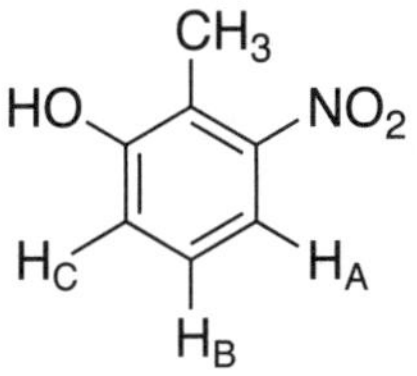

(II) For the didactic aims of this problem, please focus your attention only on the aromatic hydrogens. The proton at 7.17 ppm is coupled to the proton resonating at 7.25 ppm, which, in turn, is coupled to the proton at 7.34 ppm. Both coupling constants are about 8 Hz (J_{ortho}). The first and the second signals (7.17 and 7.25 ppm, respectively) are 24 Hz apart, while the second and the third (7.25 and 7.34 ppm, respectively) are 27 Hz apart. Thus, in both cases, the ratio $\Delta\nu/J$ is smaller than 10 and close letters in the alphabet must be used. Therefore, this spin system can be designated as ABC.

Problem 4.3

Use Pople notation to define the aromatic spin systems of the following molecules at 300 MHz. The chemical shifts (δ) are reported next to each proton in the figure.

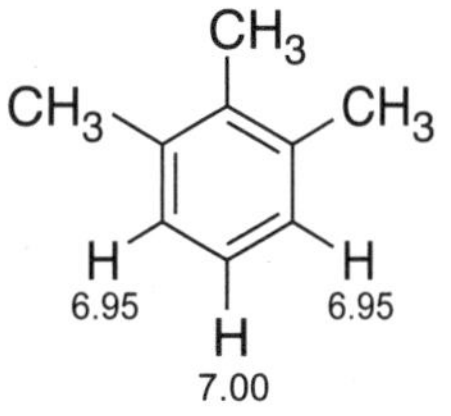

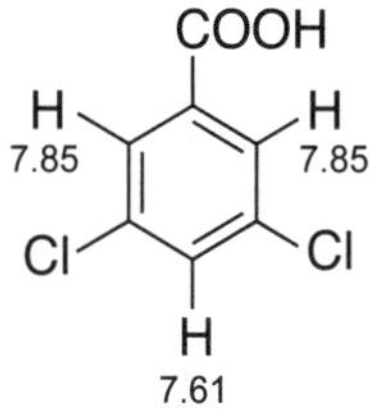

Solution

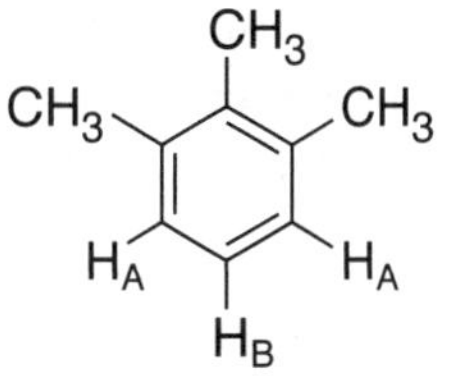

(I) This molecule is symmetric. The hydrogens at 6.95 ppm are coupled to the hydrogen at 7.00 ppm by coupling constants of about 8 Hz (J_{ortho}). At 300 MHz, the Δν is 15 Hz, so the ratio $\Delta\nu/J$ is 1.9 and close letters of the alphabet must be used. Thus, this is an A_2B system.

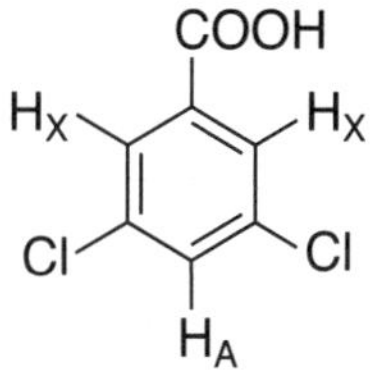

(II) Molecule II is also symmetric. The hydrogens at 7.85 ppm are coupled to the hydrogen at 7.61 ppm by coupling constants of about 3 Hz (J_{meta}). At 300 MHz, the Δν is 72 Hz and the ratio $\Delta\nu/J$ is 24, so distant letters of the alphabet must be used. Hence, this spin system can be described by the notation AX_2.

4.3 Spectral classification

NMR spectra of molecules with spin systems whose nuclei have high values of the ratio $\Delta\nu/J$ (> 10) are defined as *weakly coupled* spectra. Generally these spectra are easily interpretable according to the rationale reported in the previous chapters. These spectra are also defined as *first-order spectra*, while spectra containing only singlets are defined as *zero-order spectra* (no coupling). On the other hand, *strongly coupled* spin systems ($\Delta\nu/J \leq 10$) generate spectra defined as *second-order spectra.* These spectra are generally very difficult to interpret (except in very few cases), since very often they contain multiplets having more sub-peaks than one would expect for the corresponding first-order spectra. Second-order spectra are also characterized by an alteration in relative sub-peak intensities of the multiplets. For example, in an AB system, each doublet is distorted: the intensity of the sub-peak nearest the doublet to which it is coupled is enhanced, while the intensity of other sub-peak is decreased, forming a kind of roof (*roof effect*, or *roofing*) (Fig. 4.5).

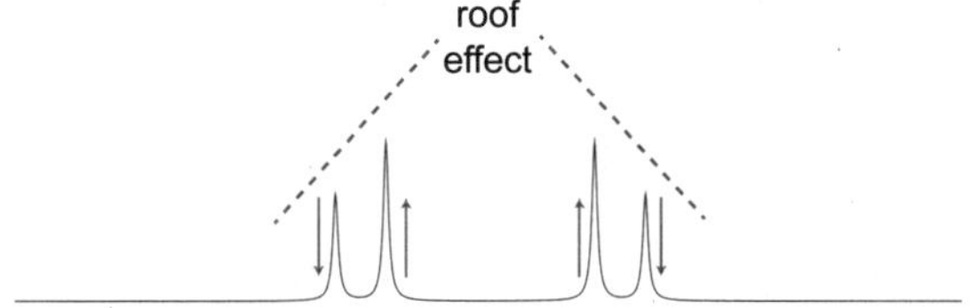

Figure 4.5 *Roof effect* in an AB system. Vertical gray arrows indicate increase (↑) and decrease (↓) of the sub-peak intensities of the doublets.

Please note that the intensification of the inner sub-peaks is achieved at the expenses of the others, that is, the integrated area under the entire signal remains constant. The closer the signals, the more intense the roof effect (Fig. 4.6).

The "roof effect" affects all type of multiplets. These distortions, which seemingly complicate the identification and interpretation of the multiplet, are instead quite informative and can aid the identification of related peaks. In particular, the multiplets that tilt to form a roof are most likely coupled to each other. For example, consider the spectrum of vinylbromide in Figure 4.7 (this spectrum was already reported in Fig. 3.18 of the previous chapter).

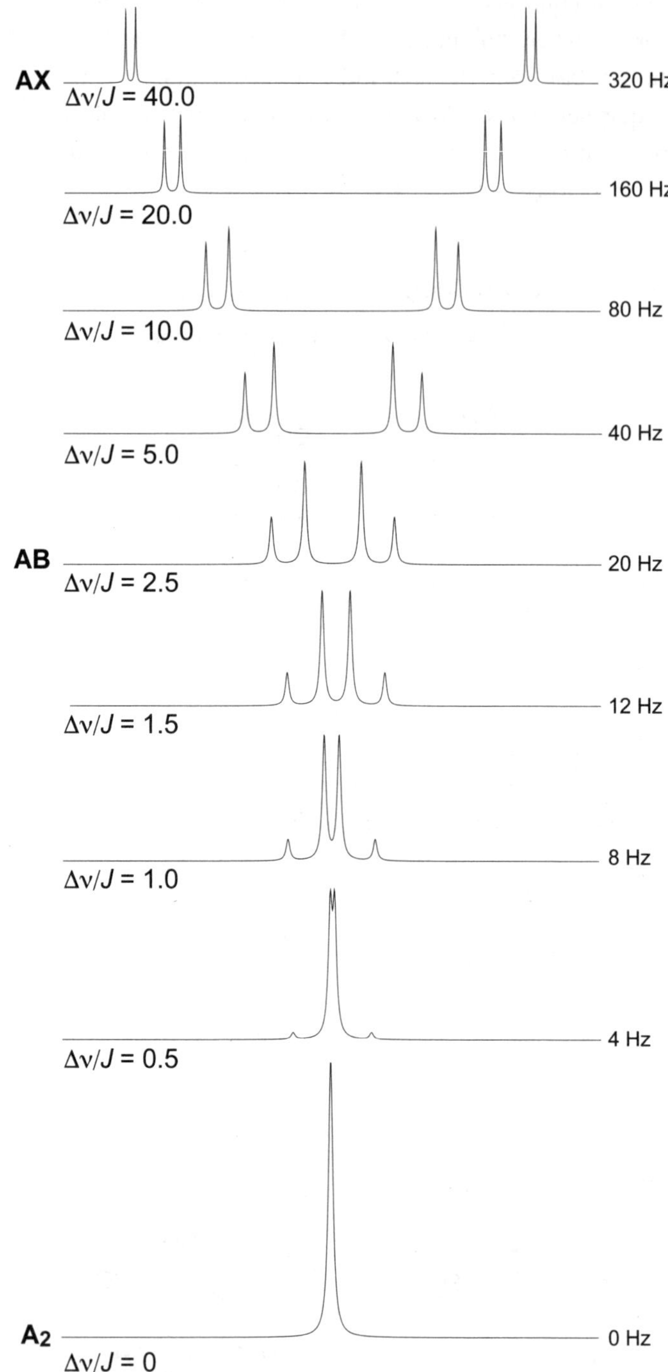

Figure 4.6 Dependence of the *roof effect* on the chemical shift frequency difference. The reported spectra are simulated at 400 MHz with a coupling constant of 8 Hz. Chemical shift frequency differences are reported on the right. If the two nuclei resonate at the same frequency ($\Delta\nu$=0), they are equivalent to each other and therefore they form an A_2 spin system. On the other hand, when resonating at frequency differences greater than 40 Hz, they can be considered to form an AX spin system (first-order spectrum), in spite of which a slight roof effect is still visible.

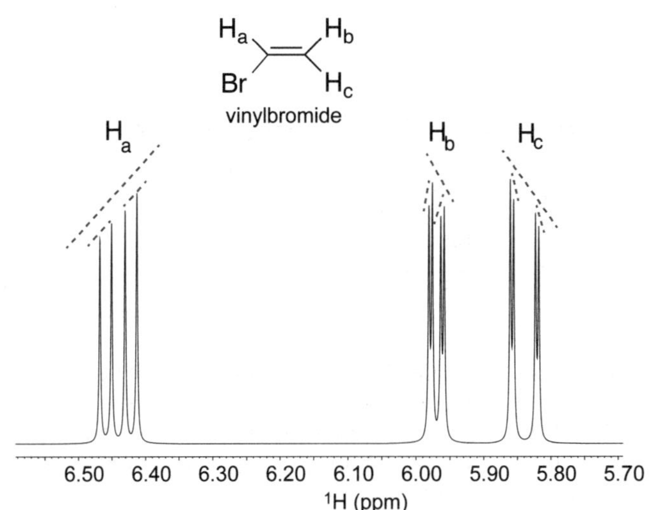

Figure 4.7 Roof effect affecting the spectrum of vinylbromide.

This spectrum is characterized by three distorted doublets of doublets. Let's first look at the signal generated by H_a (please note that the lowercase letters a, b and c are not related to Pople notation). The sub-peaks on the right side of the multiplet are more intense than those on the left side, therefore the signal is right-tilted. In contrast, the multiplet generated by H_c shows an opposite tilt, *i.e.,* to the left. Even not knowing the structure of the molecule generating the spectrum, it is possible to speculate that multiplets H_a and H_c could be coupled to each other since, together, they generate a roof effect. The case of the doublet of doublets of H_b is very interesting, in fact this signal is in the middle of the others (H_a and H_c), to which it is coupled. The splitting of the signal due to the coupling to H_c (see the small coupling constant) produces a right-tilt, while the splitting to H_a (see the large coupling constant) produces a left-tilt. Therefore, even in this case, the tilt can suggest that signal to which the other signal is coupled.

In general, the lower the value of the ratio $\Delta\nu/J$, the stronger will be the second-order effects. In the presence of strong second-order effects, the coupling constants and the chemical shifts of the multiplet could not be measured as described in the previous chapter, but these may have to be calculated. For example, the chemical shift of a doublet of an AB system cannot be taken as the midpoint between the two sub-peaks; rather, it is actually closer to the highest sub-peak. The greater the intensity difference between the two sub-peaks, the more shifted the point where the chemical shift must be measured (Fig. 4.8).

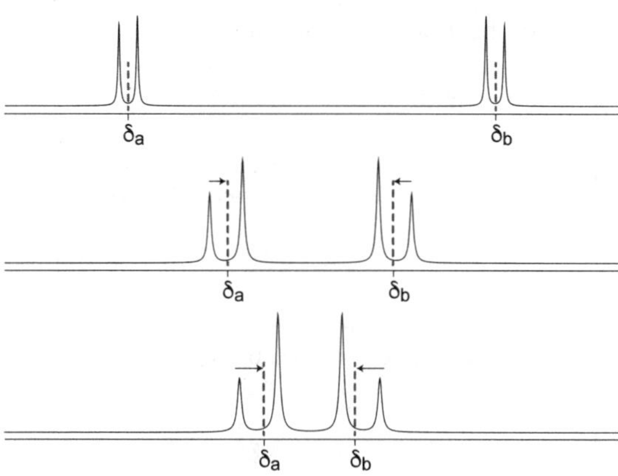

Figure 4.8 Comparison between chemical shift measurements in roofed doublets.

In general, in the case of an AB system, it is possible to refer to its chemical shifts in one of the following ways:

1. When $\Delta\nu/J_{AB} > 10$, it is possible to approximate the chemical shift of each doublet as half-way between the two sub-peaks.
2. When $\Delta\nu/J_{AB} < 10$, it is possible to refer directly to the whole AB "quartet", considering the very center of the entire AB system as the true chemical shift (Fig. 4.9).

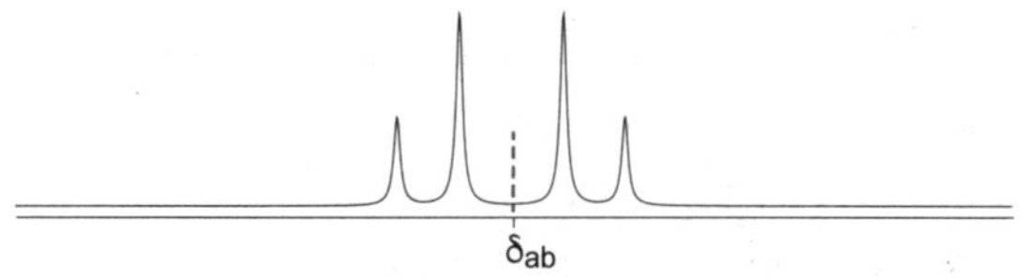

Figure 4.9 Chemical shift measurement in an AB system with $\Delta\nu/J_{AB} < 10$.

4.4 Chemical and magnetic equivalence of nuclei

As mentioned in Chapter 2, *chemically equivalent* nuclei are also generally chemical shift equivalent. However, they may not have the same NMR properties, that is, they may not be *magnetically equivalent*. Two protons are *magnetically equivalent* when they are *chemically equivalent* and they have the *same coupling constants* to any other nucleus in the molecule. In order to define whether given nuclei are magnetically equivalent or not, it is possible to follow the flowchart shown in Figure 4.10.

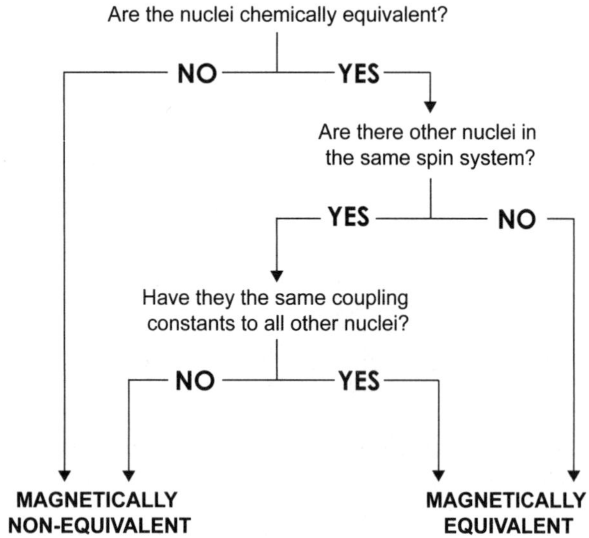

Figure 4.10 Rationale for the identification of chemically and magnetically equivalent nuclei.

Hence, magnetically equivalent protons are always chemically equivalent, and all chemically and magnetically equivalent nuclei have identical chemical shifts. Therefore, if the nuclei under consideration have different chemical shifts, then they are neither chemically nor magnetically equivalent. Please note that it could happen that two or more chemically non-equivalent nuclei accidentally share the same chemical shift. In spite of this, they remain chemically (and therefore magnetically) non-equivalent.

When a molecule possesses chemically equivalent hydrogens that are also magnetically equivalent, then the resulting NMR spectrum will be of *first-order*. On the contrary, if a set of chemically equivalent nuclei is magnetically non-equivalent, the resulting spectrum will be of *second-order*.

In summary, second-order spectra can be obtained when the studied molecule contains strongly coupled nuclei ($\Delta\nu/J \leq 10$) and/or contains magnetically non-equivalent nuclei.

Typical examples of magnetically non-equivalent protons generating second-order spectra can be found in both aromatic and non-aromatic molecules.

4.4.1 Aromatic systems

Disubstituted benzenes

1,2-homodisubstituted benzenes are characterized by a mirror plane of symmetry bisecting the 1,2 and 4,5 C-C bonds (Fig. 4.11, left panel). Therefore, the pairs of hydrogens H-3/H-6 and H-4/H-5 are chemically equivalent.

Figure 4.11 1,2-homodisubstituted benzene, where Z represents a generic substituent. Left panel: example of different coupling constants (^3J and ^4J) to the same nucleus (H-3) of chemically equivalent nuclei (H-4 and H-5). Right panel: Pople notation to describe chemically equivalent, but magnetically non-equivalent nuclei.

If you consider for example the protons H-4 and H-5, they are coupled to H-3 (and H-6) by ^3J and ^4J coupling constants, respectively, where ^3J > ^4J. Thus, although H-4 and H-5 are chemically equivalent, because they are coupled by different coupling constant to the same nucleus, they turn out to be magnetically non-equivalent. The same reasoning can be applied to H-3 and H-6, which are coupled to H-4 (and H-5) by different coupling constants. In order to represent nuclei that are chemically equivalent but magnetically non-equivalent, a different Pople notation must be used (Fig. 4.11, right panel). In particular, this spin system can be represented as AA'XX' (and not A_2X_2), where A and A' (as well as X and X') are chemically equivalent, but magnetically non-equivalent protons. This symbolism implies that $J_{AX} = J_{A'X'} \neq J_{AX'} = J_{A'X}$. Therefore, the spectrum of a 1,2-homodisubstituted benzene is of second-order. By way of example, the spectrum of the 1,2-dichlorobenzene is reported in Figure 4.12.

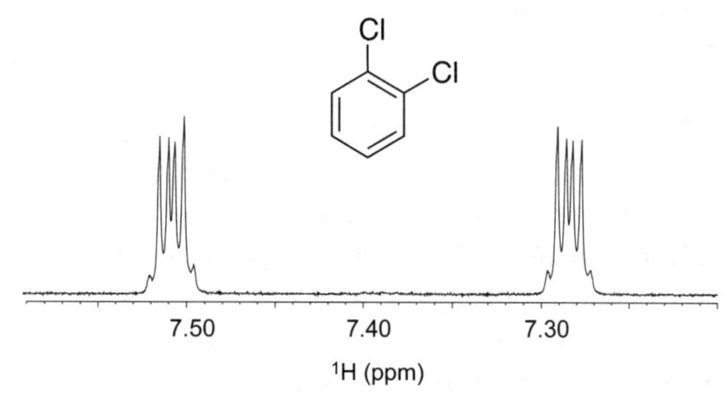

Figure 4.12 ^{1}H-NMR second-order spectrum of 1,2 dichlorobenzene (700 MHz, CD_3OD).

This is not characterized by the expected couple of doublets (not considering the long-range couplings), but by two multiplets that cannot be interpreted with the rules described in Chapter 3.

1,4-heterodisubstituted benzenes also contain magnetically non-equivalent protons. In these molecules, the pairs of hydrogens H-2/H-6 and H-3/H-5 are related by a mirror plane of symmetry passing through C-1 and C-4, and therefore they are chemically equivalent (Fig. 4.13).

Figure 4.13 1,4-heterodisubstituted benzene, where Z and Q represent generic substituents. Left panel: example of different coupling constants (^3J and ^5J) to the same nucleus (H-2) of chemically equivalent nuclei (H-3 and H-5). Right panel: Pople notation.

Nevertheless, for example, H-3 and H-5 are coupled to H-2 by two different coupling constants (^3J and ^5J, respectively). Thus, as also discussed above, the protons H-3 and H-5 (as well as H-2 and H-6) are magnetically non-equivalent, owing to their different coupling relationships. This spin system can be represented by the Pople notation AA'XX', and a second-order NMR spectrum is expected.

It is interesting to note that changing the symmetry of these systems leads to very different spectra. In fact, **1,2-heterodisubstituted** and **1,4-homodisubtituted benzenes** are characterized by all heterotopic and homotopic hydrogens, respectively (Fig. 4.14). In particular, a 1,2-heterodisubstituted benzene provides a first-order spectrum formed by two doublets and two triplets (not considering the long-range couplings), while the spectrum of a 1,4-homodisubstituted benzene, due to its symmetry, is of zero-order, having only a singlet generated by the four hydrogens.

Figure 4.14 Generic 1,2-heterodisubstituted and 1,4-homodisubstituted benzenes.

Please note also that the spectrum of a spin system predicted to be of second-order could appear as of first-order. Let's consider, for example, the aromatic spin system of tyrosine, which is classified as AA'XX'. In this case, the relatively large difference in resonance frequencies ($\Delta\nu$) of the signals and the almost null ^5J make the spectrum appears as a true A_2X_2 system (Fig. 4.15).

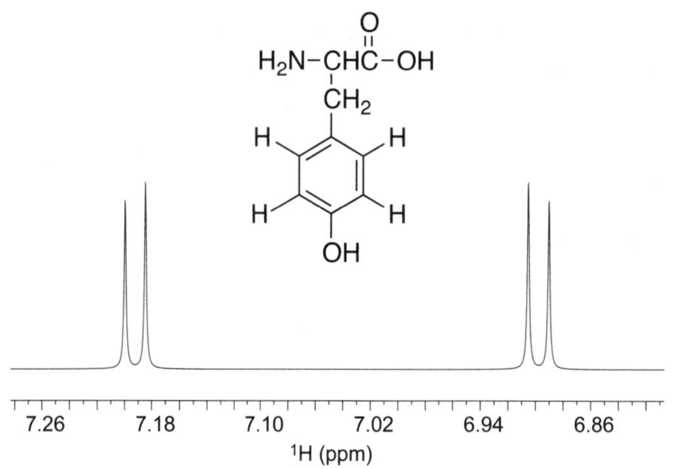

Figure 4.15 Expansion of the aromatic region of the simulated ^{1}H-NMR spectrum (400 MHz) of tyrosine. At this stage we will disregard the hydrogens bonded to the oxygen and nitrogen atoms.

Heteroaromatic and other aromatic systems

Other aromatic systems behave like 1,2-homodisubstituted (for example: diazine, pyrrole, furan, thiophene etc.) and 1,4-heterodisubstituted benzenes (for example: 4-substituted pyridine) (Fig. 4.16).

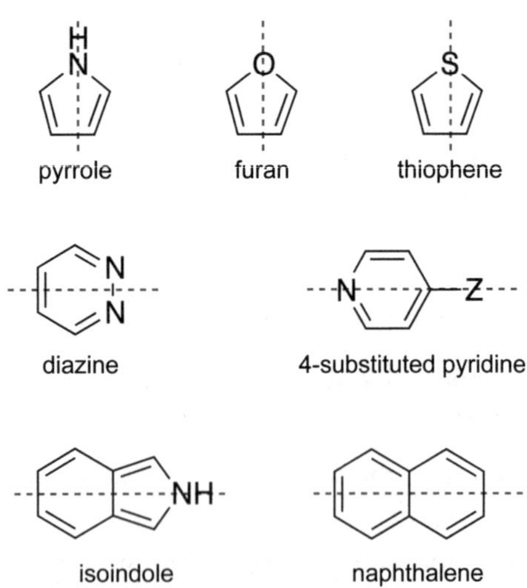

Figure 4.16 Some examples of symmetric aromatic molecules characterized by magnetic inequivalence. Z is a generic substituent.

All the cited molecules have chemically equivalent but magnetically non-equivalent pairs of protons. Furthermore, fused aromatic ring systems, such as naphthalenes and isoindoles, also show similar symmetry properties (Fig. 4.16). Please note that true A_2X_2 (or A_2B_2) systems are quite rare in aromatic molecules.

4.4.2 Aliphatic systems

A_2X_2 (or A_2B_2) systems are also difficult to find in aliphatic molecules. Generally they can be found in rigid structures, having chemically equivalent protons with identical coupling relationships. This is the case, for example, with cyclopropene (Fig. 4.17).

Figure 4.17 Cyclopropene is an example of a true A_2X_2 system.

This molecule contains a couple of elements of symmetry providing two pairs of chemically equivalent hydrogens (H_A and H_X). Hydrogens A are coupled to hydrogens X by the same coupling constants and hence they are also magnetically equivalent.

However, not all rigid aliphatic systems contain magnetically equivalent nuclei. For example, the 1,1-disubstituted cyclopropanes and 2,2-disubstituted-1,3-dioxolanes (Fig. 4.18) (and other similar structures) are classified as AA'BB', because of the presence of magnetically non-equivalent protons.

1,1- disubstituted cyclopropane

2,2-disubstituted-1,3-dioxolane

Figure 4.18 Examples of vicinal magnetically non-equivalent protons.

In fact, H_A, for example, couples to H_B and $H_{B'}$ by two different coupling constants (one geminal and another vicinal). Hence, this kind of molecule provides spectra of second-order.

Very different is the case of acyclic systems, which are characterized by conformational heterogeneity. Let's start to analyze the easy case of the ethyl group (**CH_3-CH_2-Q**, where Q is a generic substituent). Since the methyl group is always free to rotate with respect to the methylene, the coupling constants between the two groups are averaged and each hydrogen of the methylene is coupled to the hydrogens of the methyl by coupling constants of about 7 Hz. Therefore, the two sets of protons behave like they were magnetically equivalent. Hence, they can be classified either as A_3X_2 or A_3B_2. Generally, the notation A_3X_2 is the most appropriate, because the chemical shift difference between the methylene and the methyl groups is very often sufficiently large. In Figure 4.19, two spectra of molecules having ethyl groups are reported. Both molecules are symmetric and provide only a triplet and a quartet in the proton spectrum. In both cases the ratio $\Delta\nu/J$ is larger than 10, so the notation A_3X_2 must be used. However, while diethylether generates a pure first-order spectrum, the signals of tetraethylsilane are affected by a slight roof effect. This is due to the fact that the two signals in the spectrum are much closer and the ratio $\Delta\nu/J$ is only a little bit larger than 10.

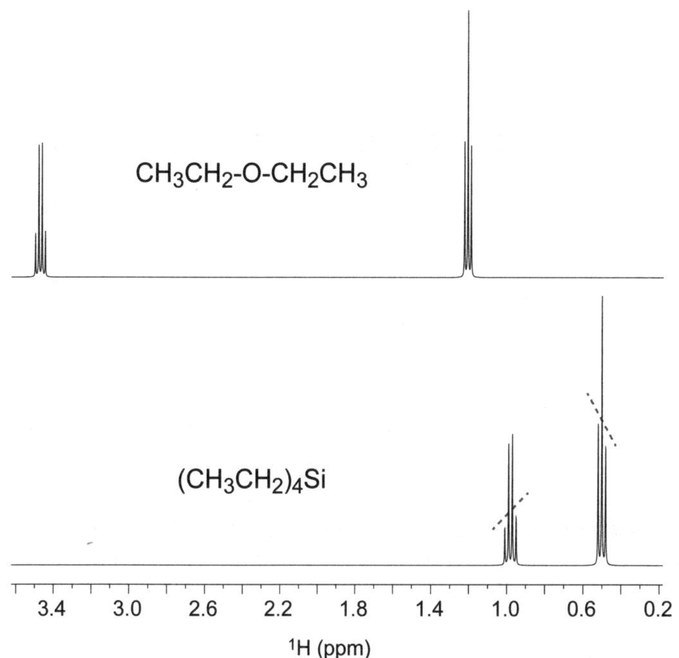

Figure 4.19 Simulated NMR spectrum (400 MHz) of diethyl ether (top) and tetraethylsilane (bottom). Gray dashed lines indicate roof effect distortion.

More attention must be paid to a spin system like **Z-CH$_2$-CH$_2$-Q** (where Z and Q are generic substituents). In this case the presence of the two substituents may hamper the free rotation of the methylenes. Therefore, some conformations could be energetically favored over others, not allowing the full averaging of the coupling constants. By way of example, some preferential conformations adopted by the generic structure Z-CH$_2$-CH$_2$-Q are depicted in Figure 4.20. In the first two conformations, H_A is coupled by two different coupling constants to H_X and $H_{X'}$. This means that, if those conformations prevail over the other, H_X and $H_{X'}$ turn out to be magnetically non-equivalent. Hence, the correct notation to describe such a system is AA'XX' (or AA'BB') and not A_2X_2 (or A_2B_2). Therefore, such moieties could provide second-order spectra.

Nevertheless, also for aliphatic systems, under certain conditions, an AA'XX' (or AA'BB') system can produce a spectrum with the appearance of a true A_2X_2 (or A_2B_2) system. The first condition is $\Delta\nu/J > 10$; the second is when the set of chemically equivalent nuclei behave like they are magnetically equivalent, that is, when the coupling constants are similar ($J_{AX} \approx J_{A'X}$ and $J_{AX} \approx J_{AX'}$), even if not necessarily identical. If one of these conditions does not occur, then the spectrum will appear as second-order.

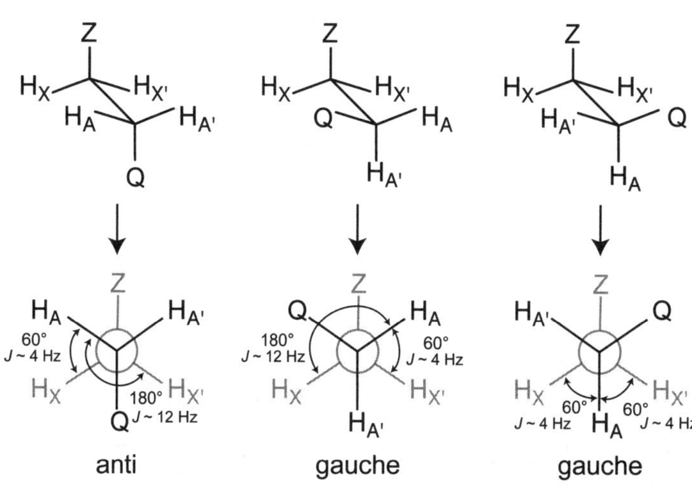

Figure 4.20 Newman projections of a generic molecule Z-CH$_2$-CH$_2$-Q. Three conformations (one anti and two gauche) are depicted indicating the different coupling patterns between H_A and H_X/$H_{X'}$. Z and Q are two generic substituents.

Let's start by analyzing a case where both conditions are met. For example, 2-(methylthio)ethanol is characterized by two methylenes bonded to groups of very different electronegativity. Therefore, they resonate at well-separated chemical shifts, and the ratio $\Delta\nu/J$ is large. Moreover, the two substituents, -SCH$_3$ and -OH, are not too bulky. This allows a free rotation of the methylenes and an almost complete averaging of the coupling constants ($J_{AX} \approx J_{A'X} \approx J_{AX'}$). *Actually, speaking about averaging of the coupling constant is not technically correct. However, for didactic reasons, this simplified explanation is sufficient.* Therefore, both conditions needed to observe a spectrum of the first-order are met and the spectrum appears as shown in Figure 4.21.

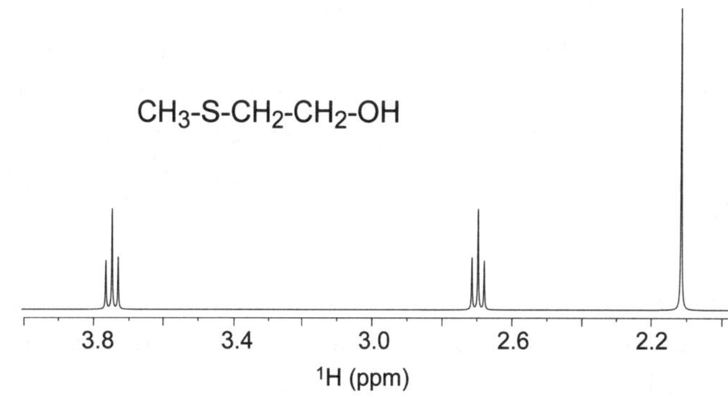

Figure 4.21 Simulated ^{1}H-NMR spectrum (400 MHz) of 2-(methylthio) ethanol. At this stage we will disregard the hydrogen bonded to the oxygen atom.

On the contrary, in 1-bromo-2-chloroethane (Fig. 4.22), both conditions needed to produce a first-order spectrum do not occur.

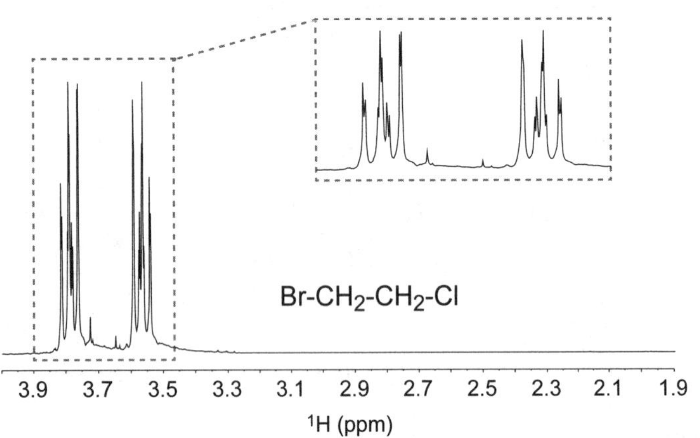

Figure 4.22 ^{1}H-NMR spectra of 1-bromo-2-chloroethane (300 MHz, $CDCl_3$).

The ratio $\Delta\nu/J$ is about 9 (at 300 MHz). Moreover, the bromine atom is surely a bulky substituent, being an element of the fourth period of the periodic table, while the chlorine atom, belonging to the third period, is moderately bulky. The close presence of these substituents hampers the free rotation of the methylenes and also the full averaging of coupling constants. In conclusion, this spin system has to be defined as AA'BB'. The spectrum appears as second-order, displaying two multiplets that are not easily interpretable (Fig. 4.22).

To summarize, even if only one of the two conditions necessary to observe a first-order spectrum does not occur, then the spectrum will appear as second-order. This is the case, for example, with 2-(trimethylsilyl)ethanol. The two methylene signals resonate at different chemical shifts, being bonded to two groups of different electronegativity [-Si(CH_3)$_3$ and -OH], so the ratio $\Delta\nu/J$ is very large. Nevertheless, the trimethylsilyl group [-Si(CH_3)$_3$] is very bulky, so it severely hampers the free rotation of the methylenes, inhibiting the process of averaging of the coupling constants. Hence, in this case, you can definitely expect that $J_{AX} \neq J_{AX'}$. So, although the chemical shift difference is large in comparison to the coupling constants, a second-order spectrum is observed (Fig 4.23). Therefore, the system has to be described as AA'XX'.

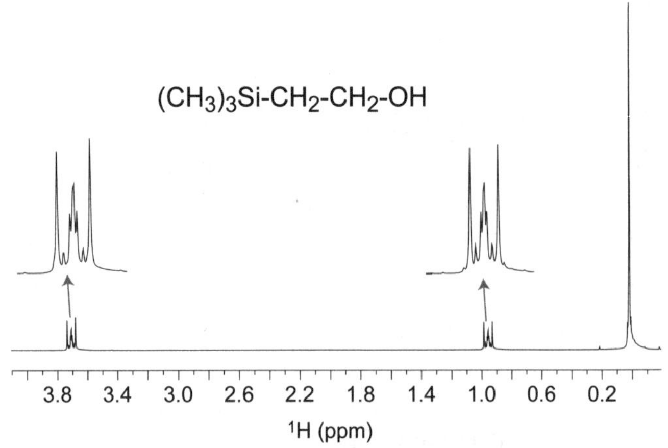

Figure 4.23 ^{1}H-NMR spectrum (300 MHz, $CDCl_3$) of 2-(trimethylsilyl) ethanol. At this stage we will disregard the hydrogen bonded to the oxygen atom.

The reasoning so far described can be extended to molecules having the generic structure **Z-CH$_2$-CH$_2$-CH$_2$-Q**. However, it should be noted that in this moiety, the first methylene and the third one are too far apart, so they do not couple to each other (Fig. 4.24).

$$Z-CH_2-CH_2-CH_2-Q \quad (J=0)$$

Figure 4.24 External methylenes do not couple to each other.

Therefore, in order to properly describe the spin system, it is necessary to evaluate the ratio $\Delta\nu/J$ for the following pairs of methylenes: Z-CH_2-CH_2- and -CH_2-CH_2-Q. In both cases, the vicinal coupling constants cannot be larger than 14 Hz. Thus, if the resonance frequency difference ($\Delta\nu$) in each pair of methylenes is smaller than 100-120 Hz, close letters of the alphabet must be used. On the contrary, if the value of $\Delta\nu$ is larger, distant letters of the alphabet are used. Furthermore, in order to determine if this moiety contains magnetically non-equivalent protons, the steric hindrance of the substituents (Z and Q) must be evaluated. By way of example, let's consider 1-bromo-3-chloropropane (Fig. 4.25) and sodium 3-(trimethylsilyl)-1-propanesulfonate (Fig. 4.26), whose $\Delta\nu$ values are sufficiently large to use the letters A, M, and X to describe their spin systems.

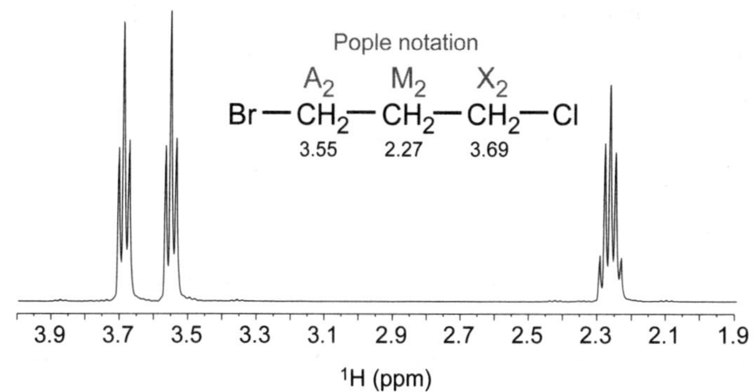

Figure 4.25 ^{1}H-NMR spectrum (300 MHz, $CDCl_3$) of 1-bromo-3-chloropropane. The chemical shifts values are reported next to each proton.

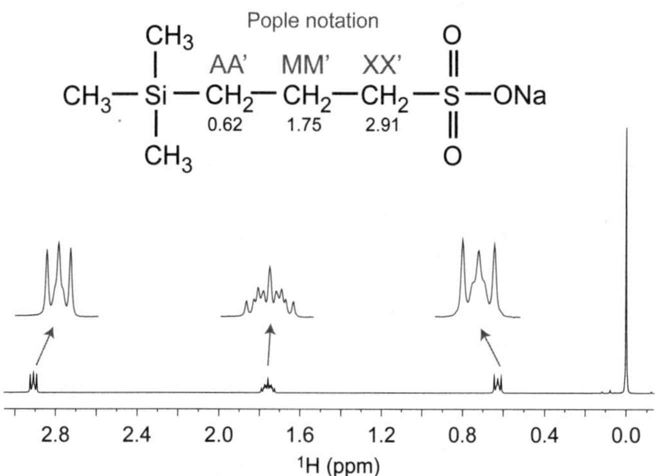

Figure 4.26 ^{1}H-NMR spectrum (300 MHz, D_2O) of sodium 3-(trimethylsilyl)-1-propanesulfonate. The chemical shifts values are reported next to each proton.

Interestingly, the two substituents of the 1-bromo-3-chloropropane (-Br and -Cl) are more distant than in 1-bromo-2-chloroethane (Fig 4.22). As a result, the methylenes are more free to rotate with respect to each other, and therefore the coupling constants are almost completely averaged. Hence, this spin system behaves more like an $A_2M_2X_2$ system. In fact, the spectrum of 1-bromo-3-chloropropane (Fig. 4.25) appears as first-order. On the contrary, the substituents in the 3-(trimethylsilyl)-1-propanesulfonate are much more bulky, and this prohibits the averaging of the coupling constants. Hence, this molecule does not meet the requirements to resemble an $A_2M_2X_2$ system, but it should be described as AA'MM'XX' and provides a second-order spectrum (Fig. 4.26).

Please note that some conformations can be favored over the others (generating nuclei that are magnetically non-equivalent) not only because of bulky substituents, but because of the formation of intramolecular hydrogen bonds. This is the case of 2-hydroxyethyl acrylate (Fig. 4.27).

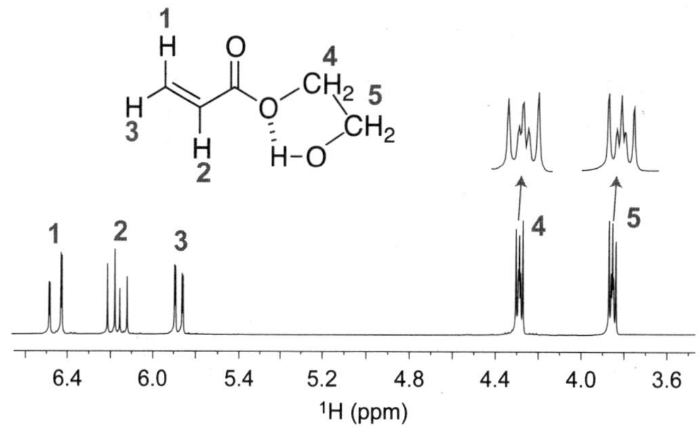

Figure 4.27 ^{1}H-NMR spectrum (300 MHz, $CDCl_3$) of 2-hydroxyethyl acrylate.

In this molecule, the -CH_2-CH_2-OH moiety is assumed to be free to rotate, which would be able to average the coupling constants of the methylenes. However, the hydroxyl group forms a hydrogen bond with the oxygen of the ester, adopting a preferential conformation that forms a pseudo five-membered ring. As a result, the proton spectrum of 2-hydroxyethyl acrylate is of the second-order (Fig. 4.27).

As mentioned at the beginning of this chapter, if the second-order effects are due to a small value of the ratio $\Delta\nu/J$, these can be reduced or cancelled using a high field spectrometer, because they increase the value of $\Delta\nu$. On the contrary, if the second-order effects are due to the presence of magnetically non-equivalent protons, the use of spectrometer operating at higher fields is not of any help, because the coupling constants are independent of the magnetic field.

Although this book does not properly explain how to interpret second-order effects, it provides useful information to alert you about the possible presence of anomalies in the proton spectra. Please note that molecules displaying second-order effects can be studied using homo- and heteronuclear 2D NMR experiments, which will be described in Chapter 7.

Problem 4.4

Use Pople notation to define the spin systems of the following molecules (assume a spectrometer frequency of 300 MHz):

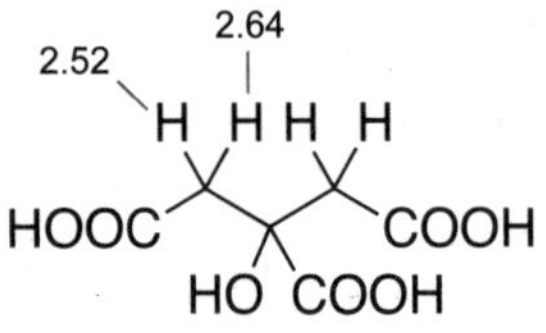

(I) 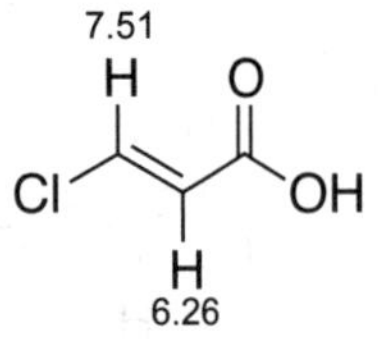 (II)

Solution

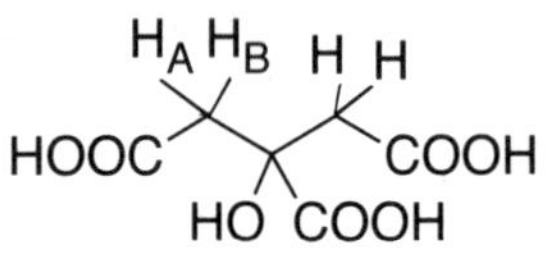

(I) 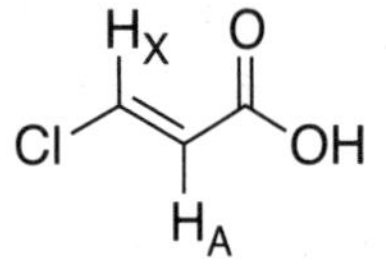 (II)

(I) This molecule is symmetric, so the two methylenes are chemically equivalent. They are 0.12 ppm apart, which corresponds to 36 Hz at 300 MHz. The two hydrogens are coupled by a geminal coupling constant (~14 Hz), their $\Delta\nu/J$ ratio is about 2.6, and close letters of the alphabet have to be used to describe the spin system (AB).

(II) In this problem you can ignore the exchangeable proton of the carboxylic acid group. The two olefinic hydrogens resonate at 1.25 ppm apart, which corresponds to 375 Hz at 300 MHz. They are coupled by a *trans* coupling constant of about 15 Hz, so the $\Delta\nu/J$ ratio is about 25 and distant letters of the alphabet must be used (AX).

Problem 4.5

Are the hydrogens H_a and H_b in 3-methylcyclopropene chemically equivalent? Are they also magnetically equivalent?

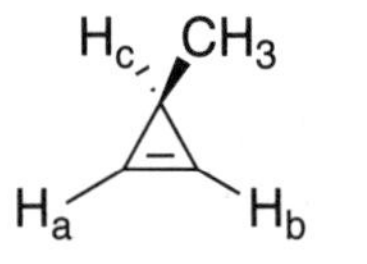

Solution

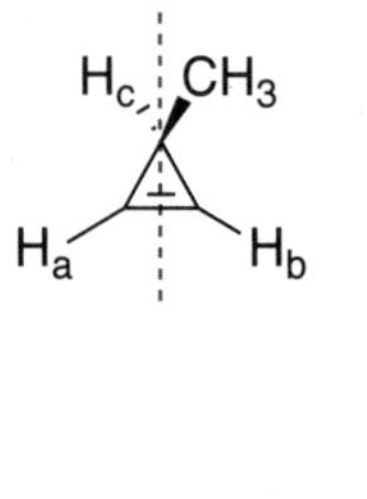

The protons H_a and H_b are chemically equivalent, because they are related by a plane of symmetry bisecting the double bond (please note that H_c and the methyl lie on the plane of symmetry). Then, they have the same coupling constant to H_c, so they are also magnetically equivalent. This is an AX_2 system.

Problem 4.6

Consider the pairs of protons H_a and H_b in 1,1-diethoxyetane, shown below.

1. Do they have they homo-, enantio-, or diastereotopic relationships?
2. Are they chemically or magnetically equivalent?
3. How many signals do they generate in a proton NMR spectrum?

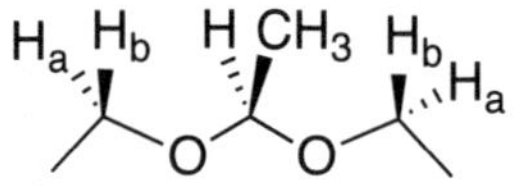

Solution

1) The molecule is symmetric and does not possess stereocenters; in fact, the central carbon possesses two identical substituents. However, if you imagine replacing hydrogen H_a or H_b with a "dummy" atom X (as in Chapter 2), you will see that two stereocenters will be automatically created (see asterisks in the figure below), and the central carbon will no longer have two identical substituents.

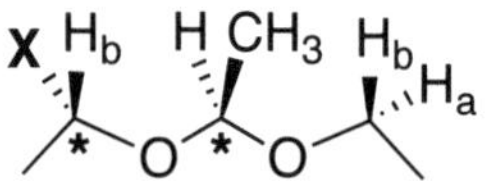

Therefore, in this molecule H_a and H_b are diastereotopic.

2) Being diastereotopic, H_a and H_b are chemically and magnetically non-equivalent.

3) H_a and H_b resonate as two distinct signals.

Problem 4.7

The proton NMR spectrum ($CDCl_3$, 300 MHz) of the 1,1-diethoxyethane (whose structure is shown in the previous problem) is reported below (numbers in parentheses indicate relative signal intensities).

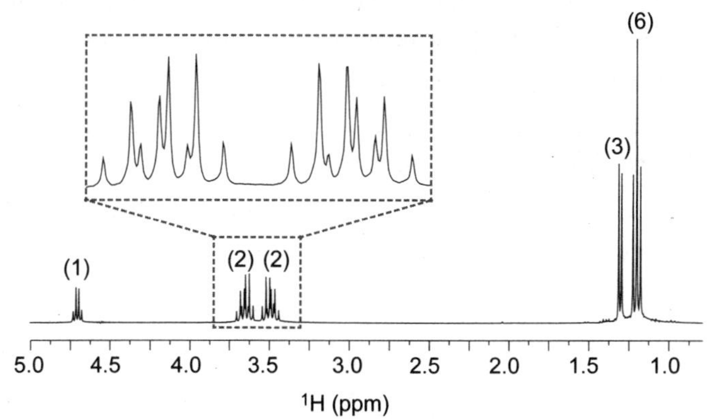

1. Assign each signal to the pertinent nucleus.
2. Explain the multiplicity of the signals expanded in the inset.

Solution

1) There are three independent spin systems in the molecule. Two of them (the ethyl groups) are symmetric. A third one is instead formed by the central methine bonded to the methyl. This last system is very easy to be assigned, because the hydrogen of the methine will resonate as a quartet at high frequency (δ 4.73 ppm), since it is bonded to two oxygen atoms. Vice versa, the methyl resonates at lower frequency as a doublet integrating for three hydrogens.
As for the ethyl groups, the methyl hydrogens are coupled by the same coupling constant to the two hydrogens of the methylene, so they resonate as a triplet (δ 1.3 ppm). Since the molecule is symmetric, the intensity of this signal will integrate for six hydrogens. On the other hand, as reported in the previous problem, the two hydrogens of the methylene are diastereotopic, and hence they generate two signals in the spectrum (each integrating for two nuclei). The methylene bonded to an oxygen atom will resonate between 3 and 4 ppm. Therefore, the correct assignment of this molecule is the following:

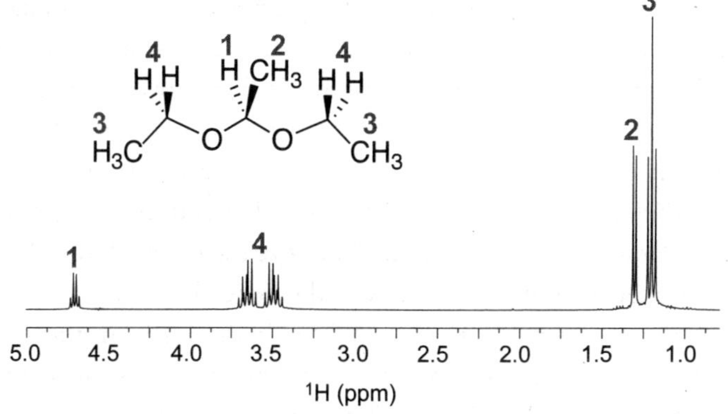

Please note that, for the methylene signals around 3.6 ppm, we cannot perform the so called *stereospecific assignment*, that is, we cannot determine which proton of the methylene (H_a or H_b) has generated which signal.
2) In order to understand the multiplicity of these signals, it is convenient first to think about the coupling between the two hydrogens of the methylene, ignoring, for the moment, that to the methyl. In this hypothesis, we should expect a classic AB system, where the two doublets are distorted by a marked roof effect.

Then, you can consider the coupling to the methyl, which splits each sub-peak of the AB system into quartets. Therefore, each multiplet appears as a doublet of quartets affected by a roof effect:

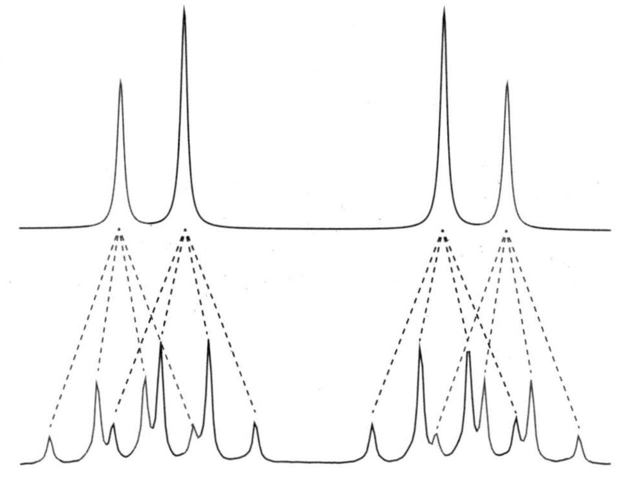

Problem 4.8

Assign each signal of the spectrum reported below ($CDCl_3$, 300 MHz) to the pertinent proton of the 3,3-dimethyl-1-butene (numbers in parentheses indicate relative signal intensities).

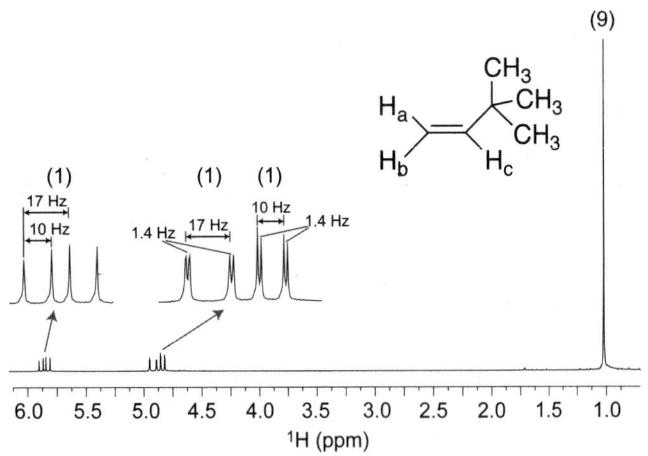

Solution

The three methyls bonded to the quaternary carbon have an identical averaged environment and therefore, altogether, they resonate as a singlet (δ 1.02 ppm). The hydrogens of the olefinic moiety (H_a, H_b and H_c), instead, are not chemically equivalent, and therefore they produce three signals in the NMR spectrum of equal relative intensities, resonating between 4.5 and 6.0 ppm. One of these resonates at 5.86 ppm. Its appearance clearly recalls the shape of a doublet of doublets (distorted by a roof effect), characterized by two large coupling constants of 10 and 17 Hz. The identification of the other two signals between 4.7 and 5.0 ppm could be more tricky for a beginner. They are very close to each other, but still separated. The first four sub-peaks on the left form a doublet of doublets characterized by large and small coupling constants (17 and 1.4 Hz, respectively). The remaining four sub-peaks on the right form another doublet of doublets, characterized by coupling constants of 10 and 1.4 Hz. Considering only their chemical shifts, it is not possible to properly assign these three signals to the respective hydrogens. Nevertheless, in this case, the analysis of the coupling constants can be of help. You should remember that the coupling pattern for an olefinic system is peculiar:

$^2J_{ab(gem)}$ = 0-3 Hz
$^3J_{bc(cis)}$ = 6-11 Hz
$^3J_{ac(trans)}$ = 12-19 Hz

Therefore, to be H_a (or H_b), a given hydrogen must have a small coupling constant to the geminal proton H_b (or H_a). This means that the doublet of doublets at 5.86 ppm, not having small coupling, can be neither H_a nor H_b. Hence, that signal must be generated by H_c. H_a must have the largest coupling constant measured for these multiplets, owing to its *trans* relationship with H_c.

Thus, the signal at 4.89 ppm (having the largest coupling constant of 17 Hz) can be assigned unambiguously to H_a. By exclusion, the signal at 4.85 ppm can be assigned to H_b.

Problem 4.9

Can you determine the Pople notation for *cis*-1,2-dichlorocyclopropane? Assume a spectrometer frequency of 300 MHz.

Solution

In order to determine the Pople notation for this spin system, you have to determine first how many chemically and magnetically equivalent protons there are in the molecule. The two hydrogens of the methines are chemically and magnetically equivalent; in fact, they are related by a plane of symmetry and they possess the same coupling patterns to all others protons in the molecule. Therefore, they generate a single signal.

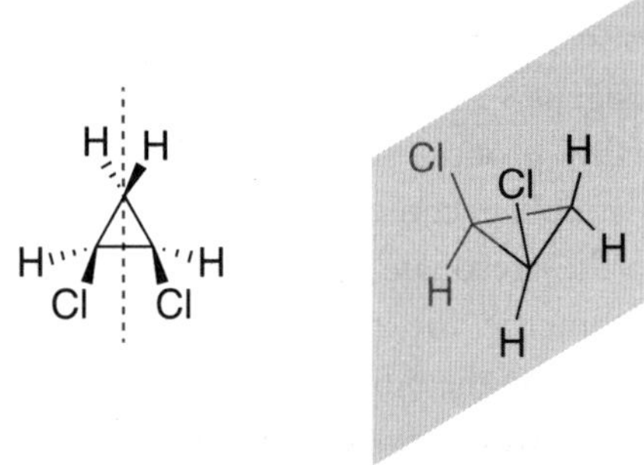

The two hydrogens of the methylene are instead chemically non-equivalent. They experience two different chemical environments arising from the two different sides of the plane in which the three-membered cycle lies. In this problem, the chemical shifts of the protons are not reported. Nevertheless, it is reasonable to suppose that the protons of the methines resonates around 3-4.5 ppm (see the scheme in the inside front cover flap), while the protons of the methylene should resonate around 2 ppm. These latter protons are expected to have a $\Delta\nu/J$ ratio < 10, since they resonate very close to each other and, being geminal, have a large coupling constant. Therefore the letters A and B can be assigned to these protons. On the other hand, the protons of the methines resonate at least 300 Hz distant from those of the methylene. This means that they have to be assigned to a distant letter in the alphabet. Therefore, this spin system can be described as ABX_2.

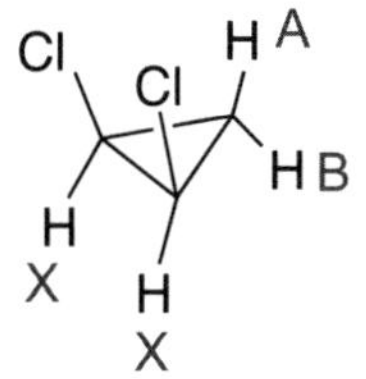

Problem 4.10

Can you determine the Pople notation for *trans*-1,2-dichlorocyclopropane? Assume a spectrometer frequency of 300 MHz.

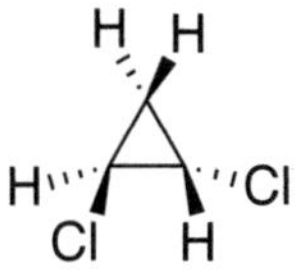

Solution

As in the previous problem, it is important to start by determining how many chemically and magnetically equivalent protons there are in the molecule. Interestingly, this molecule is characterized by a C_2 axis of symmetry, therefore both hydrogens of the two methines and those of the methylene are chemically equivalent (homotopic) to each other.

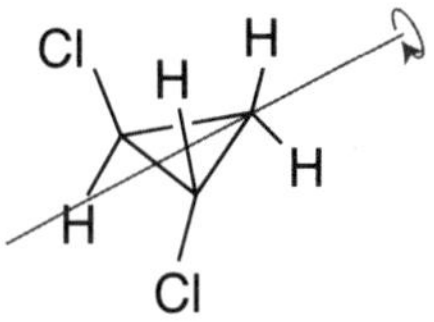

Nevertheless, these pairs of hydrogens are magnetically non-equivalent and they have different coupling patterns. Let's consider for example one methine. Its hydrogen describes different dihedral angles (and thus has different coupling constants) with the two chemically equivalent hydrogens of the methylene.

~0° dihedral angle ~120°

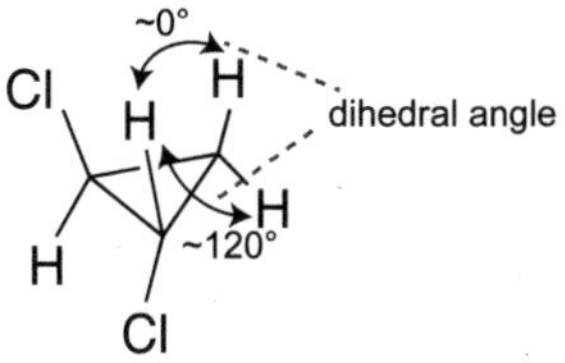

The same is also true for the other methine, and therefore the two hydrogens of the methines (as well as the two hydrogens of the methylene) are chemically equivalent, but magnetically non-equivalent. Then, the protons of the methylene are expected to resonate at least 300 Hz distant from those of the methines.

Thus, the correct Pople notation for this system is AA'XX'.

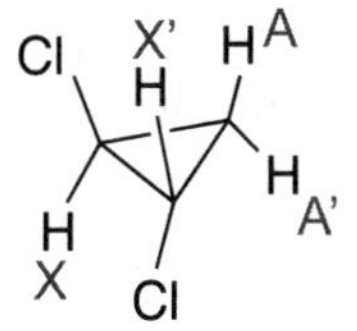

Summary Problems

Solutions to these problems are available online. Find your Activation Code and instructions on the first page of this book.

Problem 4.11

Identify the proton spin systems for the following molecules.

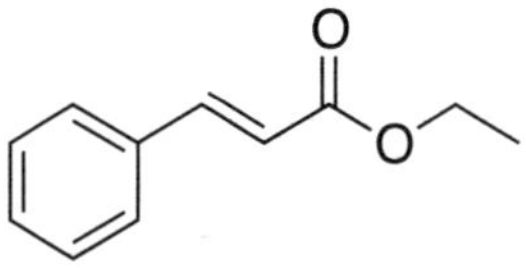

(I) 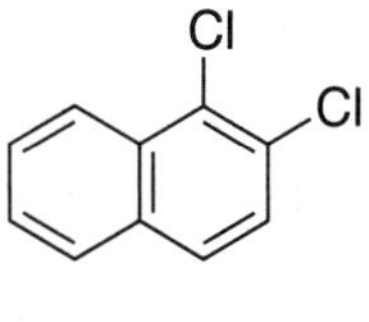 (II)

Problem 4.12

Identify the proton spin systems for the following molecules.

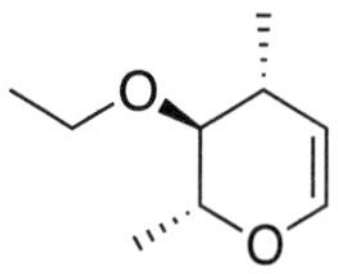

(I) 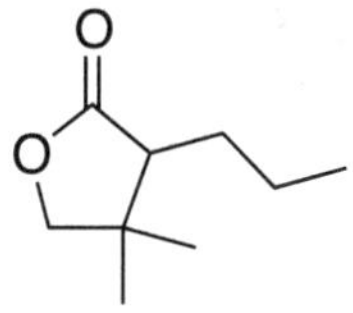 (II)

Problem 4.13

Use Pople notation to define the aromatic spin systems of the following molecules at 300 MHz. The chemical shifts (δ) are reported next to each proton in the figure.

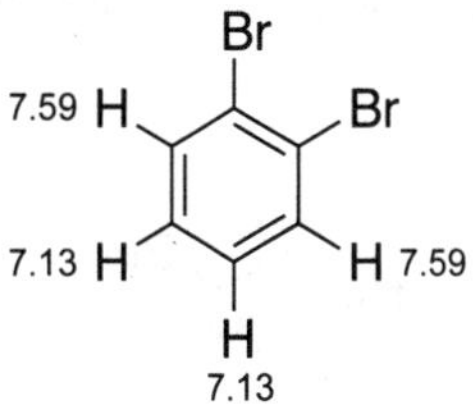

(I) 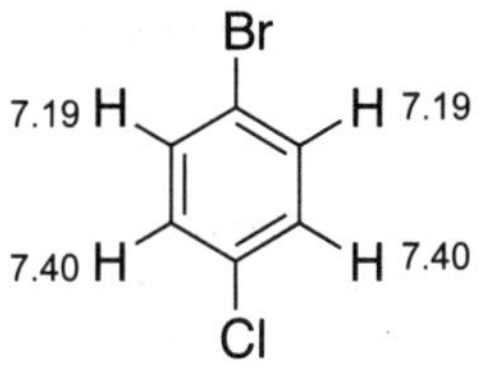 (II)

Problem 4.14

Use Pople notation to define the spin systems of the following molecules at 100 MHz. The chemical shifts of the protons are reported in the figure.

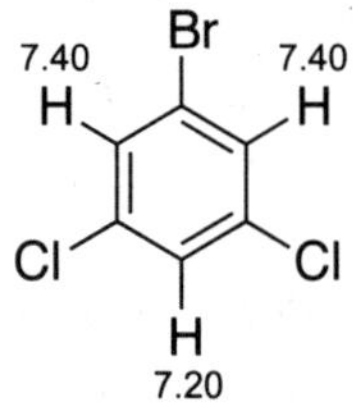

(I)

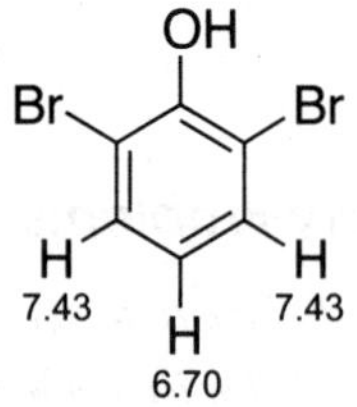

(II)

Problem 4.15

Predict if the following molecules produce second-order spectra at 300 MHz.

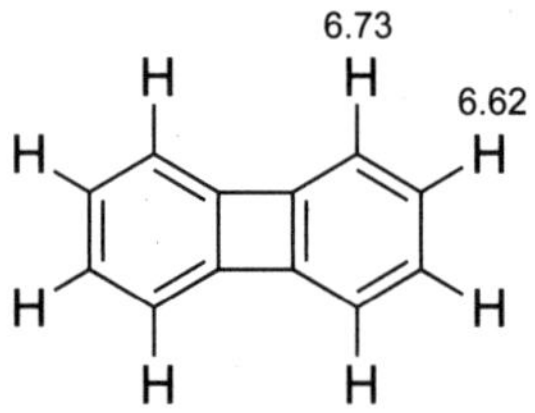

(I)

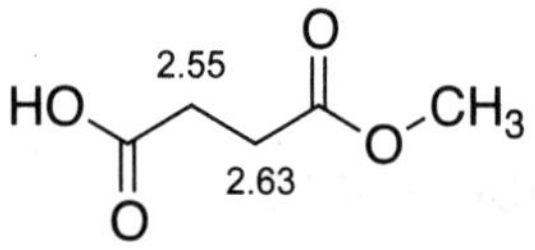

(II)

Problem 4.16

Predict if the following molecules produce second-order spectra at 300 MHz.

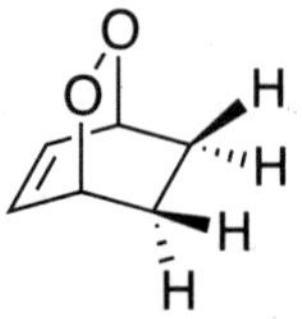

(I)

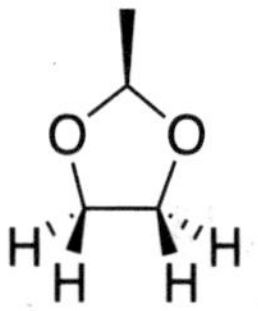

(II)

Problem 4.17

Predict if the following molecules produce second-order spectra at 300 MHz.

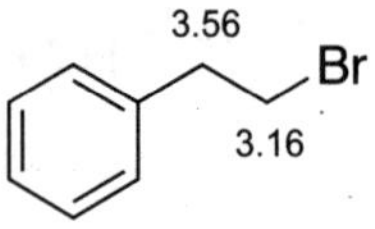

(I)

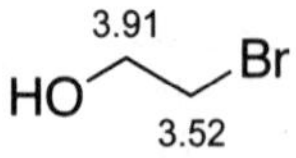

(II)

CHAPTER 5

^{13}C-NMR spectroscopy

5.1 Introduction

The most abundant isotope of carbon is ^{12}C, which accounts for 98.9% of the carbon atoms. ^{12}C does not have a nuclear magnetic moment (I=0), and therefore it is inactive from an NMR point of view. However, the ^{13}C isotope, which accounts for most of the remaining carbon atoms in nature (1.1%), has a magnetic moment just like protons (I=1/2). Therefore, ^{13}C nuclei behave similarly to protons when subjected to a magnetic field, and most of what you have learned about ^{1}H-NMR spectroscopy (shielding, coupling, etc.) is also applicable to ^{13}C-NMR.

It should be noted that the low natural abundance of ^{13}C and its low resonance frequency (1/4 compared to ^{1}H) lead this nucleus to be very insensitive to NMR (1/6000 as intense as ^{1}H), therefore it takes a great deal longer to acquire ^{13}C spectra in comparison to ^{1}H spectra. As a result, the signal to noise ratio is very unfavorable. In spite of this, ^{13}C-NMR spectroscopy plays a fundamental role in determining the structure of unknown organic molecules, in which carbon is a central element.

In the previous chapters, four main aspects of ^{1}H spectra have been analyzed: signal intensity, chemical shift, coupling (signal multiplicity), and chemical/magnetic equivalence of nuclei. These important topics will be discussed in this chapter for ^{13}C-NMR spectroscopy.

5.2 Coupling

As with ^{1}H, ^{13}C couples with all magnetically active nuclei. However, the ^{13}C-^{13}C coupling is basically negligible, because the natural abundance of ^{13}C is about 1.1% and the chances of having two ^{13}C atoms next to each other are statistically very small.

On the other hand, in organic molecules, carbon atoms are surrounded by hydrogens, so you should expect scalar couplings to those nuclei, through one, two, or three bonds. For didactic reasons, let's imagine first that carbon atoms couple only to the hydrogens that are directly bonded to them (through one bond). In this case, the carbon atoms of CH, CH_2, and CH_3 would resonate as a doublet, a triplet, and a quartet, respectively (the rationale to explain the multiplicity is the same as for ^{1}H NMR, see Chapter 3), while a quaternary carbon atom would resonate as a singlet. The coupling constants of these multiplets are huge in comparison to those observed in proton spectra (about 125 Hz for sp^3, about 160 Hz for sp^2, and about 150 Hz for sp hybridized carbon atoms). Therefore, these multiplets are "large" and the probability that they overlap to each other is high (especially for those signals having similar chemical shift). This often prevents their identification and the interpretation of their splitting patterns. All this is much more complicated considering the additional couplings of the carbon nuclei to geminal and vicinal protons (through two or three bonds), that further split the sub-peaks by coupling constants up to 10-15 Hz. For these reasons, ^{13}C spectra are always acquired with full proton decoupling (see a theory book), leading to spectra containing only singlets!

Please note that although all ^{1}H–^{13}C couplings are decoupled, couplings between carbon and other heteroatoms (if present), such as fluorine and phosphorus, will not be decoupled, so the splitting of ^{13}C signals will be still observable.

Singlets are inherently more intense than multiplets and therefore they are more visible, especially when there is a low signal-to-noise ratio. Moreover, proton decoupling further increases the ^{13}C signal intensity thanks to the nuclear Overhauser effect (NOE) generated by the decoupling of the protons directly bonded to the carbons (see a theory book for explanation). This explains why quaternary carbon atoms, which lack protons, appear less intense than those attached to protons (see next section).

5.3 Signal intensity

Different from ^{1}H-NMR, carbon signal intensities are not strictly related to the number of resonating carbons, and therefore the signals are normally not integrated. The reasons for this are related to the relaxation times of the carbon nuclei (very long) and to the NOE generated by the proton decoupling. Generally, methine, methylene, and methyl signals appear with more intensity than quaternary carbon atoms.

Problem 5.1

In the following ^{13}C spectrum there are three quaternary carbon signals. Can you mark them by asterisks?

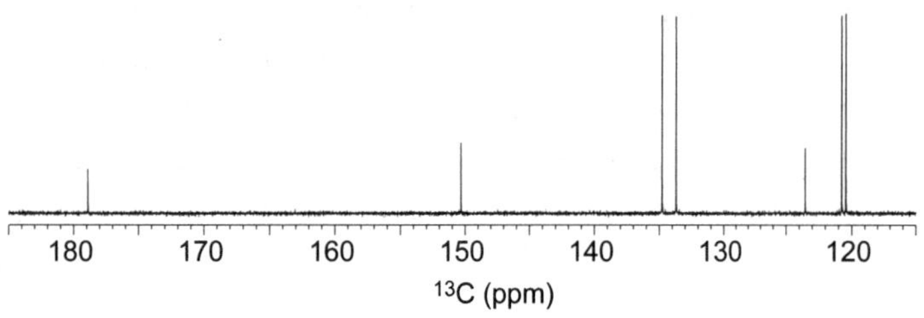

Solution

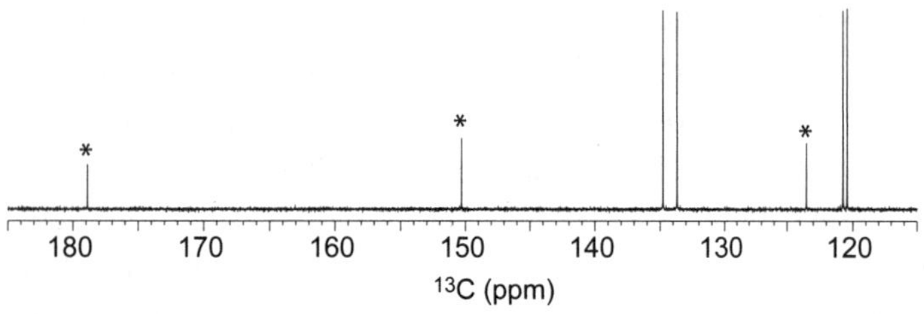

As said above, the quaternary carbon signals can be easily identified since they are usually the shortest signals in the spectrum. Please note that this is not always true. There are many cases where carbon atoms bearing hydrogens could also provide signals having low intensities. Therefore, the identification of quaternary carbon atoms looking exclusively at the intensities of signals may not be reliable, and DEPT (see Section 5.6) or 2D experiments (see Chapter 7) are necessary to unambiguously identify the quaternary carbon atoms.

Problem 5.2

In the following ^{13}C spectrum there are four quaternary carbon signals. Can you guess which ones they are?

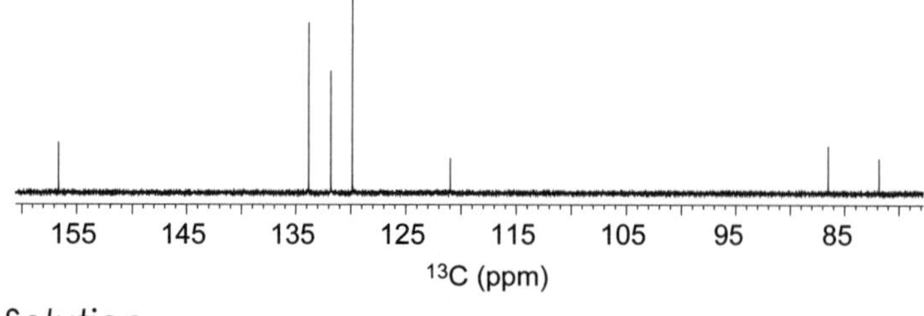

Solution

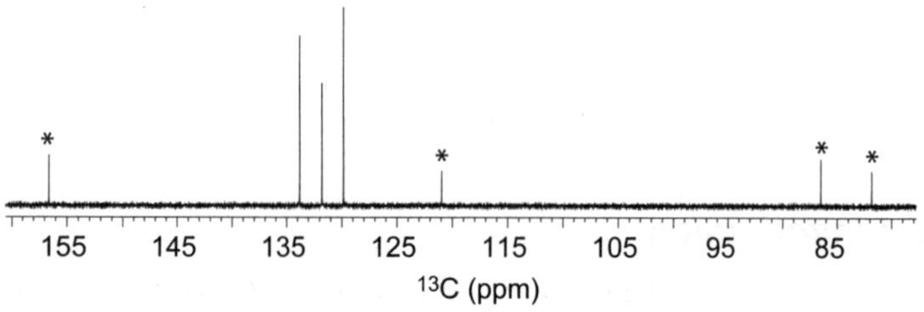

Also in this case, please note that you can only suppose which are the quaternary carbon signals.

5.4 Chemical shift

Chemical shifts of ^{13}C nuclei are spread out over a much wider range than protons; in fact, the majority of ^{13}C chemical shifts fall in the range of 0-220 ppm. Fore this reason, the signals are much more dispersed and signal overlap is not so frequent as in proton spectroscopy. Just like in ^{1}H-NMR, the compound used as a reference in ^{13}C-NMR experiments, to define the 0 ppm point, is tetramethylsilane (TMS) (at least for organic solvents). In this case the signal from the four equivalent carbon atoms serves as reference.

Generally, ^{13}C chemical shifts are reported with one digit after the decimal point (while ^{1}H chemical shift are reported two digits).

The chemical shift of a ^{13}C nucleus is influenced by essentially the same factors that influence proton chemical shifts. Therefore, highly electronegative atoms/groups and diamagnetic anisotropy effects tend to shift the signals to higher resonance frequencies. By way of example, the following table demonstrates the effects of the electronegativity of the halogens on the carbon chemical shift of methyl halides (CH_3-X):

Table 5.1 Relationship between carbon chemical shift and electronegativity of substituents in methyl halides.

X	CH_3 (ppm)	Electronegativity (Pauling units)
-F	71.6	3.98
-Cl	49.9	3.16
-Br	25.6	2.96
-I	9.6	2.66

As in proton NMR, the chemical shifts of the carbon atoms get progressively smaller as they get farther away from the deshielding substituent.

In general, ^{13}C NMR signals are distributed according to the hybridization of the carbon atoms, with sp^3 atoms at lowest frequencies (0-70 ppm); sp atoms of the acetylene type at 70-100 ppm, sp^2 atoms at 100-150 ppm, and sp^2 atoms of carbonyl groups at 160-220 ppm. A more detailed picture of the distribution of carbon chemical shifts can be found in the scheme reported in the inside back cover flap of this book. Obviously, unusual combinations of substituents can lead to shifts outside the given ranges.

Please note that, different from ^1H-NMR where aromatic and olefinic protons can be easily discriminated, sp^2 carbon atoms of the double bond type cannot be easily distinguished from those of the aromatic type.

Problem 5.3

Please indicate if the signals of the following carbon spectrum belong to sp^3, sp^2, sp^2 of carbonyl group (sp^2/C=O), or sp carbon atoms.

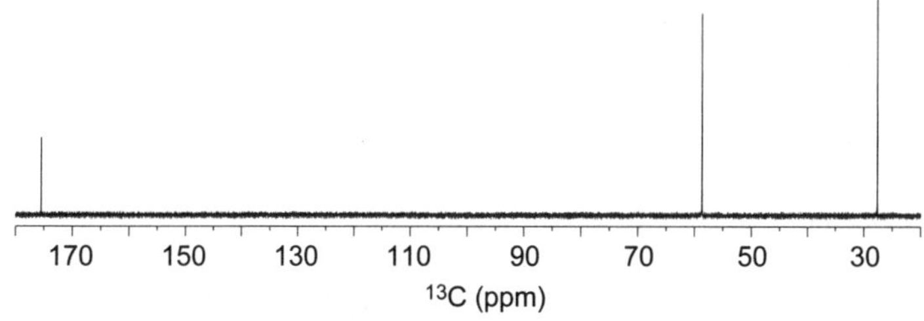

Solution

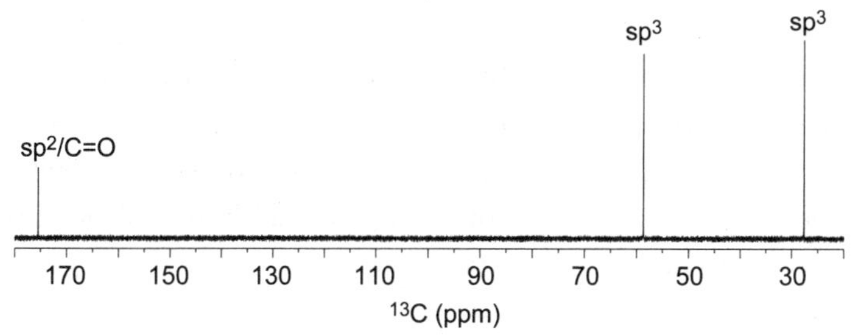

Problem 5.4

Please indicate if the signals of the following carbon spectrum belong to sp^3, sp^2, sp^2 of carbonyl group (sp^2/C=O), or sp carbon atoms.

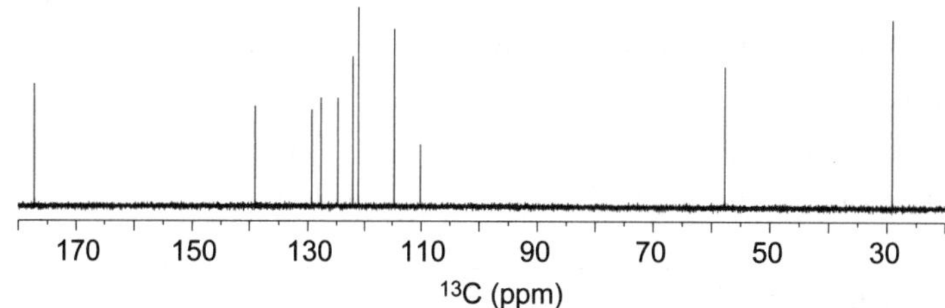

Solution

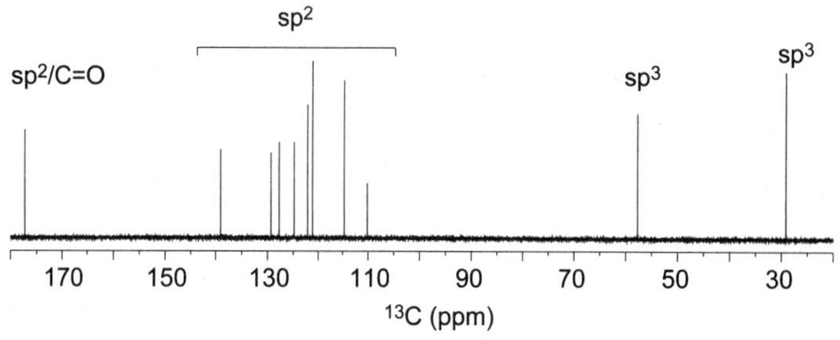

Problem 5.5

Please indicate if the signals of the following carbon spectrum belong to sp^3, sp^2, sp^2 of carbonyl group (sp^2/C=O), or sp carbon atoms.

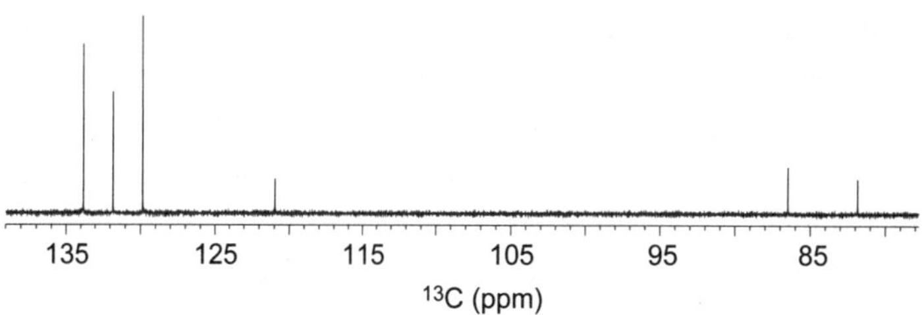

Solution

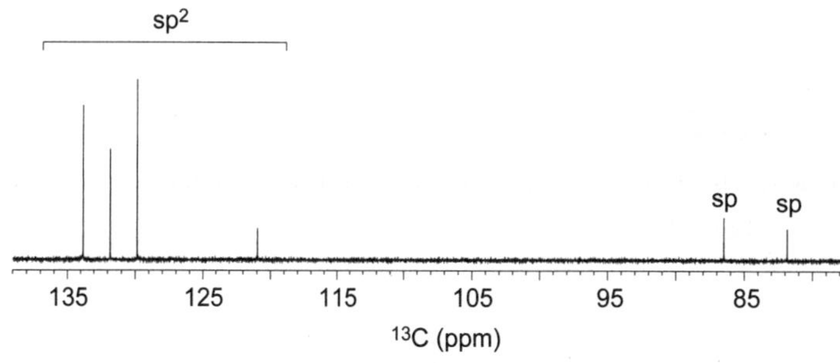

5.5 Equivalence of nuclei

Since the scalar couplings (at least those to protons) are not present in decoupled ^{13}C spectra, there is no reason to speak about magnetic equivalence. You can simply refer to the rules of symmetry described in ^{1}H-NMR spectroscopy to determine the number of chemically equivalent carbon atoms present in a molecule.

A number of problems discussed in Chapter 2 will now be discussed from the ^{13}C-NMR point of view.

Problem 5.6

How many carbon signals would you expect in the ^{13}C spectrum of the following molecules?

CH_4 (I) CH_3CH_3 (II) $CH_3CH_2CH_3$ (III) $CH_3CH_2CH_2CH_3$ (IV)

Solution

(I) There is only one carbon, so only one signal is expected.
(II) Ethane is a symmetric molecule (where the plane of symmetry bisects the C-C bond), so the two carbon atoms are chemically equivalent and resonate at the same chemical shift, generating only one signal.
(III) Propane is also symmetric, where the plane of symmetry passes through the carbon of the methylene (CH_2). Thus, the two external methyls are equivalent to each other and generate a single signal. The central carbon will generate another signal. Please note that the signal of the two methyls will be of greater intensity than the methylene's signal. Obviously, for the reasons explained above, you should not expect exactly double the intensity.
(IV) Butane is also a symmetric molecule. The plane of symmetry is placed between the two central methylenes. Thus, the two methylenes will generate a signal and the two methyls will generate another signal. So two signals will be present in the carbon spectrum.

Problem 5.7

How many signals would you expect in the ^{13}C spectrum of the following molecules?

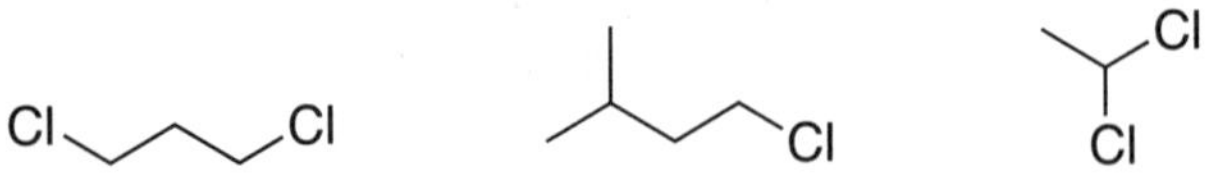

Cl Cl (I) Cl (II) Cl Cl (III)

Solution

(I) 1,3-Dichloropropane is symmetric and the plane of symmetry passes through the central methylene.

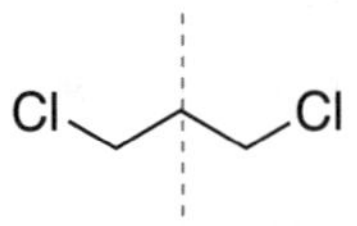

Cl Cl

Therefore, two signals are expected.
(II) 1-Chloro-3-methylbutane possesses an element of symmetry passing through the methine of the isopropyl moiety. Therefore, the two carbon atoms of the methyls will generate only one signal. The methine (bearing the two methyls) will generate another signal and the remaining two methylenes will generate two other signals.
(III) Two signals are expected in the spectrum.

Problem 5.8

How many carbon signals would you expect for the following molecules?

$H_3C{-}{\equiv}{-}CH_3$ (I) $H_3C{-}{\equiv}{-}H$ (II)

H_3C H H H (III) H_3C H H_3C H (IV)

Solution

Molecules I and IV are symmetric.

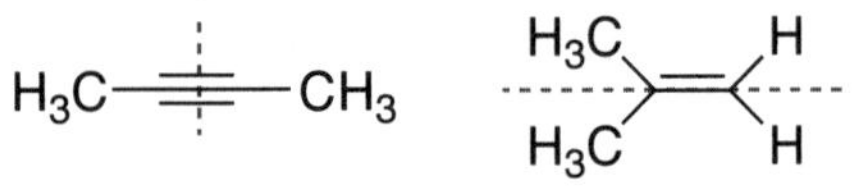

$H_3C{-}{\equiv}{-}CH_3$ (I) H_3C H H_3C H (IV)

Therefore, in the case of I, two signals are expected. As for IV, although there are four carbon atoms in the molecule, only three signals will be present in the spectrum: two from the two sp^2 carbon atoms and one arising from the two symmetric methyls. Molecules II and III are not symmetric and both will provide three signals in the carbon spectrum.

Problem 5.9

For each of the following molecules, indicate the number of signals expected in the carbon spectrum. Mark each type of equivalent carbon with a letter.

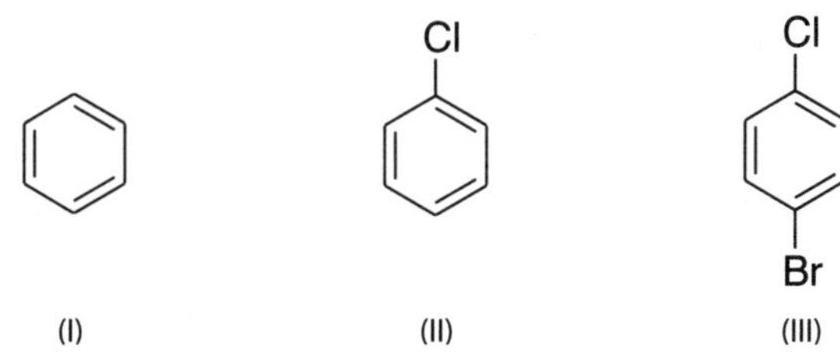

Solution

All three molecules possess elements of symmetry.

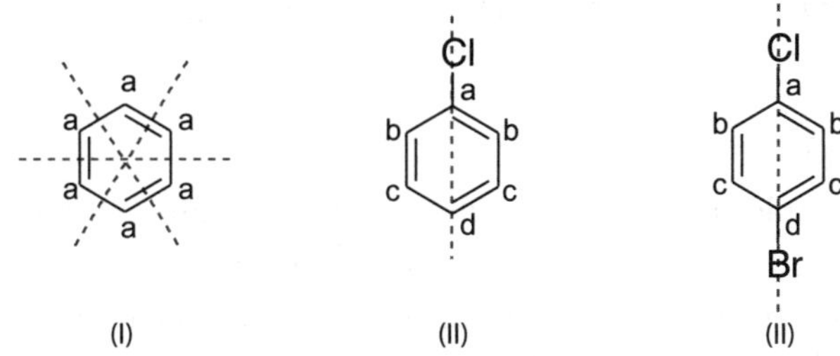

Benzene (I) is a highly symmetric molecule. All its carbon atoms are equivalent, so only one signal is expected in the spectrum. As for chlorobenzene (II), it is characterized by a plane of symmetry, and therefore four signals are present in the spectrum. The case of 1-bromo-4-chlorobenzene (III) is similar in that it has four types of carbon atoms and thus four NMR signals.

Problem 5.10

Do the following molecules provide the same ^{13}C-NMR spectrum?

Solution

As shown in the figure below, S,R-tartaric acid (I) and R,S-tartaric acid (II) are actually the same compound (*meso*), and hence they display the same spectrum.

5.6 Distortionless Enhancement by Polarization Transfer (DEPT) pulse sequence

While proton decoupling results in a much simpler spectrum, useful information about the presence of neighboring protons is lost. However, a number of NMR experiments have been optimized over the years that allow us to determine how many hydrogens are bonded to each carbon. One of the most used is the Distortionless Enhancement by Polarization Transfer (DEPT) experiment. There are a number of versions of this experiment, which can be very useful to distinguish the different types of carbon atoms within a molecule. Among these, the DEPT-135 sequence is the most useful. In this experiment, the quaternary carbon atoms are edited out of the spectrum, while methyls and methines are oppositely phased relative to the methylene signals. Generally, the spectra are plotted with methyls and methines as positive signals and methylenes as negative signals (Fig. 5.1). Usually methyls and methines can be distinguished based on their chemical shifts: methyls are much more up-field-shifted in comparison to the methines. If, in some rare cases, you encounter a signal that you cannot confidently assign to either a methyl or methine carbon, the DEPT-90 experiment may be helpful, as it contains only methine signals.

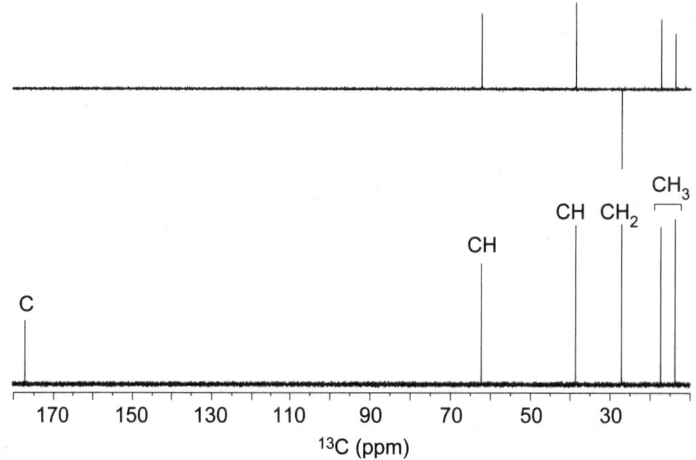

Figure 5.1 (Top) DEPT-135 spectrum. (Bottom) Regular carbon spectrum.

Therefore, it is possible to retrieve important structural information from a direct comparison between the regular carbon and DEPT spectra. In the example reported in Figure 5.1, it is clear that those spectra refer to a molecule containing a quaternary carbon at 177 ppm (attributable to a carboxylic acid, amide or ester), two methyls at 14 and 17 ppm, one methylene at 27 ppm, and two methines at 39 and 62 ppm.

Problem 5.11

^{13}C-NMR and DEPT-135 spectra of a molecule having molecular formula $C_3H_6O_2$ are reported below. Determine the structure of the molecule.

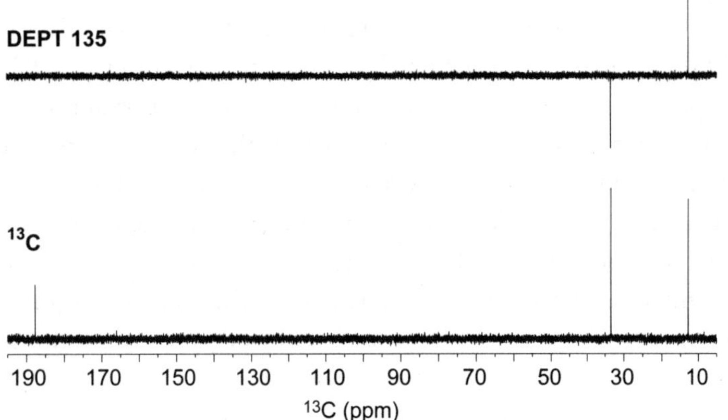

Solution

The number of signals in the spectrum is consistent with the number of carbon atoms reported in the molecular formula, suggesting that the molecule does not possess elements of symmetry. There are not olefinic or aromatic carbon atoms, but a carboxylic carbon is present. This last signal cannot be assigned to an amide, since the nitrogen atom is missing in the molecular formula. It is possible also to exclude the presence of an ester, since this would require a substituent bonded to an oxygen atom possessing a carbon resonating around 60-80 ppm. Therefore, the signal at around 187 ppm can be surely assigned to a carboxylic acid group (-COOH). The signals at 13 and 33 ppm are generated by sp^3 carbon atoms and, as suggested by the DEPT spectrum, are attributable to a CH_3 (this is in agreement with its chemical shift value) and a CH_2, respectively. Hence, you have three building blocks to be joined together. Taking into consideration the valence of the carbon atoms, there is only one molecule that can be drawn:

O
OH

Problem 5.12

^{13}C-NMR and DEPT-135 spectra of a molecule having molecular formula C_2H_7NO are reported below. Determine the structure of the molecule.

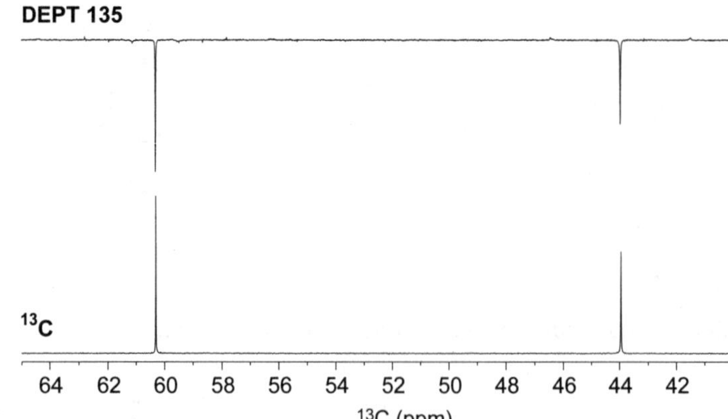

Solution

In this molecule only sp^3 carbon atoms are present. By the comparison between DEPT and ^{13}C spectra, it is clear that there are only two methylenes. As suggested by the chemical shifts, one methylene should be bonded to an oxygen atom, and the other to a nitrogen. Therefore, the structure of the molecule is:

H_2N OH

Problem 5.13

^{13}C-NMR and DEPT-135 spectra of a molecule having molecular formula $C_4H_6O_4$ are reported below. Determine the structure of the molecule.

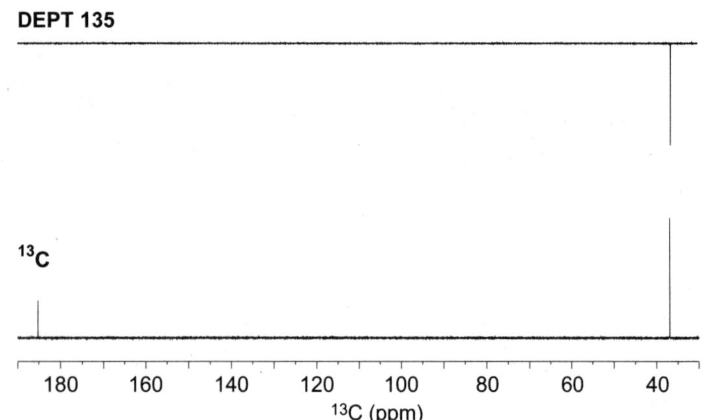

Solution

In the ^{13}C spectrum there are only two signals, while the molecular formula contains four carbon atoms. This means that the molecule is symmetric. Basically, you must determine the structure of half of the molecule, which contains the following atoms: $C_2H_3O_2$. The presence of the signal at 185 ppm is indicative of the presence of a carboxylic acid (-COOH), while the presence of ester or amide groups can be surely ruled out (see the considerations in the previous problems). The carbon signal at 37 ppm can be assigned to a methylene; therefore, it is possible to suppose that the fragment $C_2H_3O_2$ is characterized by the structure -CH_2-COOH. The structure of the entire molecule is:

Problem 5.14

^{13}C-NMR and DEPT-135 spectra of a molecule having molecular formula C_6H_6O are reported below. Determine the structure of the molecule.

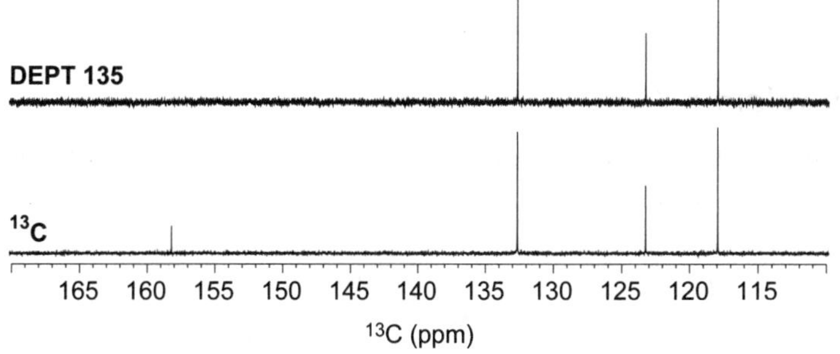

Solution

From an analysis of the chemical shifts, it is clear that it is an aromatic molecule (see the scheme in the inside back cover flap). The carbon spectrum contains only four signals out of the six expected from the molecular formula, indicating that the molecule contains an element of symmetry. There is only one quaternary carbon, suggesting that the molecule could be a monosubstituted benzene. If this hypothesis is correct, the structure would be that of phenol.

Phenol possesses a plane of symmetry that is consistent with the number of signals of the carbon spectrum. In addition, the chemical shifts are in agreement with this structure. In fact, as also discussed in Chapter 2, the mesomeric effects of the OH group strongly shield the ortho/para positions. Finally, the high chemical shift value of the quaternary carbon is in agreement with the inductive effect of the OH group that is directly bonded to it.

Problem 5.15

Which of the following molecules (both having the molecular formula $C_3H_8O_2$) has generated the attached carbon spectrum?

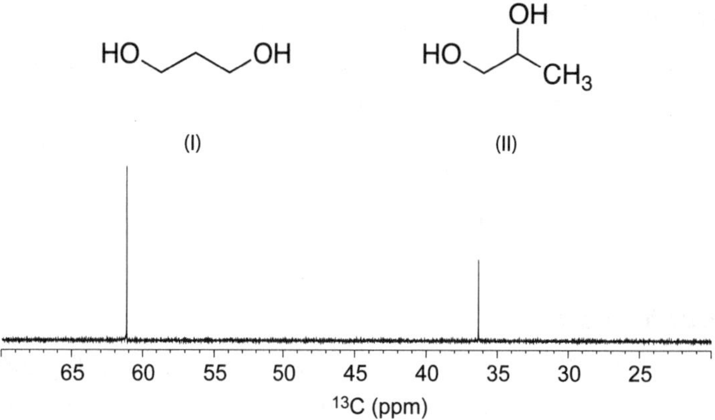

Solution

Molecule I has generated the spectrum, since it possesses a plane of symmetry and the two terminal methylenes resonate at the same chemical shift.

Summary Problems

Solutions to these problems are available online. Find your Activation Code and instructions on the first page of this book.

Problem 5.16

How many different types of carbon atoms are there in the following compounds?

$CH_3CH_2CH_2Br$ (I)

CH_3-$CH(CH_3)CH_2Br$ (II)

CH_3-$C(CH_3)_2CH_2Br$ (III)

Problem 5.17

Indicate which molecules display only two signals in the ^{13}C-NMR spectrum.

(I) (II) (III)

Problem 5.18

Which of the following molecules is characterized by a carbon spectrum having signals above 180 ppm?

CH_3CH_2CHO (I) $CH_3CH_2OCH_3$ (II)

Problem 5.19

Two molecules characterized by the same molecular formula (C_3H_6O) are propionaldehyde and acetone. Can they be distinguished by ^{13}C-NMR? Explain why.

Problem 5.20

How many different carbon signals do you expect for the following compounds?

(I) (II) (III)

Problem 5.21

How many different types of carbon atoms are there in the following compounds?

(I) (II) (III)

Problem 5.22

How many different types of carbon atoms are there in the following compounds?

(I) (II) (III)

Problem 5.23

How many different carbon signals do you expect for the following compounds?

(I) (II) (III)

Problem 5.24

How many different carbon signals do you expect for the following compounds?

(I) (II) (III)

Problem 5.25

Which of the following structures produces the carbon spectrum reported below?

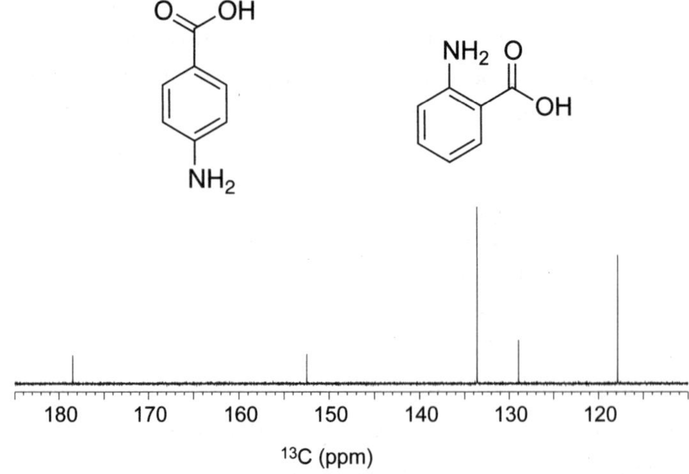

Problem 5.26

Assign each signal to the pertinent carbon of the following molecule.

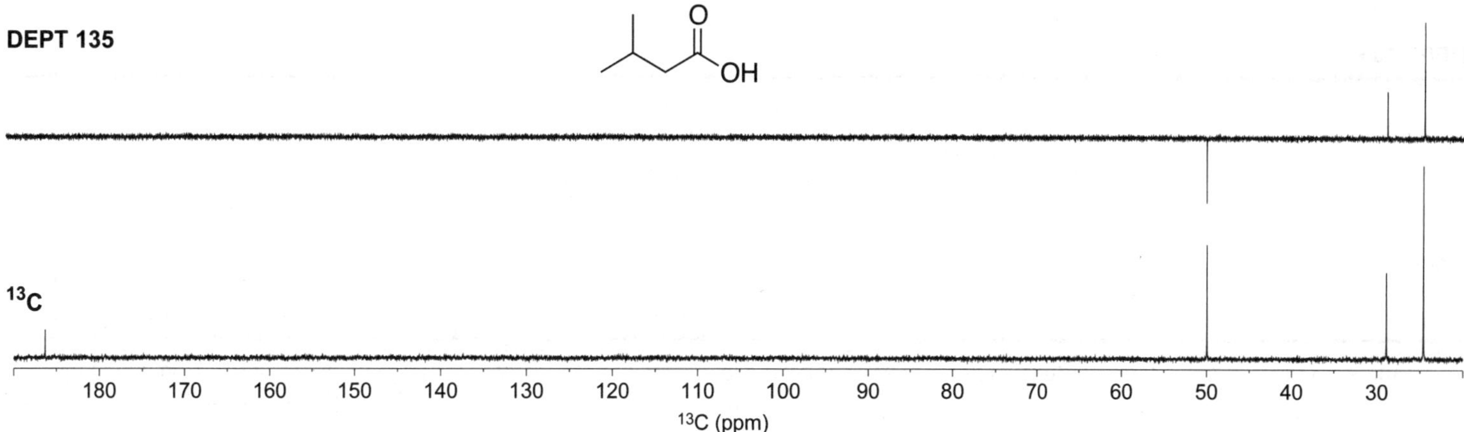

Problem 5.27

Assign each signal to the pertinent carbon of the following molecule.

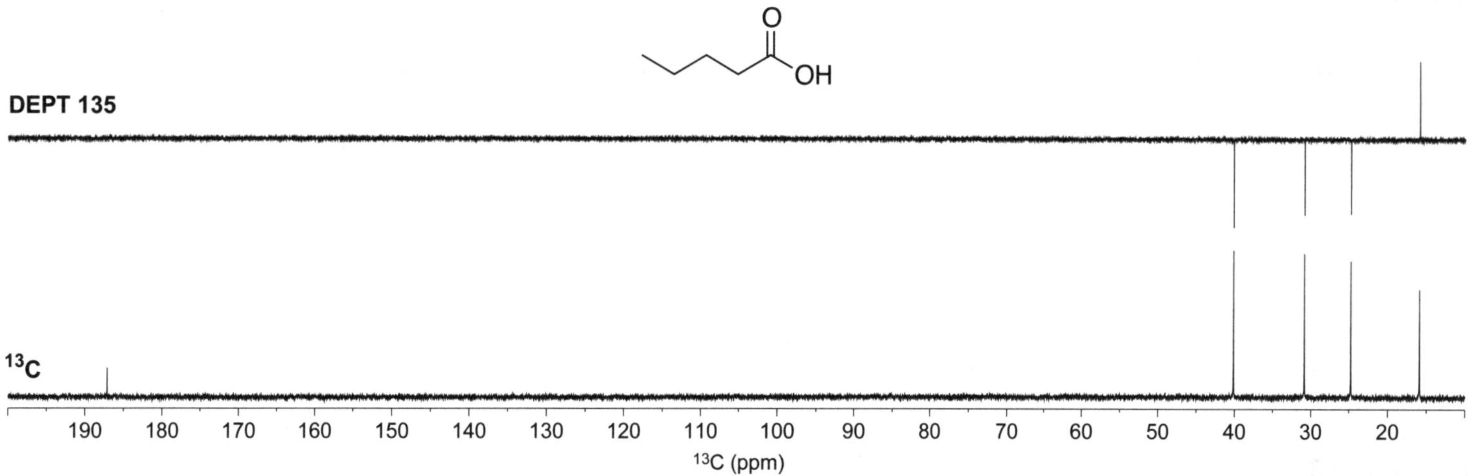

Problem 5.28

Determine the structure of the compound with molecular formula C_2H_6O, characterized by the following carbon spectrum.

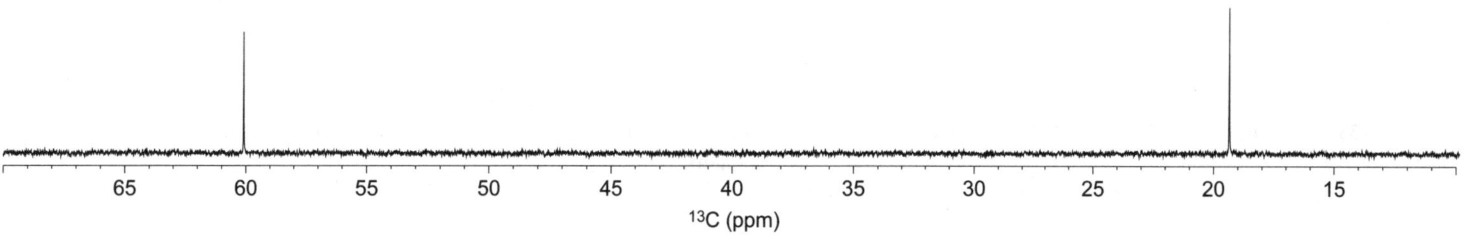

Problem 5.29

Determine the structure of the compound with molecular formula $C_4H_8O_2$, characterized by the following spectra.

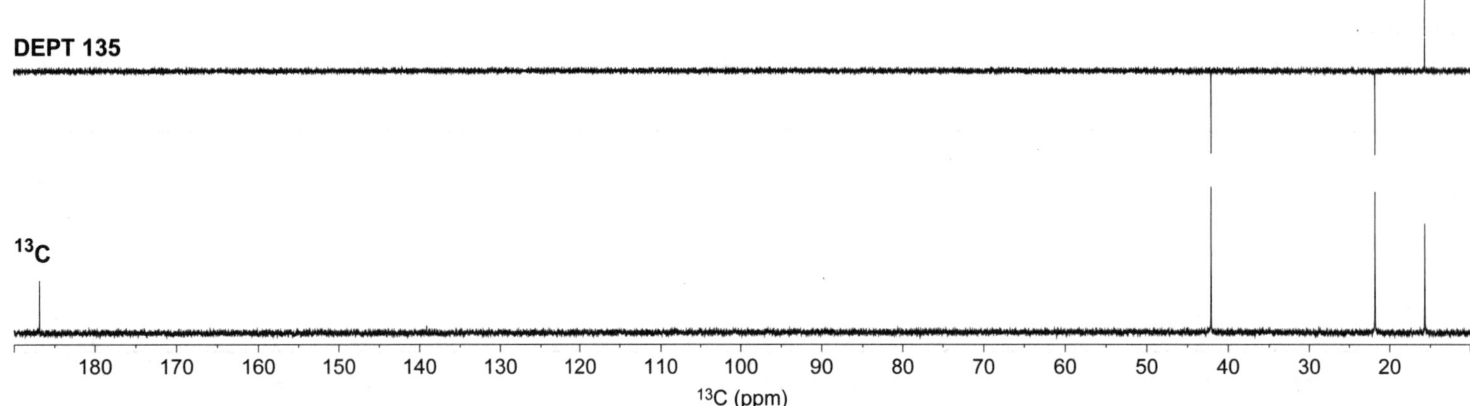

Problem 5.30

Determine the structure of the compound with molecular formula C_4H_8O, characterized by the following spectra.

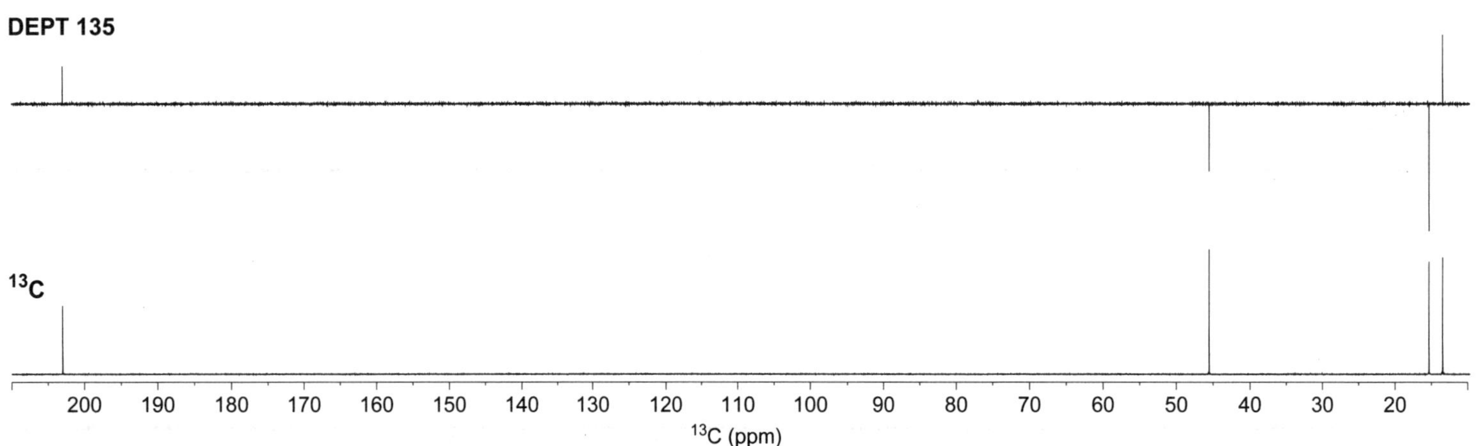

Problem 5.31

Determine the structure of dihydroxybenzene, characterized by the following carbon spectrum.

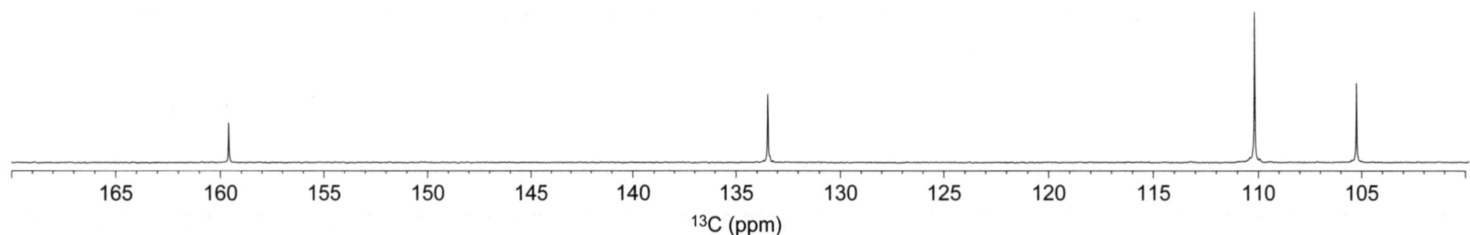

CHAPTER 6

Solvents, exchangeable protons, and hydrogen bonding in NMR

6.1 Introduction

As mentioned at the beginning of Chapter 1, this book is focused on NMR in solution, so a brief discussion about the role of the solvent is mandatory. Solvent is necessary to dissolve (if solid) or simply dilute the analyzed compound. Since most of the solvents contain proton and carbon atoms and since the solvent is generally in great excess with respect to the solute, the proton and carbon signals of the solvent can be huge in the NMR spectra in comparison to those of the solute. To avoid this problem, it is possible to use deuterated solvents (*i.e.,* $CDCl_3$ in place of $CHCl_3$; CD_3OD in place of CH_3OH, etc.) that, at least in the proton spectrum, strongly reduce the solvent signals.

6.2 Solvent signals and their multiplicity

Unfortunately, the solvent signals cannot be totally eliminated because solvent deuteration is not complete, and thus signals from the residual protons are still observable. For example, in deuterated chloroform ($CDCl_3$), a small percentage of $CHCl_3$ is still present and it produces a signal at 7.24 ppm. With methanol, the residual protonated molecules correspond to CHD_2OD and produce a signal at 3.31 ppm.

As with all NMR signals, solvent signals can exhibit multiplicity. As mentioned in Chapter 3, the general rule to predict the multiplicity of a signal is the following: n nuclei will split the NMR signal into $2nI + 1$ sub-peaks, where I is the spin quantum number of the nuclei to which a given nucleus is coupled. The spin quantum number of deuterium atoms is I = 1, so, for example, the coupling to *one* deuterium atom will split a proton signal into *three* sub-peaks ($2nI + 1 = 2{\cdot}1{\cdot}1 + 1 = 3$); the coupling to *two* deuterium atoms will split a proton signal into *five* sub-peaks ($2{\cdot}2{\cdot}1 + 1 = 5$); *three* deuterium atoms will split a proton signal into *seven* sub-peaks ($2{\cdot}3{\cdot}1 + 1 = 7$), etc. From this pattern we see that a nucleus coupled to deuterium atom(s) cannot provide even multiplets (doublet, quartet, sextet, etc.), but only odd multiplets (triplet, quintet, septet, etc.). Furthermore, the relative sub-peak intensities of the multiplets are different from those of the nuclei coupled to protons (or in general to nuclei having I = 1/2). In fact, while protons have only two orientations with respect to the applied magnetic field (m = -1/2, +1/2), the nuclei of the deuterium atoms (deuterons) have three orientations (m = -1, 0, +1). Therefore, a given nucleus coupled to only one deuteron will experience its three orientations and will generate a signal containing three sub-peaks whose relative intensities are about 1:1:1 (the rationale is very similar to that used to explain the proton scalar coupling in Chapter 3). The same rules apply also to the splitting of carbon signals in ^{13}C-NMR. For example, the carbon atom of the deuterated chloroform couples to the deuterium atom, producing the triplet shown in Figure 6.1.

What about hydrogen/deuterium and carbon/deuterium coupling constants?

Hydrogen/deuterium and carbon/deuterium coupling constants are very small in comparison to the proton/proton and carbon/proton ones. As for hydrogen/deuterium coupling constants, the vicinal ones ($^3J_{HD}$) vary from about 0 Hz (in the aromatic systems) to about 1 Hz (in aliphatic ones). Instead, the geminal constants ($^2J_{HD}$) are in a range of 1-3 Hz.

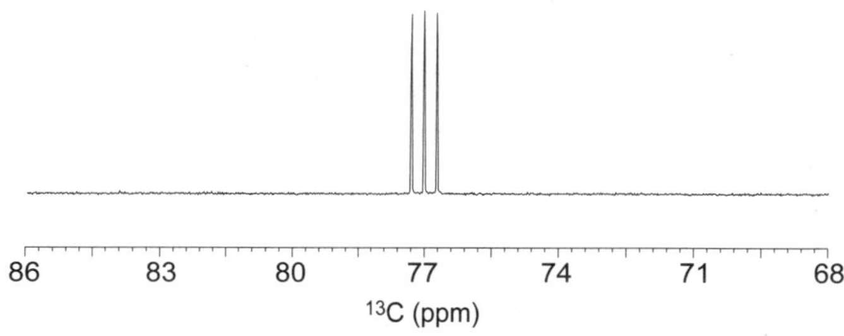

Figure 6.1 Solvent signal of $CDCl_3$ in ^{13}C-NMR spectra (75 MHz).

On the other hand, carbon/deuterium couplings between atoms directly bonded ($^1J_{CD}$) are about 20-30 Hz, while geminal couplings ($^2J_{CD}$) are basically not observable in ^{13}C-NMR spectroscopy. So, for example, the carbon signal of deuterated benzene (δ 128.4 ppm) displays coupling only to the deuteron directly bonded, thus forming a triplet having relative sub-peak intensities of 1:1:1 (like that of the $CDCl_3$), and not showing any coupling to the adjacent deuterons ($^2J_{CD}$ ~ 0) (Fig. 6.2).

$^1J_{CD}$ ~ 24 Hz $^2J_{CD}$ ~ 0 Hz

Figure 6.2 Carbon/deuterium coupling constants in deuterated benzene.

In order to predict the relative sub-peak intensities of signals coupled to nuclei having I = 1, a trinomial triangular array of coefficients must be used (Fig. 6.3, bottom). This array is similar to the binomial Pascal's triangle, already discussed in Chapter 3, and shown again here at the top of Figure 6.3. The difference between the two is that an entry in the trinomial triangle is the sum of the three (rather than the two in Pascal's triangle) entries above it.

It is also possible to use the splitting diagram, in which each splitting step divides each peak into three sub-peaks. In the example shown in Figure 6.4, a generic carbon atom couples with two deuterium atoms by $^1J_{CD}$ = 24 Hz (each square in the figure counts for 3 Hz). As you can see, as also predicted by the trinomial triangle, the relative sub-peak intensities of the resulting multiplet are 1:2:3:2:1.

Please note that the relative intensity ratios of this quintet are different from those of the quintet generated by coupling to protons (1:4:6:4:1). By the way of example, Figure 6.5 shows the comparison of two quintets (first row) and two septets (second row) generated by coupling to nuclei having I = 1/2 and 1. As you can see, although the general appearance of the multiplet is similar, the relative sub-peak intensities are different.

Binomial Pascal's triangle

N. of coupled atoms (I = 1/2)	Relative intensities of sub-peaks	Name of multiplet	Shape
0	1	*singlet*	
1	1 1	*doublet*	
2	1 2 1	*triplet*	
3	1 3 3 1	*quartet*	
4	1 4 6 4 1	*quintet*	
5	1 5 10 10 5 1	*sextet*	
...	... ...	...	...

Trinomial triangle

N. of coupled atoms (I = 1)	Relative intensities of sub-peaks	Name of multiplet	Shape
0	1	*singlet*	
1	1 1 1	*triplet*	
2	1 2 3 2 1	*quintet*	
3	1 3 6 7 6 3 1	*septet*	
...	... ...	...	...

Figure 6.3 (Top) The binomial Pascal's triangle used for couplings to nuclei having I = 1/2. (Bottom) Trinomial triangle used for couplings to nuclei having I = 1.

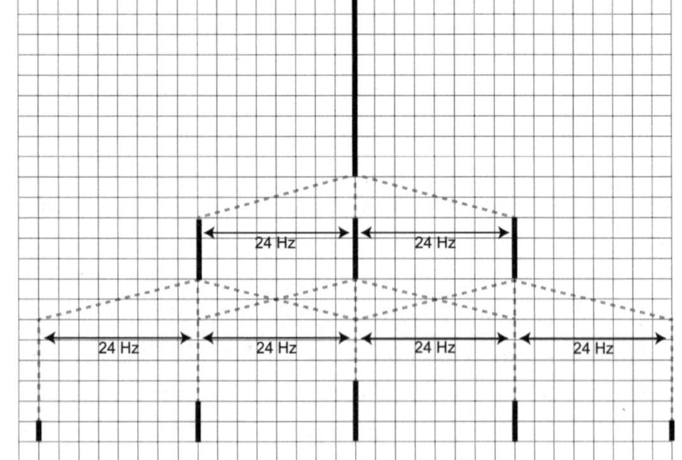

Figure 6.4 Splitting diagram for a carbon atom coupled ($^1J_{CD}$ = 24 Hz) to two deuterons.

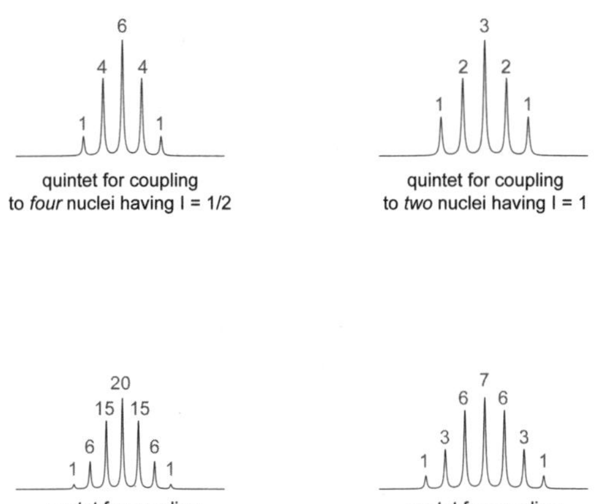

Figure 6.5 Examples of quintets and septets generated by coupling to nuclei having I = 1/2 (on the left) and I = 1 (on the right). Relative sub-peak intensities in each multiplet are reported as gray numbers.

6.3 Chemical shifts of solvent signals

Several solvents possess more than one type of hydrogen or carbon atom. The chemical shifts of the solvent signals observed in ^{1}H- and ^{13}C-NMR spectra are listed in the Table 6.1.

Table 6.1 Chemical shifts (ppm) of the solvent signals observed in ^{1}H- and ^{13}C-NMR spectra (T = 24 °C).

Solvent	Formula	^{1}H-NMR	^{13}C-NMR
Acetone	CD_3-CO-CD_3	2.05 (5)	206.3 (1) 29.8 (7)
Acetonitrile	CD_3CN	1.94 (5)	118.3 (1) 1.3 (7)
Benzene	C_6D_6	7.16 (1)	128.1 (3)
Chloroform	$CDCl_3$	7.26 (1)	77.1 (3)
Dimethyl Sulfoxide	CD_3-SO-CD_3	2.50 (5)	39.5 (7)
Methanol	CD_3OD	4.78 (1) 3.31 (5)	49.0 (7)
Water	D_2O	4.79 (1)	

Values are in ppm. The multiplicity is reported in parentheses as 1 for singlet, 3 for triplet, 5 for quintet, 7 for septet.

6.4 Physico-chemical characteristics of the solvents

Each solvent has specific physico-chemical characteristics. For example, benzene is characterized by its aromaticity, and methanol by its polarity and its ability to form hydrogen bonds. These two solvents can interact with a given solute in different ways, so we should expect that the chemical shifts of the solute signals may change depending on the solvent used.

Another physico-chemical characteristic of solvents is whether they are protic or aprotic. A protic solvent is a solvent that contains acidic hydrogens; conversely, aprotic solvents do not contain acidic hydrogens. Generally, acidic hydrogens are also defined as exchangeable. Example of exchangeable hydrogens are those bonded to oxygen atoms (as in hydroxyl groups), nitrogen atoms (as in amine groups), or sulfur atoms (as in sulphydryl groups). It is much more difficult to find exchangeable protons bonded to carbon atoms. Generally, apolar solvents are aprotic (chloroform, benzene, hexane, etc.), whereas polar solvents can be either protic or aprotic. Typical polar protic solvents are water, methanol, ethanol, etc., while typical aprotic polar solvents are dimethylsulfoxide (DMSO), acetone, *N,N*-dimethylformamide (DMF), pyridine, etc.

6.5 The appearance of the NMR signals of exchangeable protons

If a regular solvent can exchange protons, a deuterated solvent can exchange deuterons. This means that, if a molecule having exchangeable protons is dissolved in a protic deuterated solvent, all acidic hydrogens may be replaced by deuterium atoms.

The chemical exchange between hydrogen and deuterium atoms is a dynamic equilibrium. The NMR interpretation of dynamic equilibria represents a complex topic, which cannot be discussed briefly. Nevertheless, few salient points must be introduced here. A classic example used to explain dynamic processes in NMR is the analysis of the conformational equilibrium of *N,N*-dimethylformamide (Fig. 6.6).

Figure 6.6 Dynamic conformational equilibrium of *N,N*-dimethylformamide.

Due to the double bond character of the amide bond, the amide group can be considered planar and the two methyls very slowly interconvert their positions at room temperature. However, higher temperature will make this interconversion more probable and faster. Each methyl will be observable as long as it will remain in its position long enough to be sampled. As mentioned in Chapter 2, the NMR time-scale in proton NMR spectroscopy is about three seconds. Since at room temperature the two methyls interconvert at a rate much slower than every three seconds, both of them can be sampled, and, experiencing two different chemical environments, provide two distinct NMR signals (Fig. 6.7, bottom). Increasing the temperature, the interconversion rate approaches the NMR time-scale, the methyls start to appear as broad signals (Fig. 6.7). As the temperature further increases, the two methyls interconvert at an intermediate exchange rate that produces two overlapped broad signals, which successively merge into one broad signal at the so-called coalescence temperature. At this temperature, the two methyls are not distinguishable anymore. Above this point, the merged peak width decreases as temperature increases. The resulting signal appears at a weighted average position in the spectrum, which reflects the abundance of the two species in equilibrium.

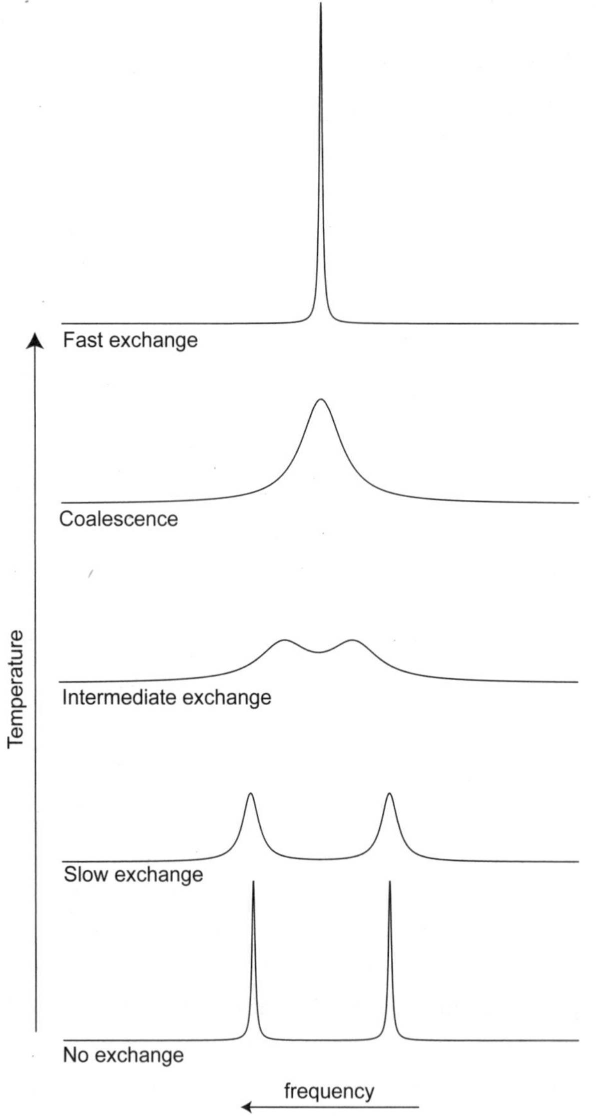

Figure 6.7 Simulated NMR spectra of *N,N*-dimethylformamide at different temperatures.

In the case of *N,N*-dimethylformamide, since the interconversion involves two equally probable conformations, the averaged signal will resonate right in the middle of the two signals representing the two species (otherwise, the signal would have been proportionally shifted towards the signal representing the most abundant species).

Proton/deuteron chemical exchange can be analyzed using the same rationale explained above, with a few important differences. First of all, this is not a conformational equilibrium, but a chemical reaction (Fig. 6.8).

$$\text{R-OH} + D_2O_{excess} \rightleftarrows \text{R-OD} + \text{HDO}$$

Figure 6.8 Example of proton/deuteron chemical exchange in a protic solvent.

Therefore, the following species are in equilibrium: R-OH, D_2O, R-OD, and HDO. Focussing on the exchangeable protons (ignoring signals from the R moiety), we expect two signals in the proton spectrum: one from the OH group and one from HDO (that resonates along with the residual solvent signal). In other words, mobile protons can reside, as it were, on R-OH or on HDO, alternatively. Clearly, the exchange rate between the two sites, which is temperature dependent, could be slow or fast on the NMR time-scale, and this will have an impact on the observed spectrum. In the slow exchange regime, the two signals will be separately observed, although, since the solvent is in great excess, the HDO signal is much more intense than the R-OH one. In the fast exchange regime, the weighted average signal will be so much shifted towards the solvent signal that in practice it will coincide with it (in other words, it is like the mobile proton spends most of its time on the solvent molecules). So, in this case, no exchangeable protons will be visible in the proton spectrum. If the exchange rate is on the NMR time scale, instead, a broader signal for the R-OH protons is expected.

Proton/deuteron chemical exchange is very fast in protic solvents (so exchangeable protons are not detected), while it is definitely slower (or not present) under aprotic conditions. Here, the chemical exchange can be sufficiently slow that the protonated species can be perfectly visible and produce signal(s) displaying also coupling(s) to non-exchangeable protons. Chemical exchange at an intermediate rate, which produces large and non-coupled signals, is generally observed in aprotic apolar solvents.

6.6 The role of hydrogen bonds

Proton/deuteron exchange rate is dependent on temperature, acidity of the solution, etc. It is also strongly dependent on the ability of the exchangeable hydrogens to form hydrogen bonds (H-bonds). An H-bond is formed when a hydrogen atom, bonded to a small strongly electronegative atom (donor) (i.e, O, N, and F), is in the vicinity of another electronegative atom (acceptor) with a lone pair of electrons (Fig. 6.9).

A:·······H—D

H-bond

Figure 6.9 General scheme of a hydrogen bond. The acceptor is reported as A, and the donor as D.

The H-bond is often described as an electrostatic dipole-dipole interaction. However, it also has some features of covalent bonding; in fact, it has a strong geometric component (the atoms are collinear) and produces interatomic distances shorter than the sum of the van der Waals radii. These features strongly reduce or inhibit the chemical exchange of the hydrogen, allowing the observation of its signal in the spectrum.

Aprotic polar solvents like DMSO, DMF, pyridine, etc. are able to interact with the exchangeable hydrogens of a solute, forming H-bonds. So, these solvents are particularly useful for observing acidic protons in the spectrum.

The exchangeable hydrogens are generally deshielded because they are bonded to electronegative atoms. However, when the same hydrogens are involved in H-bonding, they reduce their electron density (Fig 6.10), and therefore their NMR signal tends to shift further downfield (at higher chemical shift values).

For example, in dilute solutions of alcohols in non hydrogen-bonding solvents (CCl_4, $CDCl_3$, C_6D_6), the OH signal generally appears at 1-2 ppm. At higher concentrations (where the formation of H-bonds is favored), the signal moves downfield up to 5 ppm. In contrast, because of their poor propensity to form H-bonds, thiol (SH) protons typically appear in a very narrow range of chemical shift (1.2 - 2.0 ppm in $CDCl_3$).

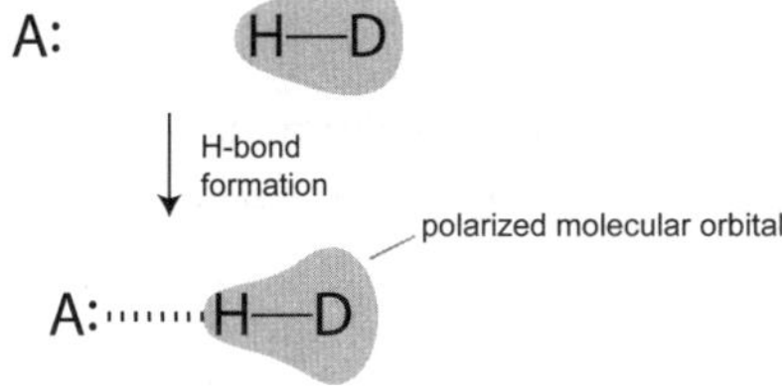

Figure 6.10 Reduction of the hydrogen electron density upon formation of the H-bond.

The formation of H-bonds is also very dependent on the acidity of the hydrogens. The more acidic the hydrogen, the stronger the propensity to form hydrogen bonds. This is the case of the carboxylic acid moieties that tend to form dimers (Fig. 6.11). These protons generally resonate very far downfield between 10 and 12 ppm.

Figure 6.11 Carboxylic acid dimer favored by the formation of intermolecular H-bonds.

The formation of H-bonds is also very much favored when they are intramolecular, especially if they form sort of five- or six-membered rings. This is the case, for example, of salicylates in which the OH group (donor) forms the H-bond with the carbonyl of the carboxylic moiety (acceptor) (Fig. 6.12, top). 1,3-Diketones, which are subject to a keto-enol tautomerization, present another interesting case (Fig. 6.12, bottom). The fraction of enol tautomer is considerably greater than the keto fraction, because it is stabilized by the internal H-bond forming a sort of six-membered ring and by the conjugation of the carbon-carbon double bond with the second carbonyl group.

Salicilate

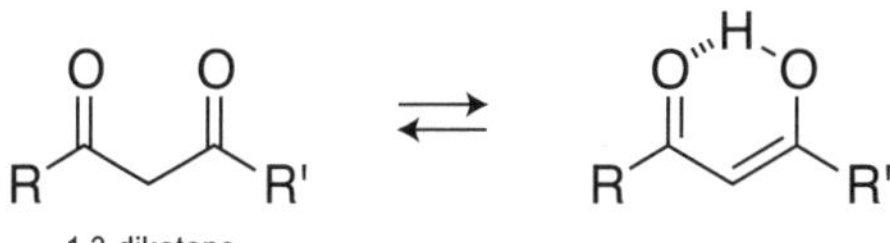

Figure 6.12 Intramolecular H-bonds forming a pseudo six-membered ring.

6.7 Water signal in different solvents

For several reasons, many solvents can contain traces of water. As mentioned above, the chemical shift values of the protons of any solute change with respect to the solvent used, and obviously this is also true for water. Actually, the signal of water varies much more than expected (Table 6.2).

Table 6.2 NMR water signal in different solvents.

Solvent	^{1}H-NMR (ppm)
Acetone	2.80
Acetonitrile	2.09
Benzene	0.40
Chloroform	1.55
Dimethyl Sulfoxide	3.31
Methanol	4.80

The explanation for this can be found in the intrinsic nature of water to form hydrogen bonds. For example, in benzene, the few water molecules dispersed in the solvent are not able to form H-bonds, so their protons resonate at 0.40 ppm. Conversely, when dissolved in DMSO, they are surrounded by solvent molecules able to form H-bonds and their proton chemical shift is 3.31 ppm.

6.8 Recognizing exchangeable protons

Generally, exchangeable protons can be recognized from their broadened appearance and, sometimes, by characteristic chemical shifts (especially when they are shifted very much downfield). However, as described above, they could also appear as sharp and coupled signals, that is, very similar to the signals produced by non-exchangeable protons. In these cases, they can be recognized by promoting their exchange by adding a source of deuterons (like a drop of D_2O), thus causing their disappearance. D_2O can be added to both water-miscible and water-immiscible solvents. In water immiscible solvents, it is important to properly shake the solution, in order to ensure that all the solute molecules get in contact with the D_2O. When the shaking is stopped, the water and the solvent separate into two distinct layers. Generally, organic solvents are less dense than aqueous solutions, and therefore they form the top layer. However, chlorinated solvents are often more dense than water, and form the bottom layer (Fig. 6.13). Since the very top and bottom of the solution in the NMR tube are not observed by the spectrometers (because they are not sampled by the receiving coils of the instruments), the thin water layer (containing also the HDO generated by the chemical exchange) will not be visible in the spectra. On the contrary, in water-miscible solvents (like acetone, DMSO, DMF, pyridine, etc.), the HDO remains in solution and will be detected. Exchangeable proton signals can also be recognized by heteronuclear 2D experiments like 2D-HSQC (see Chapter 7).

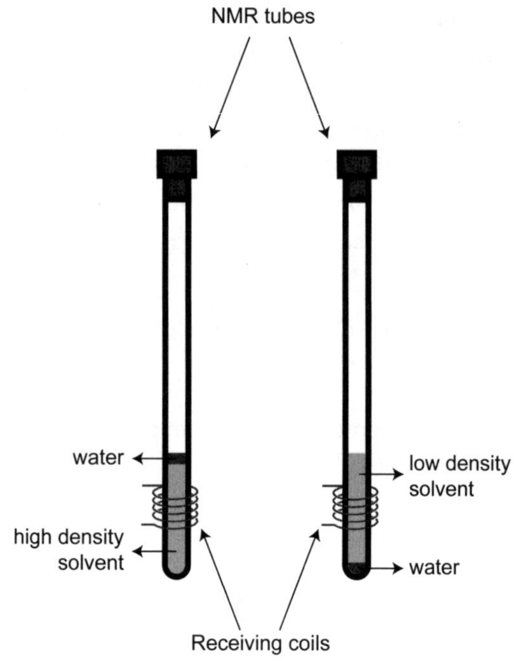

Figure 6.13 Schematic representation of organic solvent/water separation.

6.9 Solvent Selection

The NMR solvent is chosen according to the following main considerations:

1. *Solubility.* A higher solubility allows you to dissolve a larger amount of sample, and therefore the sensitivity of the experiment increases.
2. *Interference of solvent signals.* The solvent signals can overlap with the signals of the solute. Generally, it is preferable to choose a solvent whose signals do not overlap with signals from the sample.
3. *Viscosity.* High solvent viscosity leads to broad signals in NMR spectra. Generally, it is preferable to use low viscosity and low density solvents.
4. *Temperature.* To perform experiments at high temperature, a solvent with a high boiling point should be selected. On the contrary, to perform experiments at low temperature, a solvent with a low melting point should be used.

Problem 6.1

The ^{13}C-NMR spectrum (125 MHz) of (*Z*)-1,3-dichloropropene is reported below. Can you determine in which solvent the spectrum has been acquired?

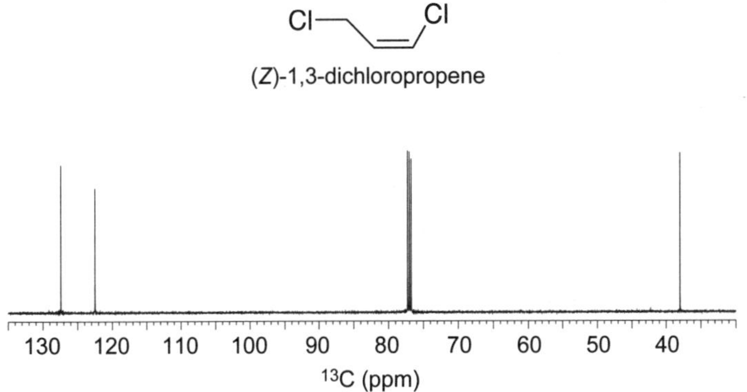

Solution

The spectrum has been acquired in deuterated chloroform, whose signal resonates at 77 ppm. The signal displays a triplet multiplicity having relative sub-peak intensities of 1:1:1, typical of a nucleus coupled to a deuteron having I = 1.

Problem 6.2

The following carbon spectrum contains two solvent signals. Has the spectrum been acquired in deuterated DMSO, methanol, ethanol, or acetone? Please explain.

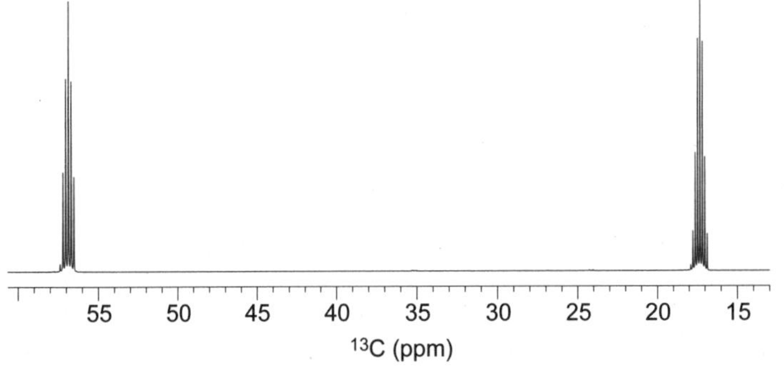

Solution

The solvent used to acquire the spectrum provides two signals. Among the solvents cited in the question, only deuterated ethanol (CD_3-CD_2-OD) and acetone (CD_3-CO-CD_3) have carbons resonating at two different frequencies. However, the chemical shift values and signals' multiplicity clearly indicate that those signals belong to the deuterated ethanol. In particular, the signal at 17.3 ppm can be assigned to the CD_3 moiety thanks to its chemical shift value and to its multiplicity (septet), which is characteristic of a nucleus coupled to three deuterons (Fig. 6.3). The signal at 57.0 ppm can be assigned to the CD_2 moiety that, being bonded to an oxygen atom, resonates at higher frequencies. Furthermore, it has a quintet multiplicity, consistent with the number of deuterium atoms bonded to the carbon atom.

Problem 6.3

The simulated proton NMR spectrum (400 MHz) of propionic acid is reported below. Can you identify the exchangeable proton signal?

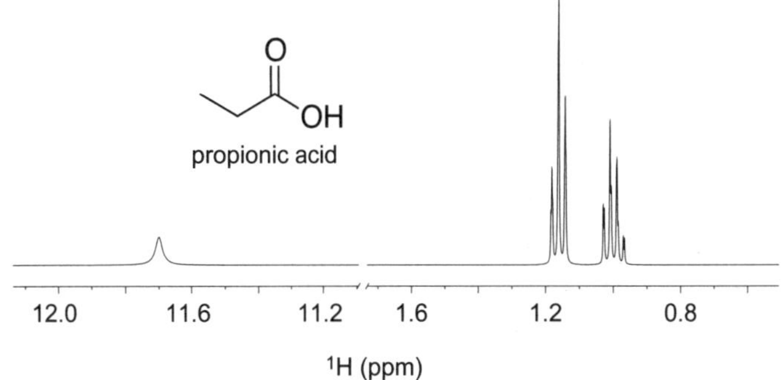

Solution

The exchangeable proton signal resonates at 11.7 ppm. It is broader than the other signals and resonates at a very high chemical shift. Remember that this value is justified by the intrinsic deshielded nature of carboxylic protons and by the tendency of carboxylic acids to form hydrogen-bonded dimers (Fig. 6.11).

Problem 6.4

Please consider deuterated pyridine.

How many signals do you expect to see in its proton and carbon spectra, respectively? What about their multiplicity?

Solution

Pyridine is a symmetric molecule. Therefore, the residual hydrogens at position 1 are chemical shift equivalents to those at position 5 (see figure below), and those at position 2 to those at position 4. Therefore three signals are expected. Since vicinal proton/deuterium coupling constants ($^3J_{HD}$) are almost null in aromatic systems, the residual solvent signals are expected to be singlet in the proton spectrum.

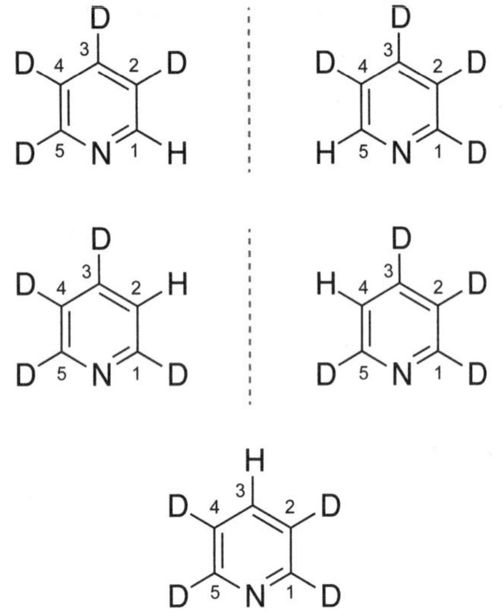

In the carbon spectrum, for the same consideration of symmetry, three signals will be observable. Since the geminal couplings ($^2J_{CD}$) are not observable, all three signals will be triplets having 1:1:1 relative sub-peak intensities.

Problem 6.5

How many signals do you expect to see for deuterated acetone in its proton and carbon spectra, respectively? What about their multiplicity?

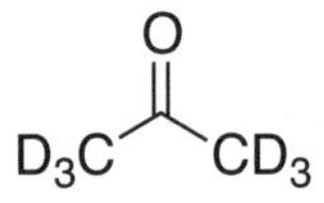

Solution

Acetone is a symmetric molecule, so only one signal is visible in the proton spectrum. The multiplicity of the proton signal will be quintet with relative sub-peak intensities of 1:2:3:2:1. In the carbon spectrum, two signals will be visible. The carbonyl carbon atom will resonate at 206.7 ppm as singlet since geminal couplings are not observable in ^{13}C-NMR spectroscopy, while the carbon atom of the -CD_3 moieties will resonate as a septet having 1:3:6:7:6:3:1 relative sub-peak intensities.

Problem 6.6

In 1-(2,5-dihydroxyphenyl)ethanone there are two exchangeable protons. They resonate at different chemical shift values. Which proton resonates at the higher frequency?

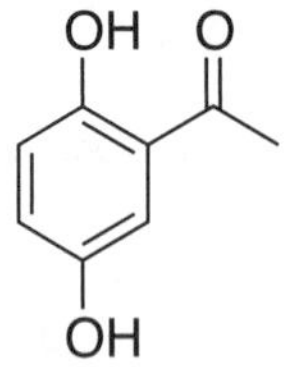

1-(2,5-dihydroxyphenyl)ethanone

Solution

The hydrogen close to the carbonyl group can form an intramolecular H-bond that is particularly stable because it forms a six-membered ring. Therefore this hydrogen will resonate at the higher frequency.

δ ~ 12 ppm H-bond

δ ~ 5 ppm

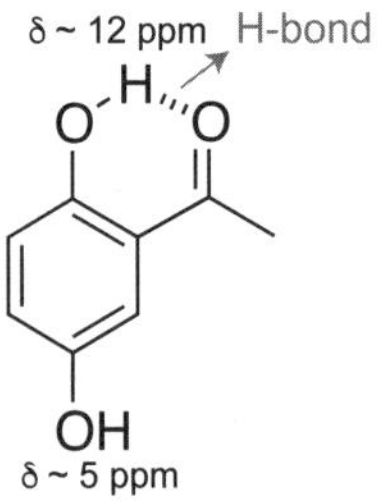

CHAPTER 7

Two-dimensional NMR spectroscopy

7.1 Introduction

The NMR spectra shown so far in this book consist of a series of signals separated (resolved) as a function of their frequencies (chemical shifts). These kinds of spectra are termed one-dimensional (1D) NMR spectra, because they correlate the signal intensities to only one set of frequencies (proton or carbon in this book).

A natural extension of this basic experiment was introduced decades ago and has been termed two-dimensional (2D) NMR, because it correlates the intensities of the signals to two sets of frequencies (F1 and F2 in Fig. 7.1).

The theory of 2D experiments is beyond the scope of this book; therefore, in this chapter, only the rationale behind their interpretation is reported.

A 2D NMR spectrum can be considered as a plot having a flat surface (lack of correlation), which is interrupted by peaks (Fig. 7.1A) that indicate a correlation between the two frequencies F1 and F2 (for this reason they are also termed "cross-peaks"). In order to easily interpret 2D spectra, they are sectioned at a given height (threshold) (Fig. 7.1B), and the intensities of the peaks are depicted by a gradient of colors, or by a contour plot (Fig. 7.1C). Each contour line (or color) joins points of the spectrum of equal intensity. 2D spectra are generally observed from the top (Fig. 7.1D). Therefore, in such a visualization, the axes of the section correspond to the two sets of frequencies (F1 and F2), and the cross-peaks are formed by almost circular concentric contour lines, whose centers are the coordinates of the correlated frequencies. These correlations have different shapes, intensities, and meanings in each different 2D experiment (see below).

2D experiments that correlate frequencies of the same nucleus are defined as *homonuclear*, while those correlating frequencies of different nuclei are defined as *heteronuclear*.

2D NMR, in all its versions, has developed into an invaluable technique, assisting the researcher with different tasks, like signal assignment and molecular structure determination. Nowadays, probably several tens of different 2D experiments have been developed and shared within the NMR community. However, here, very few of them are commented, focusing on those which even a newcomer to the field of organic chemistry can hardly ignore.

7.2 COrrelation SpectroscopY (COSY) experiment

One of the first homonuclear 2D NMR experiments developed was the COSY experiment, which gives information

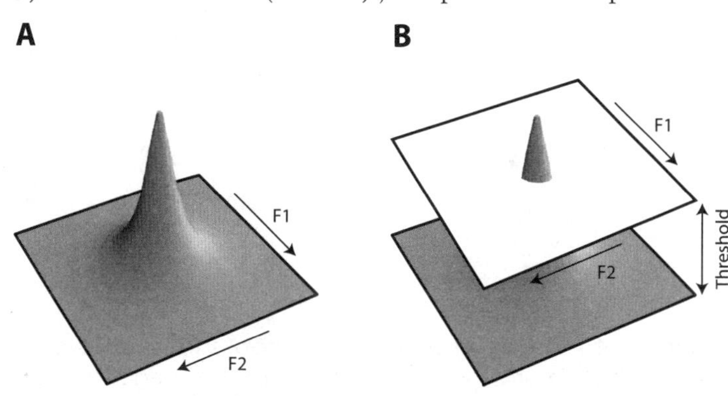

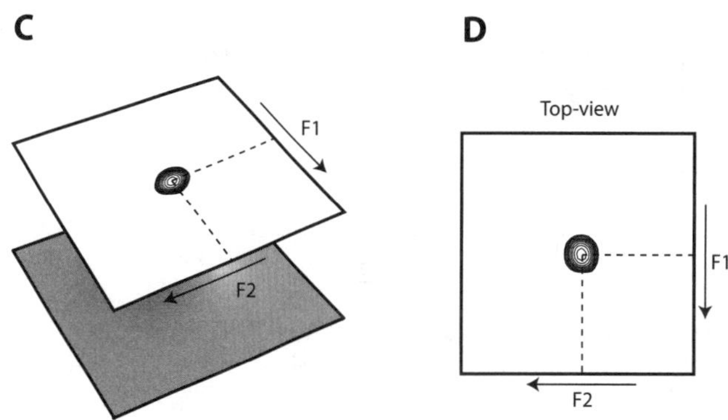

Figure 7.1 Rationale of the visualization of 2D spectra.

about protons scalarly coupled to each other. Therefore, consider a molecule containing two scalar coupled hydrogens, namely H_a and H_b, that resonate at different chemical shift values. Its COSY spectrum will appear as shown in Figure 7.2. The first characteristic of this spectrum, which is in common to all 2D homonuclear spectra, is the presence of autocorrelation peaks (H_a/H_a and H_b/H_b). These peaks show the correlation between the same chemical shift values on the two axes of the spectrum, so they do not have any structural meaning and can be ignored. They are also called diagonal peaks because they fall on the diagonal line of the spectrum, running from the lower-left corner to the upper-right corner of the spectrum (Fig. 7.2).

Interestingly, the diagonal splits the spectrum in two symmetric halves. In fact, as reported in Figure 7.2, the upper-left half contains the cross-peak correlating H_a to H_b and the lower-right half shows the same information containing the cross-peak correlating H_b to H_a. In the COSY spectrum, these correlations indicate that nucleus H_a is scalar coupled to H_b (and vice versa). Please note that signals resonating as singlets won't be coupled to any other nucleus and will produce only autocorrelation peaks.

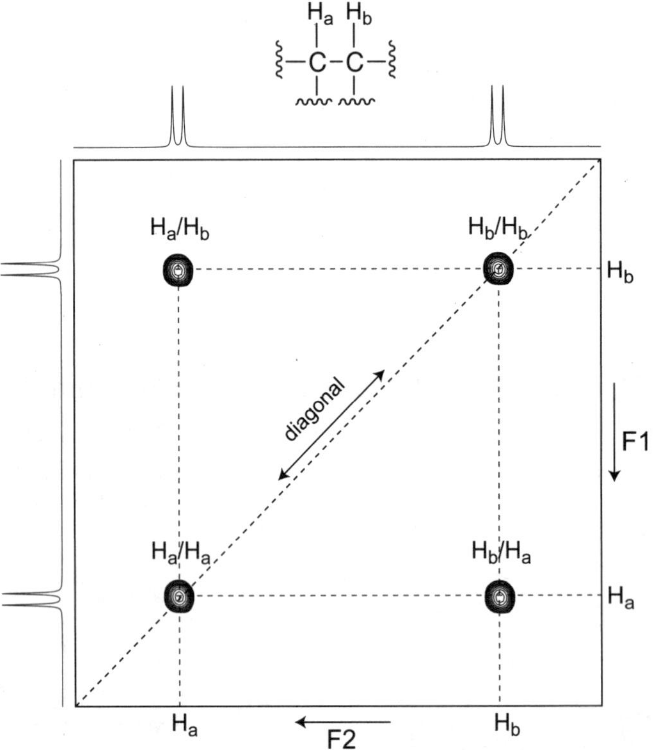

Figure 7.2 Example of a COSY experiment. The 1D proton spectrum is reported as *projections* on the two sides of the 2D plot.

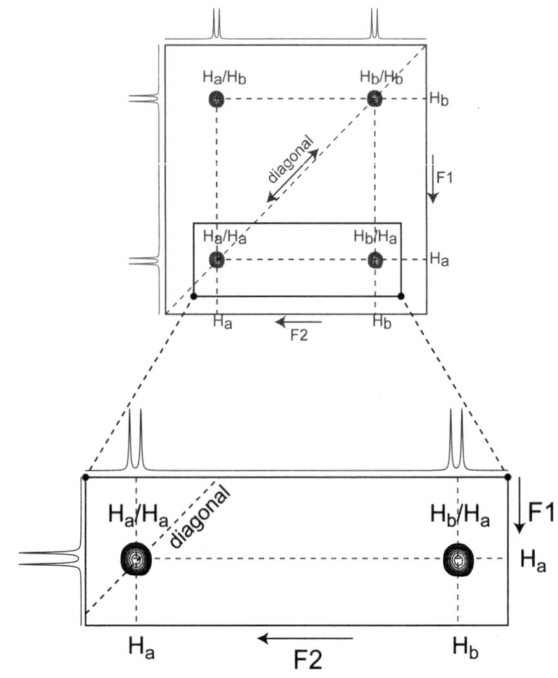

Figure 7.3 Zoomed region of the spectrum reported in Figure 7.2. In this case, the diagonal cross-peak (H_a/H_a) does not coincide with the diagonal of the rectangle in which the spectrum is enclosed. The 1D proton spectrum is shown on both sides of the 2D plot.

2D spectra are often magnified to better analyze specific spectral regions (see, for example, Fig. 7.3). In these cases, the diagonal peaks may not coincide with the geometric diagonal of the selected spectral region. However, since they correlate the same chemical shift on the two axes, they can be easily identified.

Interestingly, the shape of the cross-peaks in 2D spectra may sometimes be different from those reported in the previous figures; in particular, they may contain information about the multiplicity of the correlating signals. For example, the shape of a cross-peak correlating two doublets may look like one of those reported in Figure 7.4.

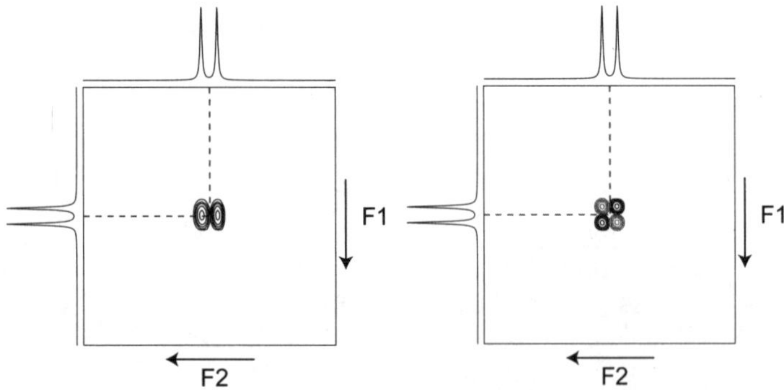

Figure 7.4 Example of shaped cross-peaks for a doublet. Sub-cross-peaks in gray on the right panel possess negative intensity with respect to the dark ones that have positive intensity (in NMR jargon, they possess opposite phase).

The shape of the cross-peaks depends on the mode of acquisition of the spectrum (pulse sequence, spectral resolution, etc.) and on the processing used to transform the spectral data. The frequencies of shaped cross-peaks are always taken at their very center (Fig. 7.4).

The number of sub-peaks in the shaped cross-peaks is related to the multiplicity of the correlating signals. For example, a triplet coupled to a doublet may have cross-peaks like those reported in Figure 7.5.

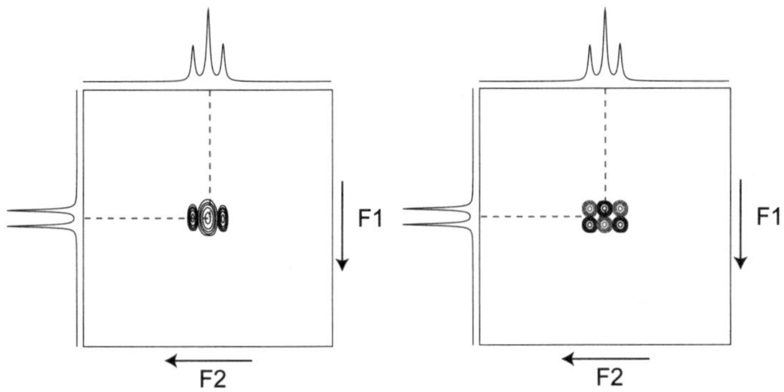

Figure 7.5 Example of shaped cross-peaks for a triplet coupled to a doublet. Sub-cross-peaks in gray on the right panel possess opposite phase with respect to the black ones.

Please note that the cross-peak intensities in COSY experiments are proportional to the magnitude of the coupling constants: small constants produce small cross-peaks and large constants produce intense cross-peaks.

The COSY experiment is very useful to analyze molecules containing large spin systems. By way of example, look at the COSY spectrum of pentanoic acid, shown in Figure 7.6. The coupling pattern can be retrieved by tracing vertical lines corresponding to each signal. Each cross-peak encountered along each line indicates a coupling to another signal, regardless of which half of the spectrum it occurs in (obviously the line will also cross the autocorrelation peak, which can be ignored). Thus, for example, it is possible to state that the signal at 2.17 ppm is coupled to the signal at 1.51 ppm (this value has to be read on the vertical scale). Then, tracing the vertical line at 1.51 ppm, obviously you will find the correlation to the signal at 2.17 ppm, but also the correlation to the signal at 1.29 ppm. You can continue to analyze the rest of the spectrum with the same rationale and determine, in this way, the sequential coupling pattern of the spin system.

It is also possible to draw on the spectrum the sequential connectivities, moving from the diagonal to the cross-peak vertically, back to the diagonal horizontally, and repeating this process till the end of the system (Fig. 7.7).

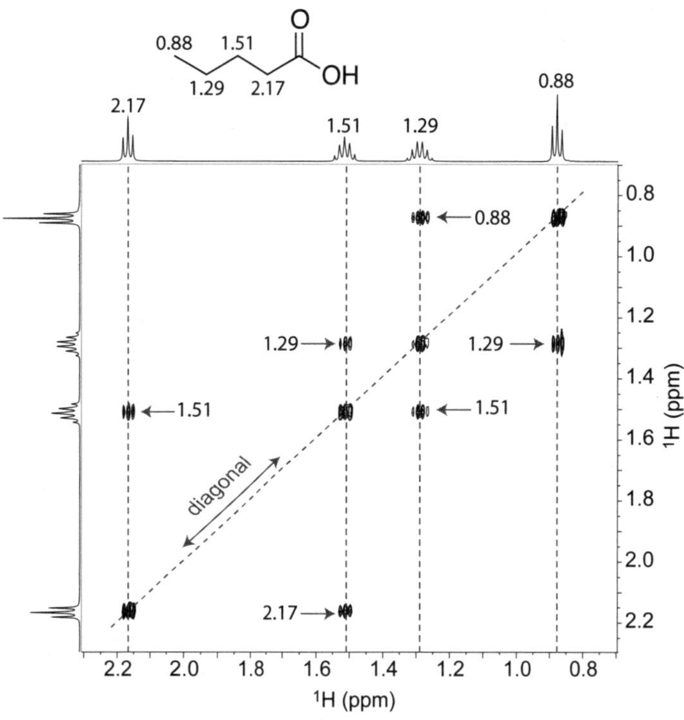

Figure 7.6 COSY spectrum (500 MHz, D_2O) of pentanoic acid.

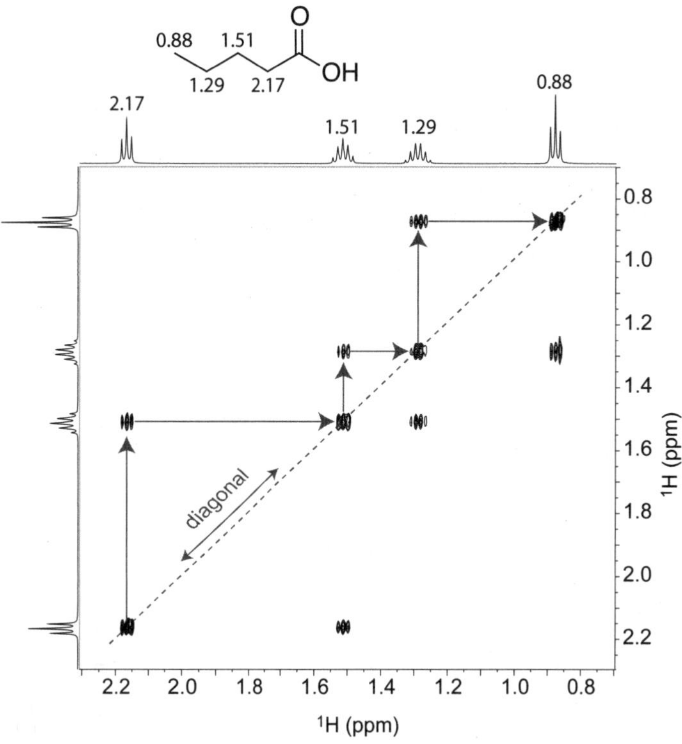

Figure 7.7 Sequential assignment of pentanoic acid in its COSY spectrum.

Problem 7.1

The 2D COSY spectrum (500 MHz, D_2O) of 4-aminobenzoic acid is reported below. Please, identify the diagonal and the correlation peaks. Please also assign each signal to the pertinent hydrogen.

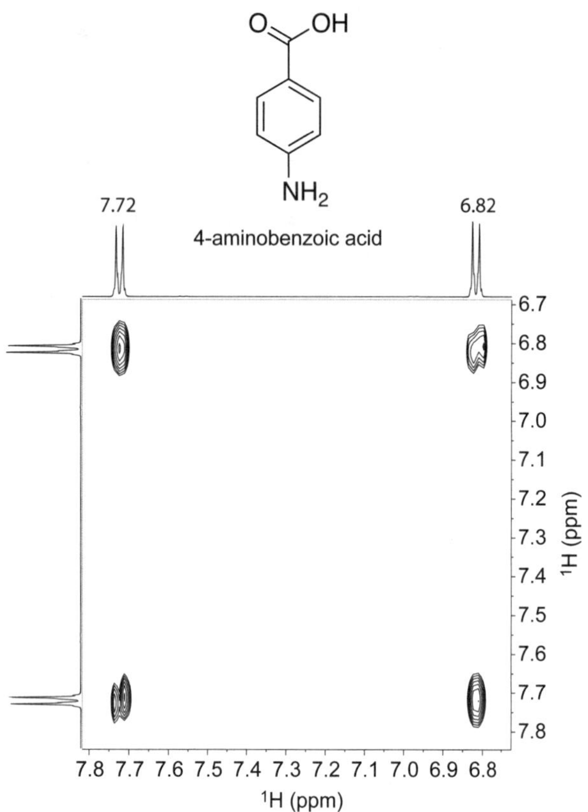

Solution

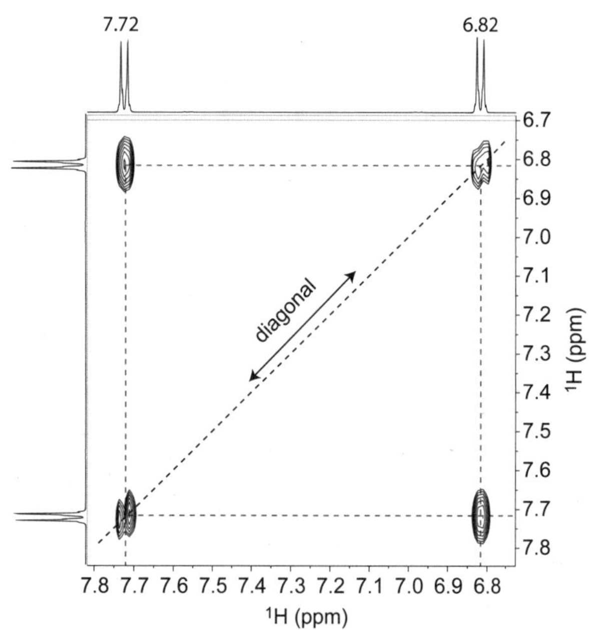

Note that, since the spectrum has been acquired in D_2O, exchangeable hydrogens are not observable.
The diagonal peaks correlate the same frequencies in both dimensions, so they are easily identified. They are joined by a dark dashed line in the figure above.
Then, there are two symmetric cross-peaks, both providing the same information: the doublet at 7.72 ppm is correlated to the doublet at 6.82 ppm. Since the molecule is symmetric (a plane of symmetry passes through C-1 and C-4), positions 2 and 3 are equivalents to positions 6 and 5, respectively.

Very often the assignment of an aromatic moiety can be performed considering the electron effects of the substituents. The mesomeric effect of the amino group in 4-aminobenzoic acid increases the electron density at positions 3 and 5, while that of the carboxylic group decreases the electron density at positions 2 and 6. Therefore, since positions 3 and 5 are the most shielded, they can be assigned to the signals resonating at the lower chemical shift value (δ 6.82 ppm). Finally, the doublet at 7.72 ppm can be assigned to the equivalent hydrogens at the positions 2 and 6.

Problem 7.2

The 2D COSY spectrum (500 MHz, D_2O) of 4-acetamido-butanoic acid is reported below. Please assign each signal to the pertinent hydrogen. Also try to draw the sequential connectivities on the spectrum.

4-acetamidobutanoic acid

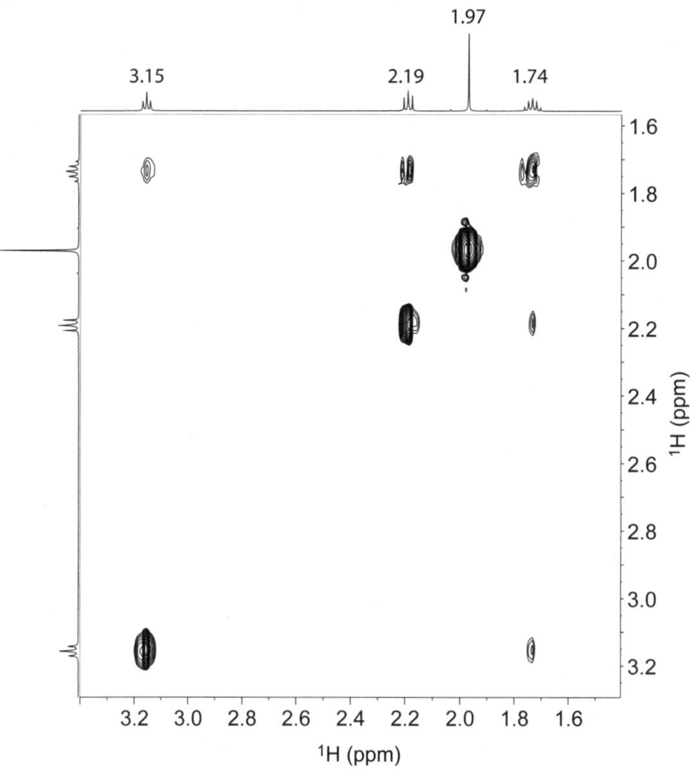

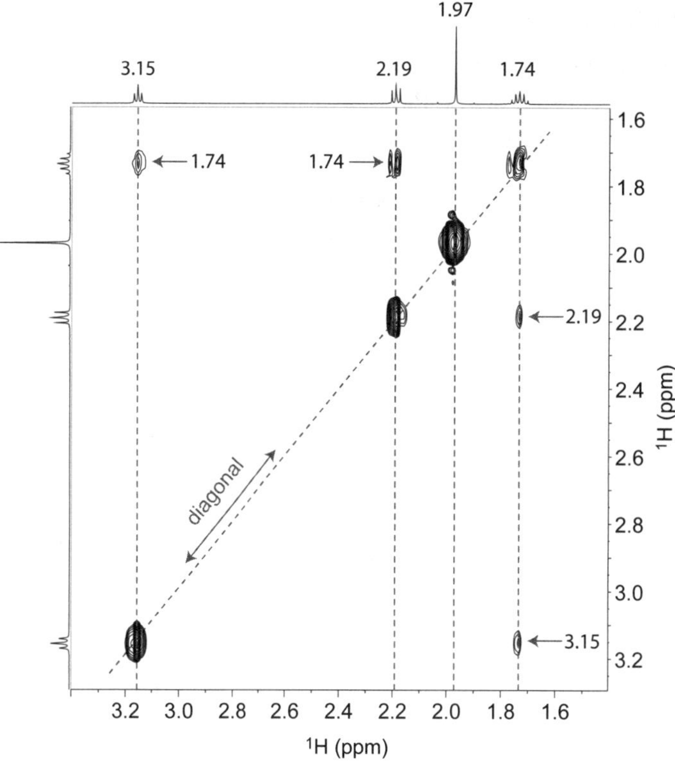

Solution

Actually, the NMR signals of this molecule can be assigned both by analyzing the proton spectrum reported in the projections of the 2D plot and by analyzing the COSY correlations. In the first case, considering the chemical shift values and multiplicity, it is possible to assign the signals of the molecule in the following way: the singlet signal at 1.97 ppm (integrating for three hydrogens) can be assigned to the methyl of the acetyl group; the triplet at 3.15 ppm to the methylene bonded to the NH group; the triplet at 2.19 ppm to the methylene in α with respect to the carbonyl group; and, by exclusion, the quintet at 1.74 ppm to the central methylene (exchangeable protons are not visible).

This assignment can be confirmed analyzing the COSY spectrum. Tracing a vertical line through each signal, it is possible to observe that the signal at 3.15 ppm correlates with signal at 1.74 ppm, which, in turn, is also correlated to the signal at 2.19 ppm. Note that the singlet at 1.97 ppm produces only an autocorrelation peak in the COSY spectrum.

Therefore the assignment and the sequential connectivities of 4-acetamidobutanoic acid are:

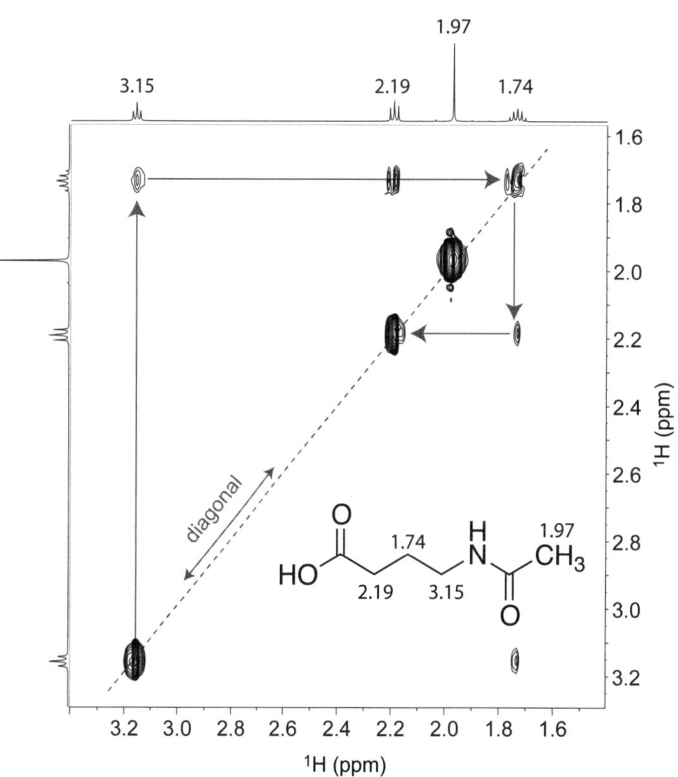

Problem 7.3

The COSY spectrum (500 MHz, D_2O) of 2-aminobenzoic acid is reported below. Please assign each signal to the pertinent hydrogen and draw the sequential connectivities on the 2D plot.

2-aminobenzoic acid

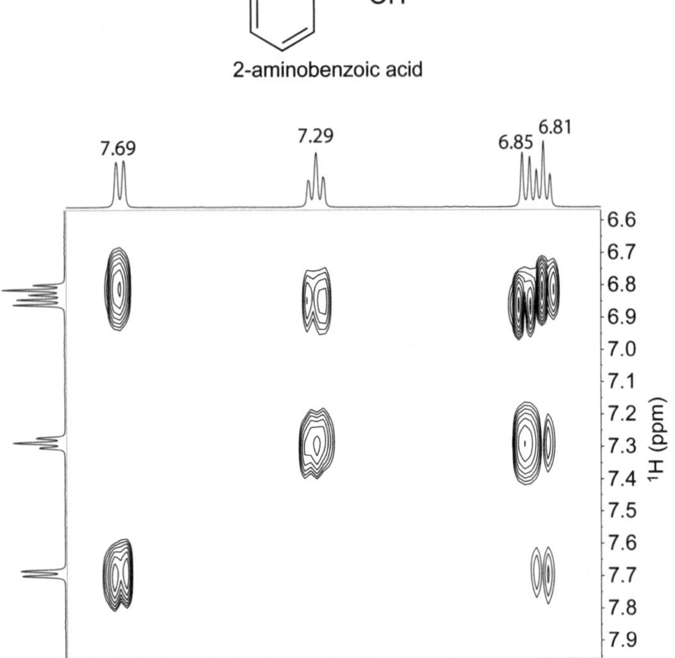

Solution

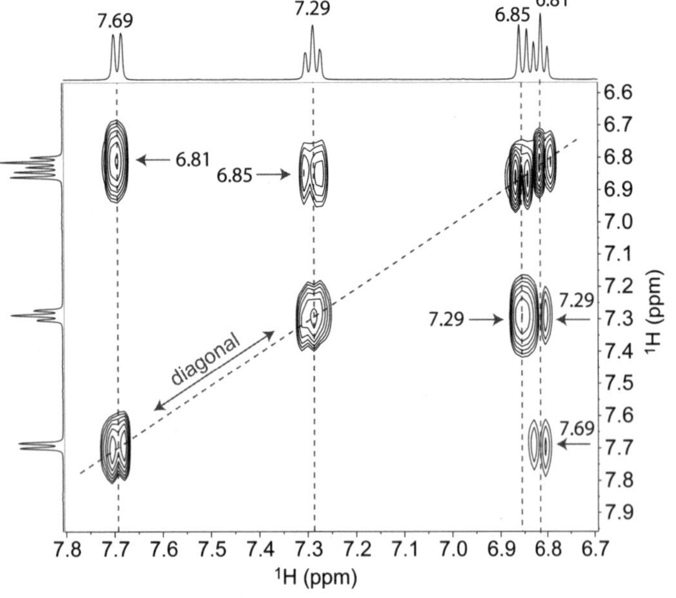

The interpretation of this spectrum is complicated by the broadness of the cross-peaks and by the partial overlapping of the cross-peaks correlating the signals at 6.81 and 6.85 ppm to the signal at 7.29 ppm. Therefore, it is very important to pay a lot of attention in tracing the lines perfectly centered on each signal. The doublet at 7.69 ppm turns out to be coupled to the triplet at 6.81 ppm. This is coupled also to the other triplet at 7.29 ppm, that, in turn, is coupled to the last doublet at 6.85 ppm. There are two different ways to place this assignment on the structure of the molecule:

However, taking into account the mesomeric effects of the two substituents, you can conclude that the correct assignment of 2-aminobenzoic acid is that on the right. Please note that exchangeable protons are not visible in this spectrum.

As for the sequential connectivities, they are difficult to draw since they change direction twice. However, if you follow the numbers in the spectrum below, you will see each coupling identified.

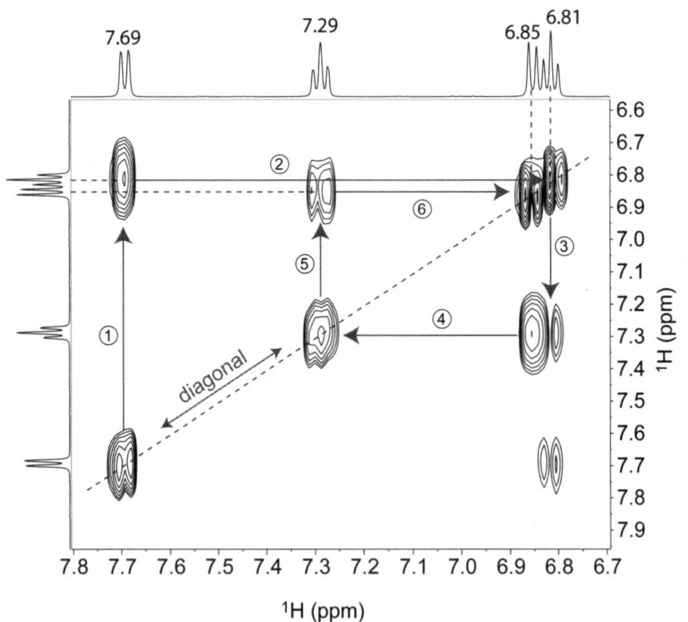

Problem 7.4

The COSY spectrum (500 MHz, D_2O) of the amino acid isoleucine is reported below. Please assign each signal to the pertinent hydrogen.

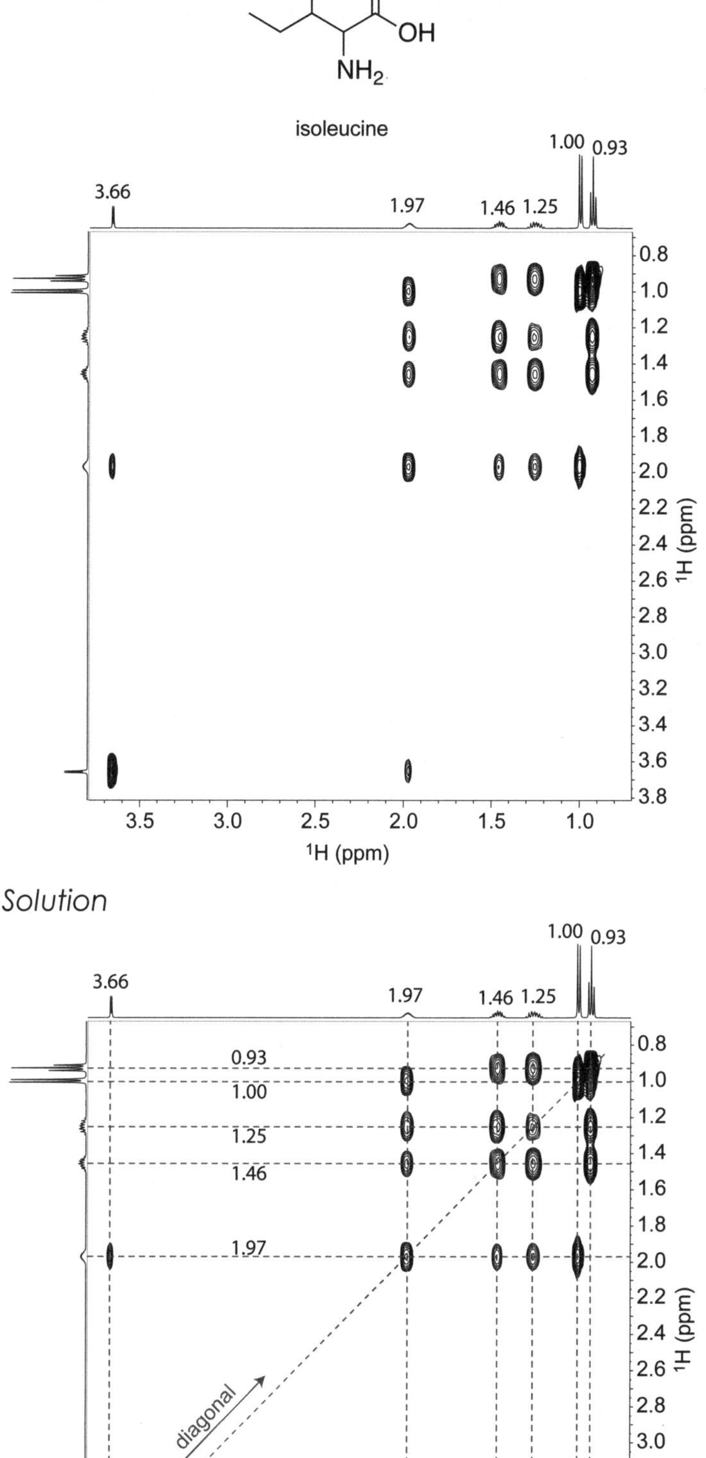

Solution

Surely, the most deshielded signal (δ 3.66 ppm) can be assigned to the Hα. It is coupled to the signal at 1.97 ppm. This signal is further coupled to other three the signals at 1.00, 1.25, and 1.46 ppm. The first signal (δ 1.00 ppm) is attributable to the methyl attached to the β position for its intensity (it integrates for three hydrogens) and multiplicity (doublet). For the same reasons, the other two signals (δ 1.25 and 1.46 ppm) can be assigned to the diastereomeric hydrogens of the γ-methylene. This methylene is then further coupled to the methyl resonating at 0.93 ppm. Note that the exchangeable protons are not visible in this spectrum. Therefore the assignment of isoleucine is:

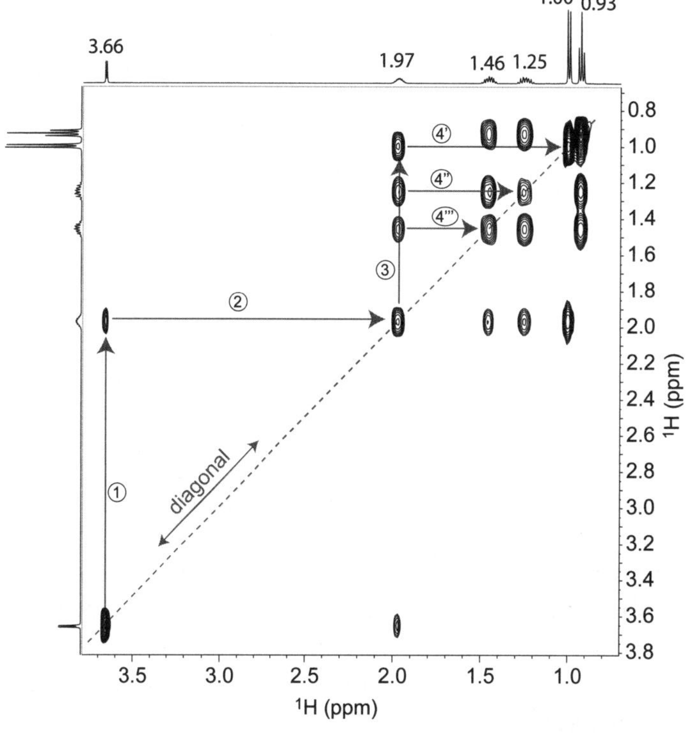

In this case, it is not convenient to draw the sequential connectivities, since it can create confusion. In fact, the signals at 1.97 ppm split the path in three:

7.3 TOtal Correlation SpectroscopY (TOCSY) experiment

The TOCSY experiment shows correlations between all protons within a given spin system, not just between protons directly coupled, as in a COSY spectrum. In fact, in this experiment, the magnetization is transferred successively over up to 5-6 bonds as long as successive protons are coupled. Transfer is interrupted by small or zero proton-proton couplings.

TOCSY is a homonuclear experiment and therefore it is characterized by autocorrelation peaks (diagonal peaks) that split the spectrum in two symmetric halves. By way of example, the TOCSY spectrum of isoleucine is shown in Figure 7.8.

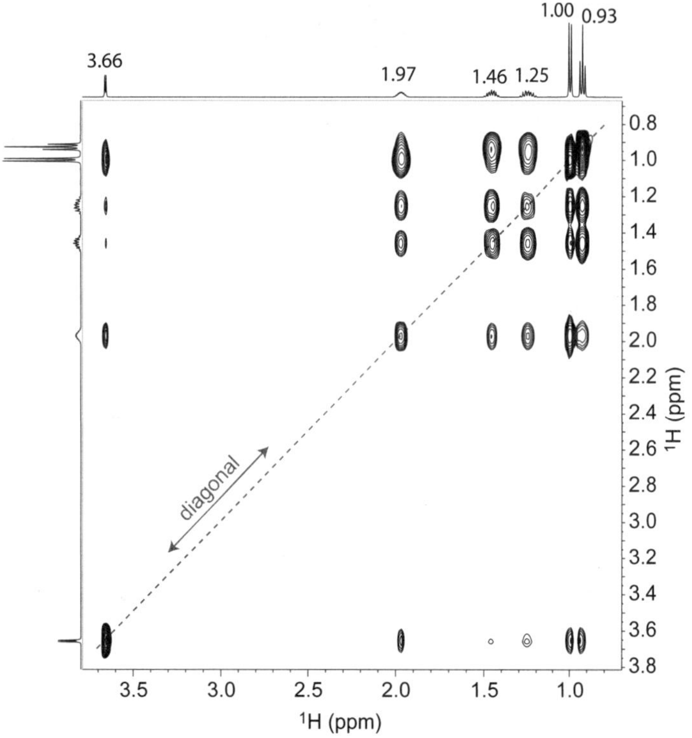

Figure 7.8 TOCSY spectrum (500 MHz, D_2O) of isoleucine.

Since isoleucine is characterized by a single spin system, all the signals are coupled to each other. It is interesting to compare the appearance of this experiment to that of the COSY acquired on the same molecule (see Problem 7.4). In the COSY spectrum, for example, the Hα (δ 3.66 ppm) turns out to be correlated only to Hβ (δ 1.97 ppm) to which it is directly coupled, while in the TOCSY experiment it is coupled to all the other signals of the spin system.

The TOCSY spectrum of isoleucine is also very useful to illustrate a phenomenon often observable in 2D spectra. Focus your attention on the top row of peaks of the spectrum (shown in expanded view in Figure 7.9). Each proton signal (δ 3.66, 1.97, 1.46 and 1.25 ppm) produces two cross-peaks (depicted as black ellipses in Fig. 7.9) correlating to the signals at 0.93 and 1.00 ppm. However, these pairs of cross-peaks are broad and close enough to each other that they form what looks like a single cross-peak.

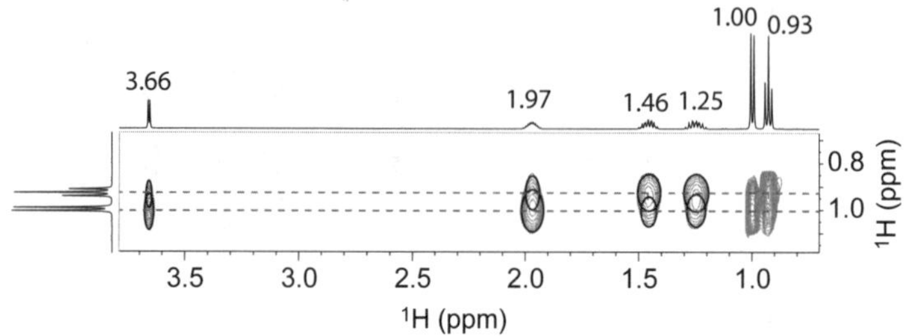

Figure 7.9 Expansion of the top row of cross-peaks of the TOCSY spectrum of isoleucine. The black ellipses indicate the original cross-peaks forming the merged peaks.

However, it is very interesting to observe that, on the contrary, the cross-peaks generated by the same correlations in the other dimension of the spectrum (look at the two column of cross-peaks on the right-most part of the spectrum) are not merged and turn out to be well resolved (Fig. 7.8 and expanded in Fig. 7.10).

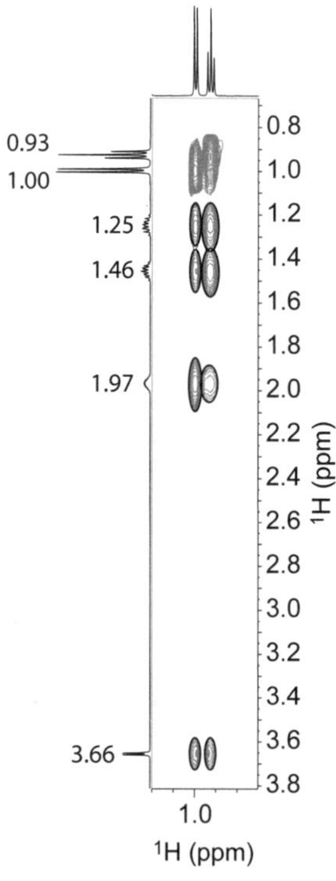

Figure 7.10 Expanded region of two right-most columns of cross-peaks from the TOCSY spectrum of isoleucine.

This is due to the fact that the second dimension (F1) in a 2D plot is generally acquired at a lower spectral resolution (see a theory book). This means that, even if the two halves of the spectrum are theoretically symmetric, they may contain minor but important differences.

TOCSY experiments are particularly useful when there is more than one spin system in the molecule. Let's consider for example the amino acid tyrosine (Fig. 7.11), which is characterized by a spin system involving the aliphatic hydrogens, namely the Hα (δ 3.42 ppm) and the two Hβs (δ 2.68-2.84 ppm), and by another spin system containing the two symmetric couples of aromatic protons at 6.59 and 6.99 ppm. In the TOCSY spectrum (Fig. 7.11), each signal couples only to the signals belonging to its own spin system, and thus two patterns of couplings are clearly distinguishable in the spectrum: one in the upper-right corner of the spectrum and another one in the lower-left corner. Note that the residual solvent signal (HDO) produces only an autocorrelation peak.

The analysis of TOCSY spectra is particularly useful when the signals of different spin systems are interspersed among each other or when there are signals overlapping.

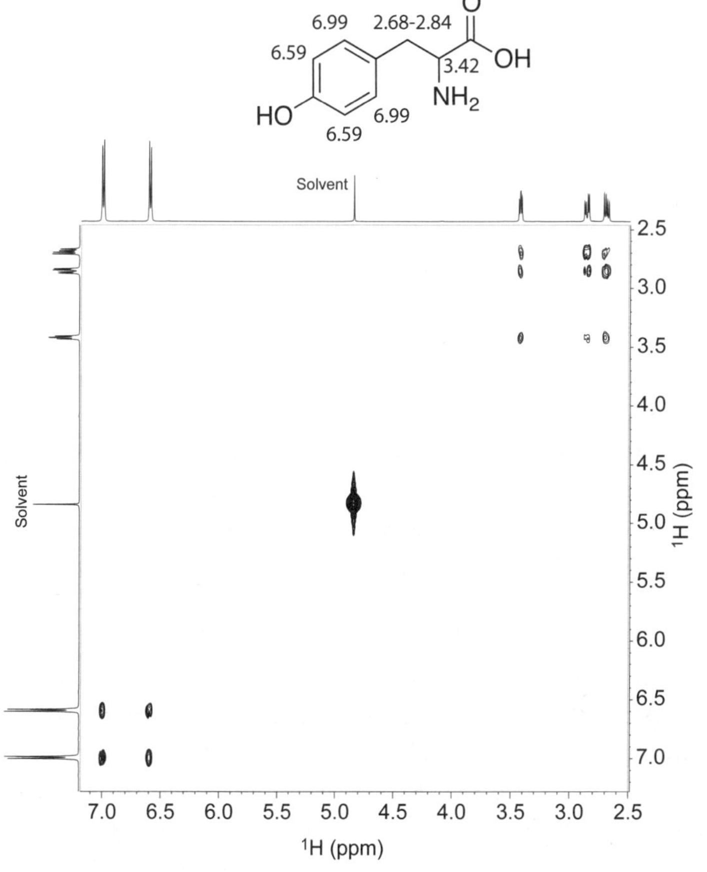

Figure 7.11 TOCSY spectrum (500 MHz, D_2O) of tyrosine.

7.4 Heteronuclear Single-Quantum Correlation (HSQC) experiment

HSQC is a heteronuclear experiment correlating directly bonded ^{1}H and X-heteronuclei (commonly ^{13}C or ^{15}N). The resulting spectrum is 2D with one axis for proton frequencies and the other one for the heteronuclear frequencies (^{13}C in this book). The interpretation of the HSQC is very simple, since there are no diagonal peaks and thus no symmetric halves. Each cross-peak in the spectrum indicates that a given hydrogen (or set of equivalent hydrogens) is attached to a given carbon. This means that if a hydrogen is bonded to a heteronuclear atom different from carbon (like oxygen, nitrogen, sulfur, etc.), it won't produce any correlation peak in the spectrum.

Interestingly, only one cross-peak will be associated with each methyl and methine carbon signal, while, depending on the nature of the hydrogens, one or two cross-peaks are aligned to each methylene carbon signal. In fact, if the hydrogens of the methylene are homotopic or enantiotopic, they resonate as a single signal in the proton spectrum and you will observe only one cross-peak aligned to the pertinent carbon signal; if the hydrogens are diastereotopic, they resonate at different chemical shift values and you will observe two cross-peaks (one for each proton signal) aligned to the same carbon signal. Obviously, there will not be any cross-peaks associated with quaternary carbons.

By way of example, the HSQC spectrum of tyrosine is shown in Figure 7.12. As you can see, it is straightforward to assign each cross-peak to a hydrogen-carbon pair. In particular, the hydrogens resonating at 6.59 and 6.99 ppm turn out to be bonded to the carbons resonating at 121.1 and 133.4 ppm, respectively. Analogously, the hydrogen at 3.42 ppm is bonded to the carbon at 60.3 ppm, and the two diastereotopic hydrogens of the methylene (2.68 and 2.84 ppm) are bonded to the carbon at 42.4 ppm. Obviously the quaternary carbon at 126.8 ppm does not show any correlation peak (please note that tyrosine possesses other two quaternary carbons resonating at 166.5 and 185.4 ppm, that are in the spectral regions omitted from the picture since they do not produce any correlation peak).

The complete assignment of the tyrosine is shown in Figure 7.13.

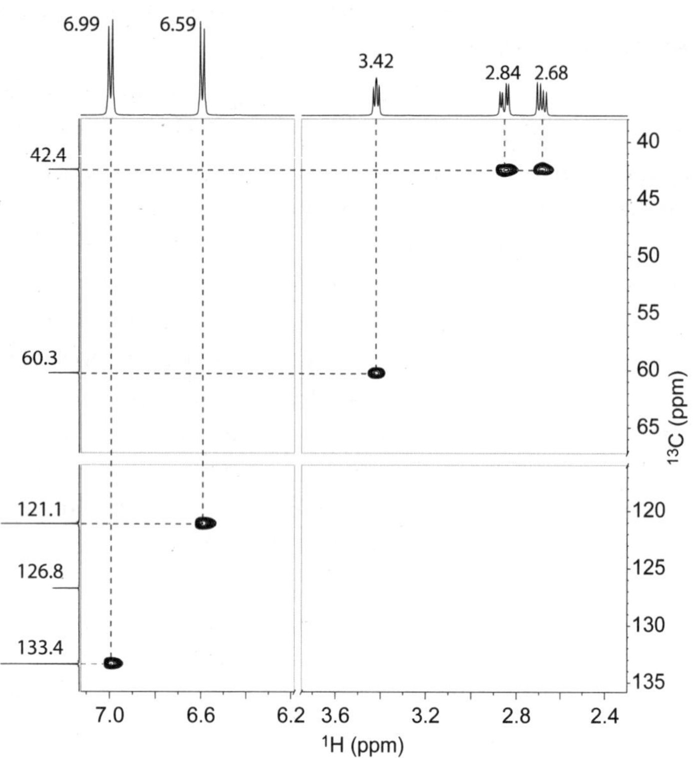

Figure 7.12 HSQC spectrum of tyrosine. Spectral regions having no cross-peaks have been omitted.

Figure 7.13 Structure and ^{1}H and ^{13}C assignments of tyrosine.

HSQC experiments are very useful when there are signals overlapping in the proton spectrum. For example, 2-hydroxyhexanoic acid (Fig. 7.14) is composed of one methine, three methylenes and one methyl.

Figure 7.14 Structure and ^{1}H and ^{13}C assignments of 2-hydroxyhexanoic acid.

Although two of the methylenes have protons resonating at the same chemical shift value (δ 1.31 ppm), they can be easily distinguished by analyzing the HSQC spectrum (Fig. 7.15). In fact, there are two distinct cross-peaks aligned to the overlapped signals at 1.31 ppm, but correlating to two carbon signals at 24.6 and 29.4 ppm.

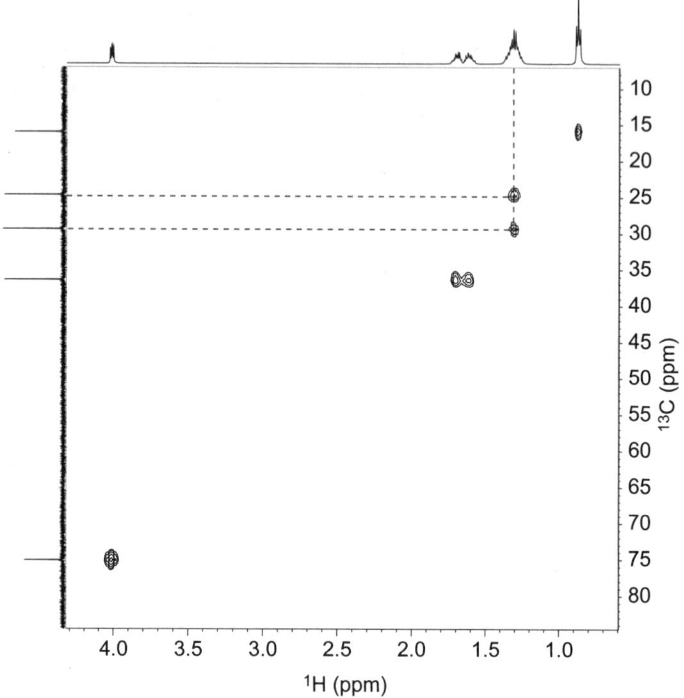

Figure 7.15 HSQC spectrum of 2-hydroxyhexanoic acid.

Another example that demonstrates the usefulness of the HSQC experiment in resolving signal overlap is the case of leucine, whose HSQC is shown in Figure 7.16.

Leucine contains two methyls, one methylene, and two methines (α and γ). The α-methine produces the most deshielded proton signal at δ 3.72 ppm, and the two methyls the most shielded signals at about 0.94-0.95 ppm. The remaining methylene and γ-methine resonate overlapped at about 1.70 ppm. First of all, the analysis of the HSQC spectrum clarifies that the signal at 0.94-0.95 ppm actually is not a triplet, but two partially superimposed doublets (Fig. 7.17). This is clear because the two HSQC cross-peaks under those proton signals are not aligned to the same proton frequency and they correlate to two carbon signals at 23.6 and 24.8 ppm. The HSQC also resolves the overlap of the signals at 1.70 ppm, which turn out to be coupled to two carbon atoms resonating at 26.8 and 42.5 ppm.

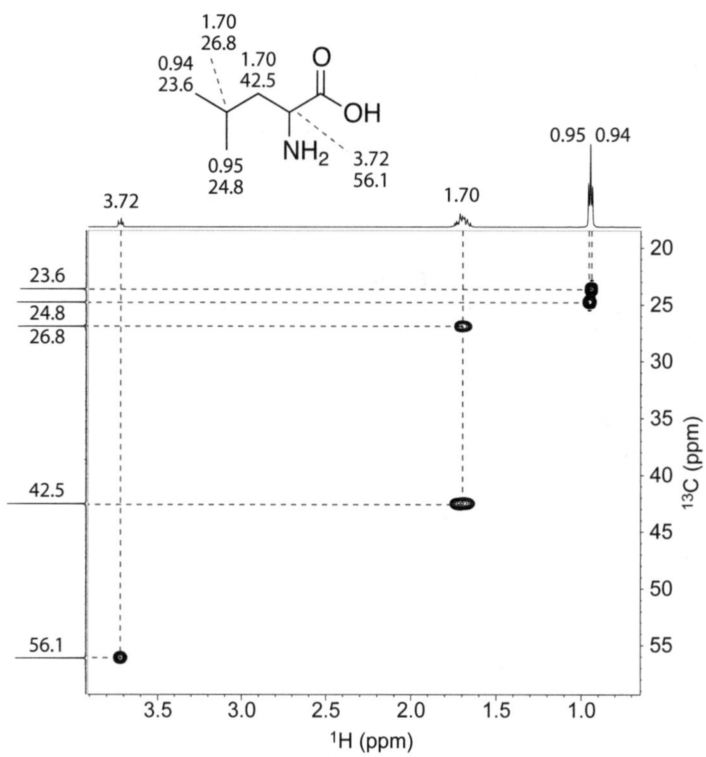

Figure 7.16 HSQC spectrum and assignment of leucine.

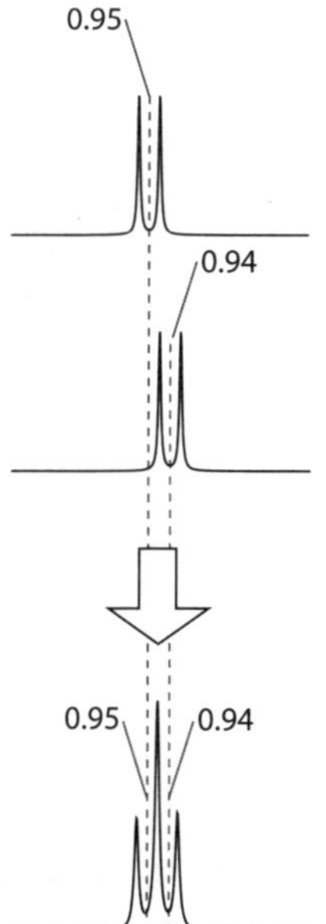

Figure 7.17 Schematic representation of the origin of the partial superimposition of the two methyl signals of leucine.

7.5 Heteronuclear Multiple Bond Correlation (HMBC) experiment

The HMBC experiment is a heteronuclear experiment correlating ^1H and X-heteronuclei (^{13}C in this book) separated from each other by 2-4 chemical bonds. HMBC is very useful to confirm the spin system determined by the COSY experiment and to connect one spin system to another. HMBC is also extremely useful to determine the position in the molecule of quaternary carbons and, when possible, to determine the position of exchangeable hydrogens (see examples below).

HMBC has no diagonal peaks and each cross-peak describes unique *long-range* proton/carbon correlations.

It is important to note that this experiment is characterized by some peculiarities. First, the magnitude of the cross-peaks does not reflect the nearness of the two coupled nuclei. For example, two nuclei that are separated by three bonds generally have a more intense cross-peak than two nuclei separated by two bonds. This is particularly evident in aromatic systems, where two-bond correlations are very weak in comparison to the three-bond ones, which are instead very intense. Second, this experiment may contain artifacts. In fact, although it is designed to get rid of the one-bond correlations, these may not be properly filtered out and may appear in the spectrum. Since HMBC experiments are acquired without ^{13}C decoupling, they appear as doublets along the proton dimension, that is, as pairs of cross-peaks aligned to a carbon signal.

Examples of these artifacts are shown in the HMBC spectrum of *n*-propanol (Fig. 7.18, bottom). The distance between the pair of cross-peaks is the direct heteronuclear coupling constant ($^1J_{CH}$); therefore, the midpoint between the two cross-peaks is centered on the signal of the hydrogen. Hence, the centers of each artifact pair are positioned at the frequencies where you find the cross-peaks in the HSQC spectrum (Fig. 7.18, top). These artifacts are easily recognizable because they are not aligned to any proton signal in the HMBC spectrum (Fig. 7.19).

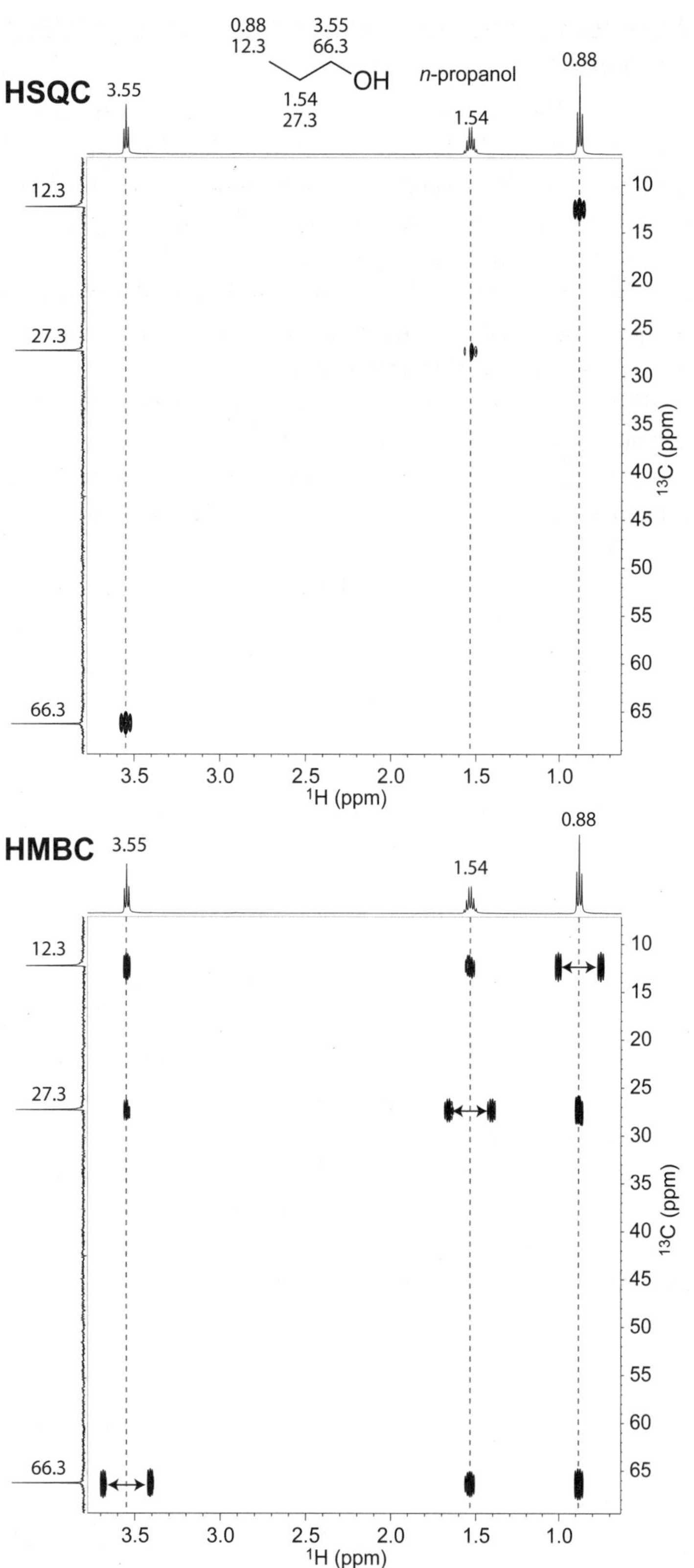

Figure 7.18 HSQC and HMBC spectra (^{1}H 500 MHz, ^{13}C 125 MHz; D_2O) of *n*-propanol. Residual one-bond couplings in the HMBC spectrum are indicated by double-headed arrows.

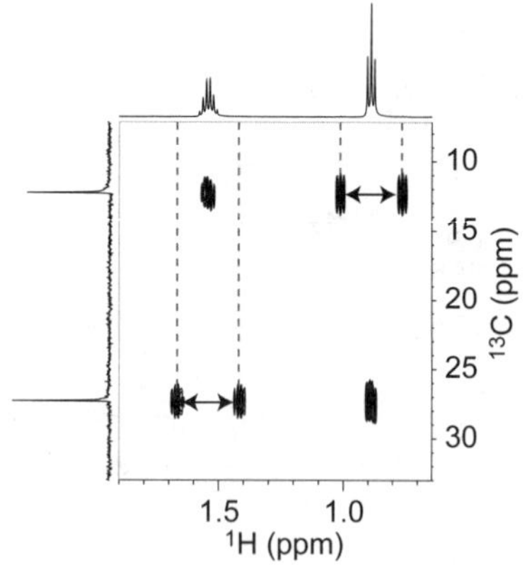

Figure 7.19 Expanded region of the HMBC spectrum of *n*-propanol. Artifacts do not correlate with any proton signal.

Long-range couplings are generally described graphically on the structure of the molecule by arrows. By convention, the arrow starts from the hydrogen and ends on the carbon atom. Figure 7.20 reports the long-range correlations observed in the HMBC spectrum of *n*-propanol.

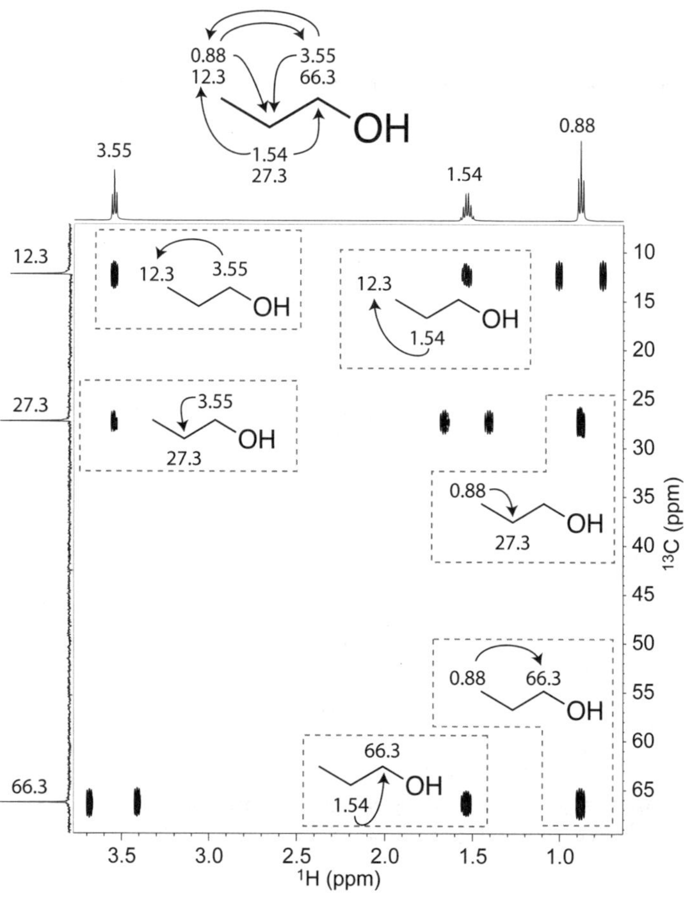

Figure 7.20 Graphical visualization of the HMBC long-range connectivities of *n*-propanol.

All the correlations are summarized in the top of the figure, while the individual correlations are highlighted in the spectrum. In particular, each correlation is reported graphically along with the pertinent cross-peak boxed in a gray dashed line.

It is important to note that, in HMBC experiments of symmetric molecules, there could be unexpected cross-peaks.

Consider for example 4-aminobenzoic acid (Fig. 7.21). This molecule is characterized by a plane of symmetry that splits the aromatic moiety in half. The protons at 7.72 ppm turn out to be directly bonded to the carbons at 133.6 ppm, while the protons at 6.82 ppm are bonded to the carbons at 117.9 ppm. However, the protons at 6.82 and 7.72 ppm are also three-bond separated from the carbons at 117.9 and 133.6 ppm, respectively, which belong to the other symmetric half of the aromatic moiety (see arrows in Figure 7.21).

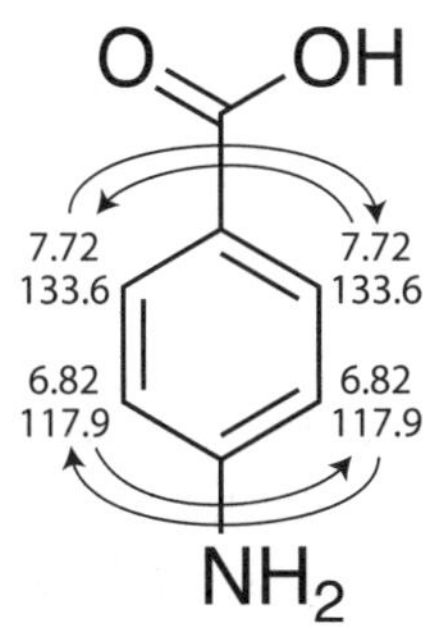

Figure 7.21 Structure of 4-aminobenzoic acid with partial proton and carbon assignments. Arrows indicate long-range couplings observed in the HMBC spectrum between symmetric moiety.

Since three-bond correlations are particularly intense, they are expected to be present in the HMBC spectrum (along with other correlations). Pay attention, because these correlations appear exactly midway between the residual one-bond couplings, so perfectly aligned to the proton signal in the 1D projection (Fig. 7.22). Generally, these cross-peaks create confusion, since they seem correlate directly bonded atoms.

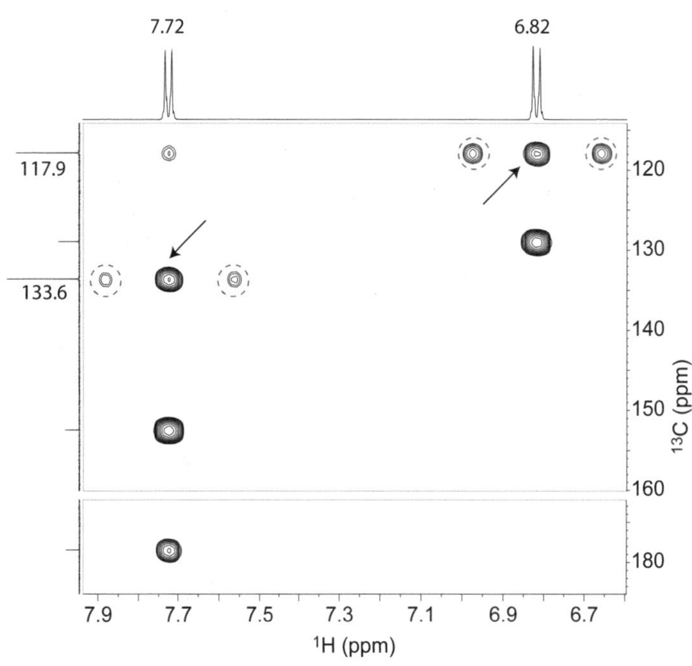

Figure 7.22 HMBC spectrum (^{1}H 500 MHz, ^{13}C 125 MHz; D_2O) of 4-aminobenzoic acid. Cross-peaks in dashed gray circles are residual one-bond couplings. Cross-peaks indicated by arrows are long-range correlations between two symmetric halves of the aromatic moiety.

Problem 7.5

Identify the residual one-bond coupling cross-peaks in the following HMBC spectrum (^{1}H 500 MHz, ^{13}C 125 MHz; D_2O).

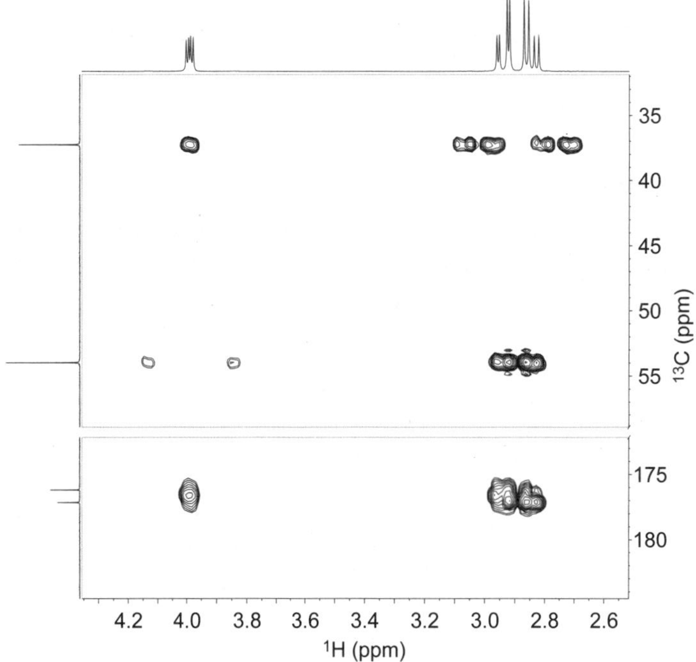

Solution

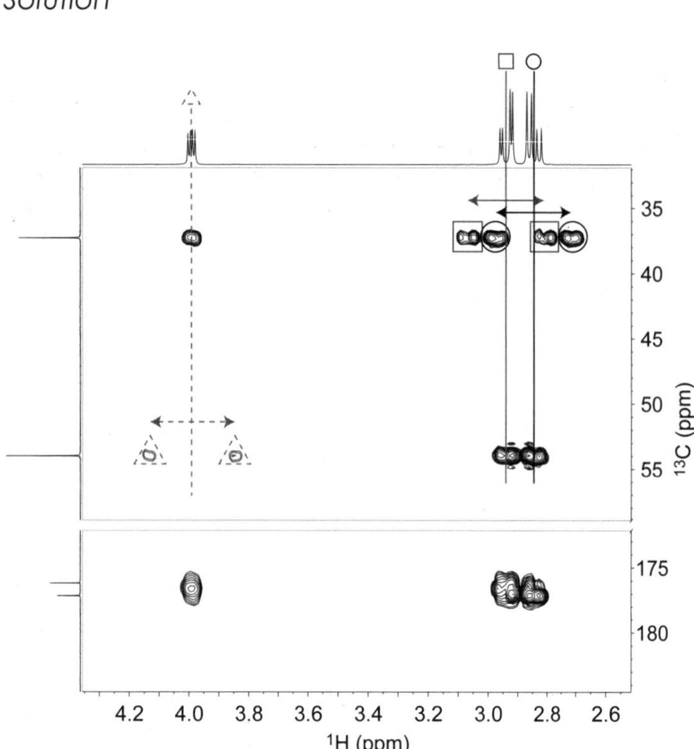

It is possible to identify the residual one-bond coupling cross-peaks by tracing vertical lines corresponding to each proton signal: the cross-peaks falling directly on the lines are long-range couplings, while all the others are presumably artifacts.
In this spectrum there are three residual one-bond correlations. The first one, very easy to identify, is aligned to the carbon at 54 ppm (dashed gray triangles). The other two are instead partially overlapped and aligned to the carbon at 37.3 ppm (gray squares and black circles).

Problem 7.6

Identify the residual one-bond coupling cross-peaks in the following HMBC spectrum (^{1}H 500 MHz, ^{13}C 125 MHz; D_2O) of the amino acid L-glutamic acid.

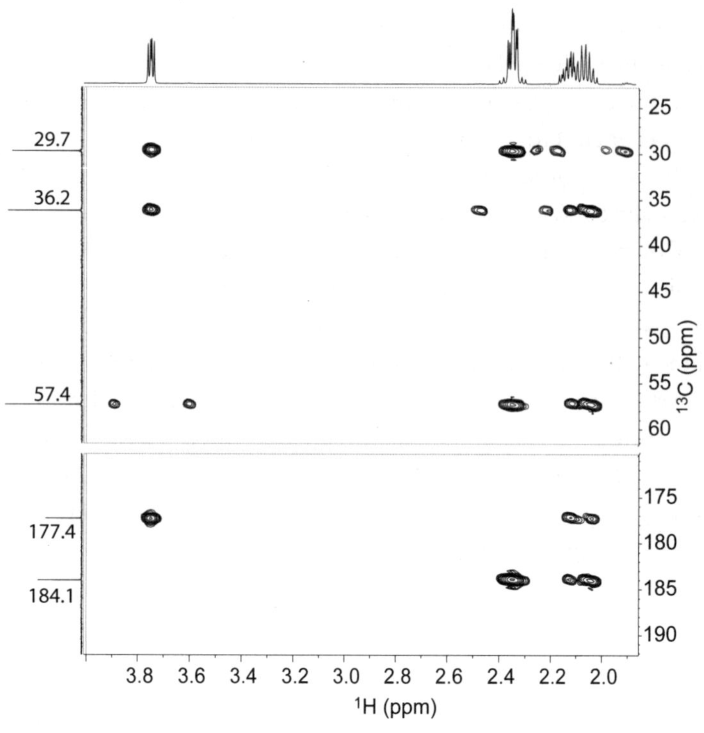

Solution

This problem can be solved using the same rationale used in the previous problem: first of all, you have to trace vertical lines corresponding to each proton signal. However, while the centers of the signals at 2.34 and 3.75 ppm are clearly identifiable, the centers of the two multiplets between 2.00 and 2.20 ppm cannot be easily determined (see figure below).

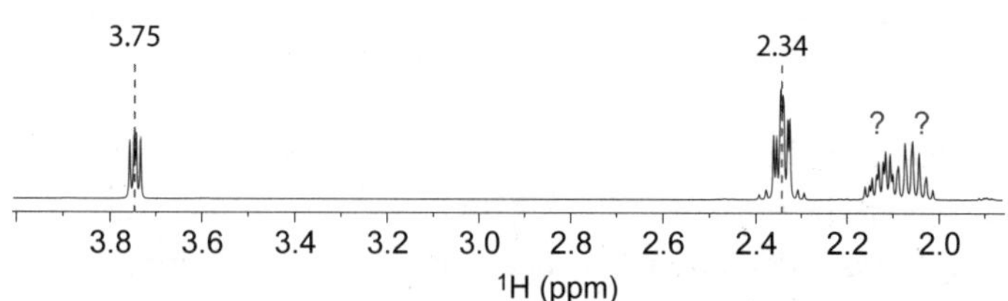

In order to find the center of those signals, you should look for the centers of the corresponding long-range correlation peaks, which must be aligned to the proton signal in the 1D projection, like in the figure below.

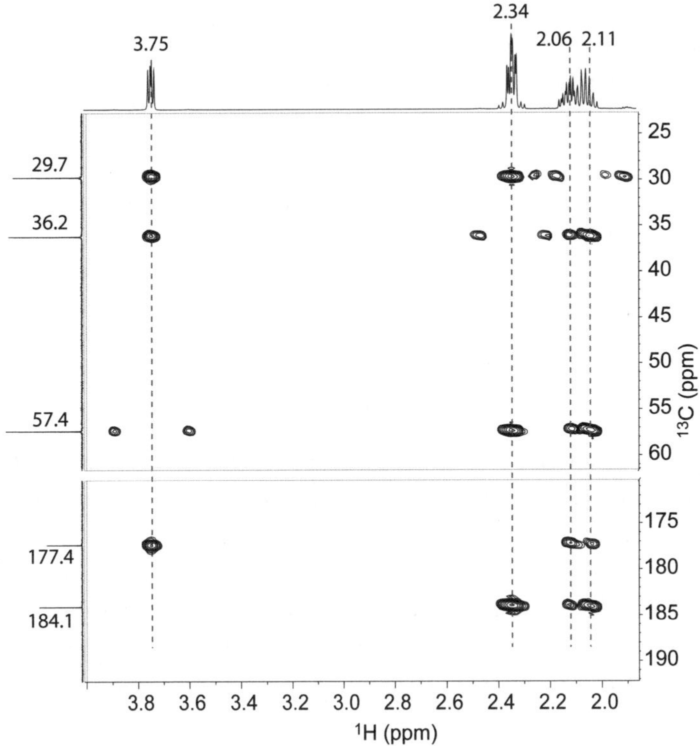

Thus, the residual one-bond correlations between the signals at 3.75 ppm (^{1}H) and 57.5 ppm (^{13}C) and between the signals at 2.34 ppm (^{1}H) and 36.2 ppm (^{13}C) are simply identified, and they are highlighted by dashed gray triangles and dashed dark hexagons, respectively (see figure below).

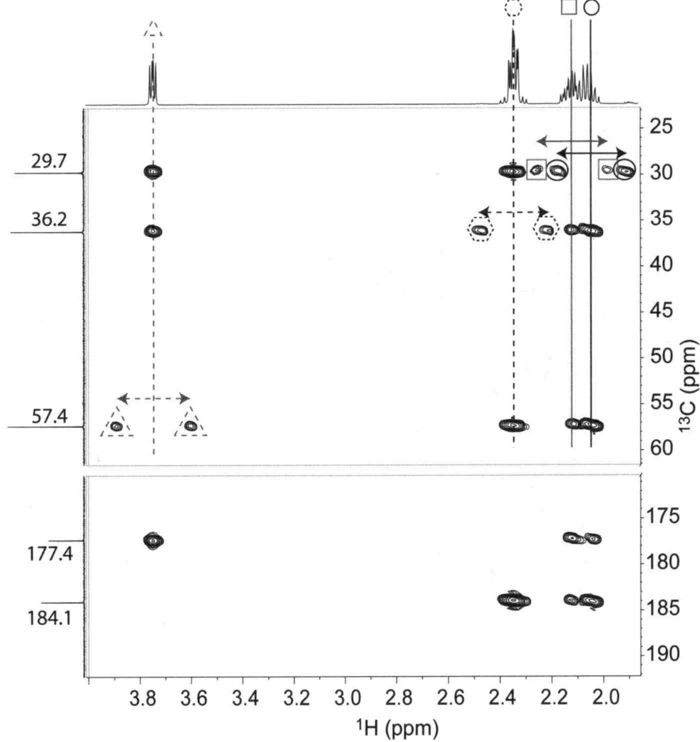

The artifacts generated by the proton signals at 2.06 and 2.11 ppm are not as easily recognized. Nevertheless, considering that the midpoint between each pair of cross-peaks must be centered on the proton resonance, it is possible to distinguish the last two pairs of artifacts, which are highlighted by gray squares and dark circles.

Problem 7.7

Imagine that, by the use of COSY and HSQC experiments, you have partially assigned the proton and the carbon resonances of the amino acid L-glutamic acid as follows:

(b) 2.06-2.11 29.7 (a)
HO 2.34 36.2 3.75 57.4 OH
NH_2

L-glutamic acid

Using the HMBC spectrum (^{1}H 500 MHz, ^{13}C 125 MHz; D_2O) reported in Problem 7.6 (which refers to the same molecule), can you assign the resonances of the two carboxylic groups (a) and (b)?

Solution

Looking at the carbon spectrum reported in the projection of the HMBC plot, you can easily identify the signals of the two carboxyl groups that resonate at 177.4 and 184.1 ppm. The problem is to determine "who's who." The strategy is to look for diagnostic long-range connectivities that may distinguish the two carbon atoms. Interestingly, the hydrogens resonating at 2.34 ppm are four bonds away from carbon (a) (no correlation should be observed), while they are only two bonds away from carbon (b). Therefore, since the proton signal at 2.34 ppm correlates only with the carbon signal at 184.1, it is possible to assign this resonance to carbon (b). By exclusion, carbon (a) can be assigned to the signal at 177.4 ppm. These assignments are further corroborated by the analysis of the connectivities of Hα (δ 3.75 ppm), which is two bonds distant from carbon (a) and four bonds distant from carbon (b). In fact, this hydrogen correlates with the signal at 177.4 ppm, and not with that at 184.1 ppm.
Please note that the long-range connectivities of the hydrogens resonating at 2.06 and 2.11 ppm are not useful to address this problem, since they are both three bonds away from the two carboxylic groups and produce similar long-range connections.

Problem 7.8

By the interpretation of COSY and HSQC experiments, a partial assignment of 4-acetamidobutyric acid has been accomplished.

4-acetamidobutyric acid

Using the HMBC spectrum (^{1}H 500 MHz, ^{13}C 125 MHz; D_2O) reported below, can you assign the resonances of the two carboxylic groups (a) and (b)?

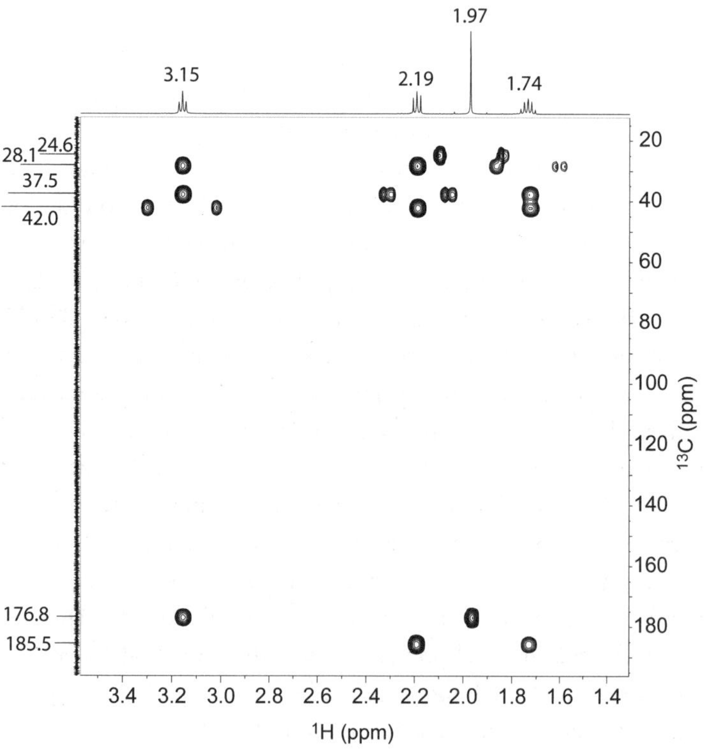

Solution

To answer to this question, you can focus your attention on the bottom part of the HMBC spectrum.

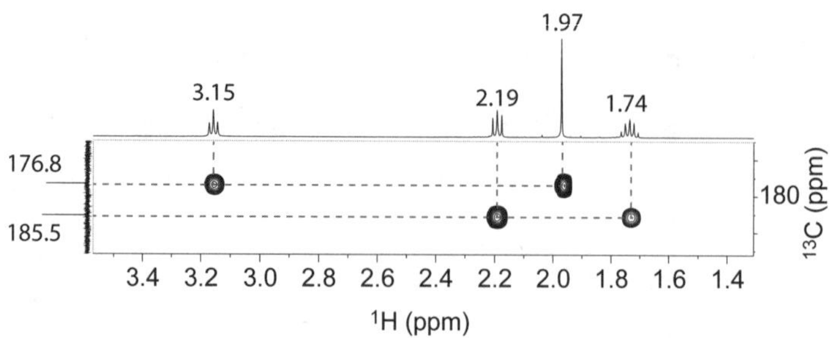

The two carbonyl groups resonate at 176.8 and 185.5 ppm. If you consider the number of intervening bonds between each hydrogen and the two carboxyl groups, you can see that only the proton signals at 2.19 and 1.74 ppm can provide correlations to carbonyl (a). On the other hand, only the hydrogens resonating at 1.97 and 3.15 ppm can provide correlations to carbonyl (b). Therefore, the complete assignment of 4-acetamidobutyric acid is:

Problem 7.9

Imagine that you are studying the structure of 3-(2'-hydroxyphenyl) propionic acid by NMR, using 1D proton and carbon spectra, DEPT, COSY, and HSQC experiments.

3-(2'-hydroxyphenyl)propionic acid

Suppose that, by the combined analysis of those experiments, you were able to define the presence of two spin systems and three quaternary carbons:

6.90 118.7	7.16 130.4	6.93 123.7	7.21 133.0
—CH—	CH—	CH—	CH—

2.48 40.4	2.84 29.0	131.5	156.1	185.6
—CH_2—	CH_2—	C	C	C

Using the HMBC experiment (^{1}H 500 MHz, ^{13}C 125 MHz; D_2O) reported below, can you assign each proton and carbon resonance to the pertinent nucleus?

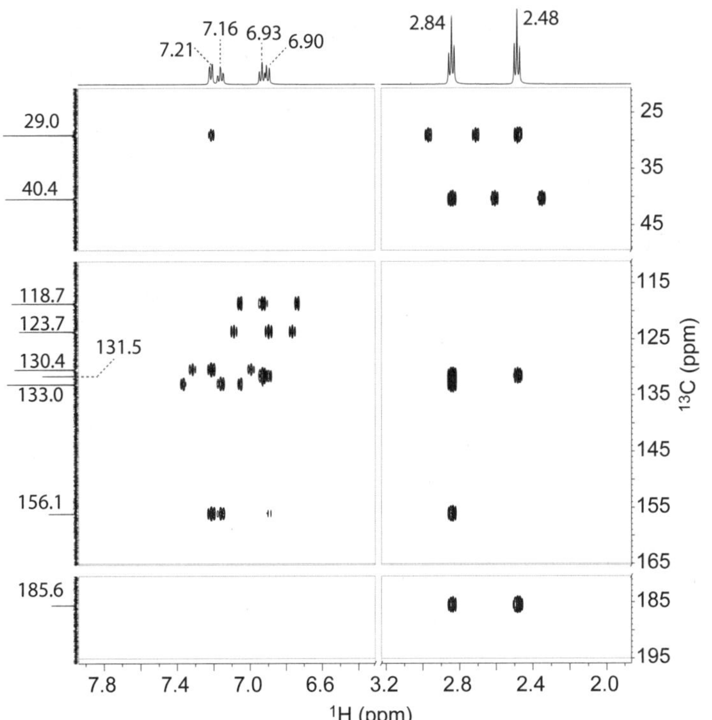

Solution

It is clear that the fragments -CH-CH-CH-CH- and -CH_2-CH_2- refer to the aromatic and aliphatic portion of the molecule, respectively. Unfortunately, you can place those assignments in different ways on the structure of the molecule. Basically, you have four possibilities:

In order to determine which is the right assignment, it is convenient first to assign the quaternary carbons. Reasoning only in terms of chemical shift values, the signal at 185.6 ppm can be definitively assigned to the carboxyl group, while the signals at 131.5 and 156.1 ppm should be assigned to the aromatic quaternary carbons (see the carbon chemical shift scheme in the inside back-cover flap).

Of these, the signal at 156.1 ppm can be assigned to the oxygen-bearing aromatic carbon (because it is more deshielded), and, by exclusion, the signal at 131.5 ppm can be assigned to the other quaternary carbon of the aromatic ring.

Let's now start to analyze the aliphatic assignments of the molecule. It is convenient to draw the structure of the molecule with the two possible assignments, showing graphically all the possible HMBC connectivities. The strategy is to evaluate which hypothesis of assignment is more consistent with the pattern of long-range correlations observed in the HMBC spectrum, paying particular attention also to the intensities of the cross-peaks.

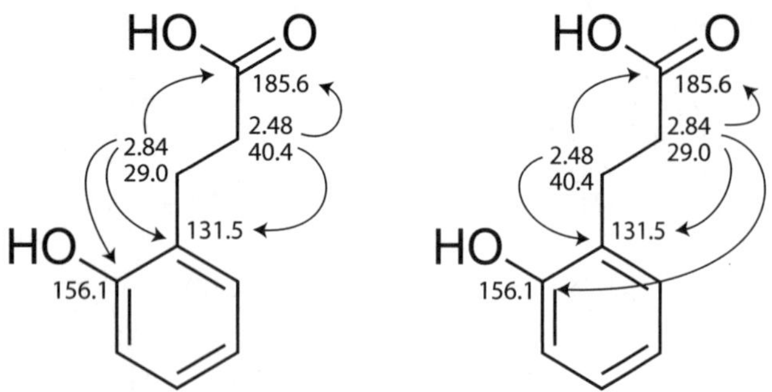

All the cross-peaks involving the aliphatic portion of the molecule are intense. All the correlations are reasonable in both the hypotheses, except the long-range coupling between the hydrogen at 2.84 ppm and the carbon at 156.1 ppm, which, in the assignment reported on the right, seems impossible, since it would be through 4 bonds (remember that four-bond correlations are very rarely seen and, if observed, have very low intensity). Therefore, the assignment reported on the right can be ruled out.

The same strategy can be used for the assignment of the aromatic moiety.

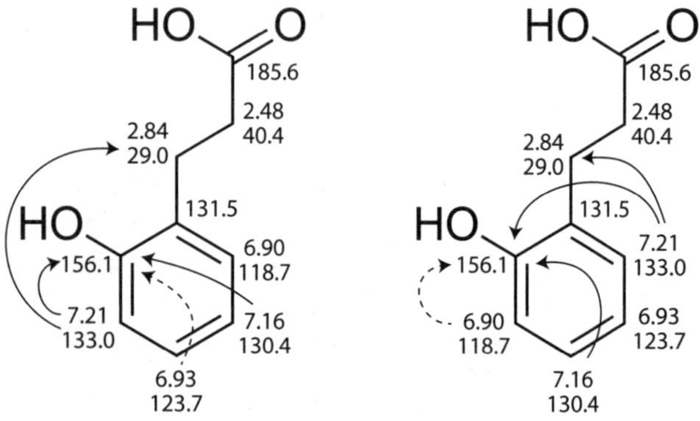

Strong and weak long-range couplings are distinguished in the figure as continuous and dashed arrows, respectively. The strong correlation between the hydrogen at 7.21 ppm to the carbon at 29.0 ppm turns out to be ${}^4J$ in the hypothesis of assignment reported on the left, while it is a ${}^3J$ connection in the hypothesis on the right. Since a ${}^4J$ coupling cannot provide an intense cross-peak in the HMBC spectrum, this correlation strongly suggests that the hypothesis on the left can be ruled out. Other correlations are also in agreement with this idea. In fact, strong correlations between the protons at 7.16 ppm and 7.21 ppm with the carbon at 156.1 ppm turn out to be ${}^4J$ and ${}^2J$ couplings in the hypothesis on the left (both should produce very low-intensity cross-peaks), while they occur through three intervening bonds in the hypothesis on the right. In addition, the coupling between the proton at 6.93 ppm and the carbon at 156.1 ppm is weak and therefore not consistent with a ${}^3J$ coupling reported in the hypothesis on the left, while it is a ${}^2J$ connection in the assignment on the right. In conclusion, it is possible to unambiguously affirm that the assignment of 3-(2'-hydroxyphenyl)propionic acid is that reported on the right.

Problem 7.10

Using the following HSQC and HMBC spectra (^{1}H 500 MHz, ^{13}C 125 MHz; D_2O), can you assign the exchangeable protons of dihydroconiferyl alcohol (the exchangeable protons are visible because the spectra have been acquired in deuterated DMSO)? The assignments of the non-exchangeable nuclei are provided.

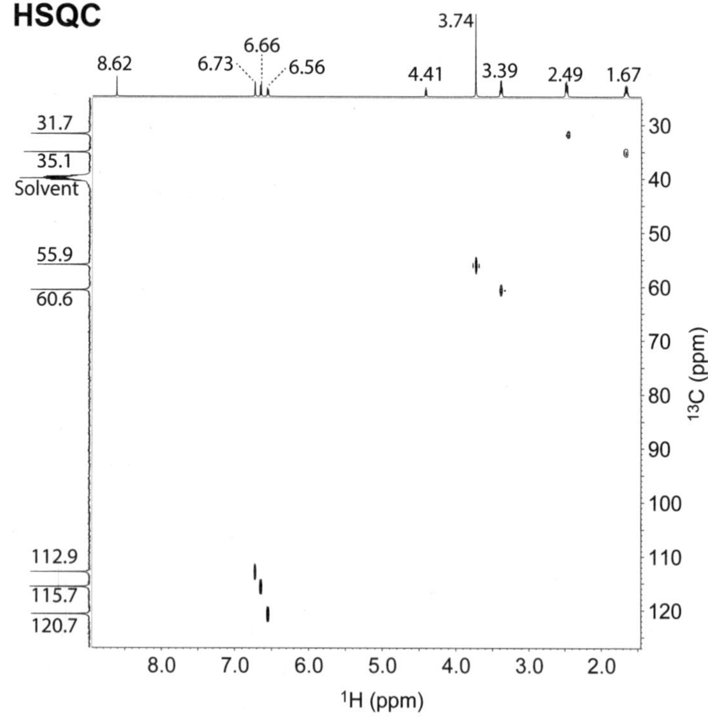

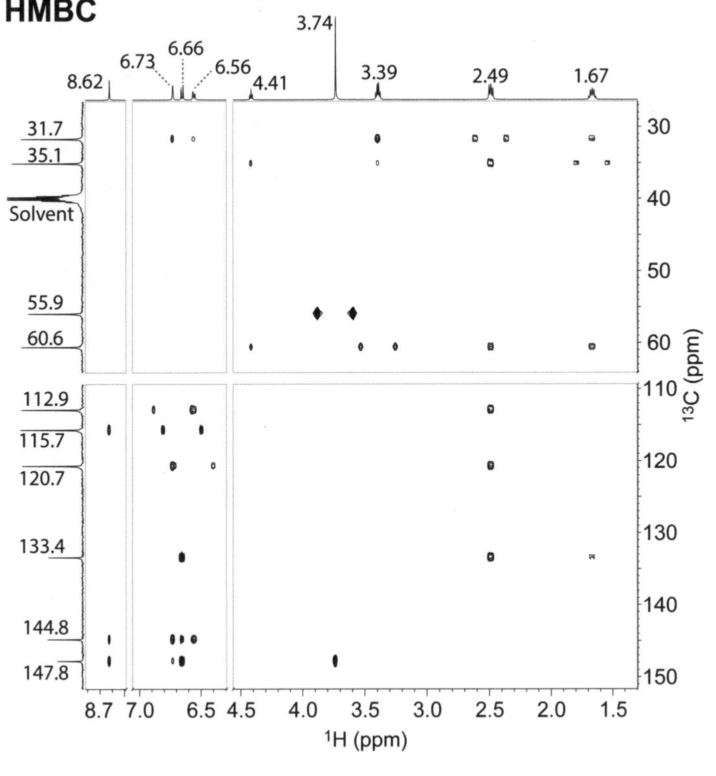

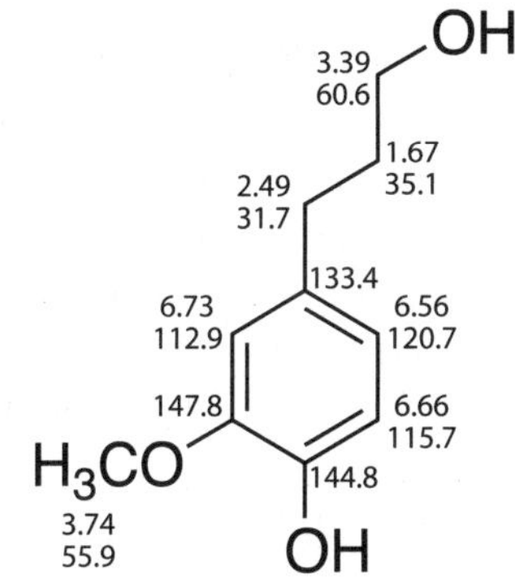

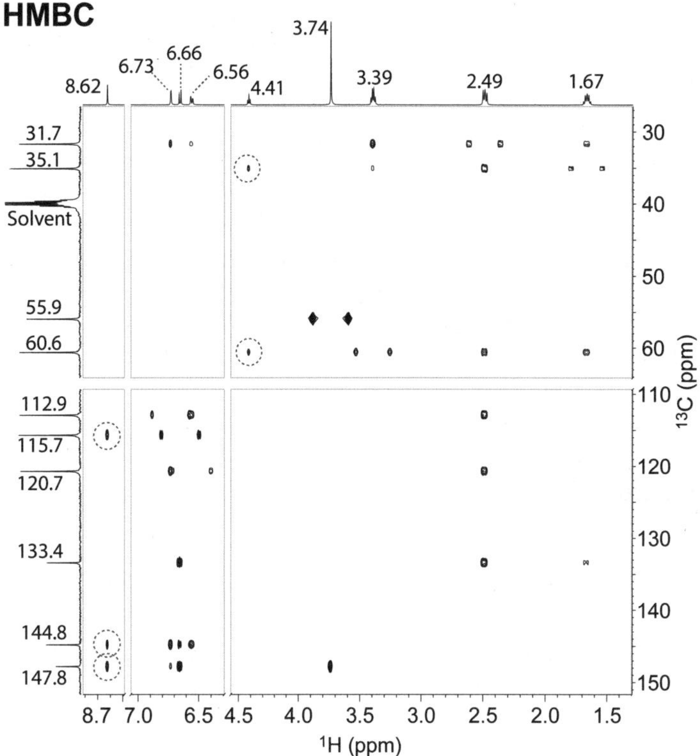

Solution

As mentioned in Section 7.4, protons bonded to heteroatoms different from carbon do not provide any cross-peak in the HSQC spectrum. In the HSQC spectrum reported above, two signals, resonating at 4.41 and 8.62 ppm, are not correlated to any carbon. To understand who's who, you can analyze the HMBC spectrum. In particular, the proton signal at 4.41 ppm can be assigned to the alcoholic hydrogen thanks to its long-range correlations to the carbon signals resonating at 35.1 and 60.6 ppm (highlighted by dashed gray circles in the figure below). Analogously, the proton signal at 8.62 ppm can be assigned to the phenolic hydrogen thanks to its long-range correlations to carbon signals at 115.7, 144.8, and 147.8 ppm (see figure below).

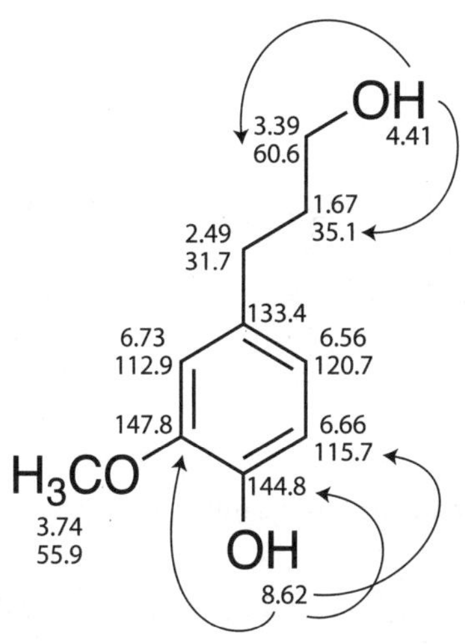

Please note that the phenolic proton resonates at higher frequencies than the alcoholic one because it is more acidic, and because it can form an intramolecular hydrogen bond with the adjacent OCH_3 group as shown in the following figure:

OH
4.41
H_3CO
H–O
8.62

Problem 7.11

The chemical structure of a disaccharide (having molecular formula $C_{12}H_{22}O_{11}$) is studied by NMR spectroscopy. In particular, by the interpretation of 1D proton and carbon, DEPT, COSY, and HSQC experiments, two sugar subunits have been determined and assigned as reported below.

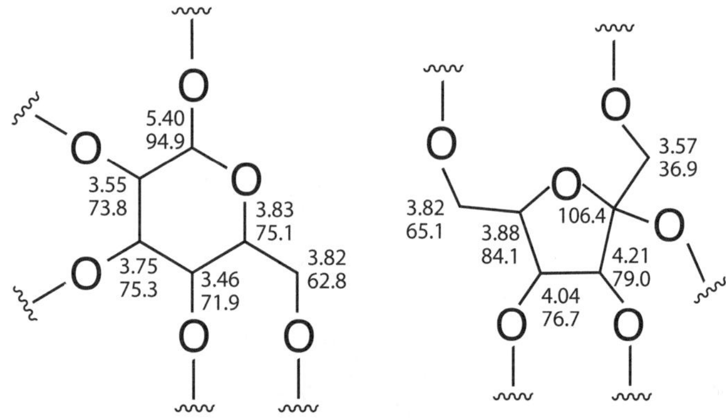

Using the HMBC spectrum (^{1}H 500 MHz, ^{13}C 125 MHz; D_2O) shown below, can you determine how the two subunits are bonded together?

Please note that one of the oxygen atoms is actually shared by the two subunits. Furthermore, in agreement with the molecular formula, the valences of the oxygen atoms not used to join the two units must be saturated by hydrogens.

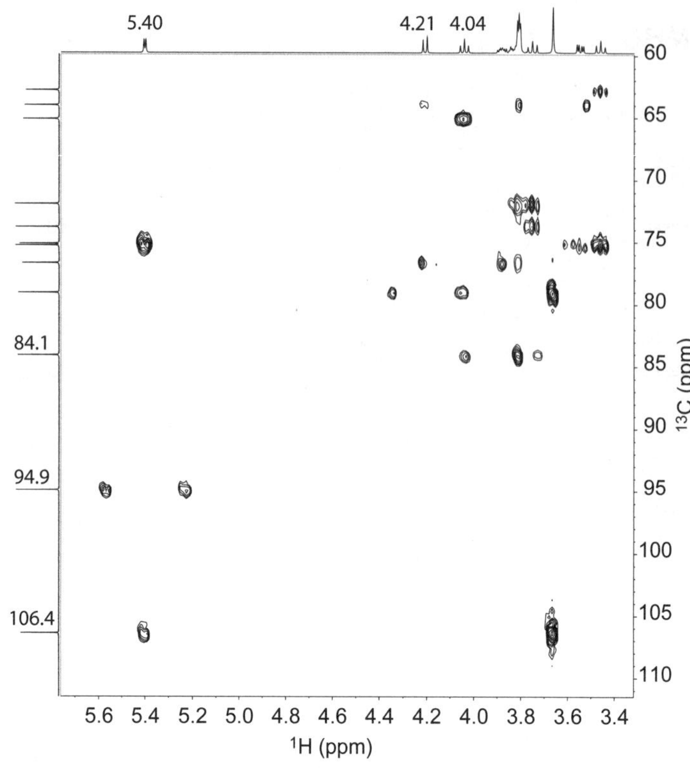

Solution

To solve this problem, you need to find at least a diagnostic long-range correlation between a proton of one subunit and a carbon of the other subunit. In this problem, the correlation between the proton at 5.40 ppm of the subunit having the six-membered ring and the carbon atom at 106.4 ppm of the subunit having the five-membered ring, clearly suggests how the two sugar moieties are bonded together.

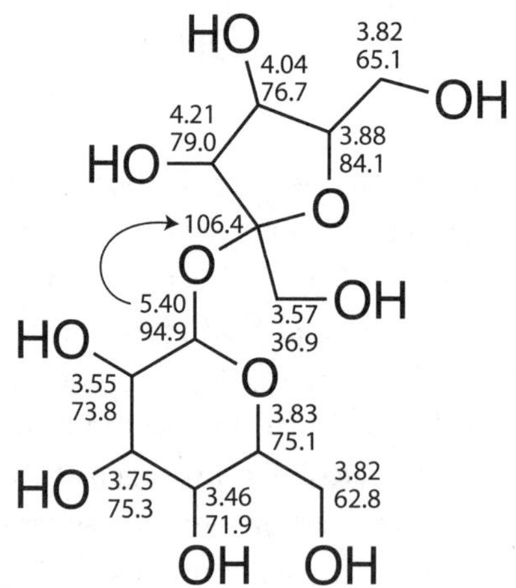

CHAPTER 8

Strategy of structural elucidation

8.1 Introduction

In the preceding chapters, very often you have been asked to assign the proton and carbon signals to the nuclei of a known chemical structure. This approach is very didactic and useful for understanding the rationale behind NMR signal assignment. In this chapter, you will instead learn how to determine the structure of an unknown compound from a set of NMR spectra. This is definitely harder and requires a suitable strategy of interpretation, which takes advantage of the large amount of information that NMR spectroscopy can provide.

Please note that there is not necessarily a 'right' or a 'wrong' way to approach a set of NMR spectra. You should choose the method that best fits your skills and perspective.

A possible strategy of interpretation is proposed here and is divided into ten steps (see below). Each step is first generally described, and then practically applied in the problems provided at the end of this chapter.

8.2 Strategy of structural elucidation

1) Calculate the number of double bond equivalents (DBEs)

Sometimes NMR spectroscopy alone is sufficient to determine the chemical structure of an unknown compound. Other times it is not, and other analytical techniques, like mass spectrometry (MS) and infrared (IR) or ultraviolet-visible (UV-Vis) spectroscopies, are necessary for the structural elucidation. In recent years, with the advent of modern NMR techniques, IR and UV-Vis spectroscopies are less frequently used to this end, and so they are not included in this book. Instead, MS continues to be (along NMR) one of the main techniques for the structural elucidation of unknown compounds. In this book, MS is only *virtually* used in order to obtain the molecular formulae of the studied molecules.

The molecular formula can be very useful to verify and/or facilitate the interpretation of NMR data. For example, it can be used to calculate the number of *double bond equivalents* (DBEs) (also termed *degree of unsaturation)*, which is the sum of π bonds and rings present in the molecule.

Some examples of molecular structures containing different combinations of π bonds and/or rings are reported in Figure 8.1.

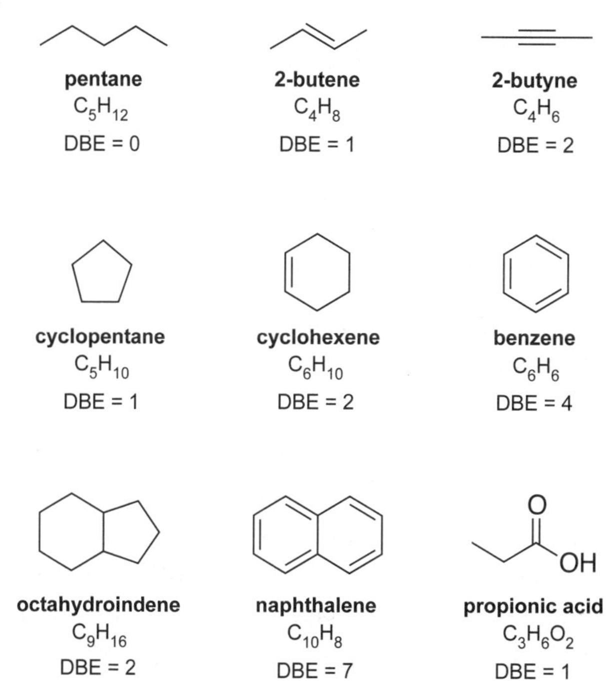

Figure 8.1 DBEs for some representative molecules.

Let's analyze the number of DBEs of each of these molecules:

- *Pentane* is a saturated acyclic alkane, so it does not have any π bonds or rings (DBE = 0).
- *2-butene* has a double bond (no ring), so DBE = 1.

- *2-butyne* contains a triple bond that is made of two π bonds (no ring), so DBE = 2.
- *Cyclopentane* is a cyclic molecule and does not contain any double bonds, so DBE = 1.
- *Cyclohexene* is a cyclic molecule and contains a double bond, so DBE = 2.
- *Benzene* is cyclic and it is formed by three π bonds, so DBE = 4.
- *Octahydroindene* is a bicyclic molecule with no double bonds, so DBE = 2.
- *Naphthalene* is a bicyclic aromatic molecule that contains 5 π bonds, so DBE = 7.
- *Propionic acid* possesses only one double bond (carbonyl), so DBE = 1.

The number of DBEs can be calculated from the molecular formula by the following equation:

$$\textbf{DBE} = C - \frac{H}{2} - \frac{X}{2} + \frac{N}{2} + 1$$

where
C = number of carbon atoms;
H = number of hydrogen atoms;
X = number of halogen atoms;
N = number of nitrogen atoms.

Thus, for example, the DBE equation for a compound having molecular formula $C_3H_6O_2$ is:

$$\text{DBE} = C - (H/2) - (X/2) + (N/2) + 1 = 3 - (6/2) - (0/2) + (0/2) + 1 = 1$$

This means that this molecule contains a ring or a π bond (it is not possible to determine which of the two).

The equation to calculate the number of DBEs is for molecules containing only carbons, hydrogens, monovalent halogens, and nitrogens (oxygen and other divalent atoms, like sulfur, do not contribute to the degree of unsaturation). However, if your molecule contains other types of atoms, they can be taken into consideration according to their valence. For example, if there is a phosphorus atom (P) that has three valences, it can be counted along with the nitrogen atoms (that have three valences as well); if there is a silicon atom (that has four valences), it can be counted along with the carbon atoms, and so on. If you do not know the valence of the atoms in the formula, you cannot use the DBE equation reported here.

Pay attention, because the DBE equation only works for neutral molecules, and only if all the non-hydrogen atoms in molecule obey the octet rule. Consider, for example, the cases of methylammonium and acetate ions (Fig. 8.2).

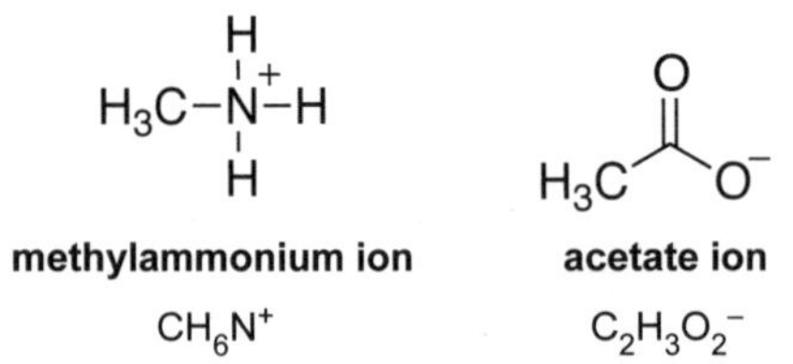

methylammonium ion CH_6N^+ **acetate ion** $C_2H_3O_2^-$

Figure 8.2 Methylammonium ion and acetate ion molecular structures.

Although methylammonium ion has all single bonds and no ring, the DBE equation indicates that this molecule has a degree of unsaturation of -1/2! [DBE = 1 - (6/2) - (0/2) + (1/2) + 1 = -1/2)].

Analogously, the calculated DBE for acetate ion is +3/2! [DBE = 2 - (3/2) - (0/2) + (0/2) + 1 = +3/2)].

In simple cases like these, you can try to calculate the number of DBE by modifying the molecular formula. In particular, you have to neutralize the charge by adding or subtracting, as necessary, a hydrogen ion (H^+), and then apply the DBE equation. Therefore, NH_4^+ should be treated as if it were NH_3, and $C_2H_3O_2^-$ as $C_2H_4O_2$. In this way, they would be correctly calculated as DBE = 0 and 1, respectively. However, this approach works only if the charge is due to protonation or deprotonation of the molecule. Unfortunately, it can be very difficult to understand from the molecular formula if the charge comes from a protonation or deprotonation.

Moreover, as said before, the DBE equation cannot be used if in the molecule there are functional groups having a higher valence state (octet expansion). The most common groups found in organic molecules that produce distortions of the value of DBE are the sulfate and sulfonate groups and the nitro group. In spite of the fact that sulfate and sulfonate groups are drawn with π bonds, they do not contribute to the number of DBE (Fig. 8.3). In contrast, the nitro group contributes one DBE.

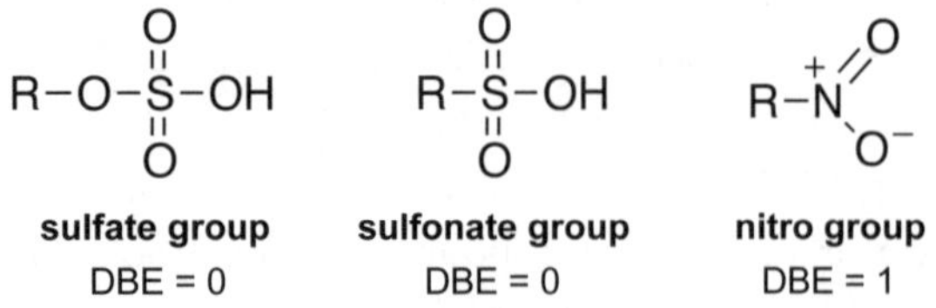

sulfate group DBE = 0 **sulfonate group** DBE = 0 **nitro group** DBE = 1

Figure 8.3 DBE for sulfate, sulfonate, and nitro groups.

Problem 8.1

Determine the number of DBEs of the chemical structures shown below.

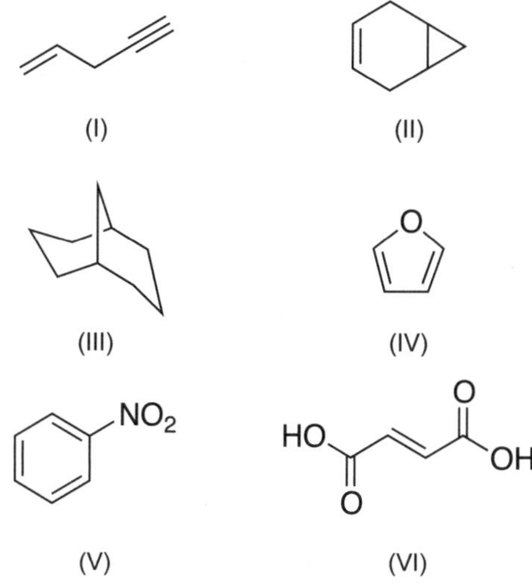

Solution

(I) One double bond (one π bond), one triple bond (two π bonds), and no ring: DBE = 3;
(II) one π bond and two rings: DBE = 3;
(III) two rings: DBE = 2;
(IV) two π bonds and one ring: DBE = 3;
(V) three π bonds, one ring, and one nitro group (that counts for one DBE): DBE = 5;
(VI) three π bonds: DBE = 3.

Problem 8.2

Determine the number of DBEs from the following molecular formulae:

(I) C_6H_{14} (II) C_6H_6O (III) C_4H_6O

Solution

(I) DBE = C - (H/2) - (X/2) + (N/2) +1 = 6 - (14/2) - (0/2) + (0/2) + 1 = 0;
(II) DBE = C - (H/2) - (X/2) + (N/2) +1 = 6 - (6/2) - (0/2) + (0/2) + 1 = 4;
(III) DBE = C - (H/2) - (X/2) + (N/2) +1 = 4 - (6/2) - (0/2) + (0/2) + 1 = 2;

Problem 8.3

Draw all possible chemical structures for a molecule having molecular formula C_3H_6.

Solution

First of all, calculate the number of DBEs:

$$DBE = C - (H/2) - (X/2) + (N/2) +1 = 3 - (6/2) - (0/2) + (0/2) + 1 = 1;$$

Therefore, there could be a ring or a π bond. So the possible structures are

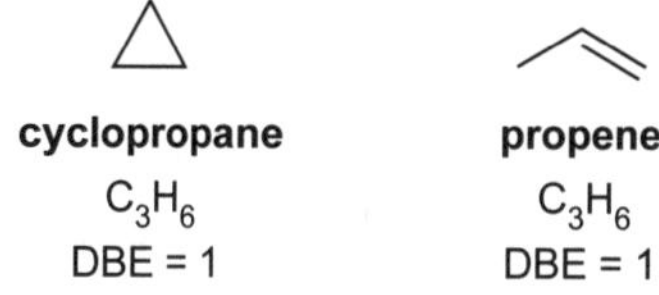

Problem 8.4

What are the possible combinations of rings, double bonds, and/or triple bonds that could exist for a compound having DBE = 3?

Solution

These are the possible combinations:

1) Three rings
2) Three double bonds
3) One triple bond and one double bond
4) One triple bond and one ring
5) Two double bonds and one ring
6) One double bond and two rings

2) Look for the solvent signals in the proton and carbon spectra

As mentioned in Chapter 6, solvent signals can appear in the proton and carbon spectra. Sometimes, they are as intense as the signals of the solute, so they could be confused with signals of the molecule. The residual signals can be easily identified using Table 6.1 (see Chapter 6), which is also reported in the inside cover of this book. The solvent used for the acquisition of the spectra is reported in each of the example problems.

3) Get preliminary information about the symmetry of the molecule

As discussed in the previous chapters, the symmetry properties of a molecule determine the number of signals observed in the proton and carbon spectra. In theory, it would be possible to determine the symmetry of a molecule by comparing the number of signals in the proton and carbon spectra to the number of hydrogen and carbon atoms in the molecular formula, respectively. Hence, if the number of proton and carbon atoms in the molecular formula were the same as the number of proton and carbon signals, the molecule would not be symmetric. In contrast, if the number of atoms in the molecular formula were larger than the number of proton and carbon signals, then the molecule would be symmetric.

Unfortunately, things are not that easy!

In fact, as you have seen in the previous chapters, it can happen that two or more non-equivalent groups of nuclei can resonate at the same chemical shift, providing false information about the symmetry of the molecule. For example, non-equivalent nuclei can be chemical shift equivalent if their chemical environment is averaged by conformational dynamic processes. Also, as mentioned in Chapter 2, two diastereotopic nuclei may resonate at the same chemical shift if they are apart from the asymmetric center. For example, let's consider the case of phenylalanine, which, possessing a chiral center, is an asymmetric molecule. It produces fewer carbon signals (7 signals) than those expected from the molecular formula (9 carbon atoms) (Fig. 8.4). This is due to the fact that the phenyl moiety is far from the chiral center and it is free to rotate averaging its chemical environment, and thus making the positions C2' and C3' equivalent to the positions C6' and C5', respectively. In other word, it is like the phenyl group experiences a *local symmetry*.

Therefore, each proton and carbon spectrum must be interpreted taking into account the possibility that part of the molecule may behave as symmetric.

In general, information about the symmetry of a molecule can be retrieved more easily from carbon than proton spectra. Carbon spectra can be grouped according to four different scenarios as discussed below. VERY IMPORTANT: the considerations discussed in scenarios 2-4 are applicable to most small organic molecules, but not to all of them. Therefore, they must not be taken as an absolute truth!

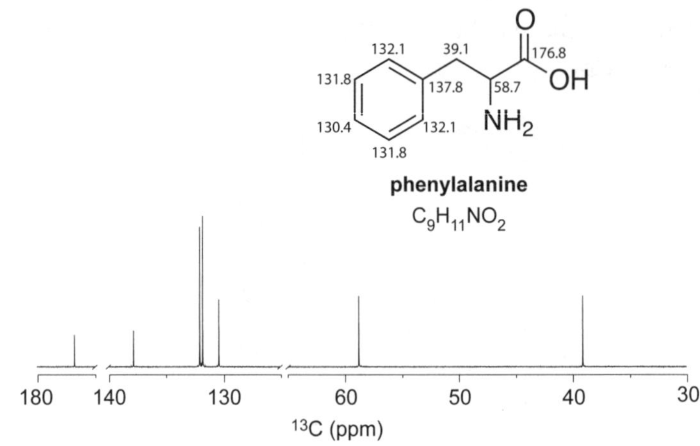

Figure 8.4 Selected regions of the carbon spectrum of phenylalanine. Numbers on the chemical structure indicate carbon assignment (values are in ppm).

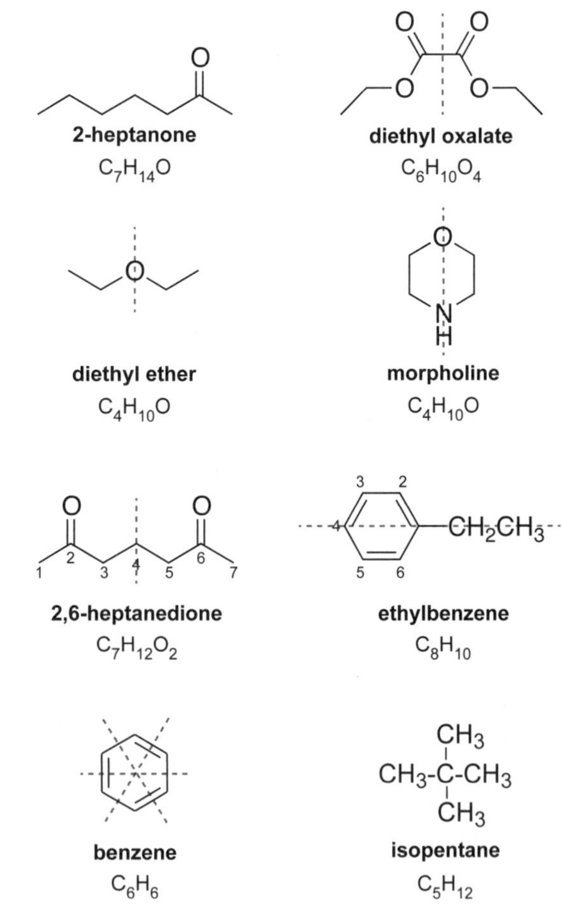

Figure 8.5 Examples of molecular structures containing different types of symmetry.

Scenario 1: The number of carbon signals is the same as the number of carbon atoms in the molecular formula.

If the number of carbon signals (be careful about any possible signal overlap) is the same as the number of carbon atoms in the molecular formula, then surely the molecule is not symmetric. For example, 2-heptanone (Fig. 8.5) is not symmetric because it produces seven signals in the carbon spectrum.

Scenario 2: The number of carbon signals is half of the number of carbon atoms in the molecular formula.

If the number of carbon signals is half that of the carbon atoms in the molecular formula, this means that the molecule may possess only one element of symmetry that does not pass through any carbon atom. This means that it bisects one or more C-C bonds or it passes through one or more heteronuclear atoms. For example, diethyl oxalate (Fig. 8.5) contains a plane of symmetry bisecting a C-C bond, and so it produces only three signals in the carbon spectrum, while its molecular formula contains six carbon atoms. Similarly, diethyl ether and morpholine have planes of symmetry passing through heteroatoms (Fig. 8.5), and so they both produce only two signals in their carbon spectrum.

Scenario 3: The number of carbon signals is less than the number of carbon atoms in the molecular formula, but larger than its half.

If the number of signals in the carbon spectrum is less than the number of carbon atoms in the molecular formula, but greater than its half, then the molecule may contain an element of symmetry passing through at least one carbon atom or may have a moiety experiencing local symmetry. For example, 2,6-heptanedione and ethylbenzene (Fig. 8.5) both have a plane of symmetry. In 2,6-heptanedione it passes through C4, so that C1, C2, and C3 turn out to be equivalent to C7, C6, and C5, respectively. In fact, this molecule produces four signals, while seven carbons are present in the molecular formula. With ethylbenzene, the plane of symmetry passes through the ethyl group and through C1 and C4 of the aromatic moiety. Therefore, C2 and C3 are equivalent to C6 and C5, respectively. The molecule produces six signals in the carbon spectrum, while there are eight carbon atoms in the molecular formula. Please note that phenylalanine (discussed above, Fig. 8.4) possesses local symmetry and it should be interpreted according to the considerations described in this scenario.

Scenario 4: The number of carbon signals is less than half of the carbon atoms in the molecular formula.

If the number of carbon signals is less than half of the carbon atoms in the molecular formula, then the molecule may be highly symmetric. By way of example, look at the structures of benzene and isopentane in Figure 8.5. Benzene produces only one signal in the carbon spectrum, while its molecular formula contains six carbon atoms. Isopentane is also highly symmetric, producing only two carbon signals.

The rationale for the determination of the symmetry of a molecule using carbon spectra is summarized in Figure 8.6.

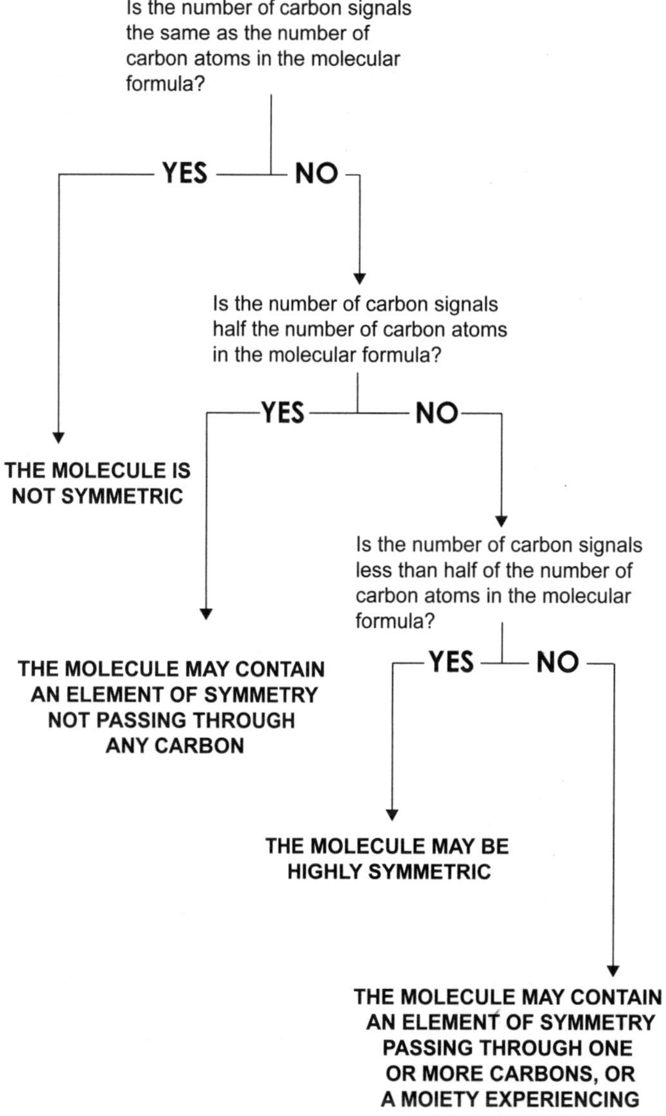

Figure 8.6 Rationale for the determination of the symmetry of a molecule using the carbon spectrum.

Problem 8.5

The carbon spectrum (125 MHz, D_2O) shown below is produced by a molecule having molecular formula $C_4H_8O_3$. Is this molecule symmetric?

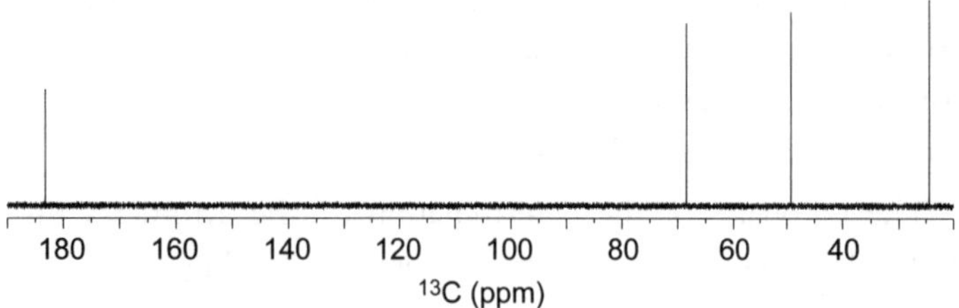

Solution

The molecule is not symmetric because the number of carbon signals is the same as the number of carbon atoms in the molecular formula.

Problem 8.6

The carbon spectrum (100 MHz, D_2O) shown below is produced by a molecule having molecular formula C_3H_9N. Is this molecule symmetric?

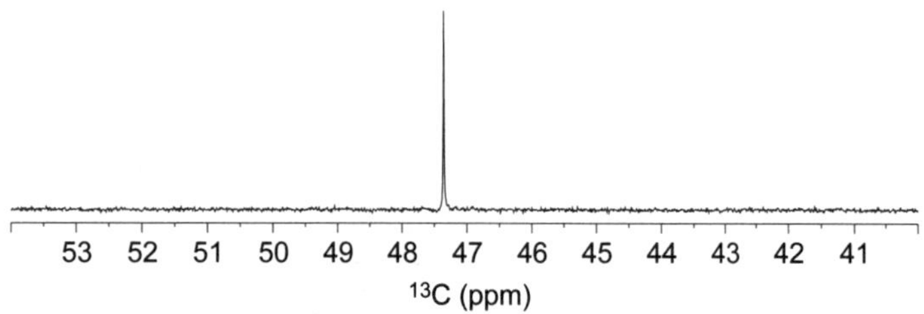

Solution

This molecule may be highly symmetric, because the ^{13}C spectrum contains only one signal, while the molecular formula contains three carbon atoms.

Problem 8.7

The carbon spectrum (100 MHz, D_2O) shown below is produced by a molecule having molecular formula $C_8H_{10}O_2$. Is the molecule symmetric?

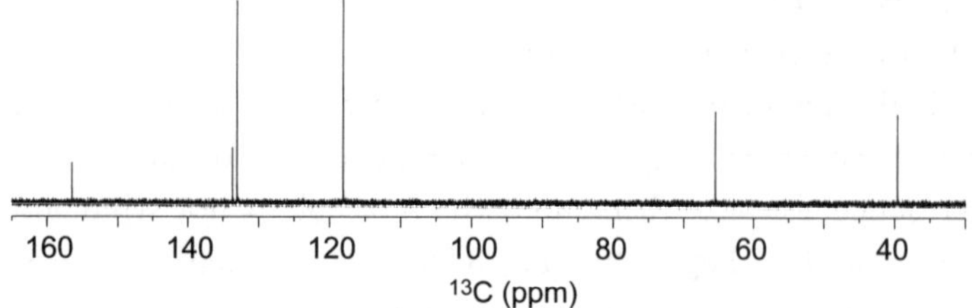

Solution

The carbon spectrum contains six signals, while the molecular formula contains eight carbon atoms. This means that the molecule may contain an element of symmetry passing through one or more carbon atoms, or it may have a moiety experiencing local symmetry. Please note that, although the carbon signal intensities are not strictly related to the number of resonating carbons, it is possible to speculate that the couples of equivalent carbons resonate at about 118 and 133 ppm. In fact, these signals appear particularly intense.

Problem 8.8

The carbon spectrum (125 MHz, D_2O) shown below is produced by a molecule having molecular formula $C_4H_6O_4$. Is this molecule symmetric?

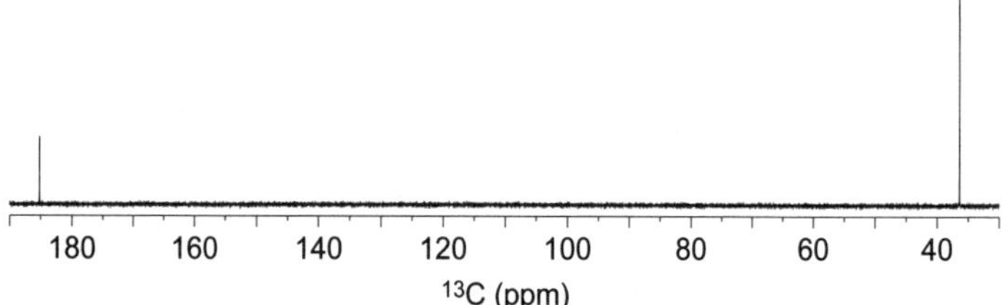

Solution

This molecule may contain an element of symmetry not passing through any carbon atom. In fact the carbon spectrum contains two signals, while the molecular formula contains four carbons.

4) Recognize the functional groups

As discussed in the previous chapters, from the analysis of the chemical shifts of the proton and carbon signals it is possible to retrieve information about the functional groups of the molecule. By way of example, the presence of aldehydes can be suggested by signals between 9.0-10.5 ppm and 190-210 ppm in proton and carbon spectra, respectively; aromatic moieties by signals between 6.0-9.0 ppm (^1H) and 90-160 ppm (^{13}C); alkenes between 3.5-8 ppm (^1H) and 60-170 ppm (^{13}C); alkynes between

1.5-3.2 (^{1}H) and 65-85 ppm (^{13}C); and so on. Please remember that aromatic and olefinic protons can be discriminated by their proton chemical shifts, but not by their carbon resonances.

^{1}H and ^{13}C chemical shift ranges for organic compounds are summarized in two schemes shown in the inside cover-flaps of this book. Obviously, unusual combinations of substituents can lead to shifts outside the given ranges.

In addition to the analysis of the chemical shifts, it is possible to analyze also the intensities of the signals in proton spectra. For example, it is possible to recognize the methyls present in a molecule, because they are generally the most intense signals. Methyls resonate at around 3.2-4 ppm when bonded to oxygen atoms (in this case they resonate as a singlet), at around 2.5-3.5 ppm when bonded to a nitrogen atom (N-methyl), and at around 1.5-2.5 ppm when bonded to sp^2 carbon atoms. Finally, when they are bonded to sp^3 carbons, they are the most shielded signals in the proton spectrum, resonating around 1 ppm.

5) Determine the number of quaternary carbons (C), methines (CH), methylenes (CH_2), and methyls (CH_3), and assign their resonances

Quaternary carbons, methines, methylenes, and methyls are the "building blocks" of small organic molecules. They can be identified by the combined analysis of the ^{13}C and DEPT spectra (Chapter 5). Then, by using the 2D HSQC experiment (Chapter 7), you can assign the proton resonances of each of these moieties.

6) Determine the number of protons that produces each proton signal

Proton signals can be accurately integrated, so it is possible to determine the number of hydrogens that produces each signal. You can simply sum the relative signal intensities reported in the spectrum (as a relative ratio with respect to the smallest integration in the spectrum), and compare the result with the number of hydrogen atoms present in the molecular formula. For example, consider the case of 1,3-dichloropropene ($C_3H_4Cl_2$), whose ^{1}H-NMR spectrum is shown in Figure 8.7. The sum of the values of the relative intensities is four (2 + 1 + 1 = 4), which is the same as the number of hydrogen atoms in the molecular formula. Therefore, you can state that both signals at 6.02 and 6.24 ppm are produced by one proton, while the signal at 4.23 is produced by two equivalent protons.

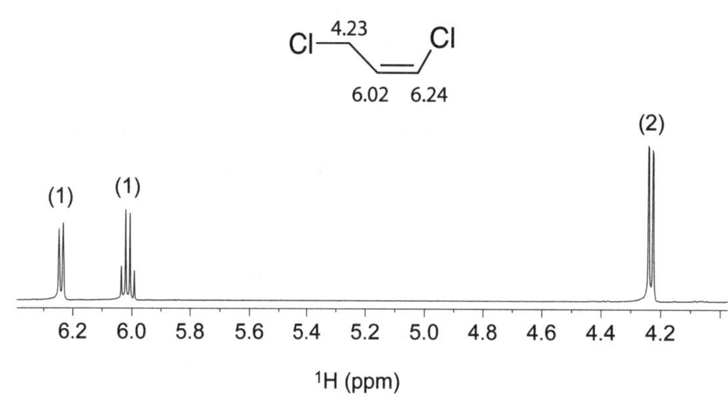

Figure 8.7 ^{1}H-NMR spectrum (500 MHz, $CDCl_3$) of *cis*-1,3-dichloropropene, having molecular formula $C_3H_4Cl_2$. Numbers on the signals indicate the relative signal intensities. Numbers on the chemical structure indicate the proton assignment (values are in ppm).

Let's now consider the proton spectrum of diethylsuccinate, having molecular formula $C_8H_{14}O_4$ (Fig. 8.8). This spectrum contains three signals having relative intensities 2:2:3. Summing the values of the relative intensities of the signals, you obtain a number that is the half of the hydrogens contained in the molecular formula (2 + 2 + 3 = 7). This means that actually each signal is produced by double the number of the protons indicated by the relative intensity ratios. Hence, the signals at 4.12 and 2.58 ppm are actually both produced by four protons (2 x 2 = 4), while the signal at 1.23 ppm is produced by six protons (3 x 2 = 6).

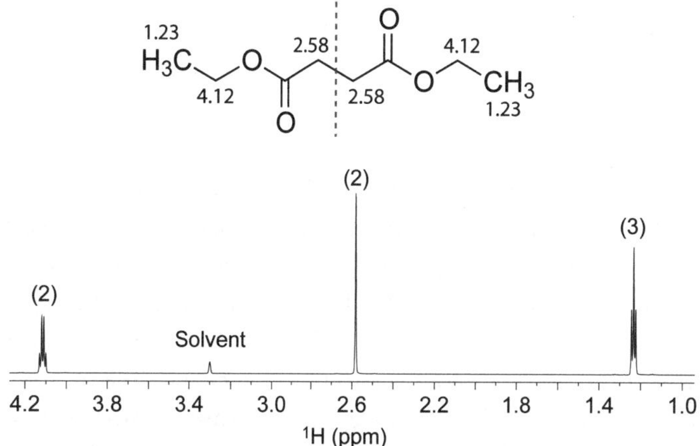

Figure 8.8 ^{1}H-NMR spectrum (700 MHz, CD_3OD), of diethylsuccinate, having molecular formula $C_8H_{14}O_4$. Numbers on the signals indicate the relative signal intensities. Numbers on the chemical structure indicate proton assignment (values are in ppm). The dashed gray line represents a plane of symmetry.

Unfortunately, there are cases where you cannot measure the relative signal intensities, because there is only one signal in the proton spectrum. For example, consider ethane and benzene (Fig. 8.9). In these cases, because only one signal is present in the spectrum, you can state that it is produced by all the hydrogens of the molecule (six, in both examples).

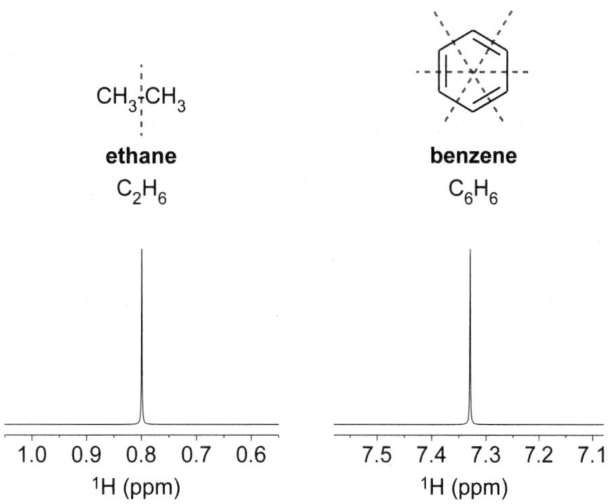

Figure 8.9 Simulated ^{1}H-NMR spectra of ethane and benzene. Dashed gray lines represent planes of symmetry.

Please also note that if a molecule contains exchangeable protons and its proton spectrum is acquired in a protic deuterated solvent, this approach does not work; in fact, the signals of the exchangeable protons are not visible in these conditions. In these cases, you can compare the sum of the relative signal intensities with the number of non-exchangeable hydrogens. This can be determined by summing all the hydrogens of the methines, methylenes, and methyls identified in the previous step. For example, a molecule of molecular formula C_2H_7NO produces the spectra shown in Figure 8.10 (please note that the spectra are acquired in D_2O).

The molecule clearly contains exchangeable protons; in fact, there is no possible combination that can explain how seven hydrogens could produce two signals of equal intensity. From the analysis of the ^{13}C and DEPT spectra you can state that the molecule is not symmetric and contains two methylenes; therefore the molecule contains four non-exchangeable protons (2 CH_2). The sum of the relative signal intensity is two, that is, half of the non-exchangeable protons in the molecule. This means that each signal actually integrates for two hydrogens.

Molecular formula: C_2H_7NO

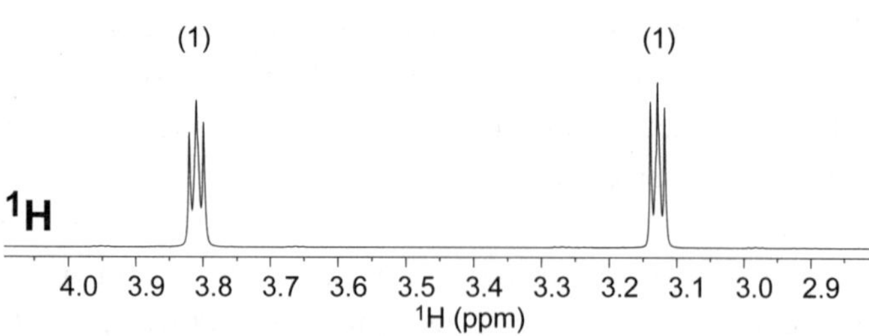

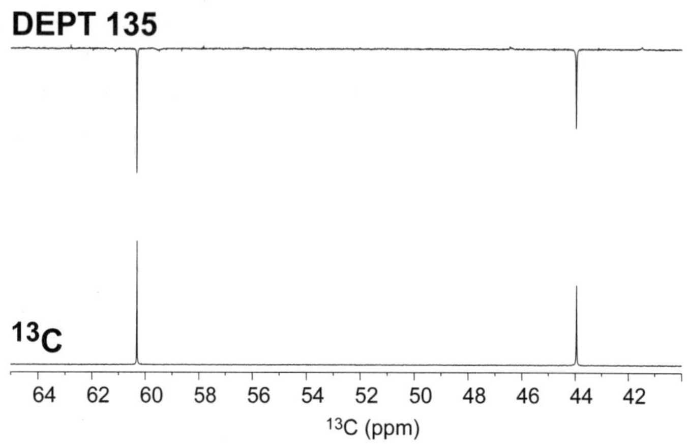

Figure 8.10 ^{1}H (500 MHz, D_2O), ^{13}C, and DEPT (125 MHz, D_2O) spectra of a molecule having molecular formula C_2H_7NO.

Let's now consider the case of a symmetric molecule having molecular formula $C_4H_{12}N_2$ (Fig. 8.11). This molecule surely contains exchangeable protons. In fact, the ^{13}C and DEPT experiments indicate that the molecule contains two couples of equivalent methylenes, counting for eight non-exchangeable protons (while the molecular formula contains twelve hydrogens). Since the sum of relative signal intensity is two, each signal integrates for four hydrogens (8/2).

7) Determine the number of exchangeable protons

If the NMR spectra are acquired in an aprotic deuterated solvent, you can determine the presence of exchangeable protons by looking at the proton signals that do not correlate to any carbon signals in the HSQC experiment.

Instead, if the NMR spectra are acquired in a protic deuterated solvent, you can determine the number of exchangeable protons by an indirect method.

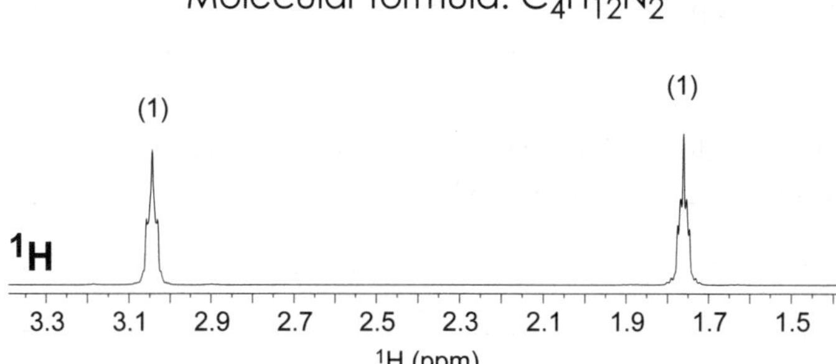

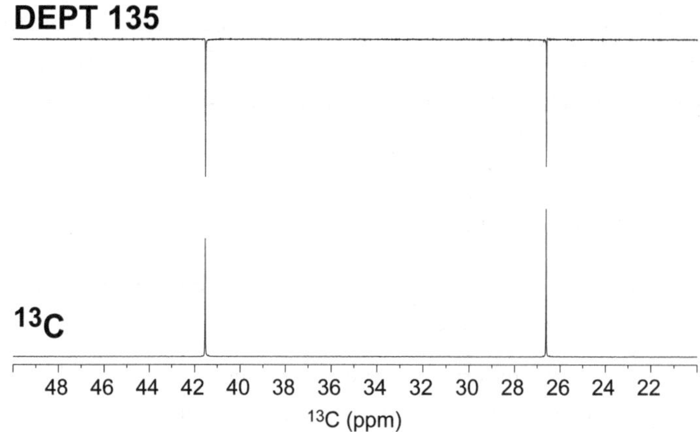

Figure 8.11 ^{1}H (500 MHz, D_2O), ^{13}C, and DEPT (125 MHz, D_2O) spectra of a molecule having molecular formula $C_4H_{12}N_2$.

Particularly, knowing the number of methines, methylenes, and methyls (see previous steps), you can first calculate the number of non-exchangeable protons, and then you can subtract this from the number of hydrogens contained in the molecular formula.

For example, consider the carbon and DEPT spectra of a molecule having molecular formula $C_4H_8O_3$ (Fig. 8.12).

The molecule contains one methyl (3 hydrogens), one methylene (2 hydrogens), and one methine (1 hydrogen), altogether counting for six hydrogens. Since the molecular formula contains eight hydrogens, two hydrogens must be exchangeable (8 - 6 = 2).

Similar reasoning can be used with molecules having chemical shift equivalent carbons (no matter if they are equivalent due to time averaging processes or to the symmetry of the molecule). For example, consider the case of a molecule having molecular formula $C_3H_{10}N_2$, whose ^{13}C and DEPT spectra are shown in Figure 8.13.

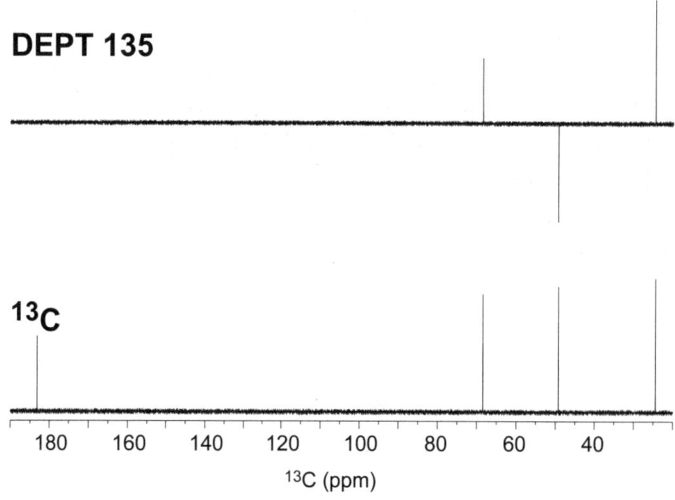

Figure 8.12 ^{13}C and DEPT spectra of a molecule having molecular formula $C_4H_8O_3$.

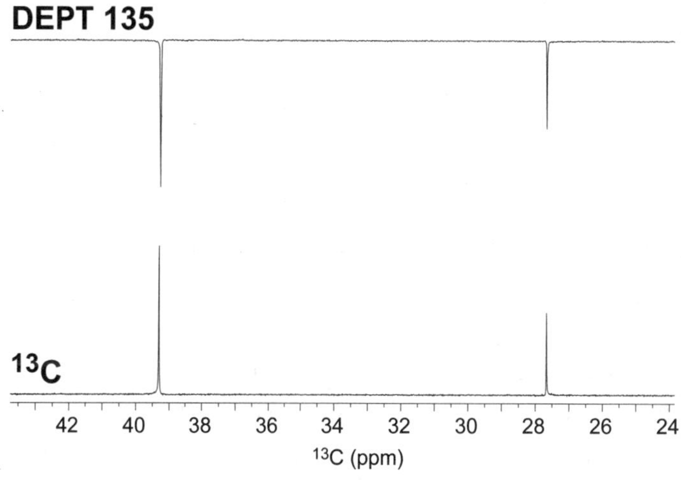

Figure 8.13 ^{13}C and DEPT spectra of a molecule having molecular formula $C_3H_{10}N_2$.

The carbon spectrum contains only two signals, meaning that two out of three carbons are chemical shift equivalent. All the signals are negative in the DEPT experiment, that is, the molecule contains three methylenes. Therefore, it is possible to conclude that there are six non-exchangeable protons (3 CH_2) and four exchangeable protons (10 - 6 = 4).

Let's now analyze the case of a molecule having molecular formula $C_9H_{11}NO_3$, whose ^{13}C and DEPT spectra are shown in Figure 8.14. The ^{13}C spectrum contains seven signals, suggesting that two pairs of carbons are chemical shift equivalent. Even though the signals in the carbon spectrum cannot be accurately integrated, it is possible to speculate that the signals at 121.1

and 133.4 ppm resonate for two carbon atoms each. Since these signals are positive in the DEPT spectrum, each signal should count for two methines. The DEPT spectrum also contains another methine and a methylene. Hence, there are seven non-exchangeable protons (5 CH and 1 CH_2), and therefore four exchangeable protons (11 - 7 = 4).

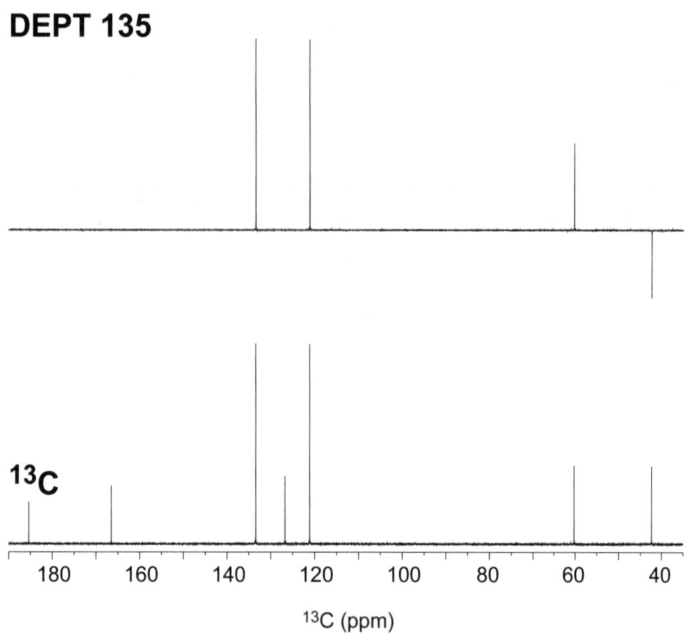

Figure 8.14 ^{13}C and DEPT spectra of a molecule having molecular formula $C_9H_{11}NO_3$.

8) Characterize all the spin systems

The methines (CH), methylenes (CH_2), and methyls (CH_3) identified in the previous step, could be bonded together, forming extended spin systems. These can be determined by the analysis of the COSY spectrum (Chapter 7) and by the analysis of the multiplicity of the proton signals (Chapter 3). The identified spin systems can be verified by the long-range correlations in the HMBC experiment (Chapter 7).

9) Connect all the fragments together

Taking into consideration the molecular formula, the chemical shifts, and the valence of each atom, it is possible to try to assemble the structural fragments identified in the previous steps. As discussed in Chapter 7, it can happen that a number of reasonable structures can be drawn. The "game" is to understand which is the right one. To this end, it is convenient to report all the HMBC correlations on all structural hypotheses and select the right one by analyzing the consistency of the correlations. For example, the presence of a cross-peak indicating a long-range coupling more than three bonds distant (for aliphatic molecules) indicates that the supposed structure is probably wrong (although in conjugated systems you can observe four-bond couplings). These concepts will be deepened in the example problems reported in the following pages.

10) Verify the consistency of all the data collected

This step serves to give you the certainty that the structure you have selected is the right one, and therefore it must not be ignored!

To this end, you have to go back to the first step of this strategy of structural elucidation to verify *a posteriori* if the selected structure is consistent with the data collected in each step. Basically you have to answer to the following questions:

- Is the molecular formula consistent with the proposed structure of the molecule?
- Does the selected structure contain the number of DBEs you have calculated?
- Is the symmetry of the molecule consistent with the number of signals in the proton and carbon spectra?
- Does the molecule contain the number of exchangeable protons you have predicted?
- Are the proton and carbon chemical shifts consistent with the chemical environment of each nucleus?
- Are the COSY correlations and the signal multiplicities consistent with the spin systems of the molecule?
- Are the HMBC correlations consistent with the supposed structure?

Some examples of application of this strategy are provided in the solved problems reported in the following pages. These are classified according to their level of difficulty (look at the top-right corner): the larger the number of filled stars, the harder the problem. For each problem, the molecular formula is provided.

Example Problems

Problem 8.9
Solvent: D_2O
Field: 500 MHz (^1H); 125 MHz (^{13}C)

Molecular formula: $C_4H_9NO_2$

Difficulty ★☆☆☆

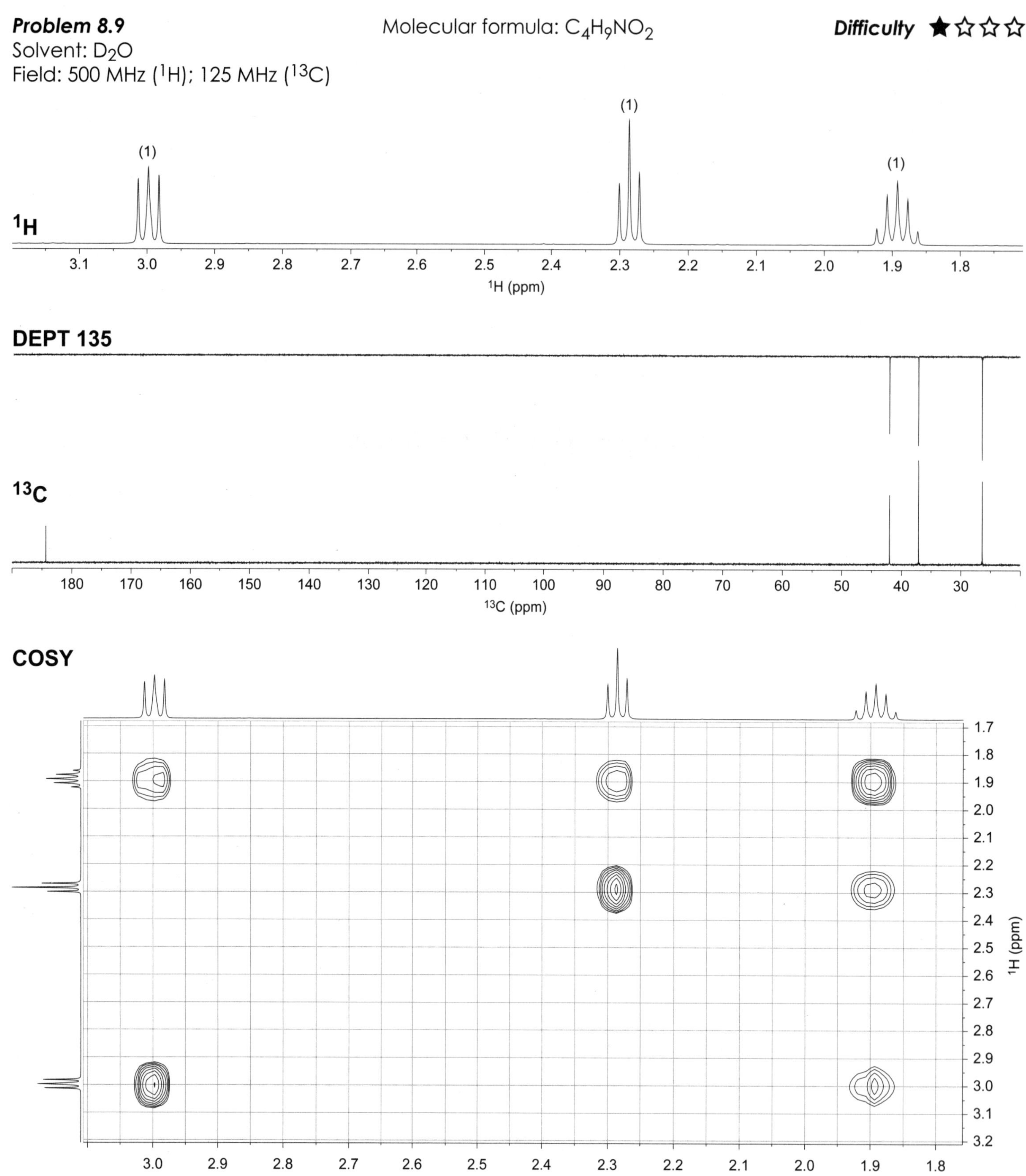

HSQC

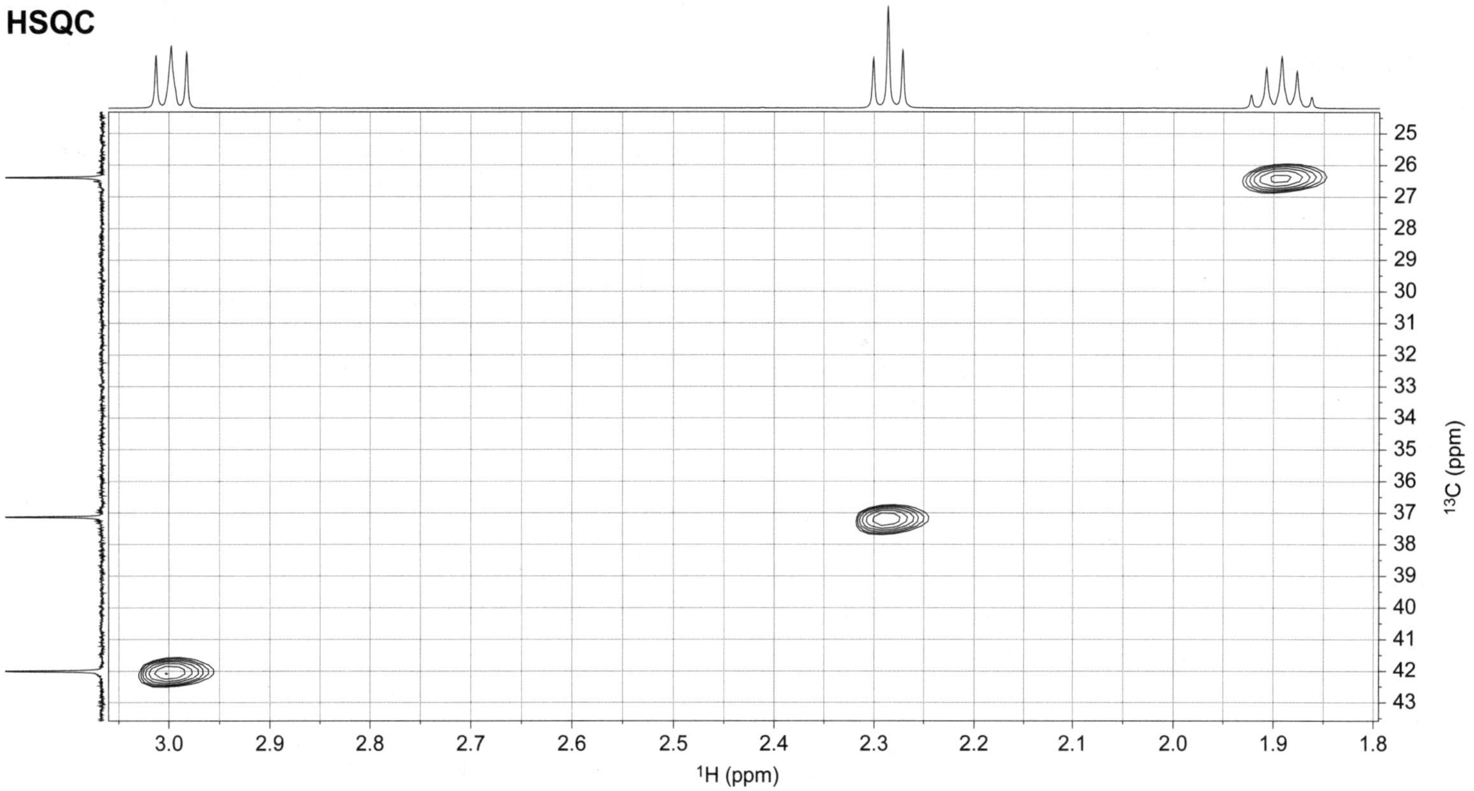
25
26
27
28
29
30
31
32
33
34
35
36
37
38
39
40
41
42
43
^{13}C (ppm)
3.0
2.9
2.8
2.7
2.6
2.5
2.4
2.3
2.2
2.1
2.0
1.9
1.8
^{1}H (ppm)

HMBC

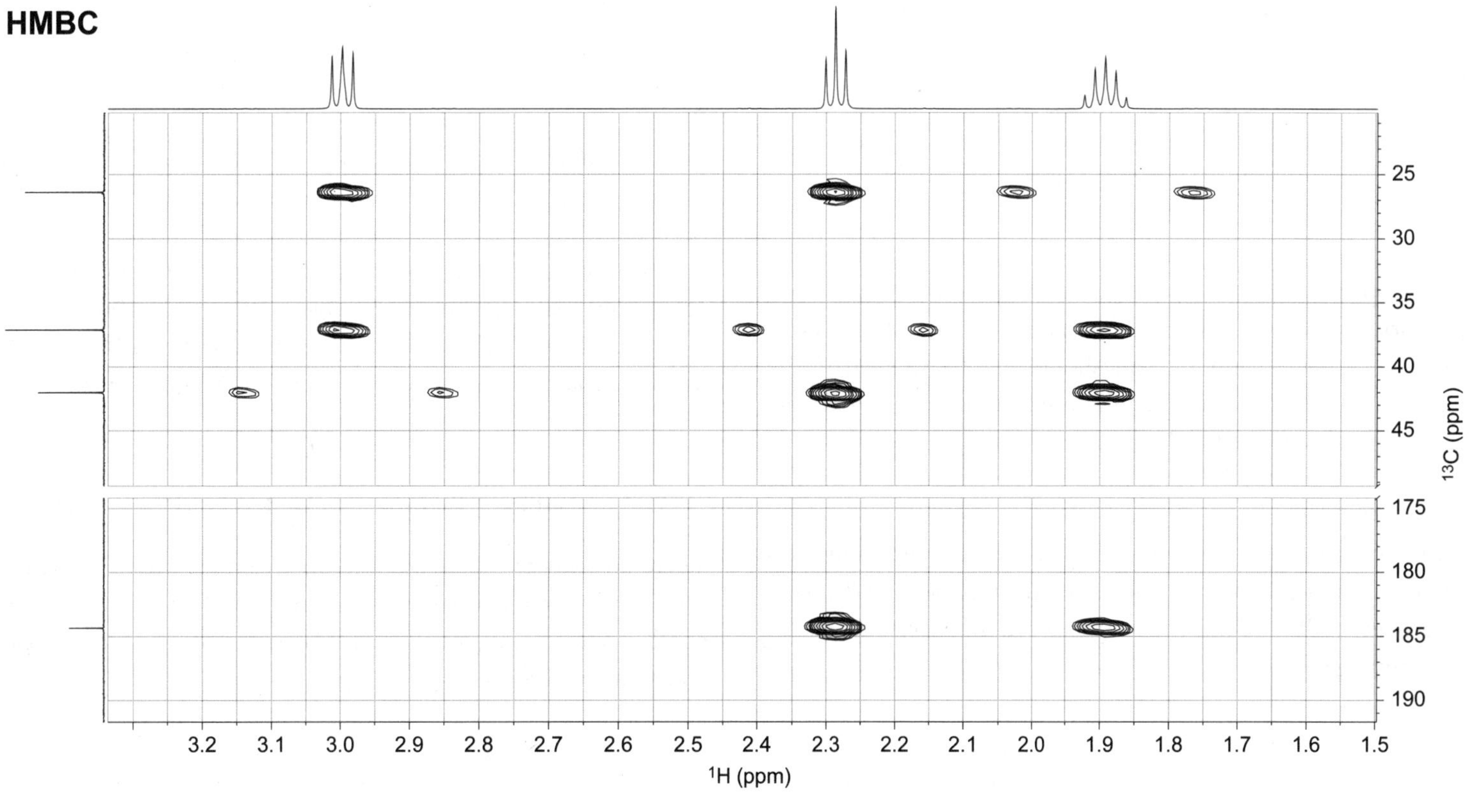
25
30
35
40
45
175
180
185
190
^{13}C (ppm)
3.2
3.1
3.0
2.9
2.8
2.7
2.6
2.5
2.4
2.3
2.2
2.1
2.0
1.9
1.8
1.7
1.6
1.5
^{1}H (ppm)

Solution

1) Calculate the number of double bond equivalents (DBEs)

DBE = C - (H/2) - (X/2) + (N/2) +1 = 4 - (9/2) - (0/2) + (1/2) + 1 = 1; this means that there is one double bond or one ring in the structure.

2) Look for the solvent signals in the proton and carbon spectra

The spectra are acquired in D_2O. The residual solvent signal is not visible because it resonates in a spectral region omitted from the figures.

3) Get preliminary information about the symmetry of the molecule

The number of carbon signals is the same as the number of carbon atoms in the molecular formula, and therefore the molecule is not symmetric.

4) Recognize the functional groups

The carbon signal at 184.4 ppm indicates the presence of a carboxylic group (or one of its derivatives). Then, the other carbon signals resonate in a chemical shift range that is consistent with the presence of an aliphatic moiety in the molecule.

5) Determine the number of quaternary carbons (C), methines (CH), methylenes (CH_2), and methyls (CH_3), and assign their resonances

The molecule contains three methylenes and one carboxyl group (see the ^{13}C and DEPT experiments). Analyzing the HSQC experiment, it is possible to assign the chemical shifts of the protons of each methylene:

3.00	2.29	1.89	
42.0	37.1	26.4	184.4
CH_2	CH_2	CH_2	C

N.B.: Due to the approximation of the chemical shift scales in the spectra reported in this and in the next problems, you may not be able to accurately determine the chemical shift value of each signal. Therefore, you should not be worried if the chemical shift values you have determined are not exactly the same as those reported in the solutions to the problems. You can improve the accuracy of your measurements by using a triangle set square. You can help yourself to trace perfect vertical and horizontal lines by placing the triangle on the chemical shift axes of the 1D and 2D spectra. The use of the triangle is particularly important in 2D experiments, especially if they contain close or overlapping cross-peaks.

6) Determine the number of protons that produces each proton signal

Since the proton spectrum has been acquired in a protic solvent (D_2O), all the signals in the ^1H spectrum are produced by non-exchangeable protons. The molecule contains three methylenes, and therefore there are six non-exchangeable protons. It is easy to suppose that each signal is produced by a methylene. This is confirmed by the analysis of the HSQC experiment. Therefore each proton signal integrates for two hydrogens.

7) Determine the number of exchangeable protons

As determined in the previous step, there are six non-exchangeable hydrogens (3 CH_2). Since the molecular formula contains nine hydrogens, there are three exchangeable protons in the molecule (9 - 6 = 3).

8) Characterize all the spin systems

By the analysis of the COSY experiment, it is possible to link together the three methylenes, forming the spin system shown below. It is consistent with the multiplicity of the proton signals and with the HMBC correlations (arrows).

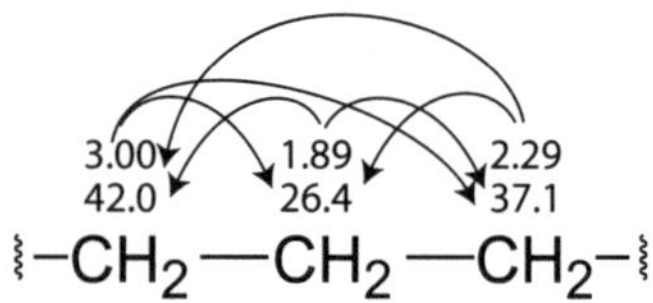

9) Connect all the fragments together

The part of the structure identified so far counts for three carbon and six hydrogen atoms. There is also a carbonyl group (C=O), which counts for another carbon and an oxygen atom. You can subtract these atoms from the molecular formula and try to understand which moieties are still to be identified. In particular three hydrogens, one nitrogen, and one oxygen atom still need to be considered. These could be an OH and an NH_2 group (this makes sense also considering that the carbonyl must be part of a carboxylic acid or one of its derivatives). Therefore, basically, you have to link together the following moieties:

–CH_2—CH_2—CH_2– –C(=O)– –NH_2 –OH

There are only two possible ways to assemble those fragments. One involves the formation of an amide and an alcohol group; another one, instead, involves the formation of a carboxylic acid and an amine group:

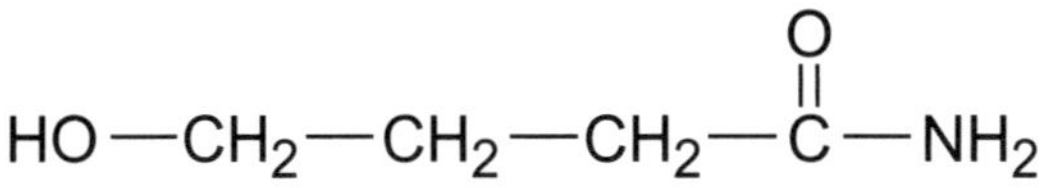

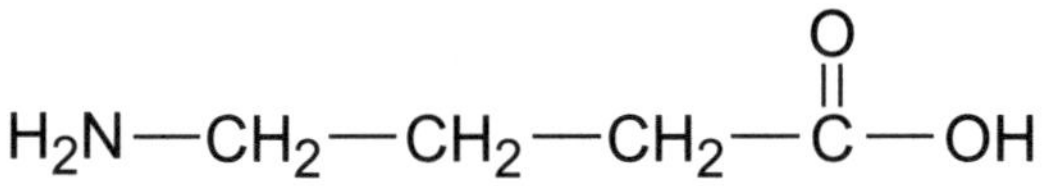

$$HO-CH_2-CH_2-CH_2-\overset{\overset{\displaystyle O}{||}}{C}-NH_2$$

$$H_2N-CH_2-CH_2-CH_2-\overset{\overset{\displaystyle O}{||}}{C}-OH$$

Which one is the correct structure?
The structure on the top contains an hydroxyl group bonded to an aliphatic carbon. This carbon should resonate at a chemical shift value larger than 60 ppm. Unfortunately, all the aliphatic carbons resonate between 26.4 and 42.0 ppm, suggesting that this structural hypothesis can be ruled out, and the correct structure is the bottom one.
Please note that there are two different ways to report the proton and carbon assignments on the correct structure:

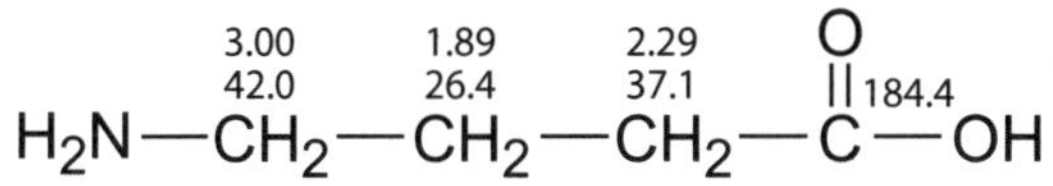

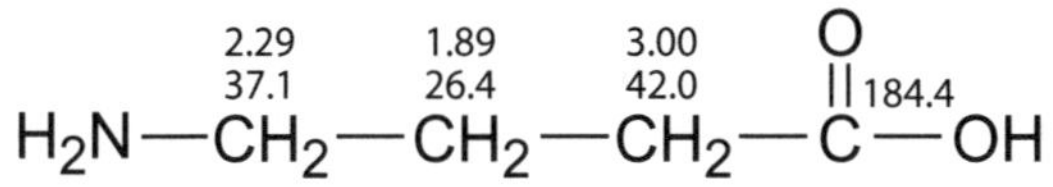

$$H_2N-\underset{}{\overset{3.00 \\ 42.0}{CH_2}}-\overset{1.89 \\ 26.4}{CH_2}-\overset{2.29 \\ 37.1}{CH_2}-\overset{\overset{\displaystyle O}{||\,184.4}}{C}-OH$$

$$H_2N-\overset{2.29 \\ 37.1}{CH_2}-\overset{1.89 \\ 26.4}{CH_2}-\overset{3.00 \\ 42.0}{CH_2}-\overset{\overset{\displaystyle O}{||\,184.4}}{C}-OH$$

In order to determine which is the right assignment, you have to evaluate the consistency of the chemical shift values with the chemical environment of each proton and carbon atom, and analyze the heteronuclear long-range correlations (HMBC). Since the hydrogens of an aliphatic carbon bonded to an amine group generally resonate around 3 ppm, it is possible to suppose that the correct assignment is that reported in the top figure. This hypothesis is also corroborated by the consistency of the chemical shift values (around 2 ppm) of the hydrogens in the α-position with respect to the carbonyl group. The HMBC correlations also agree with this; in fact, the protons at 1.89 and 2.29 ppm correlate to the carbonyl group (they are three and two bonds apart, respectively), while the hydrogens at 3.00 ppm do not show any correlations, being four bonds away.

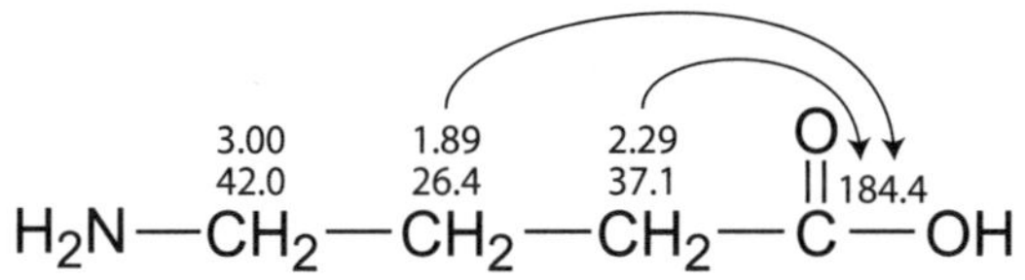

$$H_2N-\overset{3.00 \\ 42.0}{CH_2}-\overset{1.89 \\ 26.4}{CH_2}-\overset{2.29 \\ 37.1}{CH_2}-\overset{\overset{\displaystyle O}{||\,184.4}}{C}-OH$$

10) Verify the consistency of all the data collected

Is the molecular formula consistent with the structure of the molecule?
Yes, the molecule contains the same atoms contained in the molecular formula.

Does the selected structure contain the number of DBEs you have calculated?
Yes, the structure contains a carbonyl that counts for one DBE.

Is the symmetry of the molecule consistent with the number of signals in the proton and carbon spectra?
The number of signals in the proton and carbon spectra is consistent with the symmetry of the molecule.

Does the molecule contain the number of exchangeable protons you have predicted?
Yes, it does.

Are the proton and carbon chemical shifts consistent with the chemical environment of each nucleus?
Yes, they are.

Are the COSY correlations and the signal multiplicities consistent with the spin systems of the molecule?
Yes, they are.

Are the HMBC correlations consistent with the supposed structure?
Yes, they are.

THE STRUCTURE IS CORRECT!!!

Problem 8.10 Molecular formula: $C_5H_6O_4$ **Difficulty** ★☆☆☆

Solvent: D_2O

Field: 500 MHz (^{1}H); 125 MHz (^{13}C)

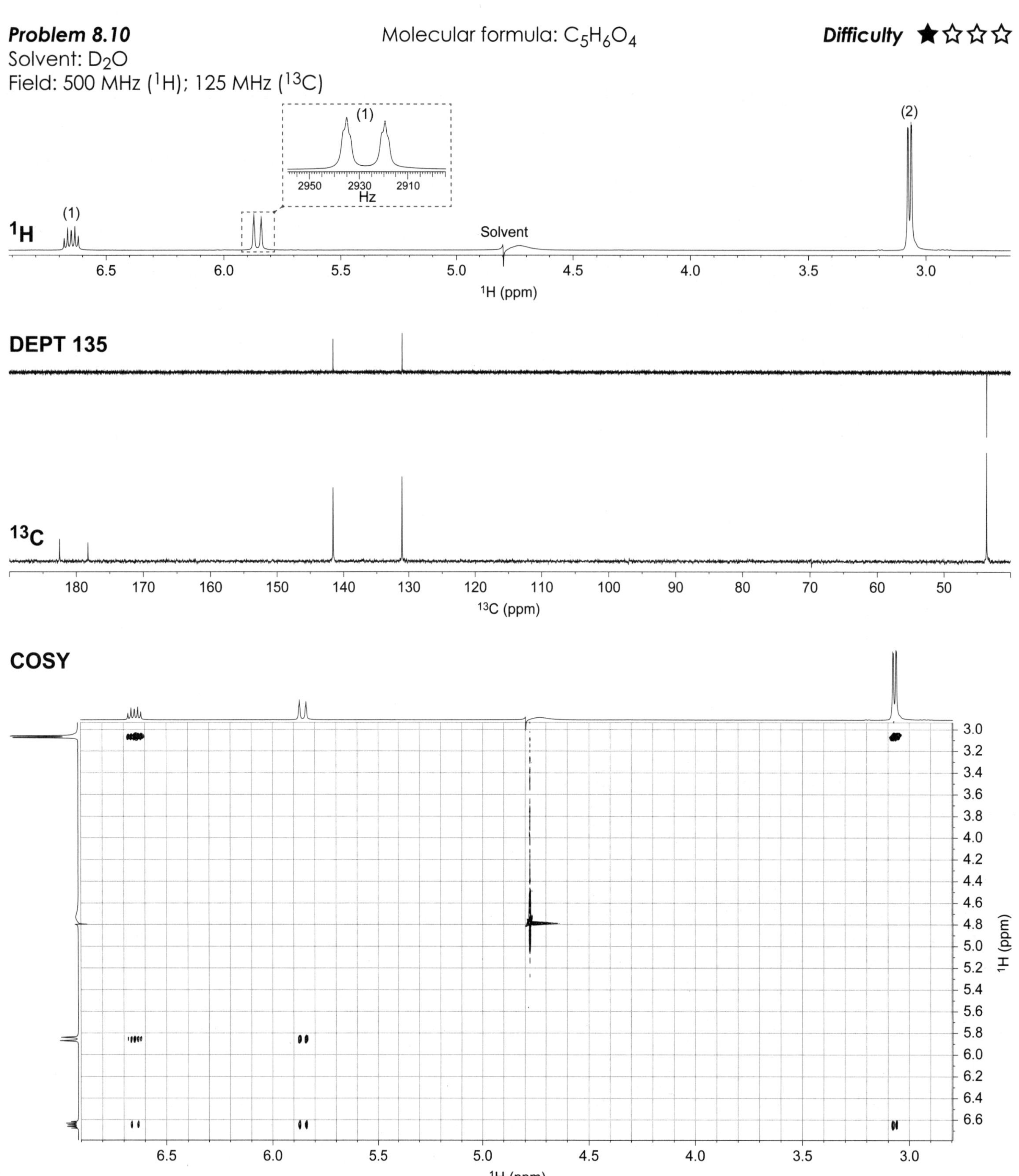

HSQC

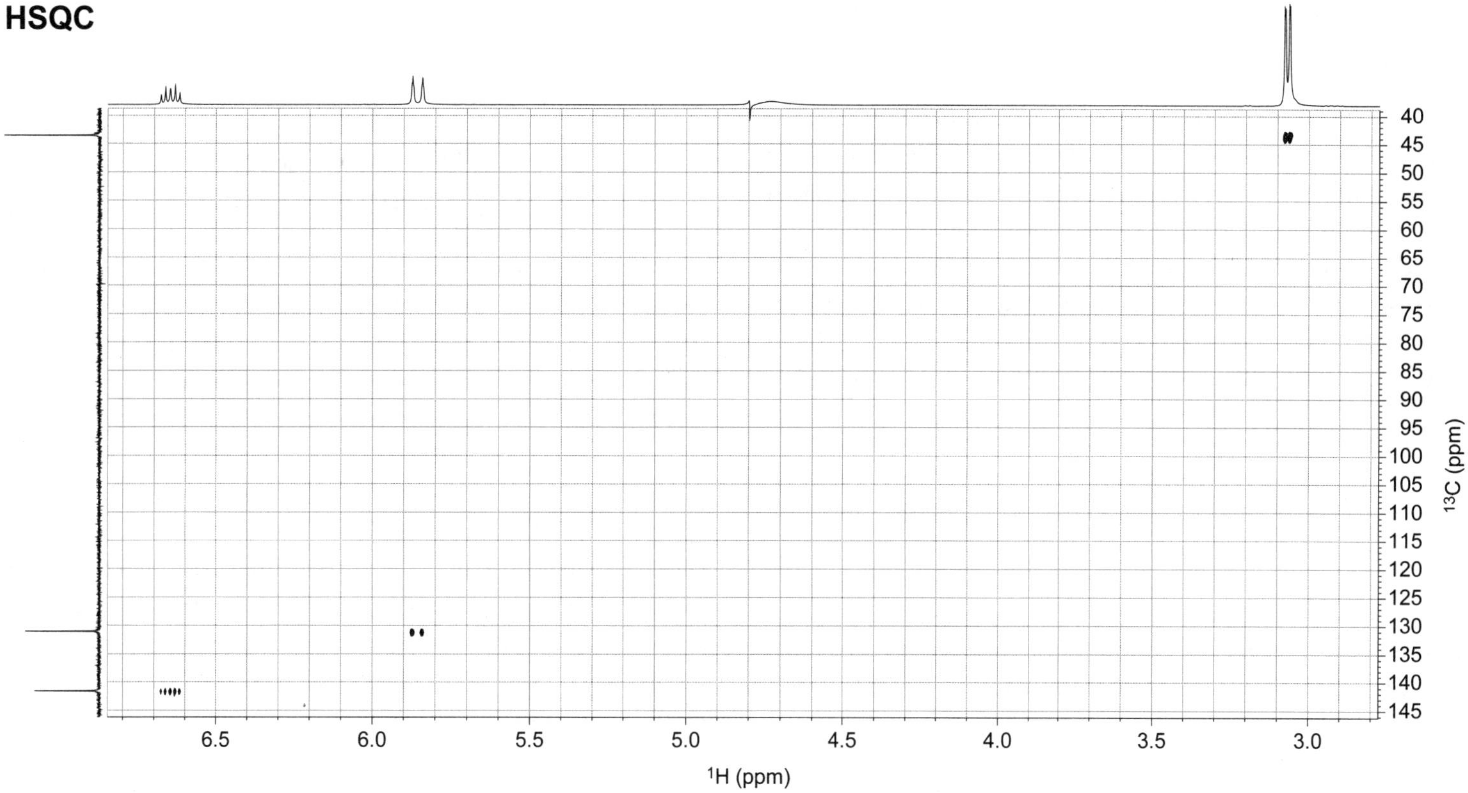
40
45
50
55
60
65
70
75
80
85
90
95
100
105
110
115
120
125
130
135
140
145
13C (ppm)
6.5
6.0
5.5
5.0
4.5
4.0
3.5
3.0
1H (ppm)

HMBC

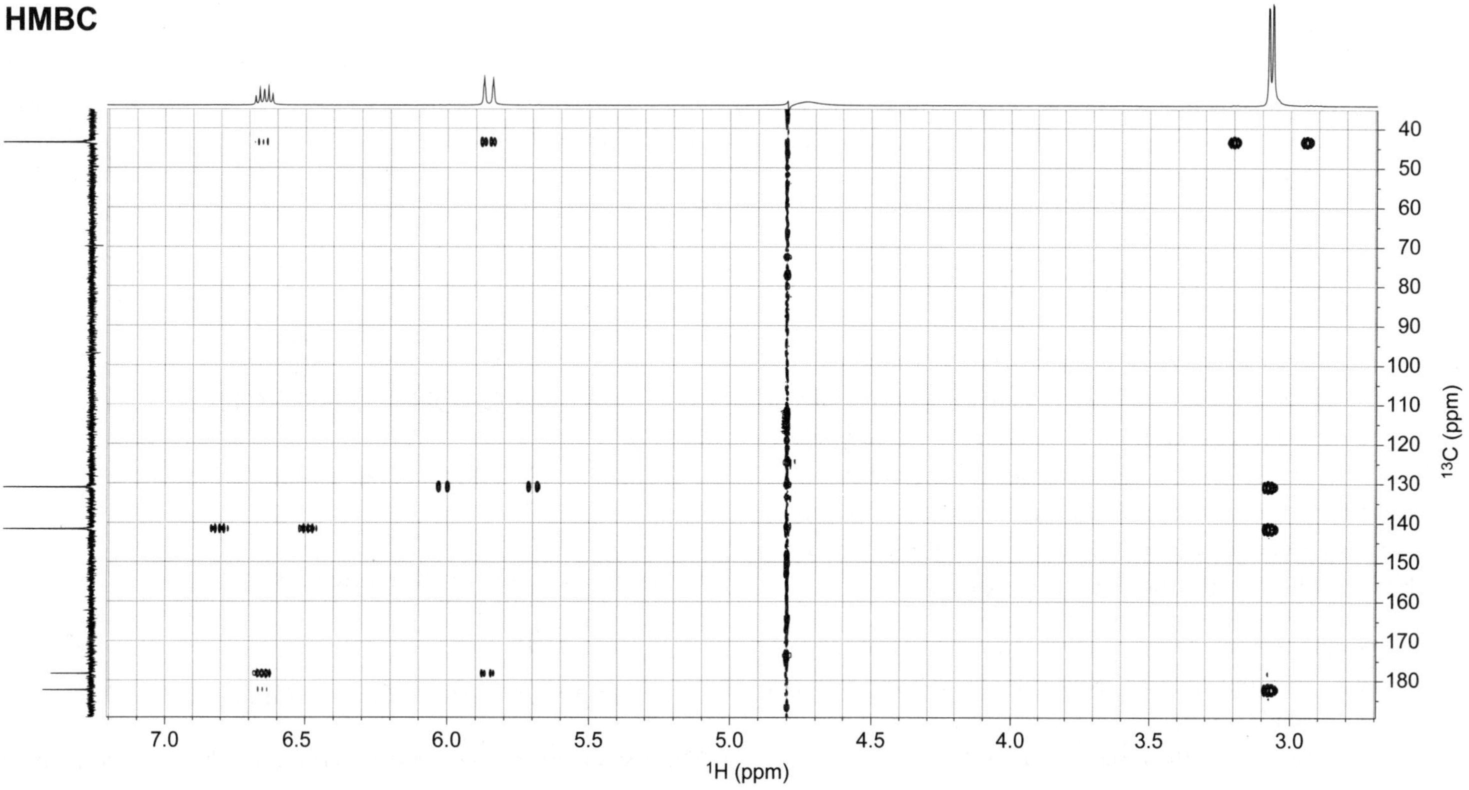
40
50
60
70
80
90
100
110
120
130
140
150
160
170
180
13C (ppm)
7.0
6.5
6.0
5.5
5.0
4.5
4.0
3.5
3.0
1H (ppm)

Solution

1) Calculate the number of double bond equivalents (DBEs)
DBE = C - (H/2) - (X/2) + (N/2) +1 = 5 - (6/2) - (0/2) + (0/2) + 1 = 3.

2) Look for the solvent signals in the proton and carbon spectra
The solvent signal resonates at 4.79 ppm (D_2O). Please note that it has an unusual shape, because the spectra have been acquired using a technique of solvent signal suppression (presaturation). Please also note that the residual solvent signal produces a stripe of small cross-peaks in the 2D experiments. This is an artifact called "t1-noise" that is particularly evident in the COSY and HSQC spectra of this problem. This artifact must be ignored.

3) Get preliminary information about the symmetry of the molecule
The number of carbon signals is the same as the number of carbon atoms in the molecular formula, and therefore the molecule is not symmetric.

4) Recognize the functional groups
The molecule contains two carboxylic carbons (178.3 and 182.5 ppm), an aliphatic carbon (43.6 ppm) and two sp^2 carbons at 131.0 and 141.6 ppm. These, along with the presence of two proton signals resonating between 5.7 and 6.7 ppm, suggest the presence of an alkene moiety.

5) Determine the number of quaternary carbons (C), methines (CH), methylenes (CH_2), and methyls (CH_3), and assign their resonances
The molecule contains one methylene, two methines and two carboxyl groups. Interpreting the HSQC experiment, it is possible to assign the proton chemical shifts of these moieties.

3.07	5.86	6.65		
43.6	131.0	141.6	178.3	182.5
CH_2	CH	CH	C	C

6) Determine the number of protons that produces each proton signal
Since the proton spectrum has been acquired in a protic solvent (D_2O), all the signals in the ^1H spectrum are produced by non-exchangeable protons. There are four non-exchangeable protons (2 CH and 1 CH_2) and the relative signal intensities are 1:1:2; therefore, the values of the relative intensities correspond to the number of hydrogens that produces each signal.

7) Determine the number of exchangeable protons
Since there are four non-exchangeable hydrogens and the molecular formula contains six hydrogens, there are two exchangeable protons in the molecule (6 - 4 = 2).

8) Characterize all the spin systems
By the analysis of the COSY experiment, it is possible to link together the three moieties. Obviously, the two sp^2 carbons turn out to be close to each other.

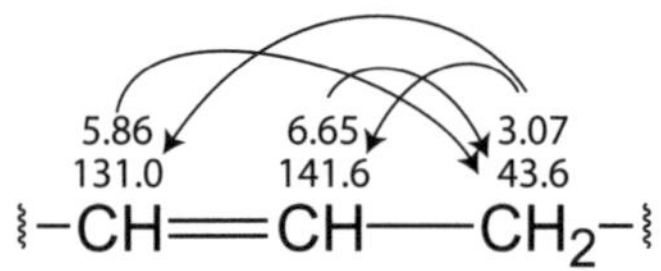

This spin system is also consistent with the HMBC correlations (arrows) and with the multiplicity of the proton signals. In particular, the signal at 3.07 ppm is a doublet of doublets. The large constant is due to the coupling to the hydrogen resonating at 6.65 ppm, while the very small constant is due to the allylic coupling to the hydrogen at 5.86 ppm. This latter signal is instead a doublet of triplets, because it is coupled by a large coupling constant to the hydrogen resonating at 6.65 ppm and by a very small coupling constants to the two hydrogens of the methylene. The signal at 6.65 ppm is also a doublet of triplets, because it is coupled by a small coupling constant to the two hydrogens of the methylene and by a large constant to the hydrogen at 5.86 ppm. Please note that two sub-peaks of the two sub-triplets are overlapped.

9) Connect all the fragments together
The part of the structure identified so far counts for three carbon and four hydrogen atoms. There are also two carbonyls (C=O), which count for other two carbon and two oxygen atoms. Subtracting these atoms from the molecular formula, you will see that two oxygen and two hydrogen atoms still need to be considered. Realistically, these are the two OH groups of the two carboxylic groups. Therefore, you have to link together the following moieties:

$$-\overset{O}{\overset{\|}{C}}-OH \qquad -\overset{O}{\overset{\|}{C}}-OH$$

$$-CH{=}CH{-}CH_2-$$

Hence, the structure of the molecule is:

$$HO-\overset{O}{\overset{\|}{C}}-CH{=}CH-CH_2-\overset{O}{\overset{\|}{C}}-OH$$

The assignment of the carboxyl carbons can be easily performed by looking at the HMBC correlations:

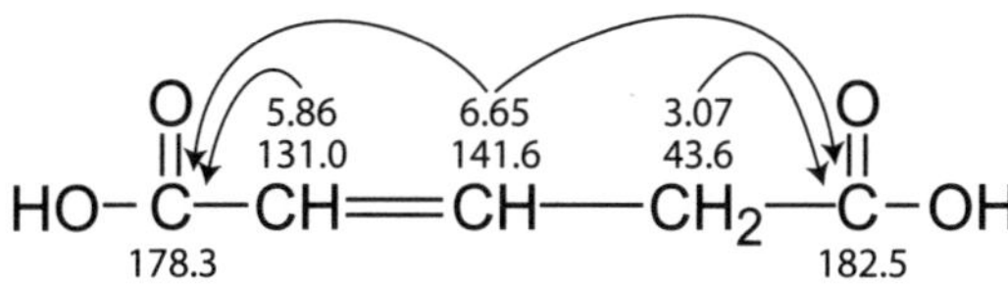

While the hydrogen resonating at 6.65 ppm correlates to both the carbonyl groups, the hydrogens at 3.07 and 5.86 ppm are actually able to discriminate between the two carbonyls.
Interestingly, you can also determine the geometry of the double bond by measuring the coupling constant(s) of one of the two olefinic hydrogens. Remember, small coupling constants (6-11 Hz) are typical of *cis* alkenes, while large couplings (12-19 Hz) are typical of *trans* alkenes. The proton spectrum contains an inset with the signal at 5.86 ppm. In order to facilitate the measurement of the coupling constant, the scale is here reported in Hz. If you measure the distance (Hz) between the two sub-triplets you will obtain the larger coupling constant:

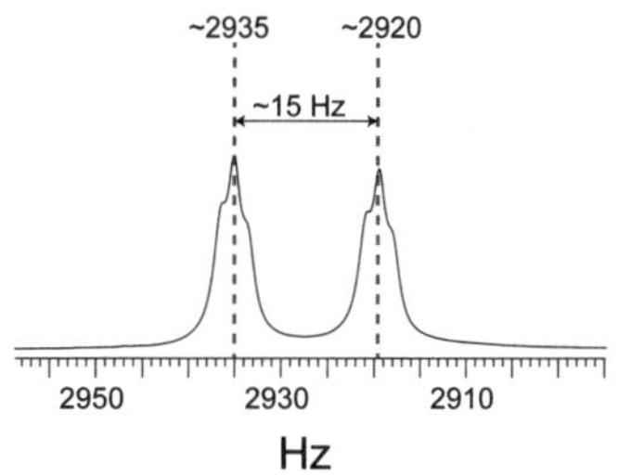

The coupling constant is about 15 Hz, so the double bond is *trans*:

10) Verify the consistency of all the data collected

Is the molecular formula consistent with the structure of the molecule?
Yes, the molecule contains the same atoms contained in the molecular formula.

Does the selected structure contain the number of DBEs you have calculated?
Yes, the molecule contains three π bonds.

Is the symmetry of the molecule consistent with the number of signals in the proton and carbon spectra?
The number of signals in the proton and carbon spectra is consistent with the symmetry of the molecule.

Does the molecule contain the number of exchangeable protons you have predicted?
Yes, it does.

Are the proton and carbon chemical shifts consistent with the chemical environment of each nucleus?
Yes, they are.

Are the COSY correlations and the signal multiplicities consistent with the spin systems of the molecule?
Yes, they are.

Are the HMBC correlations consistent with the supposed structure?
Yes, they are.

THE STRUCTURE IS CORRECT!!!

Problem 8.11 Molecular formula: $C_8H_{11}NO$ ***Difficulty*** ★★☆☆

Solvent: D_2O

Field: 500 MHz (^1H); 125 MHz (^{13}C)

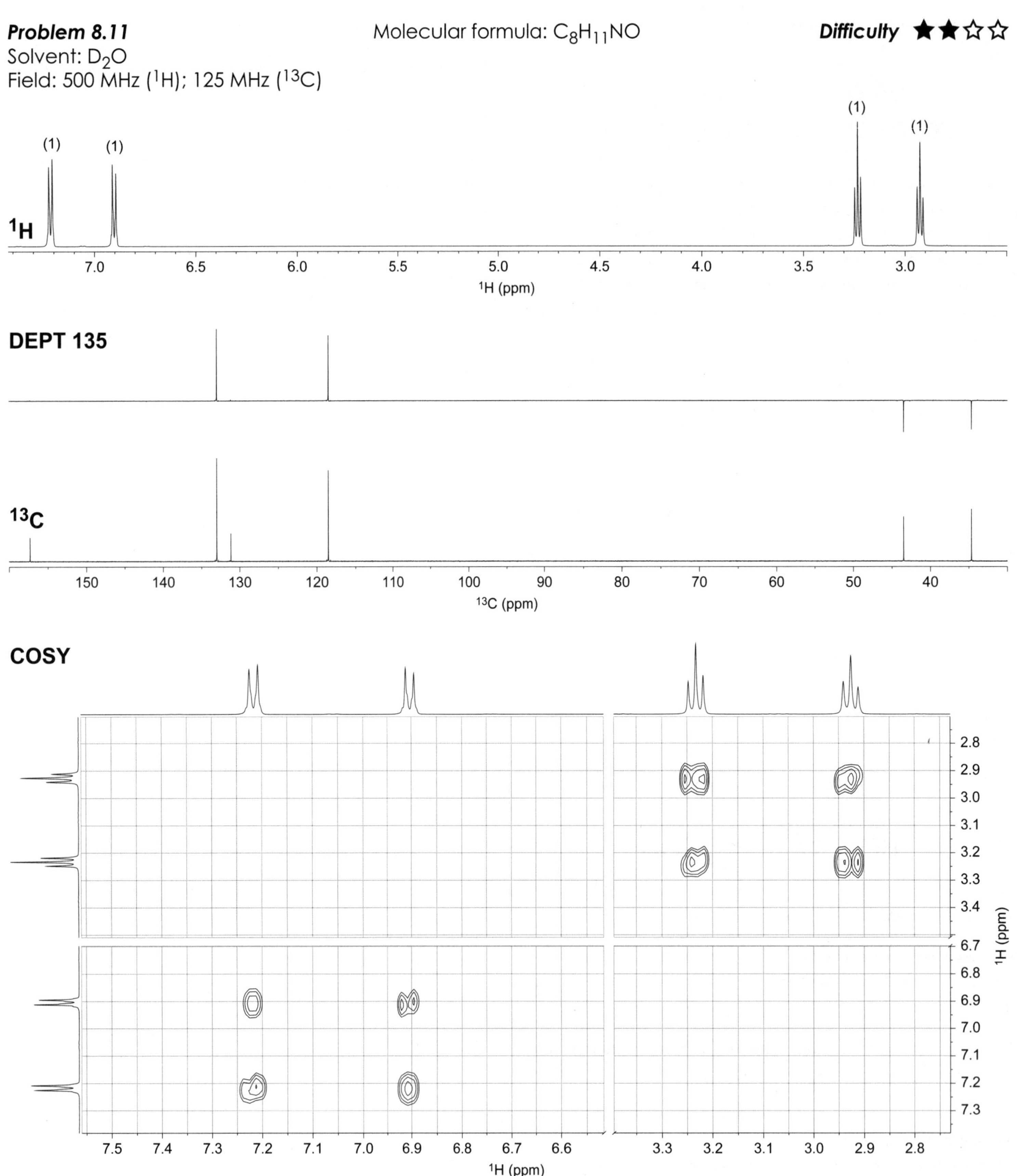

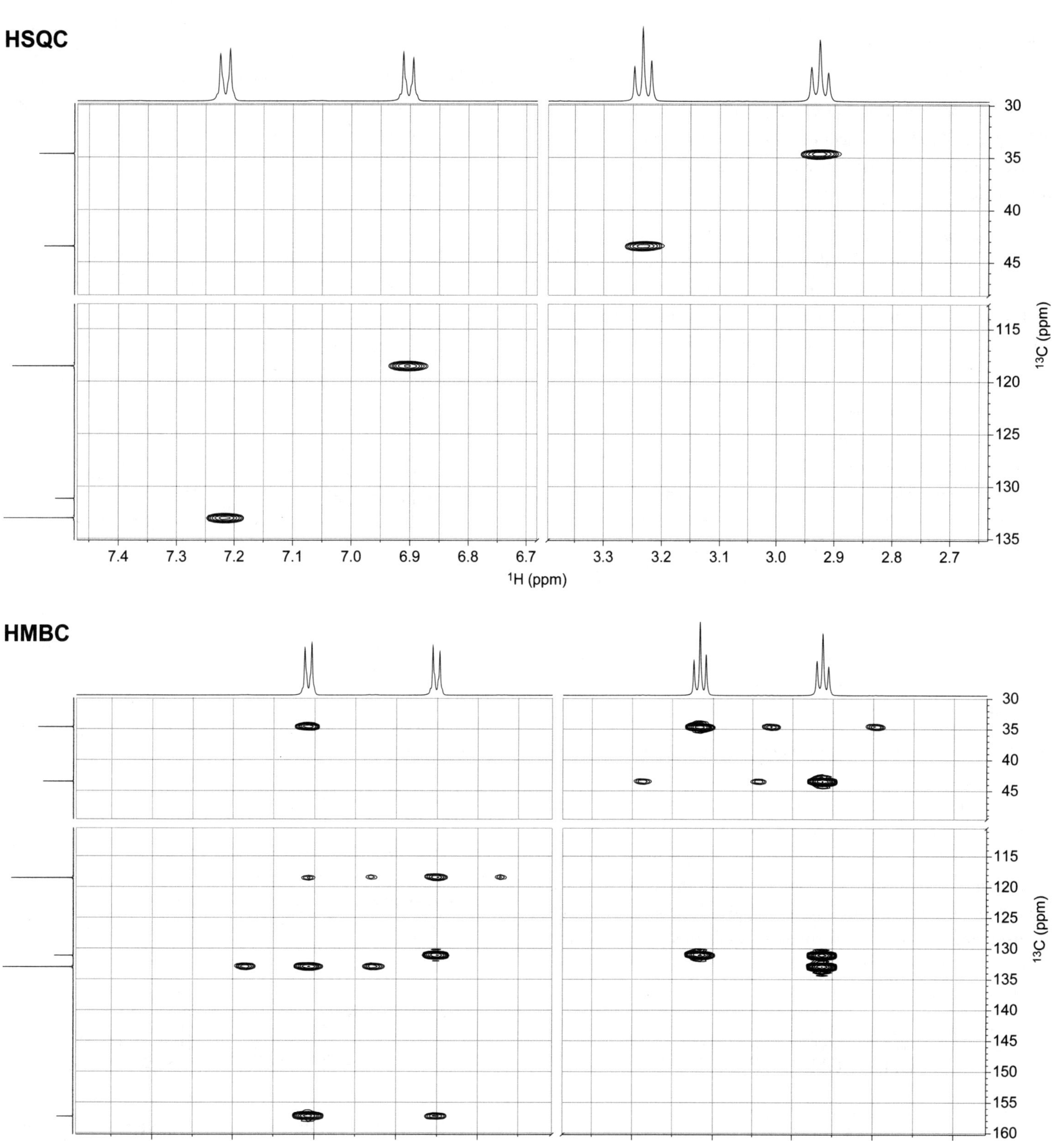
HSQC
HMBC
^{1}H (ppm)
^{13}C (ppm)

Solution

1) Calculate the number of double bond equivalents (DBEs)
DBE = C - (H/2) - (X/2) + (N/2) +1 = 8 - (11/2) - (0/2) + (1/2) + 1 = 4.

2) Look for the solvent signals in the proton and carbon spectra
The spectra have been acquired in D_2O. However, since the NMR sample is very concentrated, the residual solvent signal is not visible.

3) Get preliminary information about the symmetry of the molecule
The number of carbon signals is not the same as the number of carbon atoms in the molecular formula, but larger than its half, and therefore the molecule may have an element of symmetry passing through one or more carbon, or may be characterized by local symmetry.

4) Recognize the functional groups
The presence of two signals in the range 30-50 ppm in the carbon spectrum indicates that the molecule contains an aliphatic moiety. Moreover, the presence of other carbon signals between 115 and 160 ppm clearly indicates that the molecule contains also an aromatic moiety. Please note that two aromatic signals are particularly intense, so they may be produced by two pairs of equivalent carbons; this suggests also that the aromatic moiety may be symmetric. This hypothesis is corroborated by the presence of two doublets of equal intensity between 6.5 and 7.5 ppm in the proton spectrum, which are typical of *para*-disubstituted benzenes.

5) Determine the number of quaternary carbons (C), methines (CH), methylenes (CH_2), and methyls (CH_3), and assign their resonances
The aliphatic moiety of the molecule contains two methylenes, while the aromatic part contains two couples of equivalent methines and two quaternary carbons. Please note that one quaternary carbon signal is particularly deshielded, meaning that, probably, it is bonded to a strongly electronegative atom (considering the molecular formula, it should be an oxygen atom). The proton assignment of these fragments is accomplished by the HSQC experiment:

2.93 34.7 CH_2	3.23 43.5 CH_2	131.2 C	157.3 C

6.90 118.5 2 x CH	7.22 133.0 2 x CH

6) Determine the number of protons that produces each proton signal
Since the proton spectrum has been acquired in D_2O, the exchangeable protons are not observable, and all the signals in the ^1H spectrum are produced by non-exchangeable protons. The signals all have the same intensities and this is in agreement with the hypothesis that the aromatic moiety is symmetric. In fact, two equivalent aromatic methines produce a signal as intense as that of one methylene.

7) Determine the number of exchangeable protons
Since there are eight non-exchangeable hydrogens and the molecular formula contains eleven hydrogens, there are three exchangeable protons in the molecule (11 - 8 = 3).

8) Characterize all the spin systems
As expected, the COSY experiment indicates that the two methines are close to each other, whereas the two methylenes form another spin system.

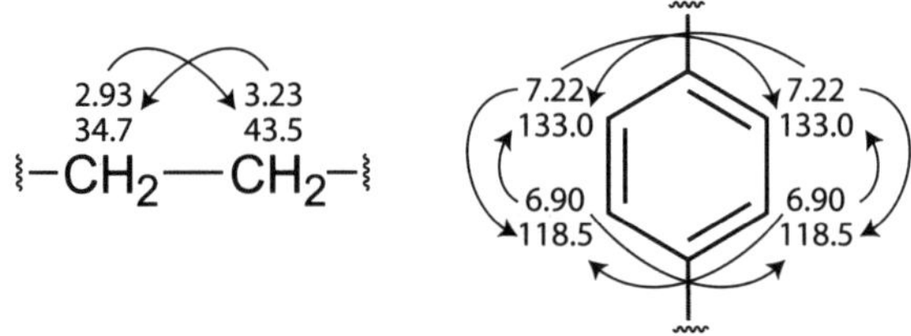

These spin systems are also consistent with the HMBC correlations (arrows), which further corroborate the hypothesis that the molecule contains a *para*-disubstituted benzene moiety. In fact, each proton correlates to the carbon of the other symmetric half of the moiety (by way of example, see the long-range correlation between the hydrogen at 7.22 ppm and the carbon at 133.0 ppm).

9) Connect all the fragments together
The part of the structure identified so far counts for eight carbon and eight hydrogen atoms. However, three hydrogens, one nitrogen, and one oxygen atom still need to be considered. These could be an OH and an NH_2 group.
Therefore, you have to link together the following moieties:

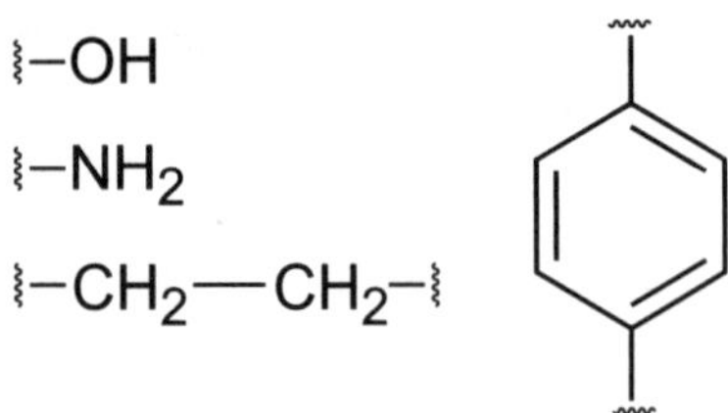

There are only two possible ways to assemble those fragments, which involve the formation of an aliphatic or an aromatic alcohol, respectively.

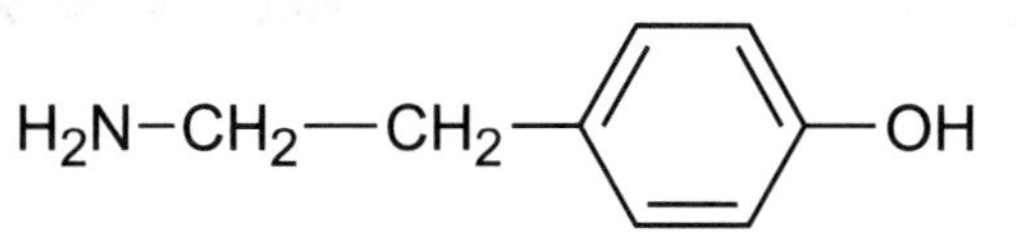

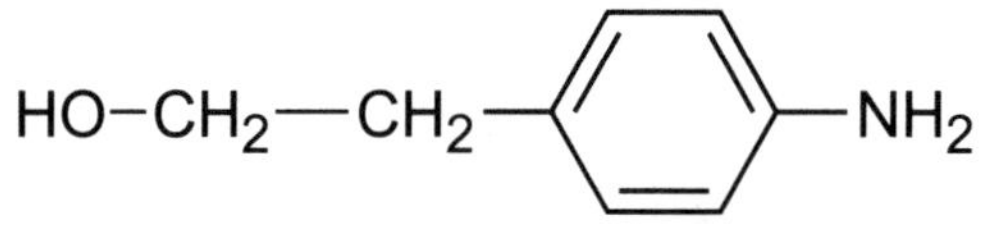

$H_2N-CH_2-CH_2-C_6H_4-OH$

$HO-CH_2-CH_2-C_6H_4-NH_2$

The structure on the bottom can be ruled out because the aliphatic alcohol would involve a carbon atom resonating at a chemical shift value larger than 60 ppm. Furthermore, the aromatic quaternary carbon at 157.3 ppm is much more consistent with the presence of an oxygen atom bonded to the aromatic ring.
It is possible to suppose that the assignment of the aromatic moiety is the following:

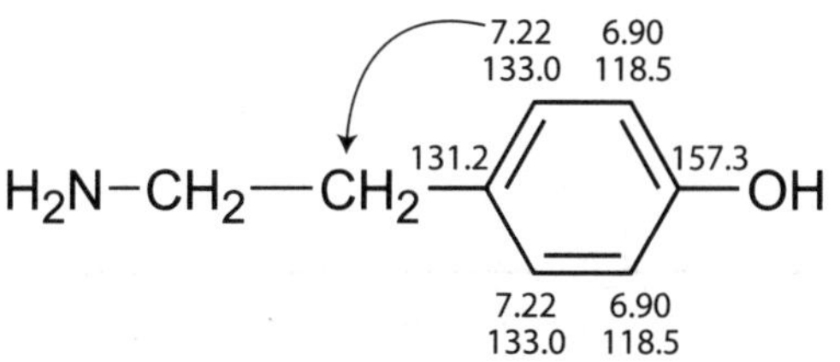

7.22 6.90
133.0 118.5
131.2 157.3
$H_2N-CH_2-CH_2-C_6H_4-OH$
7.22 6.90
133.0 118.5

The resonance at 131.2 ppm is assigned to the second quaternary carbon by exclusion. Then the proton resonances are assigned both considering the fact that the hydrogens in the *ortho* position with respect to the electron-donating group (OH) are more shielded than those in the *meta* position and considering all the long-range heteronuclear correlations (HMBC). Please remember that in aromatic systems, three-bond correlations are much more intense than those through two bonds. Finally, the correlation of the signal at 7.22 ppm to the aliphatic carbon at 34.7 ppm further confirms the assignment of the aromatic moiety. This last correlation suggests also the assignment of the aliphatic part of the molecule, so the complete assignment of the molecule is:

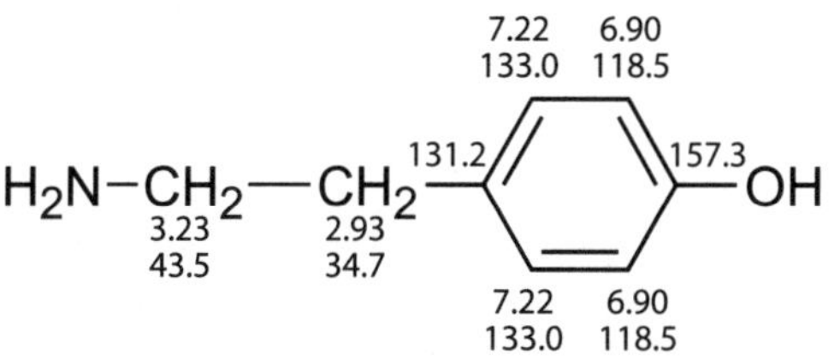

7.22 6.90
133.0 118.5
131.2 157.3
$H_2N-CH_2-CH_2-C_6H_4-OH$
3.23 2.93
43.5 34.7
7.22 6.90
133.0 118.5

10) Verify the consistency of all the data collected

Is the molecular formula consistent with the structure of the molecule?
Yes, the molecule contains the same atoms contained in the molecular formula.

Does the selected structure contain the number of DBEs you have calculated?
Yes, the molecule contains an aromatic ring counting for four DBEs (three π bonds and one ring).

Is the symmetry of the molecule consistent with the number of signals in the proton and carbon spectra?
The number of signals in the proton and carbon spectra is consistent with the symmetry of the molecule.

Does the molecule contain the number of exchangeable protons you have predicted?
Yes, it does.

Are the proton and carbon chemical shifts consistent with the chemical environment of each nucleus?
Yes, they are.

Are the COSY correlations and the signal multiplicities consistent with the spin systems of the molecule?
Yes, they are.

Are the HMBC correlations consistent with the supposed structure?
Yes, they are.

THE STRUCTURE IS CORRECT!!!

Problem 8.12 Molecular formula: $C_6H_{12}O_3$ ***Difficulty*** ★★☆☆
Solvent: D_2O
Field: 500 MHz (^{1}H); 125 MHz (^{13}C)

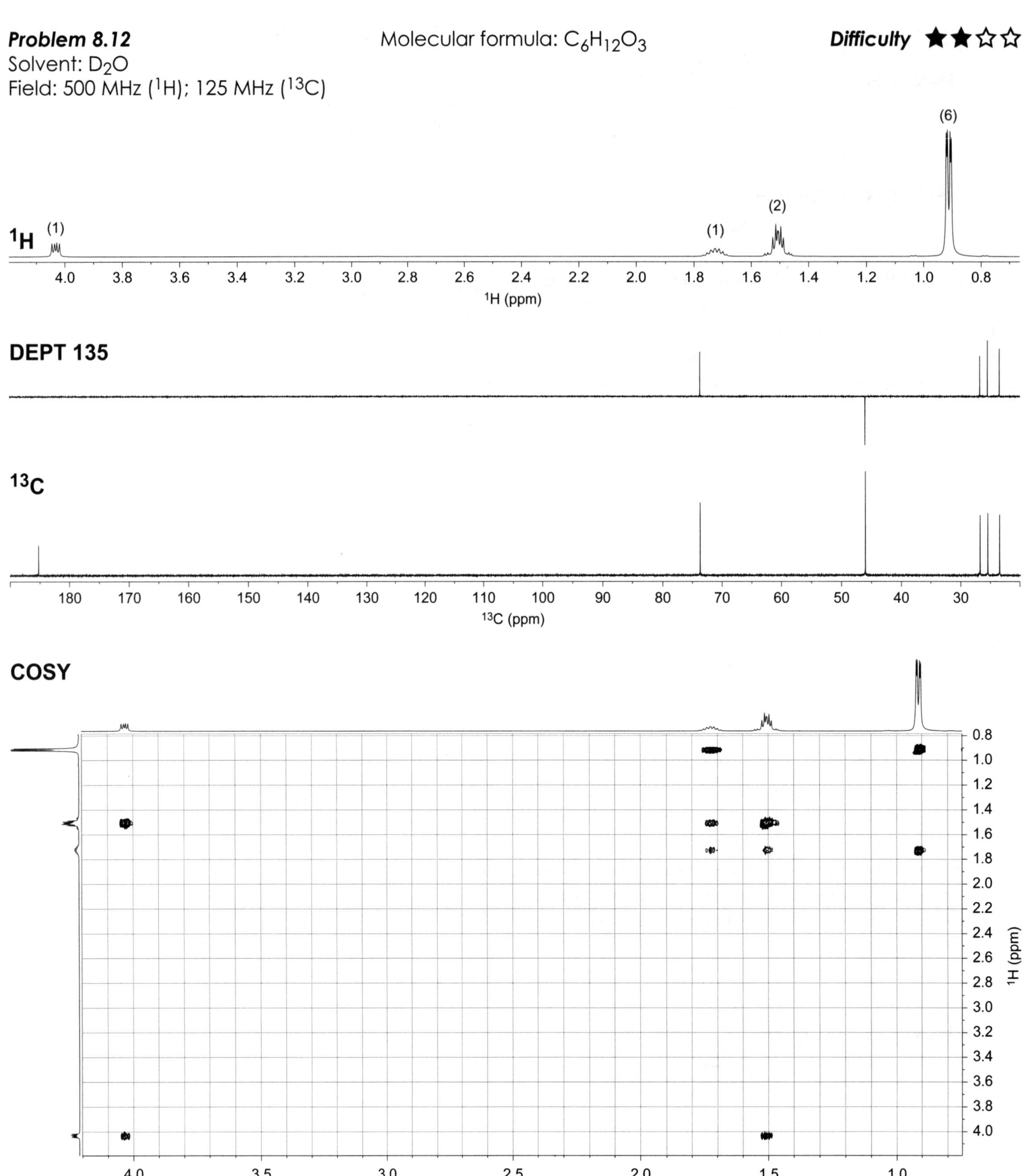

HSQC

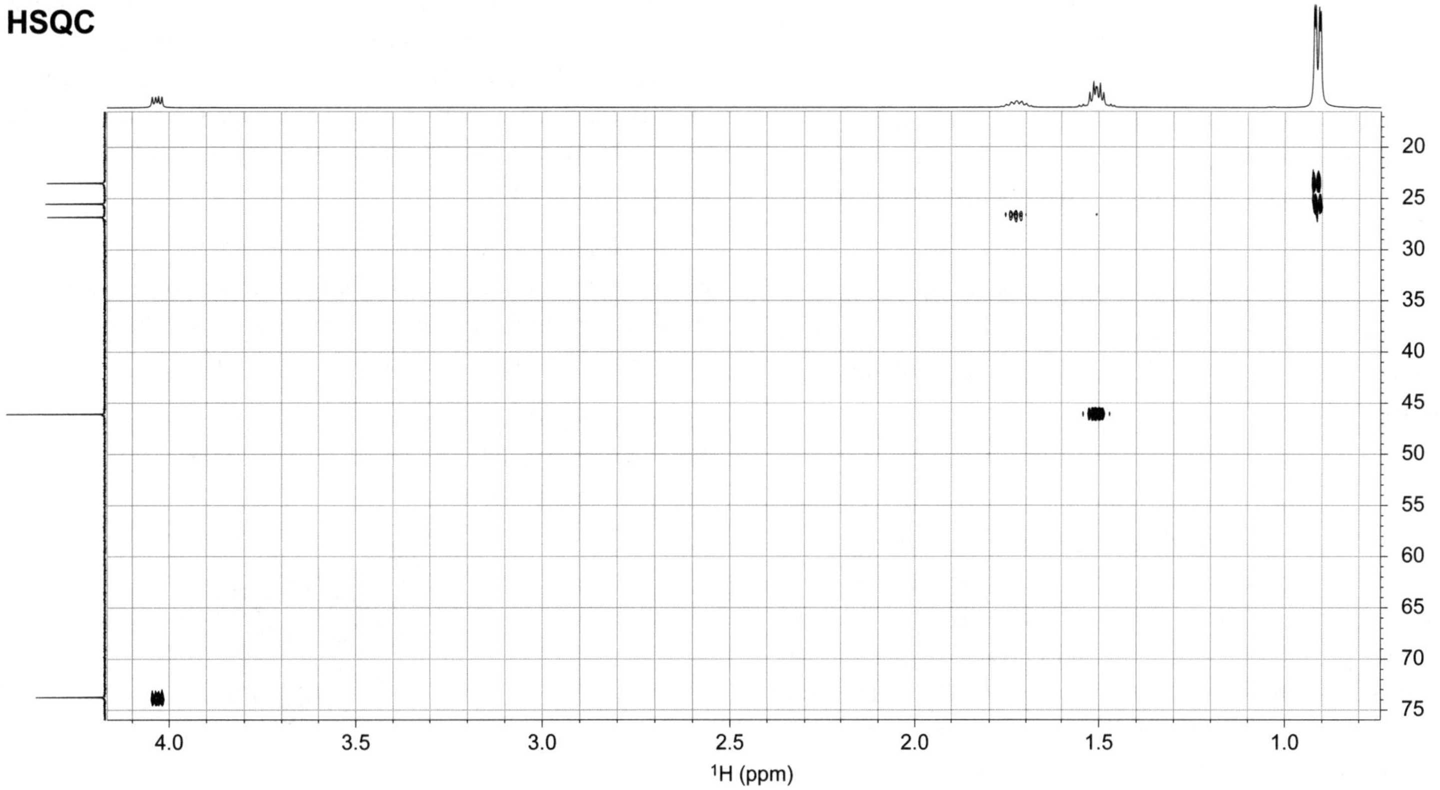

1H (ppm): 4.0, 3.5, 3.0, 2.5, 2.0, 1.5, 1.0
13C (ppm): 20, 25, 30, 35, 40, 45, 50, 55, 60, 65, 70, 75

HMBC

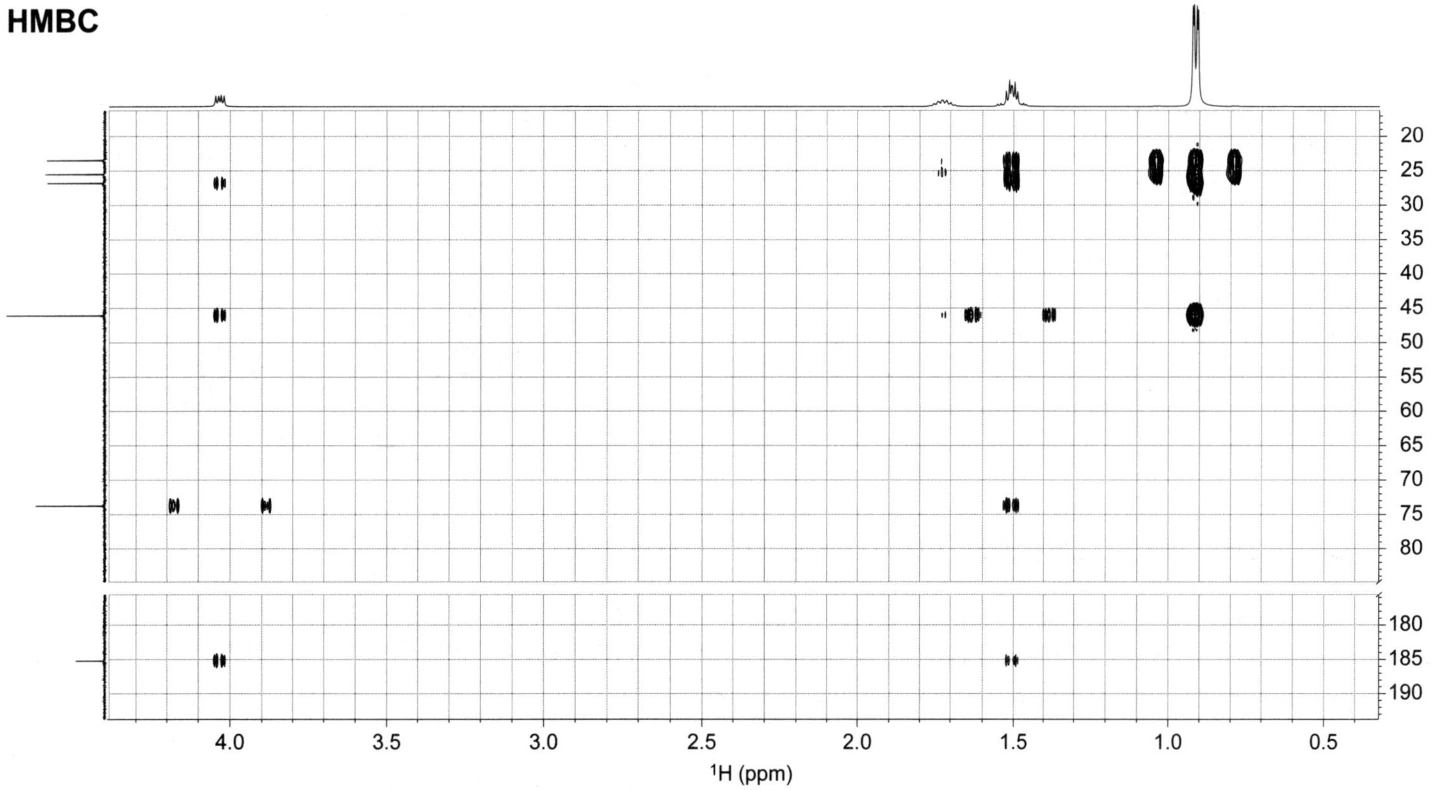

1H (ppm): 4.0, 3.5, 3.0, 2.5, 2.0, 1.5, 1.0, 0.5
13C (ppm): 20, 25, 30, 35, 40, 45, 50, 55, 60, 65, 70, 75, 80, 180, 185, 190

Solution

1) Calculate the number of double bond equivalents (DBEs)

DBE = C - (H/2) - (X/2) + (N/2) +1 = 6 - (12/2) - (0/2) + (0/2) + 1 = 1.

2) Look for the solvent signals in the proton and carbon spectra

The spectra have been acquired in D_2O. However, since the NMR sample is very concentrated, the residual solvent signal is not visible.

3) Get preliminary information about the symmetry of the molecule

The number of carbon signals is the same as the number of carbon atoms in the molecular formula, so the molecule is not symmetric.

4) Recognize the functional groups

The carbon signal at 185.2 ppm indicates the presence of a carboxylic acid or an ester (it is possible to rule out the presence of an amide group for the lack of nitrogen atoms in the molecular formula). Then, there is a carbon signal resonating at 73.8 ppm, probably because it is bonded to an oxygen atom. The remaining carbon signals resonate at very small chemical shift values, consistent with the presence of an aliphatic moiety in the molecule. Two of these signals should be produced by two methyls, as suggested by the presence a multiplet integrating for six hydrogens at about 0.9 ppm in the proton spectrum.

5) Determine the number of quaternary carbons (C), methines (CH), methylenes (CH_2), and methyls (CH_3), and assign their resonances

There is a carboxylic carbon (185.2 ppm), a methylene (46.1 ppm), one oxygen bearing methine (73.8 ppm) and three other carbon signals (23.5, 25.5, and 26.8 ppm) that can be attributed both to methyls and to methines. By analysis of the HSQC spectrum, it is clear that the two most shielded carbon signals (23.5 and 25.5 ppm) correlate with the intense proton signal at 0.90 ppm (attributable to two methyls), while the carbon signal at 26.8 ppm is surely a methine because it correlates to the proton signal at 1.73 ppm that integrates for one hydrogen:

0.92	0.91	1.51	1.73	4.03	
23.5	25.5	46.1	26.8	73.8	185.2
CH_3	CH_3	CH_2	CH	CH	C

Please note that the analysis of the HSQC indicates also that the signal at 0.9 ppm is actually not a doublet of doublets, but rather two overlapped doublets slightly shifted relative to each other.

6) Determine the number of protons that produces each proton signal

The proton spectrum has been acquired in D_2O, and so all the signals in the ^1H spectrum are produced by non-exchangeable protons. The molecule contains ten non-exchangeable protons (2 CH_3, 1 CH_2, and 2 CH), and the relative signal intensities are 1:1:2:6 (whose sum is 10). Therefore, the values of the relative intensities correspond to the number of hydrogens that produces each signal.

7) Determine the number of exchangeable protons

Since there are ten non-exchangeable hydrogens and the molecular formula contains twelve hydrogens, there are two exchangeable protons in the molecule (12 - 10 = 2).

8) Characterize all the spin systems

From the analysis of the COSY spectrum it is possible to define the following spin system:

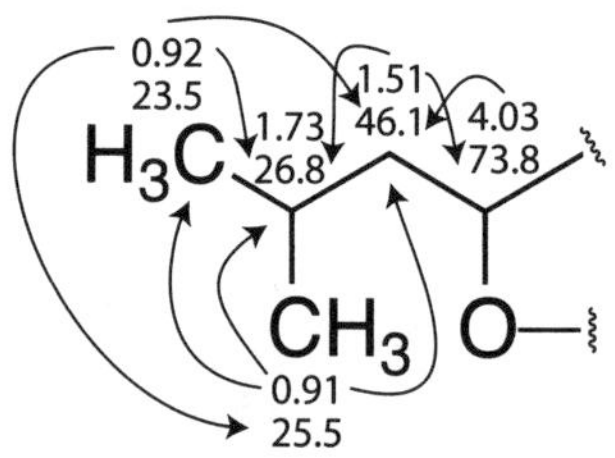

This system is also consistent with the HMBC correlations (only some correlations are shown as arrows in the figure).

9) Connect all the fragments together

The structure of the molecule identified so far already has all the carbon atoms contained in the molecular formula, so it is possible to exclude the presence of an ester group. Therefore, the structure of the molecule can be completed as shown below:

0.92
23.5
H_3C
1.73
26.8
1.51
46.1
4.03
73.8
O
185.2
OH
CH_3
OH
0.91
25.5

10) Verify the consistency of all the data collected

Is the molecular formula consistent with the structure of the molecule?
Yes, the molecule has the same atoms contained in the molecular formula.

Does the selected structure contain the number of DBEs you have calculated?
Yes, the molecule contains one DBE.

Is the symmetry of the molecule consistent with the number of signals in the proton and carbon spectra?
The number of signals in the proton and carbon spectra is consistent with the symmetry of the molecule.

Does the molecule contain the number of exchangeable protons you have predicted?
Yes, it does.

Are the proton and carbon chemical shifts consistent with the chemical environment of each nucleus?
Yes, they are.

Are the COSY correlations and the signal multiplicities consistent with the spin systems of the molecule?
Yes, they are.

Are the HMBC correlations consistent with the supposed structure?
Yes, they are.

THE STRUCTURE IS CORRECT!!!

Problem 8.13 Molecular formula: $C_8H_{16}O$ **Difficulty** ★★★☆

Solvent: $CDCl_3$

Field: 500 MHz (^1H); 125 MHz (^{13}C)

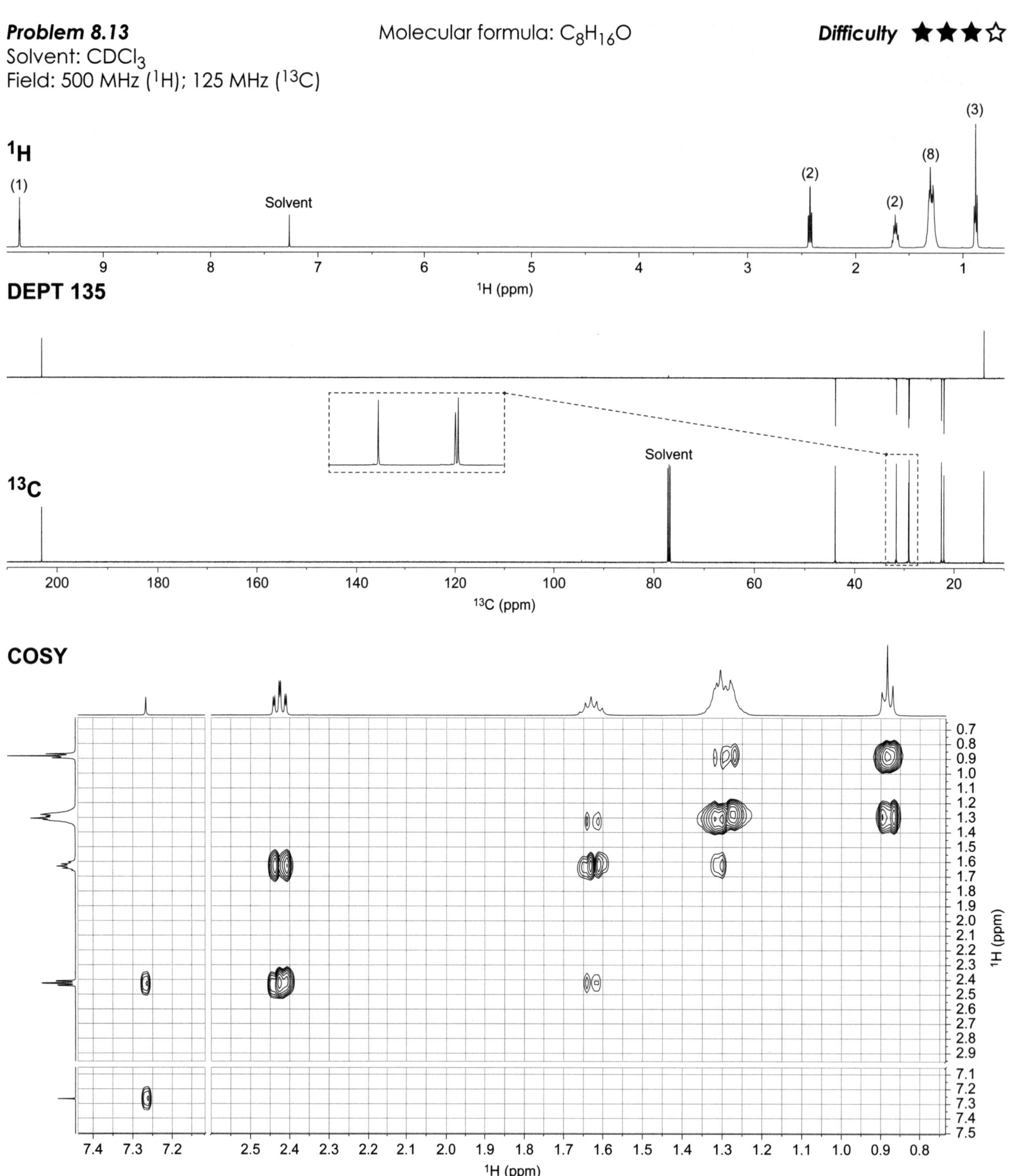

HSQC

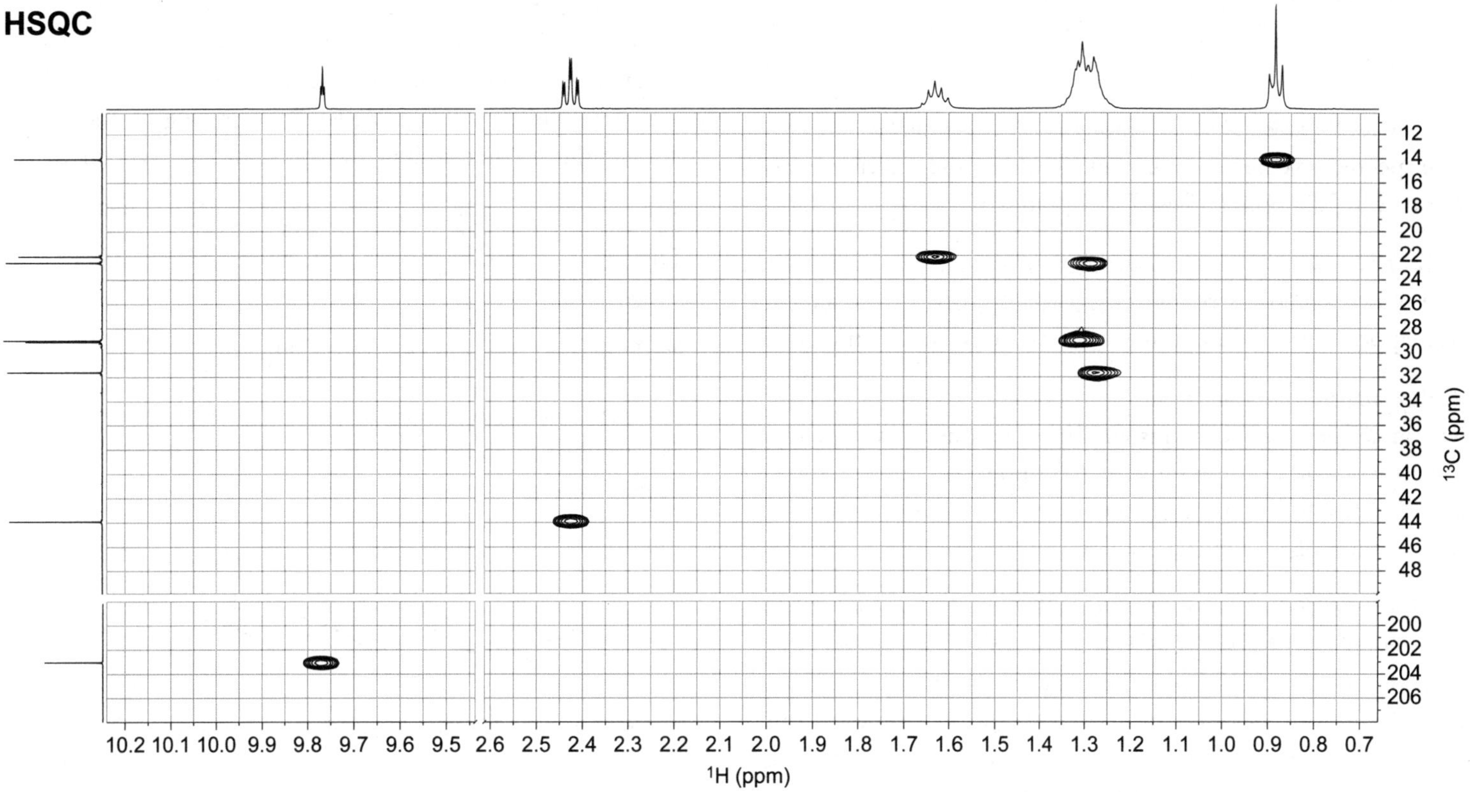

10.2 10.1 10.0 9.9 9.8 9.7 9.6 9.5 2.6 2.5 2.4 2.3 2.2 2.1 2.0 1.9 1.8 1.7 1.6 1.5 1.4 1.3 1.2 1.1 1.0 0.9 0.8 0.7
^1H (ppm)

12 14 16 18 20 22 24 26 28 30 32 34 36 38 40 42 44 46 48 200 202 204 206
^{13}C (ppm)

HMBC

10.3 10.2 10.1 10.0 9.9 9.8 9.7 9.6 2.6 2.5 2.4 2.3 2.2 2.1 2.0 1.9 1.8 1.7 1.6 1.5 1.4 1.3 1.2 1.1 1.0 0.9 0.8 0.7
^1H (ppm)

10 12 14 16 18 20 22 24 26 28 30 32 34 36 38 40 42 44 46 48 198 200 202 204 206 208
^{13}C (ppm)

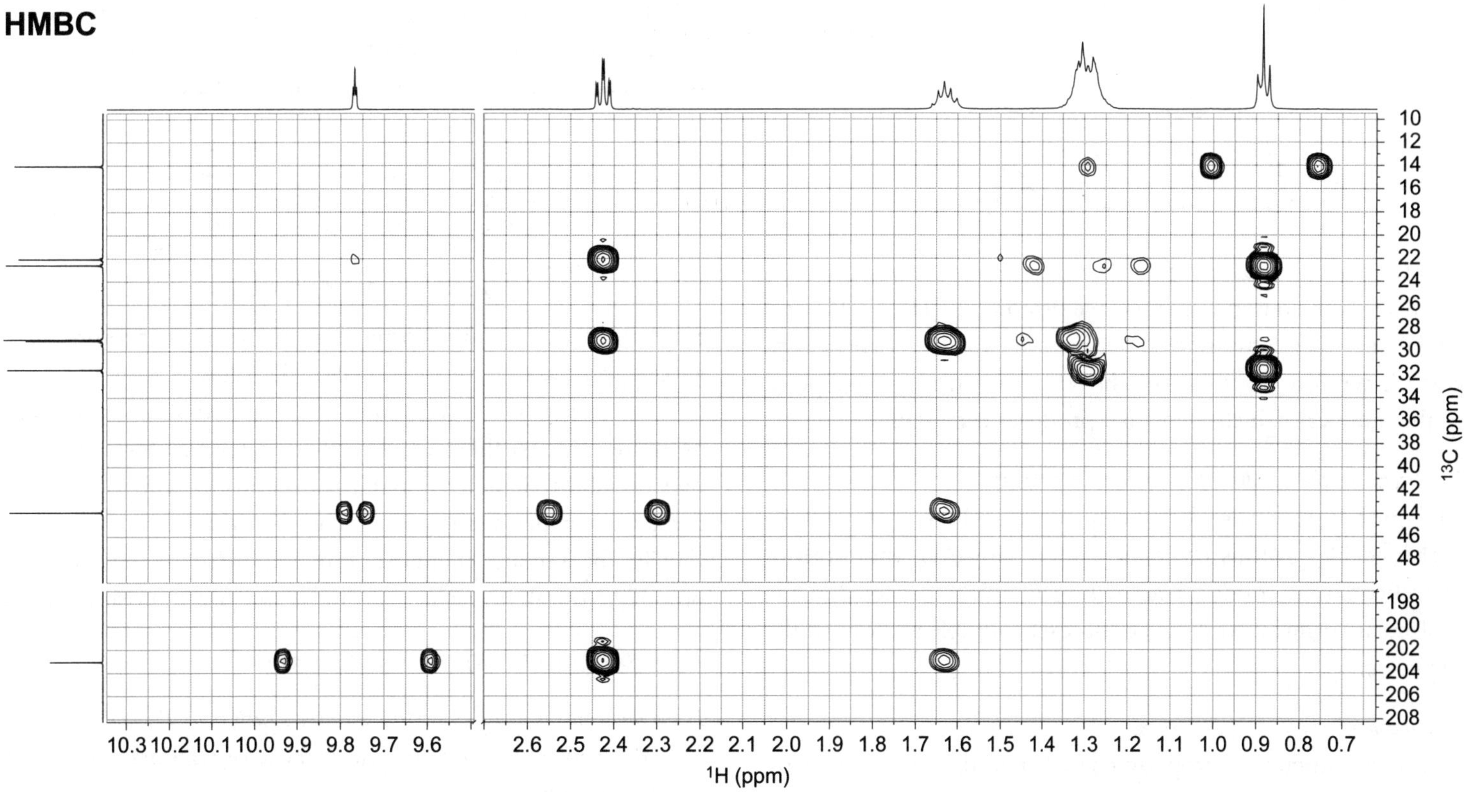

Solution

1) Calculate the number of double bond equivalents (DBEs)
DBE = C - (H/2) - (X/2) + (N/2) +1 = 8 - (16/2) - (0/2) + (0/2) + 1 = 1.

2) Look for the solvent signals in the proton and carbon spectra
The spectra have been acquired in $CDCl_3$, so the solvent signals resonate at 7.26 ppm (^{1}H) and 77.1 ppm (^{13}C).

3) Get preliminary information about the symmetry of the molecule
The carbon spectrum contains eight signals (two of them almost overlapped at around 29 ppm), so the molecule is not symmetric.

4) Recognize the functional groups
The molecule clearly contains an aldehyde group as indicated by the presence of a carbon signal at 203.1 ppm and a proton signal at 9.77 ppm. The remaining signals are instead consistent with the presence of an aliphatic moiety.

5) Determine the number of quaternary carbons (C), methines (CH), methylenes (CH_2), and methyls (CH_3), and assign their resonances
In addition to the aldehyde, there are six methylenes and one methyl. As you can see from the HSQC experiment, four methylenes have very similar proton chemical shifts, resonating around 1.30 ppm. In particular, two of these have the same proton and carbon chemical shifts:

0.88 14.1 CH_3	1.62 22.2 CH_2	1.29 22.6 CH_2	9.77 203.1 CH=O
1.28 31.7 CH_2	1.31 29.1 CH_2	1.31 29.1 CH_2	2.42 44.0 CH_2

6) Determine the number of protons that produces each proton signal
There are sixteen non-exchangeable protons (1 CH_3, 6 CH_2, and 1 CH) and the relative signal intensities are 1:2:2:8:3 (whose sum is 16). Therefore, the values of the relative intensities correspond to the number of hydrogens that produces each signal.

7) Determine the number of exchangeable protons
The proton spectrum has been acquired in $CDCl_3$ and so exchangeable protons could be visible. However, the moieties identified so far count for all hydrogen atoms contained in the molecular formula, so there are no exchangeable protons in the molecule.

8) Characterize all the spin systems
It is very difficult to properly characterize the spin system of this molecule, because there are many chemical shift equivalent methylenes coupled to each other. You can follow the spin system only partially, starting from the signal resonating at 9.77 ppm or from the signal at 0.88 ppm. In both cases you will stop the characterization of the spin system when trying to find the correlations between the methylenes resonating at about 1.30 ppm. Definitely, you can draw the following sub-structures:

0.88 14.1 CH_3–	1.29 22.6 –CH_2–	1.62 2.42 9.77 22.2 44.0 203.1 –CH_2–CH_2–CH=O
1.31 29.1 –CH_2–	1.31 29.1 –CH_2–	1.28 31.7 –CH_2–

9) Connect all the fragments together
It is very easy to connect all the moieties together; in fact, the molecule is octanal:

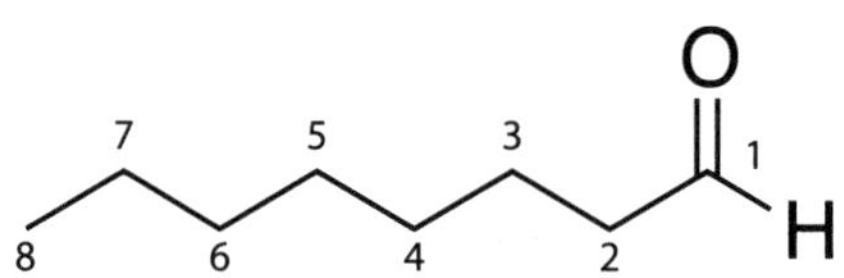

The problem now is to assign all the resonances. This can be performed using the long-range correlations in the HMBC experiment. In particular, the correlation between the signal at 2.42 ppm and the carbon at 29.1 ppm indicates that one of the two chemical shift equivalent methylenes (1.31 ppm/29.1 ppm) is attached to the methylene resonating at 1.62 ppm/22.2 ppm.

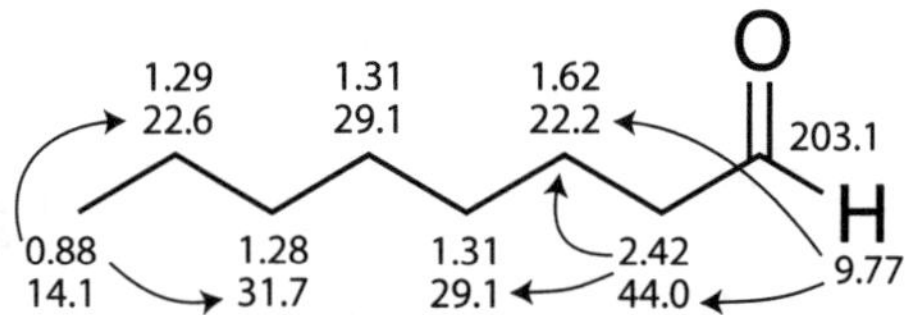

Similarly, the long-range correlations between the proton signal at 0.88 ppm and the carbon signals at 22.6 and 31.7 ppm indicate that the methylenes at 1.28 ppm/31.7 ppm and 1.29 ppm/22.6 ppm are close to the methyl. Unfortunately, at this stage, it is only possible to tentatively assign the methylene with the smallest carbon resonance as the one attached to the methyl.
Finally, the second chemical shift equivalent methylene (1.31 ppm/29.1 ppm), by exclusion, can be assigned to position 5.

10) Verify the consistency of all the data collected

Is the molecular formula consistent with the structure of the molecule?
Yes, the molecule contains the same atoms contained in the molecular formula.

Does the selected structure contain the number of DBEs you have calculated?
Yes, the molecule contains one DBE.

Is the symmetry of the molecule consistent with the number of signals in the proton and carbon spectra?
The number of signals in the proton and carbon spectra is consistent with the symmetry of the molecule.

Does the molecule contain the number of exchangeable protons you have predicted?
Yes, the molecule does not contains exchangeable protons.

Are the proton and carbon chemical shifts consistent with the chemical environment of each nucleus?
Yes, they are.

Are the COSY correlations and the signal multiplicities consistent with the spin systems of the molecule?
Yes, they are.

Are the HMBC correlations consistent with the supposed structure?
Yes, they are.

THE STRUCTURE IS CORRECT!!!

Problem 8.14 Molecular formula: $C_8H_8O_2$ **Difficulty** ★★★☆

Solvent: $CDCl_3$

Field: 400 MHz (^1H); 100 MHz (^{13}C)

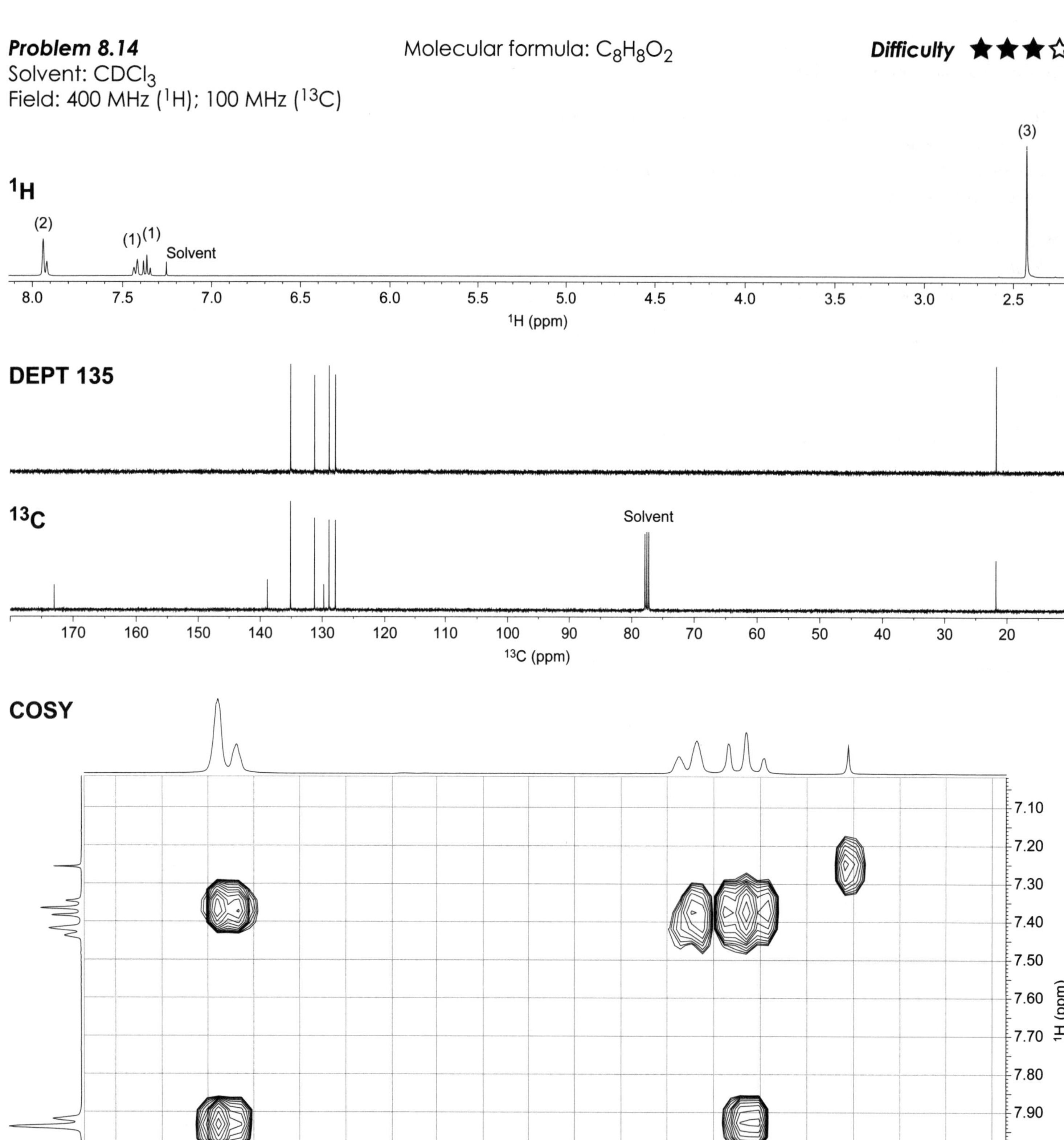

HSQC

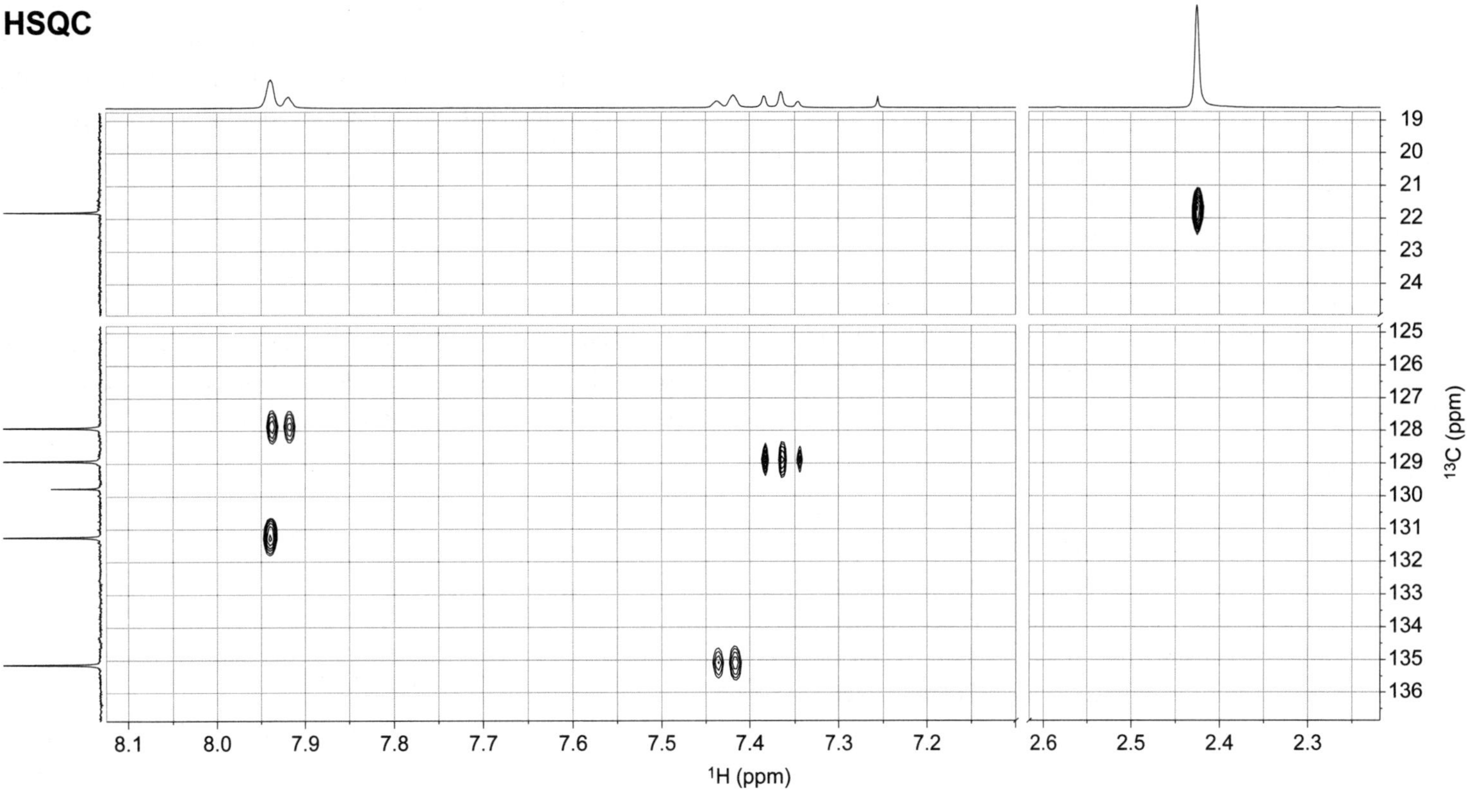
19
20
21
22
23
24
125
126
127
128
129
130
131
132
133
134
135
136
^{13}C (ppm)
8.1
8.0
7.9
7.8
7.7
7.6
7.5
7.4
7.3
7.2
2.6
2.5
2.4
2.3
^{1}H (ppm)

HMBC

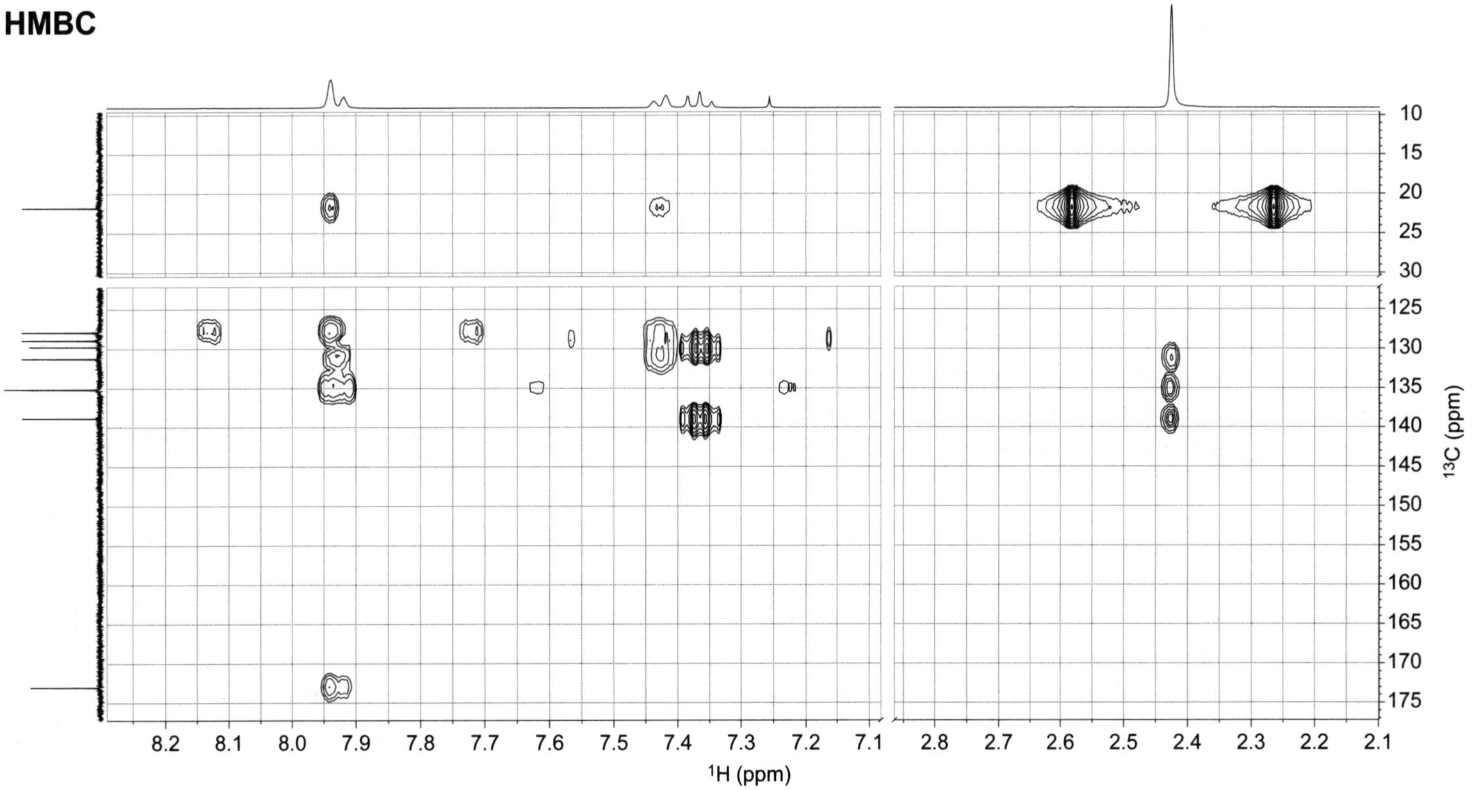
10
15
20
25
30
125
130
135
140
145
150
155
160
165
170
175
^{13}C (ppm)
8.2
8.1
8.0
7.9
7.8
7.7
7.6
7.5
7.4
7.3
7.2
7.1
2.8
2.7
2.6
2.5
2.4
2.3
2.2
2.1
^{1}H (ppm)

Solution

1) Calculate the number of double bond equivalents (DBEs)

DBE = C - (H/2) - (X/2) + (N/2) +1 = 8 - (8/2) - (0/2) + (0/2) + 1 = 5.

2) Look for the solvent signals in the proton and carbon spectra

The spectra have been acquired in $CDCl_3$, so the solvent signals resonate at 7.26 ppm (very low intensity) and 77.1 ppm in the proton and carbon spectra, respectively.

3) Get preliminary information about the symmetry of the molecule

The number of carbon signals is the same as the number of carbon atoms in the molecular formula, so the molecule is not symmetric.

4) Recognize the functional groups

The molecule clearly contains an aromatic moiety and a carboxylic group or an ester (an amide group can be ruled out because there are no nitrogen atoms in the molecular formula). Furthermore, as suggested also by the presence of an intense singlet at 2.43 ppm in the proton spectrum, there is a methyl that probably (see its proton chemical shift) is bonded to the aromatic moiety.

5) Determine the number of quaternary carbons (C), methines (CH), methylenes (CH_2), and methyls (CH_3), and assign their resonances

The molecule contains four aromatic methines, one methyl, two aromatic quaternary carbon atoms, and a carboxylic group:

2.43 21.8 CH_3	129.8 C	138.9 C	173.0 C
7.93 127.9 CH	7.36 128.9 CH	7.94 131.2 CH	7.43 135.1 CH

6) Determine the number of protons that produces each proton signal

The molecule contains seven non-exchangeable protons (4 CH and 1 CH_3) and the sum of the relative signal intensities (2:1:1:3) is seven as well. Therefore, the values of the relative intensities correspond to the number of hydrogens that produces each signal.

7) Determine the number of exchangeable protons

The moieties identified so far count for seven hydrogen atoms, while the molecular formula contains eight hydrogens, therefore there is one exchangeable proton in the molecule. In spite of the proton spectrum having been acquired in $CDCl_3$, this proton is not visible in the spectra.

8) Characterize all the spin systems

The COSY spectrum and the multiplicity of the proton signals are difficult to interpret. In particular, from the analysis of the COSY and HSQC spectra it is clear that the signal at around 7.93 ppm is actually the result of the superimposition of two signals, realistically a singlet at 7.94 ppm and a doublet at 7.93 ppm.

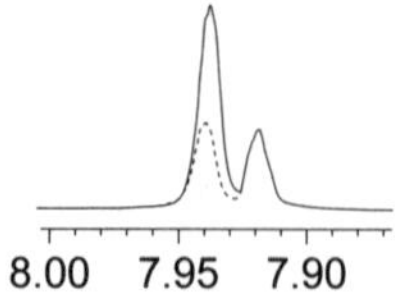

Then, there are a doublet and a triplet at 7.43 and 7.36 ppm, respectively. These signals should be coupled to each other; in fact, they are related by a marked roof effect.

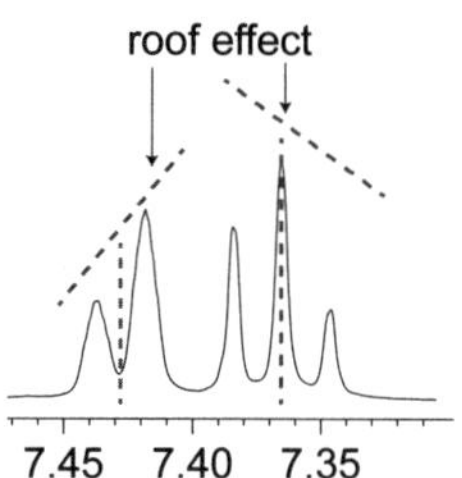

Please note the doublet is characterized by a line-width larger than that of the triplet. This is probably due to a very small long-range coupling, which is quite common in aromatic systems. Therefore, in summary, there are four aromatic proton signals: two doublets, a triplet, and a singlet. This pattern is generally consistent with a *meta*-disubstituted benzene, as shown below.

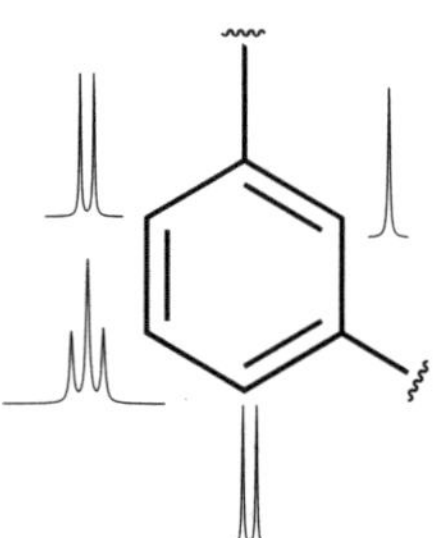

Unfortunately, the COSY spectrum is affected by the superimposition of several cross-peaks, and you can only unambiguously determine the correlation between the signals at 7.93 and 7.36 ppm. However, keeping in mind the hypothesis of the *meta*-disubstituted benzene, it is possible to "unearth" also the cross-peaks between the triplet at 7.36 ppm and the doublet at 7.43 ppm (indicated as black circle in the figure below), which are basically merged with the diagonal peaks.

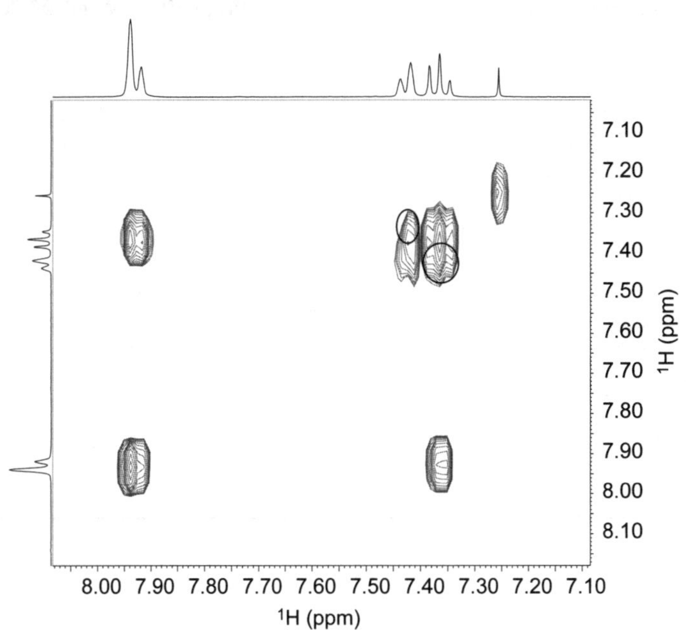

Hence, for the time being, it is possible to tentatively define the following spin systems:

7.93
127.9
7.94
131.2
7.36
128.9
7.43
135.1
2.43
21.8
$-CH_3$

9) Connect all the fragments together

The moieties shown in the previous figure count for seven carbon atoms and seven hydrogens, meaning that a carboxylic acid group (COOH) still has to be positioned. Clearly, this group and the methyl are the substituents of the aromatic moiety.

They can be positioned thanks to the diagnostic HMBC correlations. Some of them are shown as arrows in the following figure:

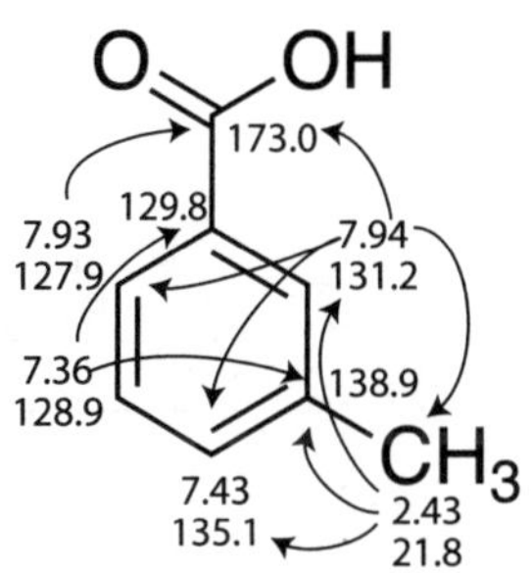

10) Verify the consistency of all the data collected

Is the molecular formula consistent with the structure of the molecule?
Yes, the molecule contains the same atoms contained in the molecular formula.

Does the selected structure contain the number of DBEs you have calculated?
Yes, the molecule contains five DBEs: four for the aromatic moiety and one for the carboxylic acid group.

Is the symmetry of the molecule consistent with the number of signals in the proton and carbon spectra?
The number of signals in the proton and carbon spectra is consistent with the symmetry of the molecule.

Does the molecule contain the number of exchangeable protons you have predicted?
Yes, the molecule contains only one exchangeable proton.

Are the proton and carbon chemical shifts consistent with the chemical environment of each nucleus?
Yes, they are.

Are the COSY correlations and the signal multiplicities consistent with the spin systems of the molecule?
Yes, they are.

Are the HMBC correlations consistent with the supposed structure?
Yes, they are.

THE STRUCTURE IS CORRECT!!!

Problem 8.15
Solvent: DMSO
Field: 500 MHz (^{1}H); 125 MHz (^{13}C)

Molecular formula: $C_{10}H_{12}O_3$

Difficulty ★★★★

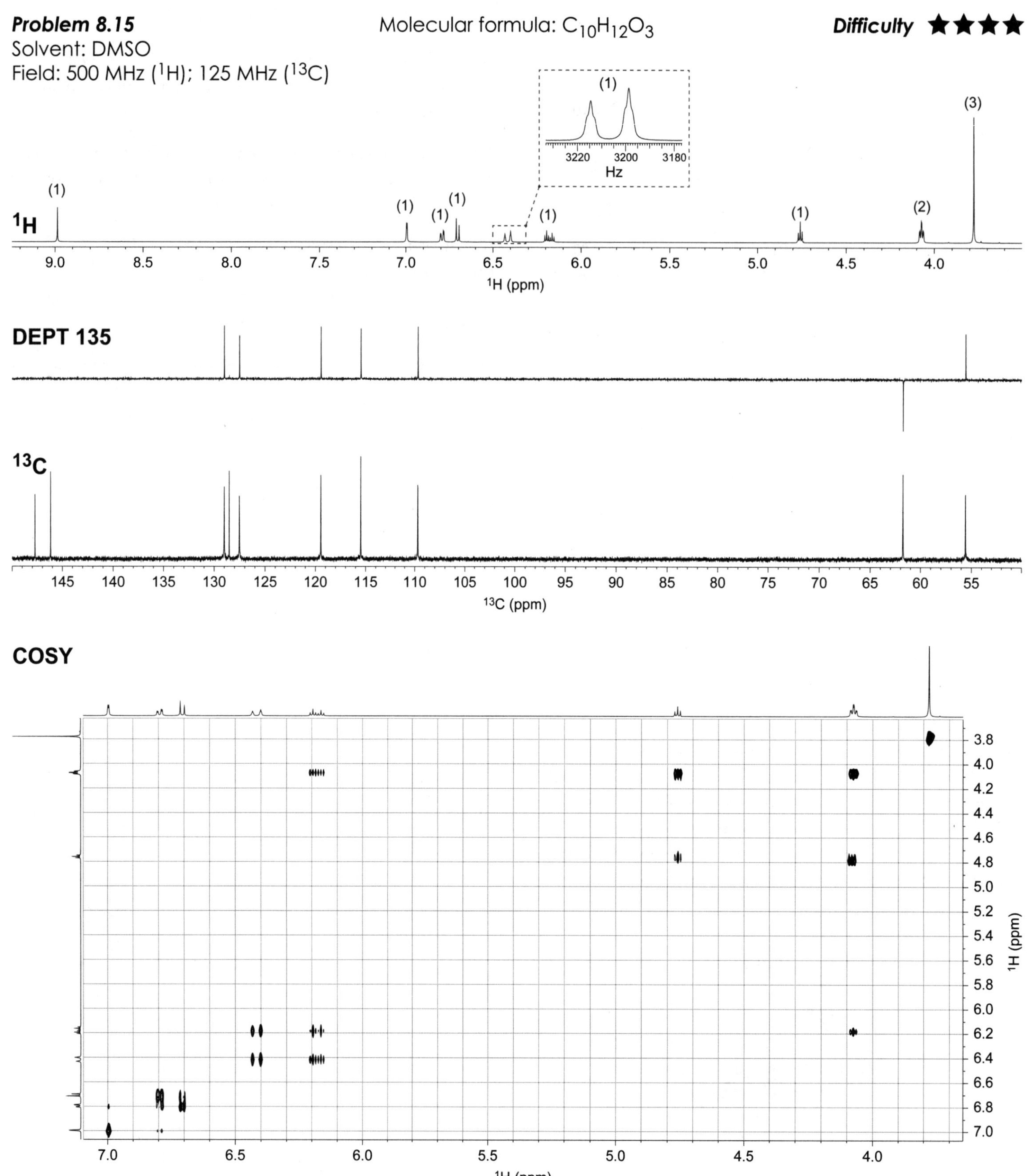

HSQC

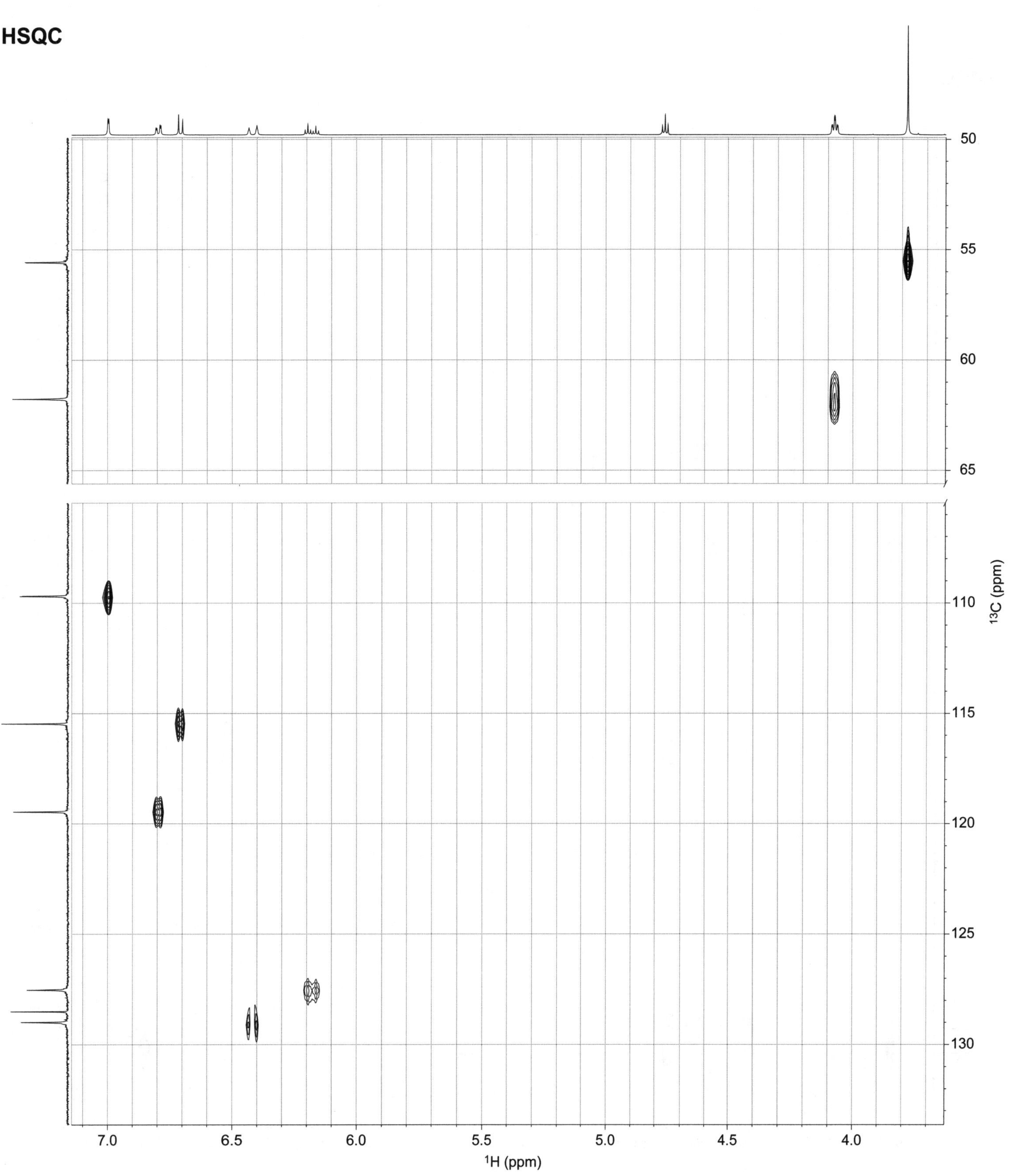

HMBC

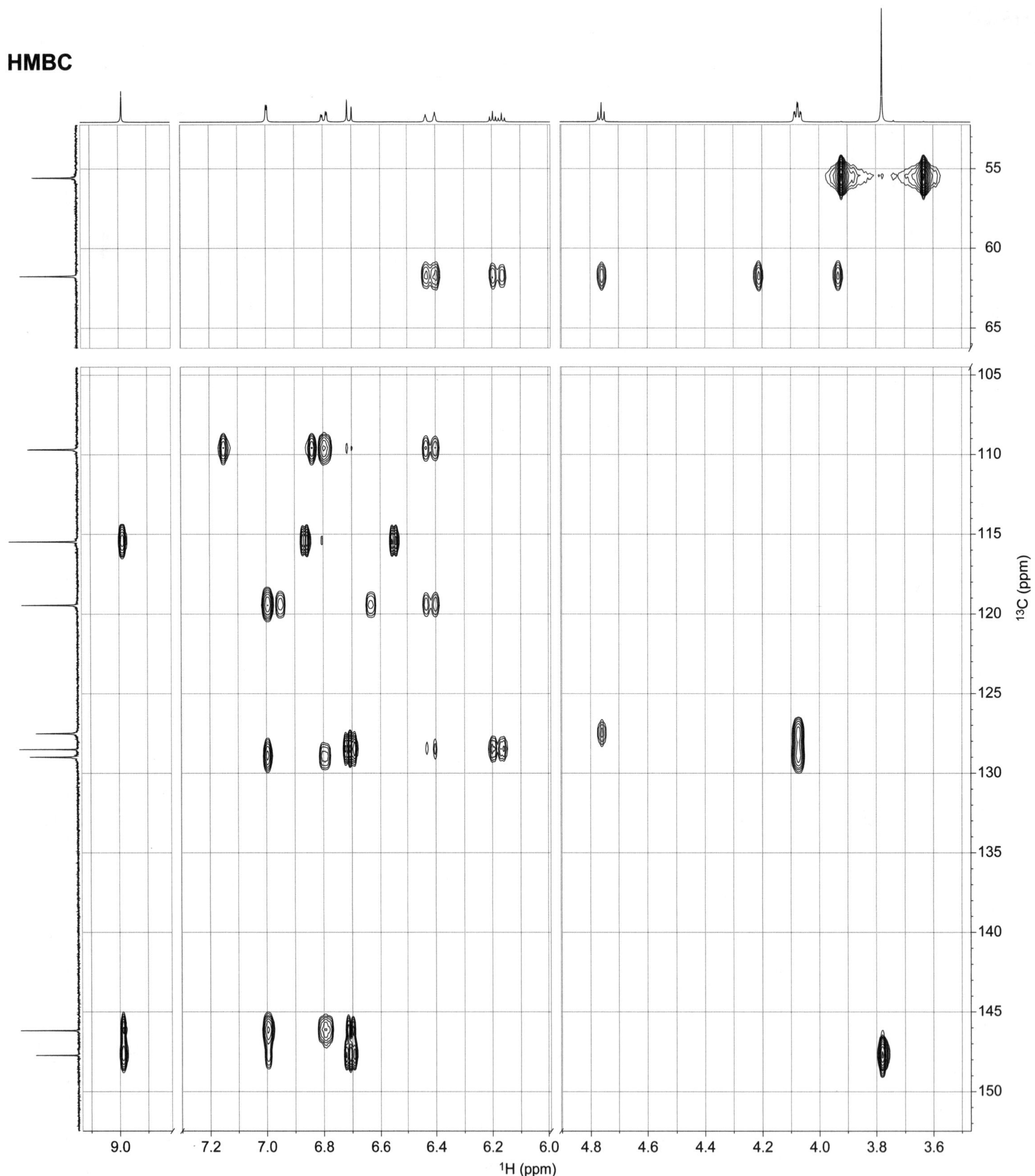

55
60
65
105
110
115
120
125
130
135
140
145
150
^{13}C (ppm)
9.0
7.2
7.0
6.8
6.6
6.4
6.2
6.0
4.8
4.6
4.4
4.2
4.0
3.8
3.6
^{1}H (ppm)

Solution

1) Calculate the number of double bond equivalents (DBEs)
DBE = C - (H/2) - (X/2) + (N/2) +1 = 10 - (12/2) - (0/2) + (0/2) + 1 = 5.

2) Look for the solvent signals in the proton and carbon spectra
The spectra have been acquired in deuterated DMSO. The residual solvent signals are not visible because they resonate in spectral regions omitted from the figures.

3) Get preliminary information about the symmetry of the molecule
The number of carbon signals is the same as the number of carbon atoms in the molecular formula, so the molecule is not symmetric.

4) Recognize the functional groups
The molecule clearly contains an aromatic and an olefinic moiety. It also possesses a methyl (3.88 ppm/58.6 ppm) bonded to an oxygen atom (methoxy group). Please note that, although there is a signal at 9.98 ppm, there is no aldehydic group; in fact, there is no signal around 200 ppm in the carbon spectrum. This signal is actually produced by an exchangeable proton, because it does not correlate to any carbon signal in the HSQC experiment.

5) Determine the number of quaternary carbons (C), methines (CH), methylenes (CH_2), and methyls (CH_3), and assign their resonances
The molecule contains three aromatic and two olefinic methines, one methylene, one methyl, and three aromatic quaternary carbon atoms:

3.88 58.6 CH_3	147.7 C	146.2 C	128.5 C
7.00 109.7 CH	6.71 115.5 CH	6.80 119.5 CH	
6.18 127.5 CH	6.42 129.0 CH	4.07 61.8 CH_2	

6) Determine the number of protons that produces each proton signal
There are ten non-exchangeable protons (5 CH, 1 CH_2, and 1 CH_3), while the sum of the relative signal intensities (1:1:1:1:1:1:1:2:3) is 12! This means that there are a couple of signals that actually can be attributed to exchangeable protons (see next step) and that the values of the relative intensities correspond to the number of hydrogens that produces each signal.

7) Determine the number of exchangeable protons
The proton spectrum has been acquired in DMSO, one of the best solvents to observe exchangeable protons. As described in the previous steps, there are two signals that can be attributed to exchangeable protons. The first one is that resonating at 8.98 ppm, and the second one is that resonating at 4.76 ppm. In fact, neither signal correlates to any carbon signal in the HSQC experiment.

8) Characterize all the spin systems
From the analysis of the COSY spectrum it is clear that there are three spin systems in the molecule. The first one is the methoxy group. Another one can also be easily identified and involves the olefinic part of the molecule:

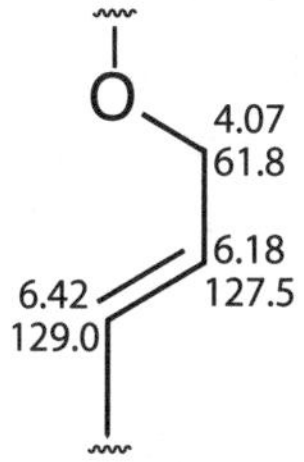

The last spin system involves the aromatic protons; however, its identification requires some additional reasonings. In particular, the molecule is composed of eight sp^2 carbons, two of them involving the formation of the alkene group described above. This means that the aromatic moiety is composed of six carbon atoms, suggesting that it is realistically a benzene derivative. The fact that there are only three aromatic methines indicates that the benzene is trisubstituted. The methines produce three signals in the proton spectrum. The first one is the doublet at 7.00 ppm, having a small coupling constant probably due to a long-range coupling. Another signal is the doublet of doublets that resonates at 6.80 ppm, characterized by a small and a large coupling constant. Realistically the small coupling constant is due to a long-range proton coupling, while the large one is a vicinal coupling. Finally, the last aromatic signal resonates as a doublet at 6.71 ppm, having a large (vicinal) coupling constant.
A trisubstituted benzene can have one of the following three substitution patterns:

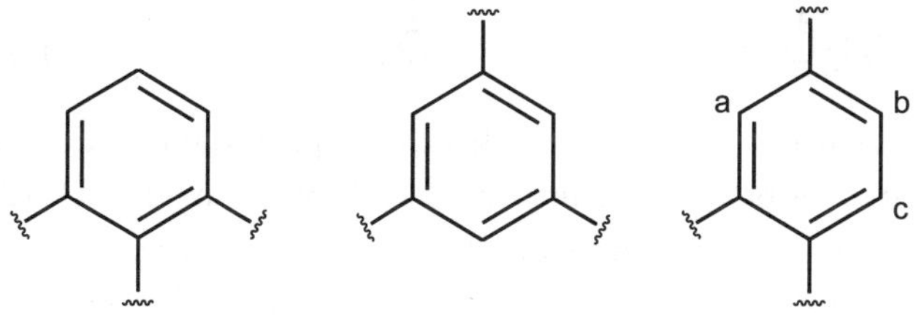

The substitution pattern that fits best with the multiplicity of the aromatic signals is the third one. In fact, the doublet at 7.00 ppm can be attributed to the hydrogen in position "a". Its small coupling constant ($^4J_{CH}$ ~ 2 Hz) is due to the *meta*-coupling with the hydrogen in position "b". This hydrogen resonates at 6.80 ppm as a doublet of doublets, for its coupling to the hydrogens in positions "a" ($^4J_{CH}$ ~ 2 Hz) and "c" ($^3J_{CH}$ ~8 Hz). The hydrogen in position "c", instead, resonates at 6.71 ppm as a doublet ($^3J_{CH}$ ~8 Hz), because it is coupled only to the hydrogen in position "b".
Keeping all this in mind, you can go back to the COSY spectrum and verify that this hypothesis is indeed correct.

9) Connect all the fragments together
The moieties identified so far are: the trisubstituted benzene, the methoxy group, and the olefinic fragment. The parts of the structure identified so far count for ten carbon, eleven hydrogen, and two oxygen atoms. Therefore, an OH group and a hydrogen still need to be considered.

Taking into account the multiplicity of the proton signals, and the chemical shifts of the proton and carbon signals, it is possible to propose three structures.

Since the hydrogens of the methyl are coupled to the quaternary aromatic carbon resonating at 147.7 ppm in the HMBC experiment, the structure on the right can be ruled out. In order to decide between the first two structures, it is convenient to assign as many resonances as possible. Thus, the signal of the exchangeable proton at 4.76 ppm can be assigned to the alcoholic hydrogen, thanks to its long-range correlation to the carbon resonating at 61.8 ppm (HMBC) and to its correlation to the signal at 4.07 ppm in the COSY spectrum. The signal of the exchangeable proton at 8.98 ppm can be assigned instead to the phenolic hydrogen, thanks to its long-range correlations to the carbons resonating at 109.7, 146.2, and 147.7 ppm. Thanks to the long-range correlations of the olefinic hydrogens at 6.18 and 6.42 ppm to the carbon signal at 128.5 ppm, it is possible to assign the quaternary aromatic carbon to where the alkene is bonded. Moreover, the proton signal at 6.42 ppm is also coupled to the carbons at 109.7 and 119.5 ppm, indicating that these two methines are adjacent to the carbon at 128.5 ppm. Please note that the phenolic proton at 8.98 is not correlated to any of these carbons, and this means that the structure on the left can be ruled out, leaving the central structure as the correct one.

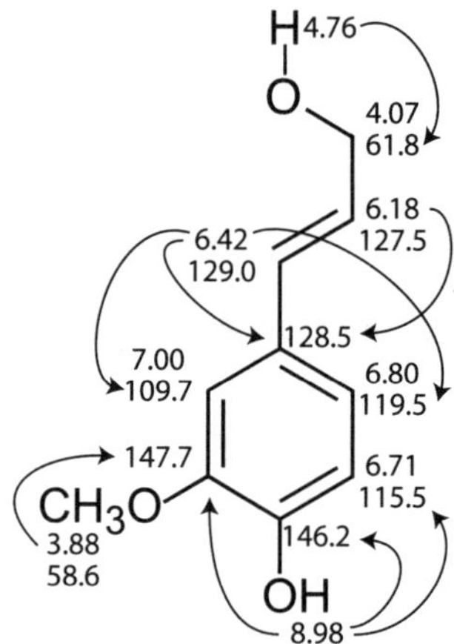

There are also other HMBC correlations involving the aromatic proton and carbon atoms that further corroborate the proposed assignment and confirm once again the structure of the molecule. Finally, it is possible to define the geometry of the alkene group. If you look at the inset of the signal at 6.42 ppm, you will see that the two sub-peaks are about 16 Hz apart, indicating that the alkene adopts a *trans* geometry (see Chapter 3).

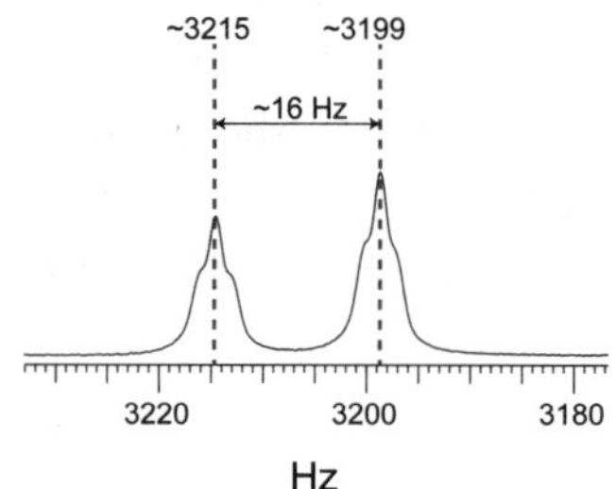

10) Verify the consistency of all the data collected

Is the molecular formula consistent with the structure of the molecule?
Yes, the molecule contains the same atoms contained in the molecular formula.

Does the selected structure contain the number of DBEs you have calculated?
Yes, the molecule contains five DBEs: four for the aromatic moiety and one for the alkene group.

Is the symmetry of the molecule consistent with the number of signals in the proton and carbon spectra?
The number of signals in the proton and carbon spectra is consistent with the symmetry of the molecule.

Does the molecule contain the number of exchangeable protons you have predicted?
Yes, the molecule contains two exchangeable protons (4.76 and 8.98 ppm), which are visible in the proton spectrum.

Are the proton and carbon chemical shifts consistent with the chemical environment of each nucleus?
Yes, they are.

Are the COSY correlations and the signal multiplicities consistent with the spin systems of the molecule?
Yes, they are.

Are the HMBC correlations consistent with the supposed structure?
Yes, they are.

THE STRUCTURE IS CORRECT!!!

Problem 8.16 Molecular formula: $C_{11}H_{12}N_2O_2$ **Difficulty** ★★★★

Solvent: D_2O

Field: 500 MHz (^1H); 125 MHz (^{13}C)

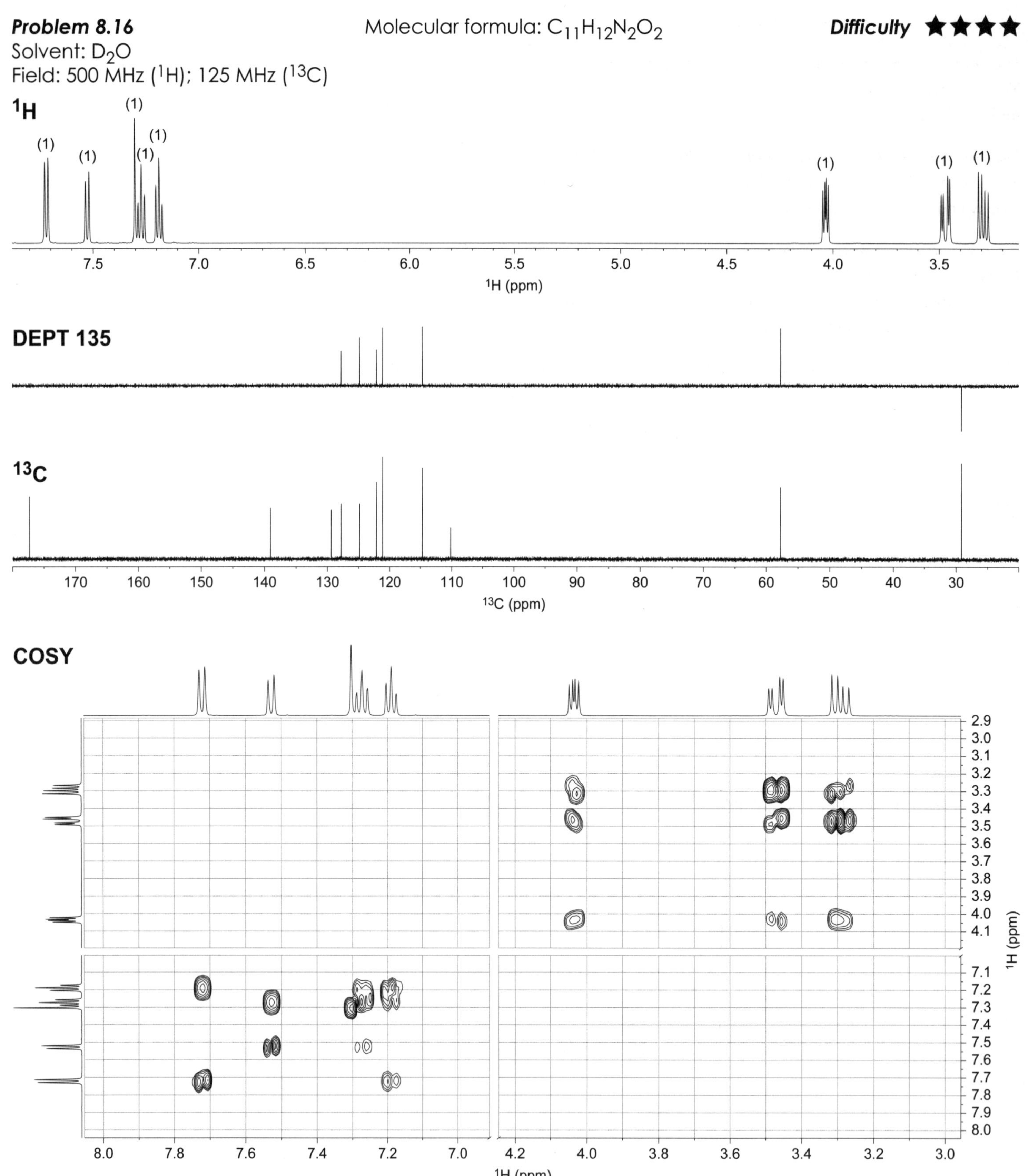

HSQC

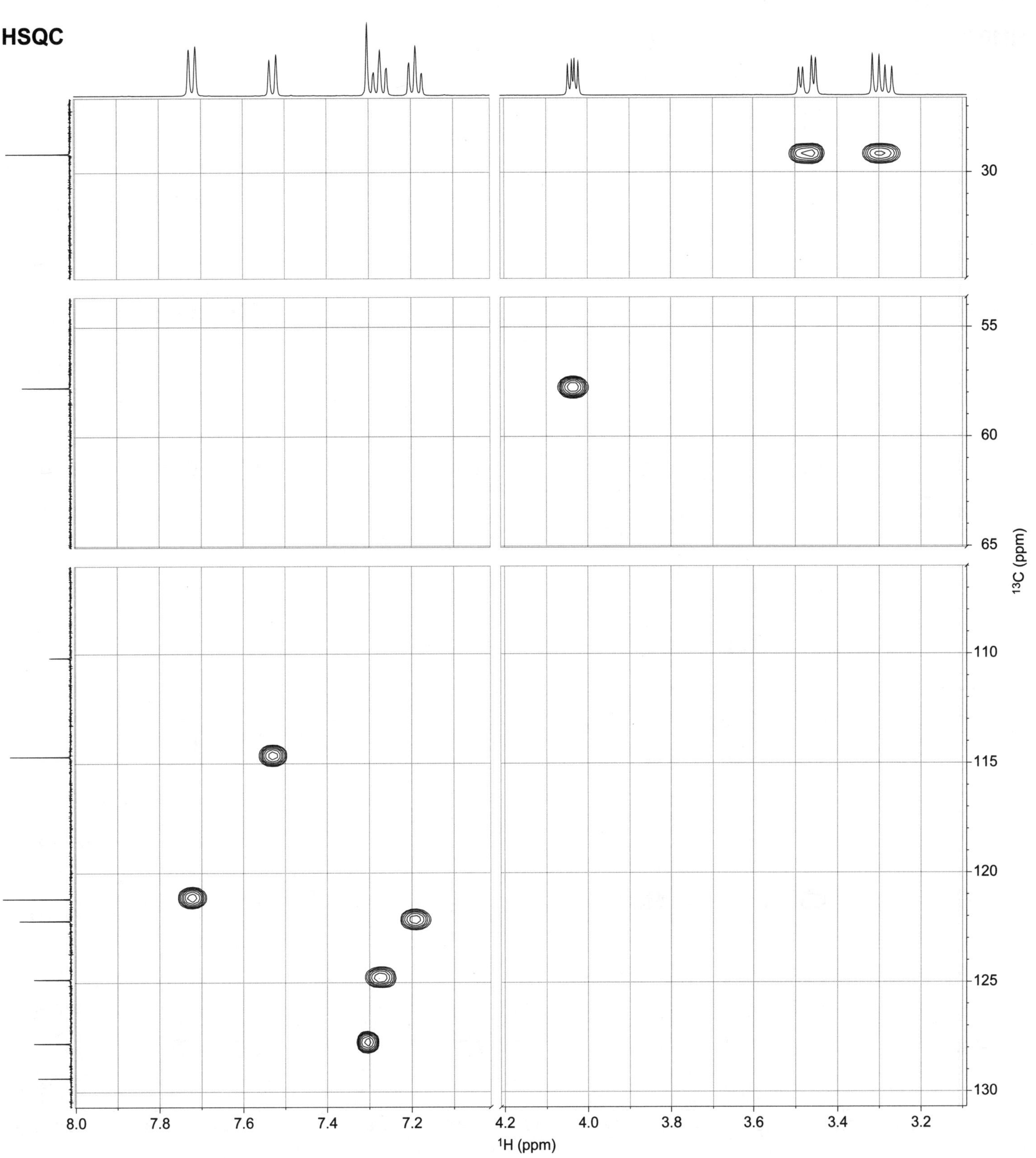

30
55
60
65
^{13}C (ppm)
110
115
120
125
130
8.0
7.8
7.6
7.4
7.2
4.2
4.0
3.8
3.6
3.4
3.2
^1H (ppm)

HMBC

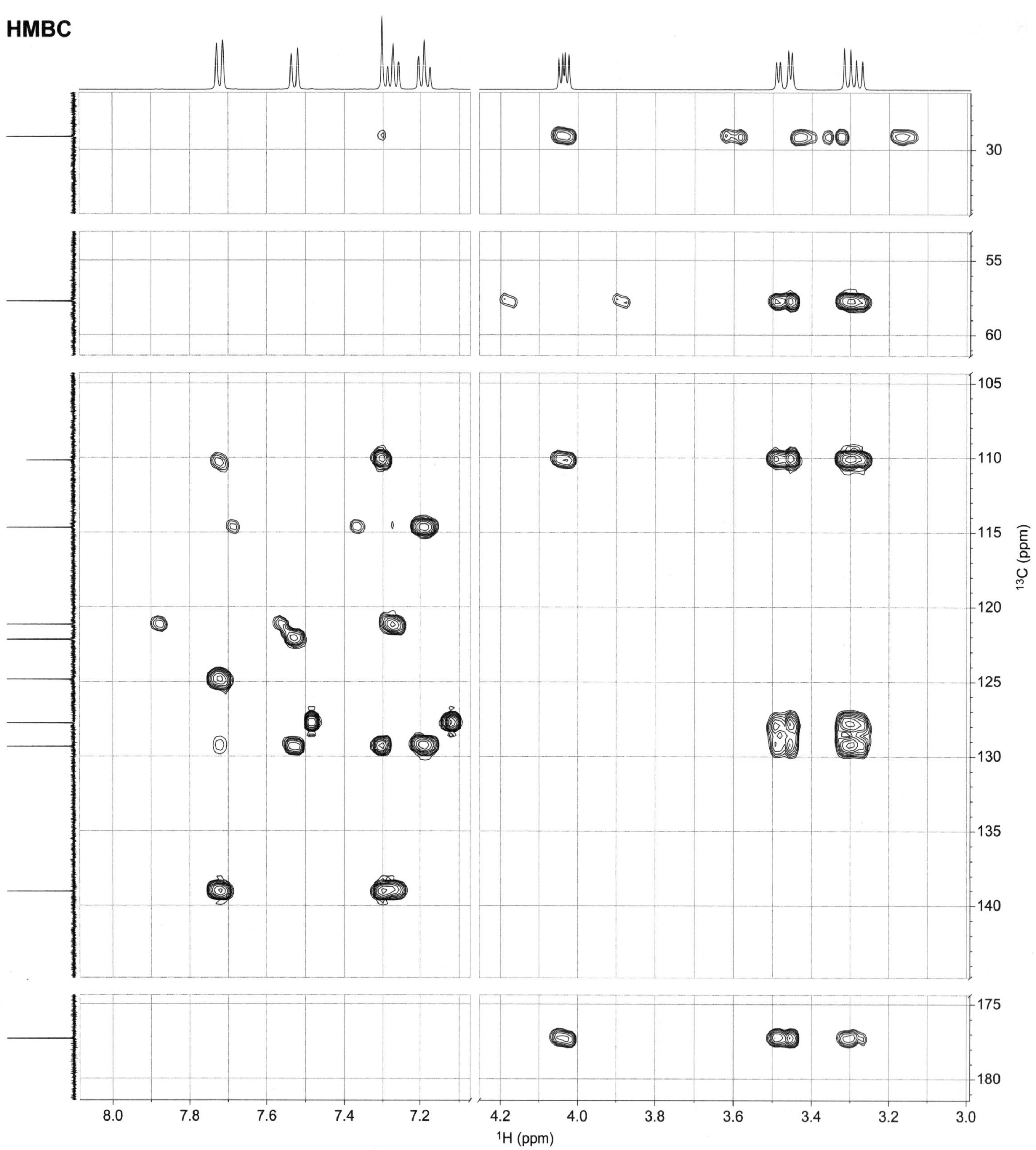

Solution

1) Calculate the number of double bond equivalents (DBEs)
DBE = C - (H/2) - (X/2) + (N/2) +1 = 11 - (12/2) - (0/2) + (2/2) + 1 = 7.

2) Look for the solvent signals in the proton and carbon spectra
The spectra have been acquired in D_2O. However, since the NMR sample is very concentrated, the residual solvent signal is not visible.

3) Get preliminary information about the symmetry of the molecule
The number of carbon signals is the same as the number of carbon atoms in the molecular formula, so the molecule is not symmetric.

4) Recognize the functional groups
There is a carboxylic group (or one of its derivatives), and an aliphatic and an aromatic moiety. The latter involves eight carbon atoms!

5) Determine the number of quaternary carbons (C), methines (CH), methylenes (CH_2), and methyls (CH_3), and assign their resonances
The molecule contains one aliphatic methylene, one aliphatic methine, five aromatic methines, three quaternary aromatic carbons, and a carboxylic group (or derivative):

110.2	129.4	139.1	177.3
C	C	C	C

7.53 114.7	7.72 121.2	7.19 122.2	7.28 124.9
CH	CH	CH	CH

7.30 127.8	4.04 57.8	3.29-3.47 29.2
CH	CH	CH_2

6) Determine the number of protons that produces each proton signal
There are eight non-exchangeable protons (6 CH and 1 CH_2) and there are eight signals in the proton spectrum, all having the same intensity. This means that each signal is produced by one hydrogen.

7) Determine the number of exchangeable protons
Since there are eight non-exchangeable hydrogens and the molecular formula contains twelve hydrogens, there are four exchangeable protons in the molecule (12 - 8 = 4).

8) Characterize all the spin systems
From the analysis of the COSY spectrum it is clear that there are three spin systems in the molecule. One is formed by the non-coupled aromatic methine resonating as a singlet at 7.30 ppm. The two other spin systems are also easily identified and involve the aliphatic and aromatic parts of the molecule. In particular, the first is formed by a methine bonded to a methylene and the second by four aromatic methines bonded in a row:

7.30 / 127.8: –CH–

4.04 / 57.8, 3.29-3.47 / 29.2: –CH—CH_2–

7.53 / 114.7, 7.19 / 122.2, 7.28 / 124.9, 7.72 / 121.2: –CH—CH—CH—CH–

The HMBC correlations further confirm these findings.

9) Connect all the fragments together
The spin systems described above count for seven carbon and eight hydrogen atoms. As mentioned above, the molecule also contains three aromatic quaternary carbons and a carbonyl. Therefore, two nitrogen, one oxygen, and four exchangeable hydrogen atoms still need to be considered. From the analysis of carbon chemical shifts it is possible to state that no methine or methylene or quaternary carbon is bonded to an oxygen atom (aliphatic carbons should resonate above 60 ppm and aromatic quaternary carbons above 140 ppm when bonded to an oxygen atom). Therefore, the oxygen atom must be bonded to the carbonyl group. For the same reason, the presence of an ester can be also ruled out, so the molecule contains a carboxylic acid group. Therefore, it remains to assign two nitrogens and three exchangeable hydrogens that, realistically, should form an NH and an NH_2 group. In summary, the following fragments must be bonded together.

7.30 / 127.8: –CH–

4.04 / 57.8, 3.29-3.47 / 29.2: –CH(–)—CH_2–

7.53 / 114.7, 7.19 / 122.2, 7.28 / 124.9, 7.72 / 121.2: –CH—CH—CH—CH–

–NH–   –NH_2 –C(=O)–OH (177.3)

110.2 C, 129.4 C, 139.1 C

Let's start by analyzing the aromatic part of the molecule. Remember that in aromatic rings the ${}^3J_{CH}$ HMBC correlations are much more intense than the ${}^2J_{CH}$ or ${}^4J_{CH}$ ones. Thus, the different intensities of the cross-peaks can be used for the structural elucidation. For example, the doublet at 7.53 ppm displays two intense HMBC correlations to the carbon signals at 122.2 and 129.4 ppm. The correlation to the signal at 122.2 ppm does not provide any additional structural information; in fact, you already know that these two atoms are three bonds apart (see the spin systems defined in the previous step). However, the correlation to the carbon at 129.4 ppm indicates that these atoms are also three bonds apart. Analogously, the proton signal at 7.72 ppm displays, as expected, an intense HMBC correlation to the carbon at 124.9 ppm, and a diagnostic correlation to the quaternary carbon at 139.1 ppm.

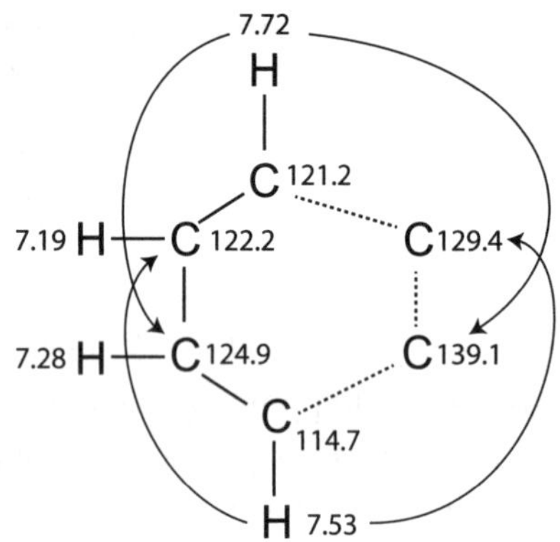

Therefore, you can suppose that these atoms take part in the formation of a six-membered aromatic ring as shown in the figure below. This hypothesis is confirmed by also looking at the three-bond HMBC correlations of the protons at 7.19 and 7.28 ppm.

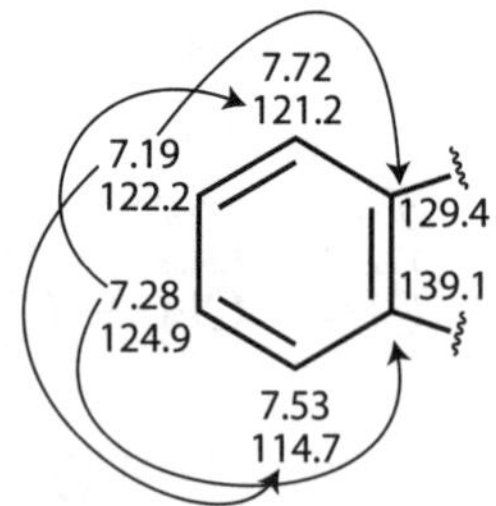

The structural characterization of the aromatic moiety is not yet complete; in fact, a quaternary carbon (110.2 ppm) and a methine (7.30 ppm/127.8 ppm) have still to be placed. Interestingly, the long-range correlation between the proton resonating at 7.72 ppm and the carbon at 110.2 ppm indicates that this carbon is adjacent to the quaternary carbon at 129.4 ppm.

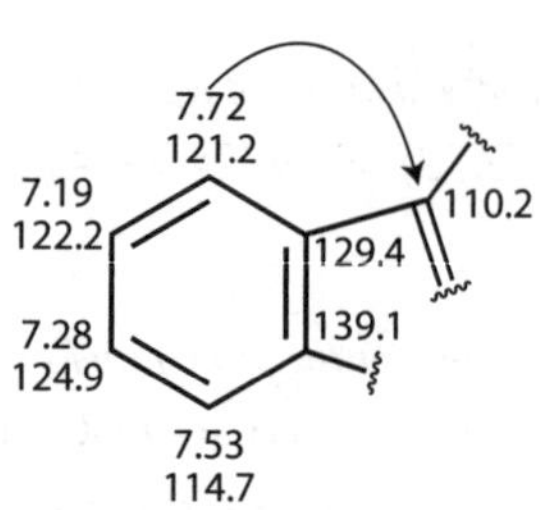

Finally, the last methine can be placed next to this quaternary carbon, as suggested by the long-range coupling between the hydrogen resonating as a singlet at 7.30 ppm and the carbon signals at 110.2, 129.4, and 139.1 ppm.

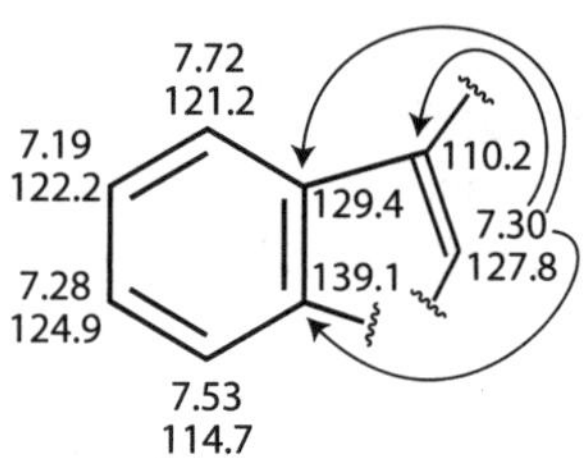

Interestingly, the correlation between the proton signal at 7.30 ppm and the carbon signal at 139.1 ppm is intense and it can be explained assuming that these nuclei are closer than four intervening bonds. Therefore, it is easy to suppose that a single atom joins the carbons at 127.8 and 139.1 ppm, closing a second aromatic ring. Since no additional aromatic carbon atom is available, it is realistic to suppose that the NH group is actually part of the ring, forming an indole moiety.

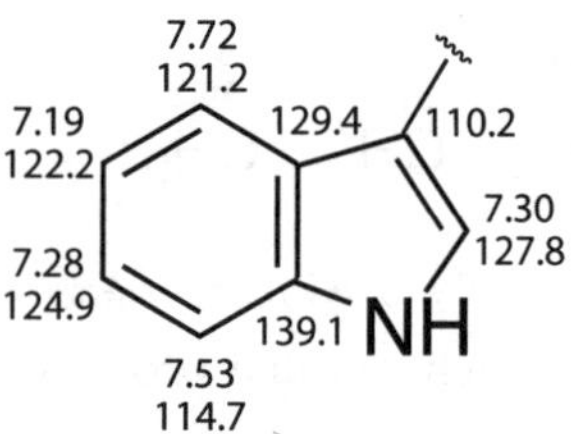

Let's now analyze the HMBC correlations involving the aliphatic part of the molecule. Note that both the methylene and the methine are connected to the aromatic quaternary carbon resonating at 110.2 ppm, and to the carbonyl at 177.3 ppm. However, the methylene is also connected to the other aromatic carbons resonating at 124.9 and 127.8 ppm. This means that the methylene is actually bonded to the aromatic moiety of the molecule.

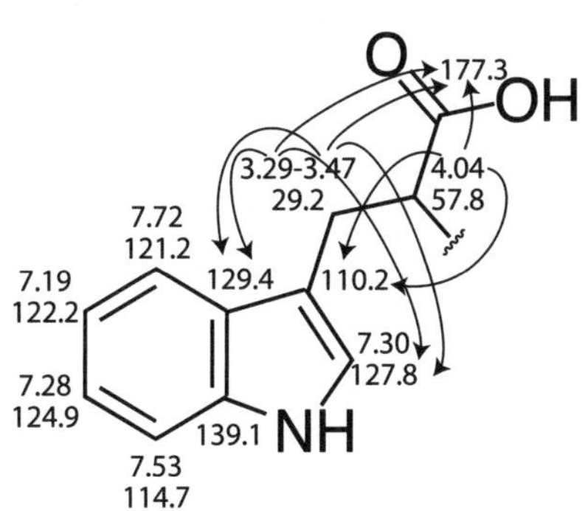

It is clear, at this point, that the structure is that of tryptophan and thus the remaining amino group has to be bonded to the α-methine.

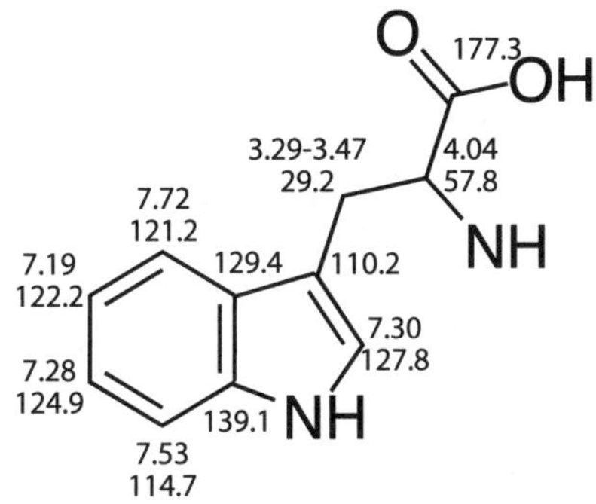

10) Verify the consistency of all the data collected

Is the molecular formula consistent with the structure of the molecule?
Yes, the molecule contains the same atoms contained in the molecular formula.

Does the selected structure contain the number of DBEs you have calculated?
Yes, the molecule contains seven DBEs: six for the aromatic moiety and one for the carboxylic acid group.

Is the symmetry of the molecule consistent with the number of signals in the proton and carbon spectra?
The number of signals in the proton and carbon spectra is consistent with the symmetry of the molecule.

Does the molecule contain the number of exchangeable protons you have predicted?
Yes, the molecule contains four exchangeable protons.

Are the proton and carbon chemical shifts consistent with the chemical environment of each nucleus?
Yes, they are.

Are the COSY correlations and the signal multiplicities consistent with the spin systems of the molecule?
Yes, they are.

Are the HMBC correlations consistent with the supposed structure?
Yes, they are.

THE STRUCTURE IS CORRECT!!!

CHAPTER 9

Summary problems

This chapter contains summary problems, the solutions of which are available online (find your Activation Code and instructions on the first page of this book).

The problems are classified by level of difficulty (look at the top-right corner of the page: the larger the number of dark stars, the harder the problem), and most contain a full set of NMR spectra (^{1}H, ^{13}C, DEPT, COSY, HSQC, and HMBC).

As mentioned in the previous chapter, due to the approximation of the chemical shift scales in the spectra of the problems, you may not be able to accurately determine the chemical shift value of each signal. Therefore, you should not be worried if the chemical shift values you have determined are not exactly the same as those reported in the solutions to the problems. You can improve the accuracy of your measurements using a triangle set square. You can help yourself to trace perfect vertical and horizontal lines by placing the triangle on the chemical shift axes of the 1D and 2D spectra. The use of the triangle is particularly important in 2D experiments, especially if they contain close or overlapping cross-peaks.

Problem 9.1
Solvent: D_2O
Field: 500 MHz (^1H); 125 MHz (^{13}C)

Molecular formula: $C_3H_{10}N_2$

Difficulty ★☆☆☆

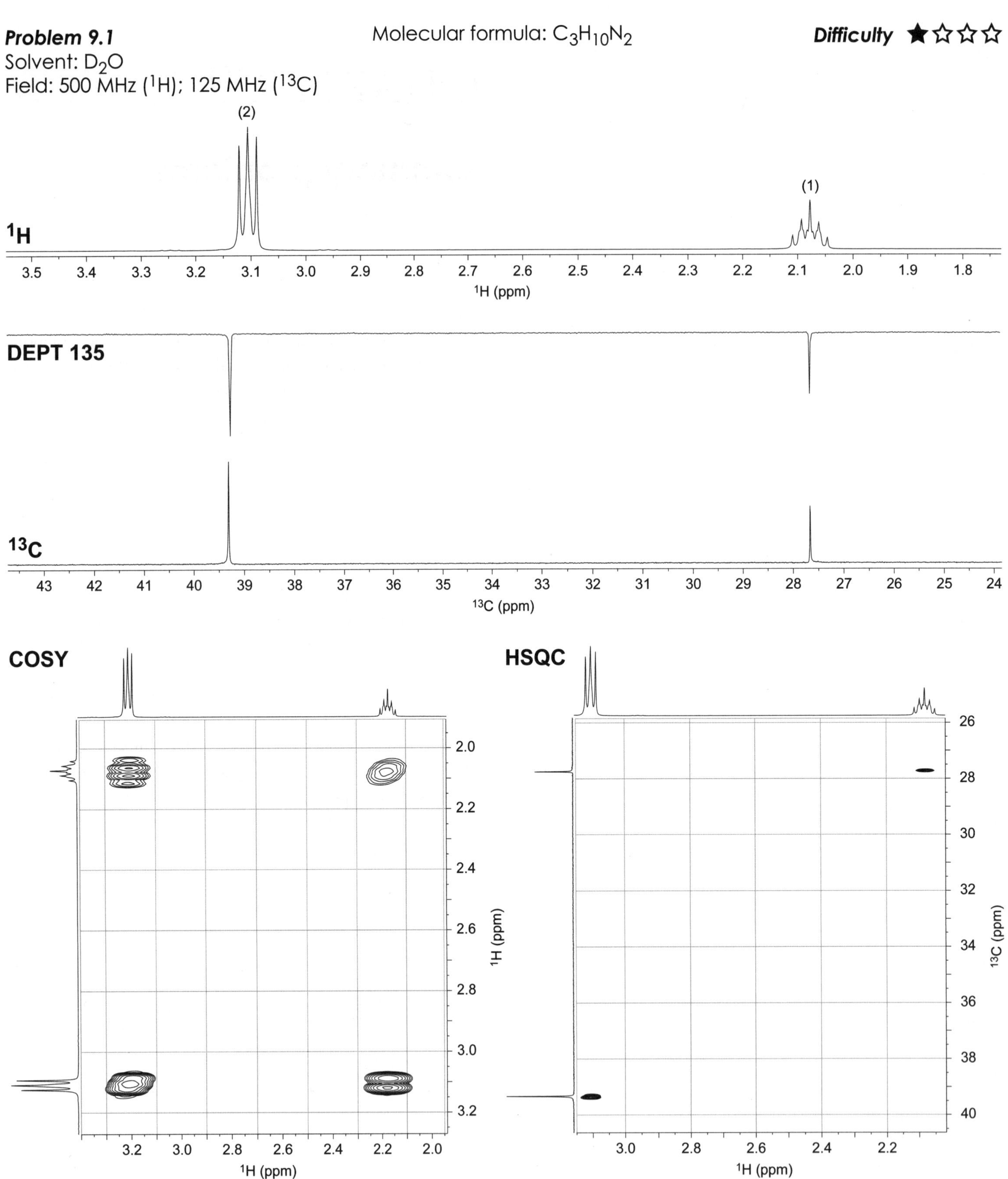

Problem 9.2 Molecular formula: $C_4H_{10}O$ ***Difficulty*** ★☆☆☆

Solvent: D_2O

Field: 500 MHz (^{1}H); 125 MHz (^{13}C)

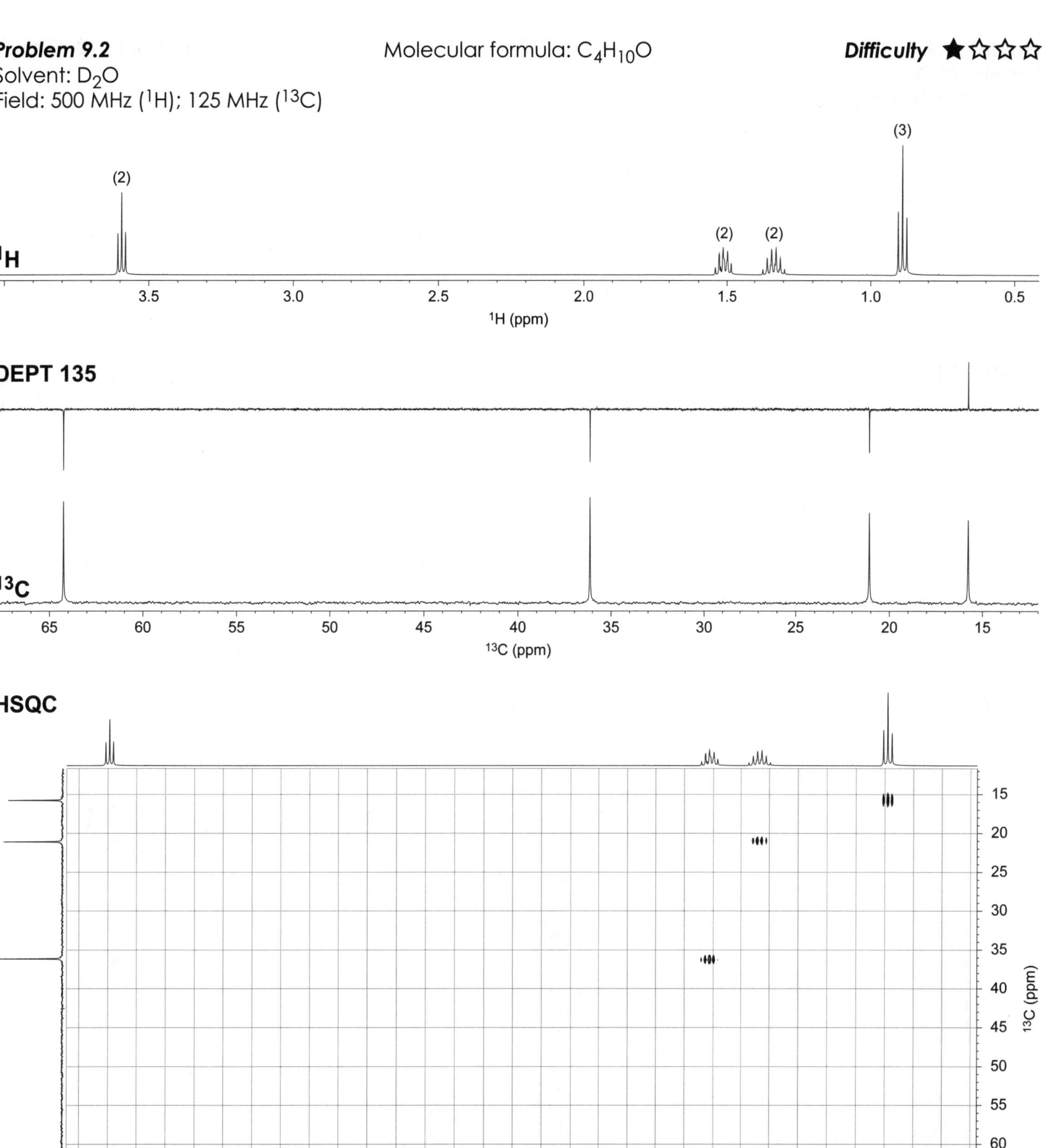

Problem 9.3
Solvent: D_2O
Field: 500 MHz (^1H); 125 MHz (^{13}C)

Molecular formula: $C_4H_8O_2$

Difficulty ★☆☆☆

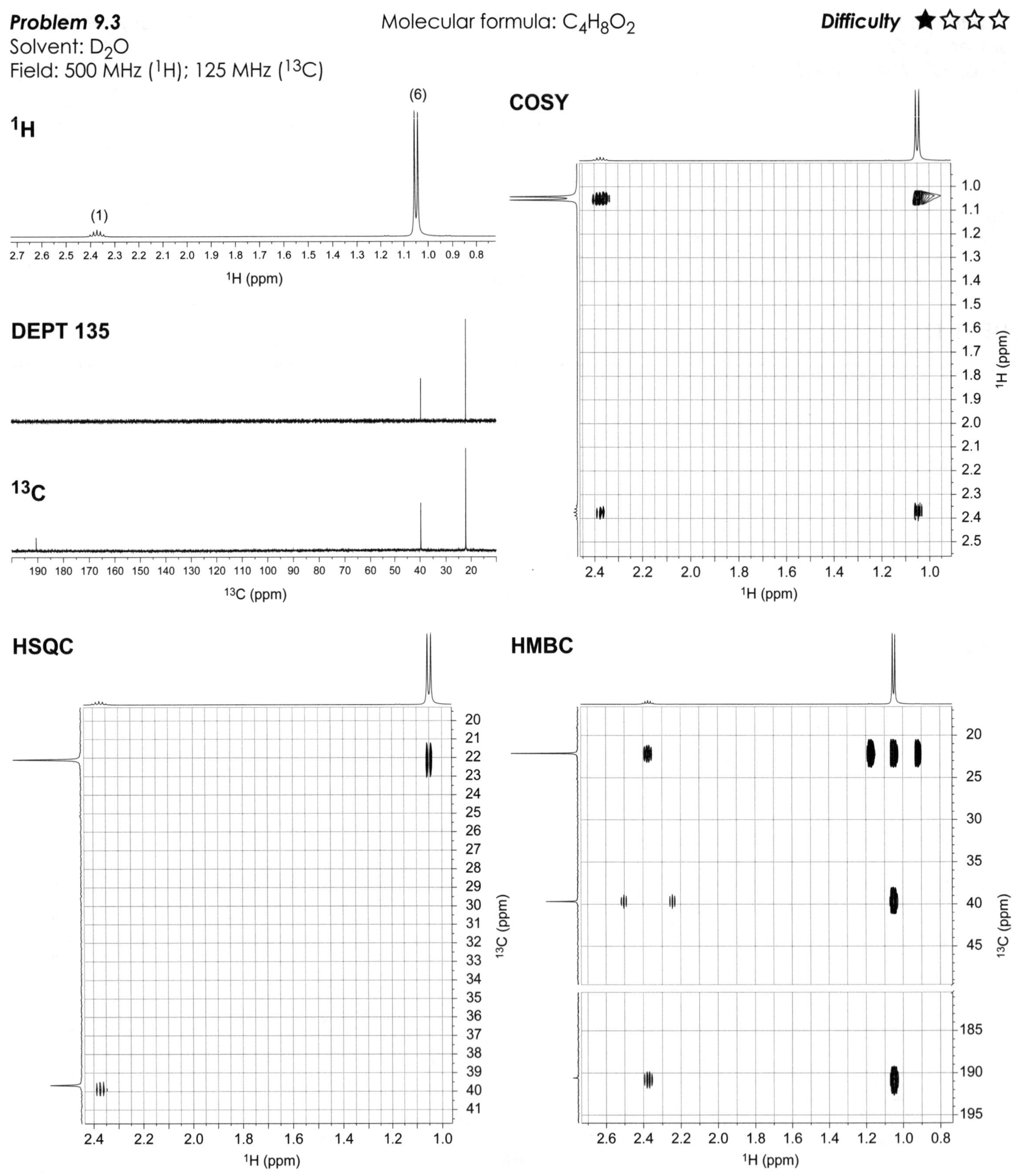

Problem 9.4 Molecular formula: $C_3H_8O_2$ ***Difficulty*** ★☆☆☆

Solvent: D_2O

Field: 500 MHz (^1H); 125 MHz (^{13}C)

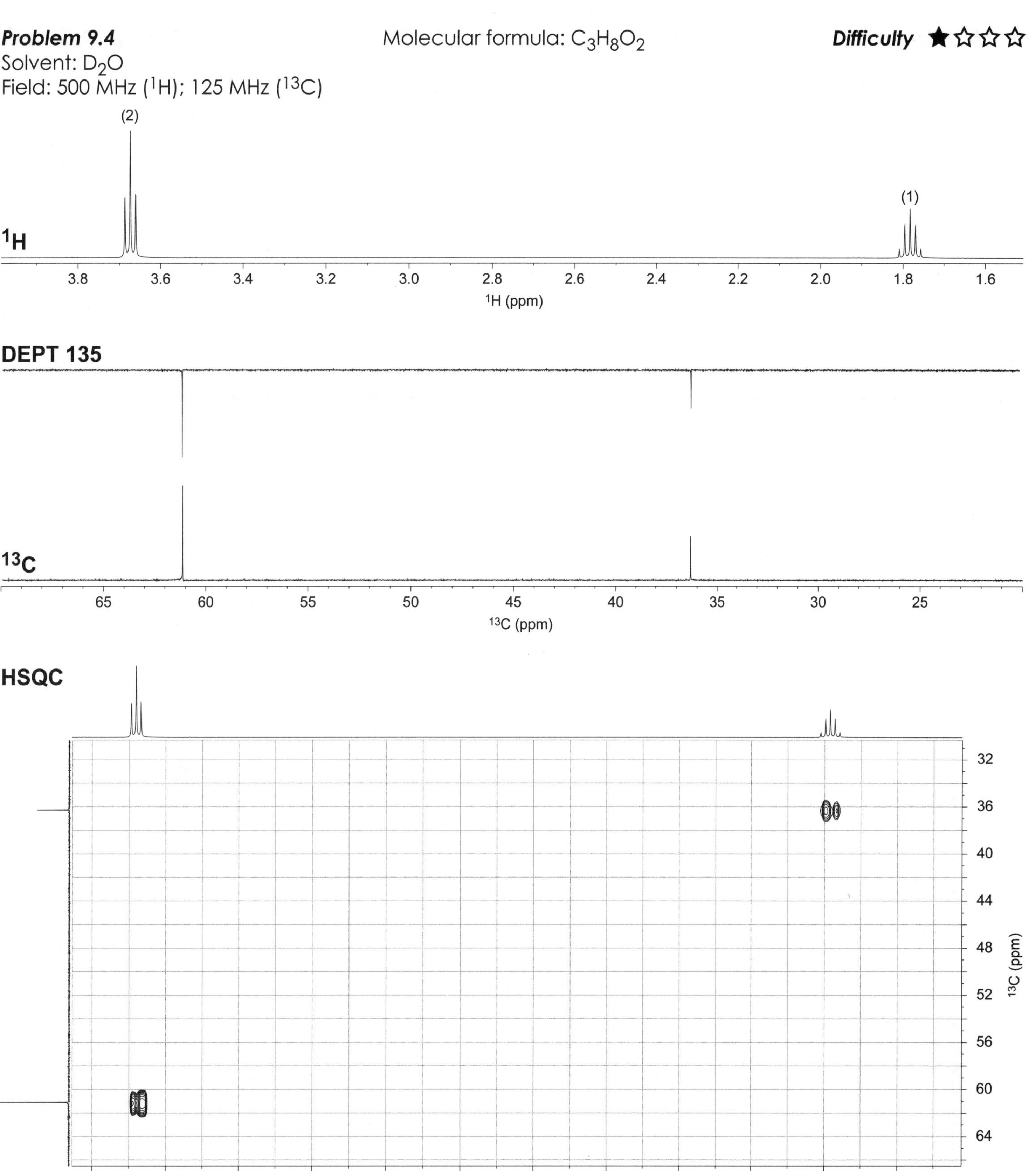

Problem 9.5
Solvent: D_2O
Field: 400 MHz (^1H); 100 MHz (^{13}C)

Molecular formula: $C_4H_8O_3$

Difficulty ★☆☆☆

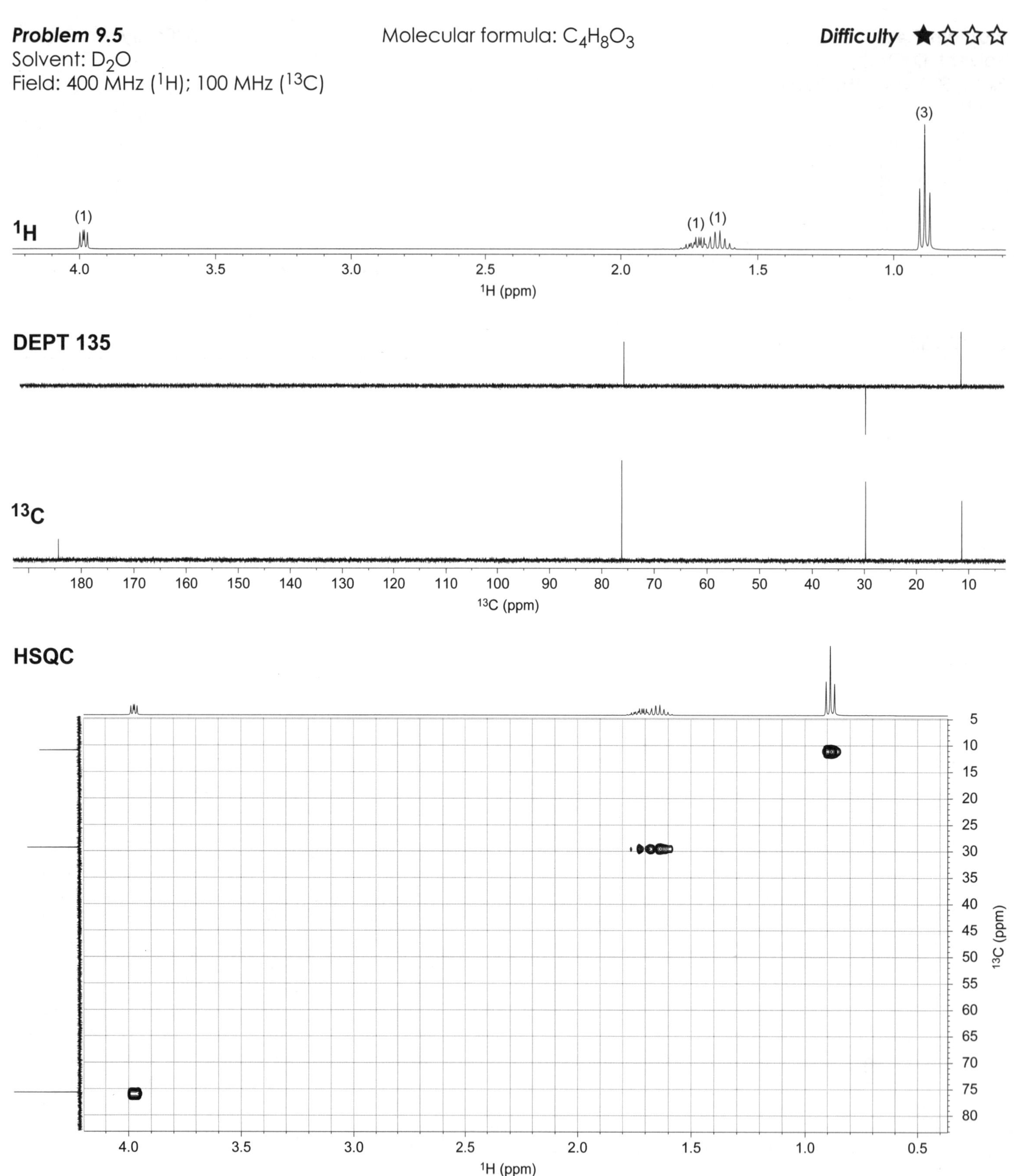

Problem 9.6 Molecular formula: $C_5H_{10}O_2$ ***Difficulty*** ★☆☆☆

Solvent: D_2O

Field: 500 MHz (^{1}H); 125 MHz (^{13}C)

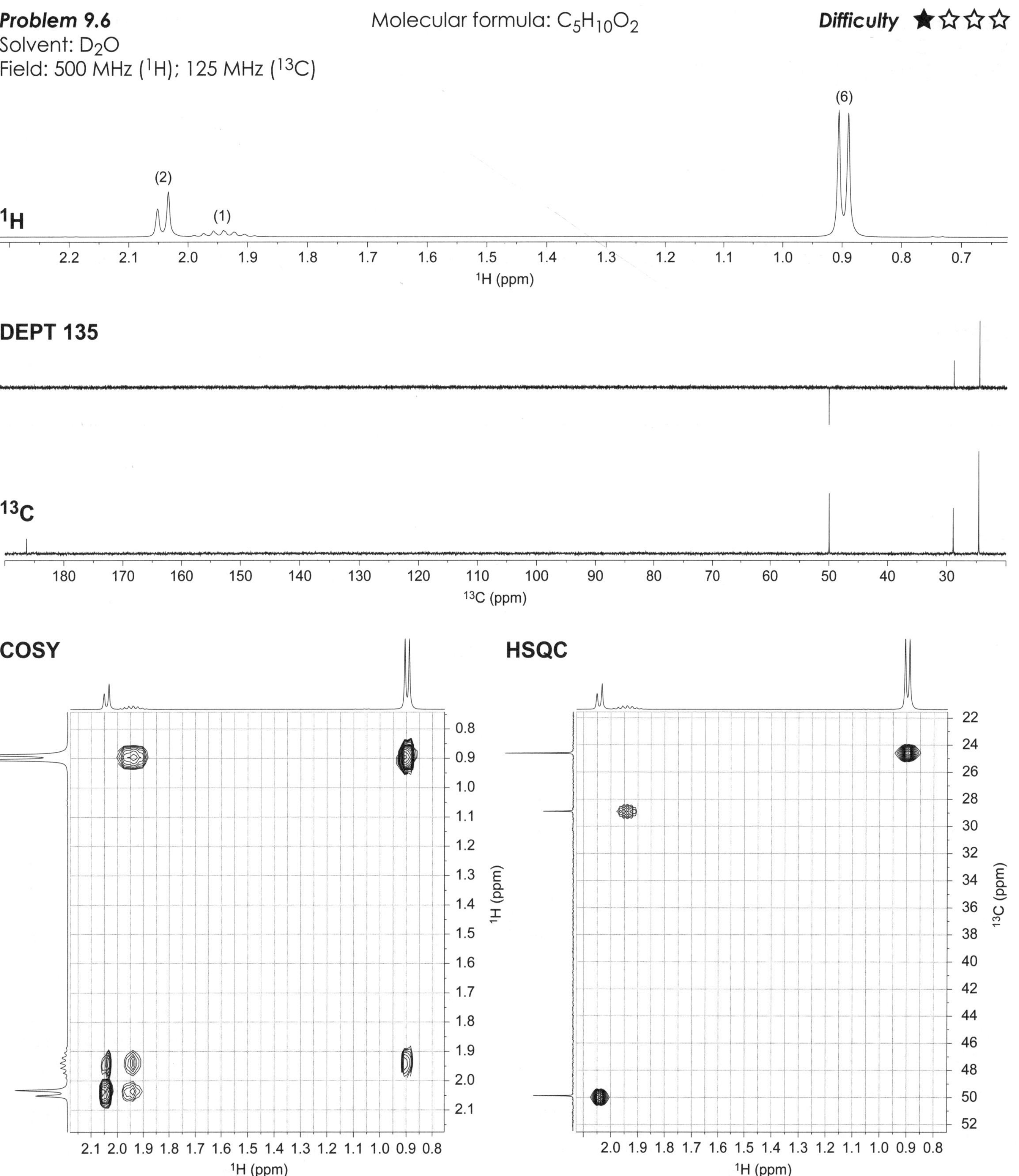

Problem 9.7 Molecular formula: C_6H_6O ***Difficulty*** ★☆☆☆

Solvent: D_2O

Field: 400 MHz (^{1}H); 100 MHz (^{13}C)

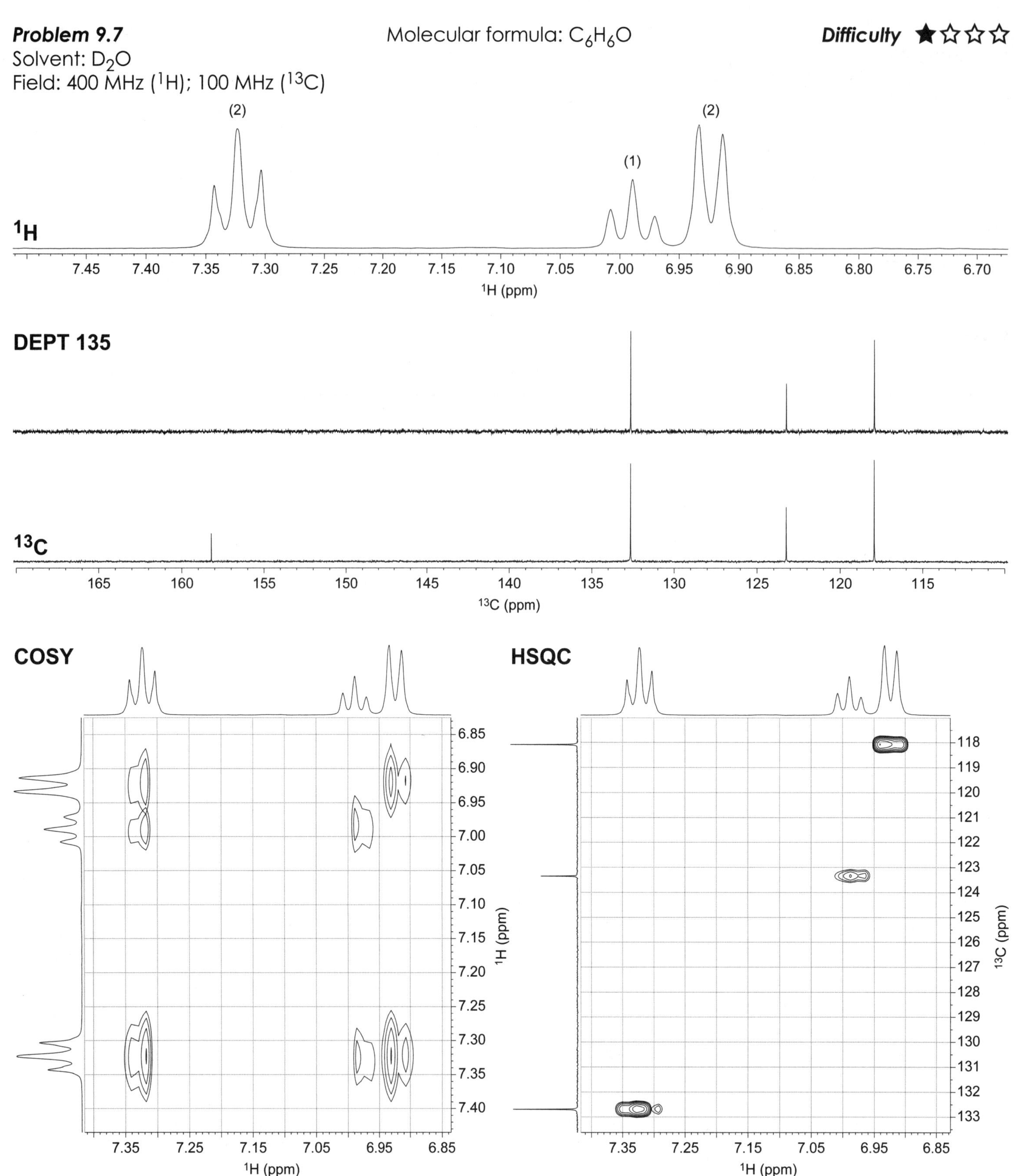

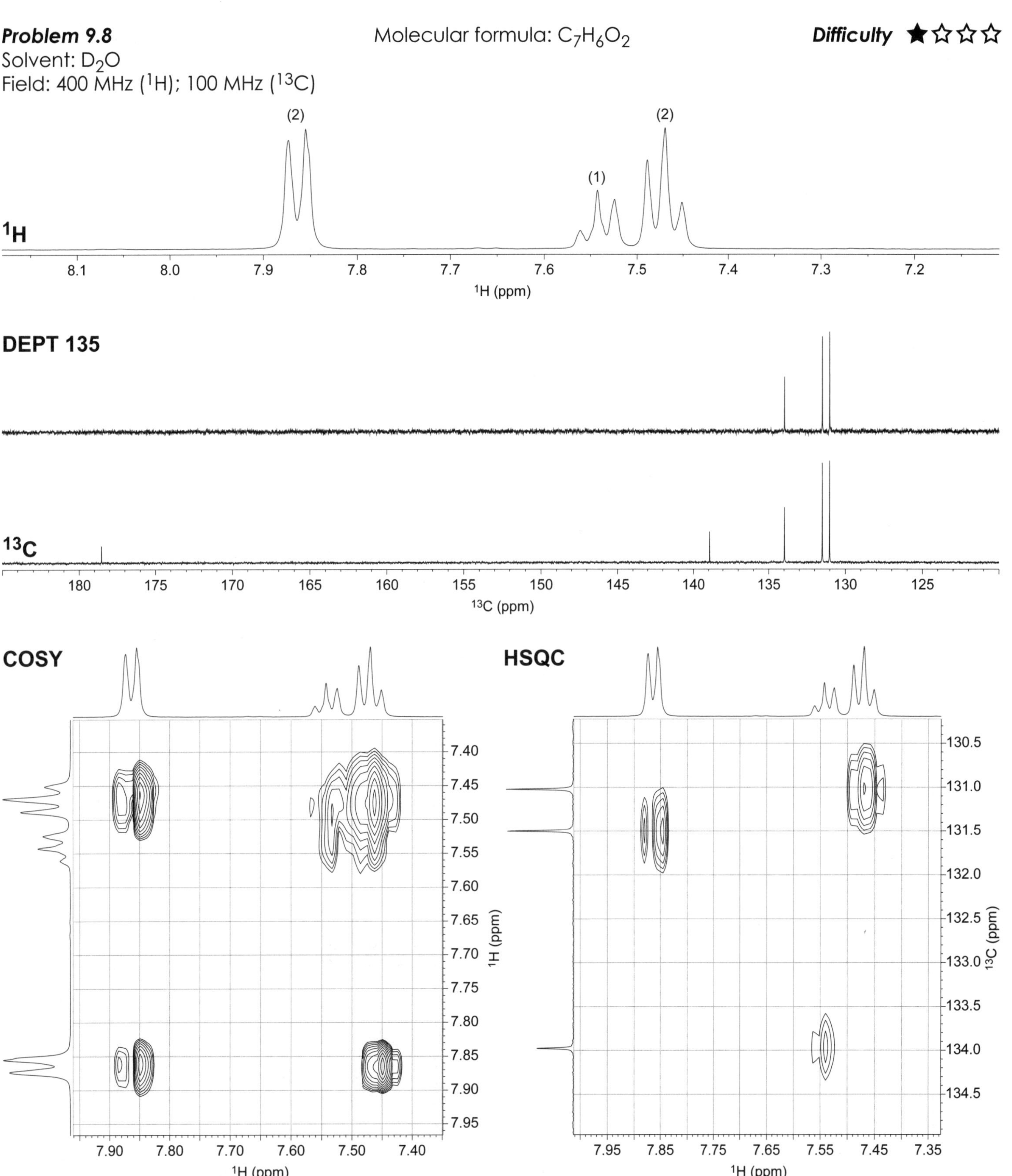

Problem 9.8
Solvent: D2O
Field: 400 MHz (1H); 100 MHz (13C)
Molecular formula: C7H6O2
Difficulty
1H
(2)
(1)
(2)
1H (ppm)
DEPT 135
13C
13C (ppm)
COSY
1H (ppm)
1H (ppm)
HSQC
13C (ppm)
1H (ppm)

Problem 9.9
Solvent: D_2O
Field: 500 MHz (^1H); 125 MHz (^{13}C)

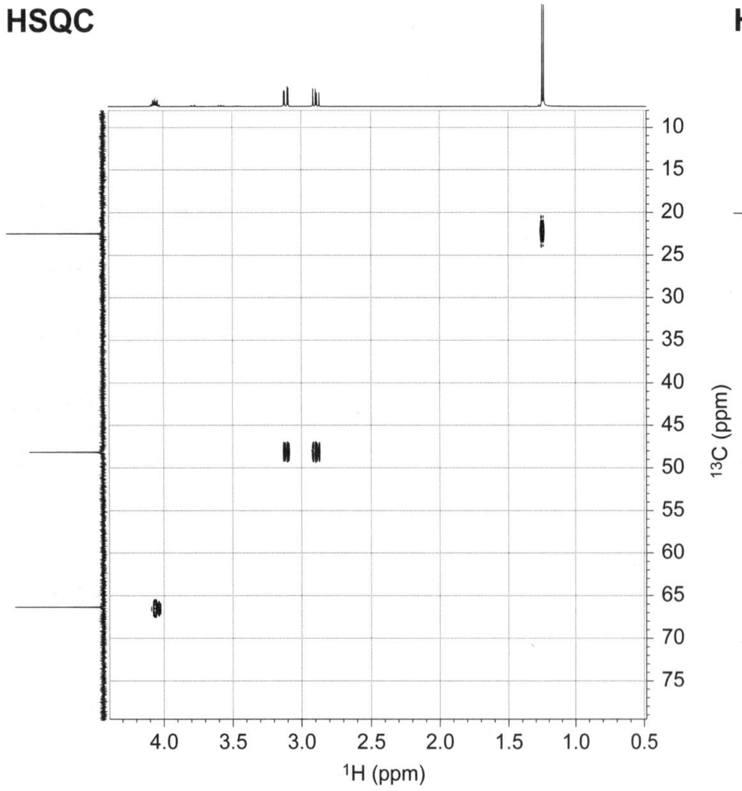

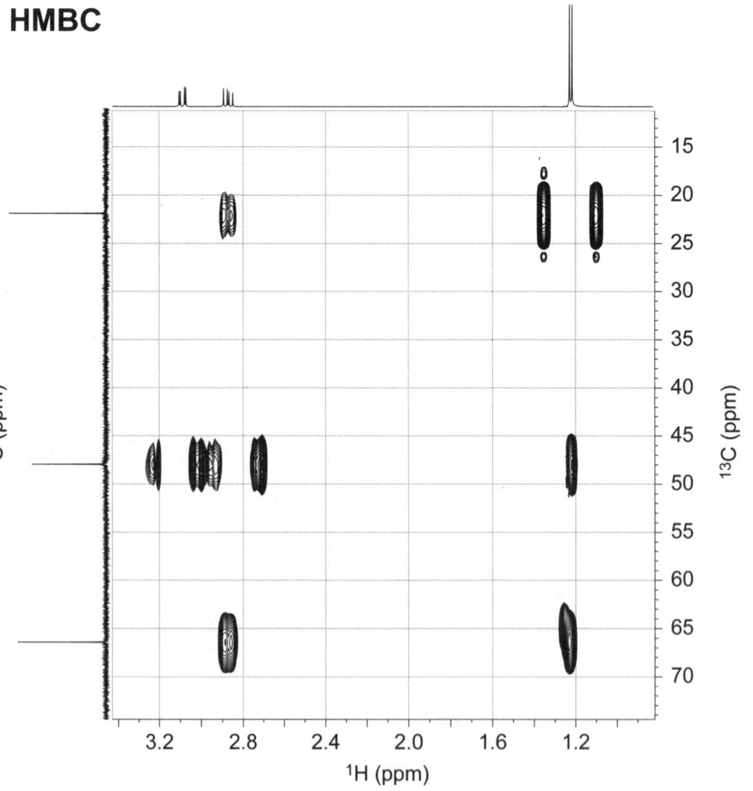

Molecular formula: C_3H_9NO

Difficulty ★☆☆☆

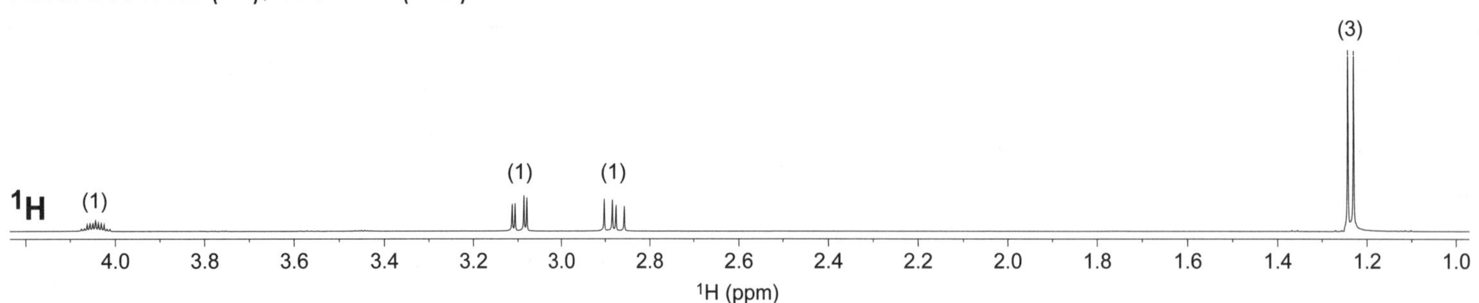

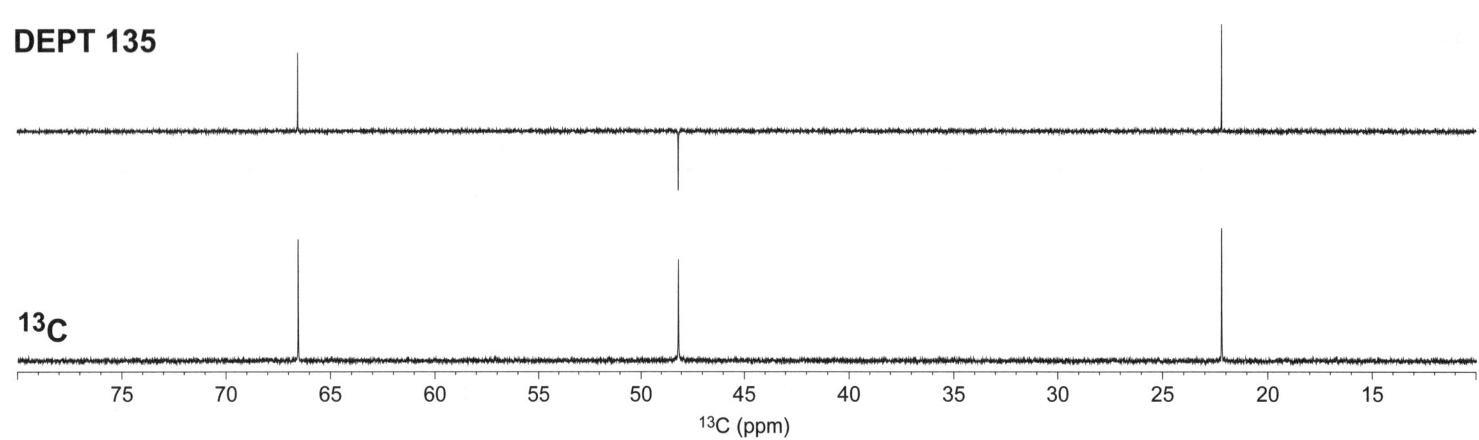

HSQC

HMBC

Problem 9.10 Molecular formula: $C_7H_5IO_2$ **Difficulty** ★☆☆☆

Solvent: CD_3OD

Field: 700 MHz (^1H); 175 MHz (^{13}C)

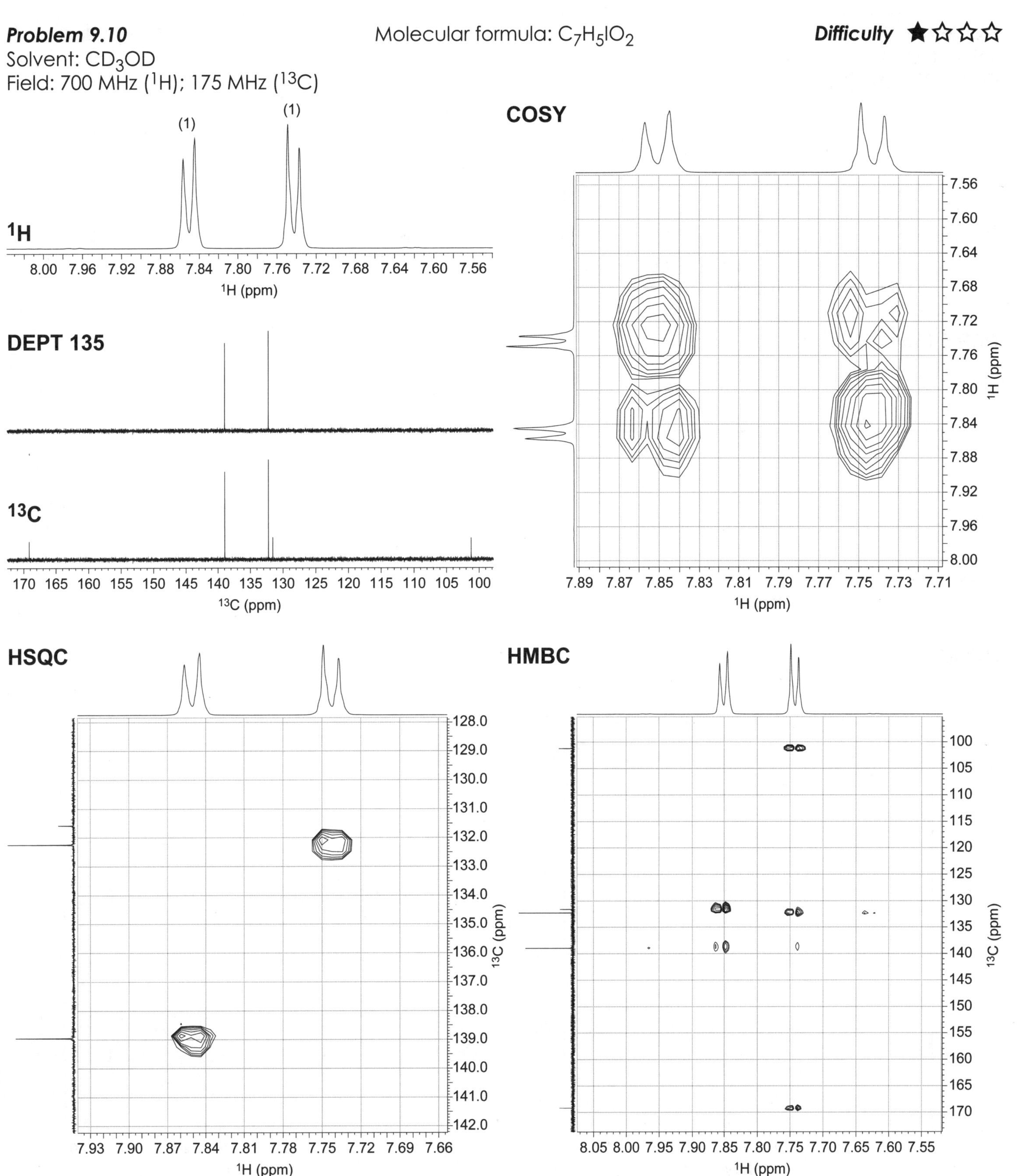

Problem 9.11
Solvent: D_2O
Field: 500 MHz (^1H); 125 MHz (^{13}C)

Molecular formula: $C_3H_6O_3$

Difficulty ★☆☆☆

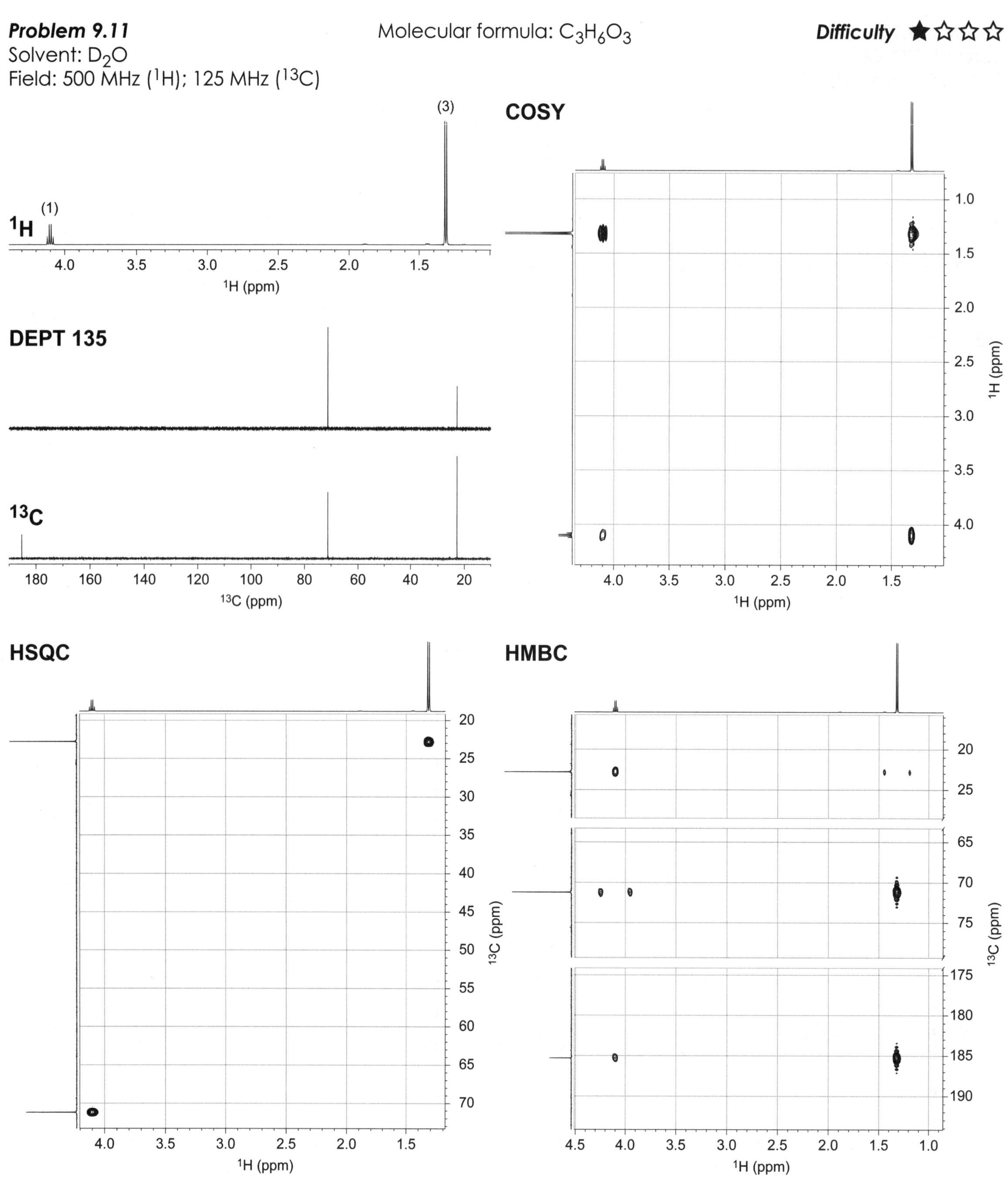

Problem 9.12 Molecular formula: $C_3H_3O_2Cl$ **Difficulty** ★☆☆☆

Solvent: D_2O

Field: 400 MHz (^1H); 100 MHz (^{13}C)

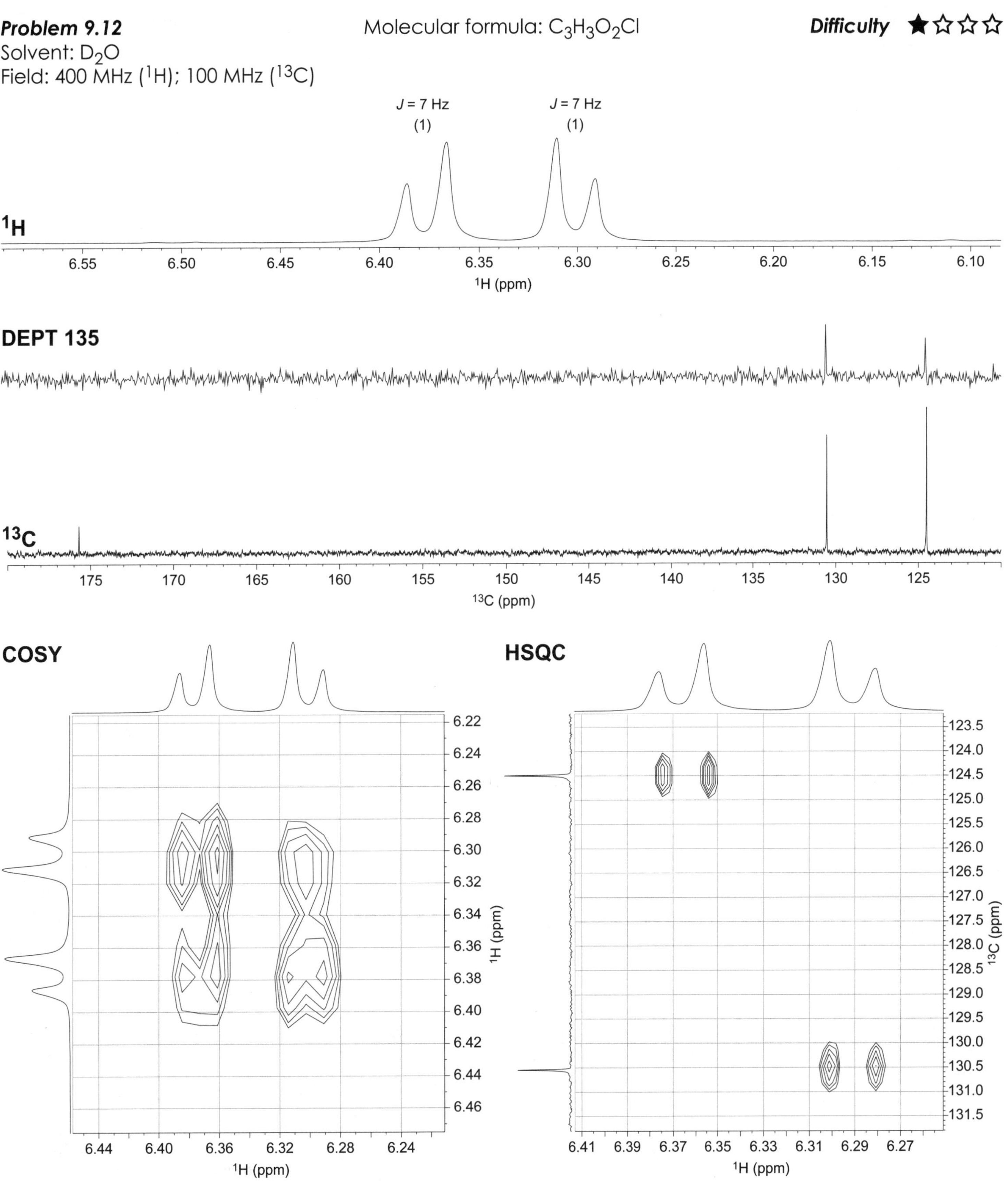

Problem 9.13
Solvent: D_2O
Field: 400 MHz (^1H); 100 MHz (^{13}C)

Molecular formula: $C_6H_5NO_2$
Hint: It is a pyridine derivative

Difficulty ★☆☆☆

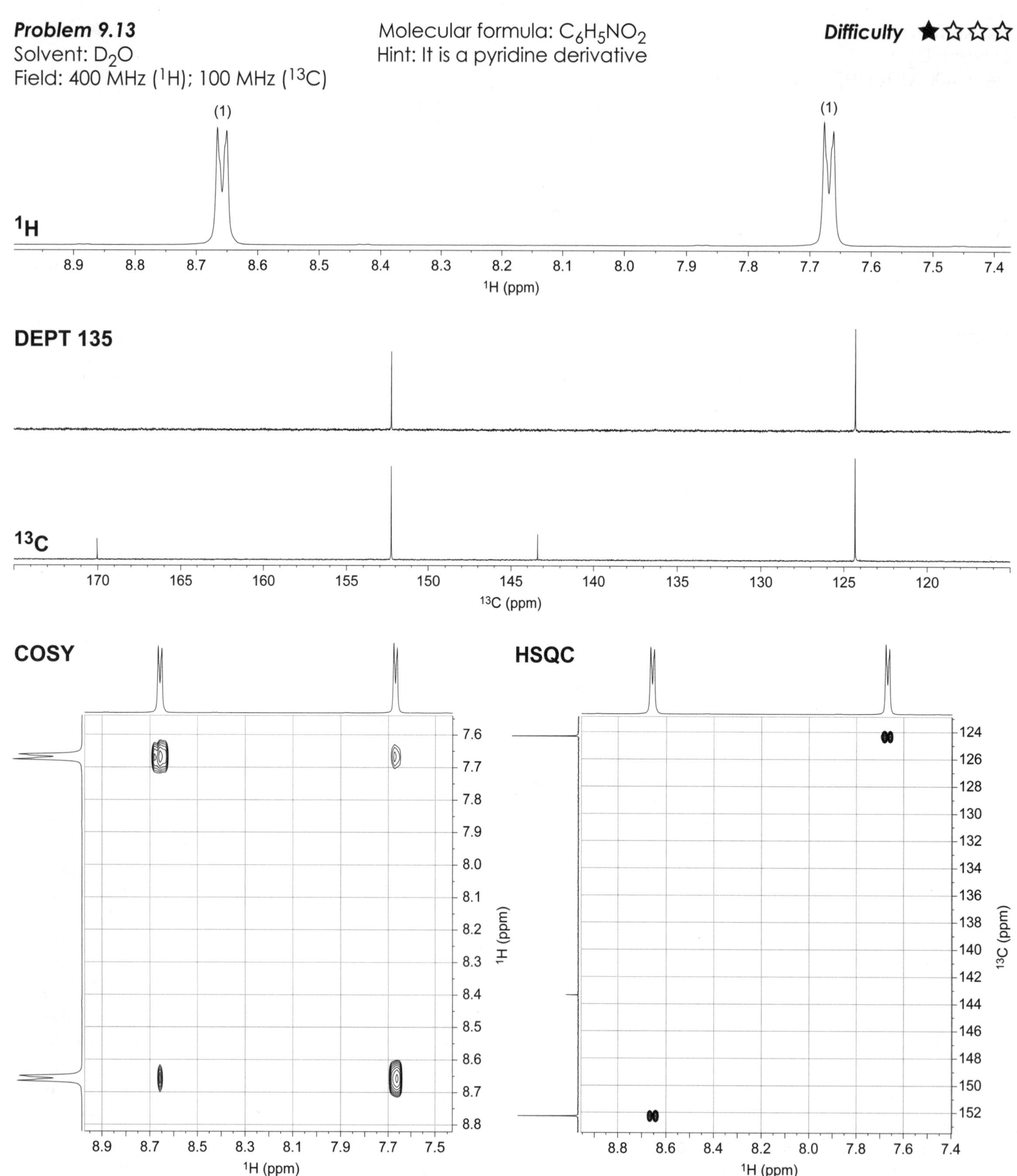

Problem 9.14 Molecular formula: $C_5H_{11}NO_2$ ***Difficulty*** ★☆☆☆

Solvent: D_2O

Field: 400 MHz (^1H); 100 MHz (^{13}C)

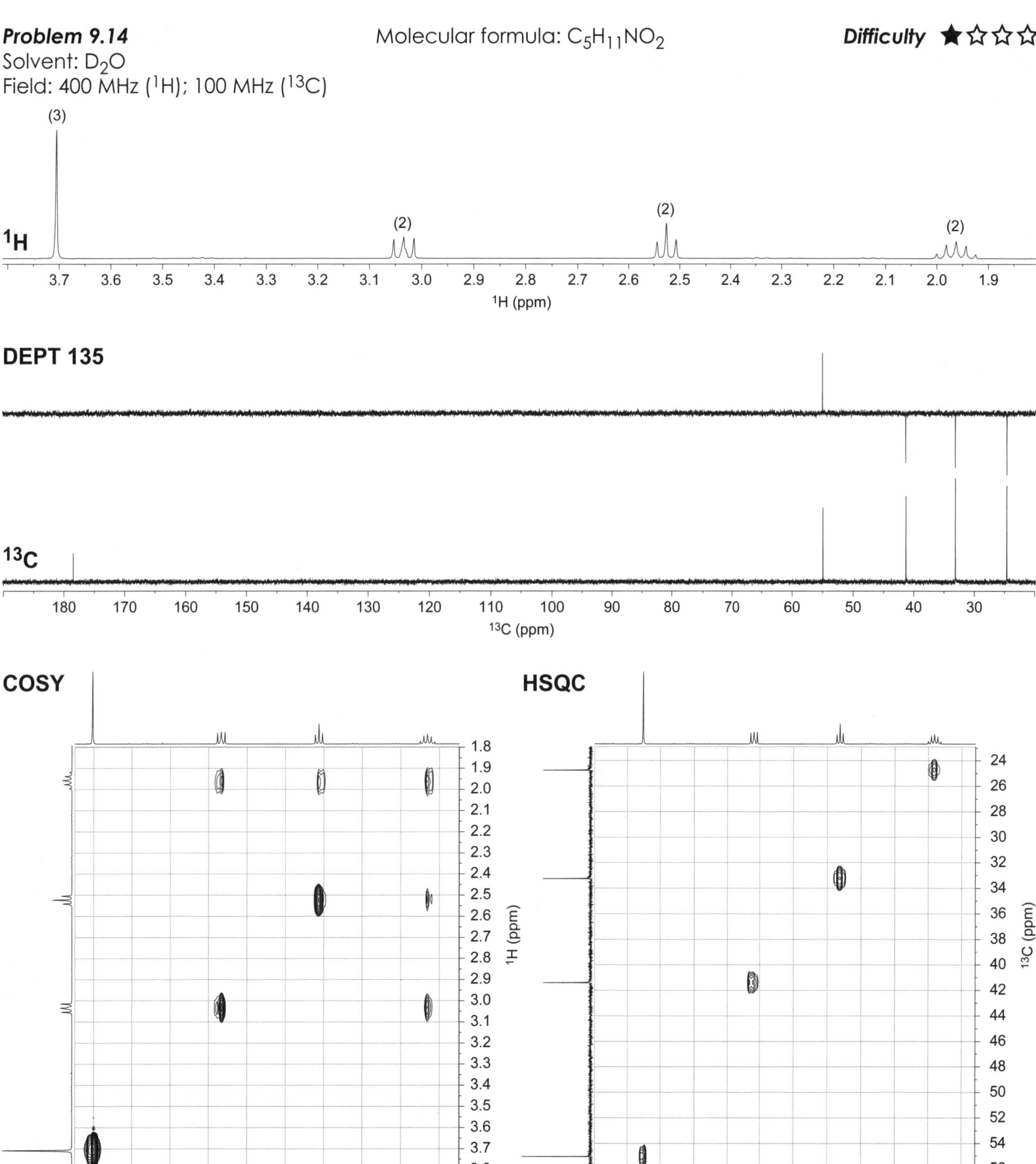

Problem 9.15
Solvent: D_2O
Field: 500 MHz (^1H); 125 MHz (^{13}C)

Molecular formula: $C_3H_7NO_2S$
Hint: It is a natural amino acid

Difficulty ★☆☆☆

^1H

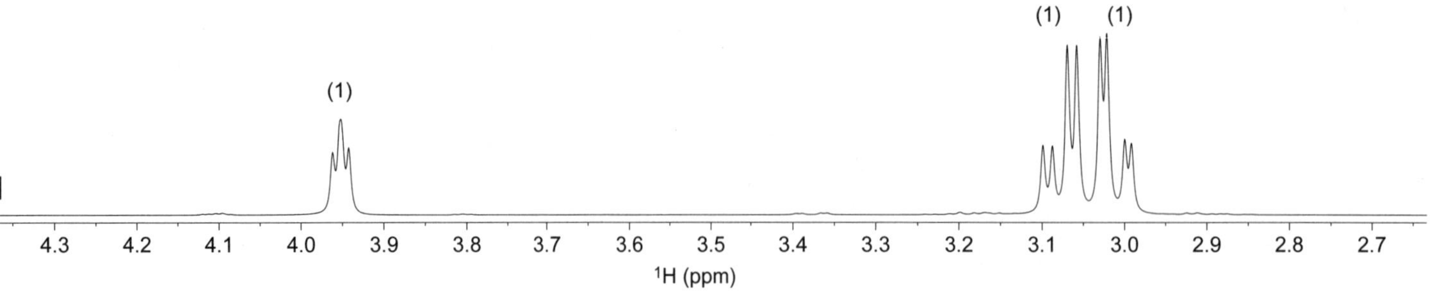

(1) (1) (1)

4.3 4.2 4.1 4.0 3.9 3.8 3.7 3.6 3.5 3.4 3.3 3.2 3.1 3.0 2.9 2.8 2.7
^1H (ppm)

DEPT 135

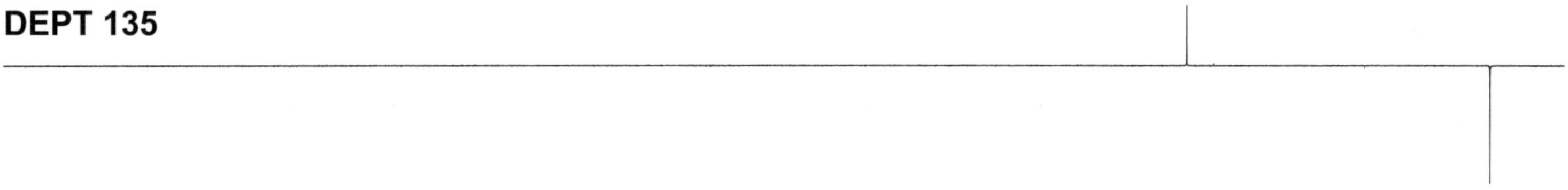

^{13}C

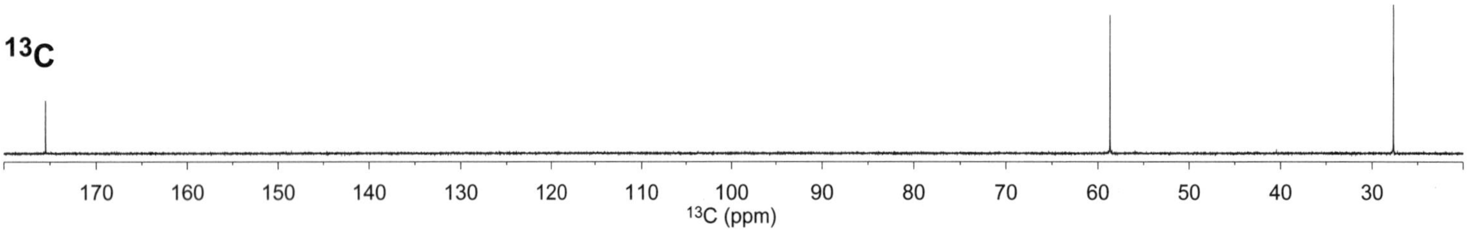

170 160 150 140 130 120 110 100 90 80 70 60 50 40 30
^{13}C (ppm)

COSY

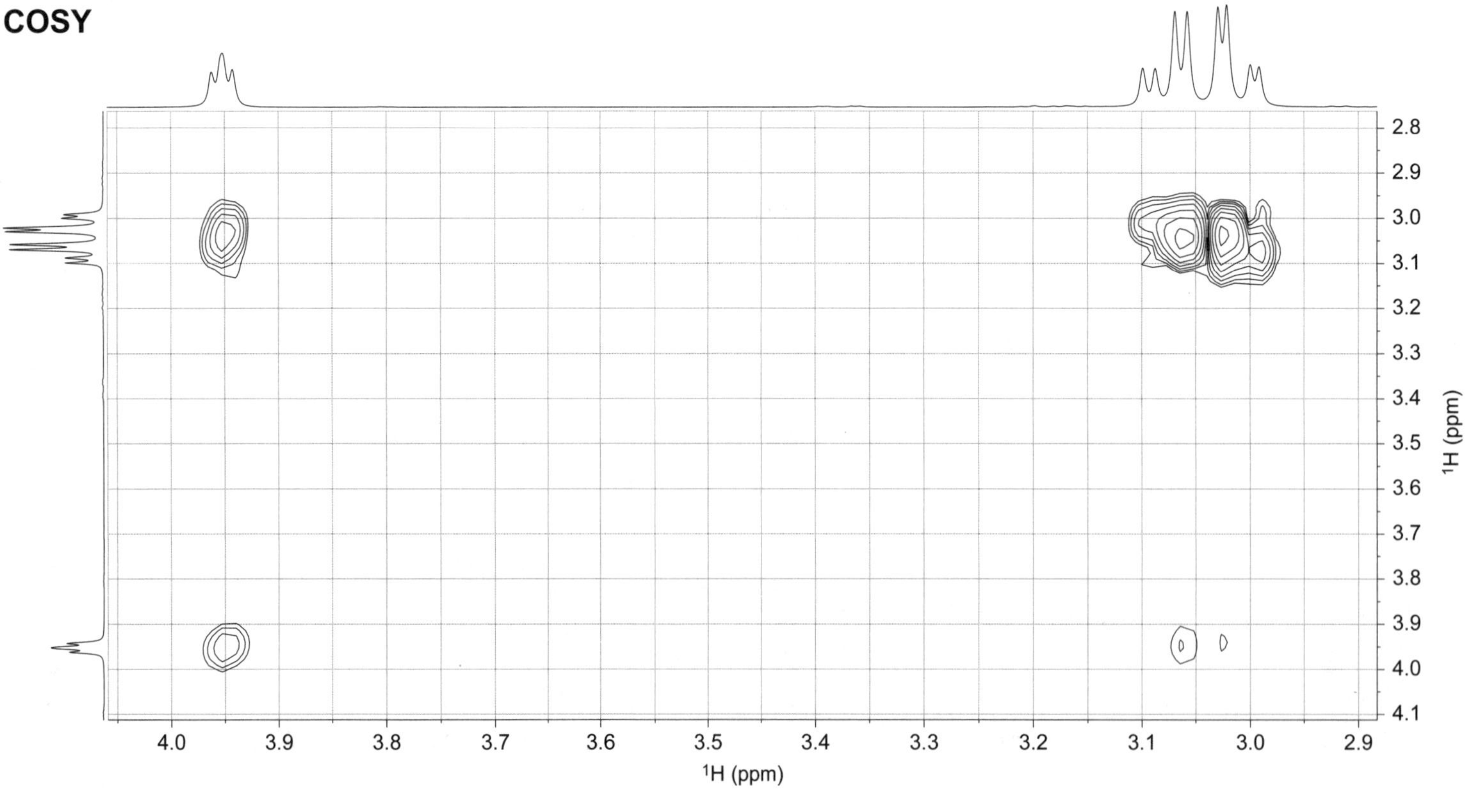

4.0 3.9 3.8 3.7 3.6 3.5 3.4 3.3 3.2 3.1 3.0 2.9
^1H (ppm)

2.8 2.9 3.0 3.1 3.2 3.3 3.4 3.5 3.6 3.7 3.8 3.9 4.0 4.1
^1H (ppm)

HSQC

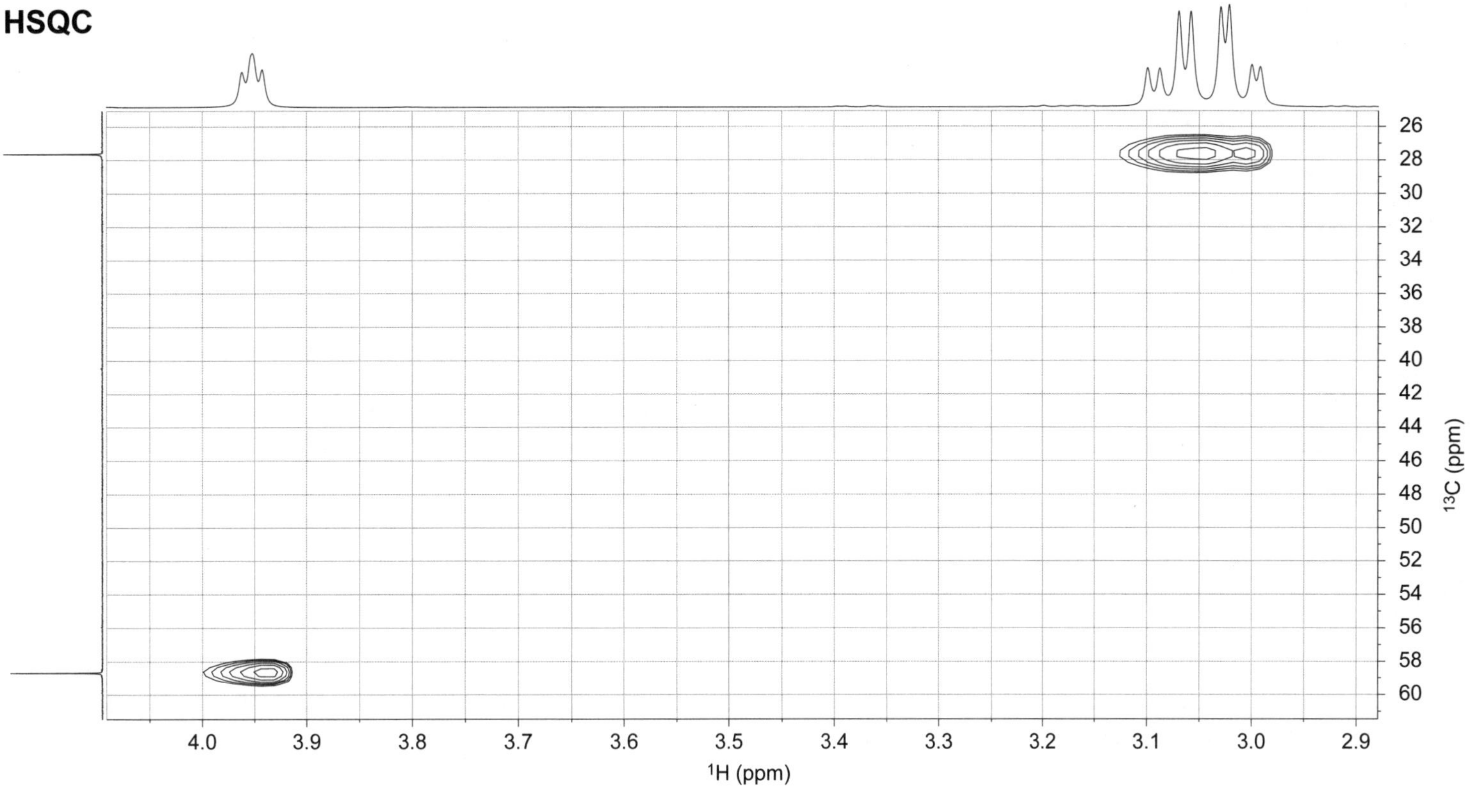
26
28
30
32
34
36
38
40
42
44
46
48
50
52
54
56
58
60
^{13}C (ppm)
4.0
3.9
3.8
3.7
3.6
3.5
3.4
3.3
3.2
3.1
3.0
2.9
^{1}H (ppm)

HMBC

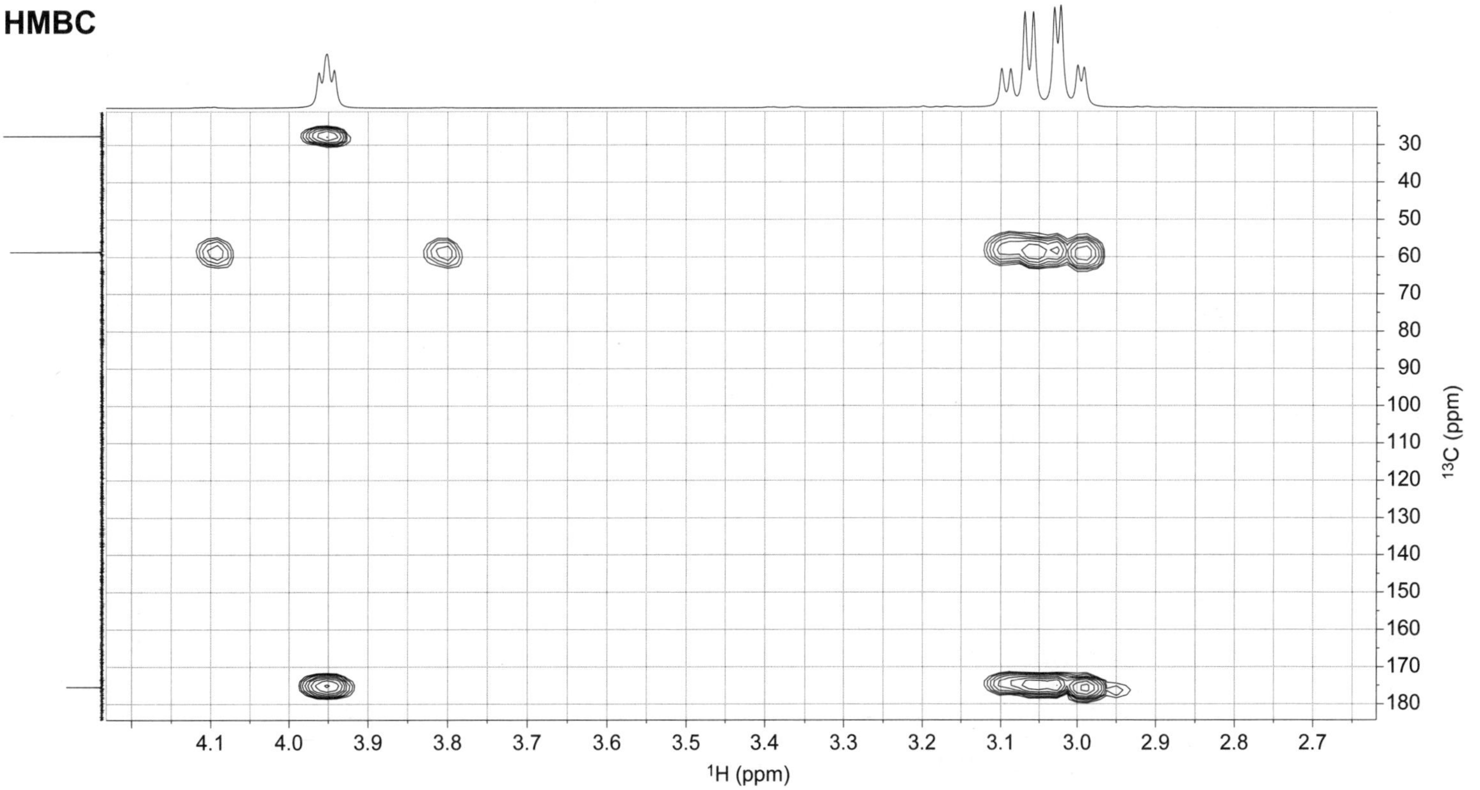
30
40
50
60
70
80
90
100
110
120
130
140
150
160
170
180
^{13}C (ppm)
4.1
4.0
3.9
3.8
3.7
3.6
3.5
3.4
3.3
3.2
3.1
3.0
2.9
2.8
2.7
^{1}H (ppm)

Problem 9.16 Molecular formula: $C_5H_9NO_4$ **Difficulty** ★☆☆☆

Solvent: D_2O

Field: 500 MHz (^1H); 125 MHz (^{13}C)

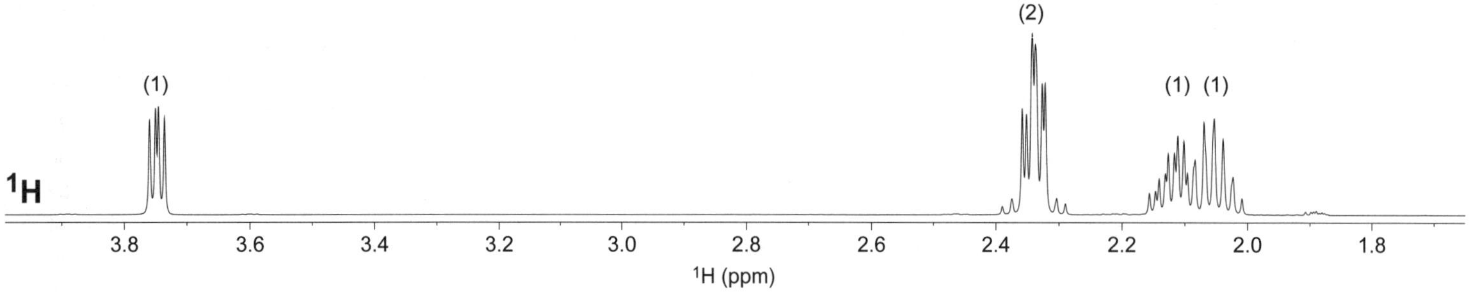

^1H

(1) (2) (1) (1)

3.8 3.6 3.4 3.2 3.0 2.8 2.6 2.4 2.2 2.0 1.8

^1H (ppm)

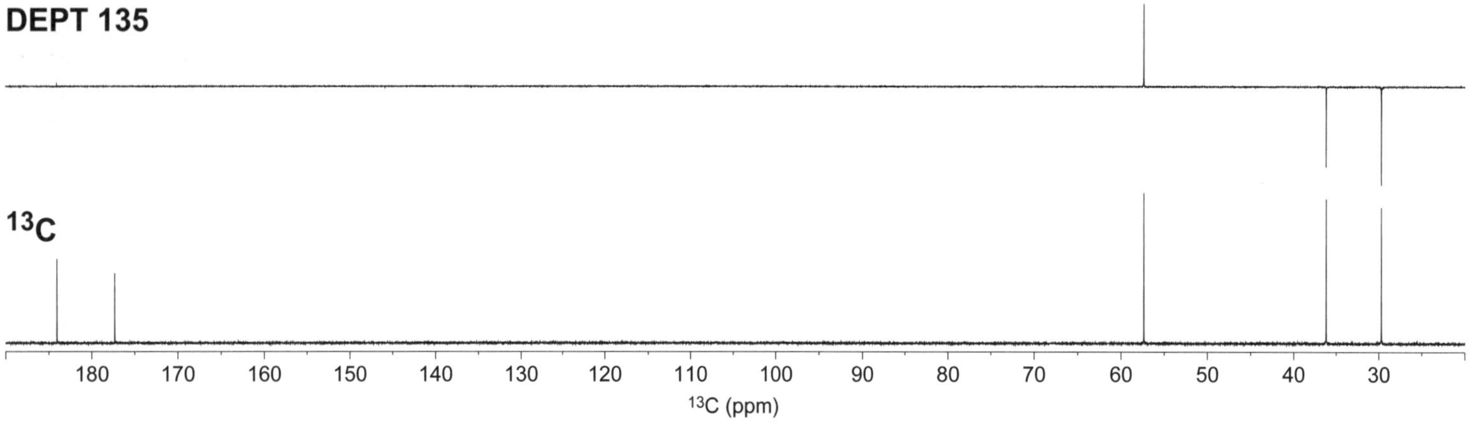

DEPT 135

^{13}C

180 170 160 150 140 130 120 110 100 90 80 70 60 50 40 30

^{13}C (ppm)

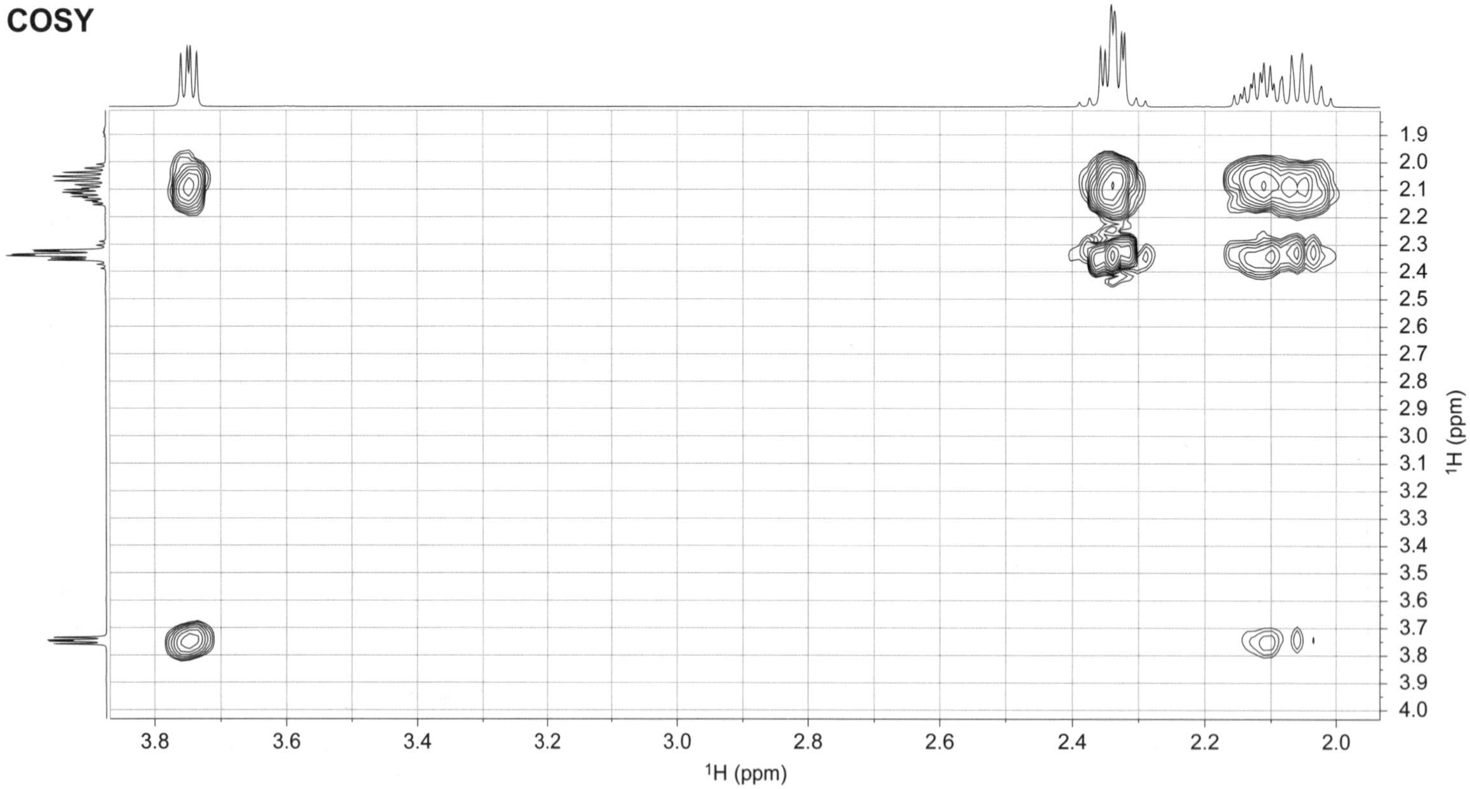

COSY

1.9 2.0 2.1 2.2 2.3 2.4 2.5 2.6 2.7 2.8 2.9 3.0 3.1 3.2 3.3 3.4 3.5 3.6 3.7 3.8 3.9 4.0

^1H (ppm)

3.8 3.6 3.4 3.2 3.0 2.8 2.6 2.4 2.2 2.0

^1H (ppm)

HSQC

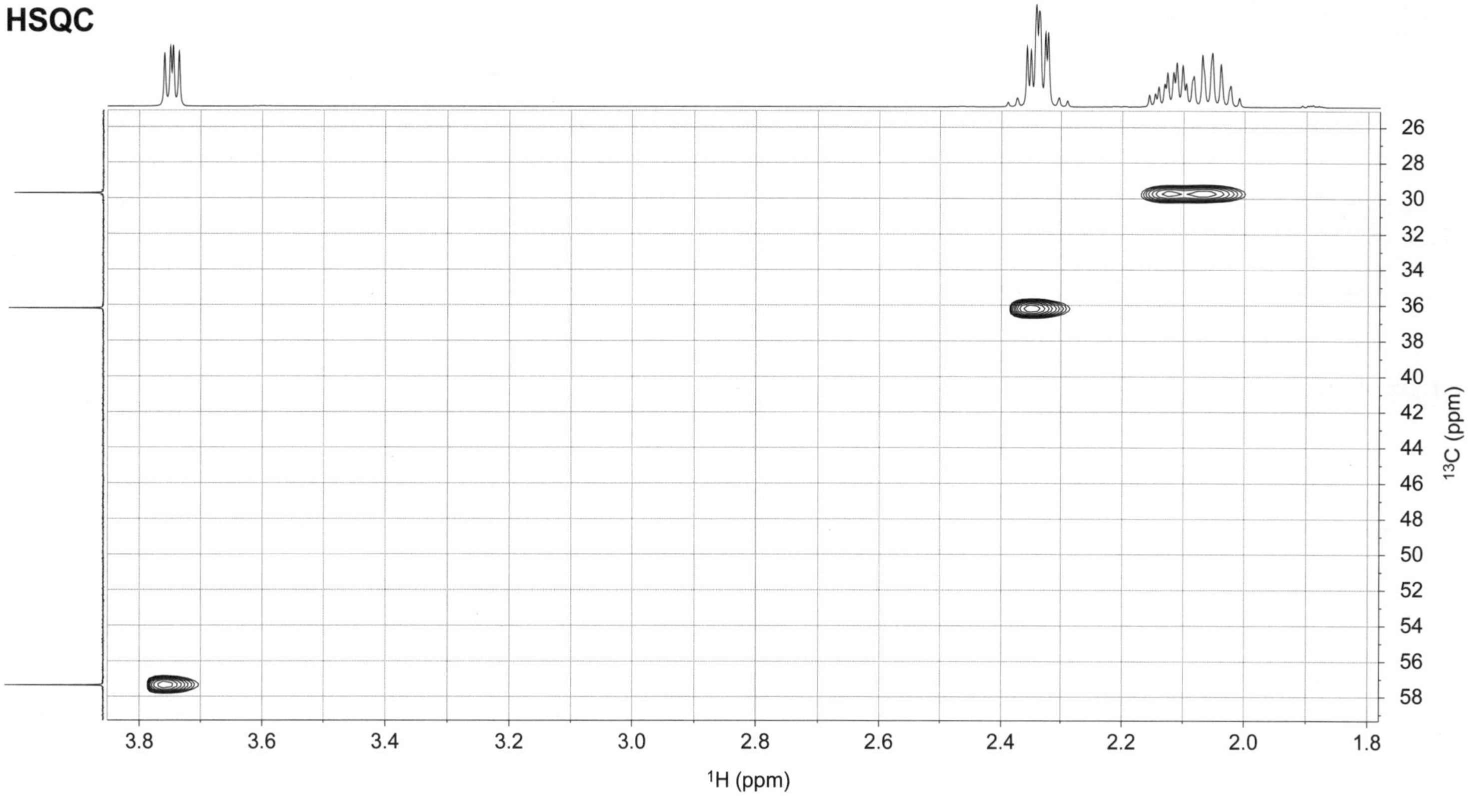

HMBC

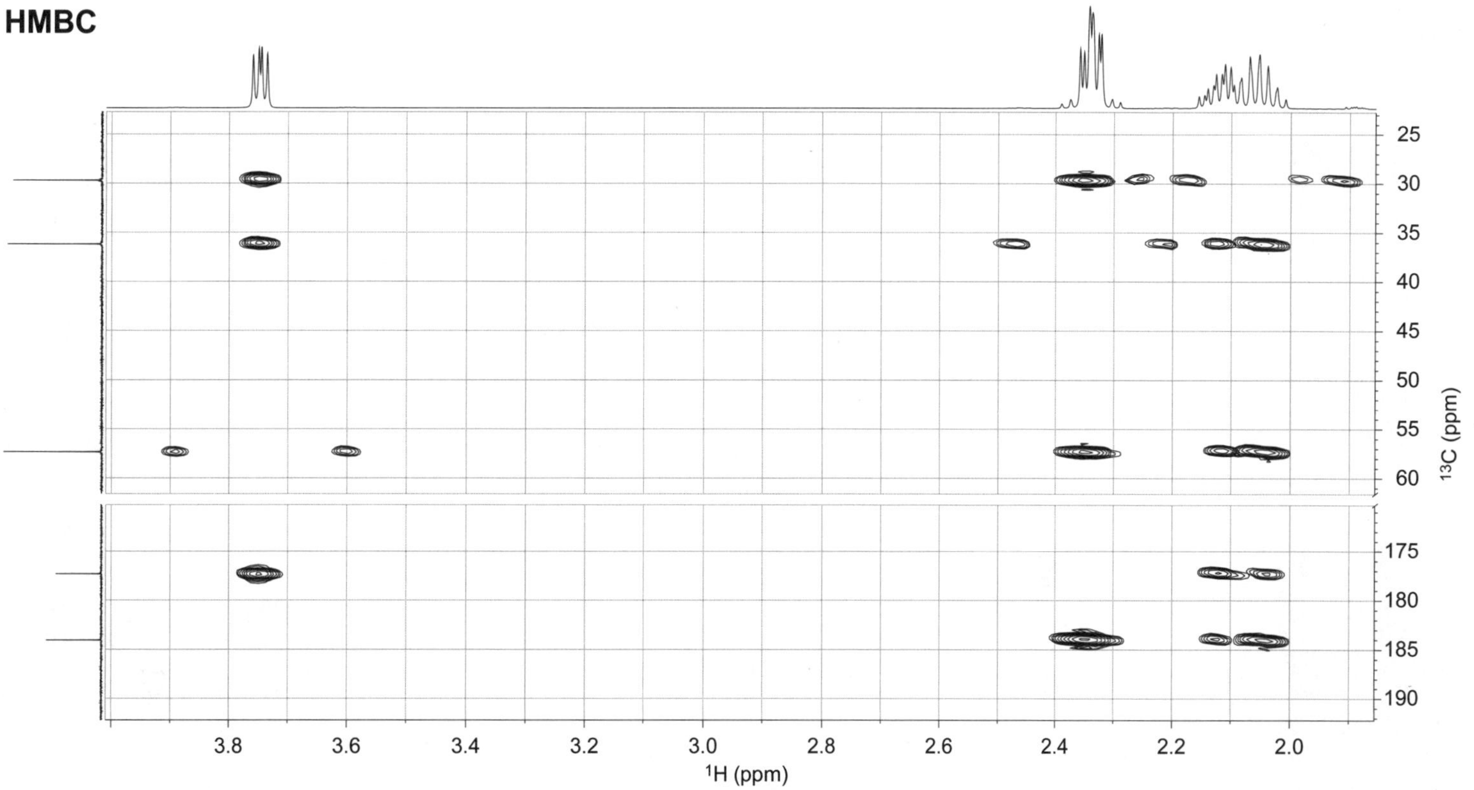

Problem 9.17
Solvent: D_2O
Field: 500 MHz (^1H); 125 MHz (^{13}C)

Molecular formula: $C_4H_9NO_3$

Difficulty ★☆☆☆

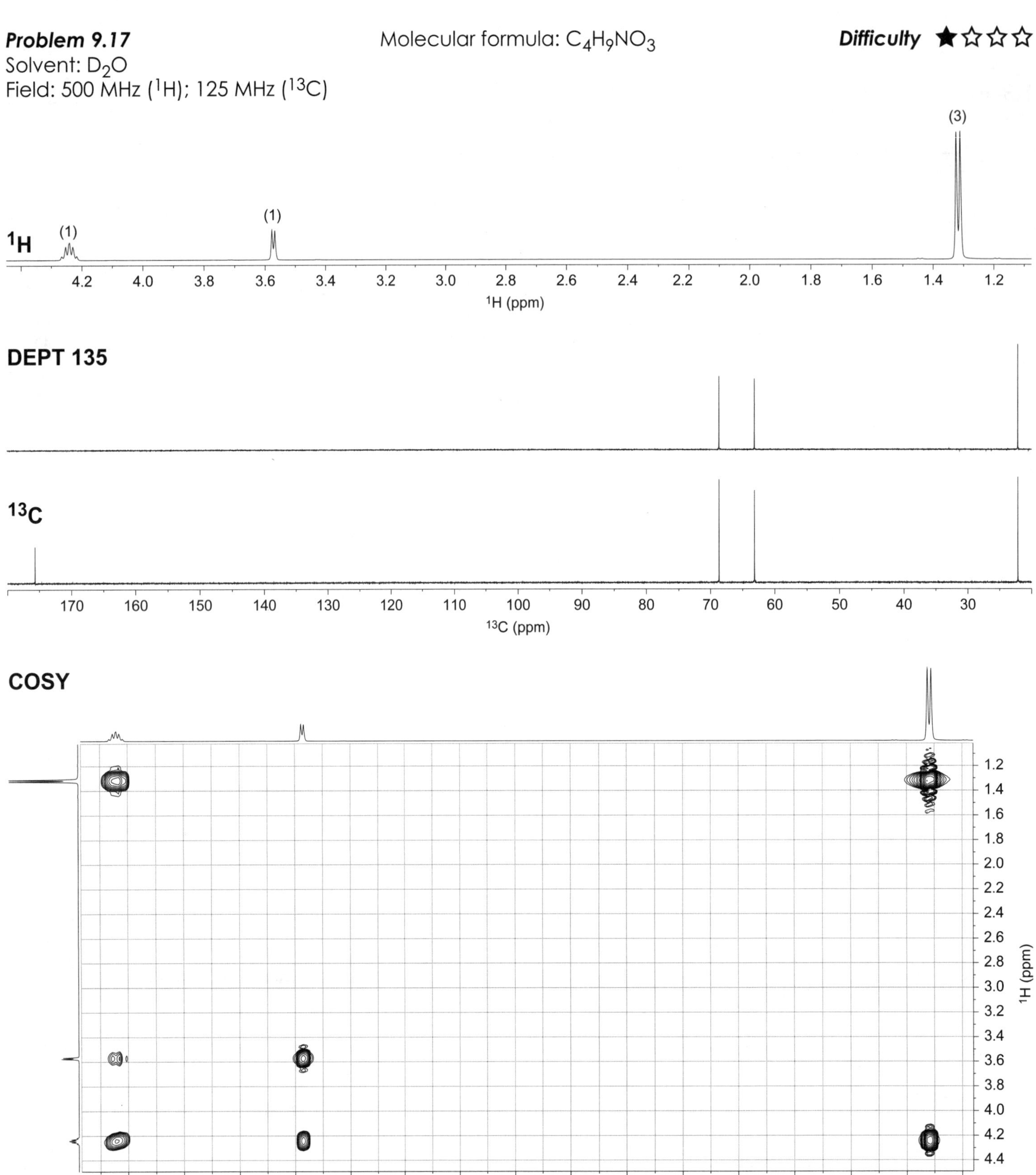

HSQC

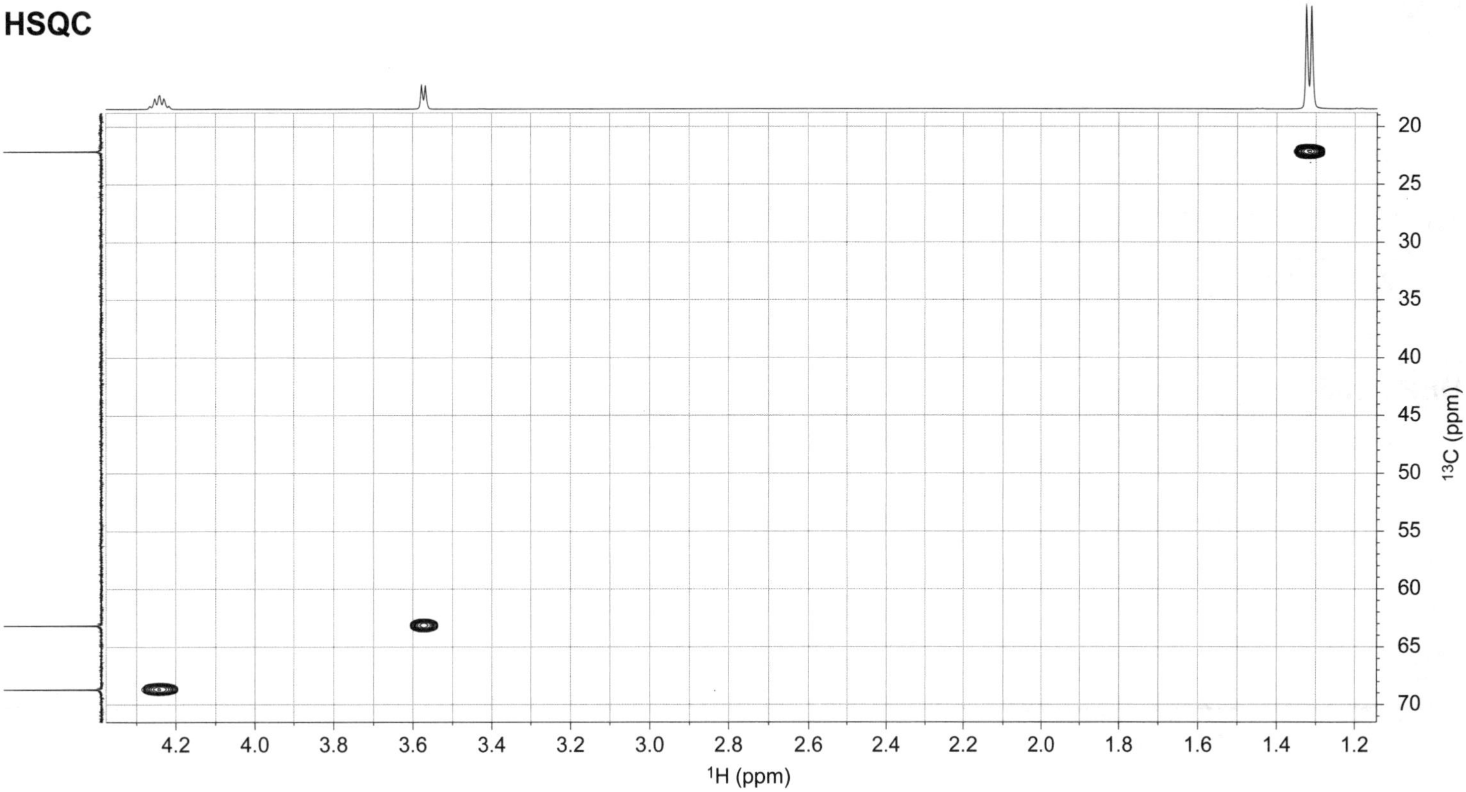
20
25
30
35
40
45
50
55
60
65
70
^{13}C (ppm)
4.2 4.0 3.8 3.6 3.4 3.2 3.0 2.8 2.6 2.4 2.2 2.0 1.8 1.6 1.4 1.2
^{1}H (ppm)

HMBC

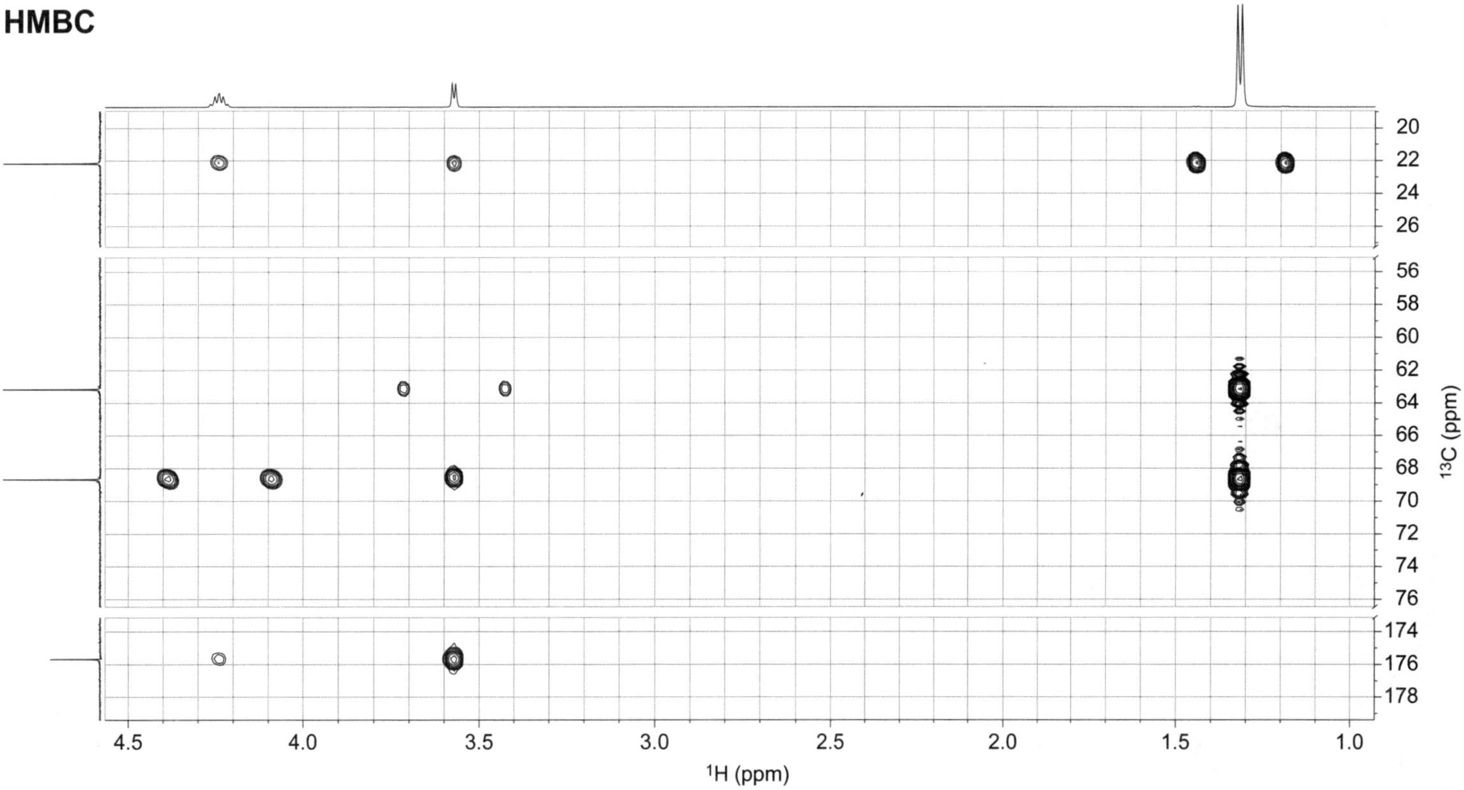
20
22
24
26
56
58
60
62
64
66
68
70
72
74
76
174
176
178
^{13}C (ppm)
4.5 4.0 3.5 3.0 2.5 2.0 1.5 1.0
^{1}H (ppm)

Problem 9.18 Molecular formula: $C_5H_{11}NO_2$ ***Difficulty*** ★☆☆☆

Solvent: D_2O

Field: 500 MHz (^1H); 125 MHz (^{13}C)

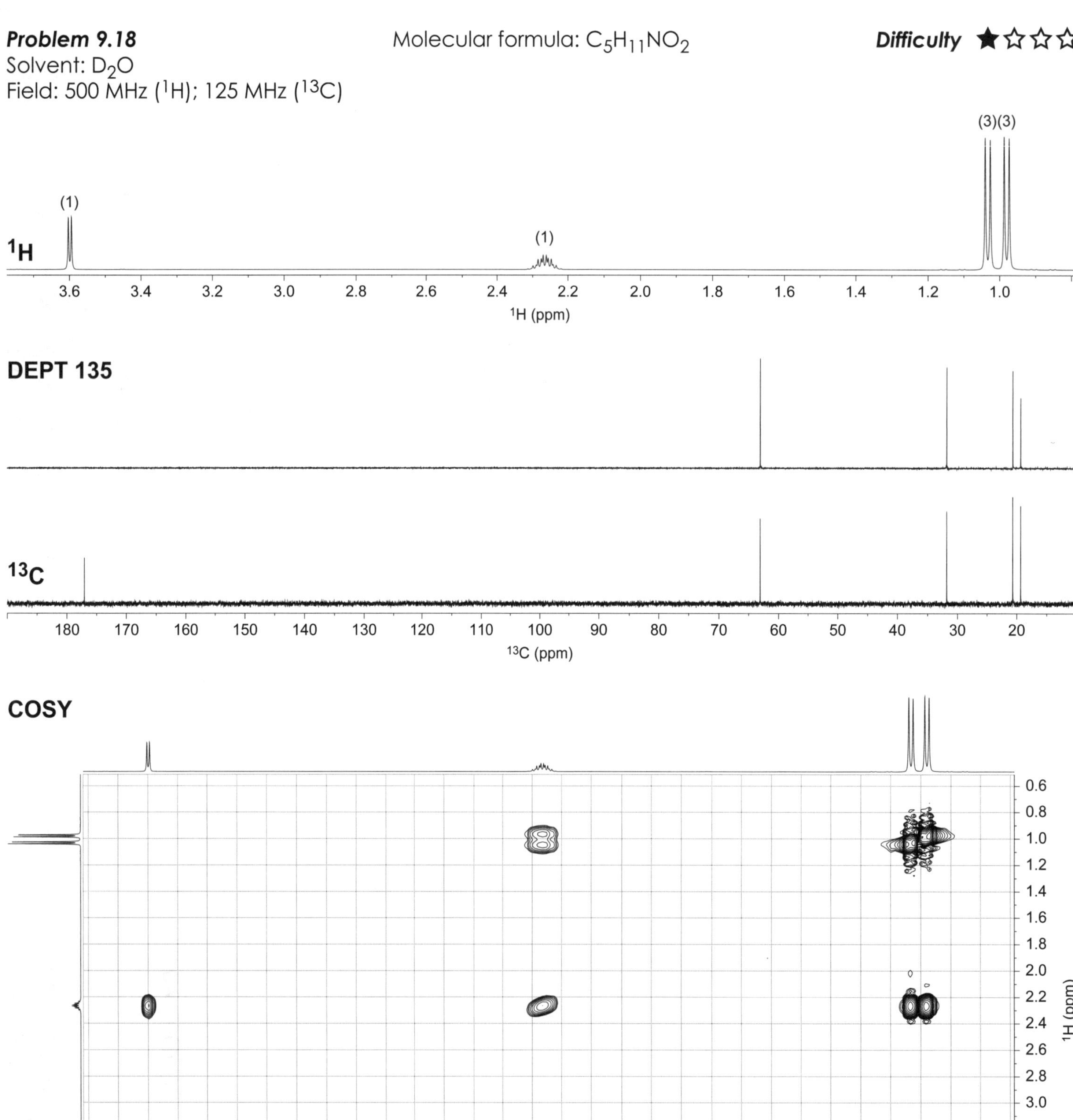

HSQC

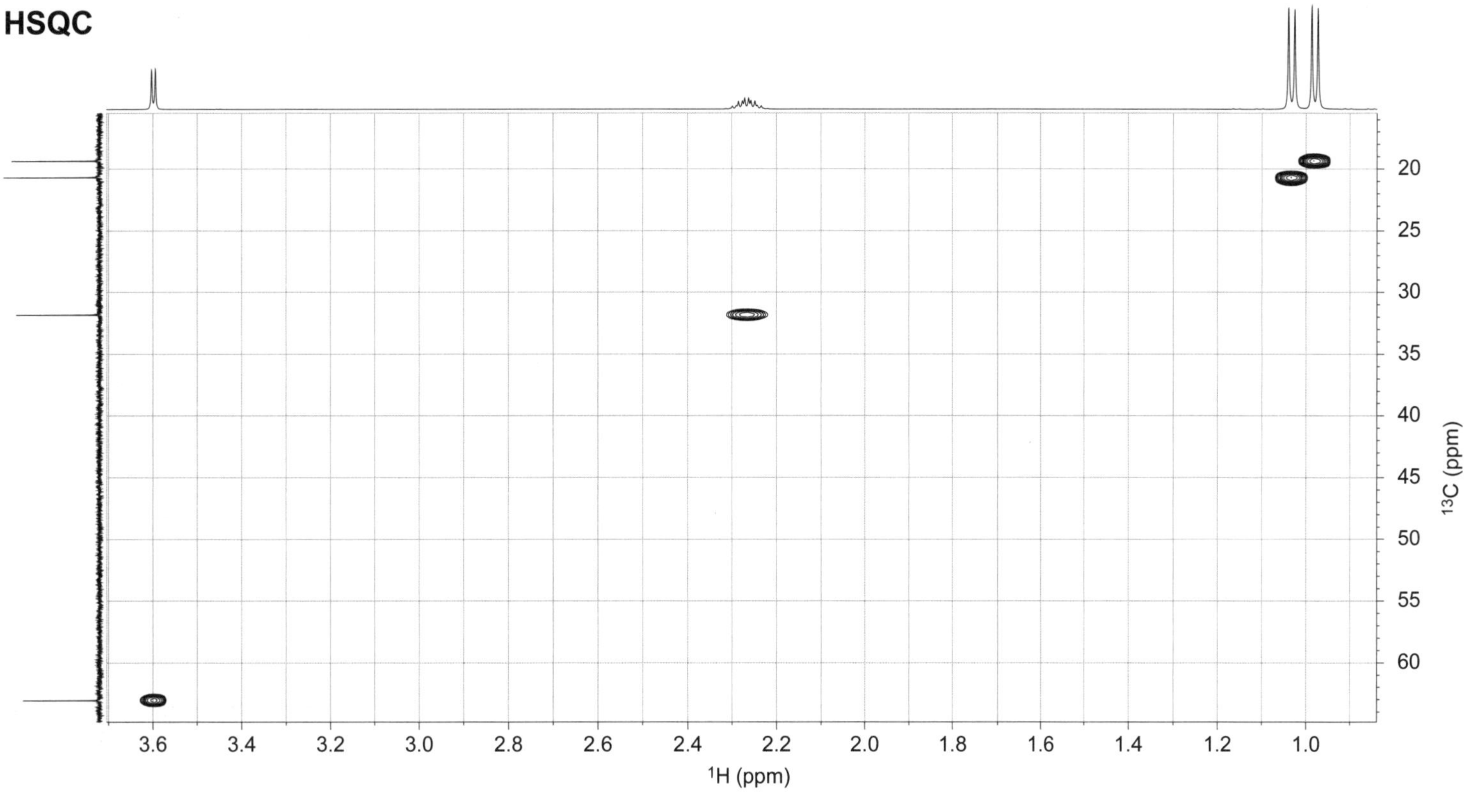
3.6
3.4
3.2
3.0
2.8
2.6
2.4
2.2
2.0
1.8
1.6
1.4
1.2
1.0
¹H (ppm)
20
25
30
35
40
45
50
55
60
¹³C (ppm)

HMBC

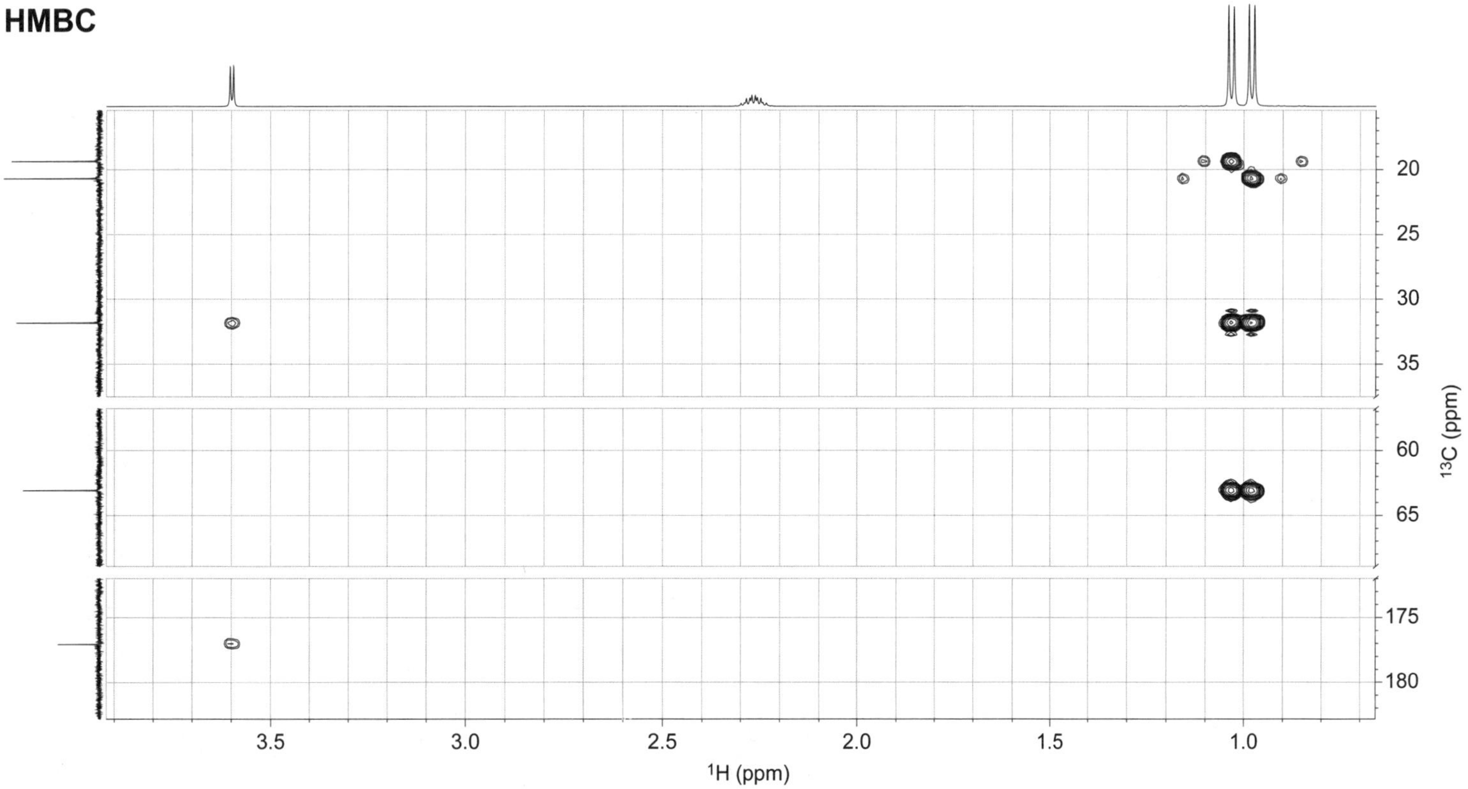
3.5
3.0
2.5
2.0
1.5
1.0
¹H (ppm)
20
25
30
35
60
65
175
180
¹³C (ppm)

Problem 9.19
Solvent: D_2O
Field: 500 MHz (^1H); 125 MHz (^{13}C)

Molecular formula: $C_7H_7NO_2$
Hint: there is an amine group

Difficulty ★☆☆☆

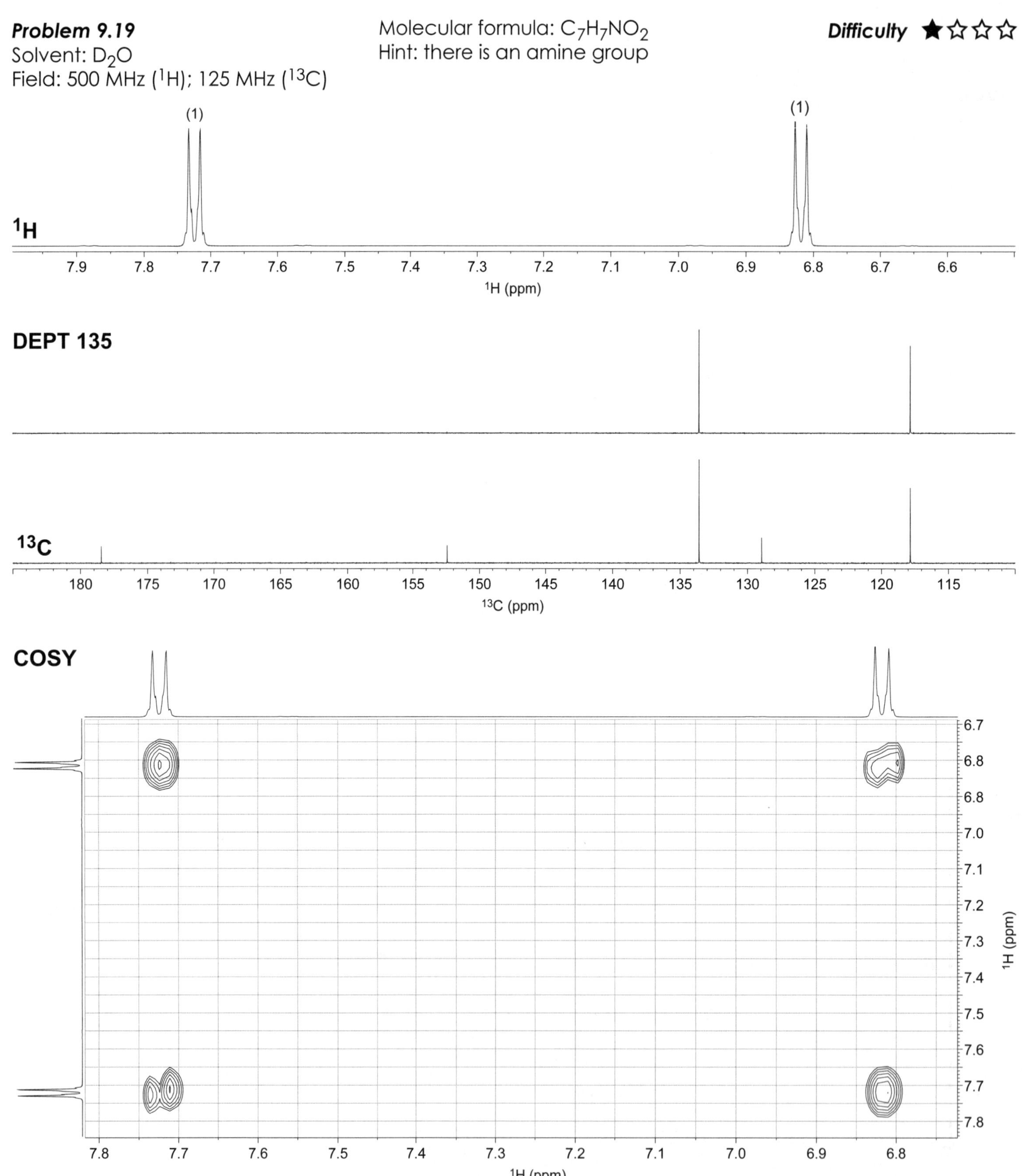

HSQC

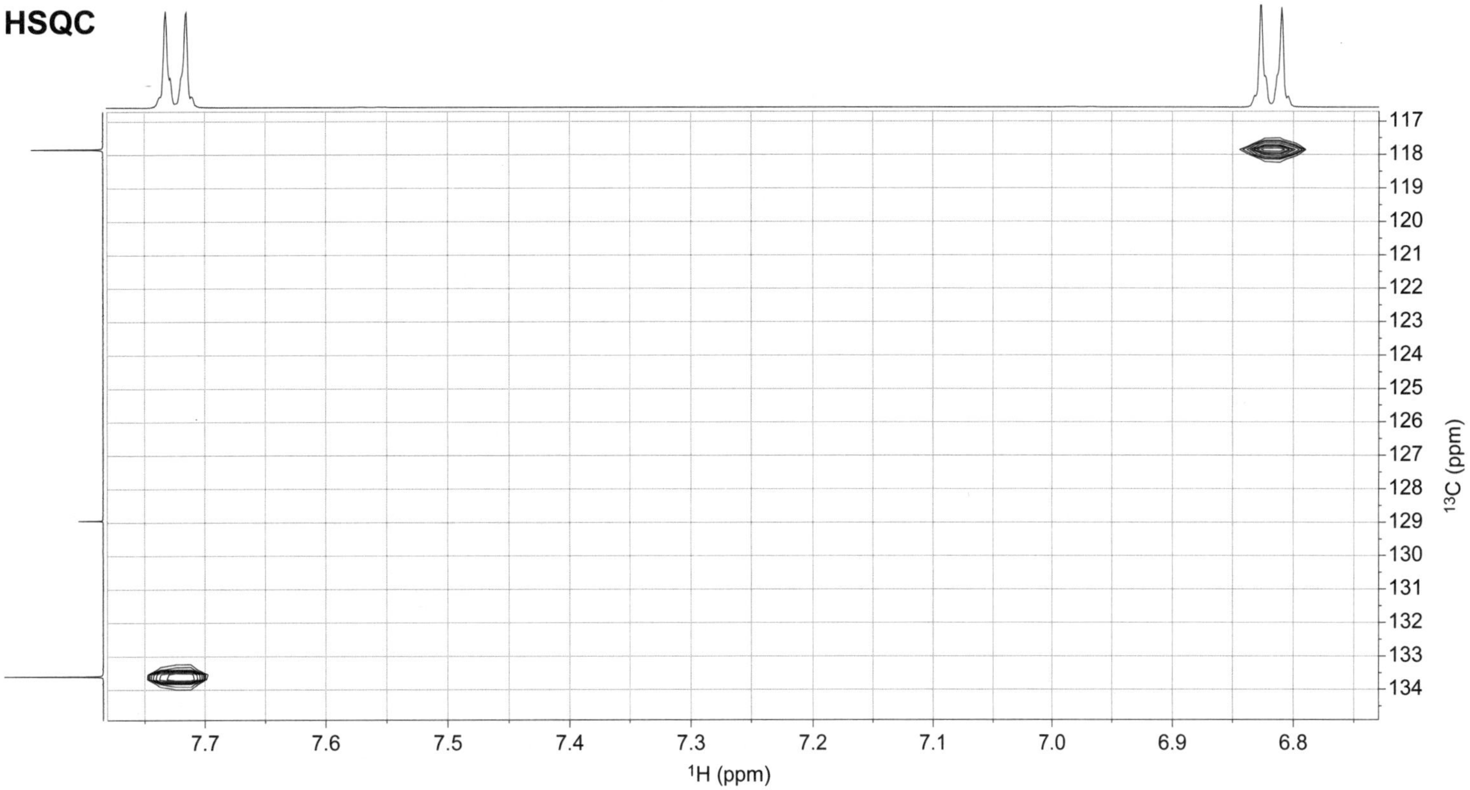
117
118
119
120
121
122
123
124
125
126
127
128
129
130
131
132
133
134
^{13}C (ppm)
7.7
7.6
7.5
7.4
7.3
7.2
7.1
7.0
6.9
6.8
^{1}H (ppm)

HMBC

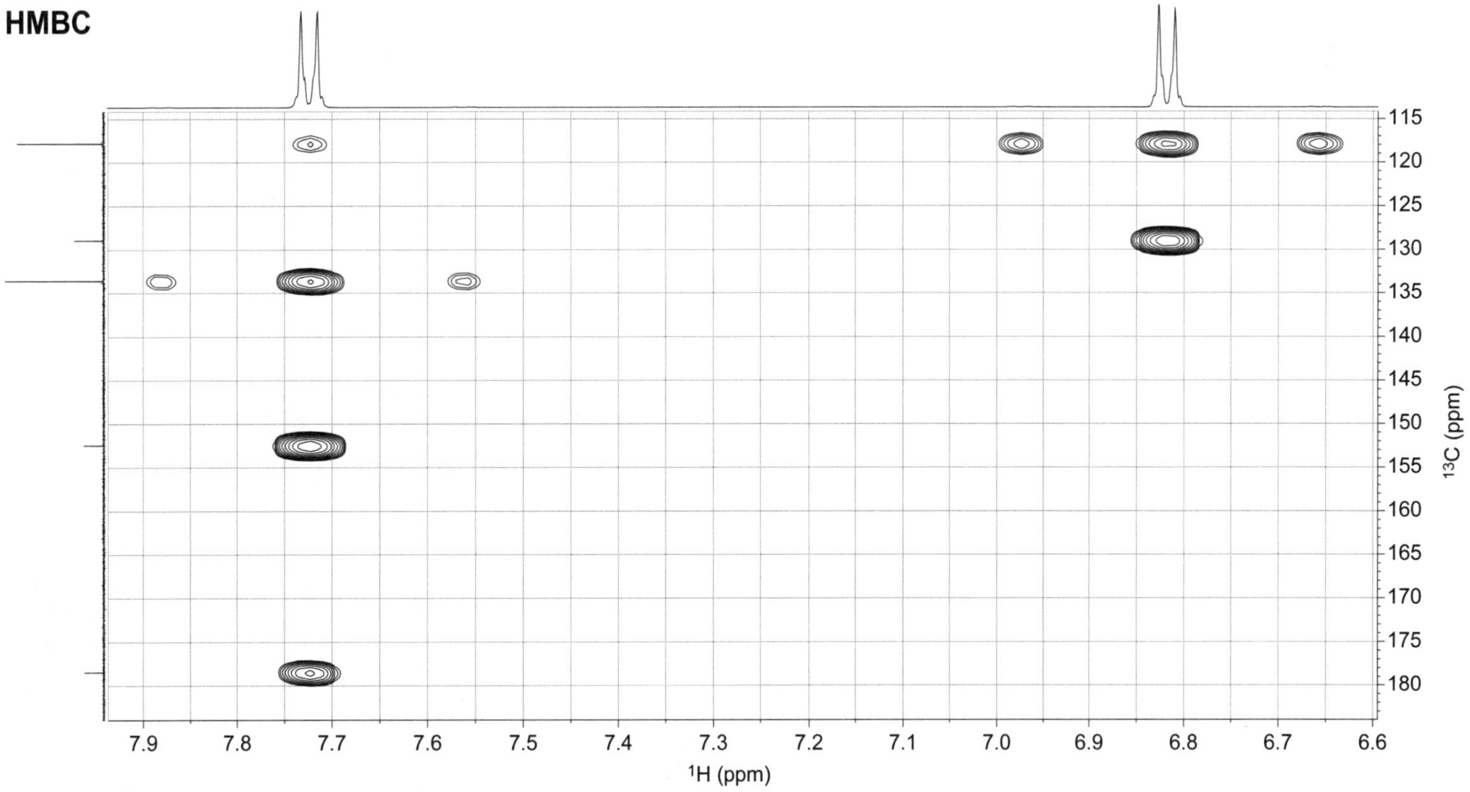
115
120
125
130
135
140
145
150
155
160
165
170
175
180
^{13}C (ppm)
7.9
7.8
7.7
7.6
7.5
7.4
7.3
7.2
7.1
7.0
6.9
6.8
6.7
6.6
^{1}H (ppm)

Problem 9.20
Solvent: D_2O
Field: 500 MHz (^1H); 125 MHz (^{13}C)

Molecular formula: $C_7H_7NO_2$
Hint: It is a carboxylic acid

Difficulty ★☆☆☆

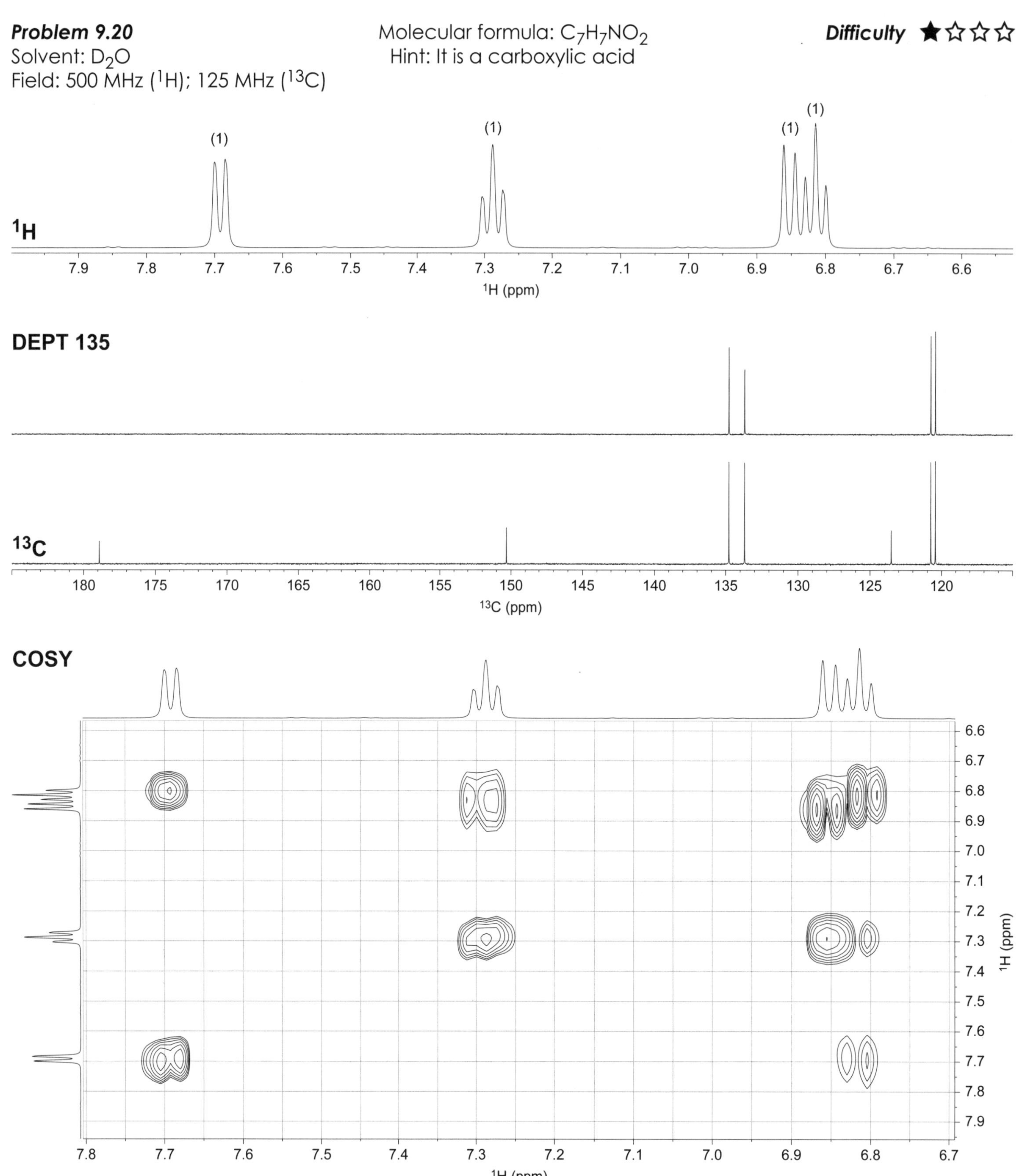

HSQC

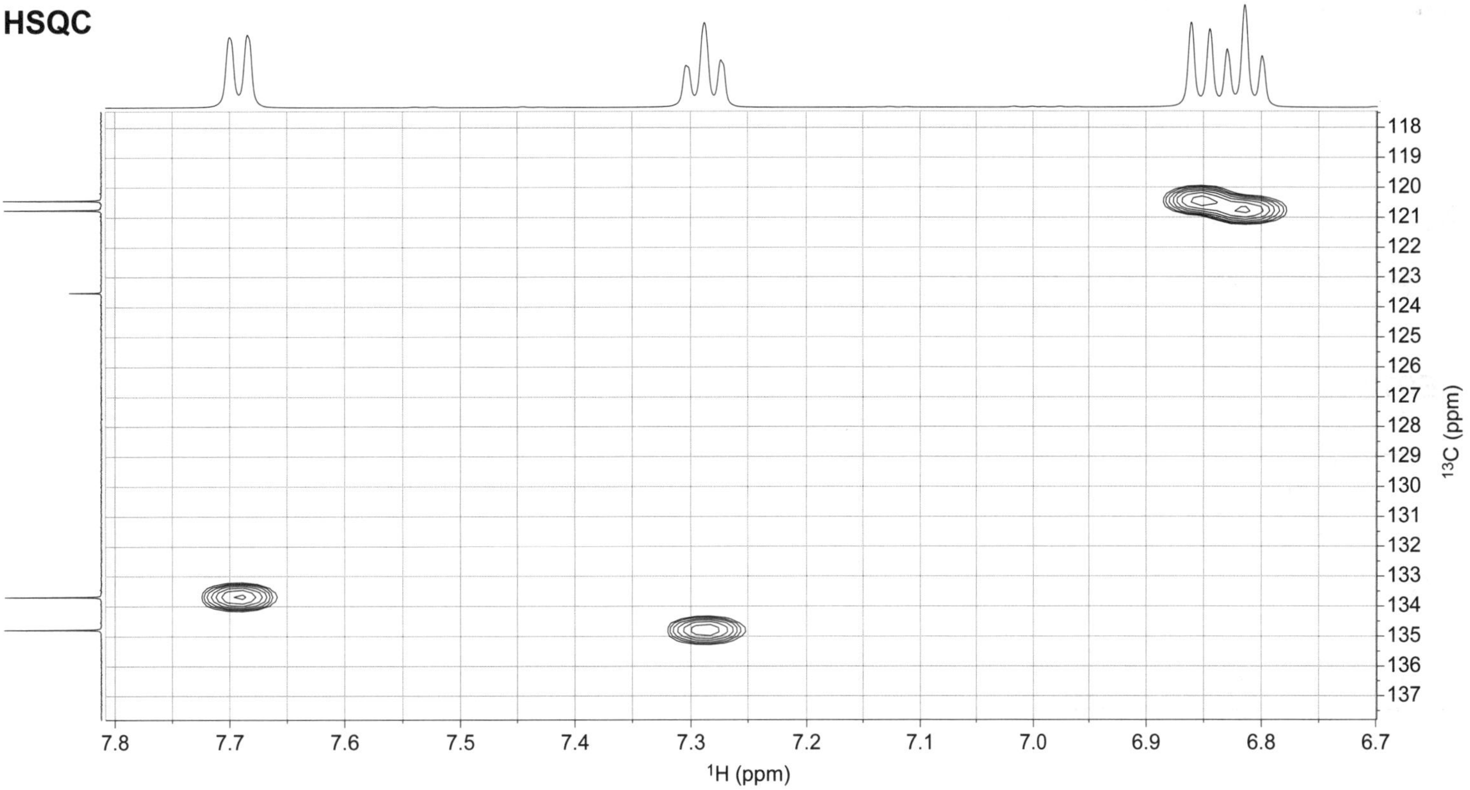

HMBC

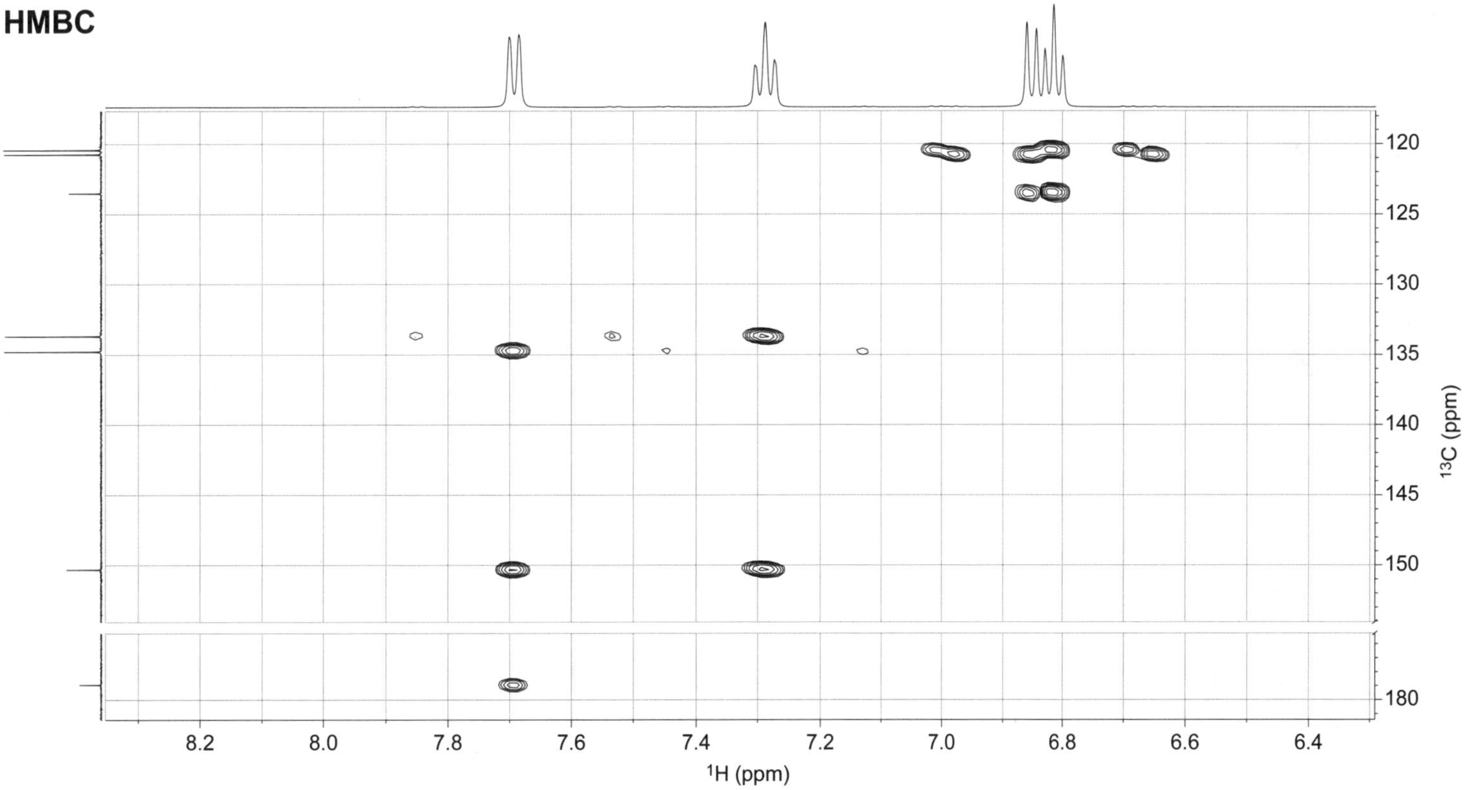

Problem 9.21 Molecular formula: $C_4H_8O_3$ **Difficulty** ★☆☆☆

Solvent: D_2O

Field: 500 MHz (^{1}H); 125 MHz (^{13}C)

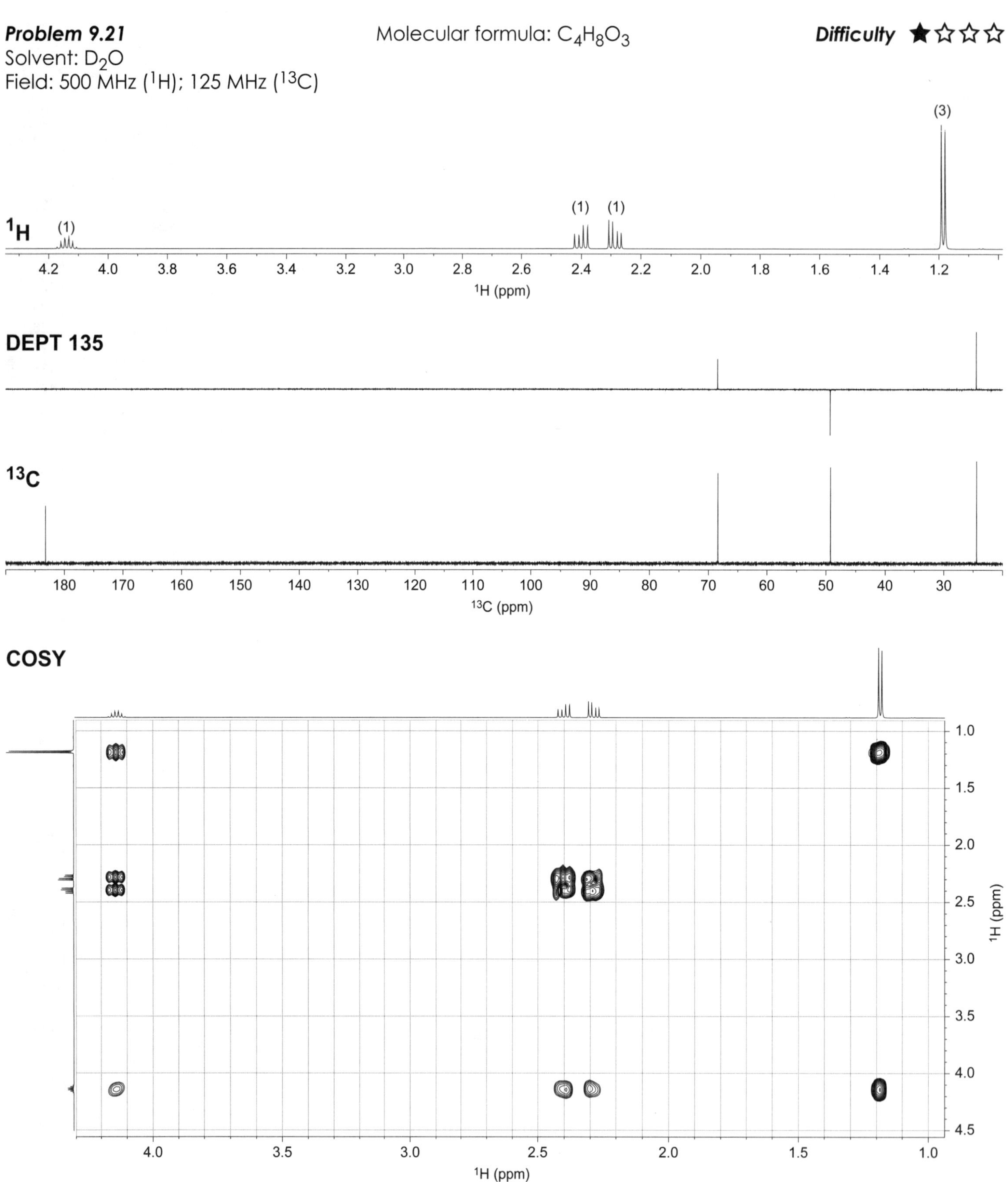

HSQC

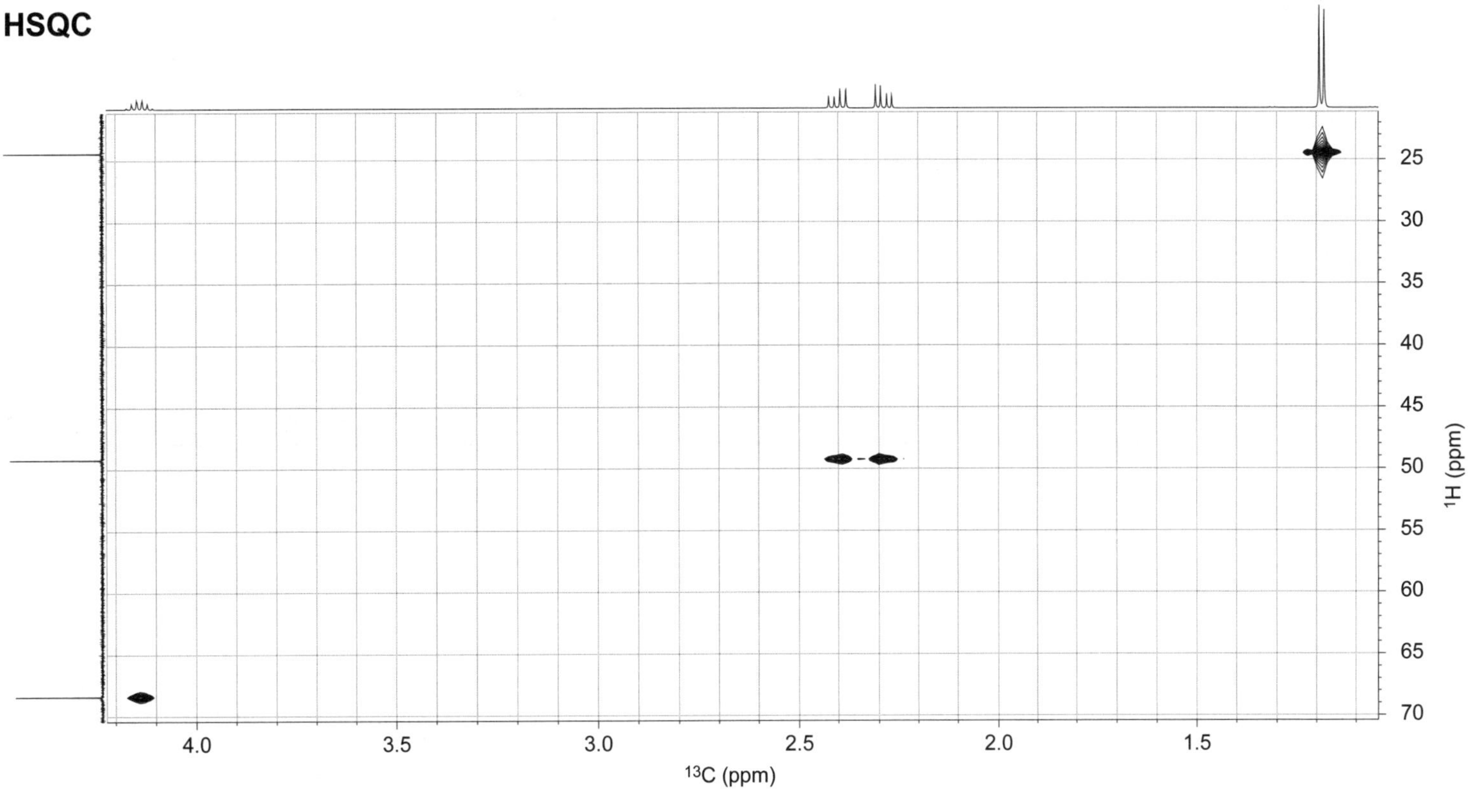
25
30
35
40
45
50
55
60
65
70
4.0
3.5
3.0
2.5
2.0
1.5
13C (ppm)
1H (ppm)

HMBC

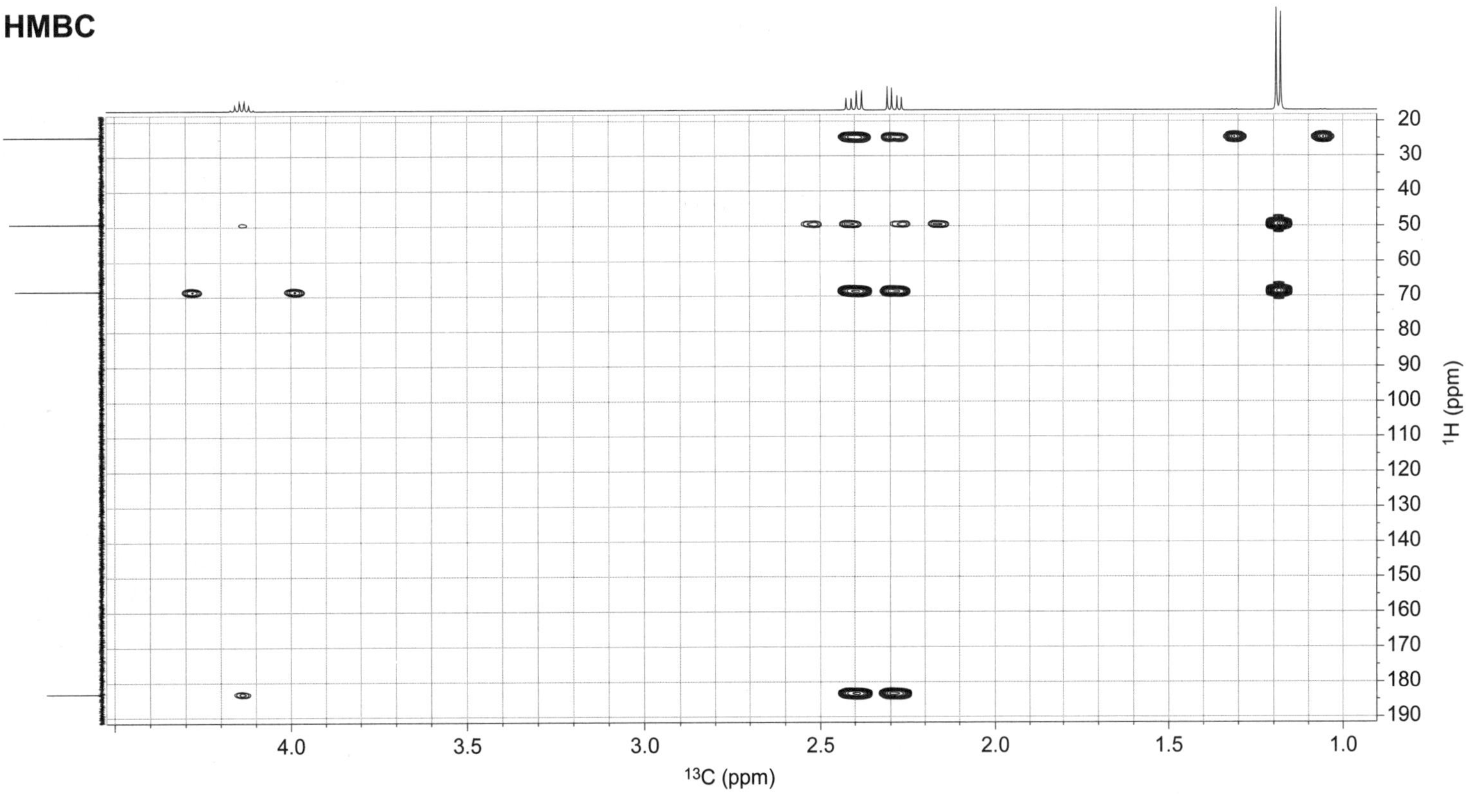
20
30
40
50
60
70
80
90
100
110
120
130
140
150
160
170
180
190
4.0
3.5
3.0
2.5
2.0
1.5
1.0
13C (ppm)
1H (ppm)

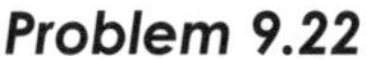

Problem 9.22 Molecular formula: $C_3H_8O_2$ ***Difficulty*** ★☆☆☆

Solvent: D_2O

Field: 500 MHz (^1H); 125 MHz (^{13}C)

^{1}H

(3)
(1) (1)
(1)
4.0 3.8 3.6 3.4 3.2 3.0 2.8 2.6 2.4 2.2 2.0 1.8 1.6 1.4 1.2 1.0 0.8
¹H (ppm)

DEPT 135

^{13}C

75 70 65 60 55 50 45 40 35 30 25 20 15
¹³C (ppm)

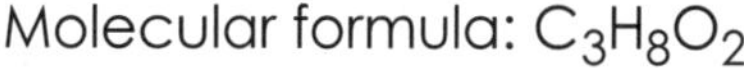

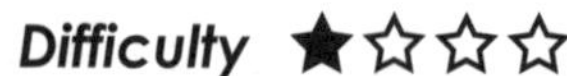

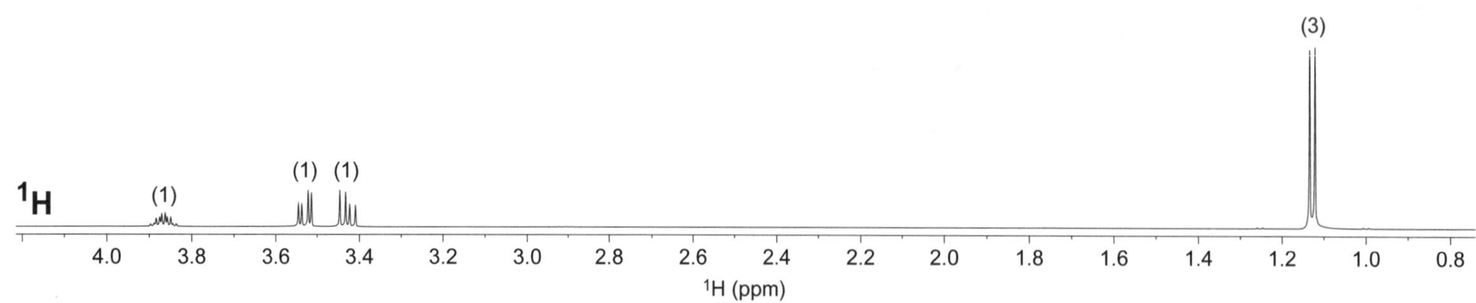

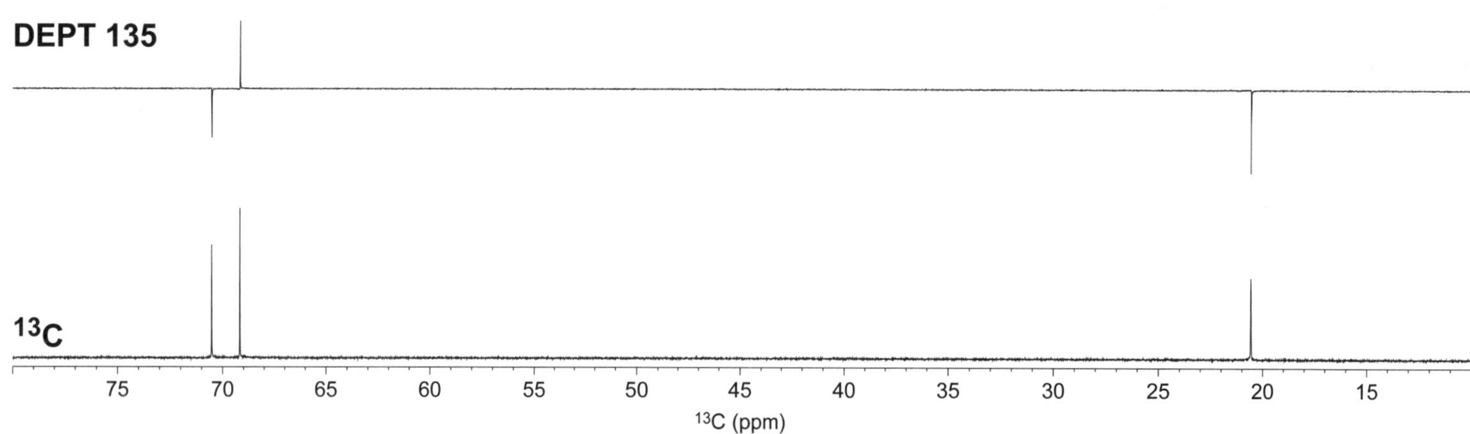

HSQC

15 20 25 30 35 40 45 50 55 60 65 70 75
¹³C (ppm)
4.0 3.5 3.0 2.5 2.0 1.5 1.0
¹H (ppm)

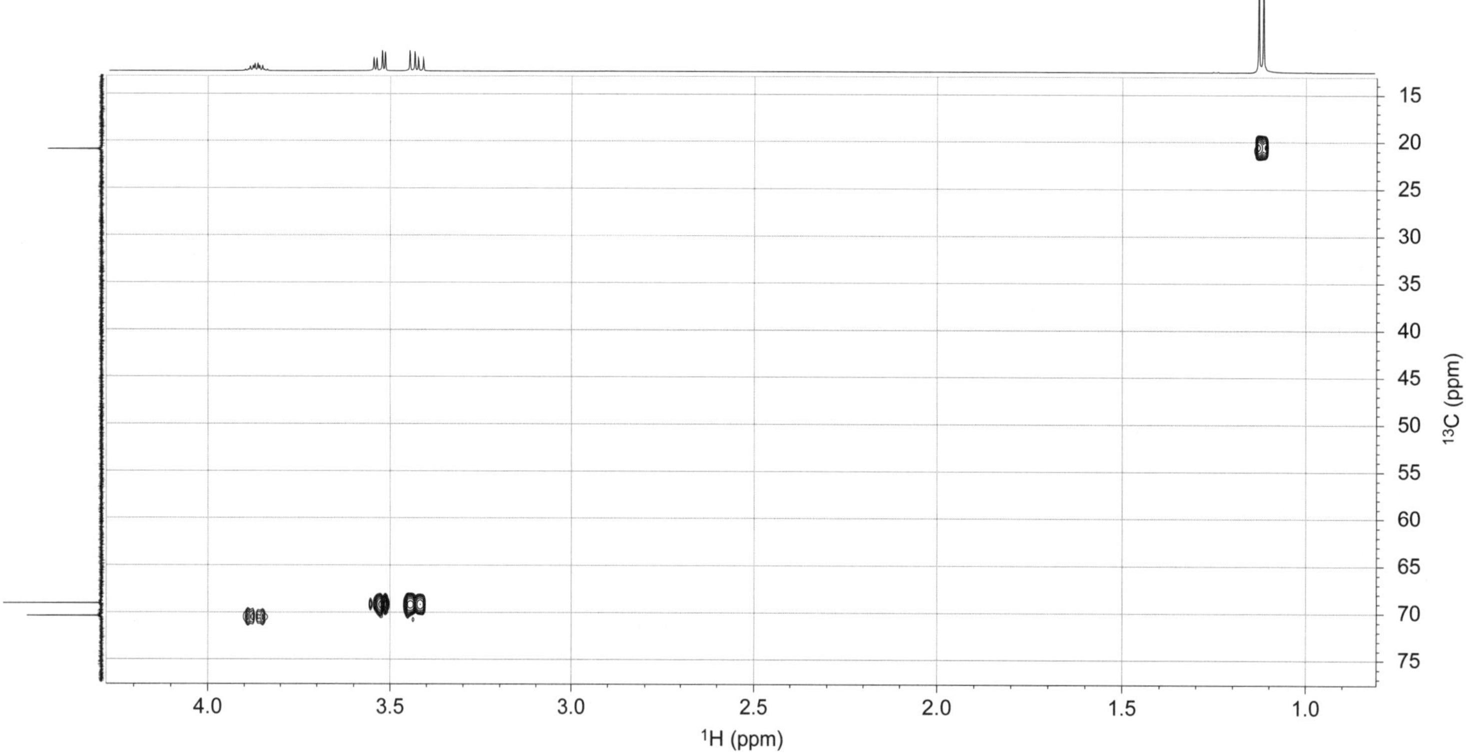

HMBC

4.0
3.5
3.0
2.5
2.0
1.5
1.0
0.5
^{1}H (ppm)
15
20
25
30
35
40
45
50
55
60
65
70
75
80
^{13}C (ppm)

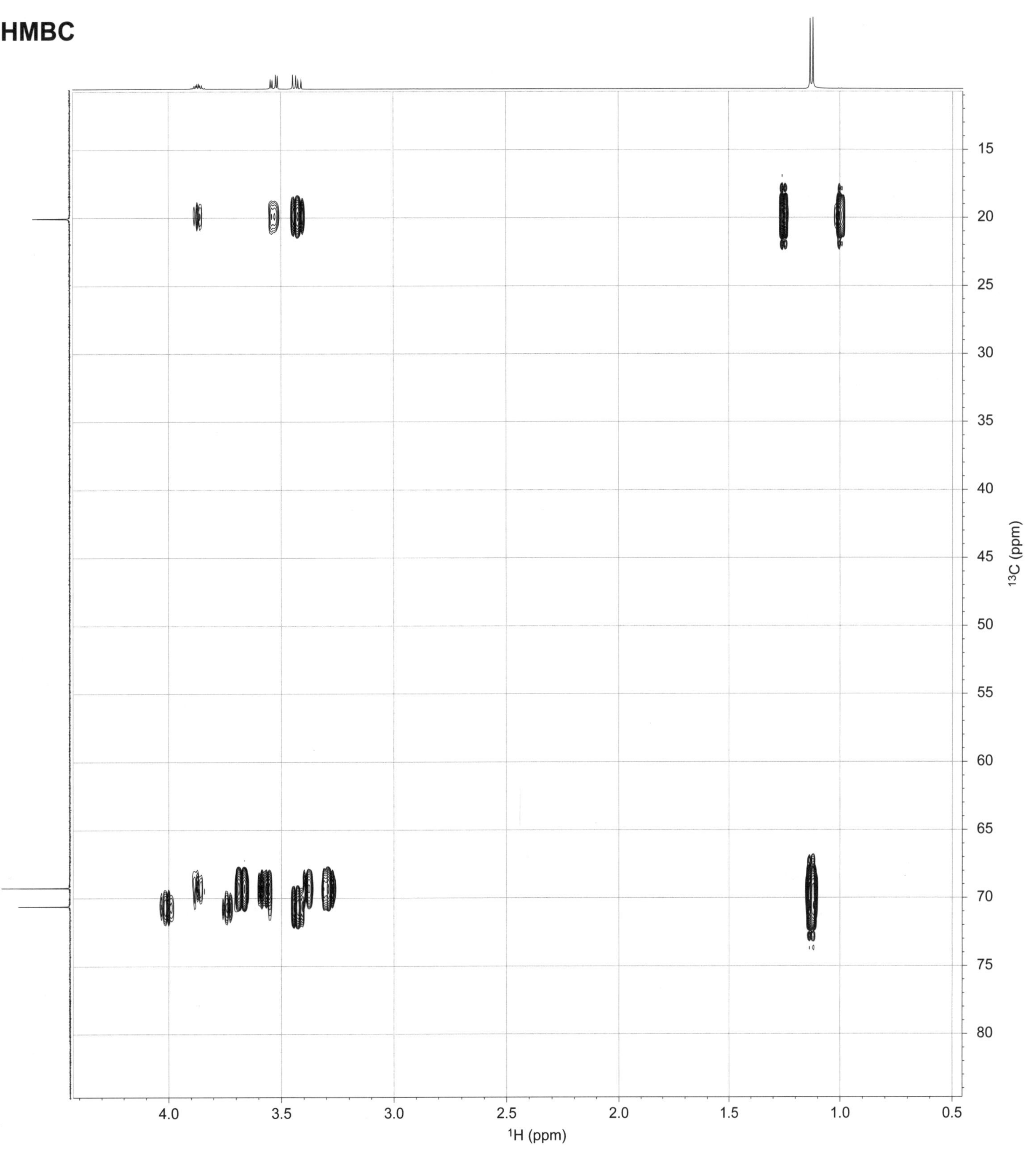

Problem 9.23 Molecular formula: $C_5H_9NO_4$ ***Difficulty*** ★☆☆☆

Solvent: D_2O

Field: 500 MHz (^{1}H); 125 MHz (^{13}C)

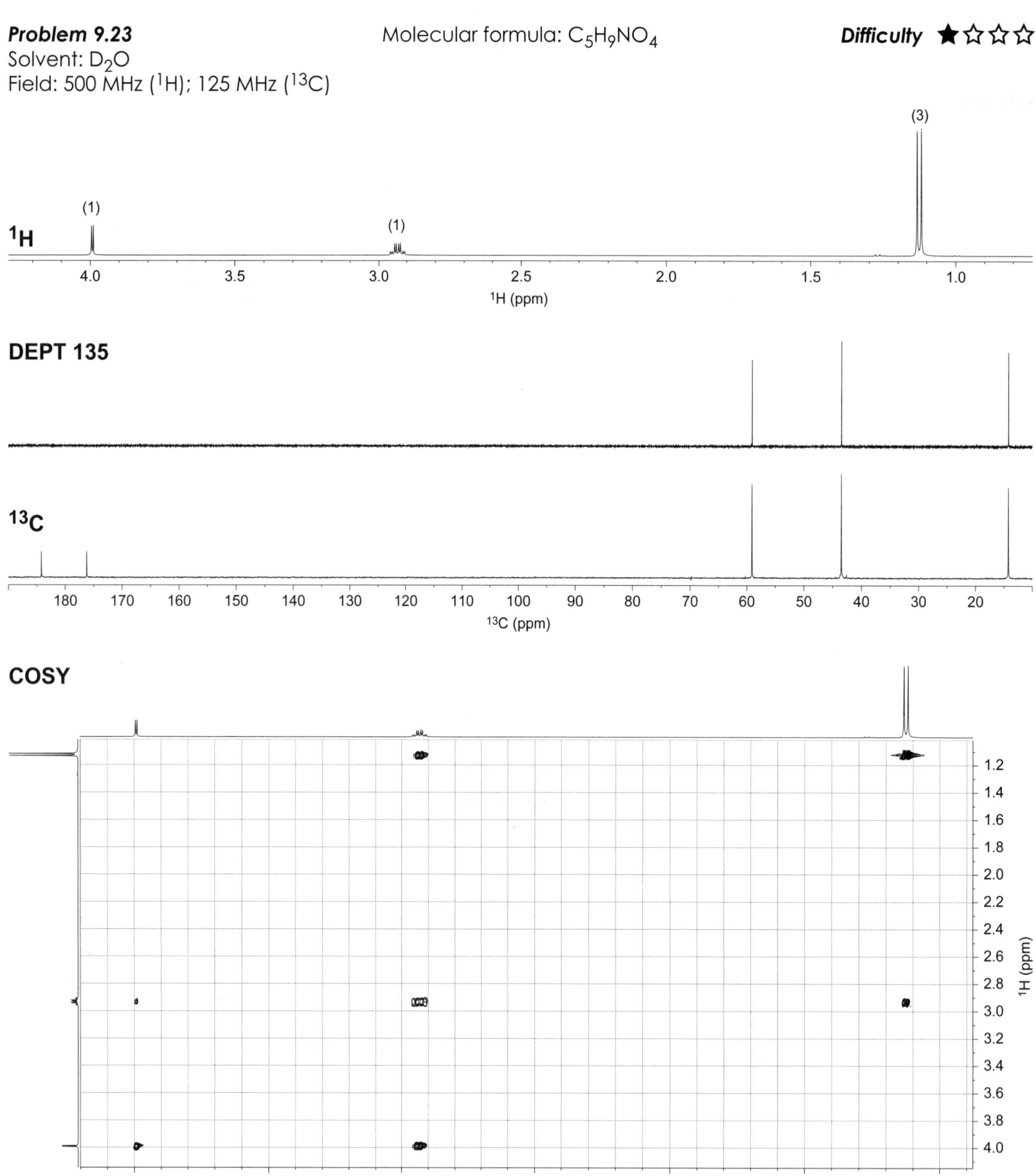

HSQC

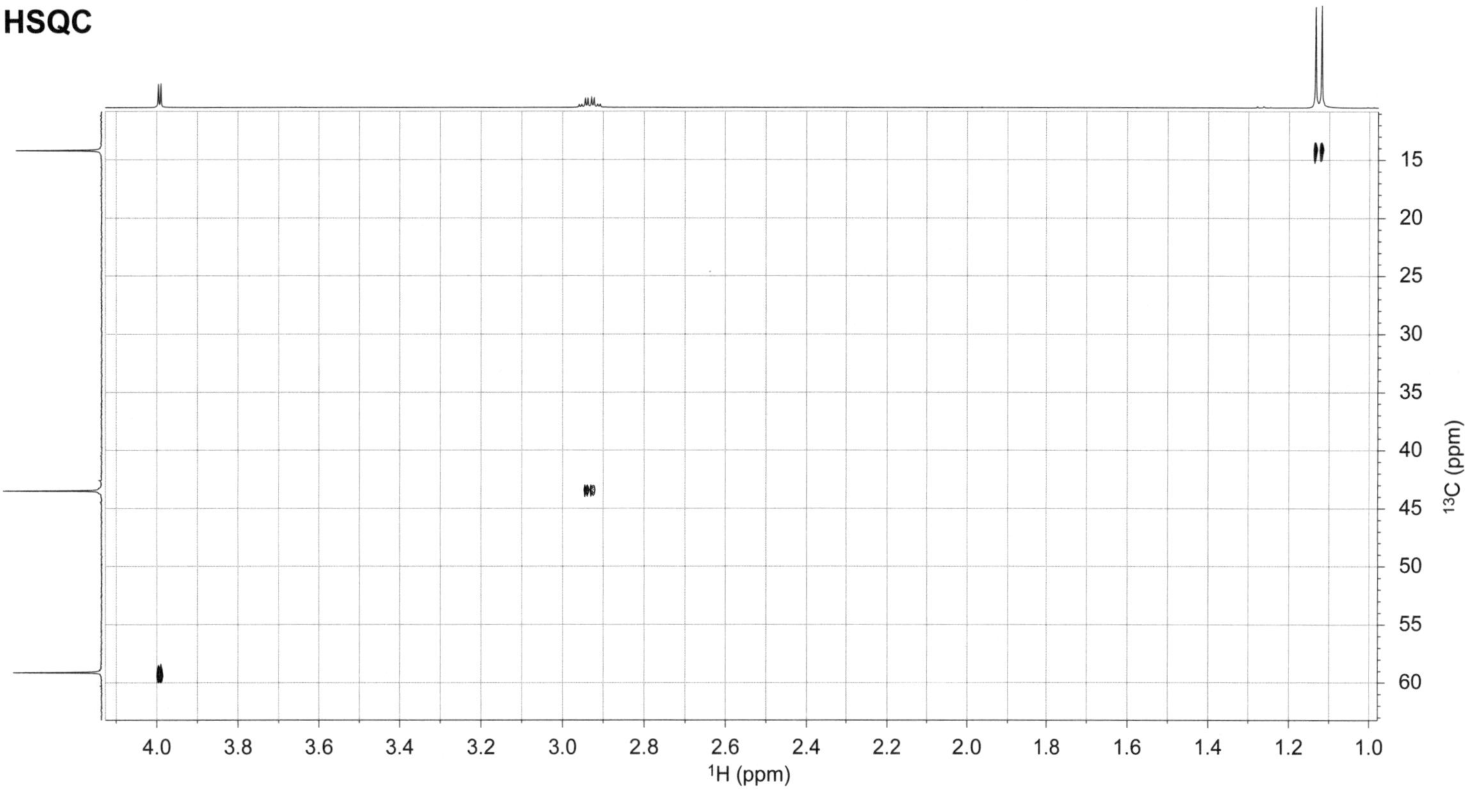

HMBC

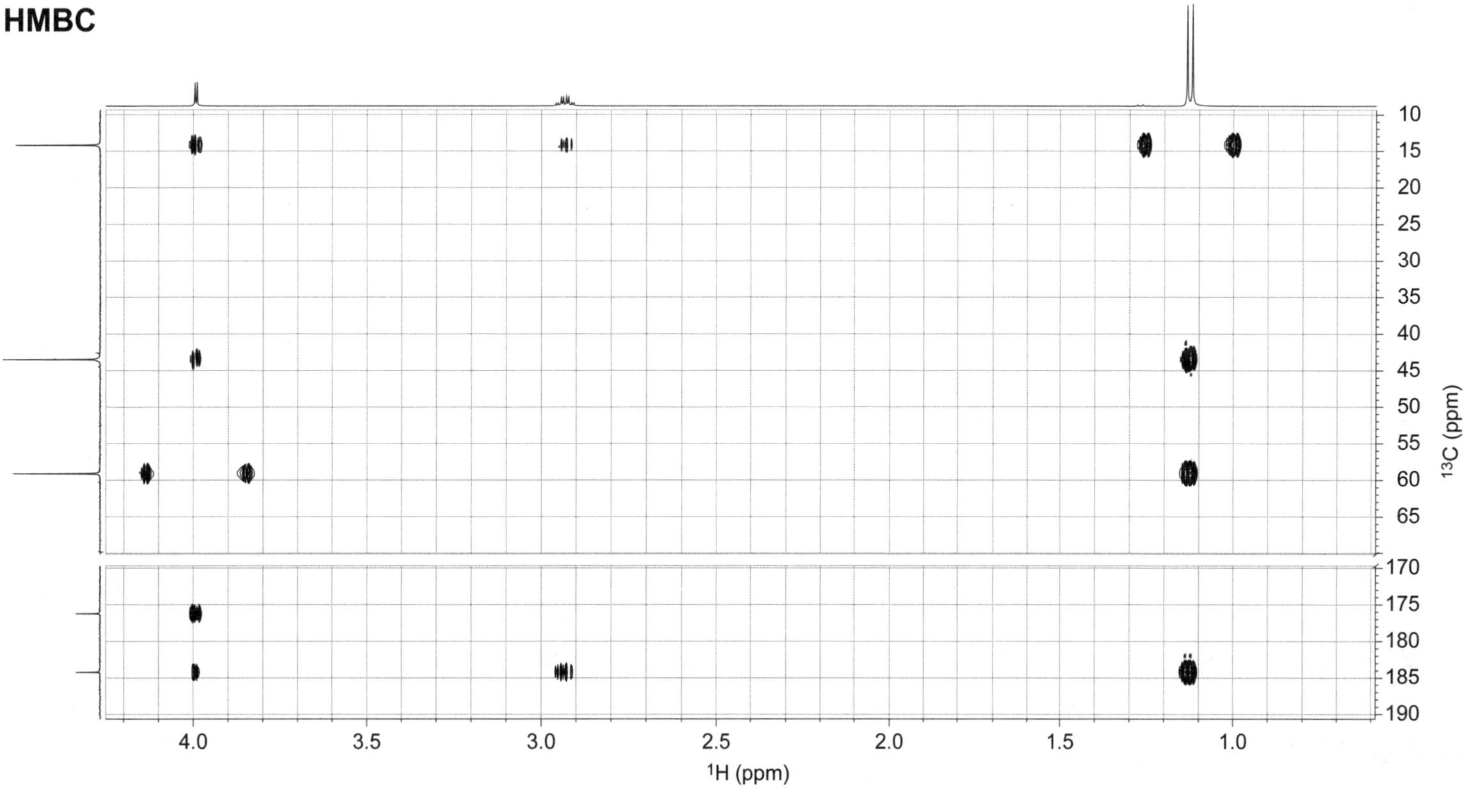

Problem 9.24 Molecular formula: $C_7H_{12}O_4$ ***Difficulty*** ★☆☆☆

Solvent: D_2O

Field: 500 MHz (^1H); 125 MHz (^{13}C)

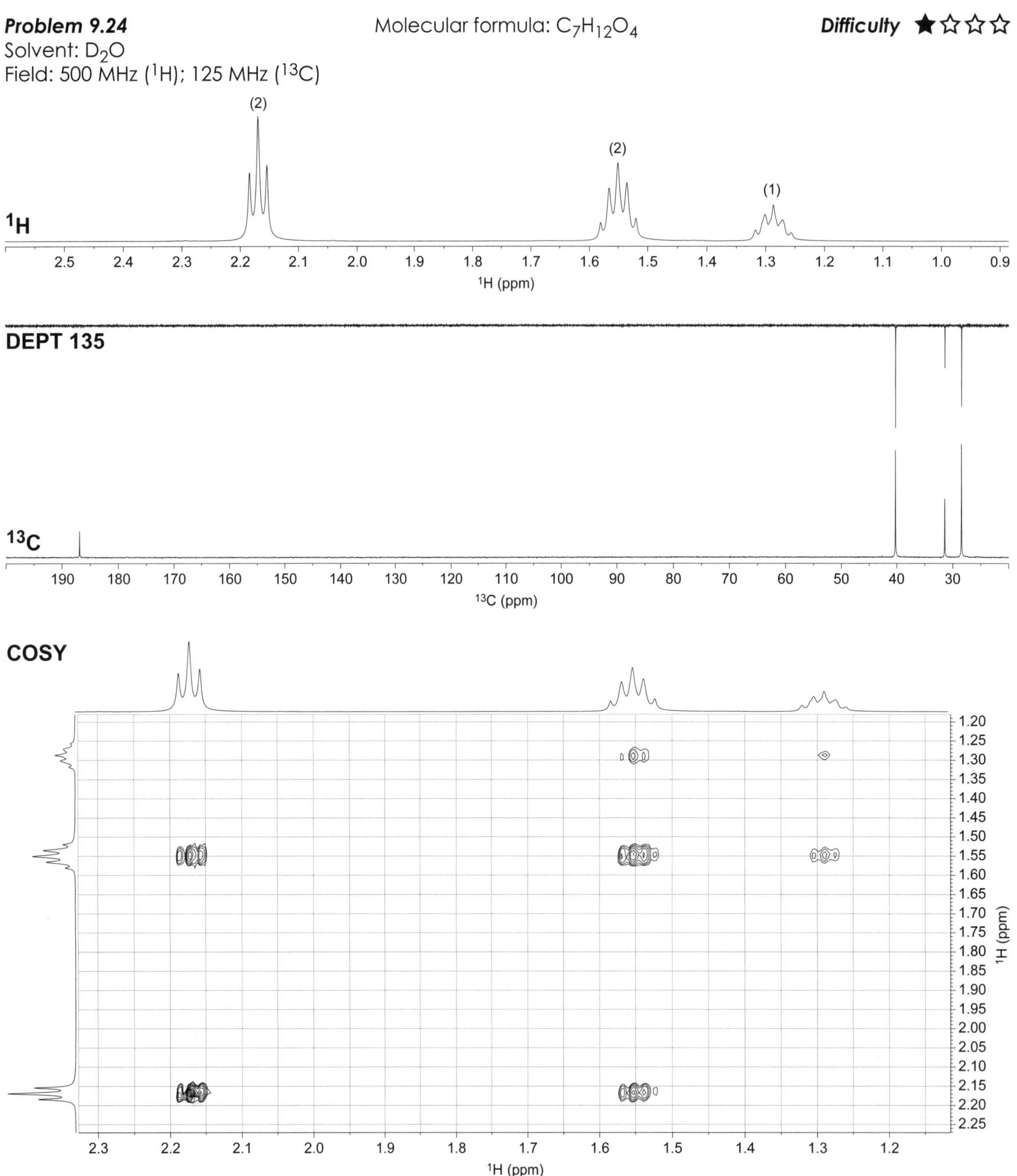

HSQC

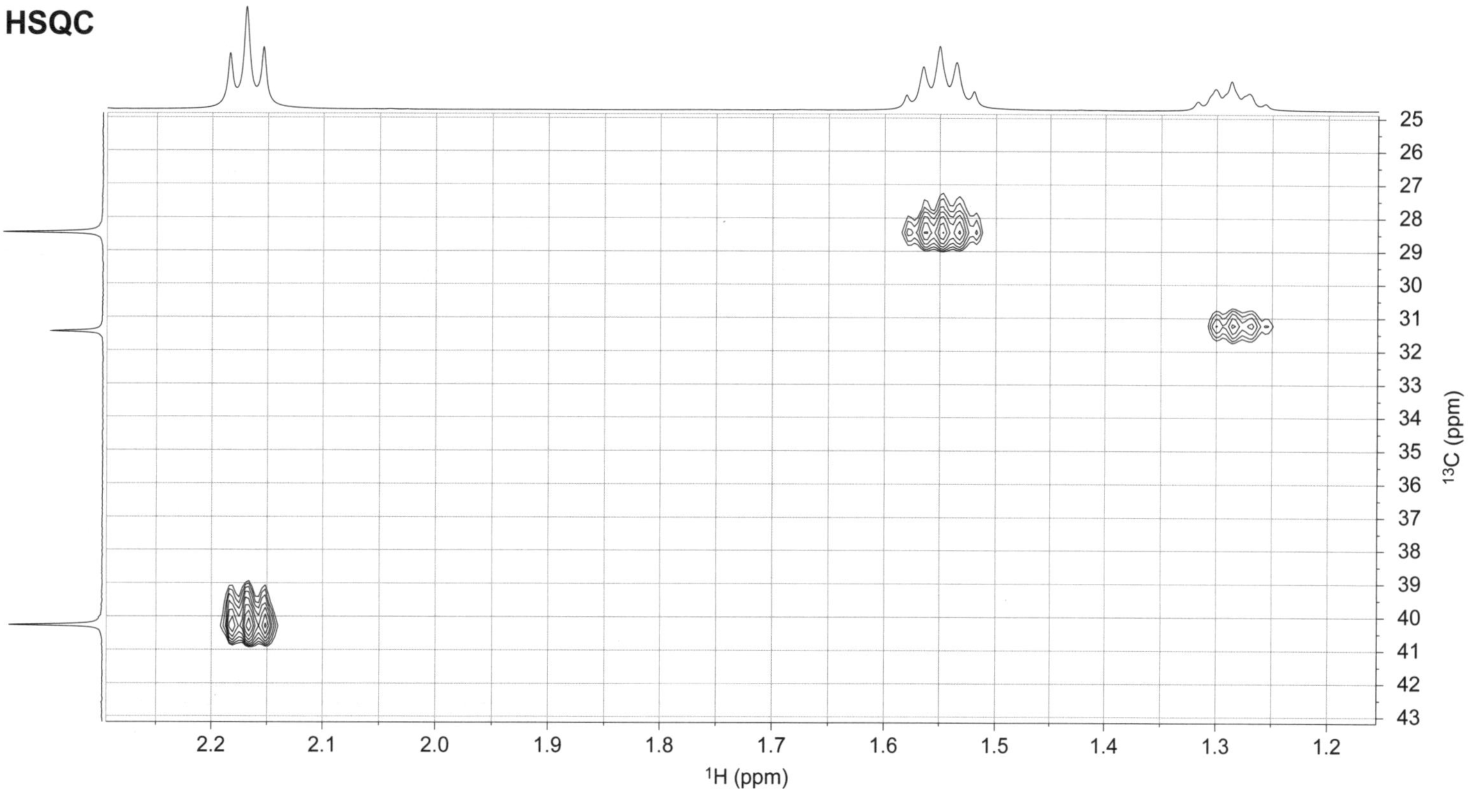

25
26
27
28
29
30
31
32
33
34
35
36
37
38
39
40
41
42
43
^{13}C (ppm)
2.2
2.1
2.0
1.9
1.8
1.7
1.6
1.5
1.4
1.3
1.2
^{1}H (ppm)

HMBC

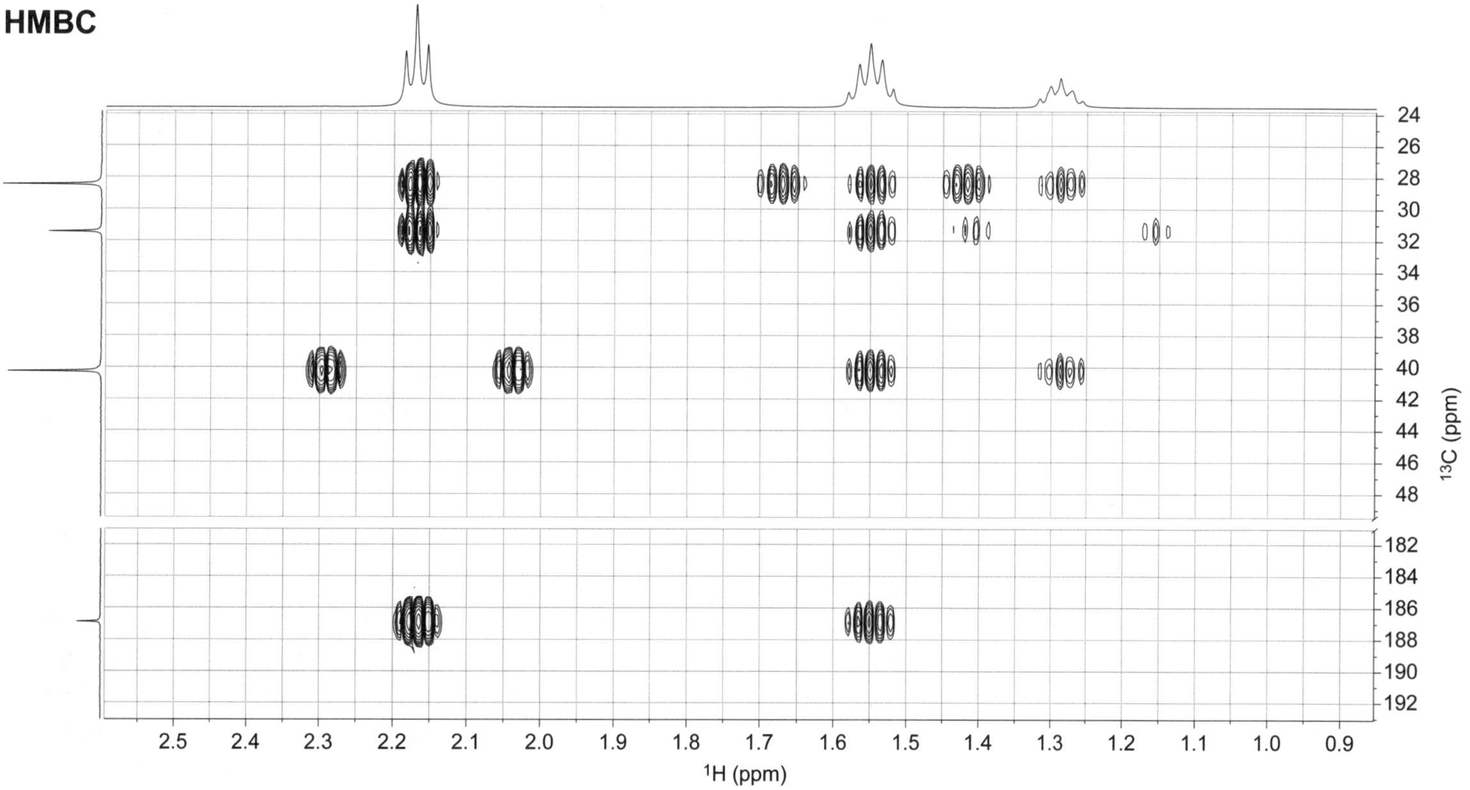

24
26
28
30
32
34
36
38
40
42
44
46
48
182
184
186
188
190
192
^{13}C (ppm)
2.5
2.4
2.3
2.2
2.1
2.0
1.9
1.8
1.7
1.6
1.5
1.4
1.3
1.2
1.1
1.0
0.9
^{1}H (ppm)

Problem 9.25 Molecular formula: $C_4H_9NO_2$ ***Difficulty*** ★☆☆☆

Solvent: D_2O

Field: 500 MHz (^1H); 125 MHz (^{13}C)

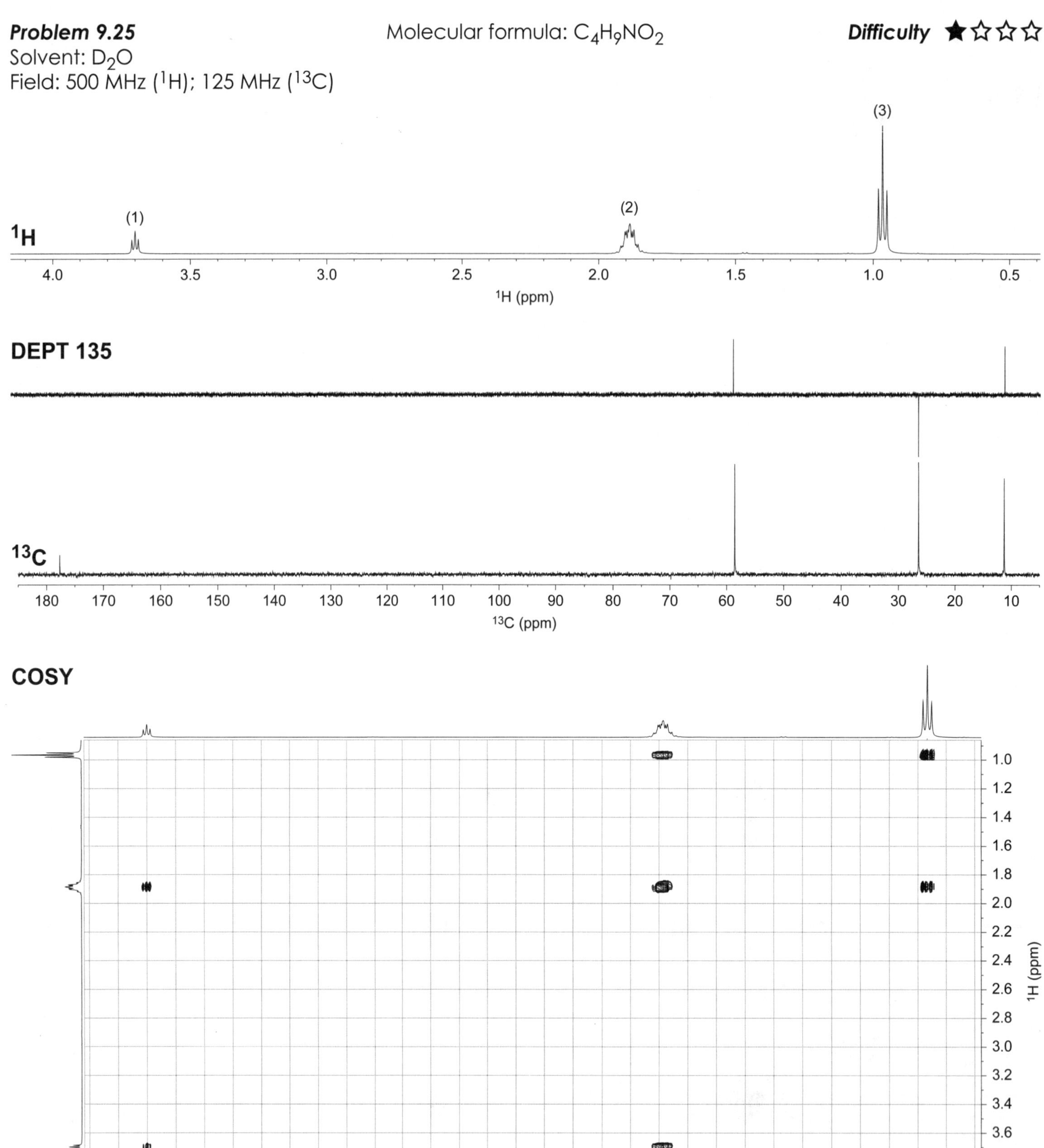

HSQC

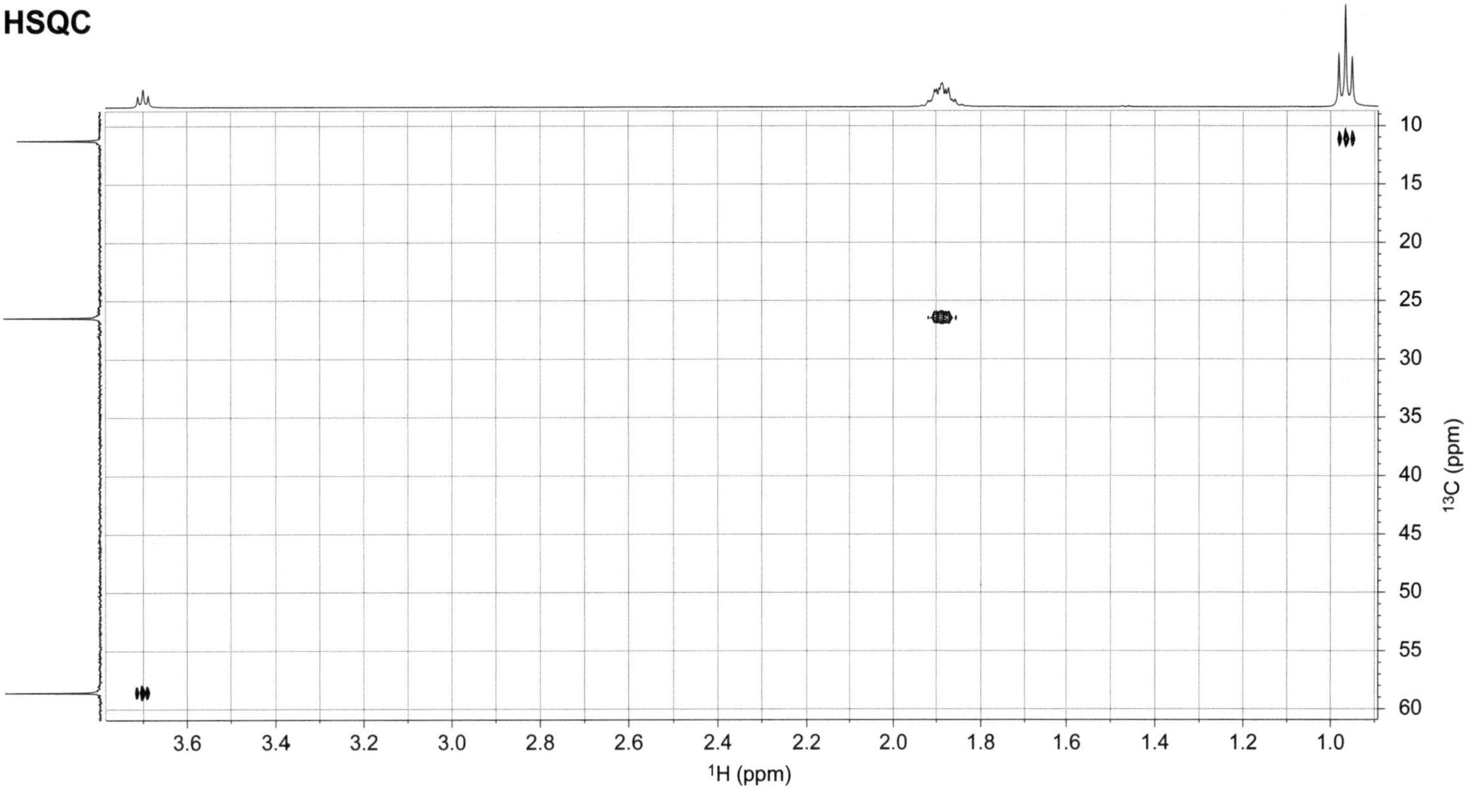

^1H (ppm) axis: 3.6, 3.4, 3.2, 3.0, 2.8, 2.6, 2.4, 2.2, 2.0, 1.8, 1.6, 1.4, 1.2, 1.0

^{13}C (ppm) axis: 10, 15, 20, 25, 30, 35, 40, 45, 50, 55, 60

HMBC

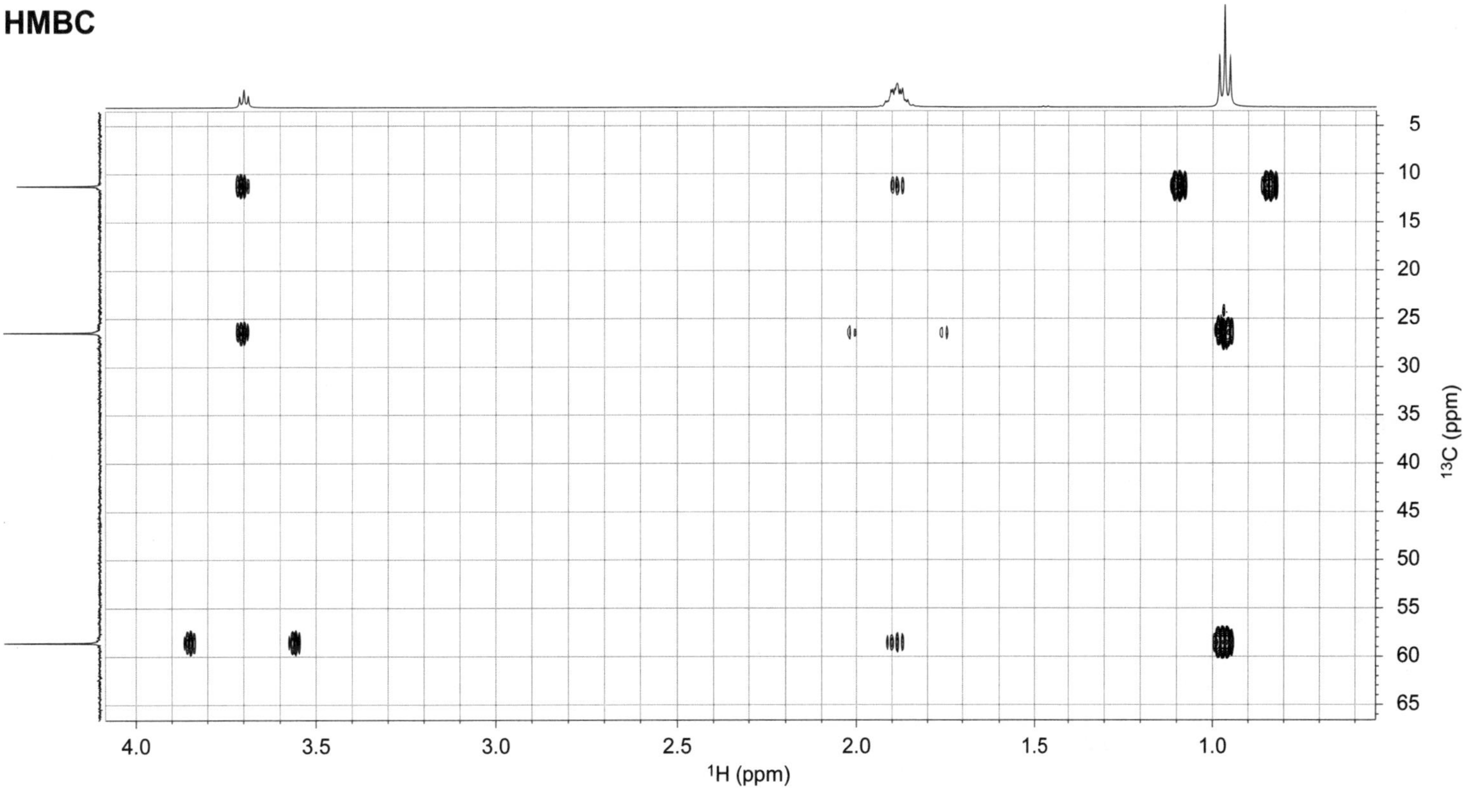

^1H (ppm) axis: 4.0, 3.5, 3.0, 2.5, 2.0, 1.5, 1.0

^{13}C (ppm) axis: 5, 10, 15, 20, 25, 30, 35, 40, 45, 50, 55, 60, 65

Problem 9.26
Solvent: D_2O
Field: 500 MHz (^1H); 125 MHz (^{13}C)

Molecular formula: $C_6H_{13}NO_2$

Difficulty ★☆☆☆

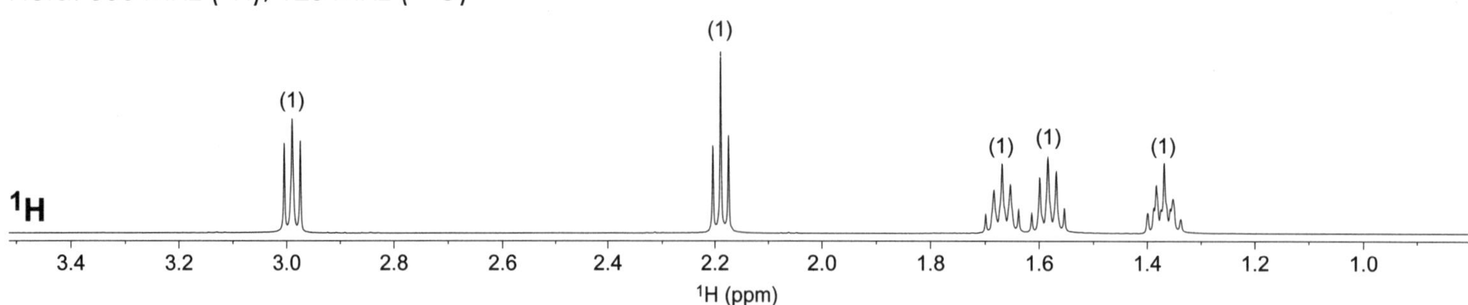

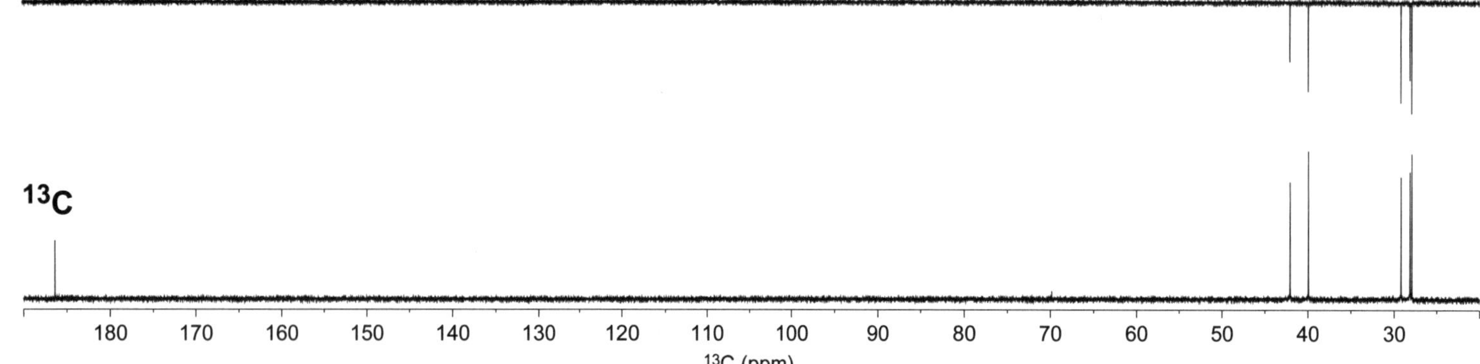

COSY

3.2 3.0 2.8 2.6 2.4 2.2 2.0 1.8 1.6 1.4 1.2

^1H (ppm)

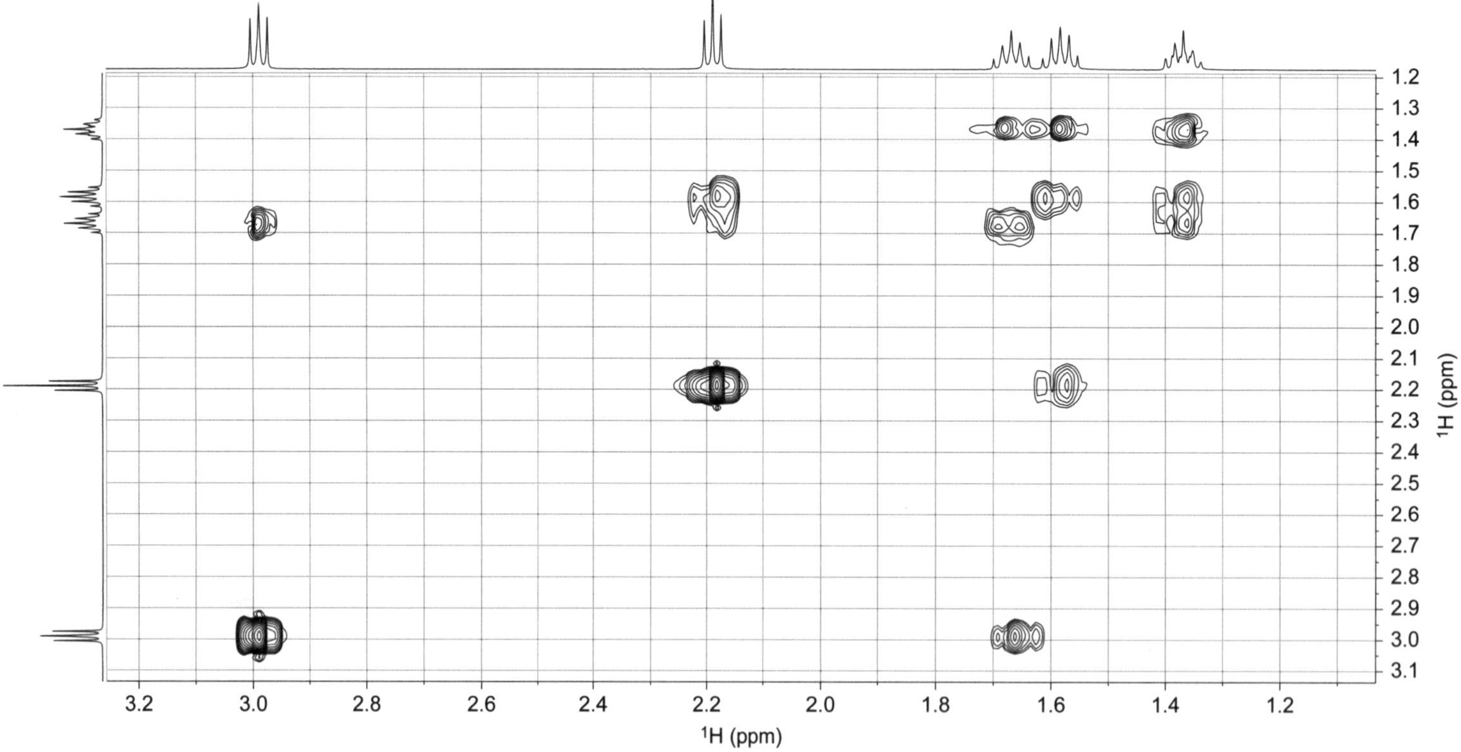

HSQC

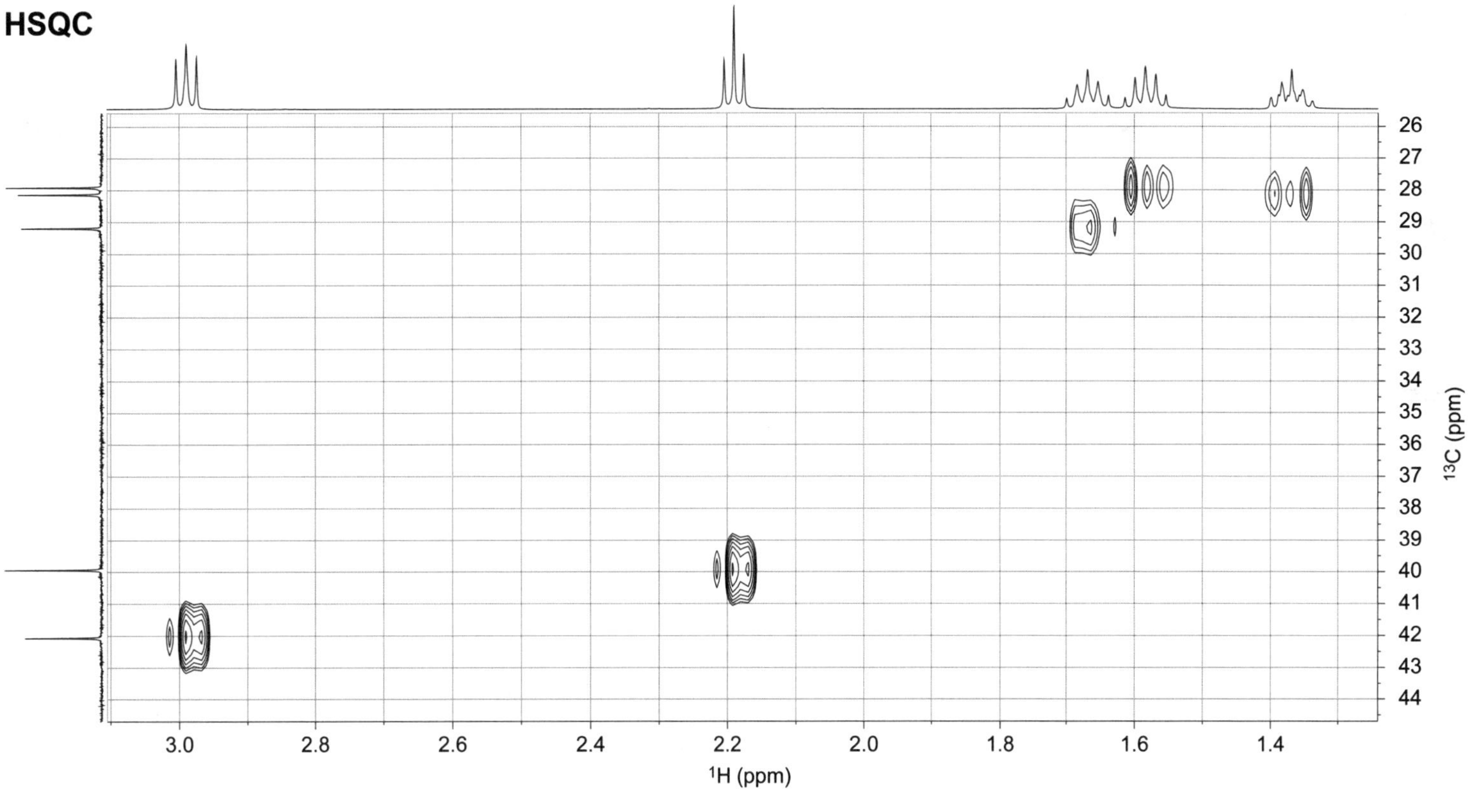

26 27 28 29 30 31 32 33 34 35 36 37 38 39 40 41 42 43 44

^{13}C (ppm)

3.0 2.8 2.6 2.4 2.2 2.0 1.8 1.6 1.4

^{1}H (ppm)

HMBC

22 24 26 28 30 32 34 36 38 40 42 44 46

184 186 188 190 192

^{13}C (ppm)

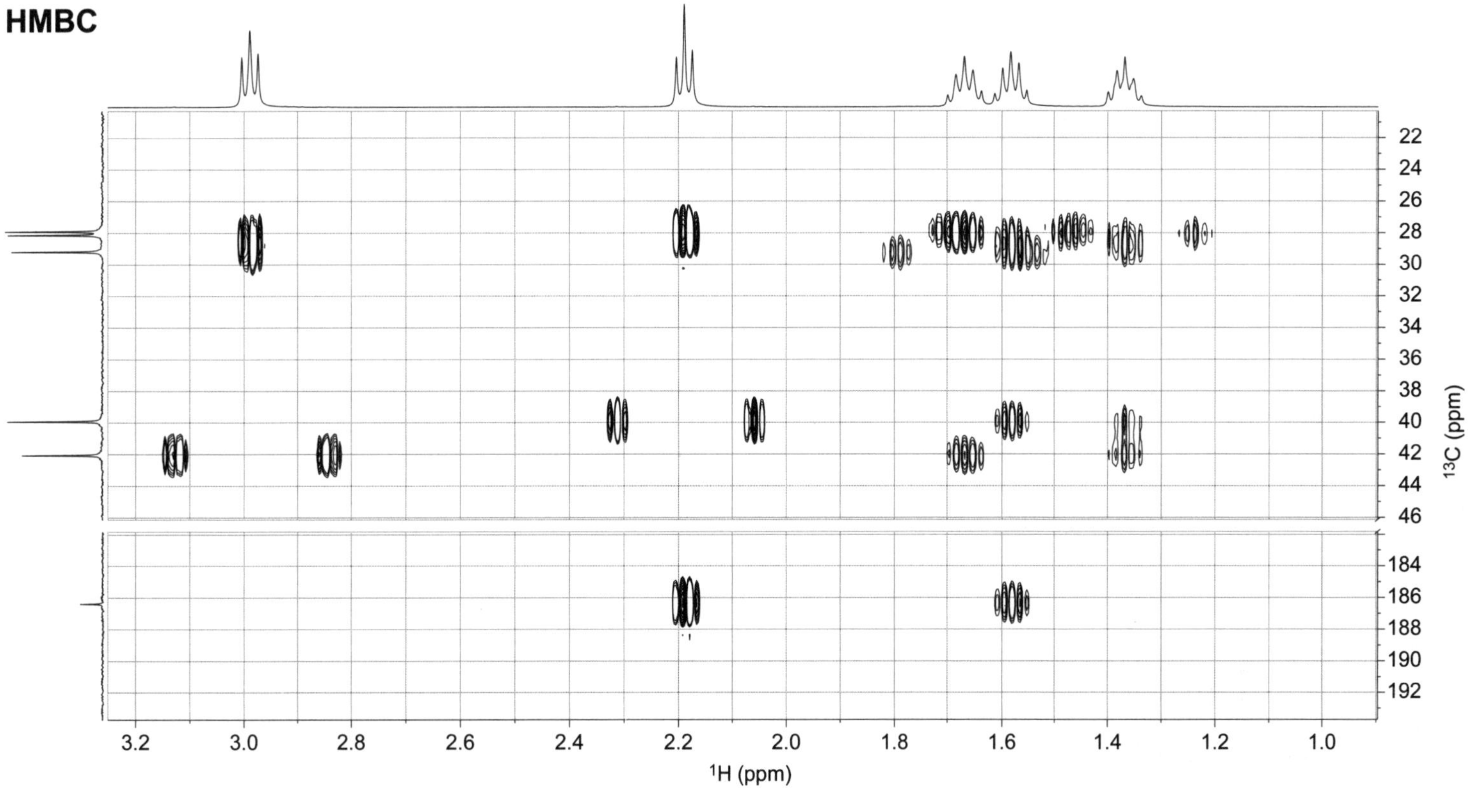

3.2 3.0 2.8 2.6 2.4 2.2 2.0 1.8 1.6 1.4 1.2 1.0

^{1}H (ppm)

Problem 9.27
Solvent: D_2O
Field: 500 MHz (^1H); 125 MHz (^{13}C)

Molecular formula: $C_8H_7O_2Cl$

Difficulty ★☆☆☆

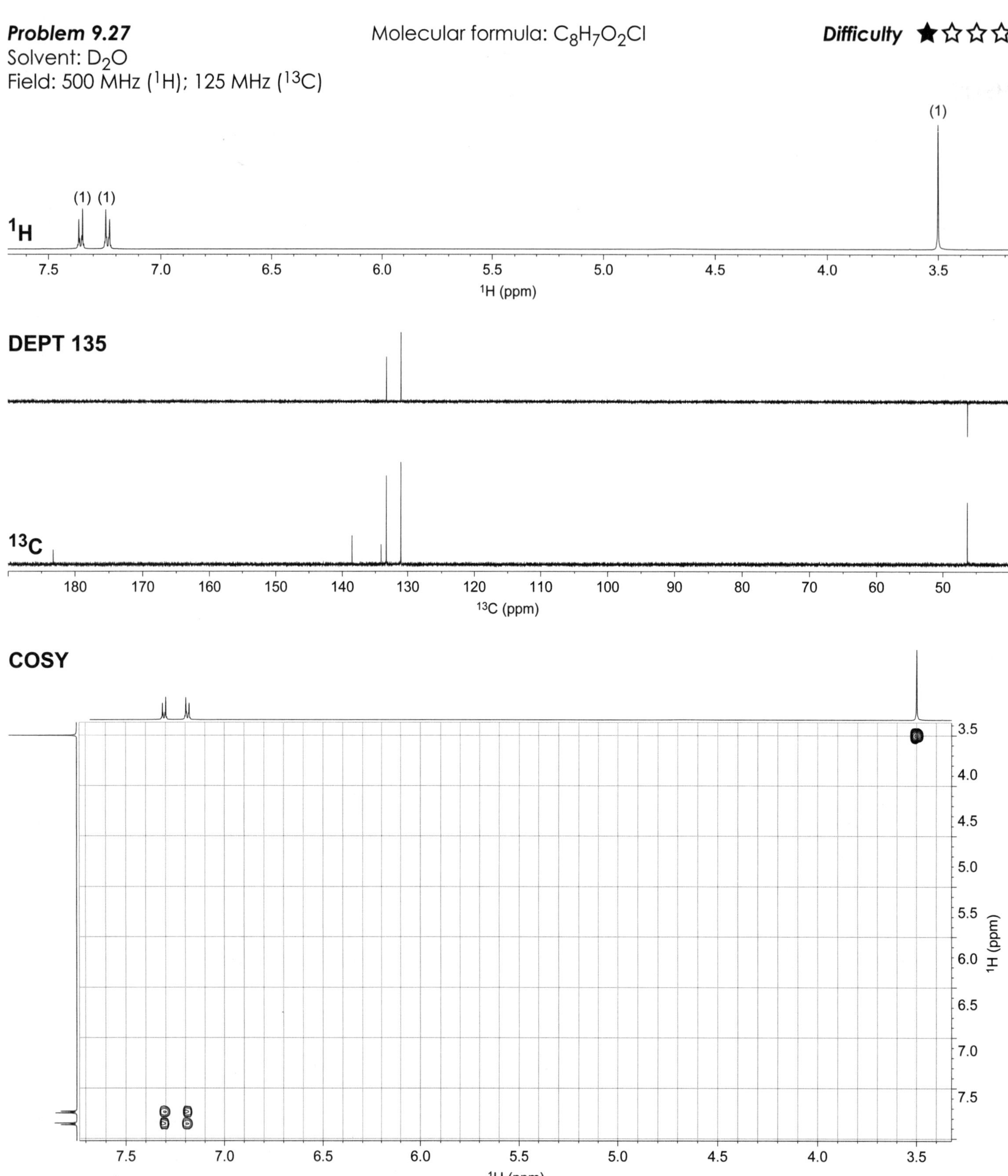

HSQC

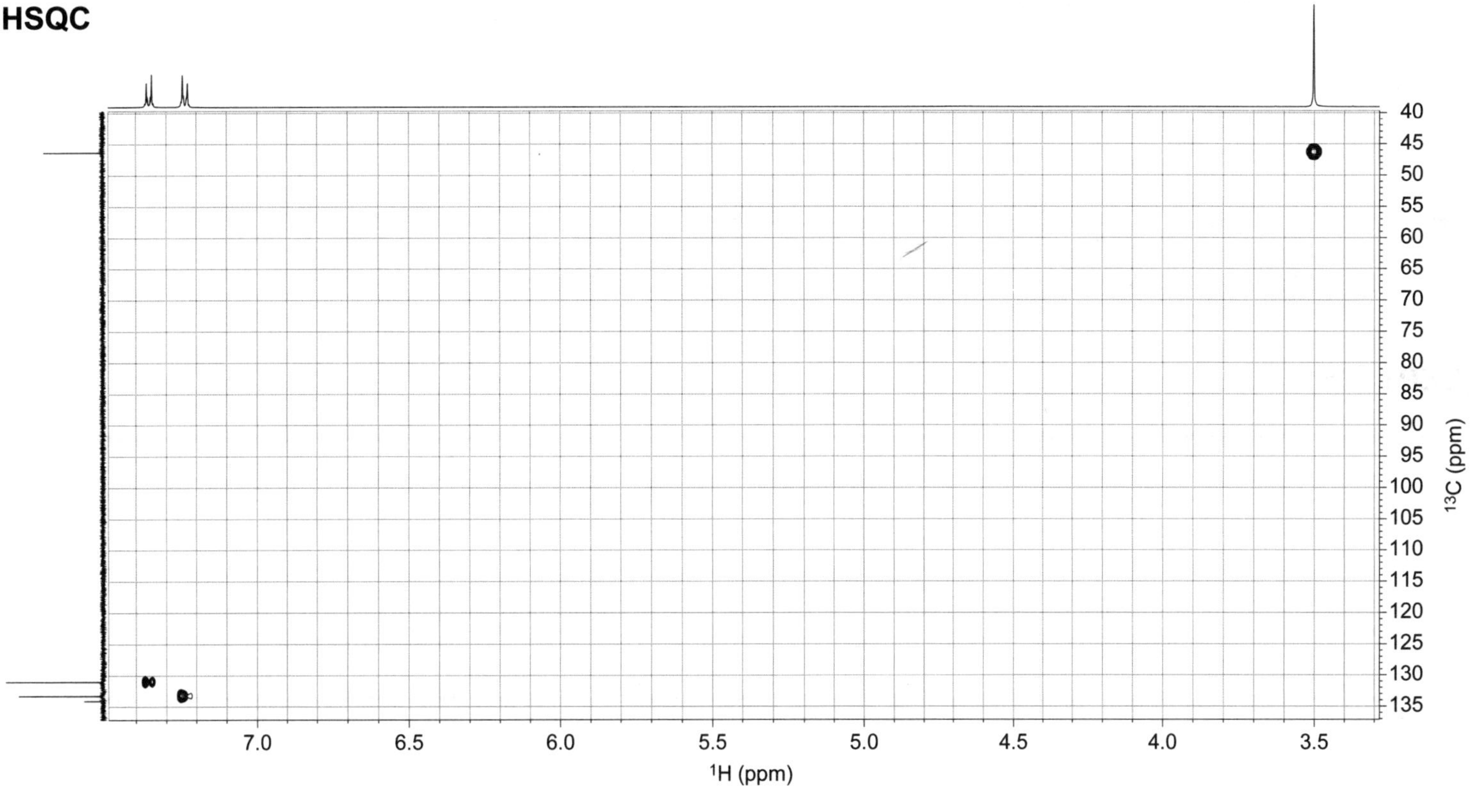
40
45
50
55
60
65
70
75
80
85
90
95
100
105
110
115
120
125
130
135
13C (ppm)
7.0
6.5
6.0
5.5
5.0
4.5
4.0
3.5
1H (ppm)

HMBC

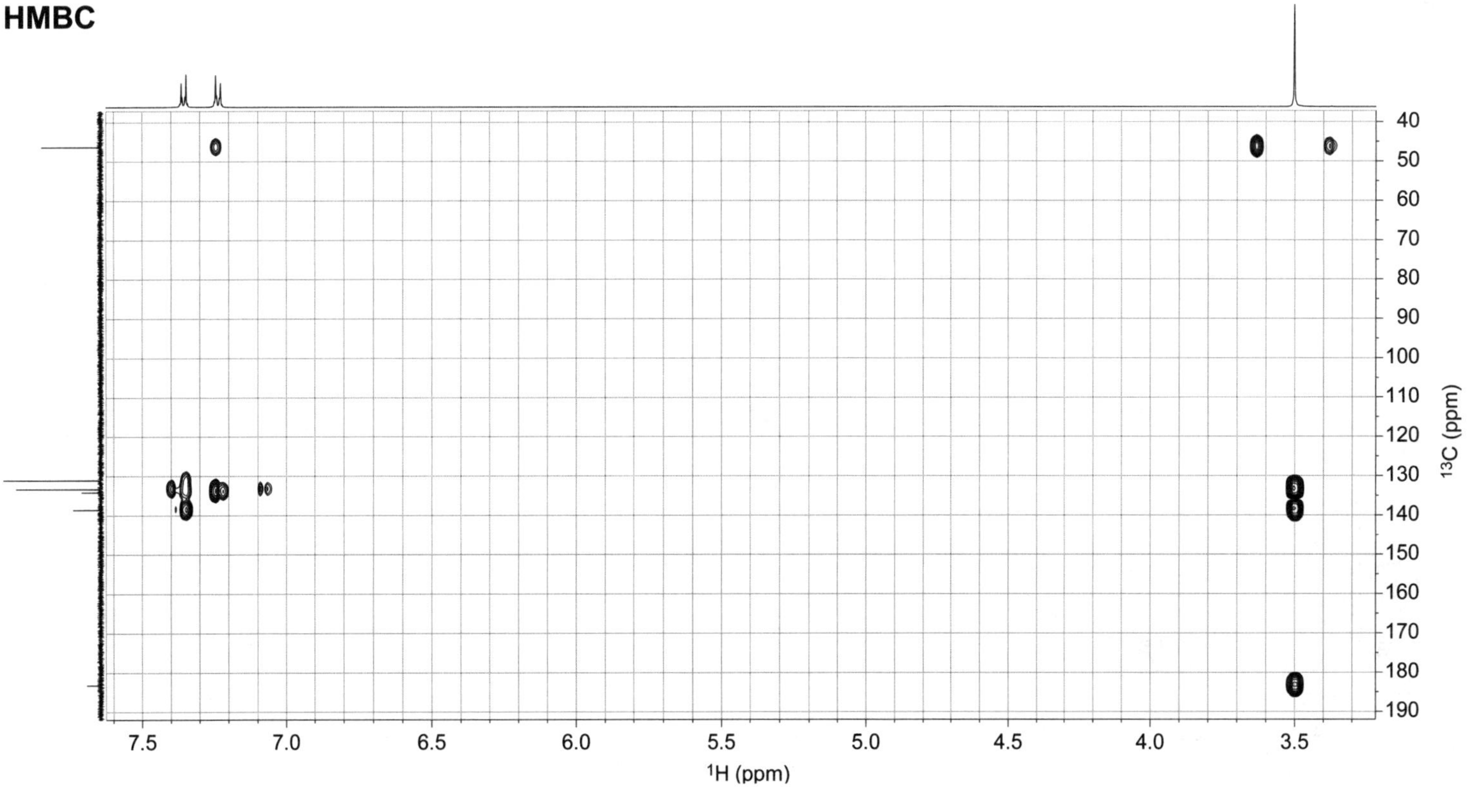
40
50
60
70
80
90
100
110
120
130
140
150
160
170
180
190
13C (ppm)
7.5
7.0
6.5
6.0
5.5
5.0
4.5
4.0
3.5
1H (ppm)

Problem 9.28 Molecular formula: $C_4H_8O_2$ ***Difficulty*** ★☆☆☆

Solvent: D_2O

Field: 500 MHz (^1H); 125 MHz (^{13}C)

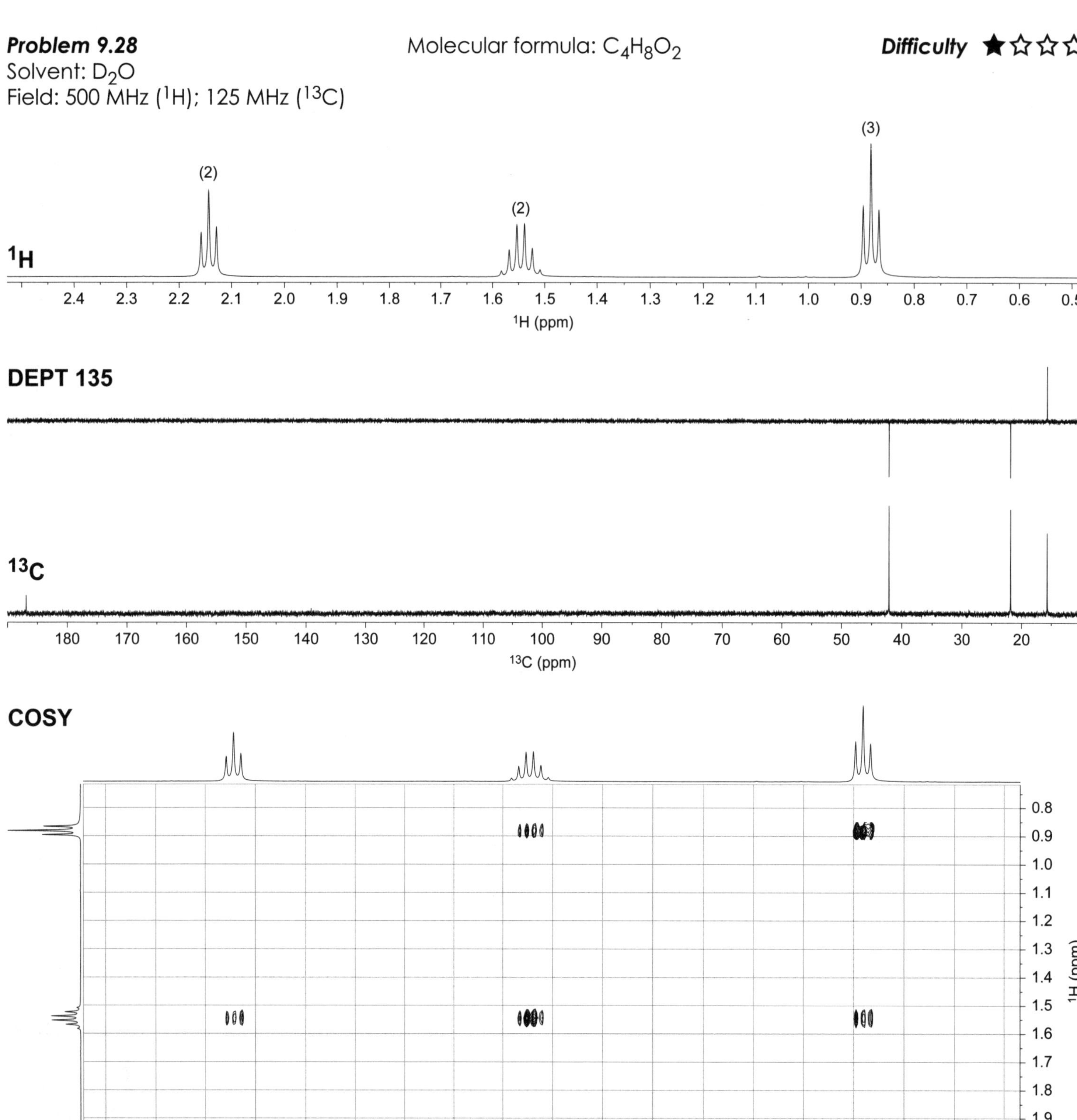

HSQC

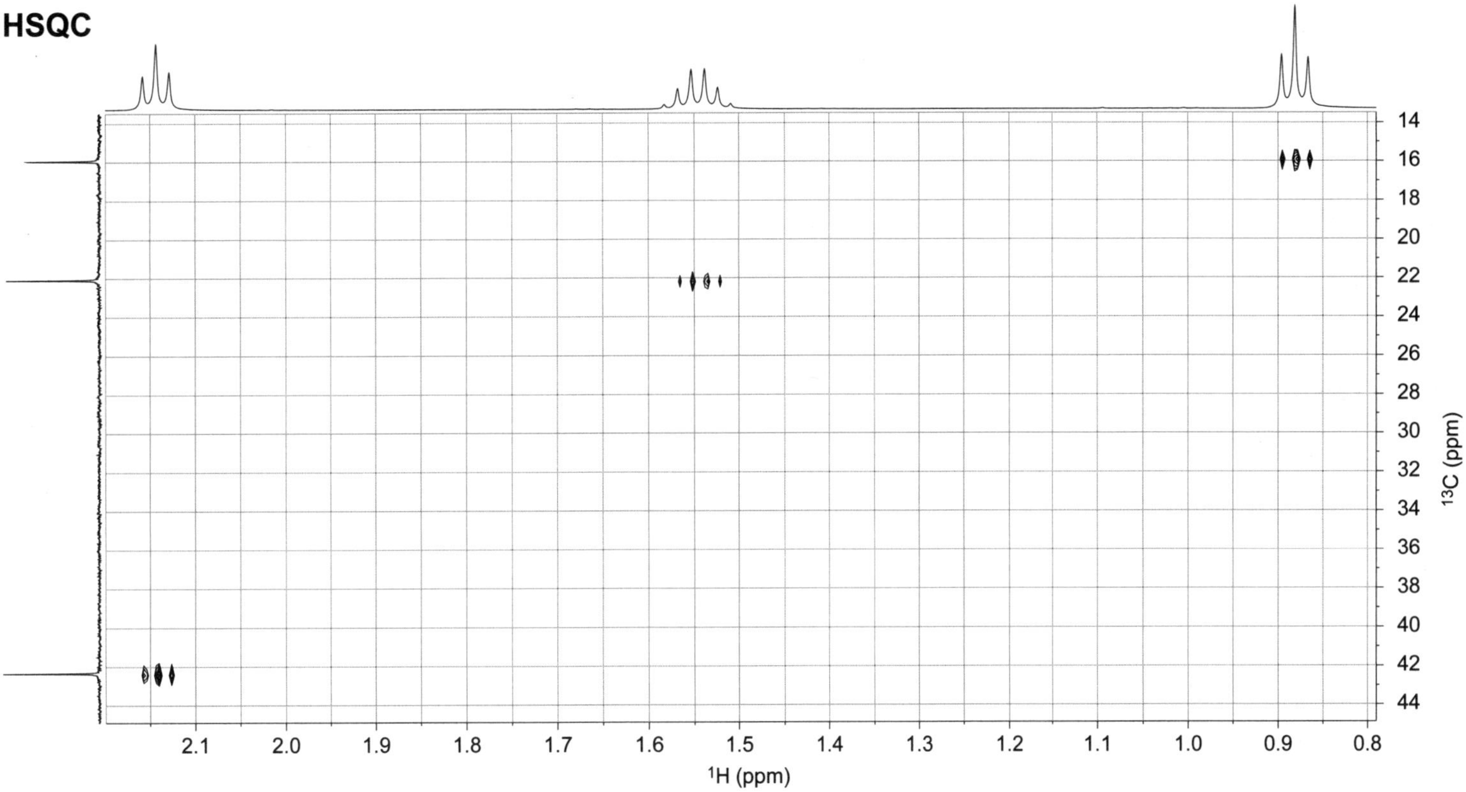
14
16
18
20
22
24
26
28
30
32
34
36
38
40
42
44
2.1
2.0
1.9
1.8
1.7
1.6
1.5
1.4
1.3
1.2
1.1
1.0
0.9
0.8
1H (ppm)
13C (ppm)

HMBC

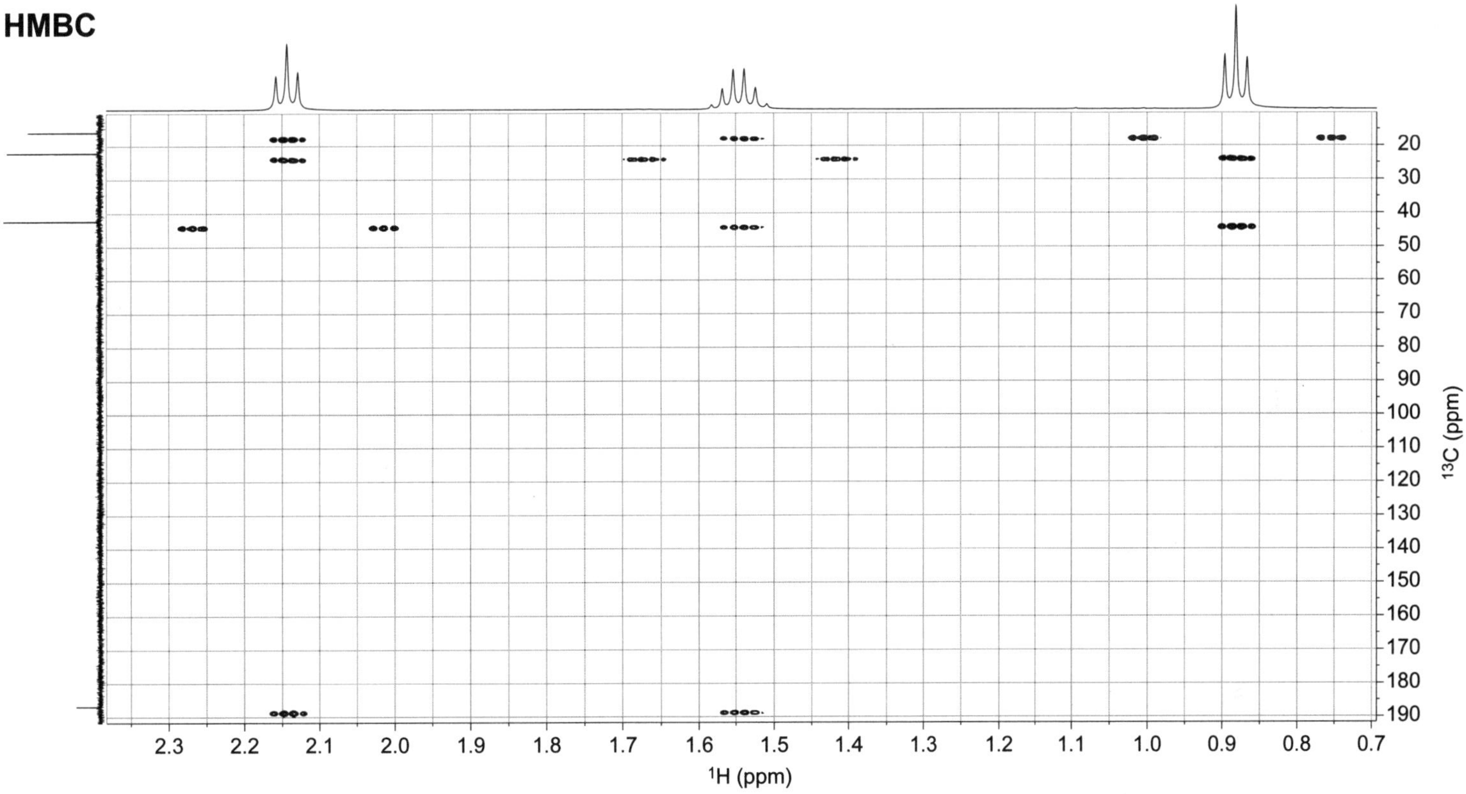
20
30
40
50
60
70
80
90
100
110
120
130
140
150
160
170
180
190
2.3
2.2
2.1
2.0
1.9
1.8
1.7
1.6
1.5
1.4
1.3
1.2
1.1
1.0
0.9
0.8
0.7
1H (ppm)
13C (ppm)

Problem 9.29

Difficulty ★☆☆☆

Solvent: D_2O
Field: 500 MHz (^1H); 125 MHz (^{13}C)

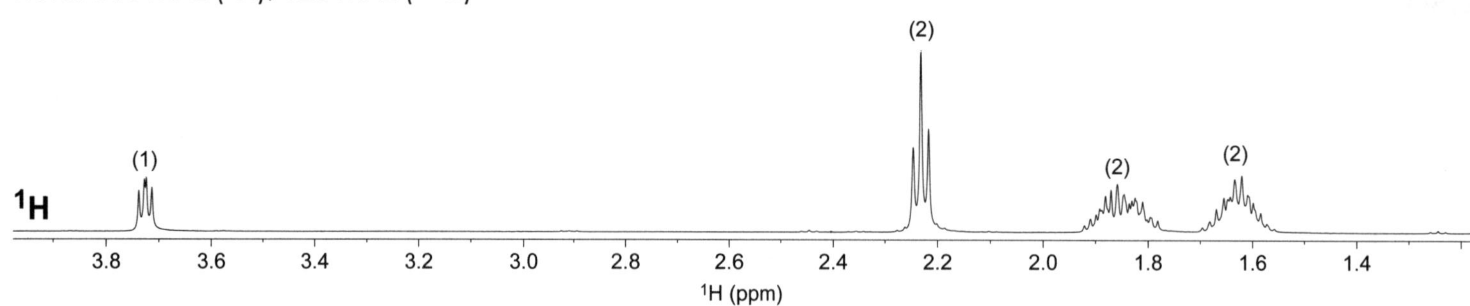

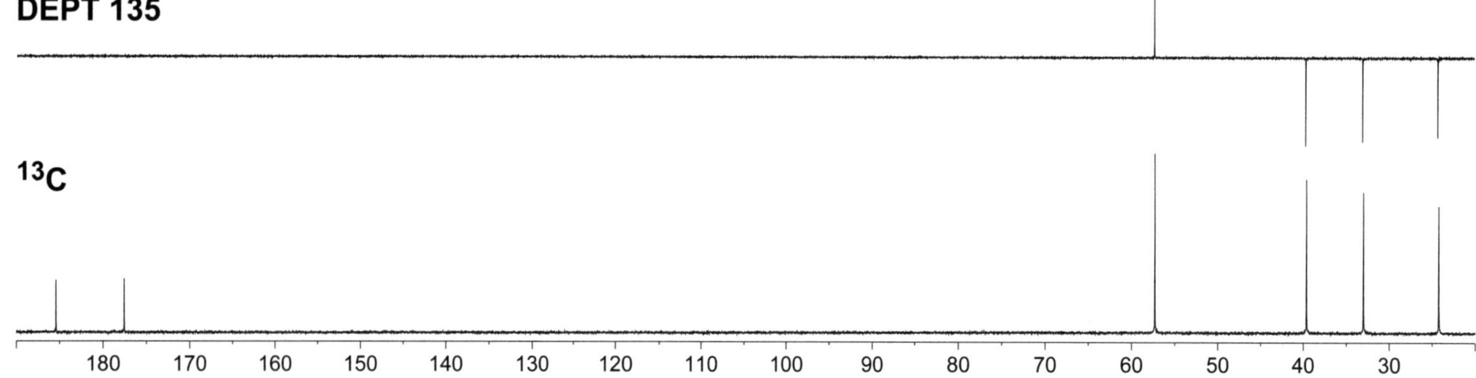

COSY

3.8 3.6 3.4 3.2 3.0 2.8 2.6 2.4 2.2 2.0 1.8 1.6 1.4

^1H (ppm)

1.6 1.7 1.8 1.9 2.0 2.1 2.2 2.3 2.4 2.5 2.6 2.7 2.8 2.9 3.0 3.1 3.2 3.3 3.4 3.5 3.6 3.7 3.8

^1H (ppm)

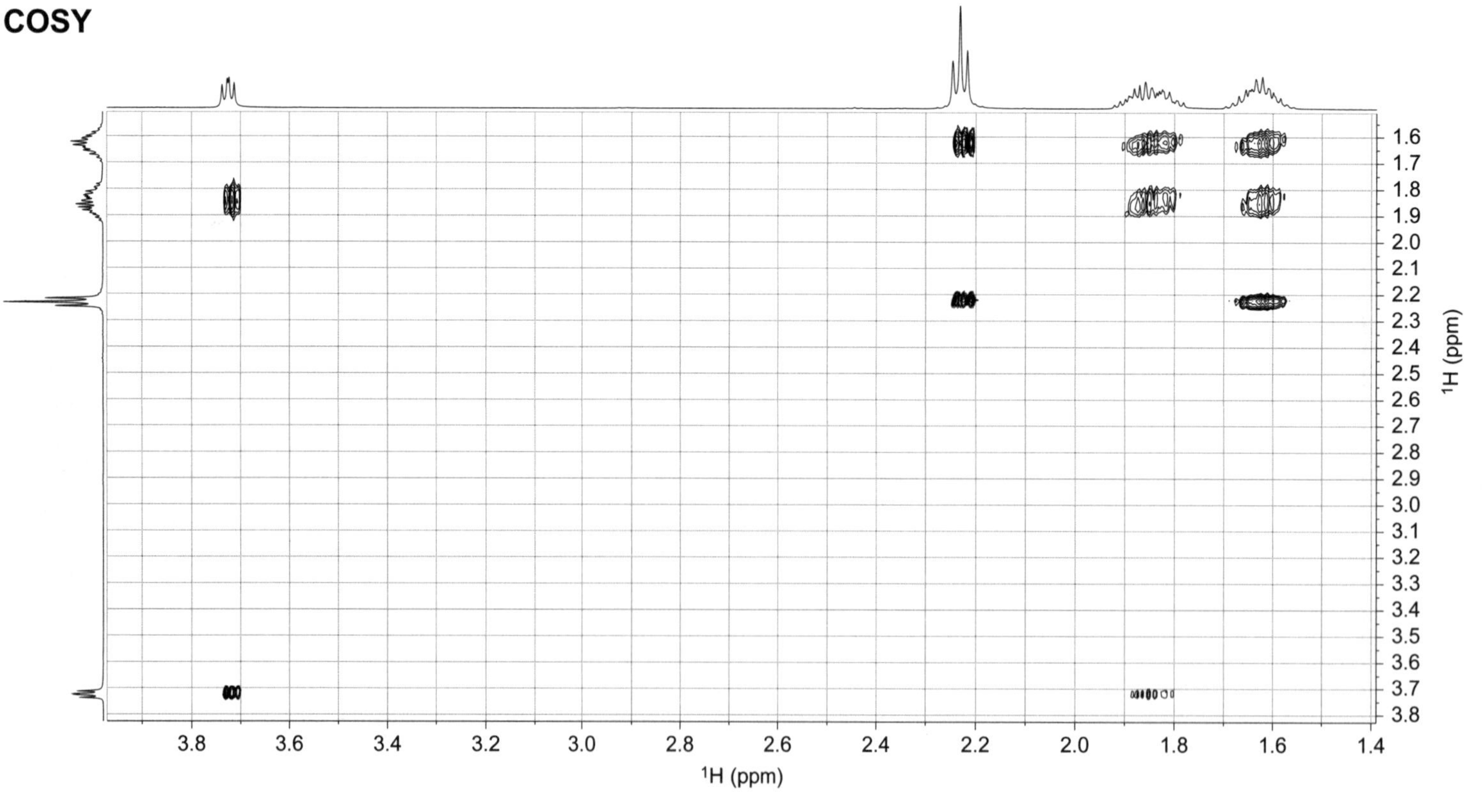

HSQC

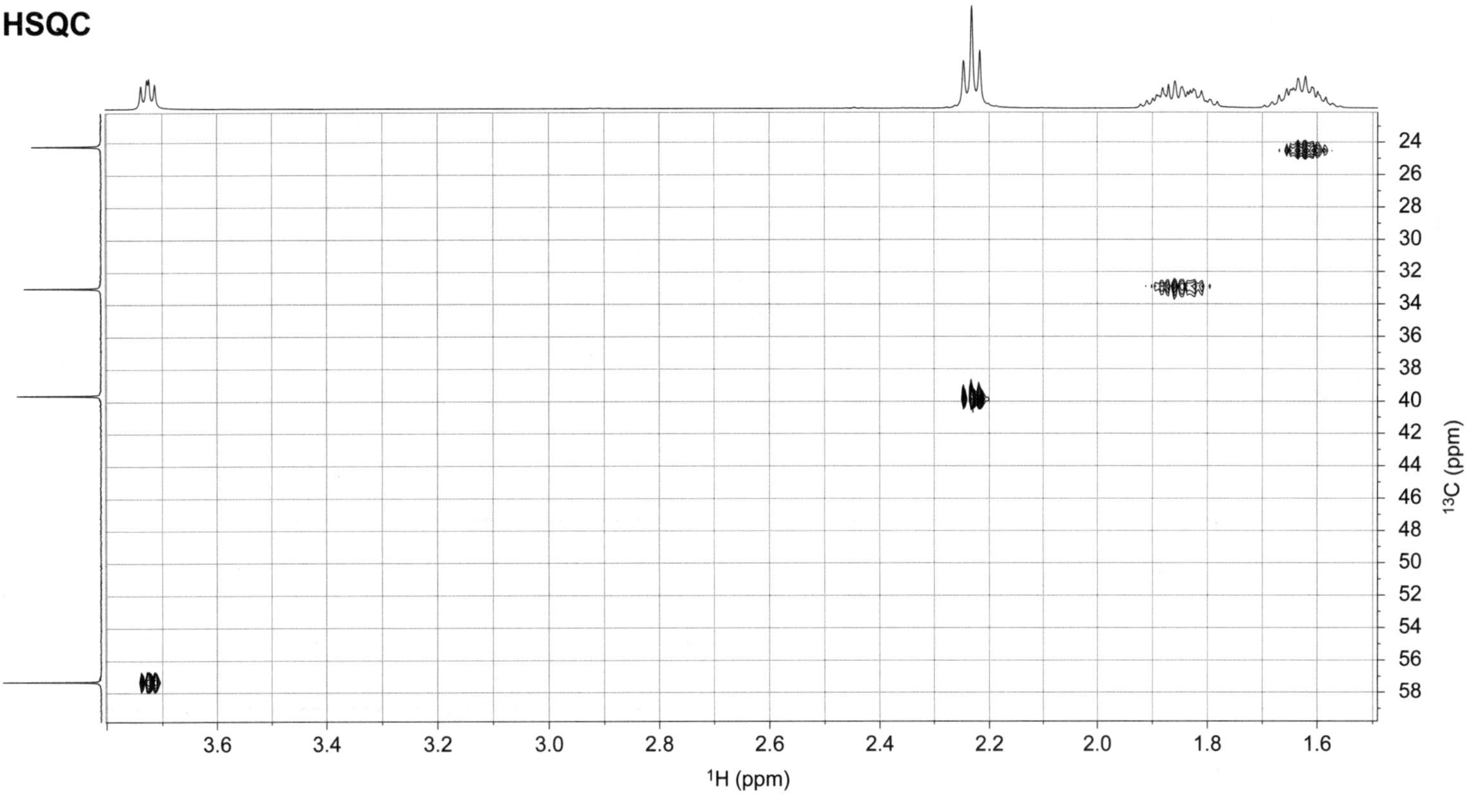

24
26
28
30
32
34
36
38
40
42
44
46
48
50
52
54
56
58
¹³C (ppm)
3.6
3.4
3.2
3.0
2.8
2.6
2.4
2.2
2.0
1.8
1.6
¹H (ppm)

HMBC

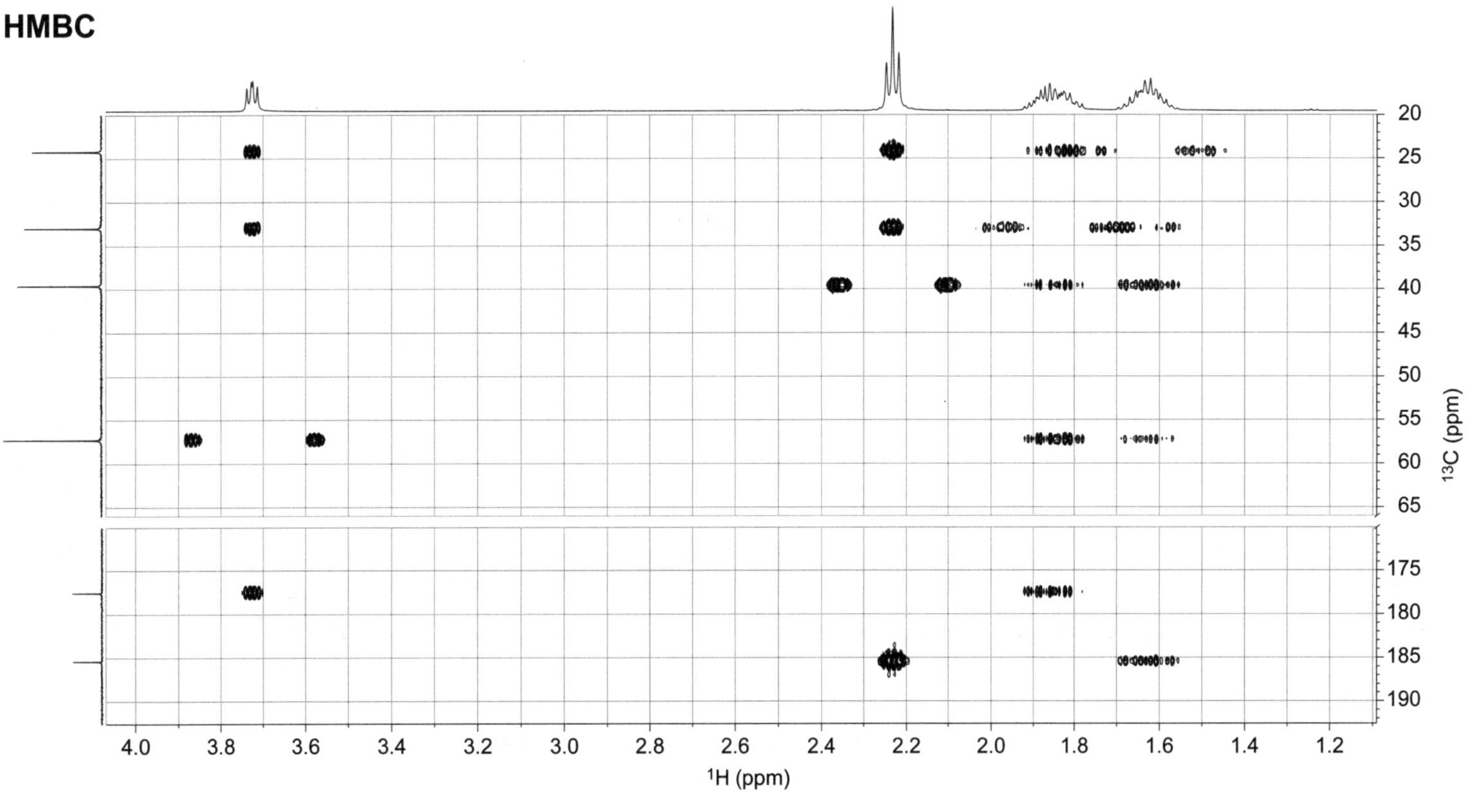

20
25
30
35
40
45
50
55
60
65
175
180
185
190
¹³C (ppm)
4.0
3.8
3.6
3.4
3.2
3.0
2.8
2.6
2.4
2.2
2.0
1.8
1.6
1.4
1.2
¹H (ppm)

Problem 9.30 Molecular formula: $C_7H_5O_2Cl$ ***Difficulty*** ★☆☆☆

Solvent: D_2O

Field: 500 MHz (^1H); 125 MHz (^{13}C)

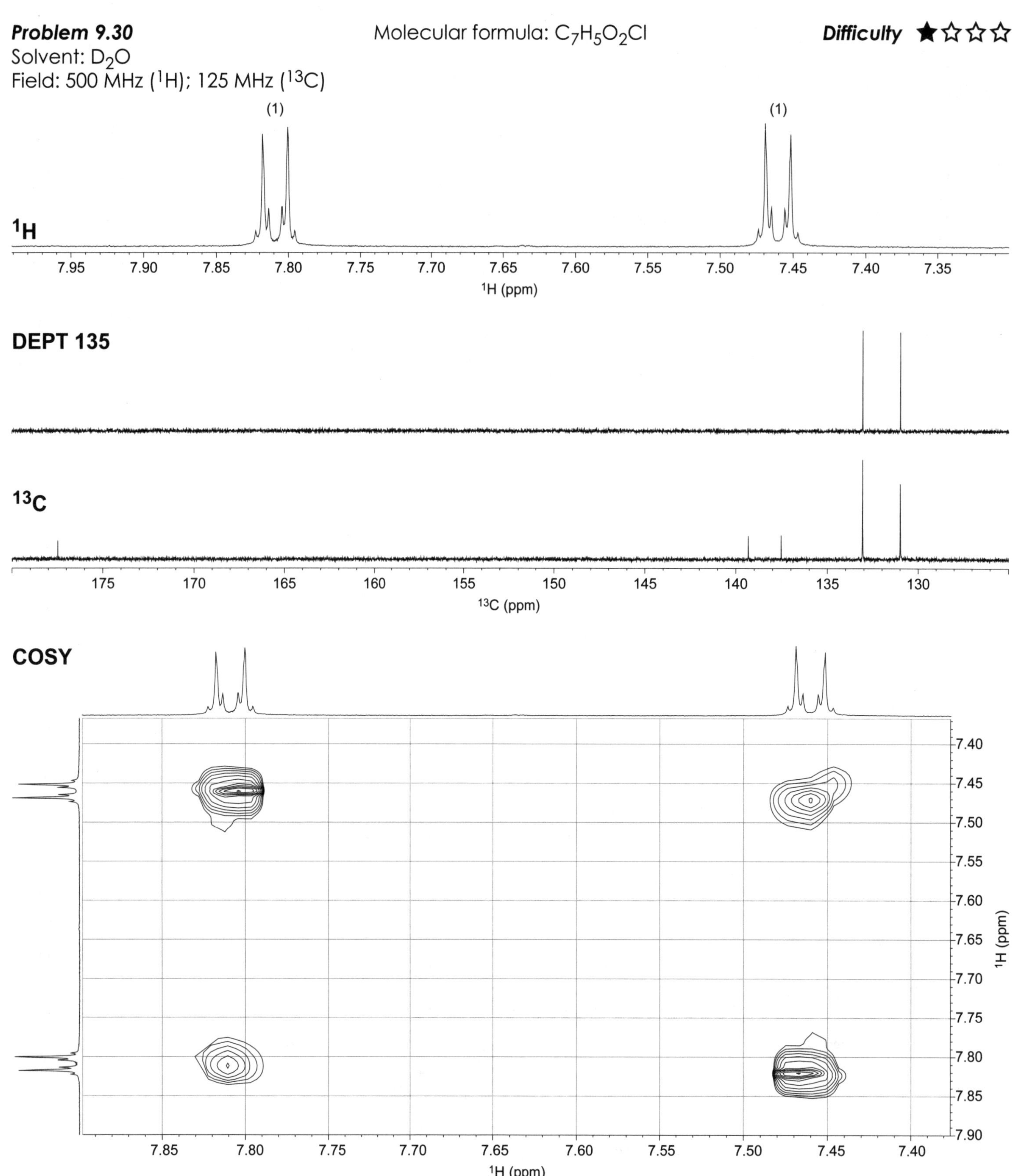

HSQC

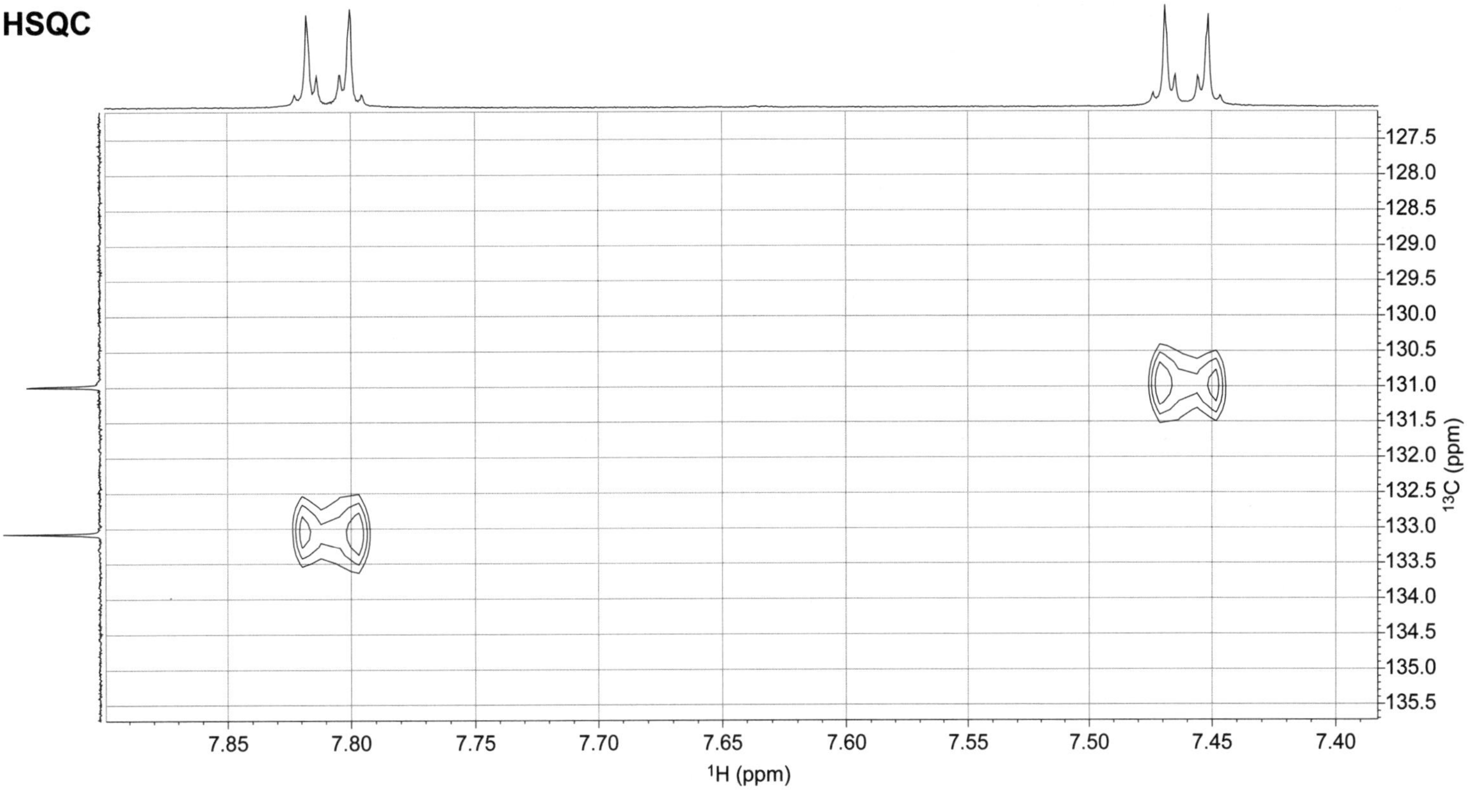
127.5
128.0
128.5
129.0
129.5
130.0
130.5
131.0
131.5
132.0
132.5
133.0
133.5
134.0
134.5
135.0
135.5
^{13}C (ppm)
7.85
7.80
7.75
7.70
7.65
7.60
7.55
7.50
7.45
7.40
^{1}H (ppm)

HMBC

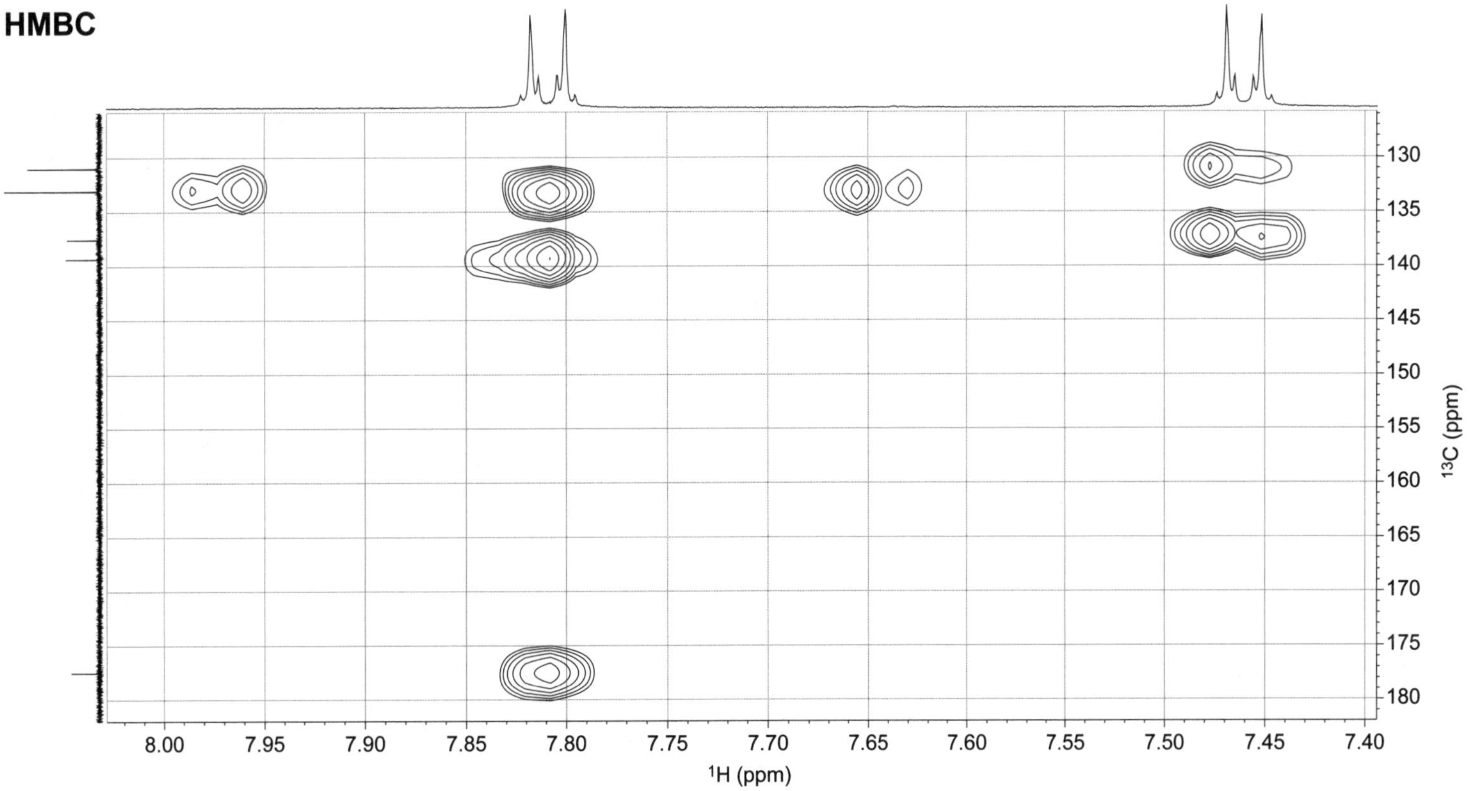
130
135
140
145
150
155
160
165
170
175
180
^{13}C (ppm)
8.00
7.95
7.90
7.85
7.80
7.75
7.70
7.65
7.60
7.55
7.50
7.45
7.40
^{1}H (ppm)

Problem 9.31
Solvent: D_2O
Field: 500 MHz (^1H); 125 MHz (^{13}C)

Molecular formula: C_3H_8O

Difficulty ★☆☆☆

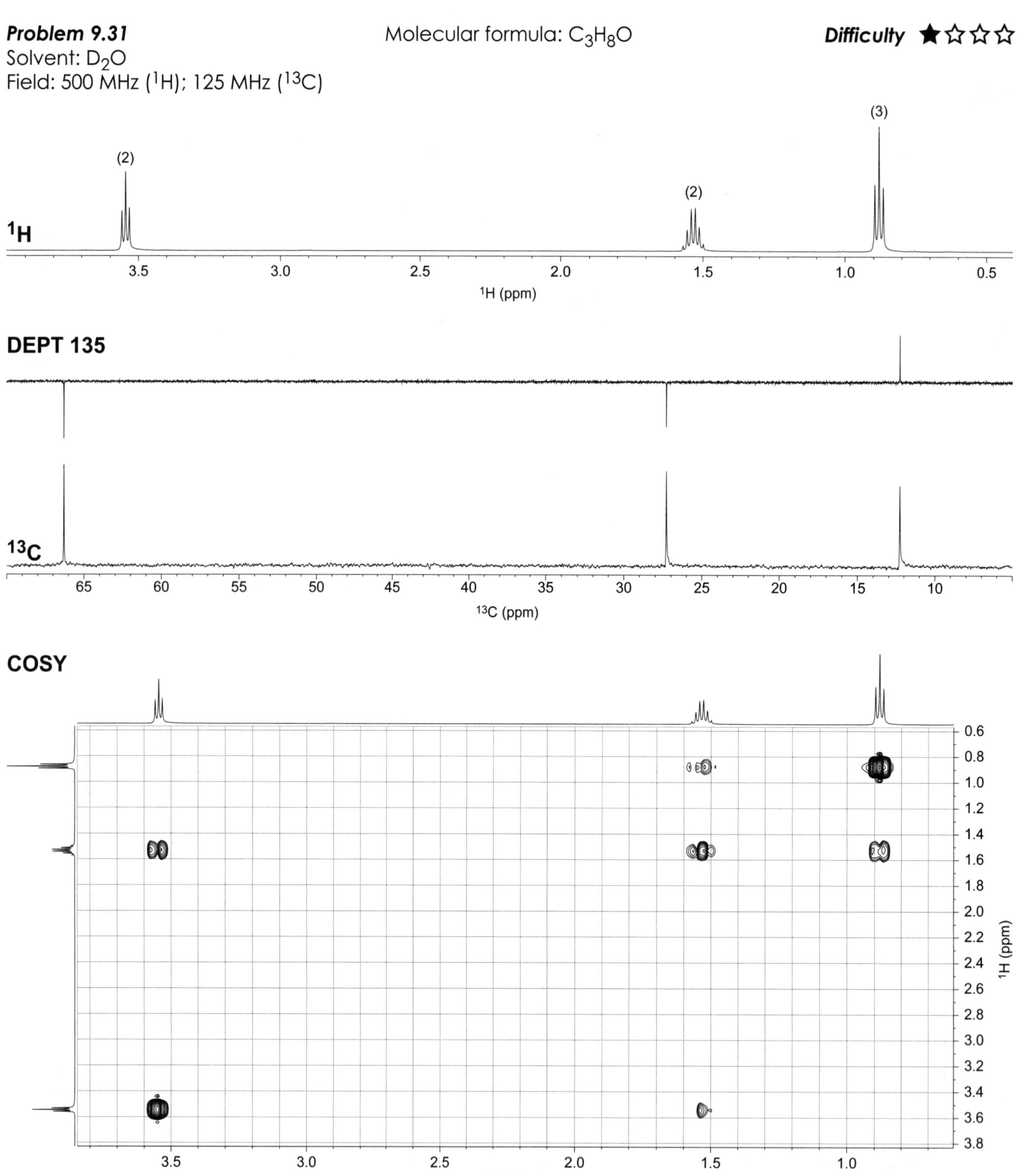

HSQC

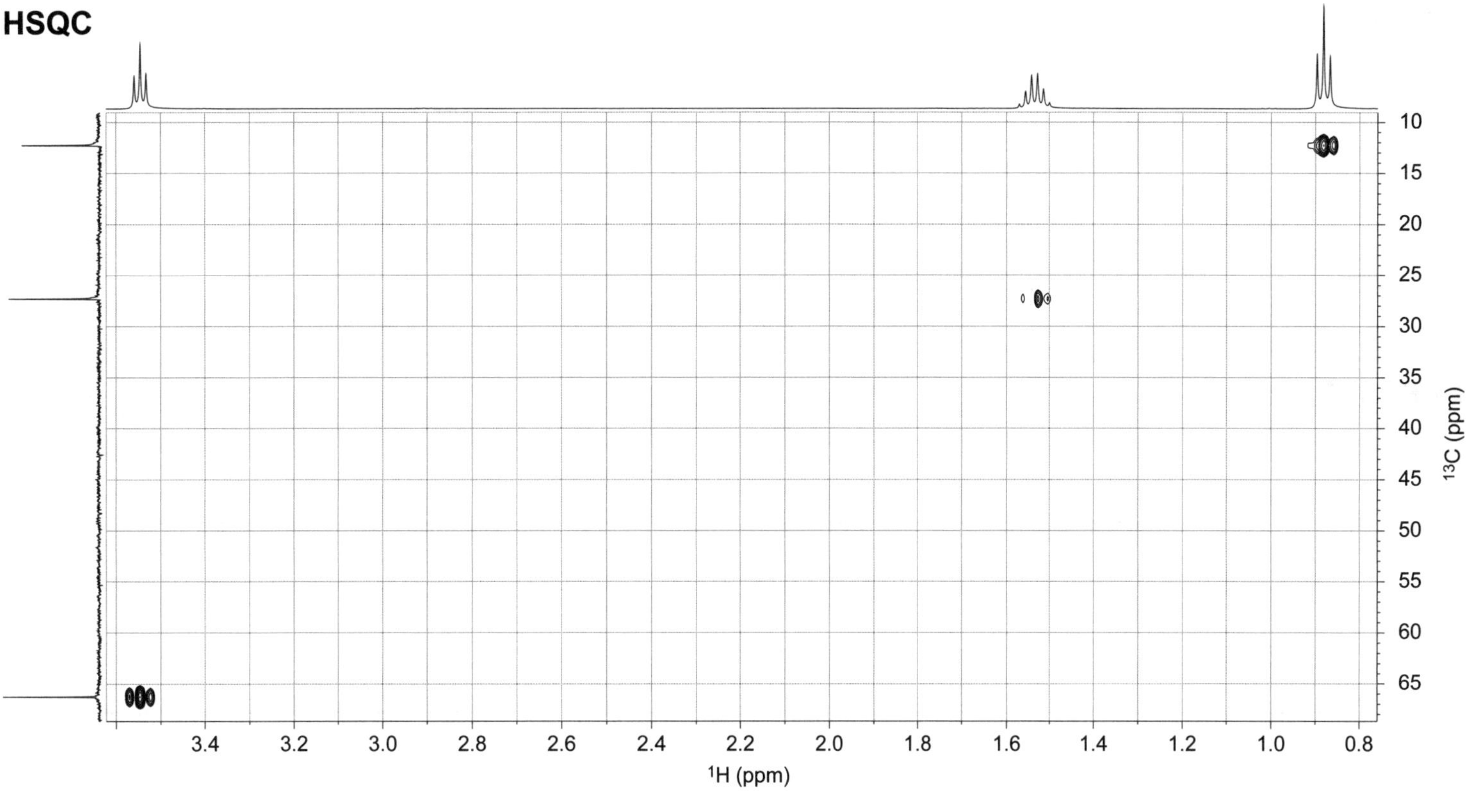

HMBC

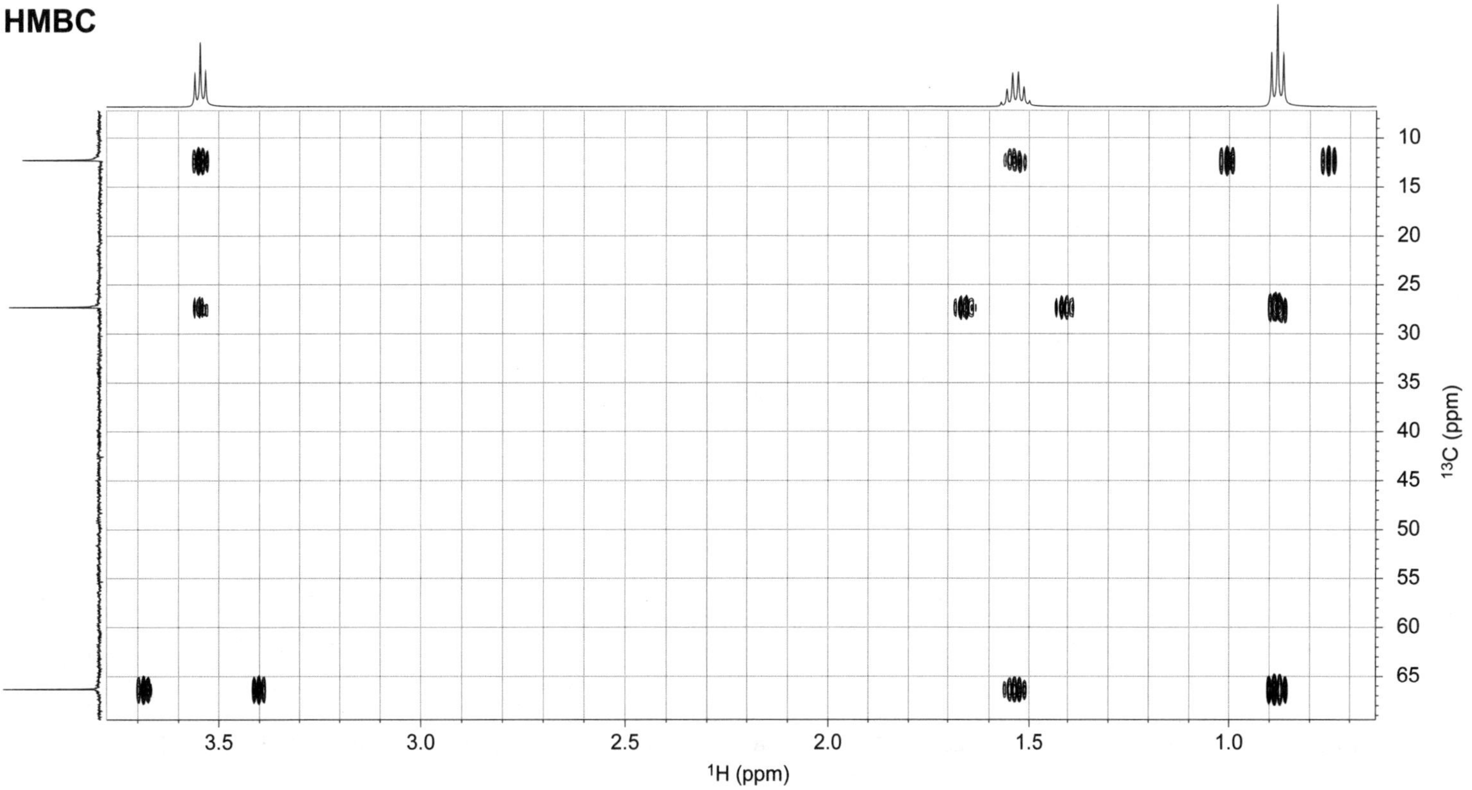

Problem 9.32
Solvent: D_2O
Field: 500 MHz (^1H); 125 MHz (^{13}C)

Molecular formula: C_6H_5OCl

Difficulty ★☆☆☆

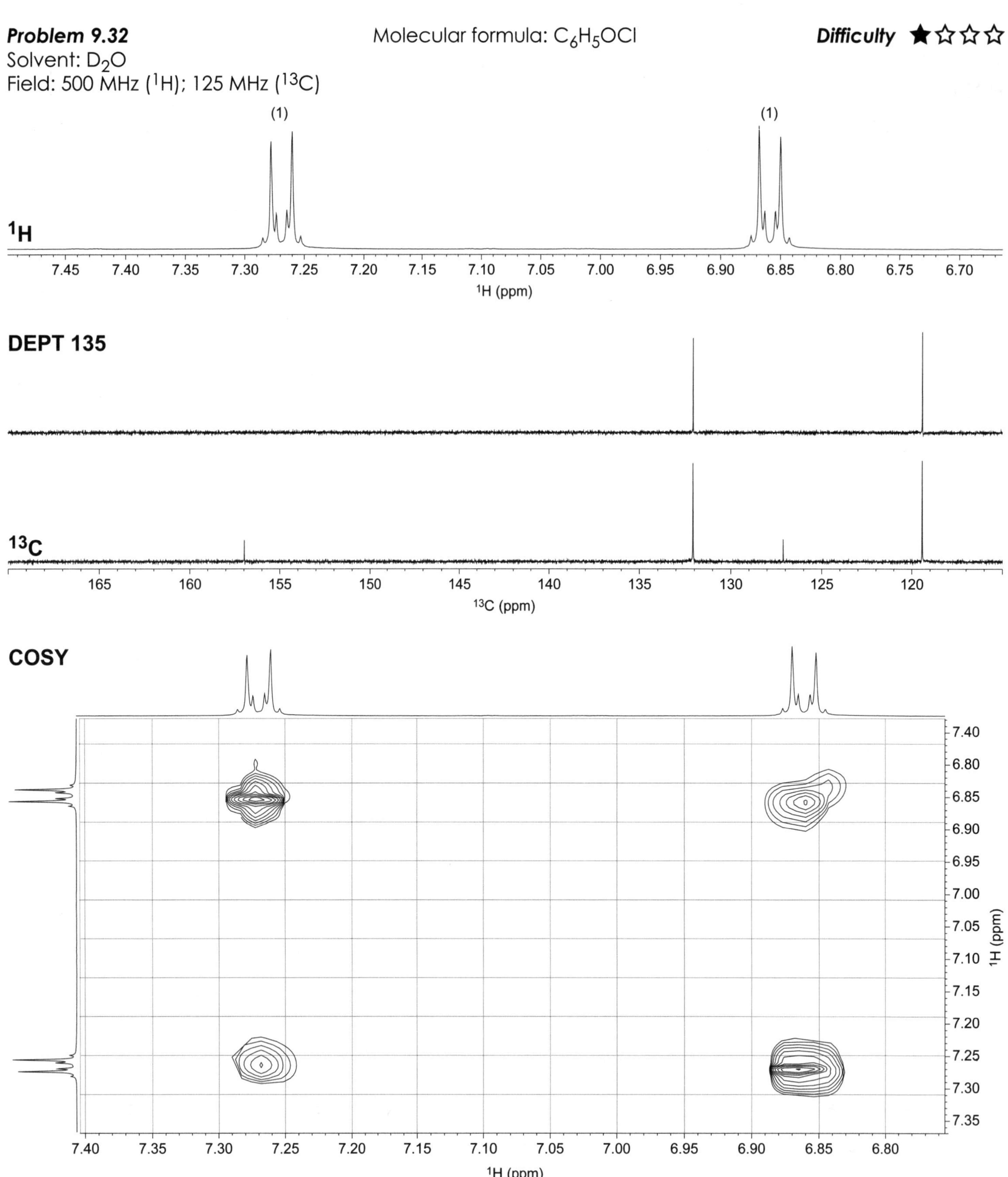

HSQC

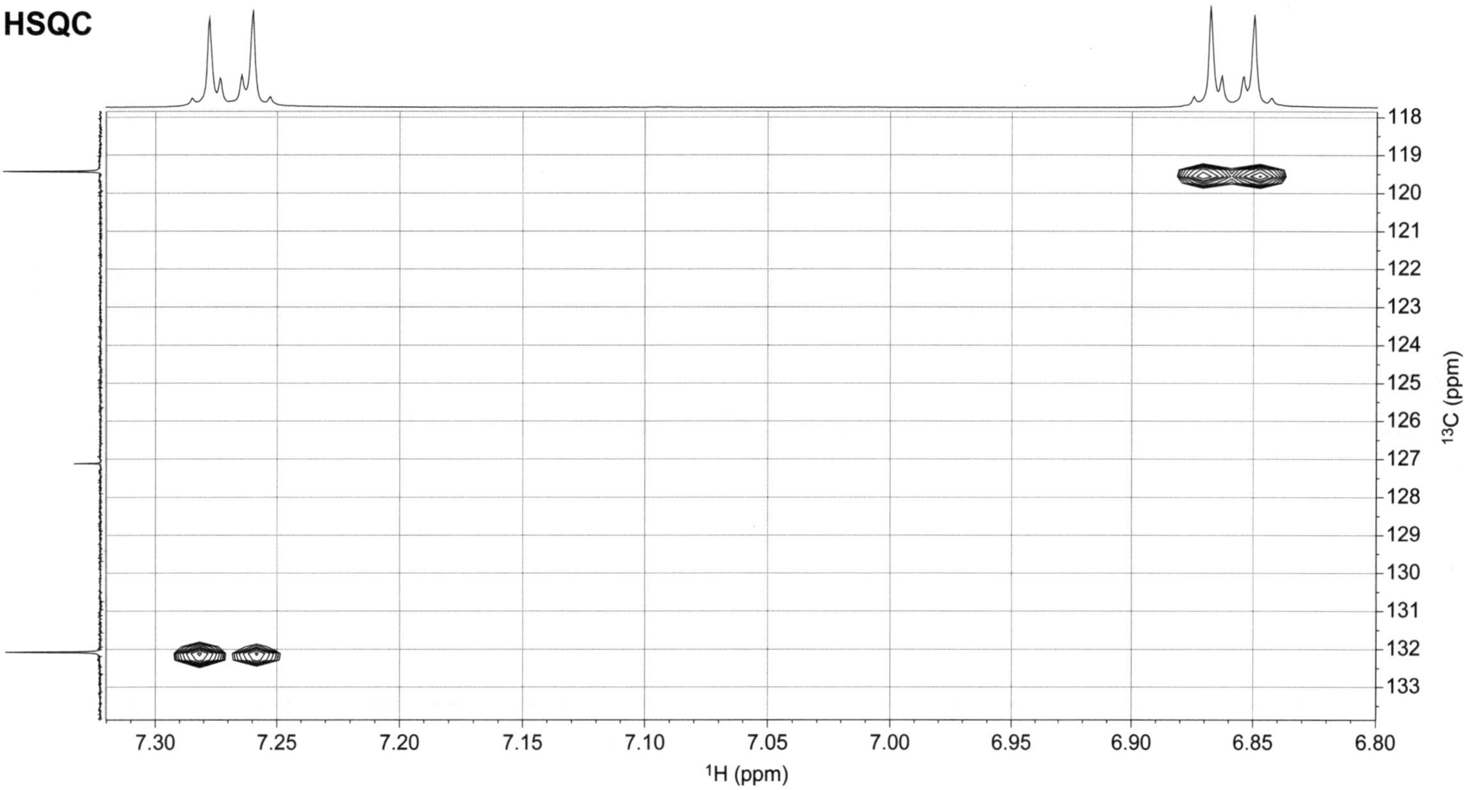
118
119
120
121
122
123
124
125
126
127
128
129
130
131
132
133
13C (ppm)
7.30
7.25
7.20
7.15
7.10
7.05
7.00
6.95
6.90
6.85
6.80
1H (ppm)

HMBC

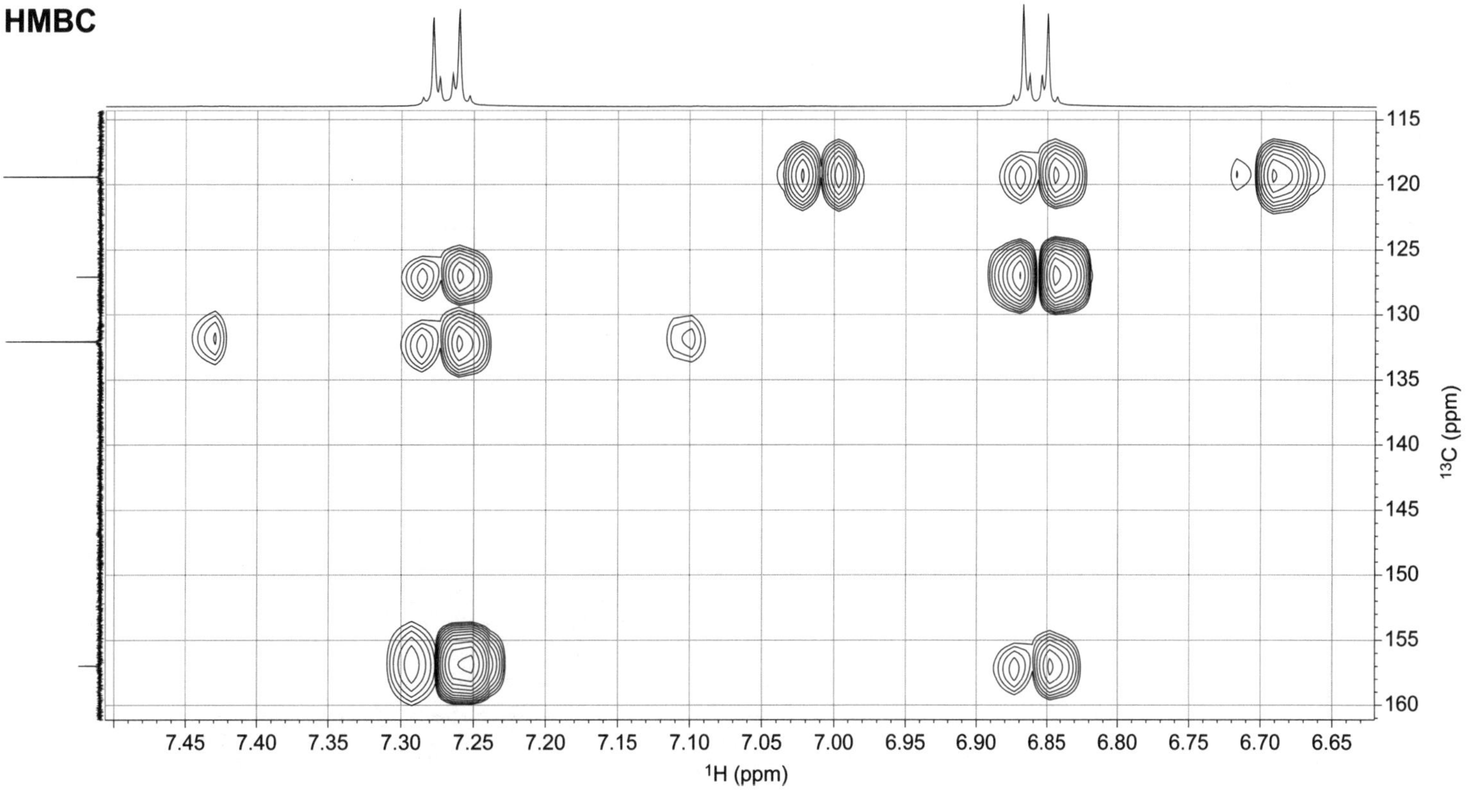
115
120
125
130
135
140
145
150
155
160
13C (ppm)
7.45
7.40
7.35
7.30
7.25
7.20
7.15
7.10
7.05
7.00
6.95
6.90
6.85
6.80
6.75
6.70
6.65
1H (ppm)

Problem 9.33 Molecular formula: $C_6H_{11}NO_3$ ***Difficulty*** ★☆☆☆

Solvent: D_2O

Field: 500 MHz (^1H); 125 MHz (^{13}C)

^{1}H

DEPT 135

^{13}C

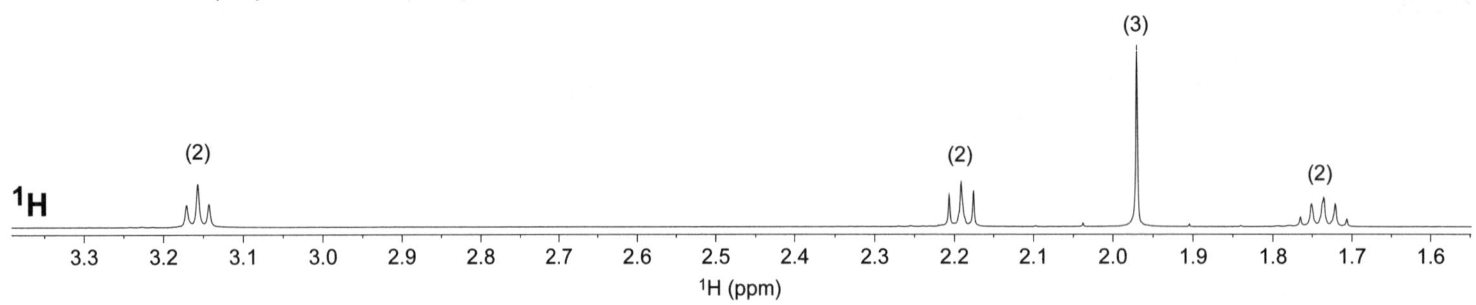

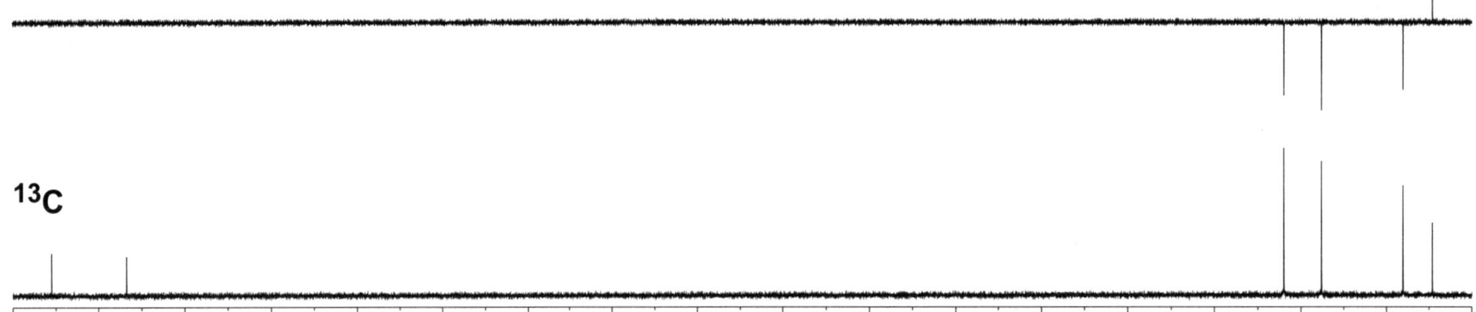

COSY

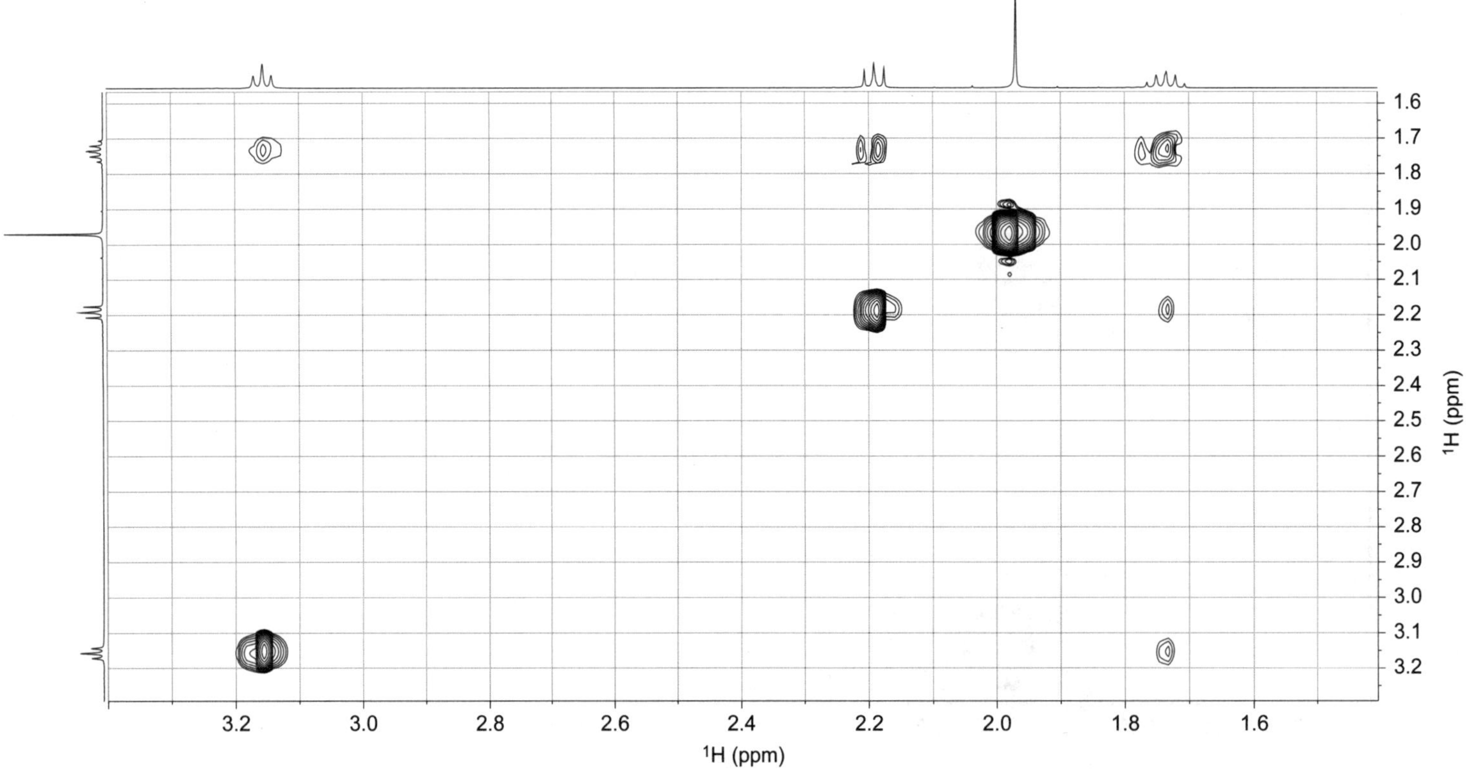

HSQC

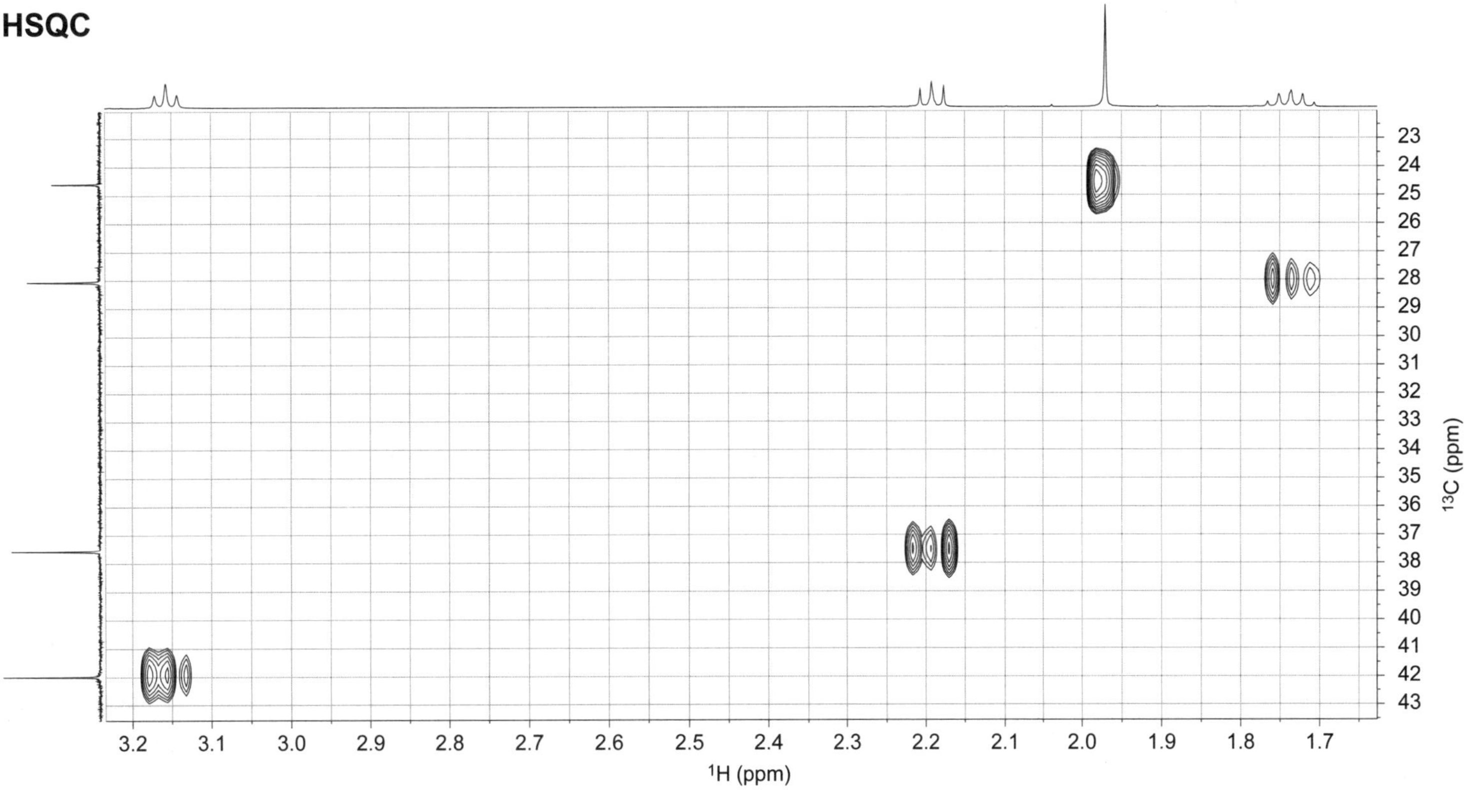

3.2 3.1 3.0 2.9 2.8 2.7 2.6 2.5 2.4 2.3 2.2 2.1 2.0 1.9 1.8 1.7

^{1}H (ppm)

23 24 25 26 27 28 29 30 31 32 33 34 35 36 37 38 39 40 41 42 43

^{13}C (ppm)

HMBC

3.4 3.2 3.0 2.8 2.6 2.4 2.2 2.0 1.8 1.6 1.4

^{1}H (ppm)

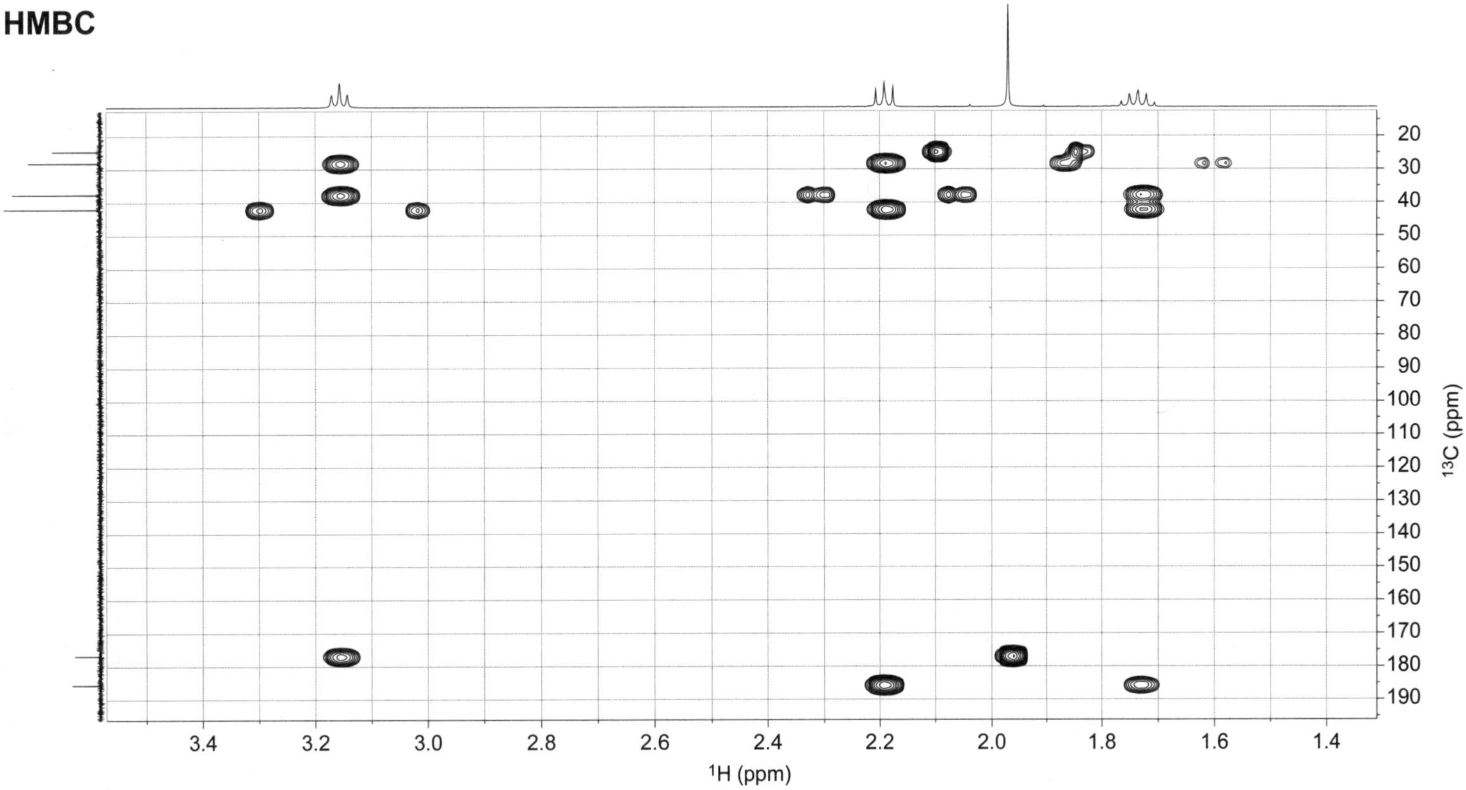

20 30 40 50 60 70 80 90 100 110 120 130 140 150 160 170 180 190

^{13}C (ppm)

Problem 9.34 Molecular formula: $C_6H_{12}O_3$ **Difficulty** ★☆☆☆

Solvent: D_2O

Field: 500 MHz (^1H); 125 MHz (^{13}C)

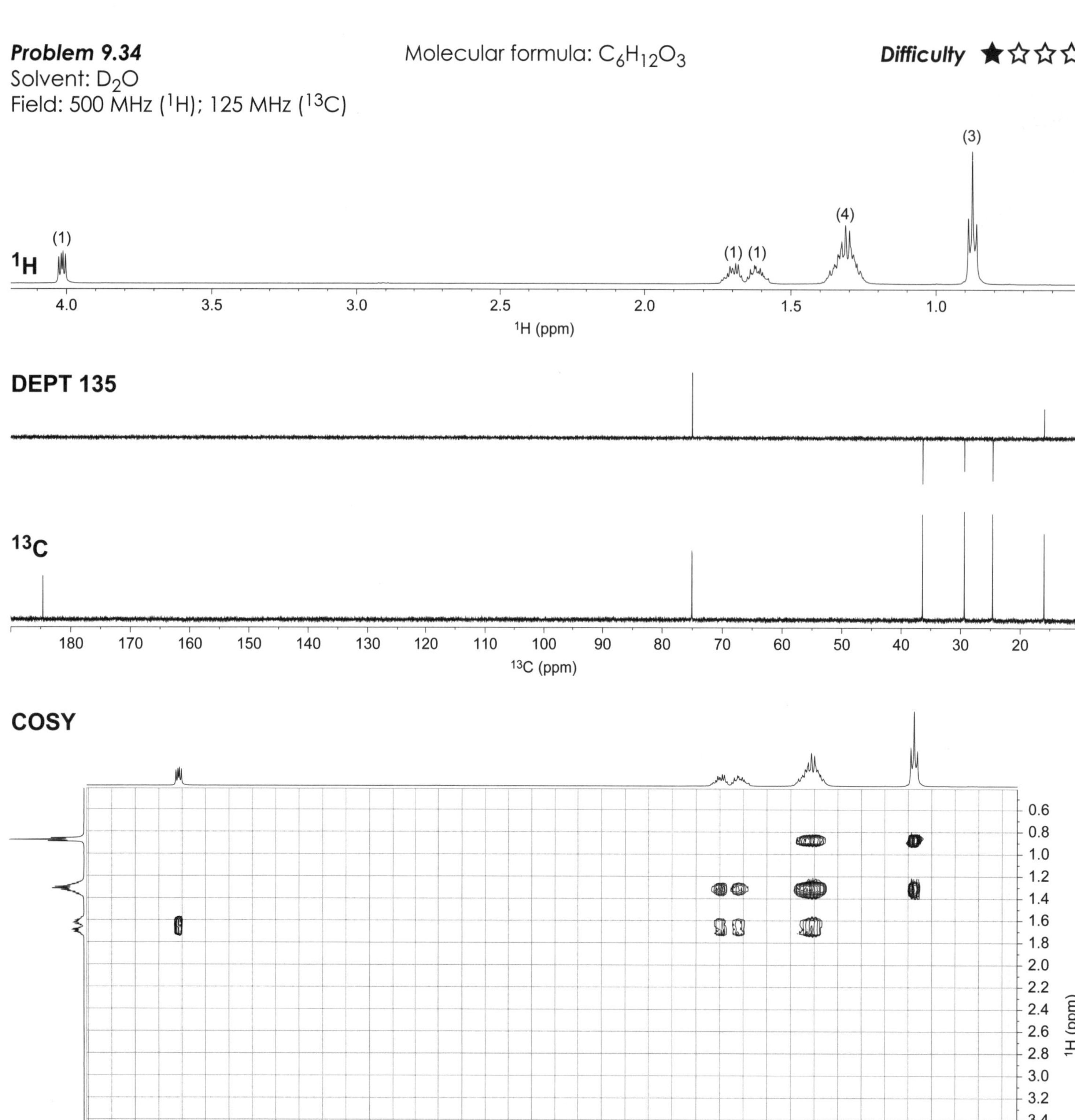

HSQC

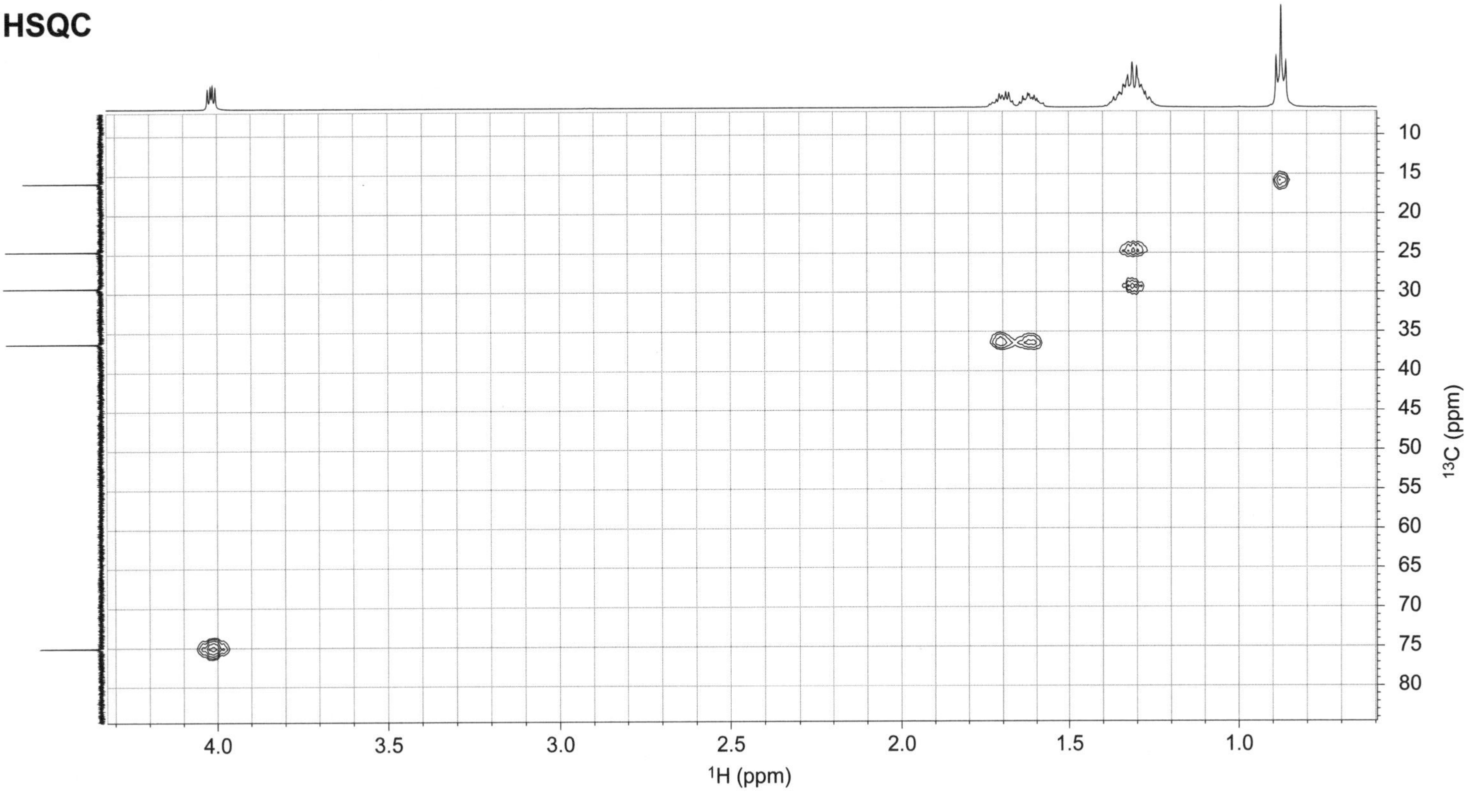
4.0
3.5
3.0
2.5
2.0
1.5
1.0
¹H (ppm)
10
15
20
25
30
35
40
45
50
55
60
65
70
75
80
¹³C (ppm)

HMBC

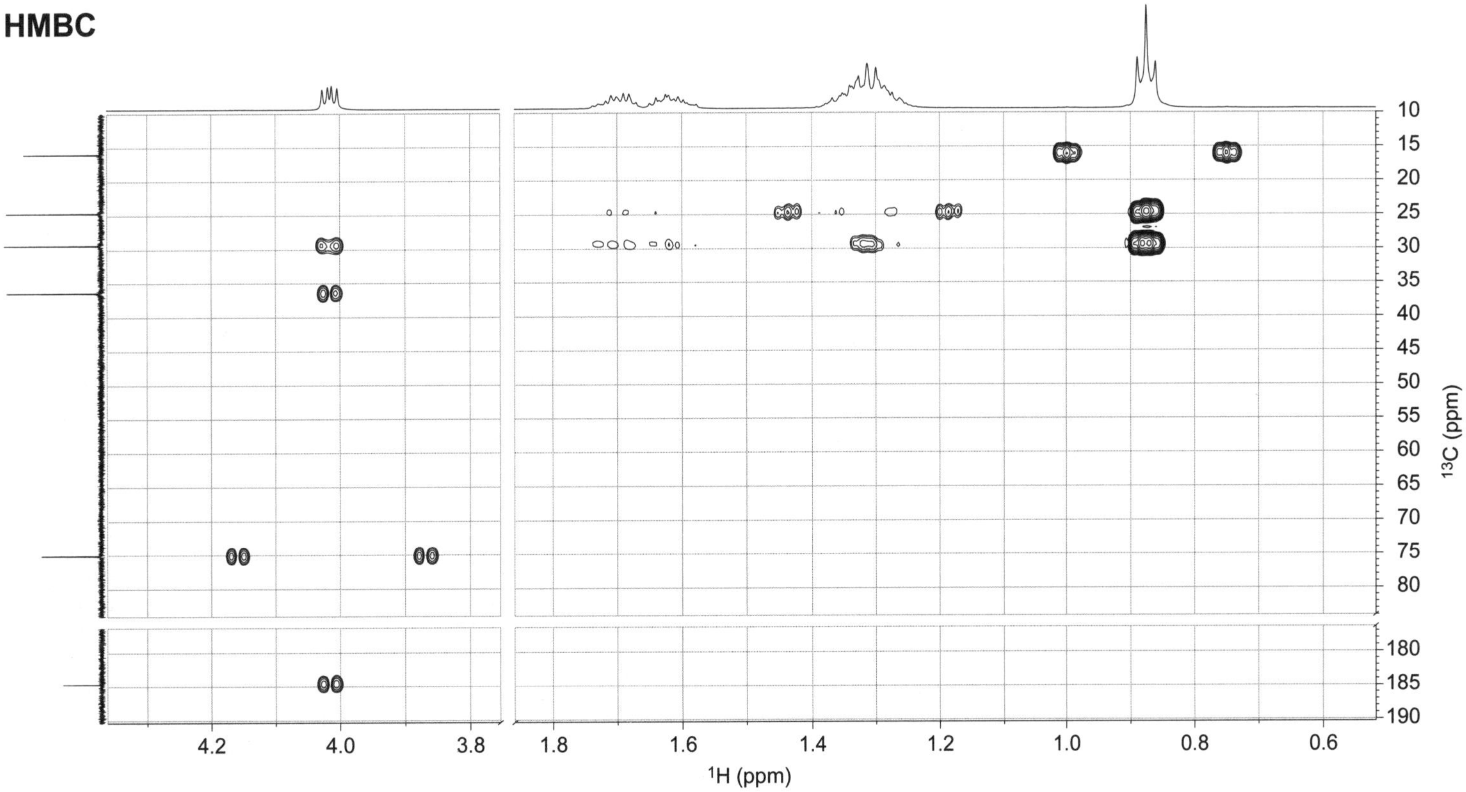
4.2
4.0
3.8
1.8
1.6
1.4
1.2
1.0
0.8
0.6
¹H (ppm)
10
15
20
25
30
35
40
45
50
55
60
65
70
75
80
180
185
190
¹³C (ppm)

Problem 9.35 Molecular formula: $C_7H_6O_2$ **Difficulty** ★☆☆☆

Solvent: CD_3-SO-CD_3

Field: 500 MHz (^{1}H); 125 MHz (^{13}C)

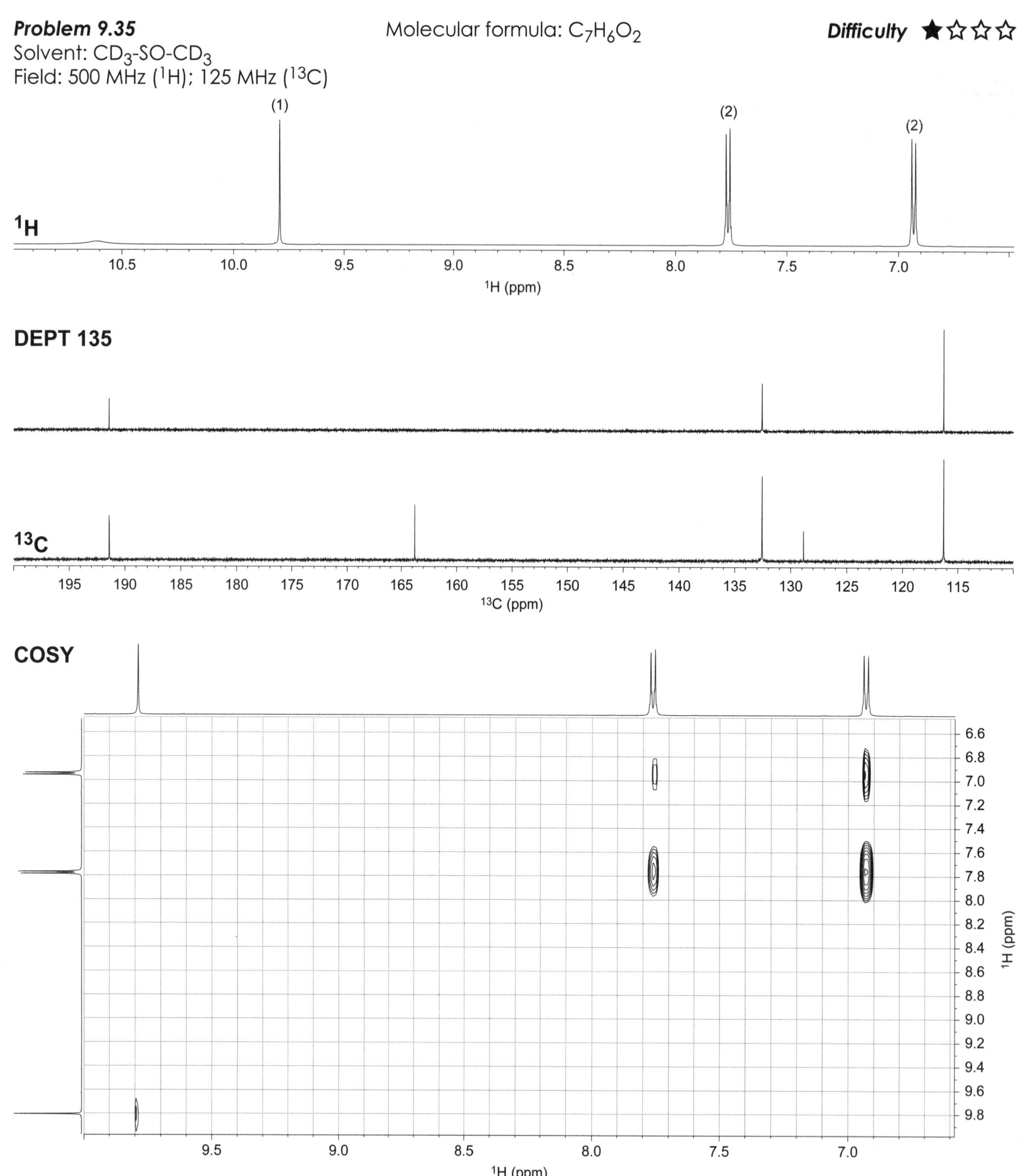

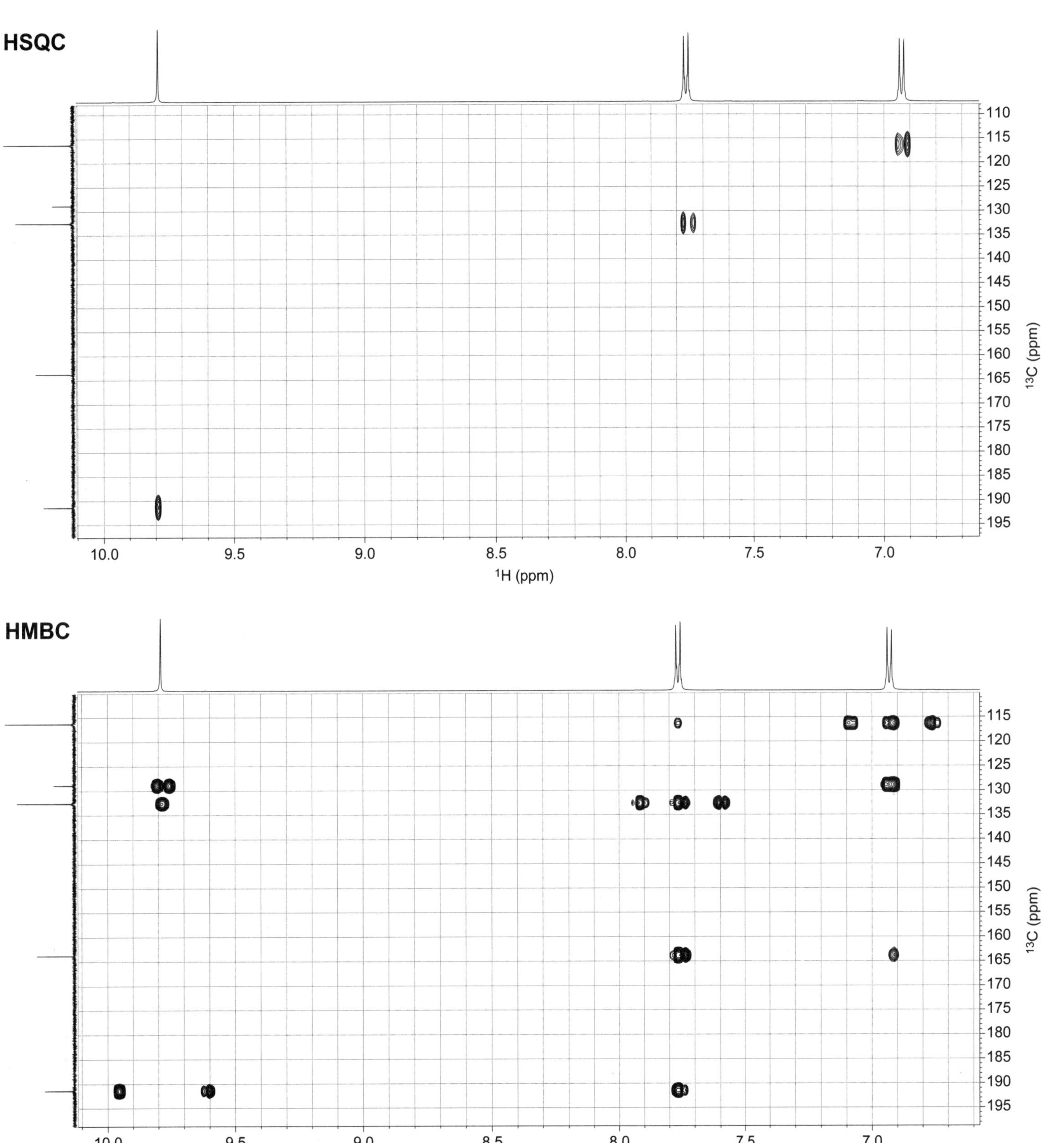

HSQC
HMBC
10.0
9.5
9.0
8.5
8.0
7.5
7.0
^{1}H (ppm)
^{13}C (ppm)
110
115
120
125
130
135
140
145
150
155
160
165
170
175
180
185
190
195

Problem 9.36 Molecular formula: $C_7H_8O_2$ ***Difficulty*** ★☆☆☆

Solvent: D_2O

Field: 500 MHz (^1H); 125 MHz (^{13}C)

DEPT 135

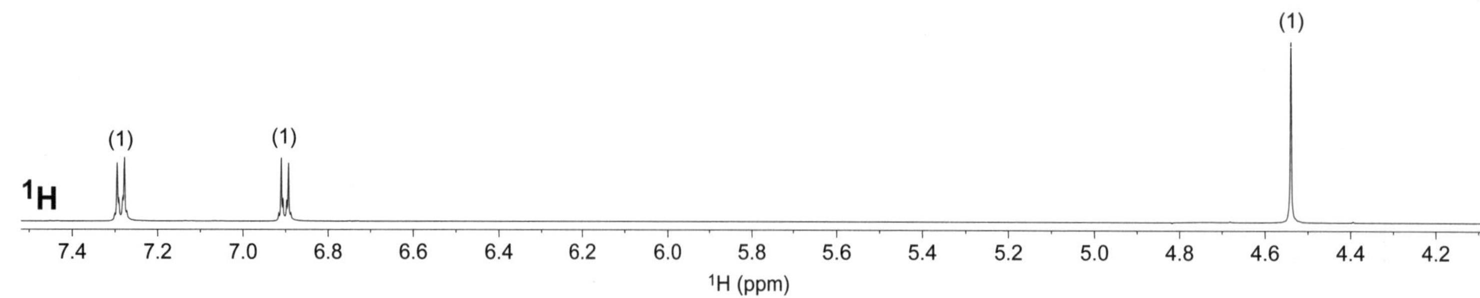

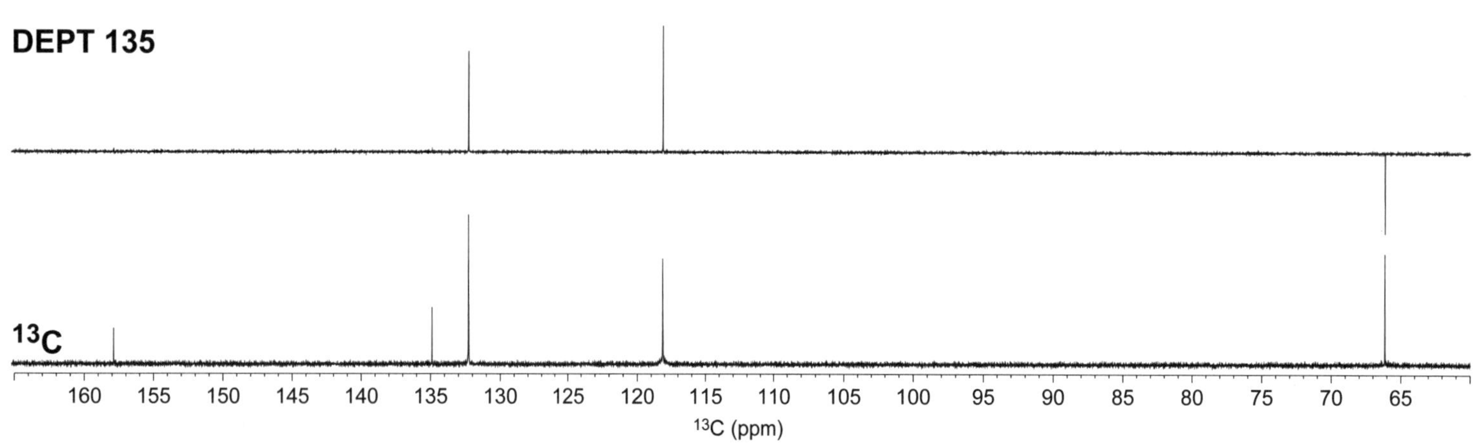

HSQC

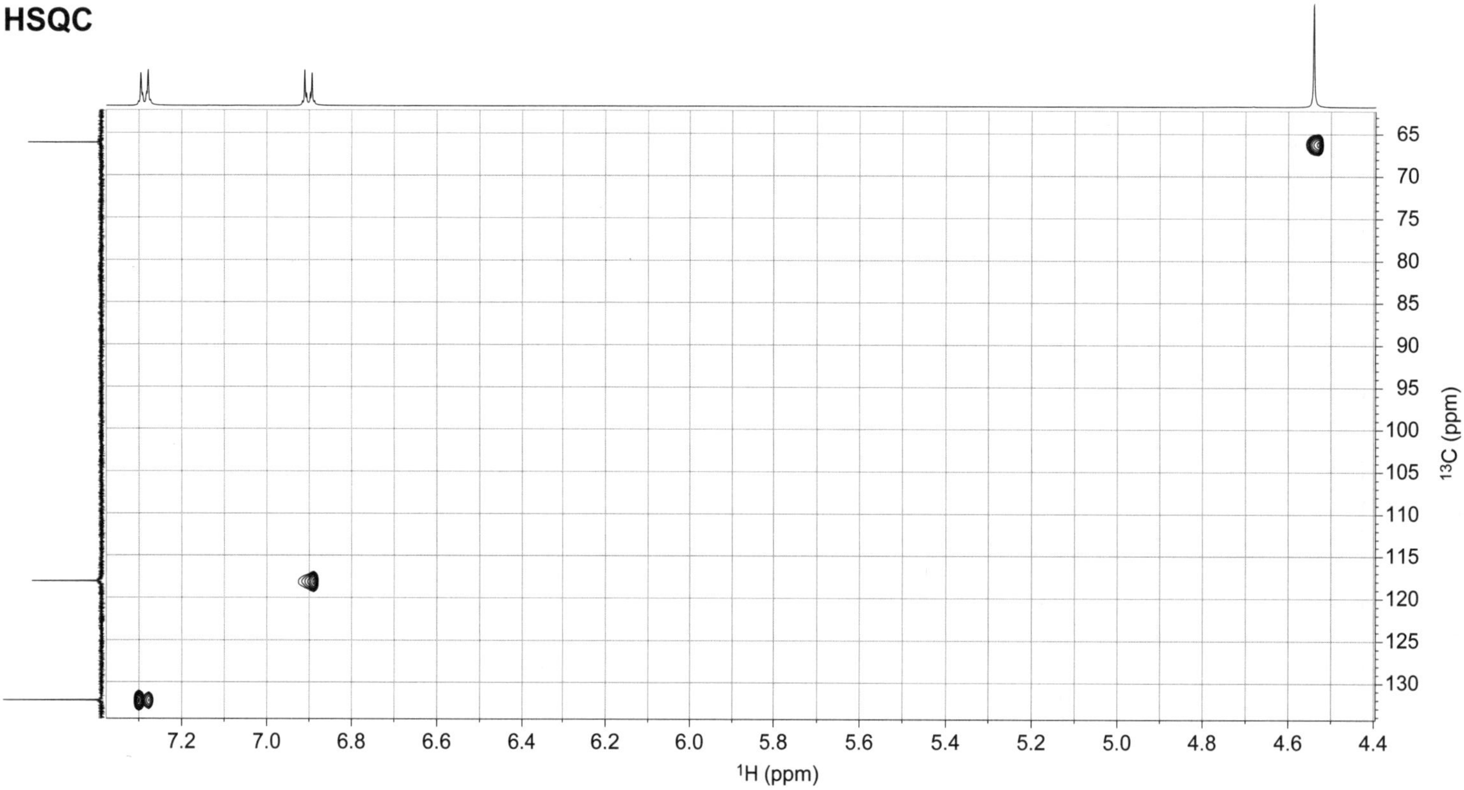

HMBC

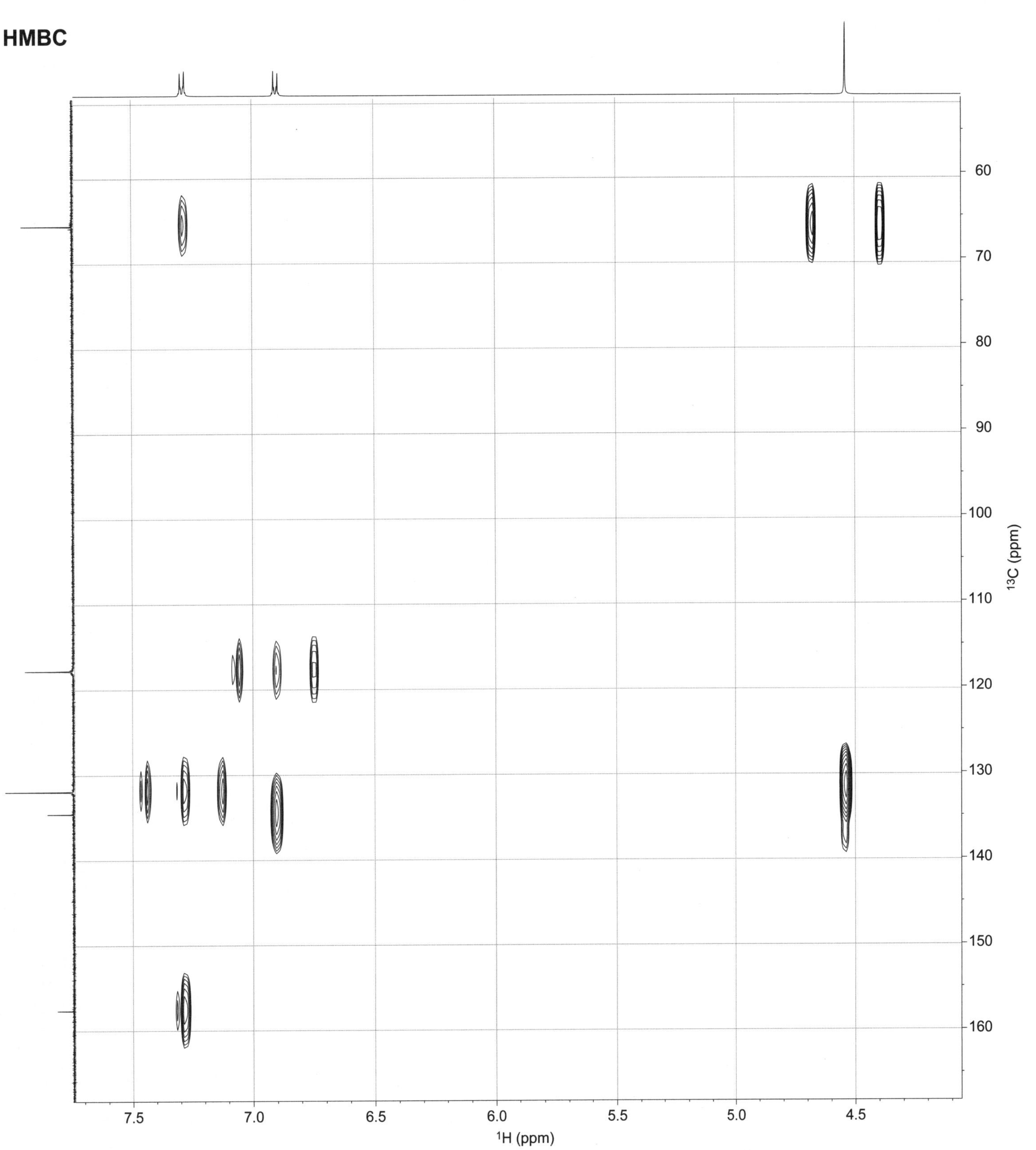

Problem 9.37 Molecular formula: C_7H_7NO **Difficulty** ★☆☆☆

Solvent: CD_3OD

Field: 500 MHz (^1H); 125 MHz (^{13}C)

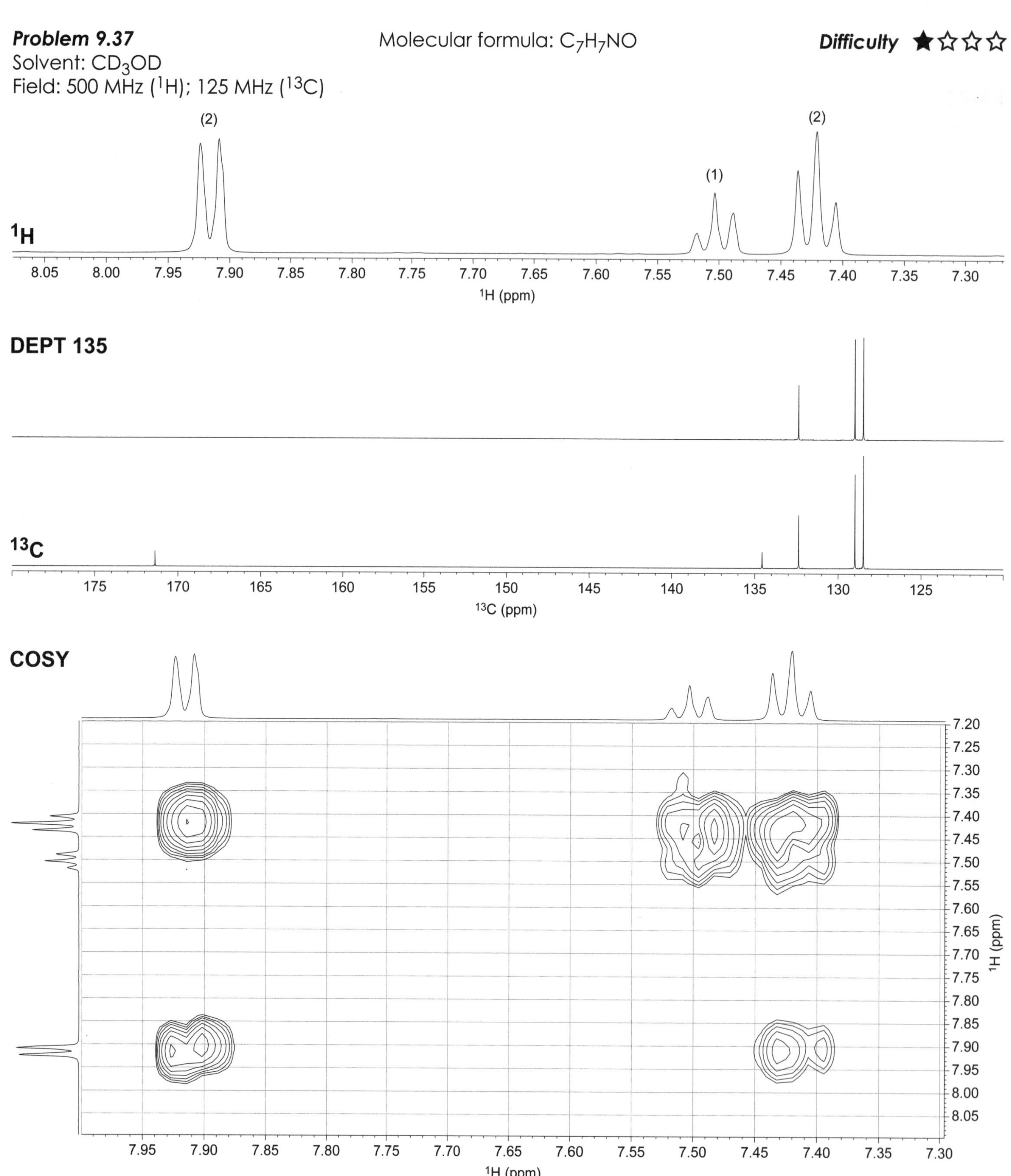

HSQC

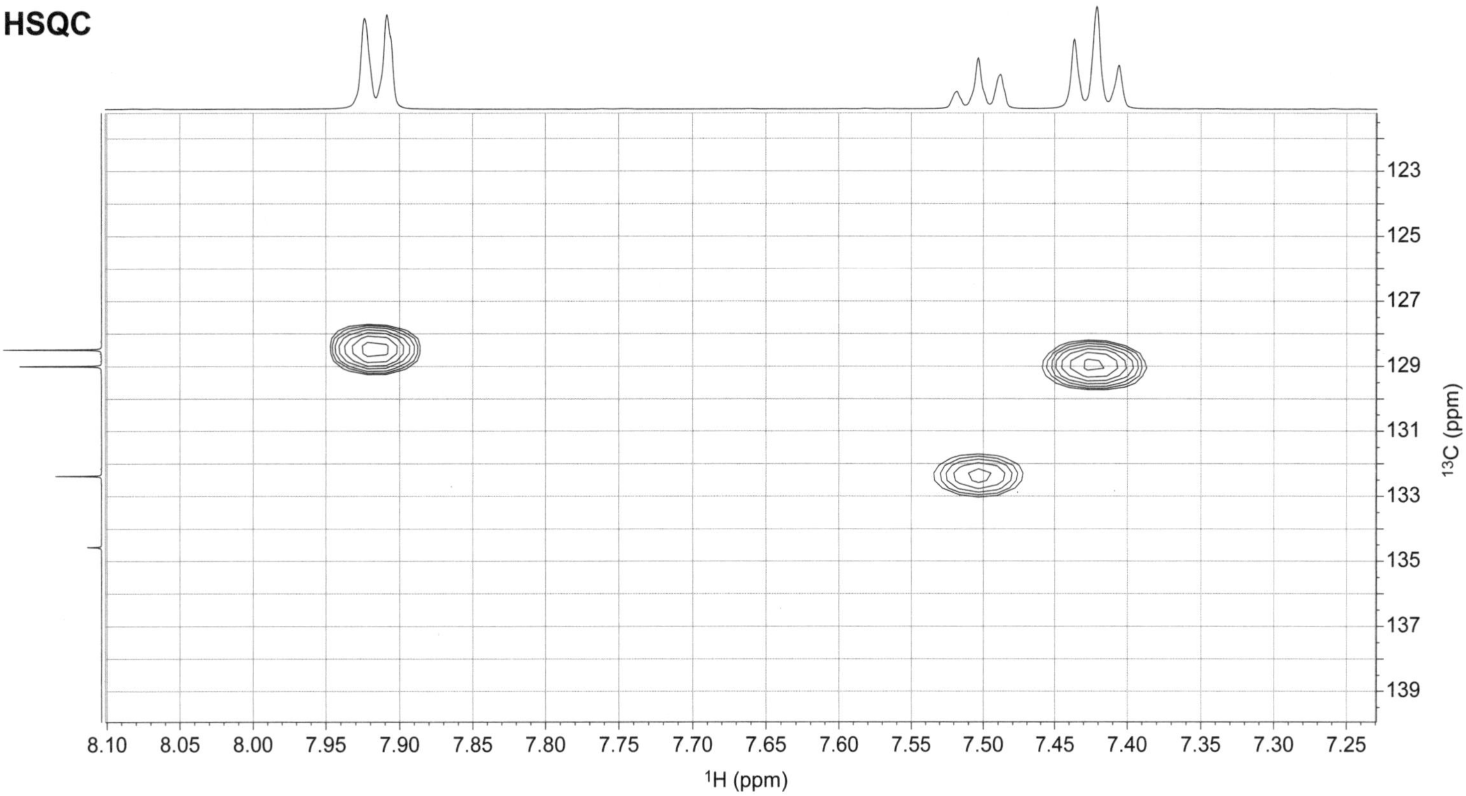

123
125
127
129
131
133
135
137
139

^{13}C (ppm)

8.10 8.05 8.00 7.95 7.90 7.85 7.80 7.75 7.70 7.65 7.60 7.55 7.50 7.45 7.40 7.35 7.30 7.25

^{1}H (ppm)

HMBC

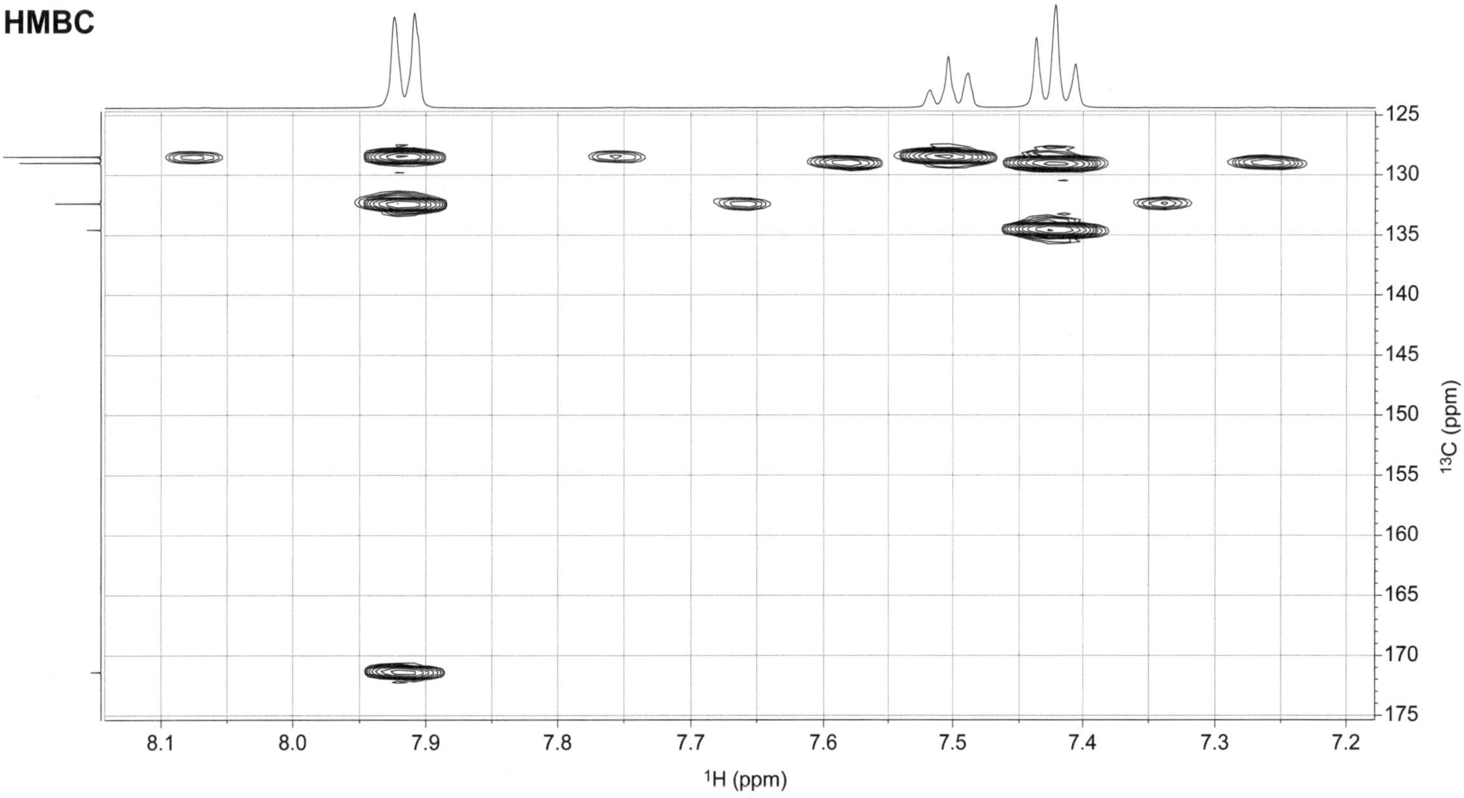

125
130
135
140
145
150
155
160
165
170
175

^{13}C (ppm)

8.1 8.0 7.9 7.8 7.7 7.6 7.5 7.4 7.3 7.2

^{1}H (ppm)

Problem 9.38 Molecular formula: $C_6H_5NO_2$ ***Difficulty*** ★☆☆☆

Solvent: CD_3-CO-CD_3

Field: 500 MHz (^1H); 125 MHz (^{13}C)

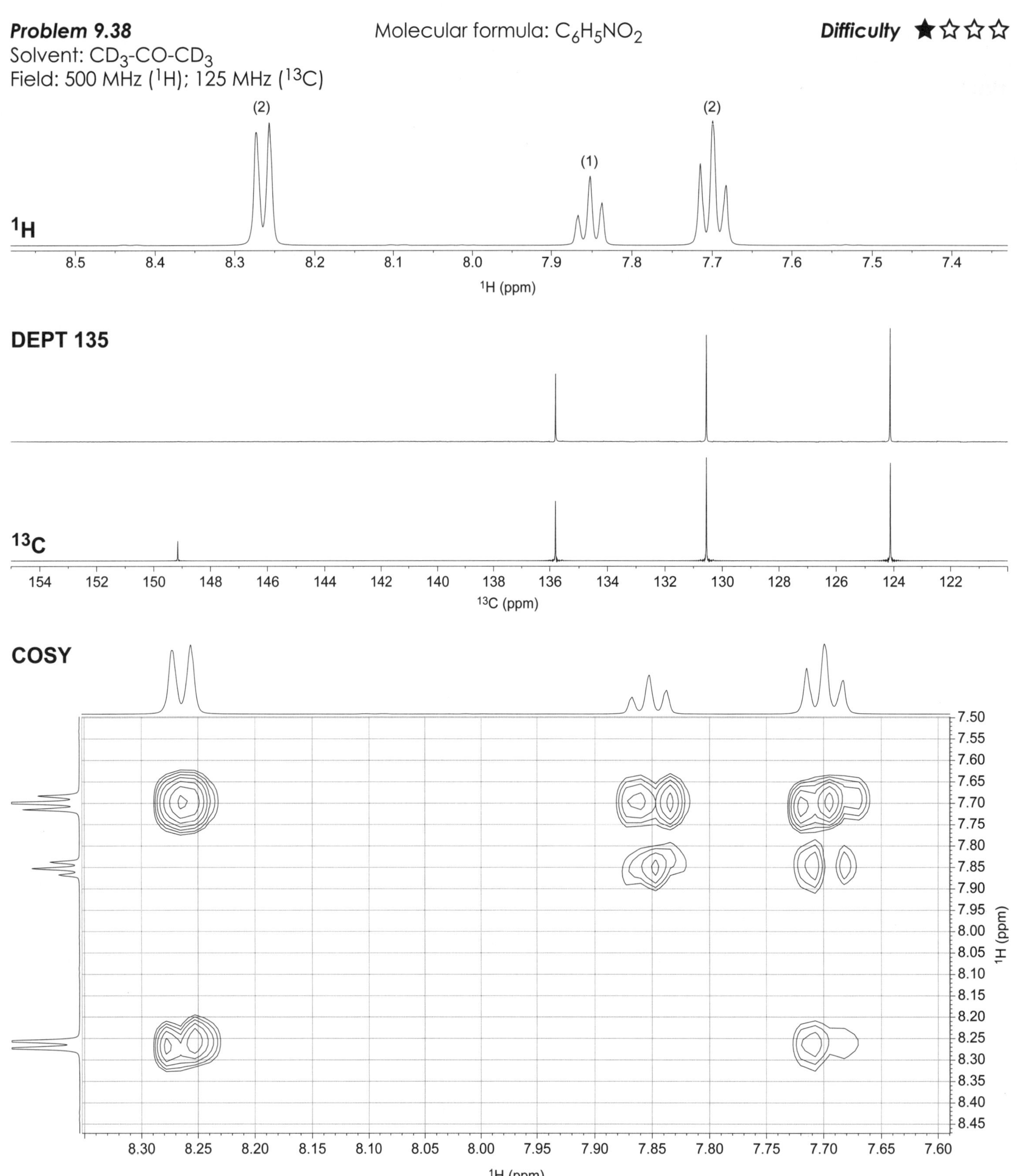

HSQC

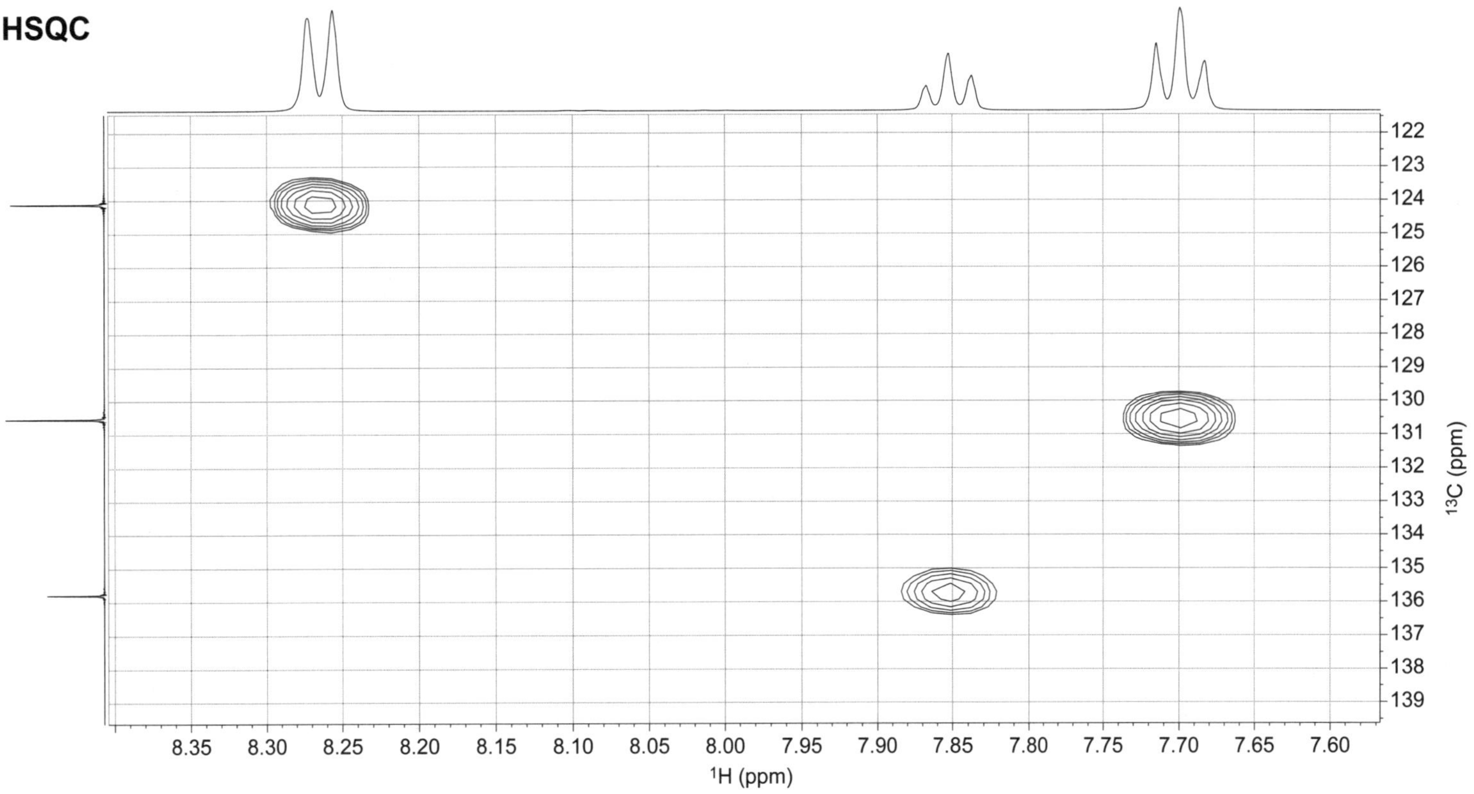

122
123
124
125
126
127
128
129
130
131
132
133
134
135
136
137
138
139
13C (ppm)
8.35
8.30
8.25
8.20
8.15
8.10
8.05
8.00
7.95
7.90
7.85
7.80
7.75
7.70
7.65
7.60
1H (ppm)

HMBC

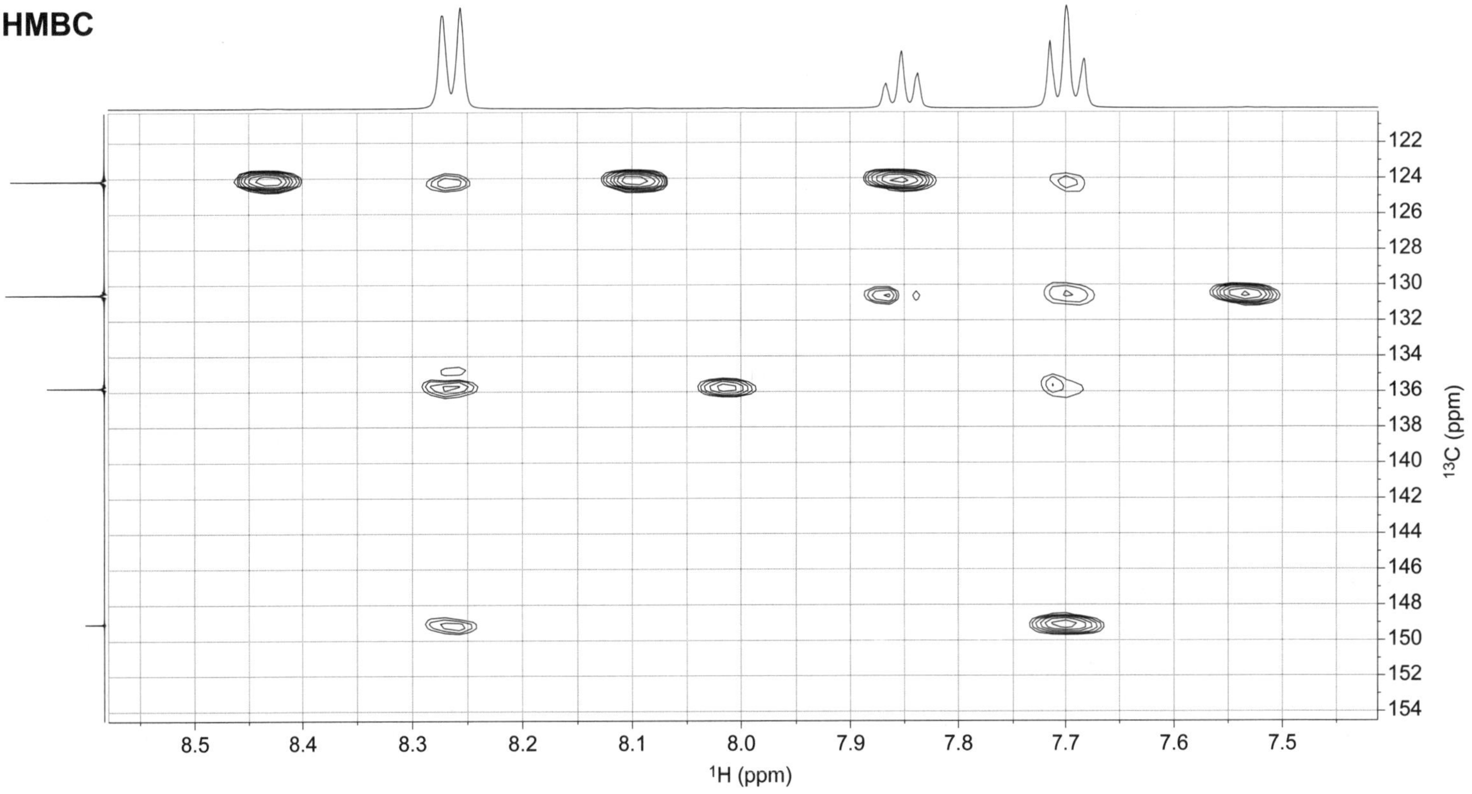

122
124
126
128
130
132
134
136
138
140
142
144
146
148
150
152
154
13C (ppm)
8.5
8.4
8.3
8.2
8.1
8.0
7.9
7.8
7.7
7.6
7.5
1H (ppm)

Problem 9.39 Molecular formula: $C_3H_4Cl_2$ ***Difficulty*** ★☆☆☆

Solvent: $CDCl_3$

Field: 500 MHz (^1H); 125 MHz (^{13}C)

J = 7 Hz
(1)
J = 7 Hz
(1)
J = 7 Hz
(2)
1H
6.2
6.0
5.8
5.6
5.4
5.2
5.0
4.8
4.6
4.4
4.2
1H (ppm)

DEPT 135
Solvent
13C
130
125
120
115
110
105
100
95
90
85
80
75
70
65
60
55
50
45
40
35
13C (ppm)

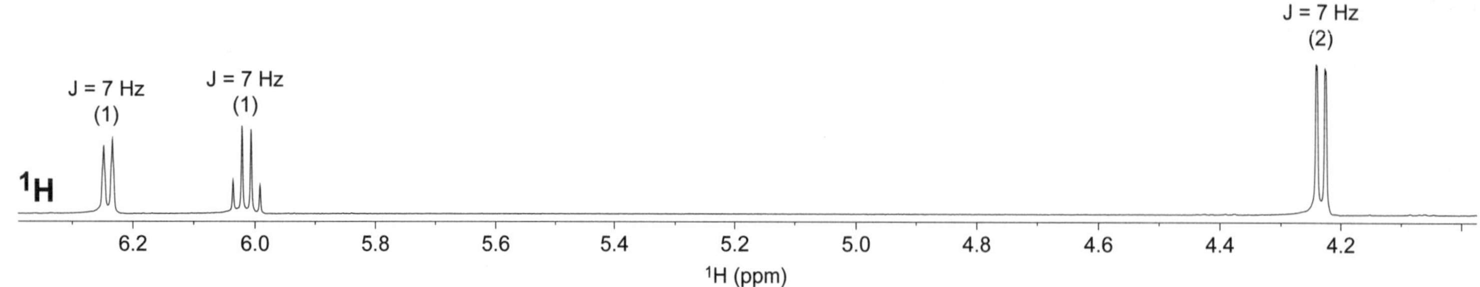

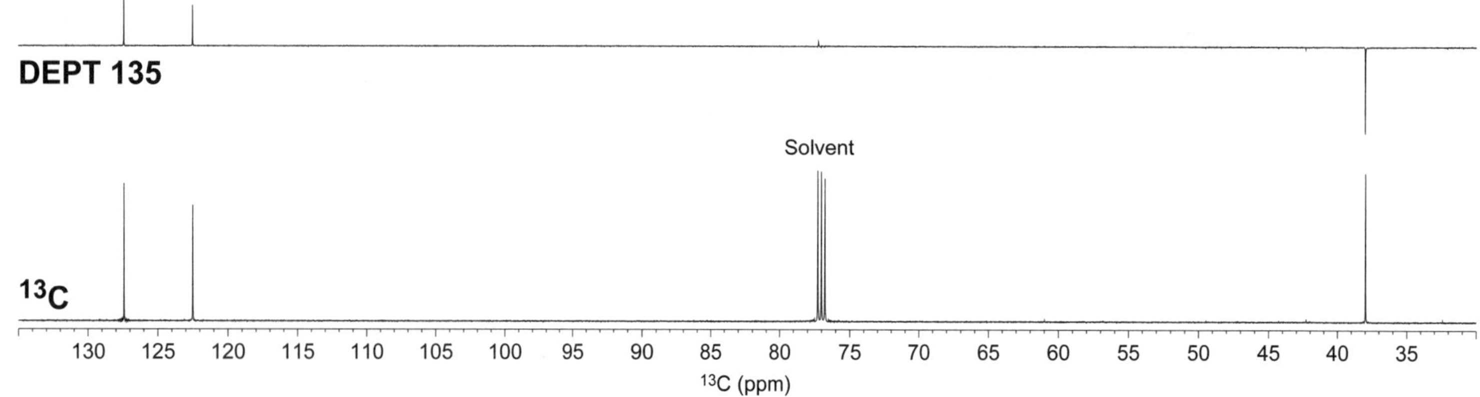

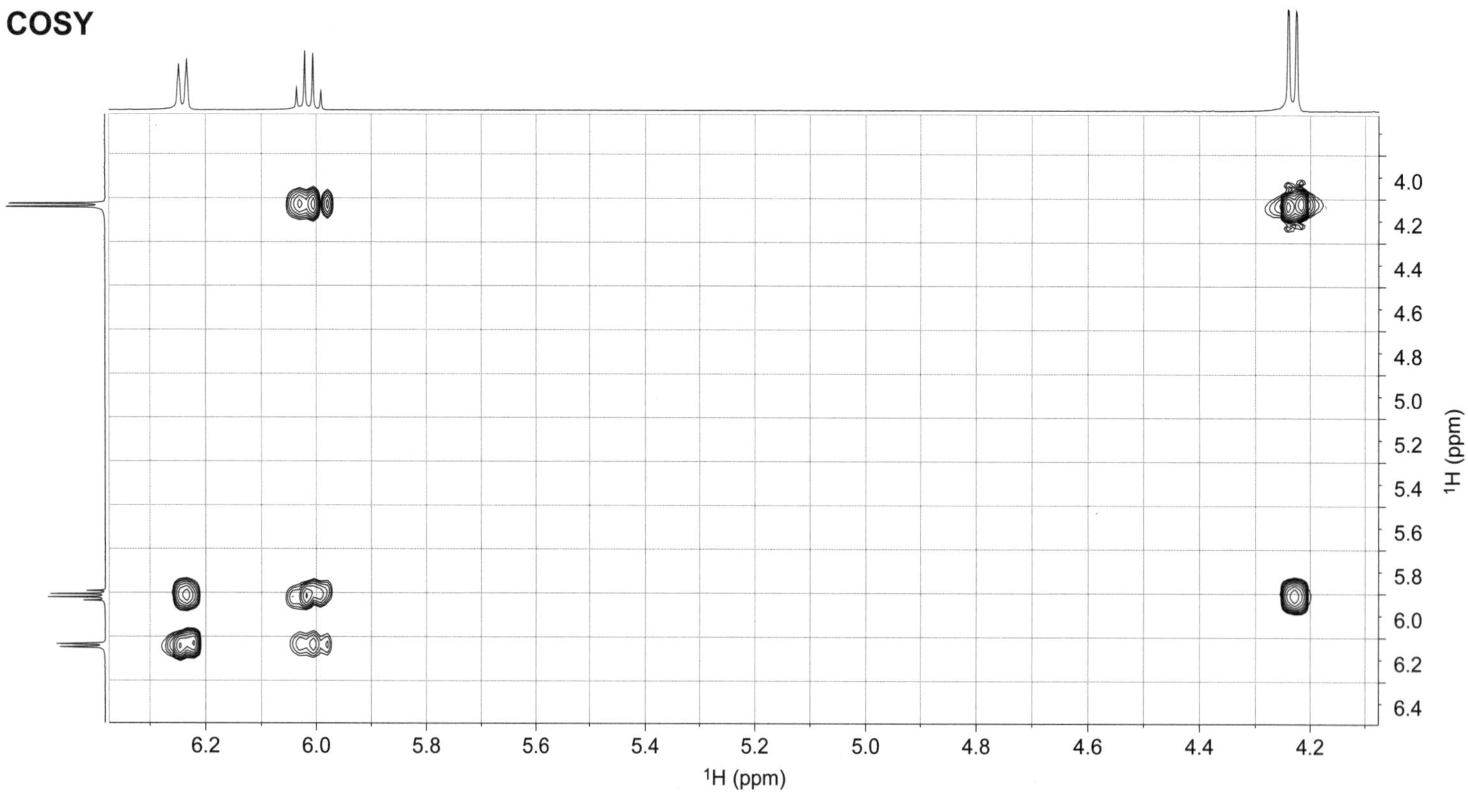

COSY
6.2
6.0
5.8
5.6
5.4
5.2
5.0
4.8
4.6
4.4
4.2
1H (ppm)
4.0
4.2
4.4
4.6
4.8
5.0
5.2
5.4
5.6
5.8
6.0
6.2
6.4
1H (ppm)

HSQC

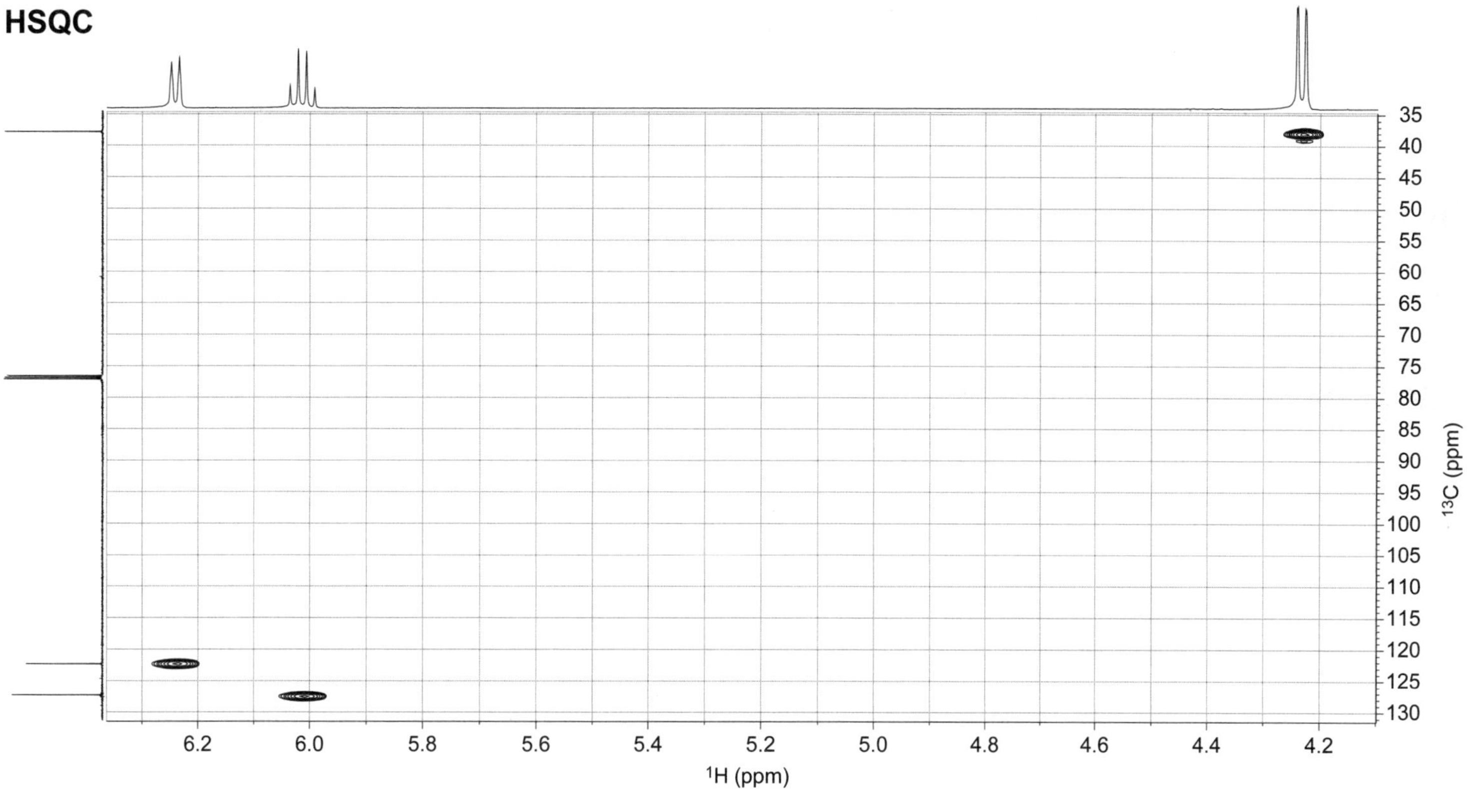

6.2 6.0 5.8 5.6 5.4 5.2 5.0 4.8 4.6 4.4 4.2

^{1}H (ppm)

35 40 45 50 55 60 65 70 75 80 85 90 95 100 105 110 115 120 125 130

^{13}C (ppm)

HMBC

6.4 6.2 6.0 5.8 5.6 5.4 5.2 5.0 4.8 4.6 4.4 4.2 4.0

^{1}H (ppm)

25 30 35 40 45 50 55 60 65 70 75 80 85 90 95 100 105 110 115 120 125 130 135

^{13}C (ppm)

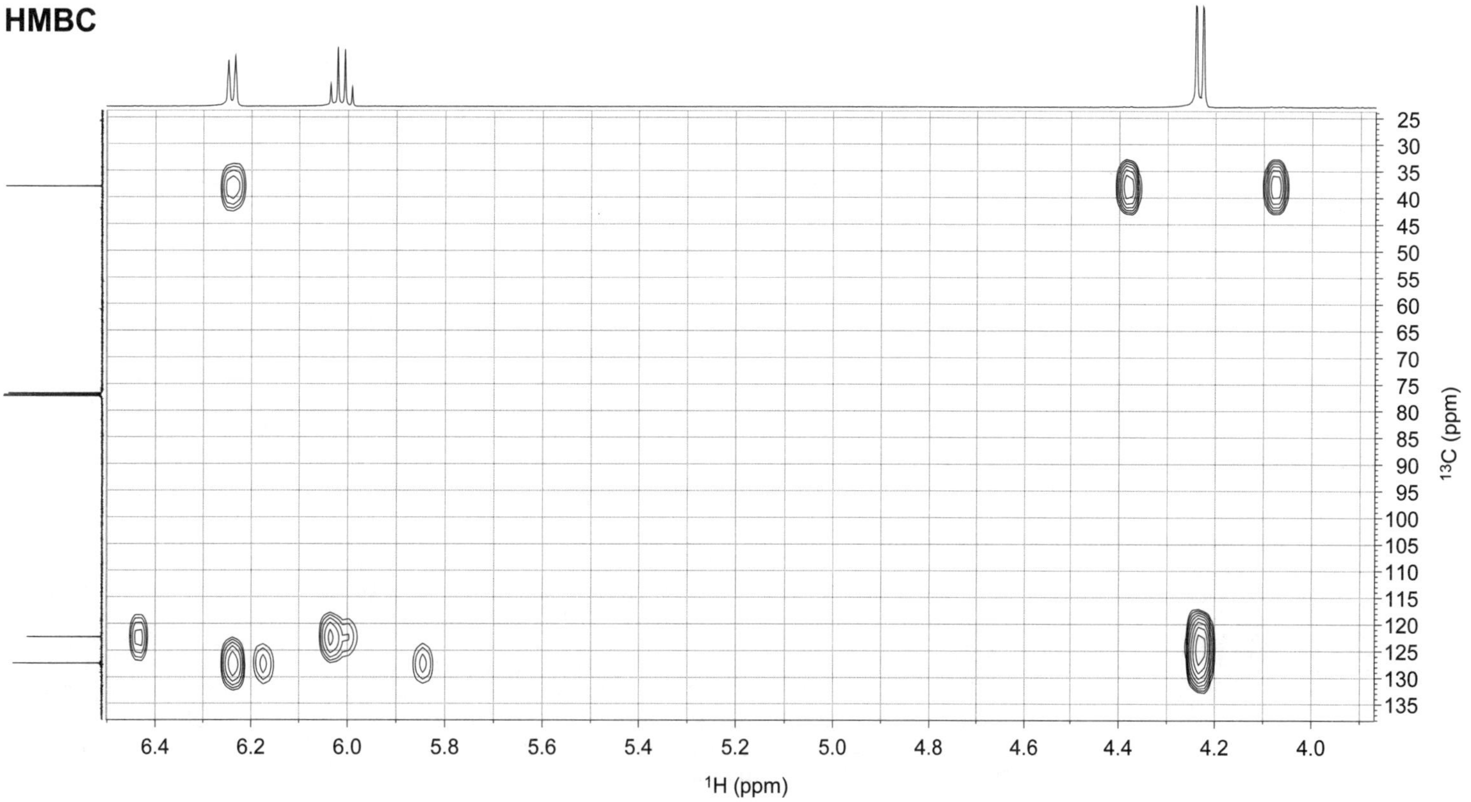

Problem 9.40 Molecular formula: $C_4H_9NO_2$ ***Difficulty*** ★☆☆☆

Solvent: D_2O

Field: 500 MHz (^1H); 125 MHz (^{13}C)

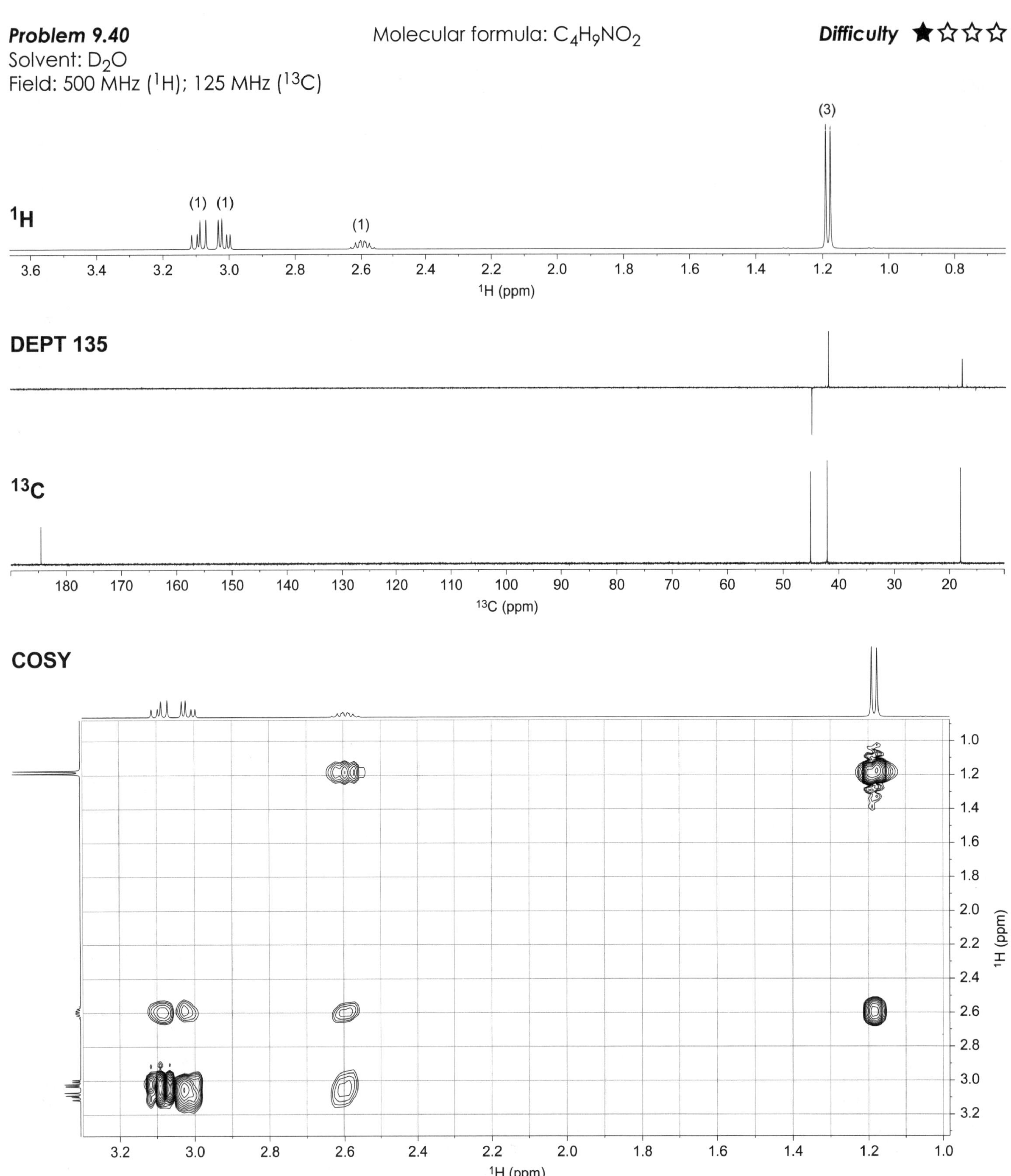

HSQC

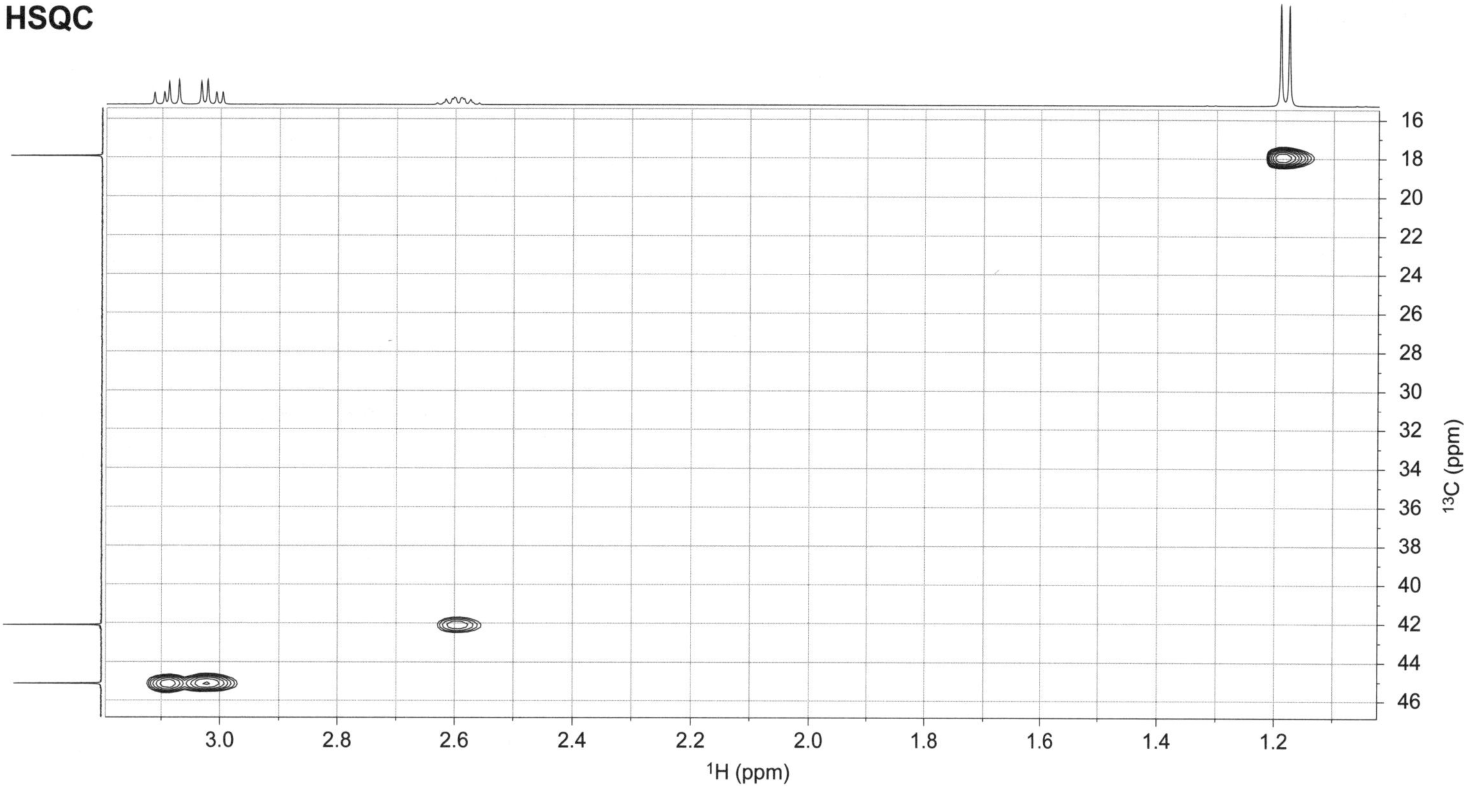
16
18
20
22
24
26
28
30
32
34
36
38
40
42
44
46
13C (ppm)
3.0
2.8
2.6
2.4
2.2
2.0
1.8
1.6
1.4
1.2
1H (ppm)

HMBC

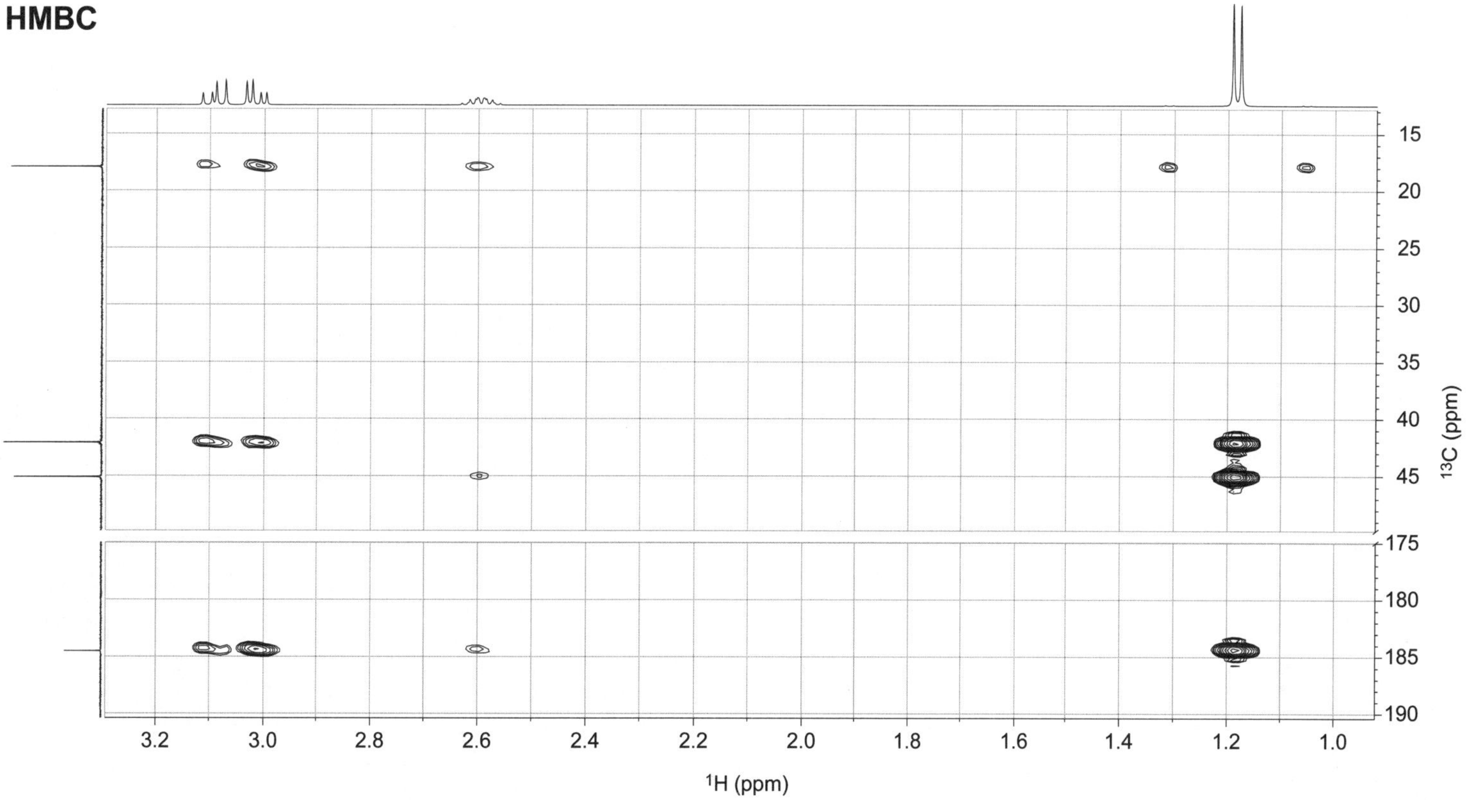
15
20
25
30
35
40
45
13C (ppm)
175
180
185
190
3.2
3.0
2.8
2.6
2.4
2.2
2.0
1.8
1.6
1.4
1.2
1.0
1H (ppm)

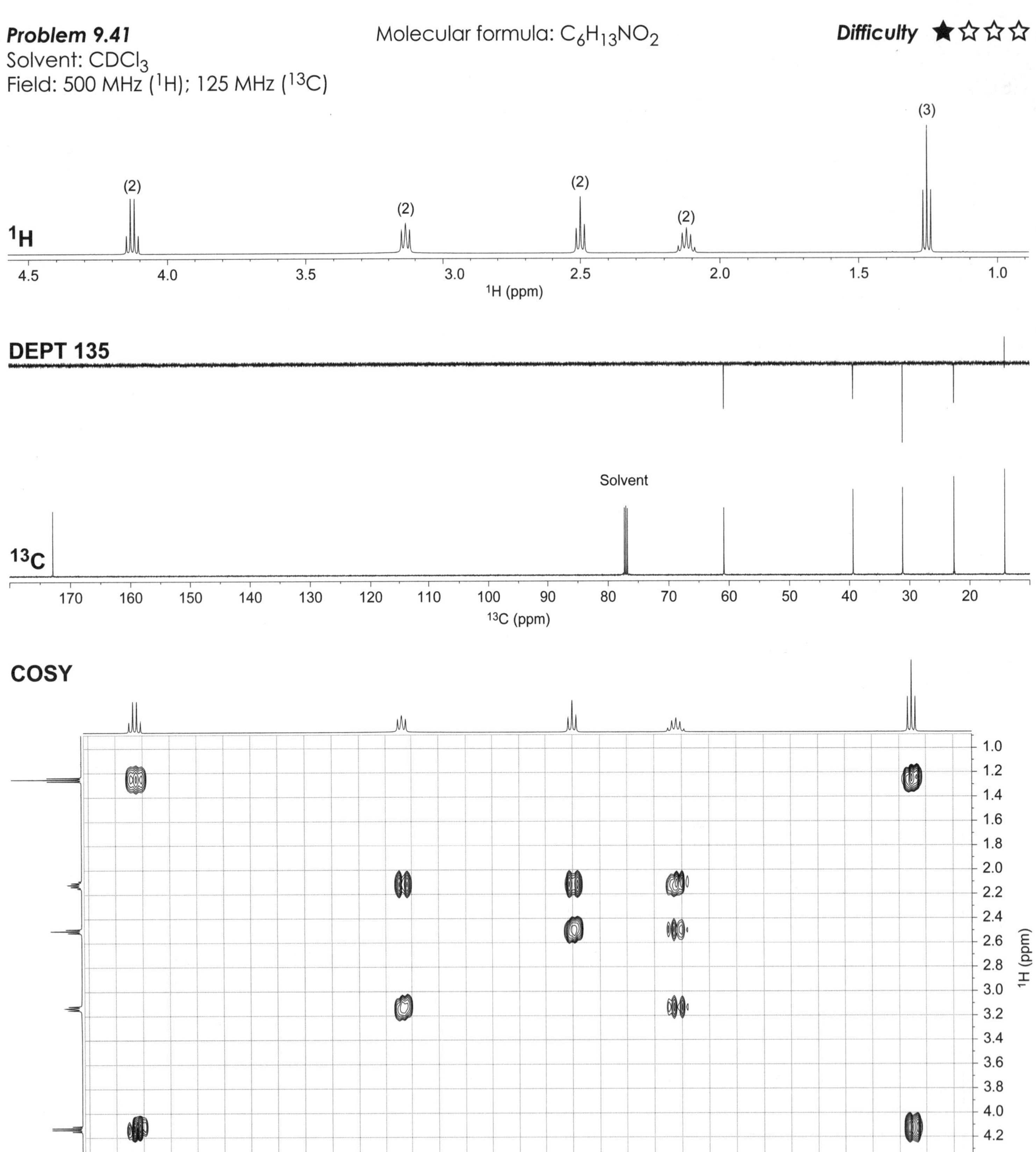
Problem 9.41
Molecular formula: $C_6H_{13}NO_2$
Difficulty
Solvent: $CDCl_3$
Field: 500 MHz (^1H); 125 MHz (^{13}C)
(2)
(2)
(2)
(2)
(3)
^1H
4.5 4.0 3.5 3.0 2.5 2.0 1.5 1.0
^1H (ppm)
DEPT 135
Solvent
^{13}C
170 160 150 140 130 120 110 100 90 80 70 60 50 40 30 20
^{13}C (ppm)
COSY
1.0 1.2 1.4 1.6 1.8 2.0 2.2 2.4 2.6 2.8 3.0 3.2 3.4 3.6 3.8 4.0 4.2
^1H (ppm)
4.0 3.5 3.0 2.5 2.0 1.5
^1H (ppm)

HSQC

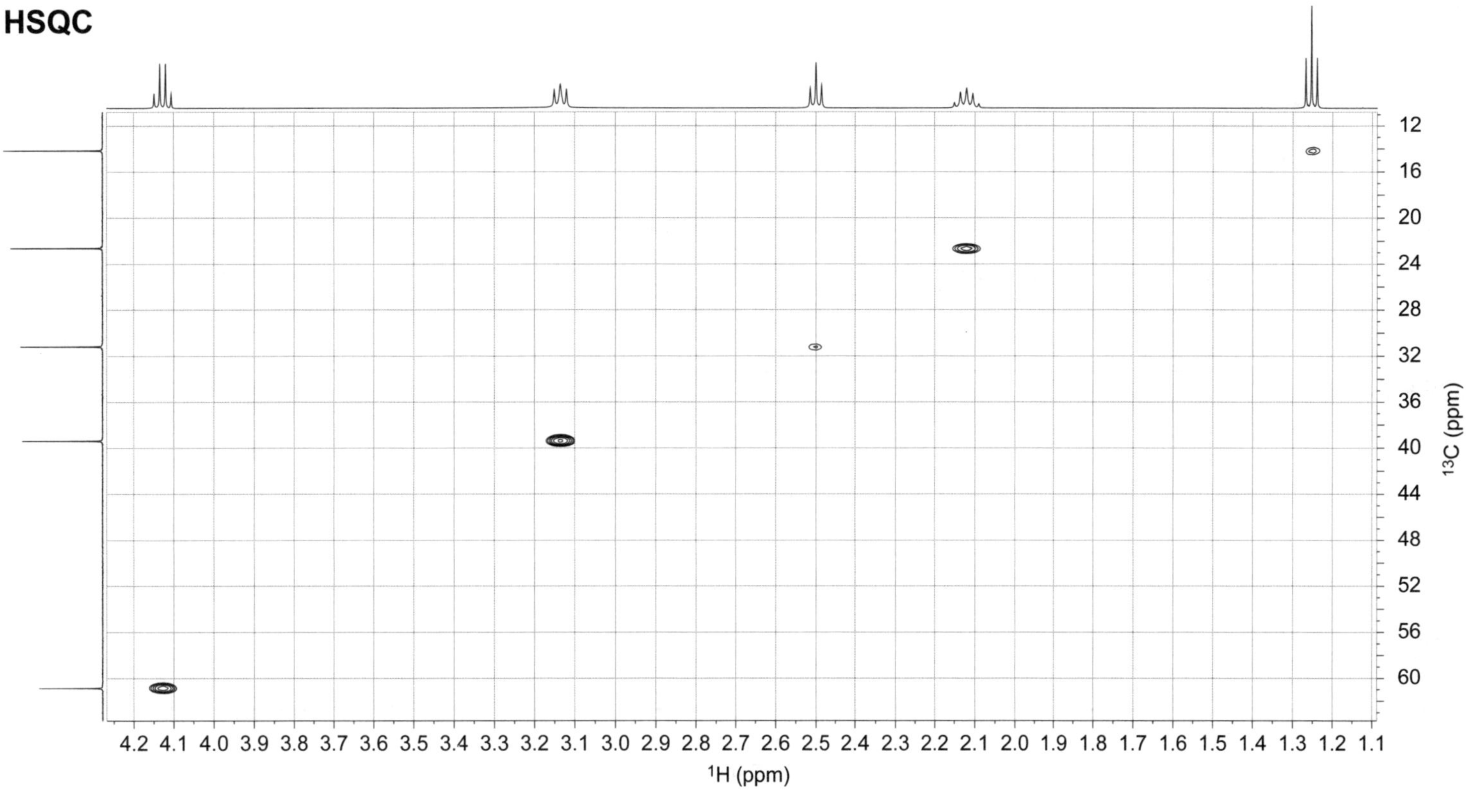
12
16
20
24
28
32
36
40
44
48
52
56
60
^{13}C (ppm)
4.2 4.1 4.0 3.9 3.8 3.7 3.6 3.5 3.4 3.3 3.2 3.1 3.0 2.9 2.8 2.7 2.6 2.5 2.4 2.3 2.2 2.1 2.0 1.9 1.8 1.7 1.6 1.5 1.4 1.3 1.2 1.1
^{1}H (ppm)

HMBC

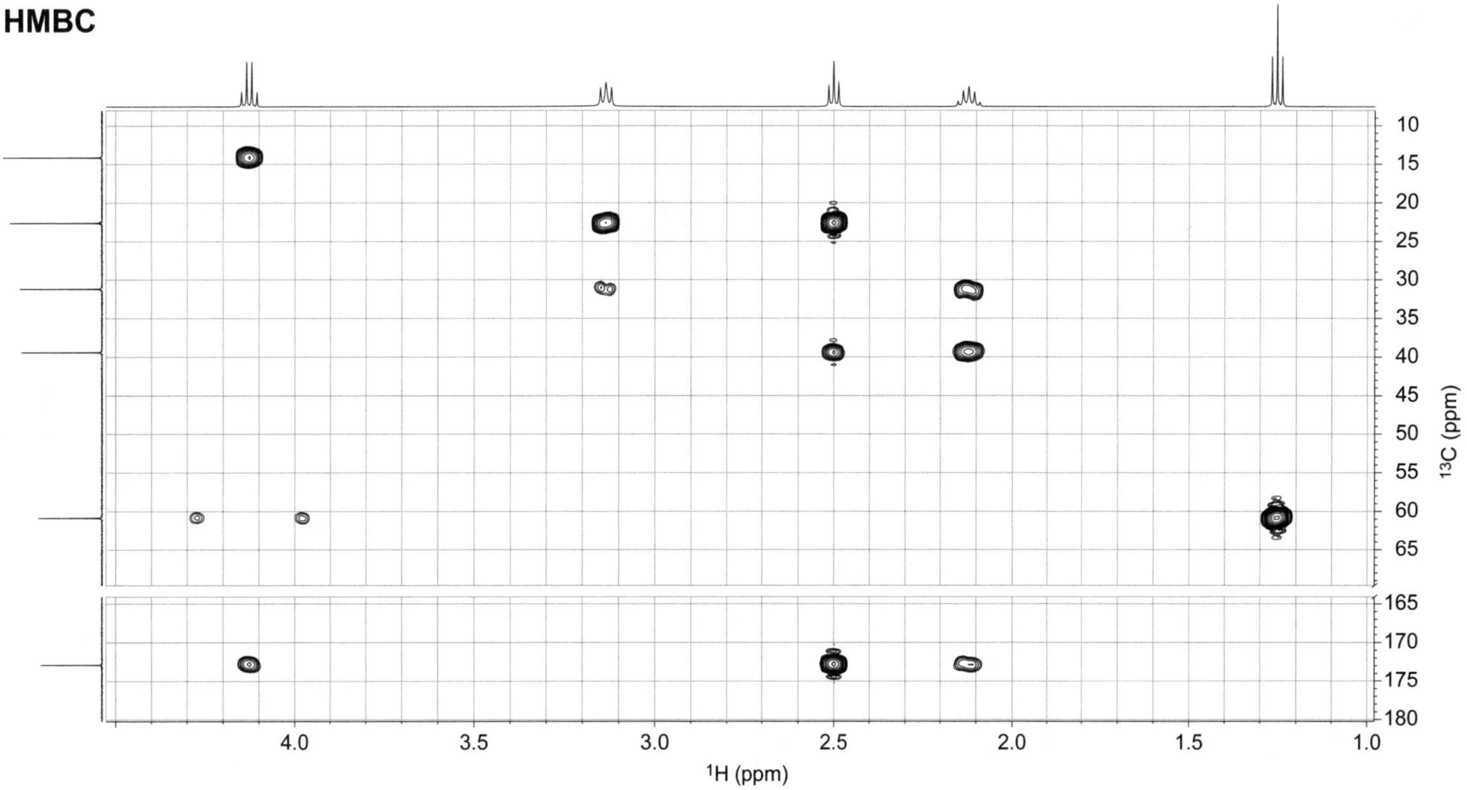
10
15
20
25
30
35
40
45
50
55
60
65
165
170
175
180
^{13}C (ppm)
4.0
3.5
3.0
2.5
2.0
1.5
1.0
^{1}H (ppm)

Problem 9.42 Molecular formula: $C_5H_{11}NO_2$ **Difficulty** ★☆☆☆

Solvent: D_2O

Field: 500 MHz (^{1}H); 125 MHz (^{13}C)

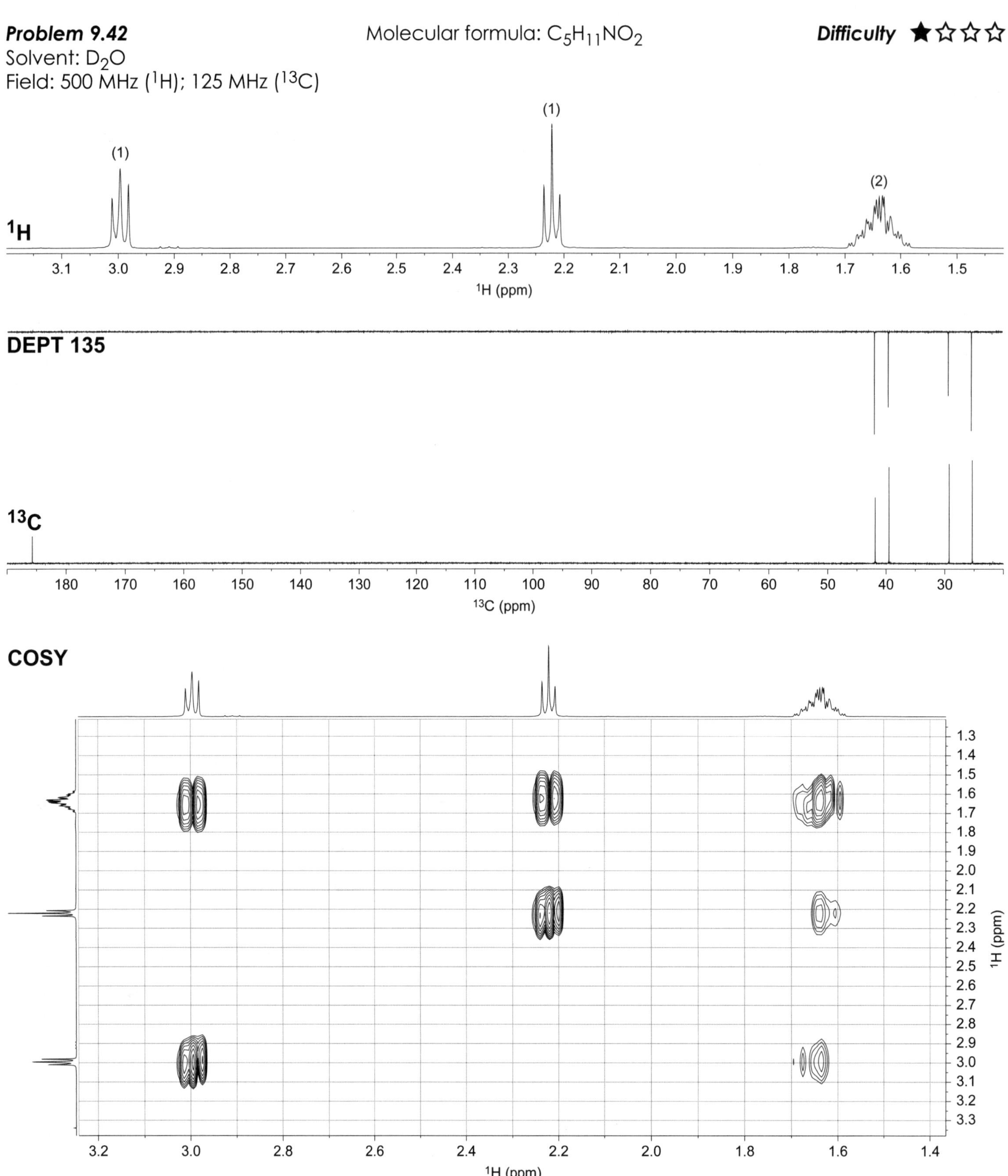

HSQC

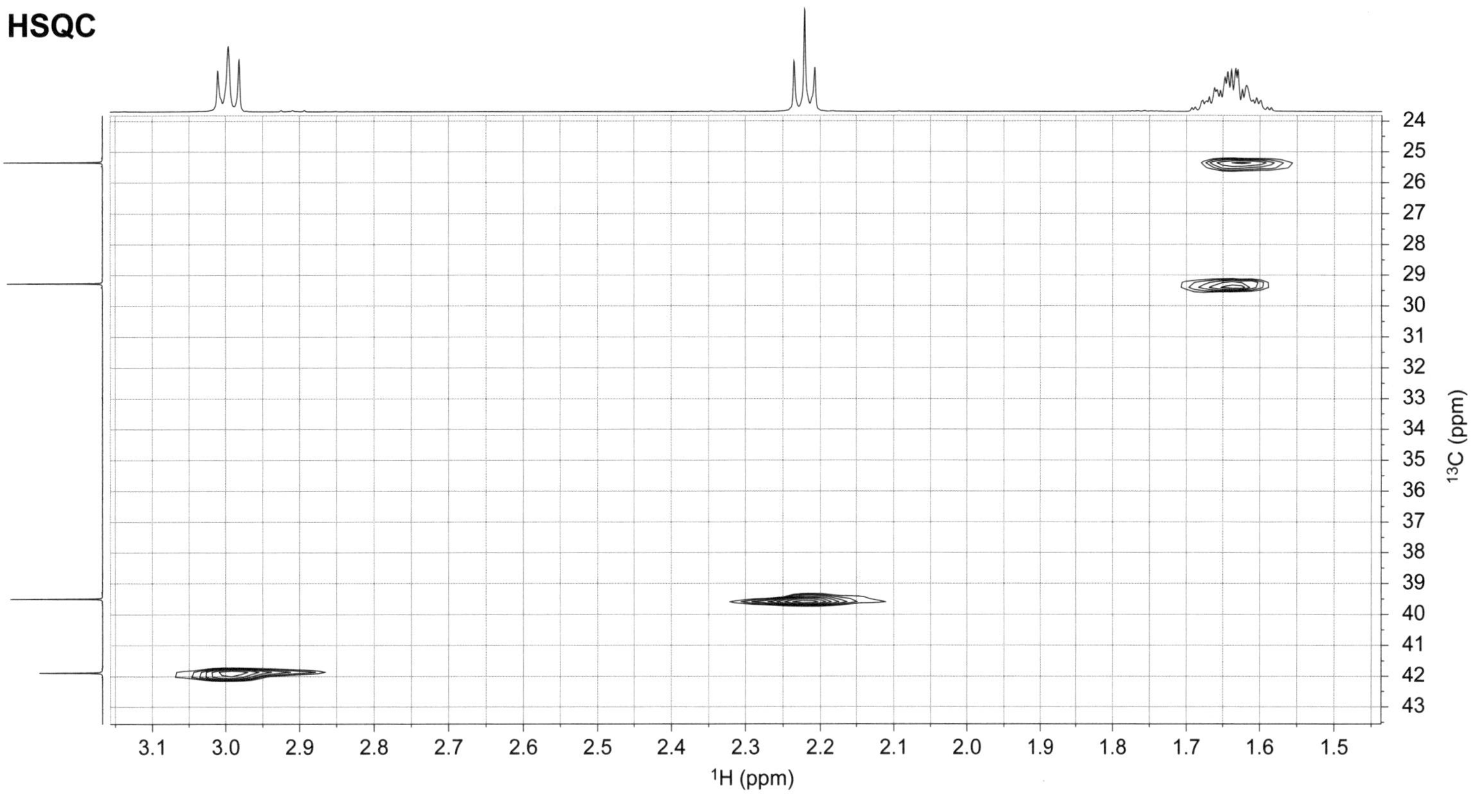
24
25
26
27
28
29
30
31
32
33
34
35
36
37
38
39
40
41
42
43
^{13}C (ppm)
3.1
3.0
2.9
2.8
2.7
2.6
2.5
2.4
2.3
2.2
2.1
2.0
1.9
1.8
1.7
1.6
1.5
^{1}H (ppm)

HMBC

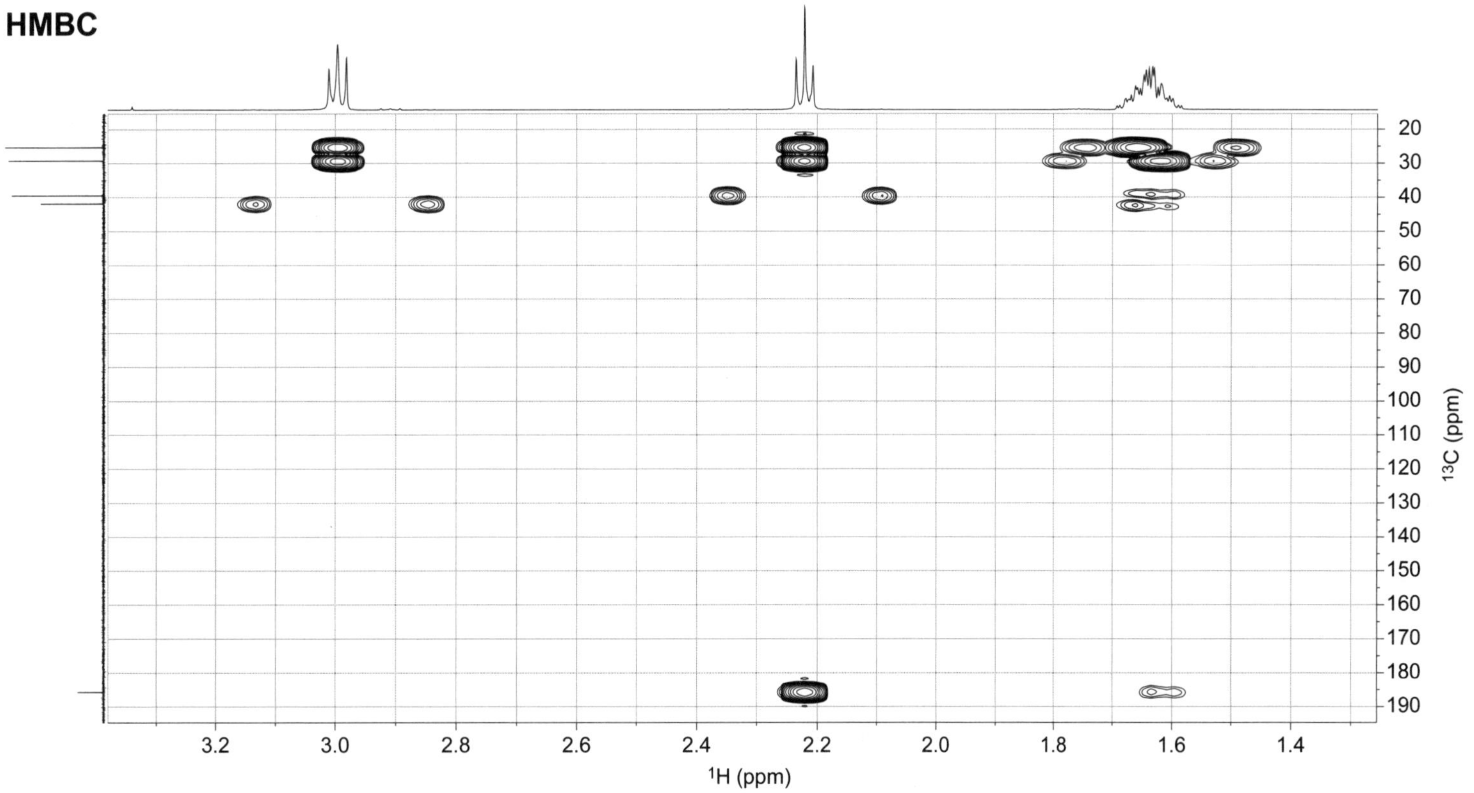
20
30
40
50
60
70
80
90
100
110
120
130
140
150
160
170
180
190
^{13}C (ppm)
3.2
3.0
2.8
2.6
2.4
2.2
2.0
1.8
1.6
1.4
^{1}H (ppm)

Problem 9.43 Molecular formula: $C_7H_5IO_2$ **Difficulty** ★☆☆☆

Solvent: CD_3OD

Field: 700 MHz (^1H); 175 MHz (^{13}C)

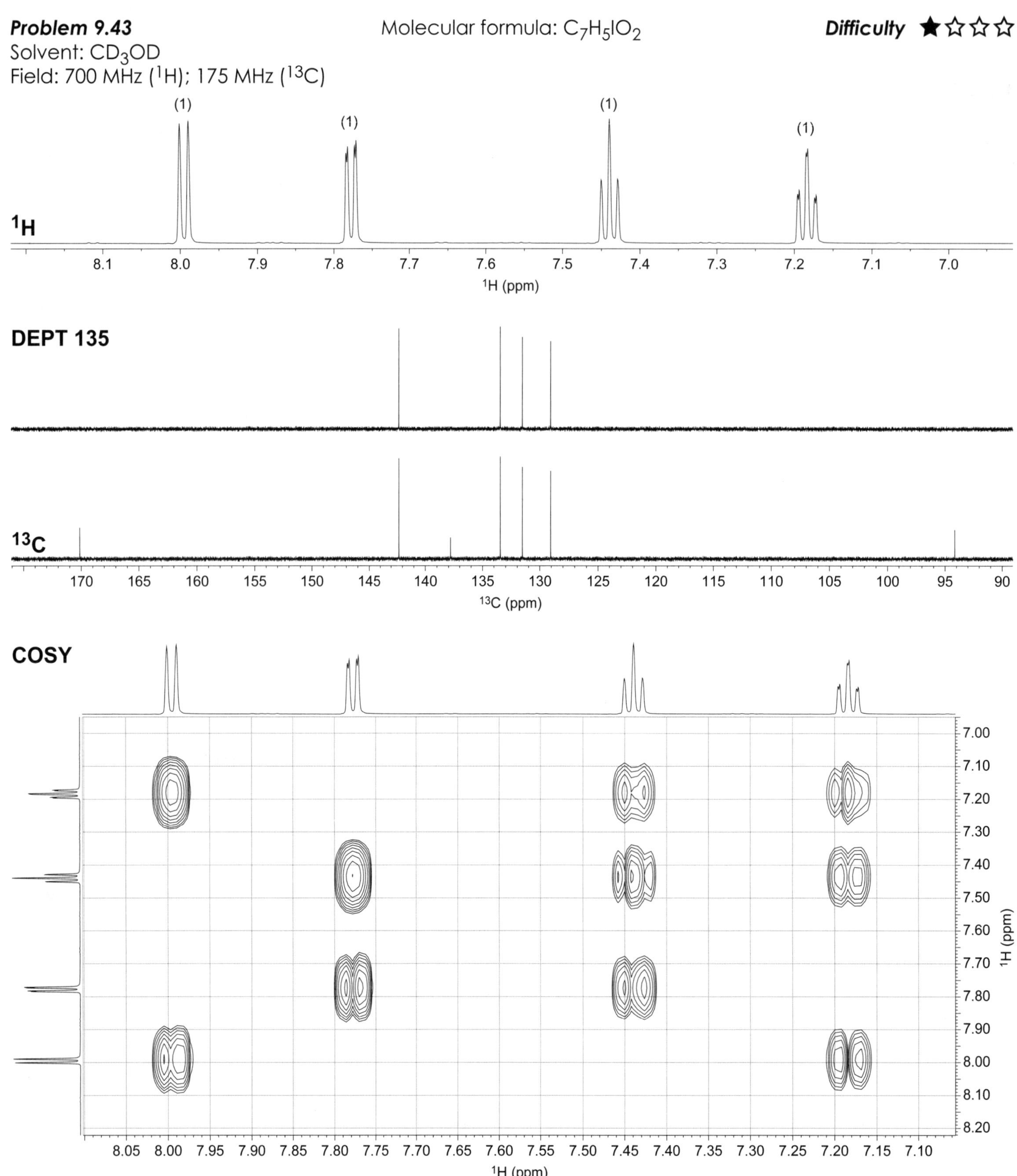

HSQC

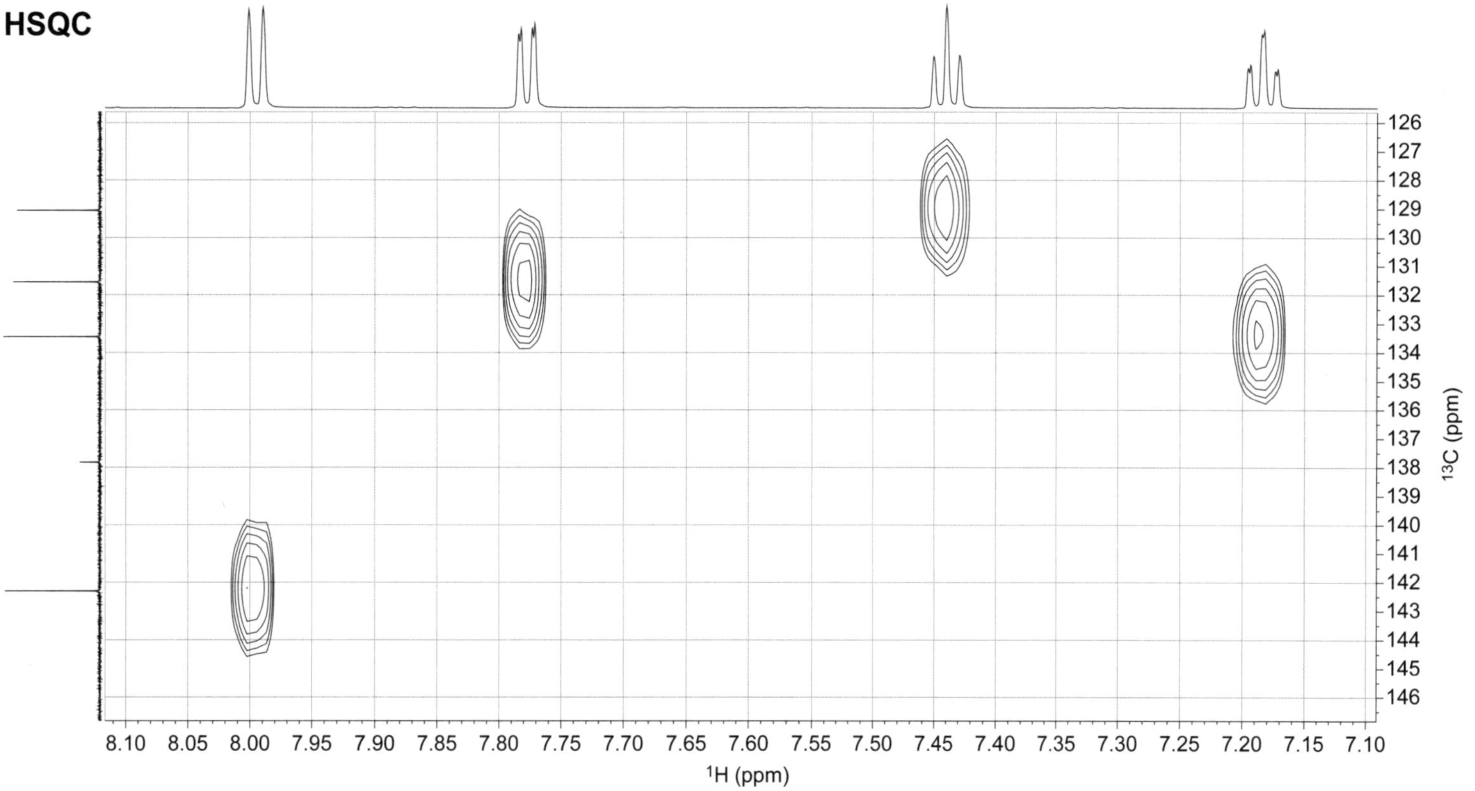

126 127 128 129 130 131 132 133 134 135 136 137 138 139 140 141 142 143 144 145 146

^{13}C (ppm)

8.10 8.05 8.00 7.95 7.90 7.85 7.80 7.75 7.70 7.65 7.60 7.55 7.50 7.45 7.40 7.35 7.30 7.25 7.20 7.15 7.10

^{1}H (ppm)

HMBC

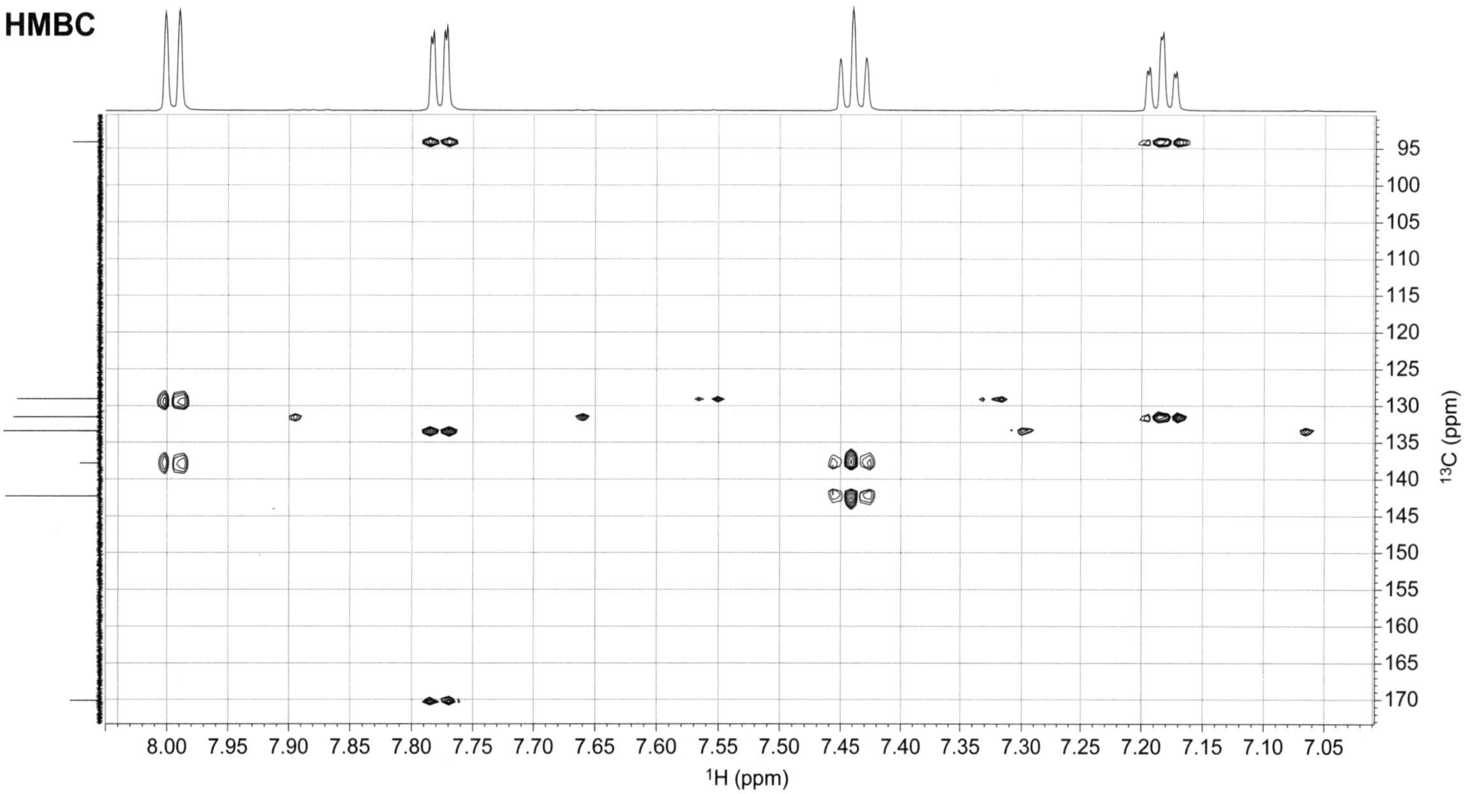

95 100 105 110 115 120 125 130 135 140 145 150 155 160 165 170

^{13}C (ppm)

8.00 7.95 7.90 7.85 7.80 7.75 7.70 7.65 7.60 7.55 7.50 7.45 7.40 7.35 7.30 7.25 7.20 7.15 7.10 7.05

^{1}H (ppm)

Problem 9.44
Solvent: C_6D_6
Field: 500 MHz (^1H); 125 MHz (^{13}C)

Molecular formula: $C_8H_{10}O$

Difficulty ★★☆☆

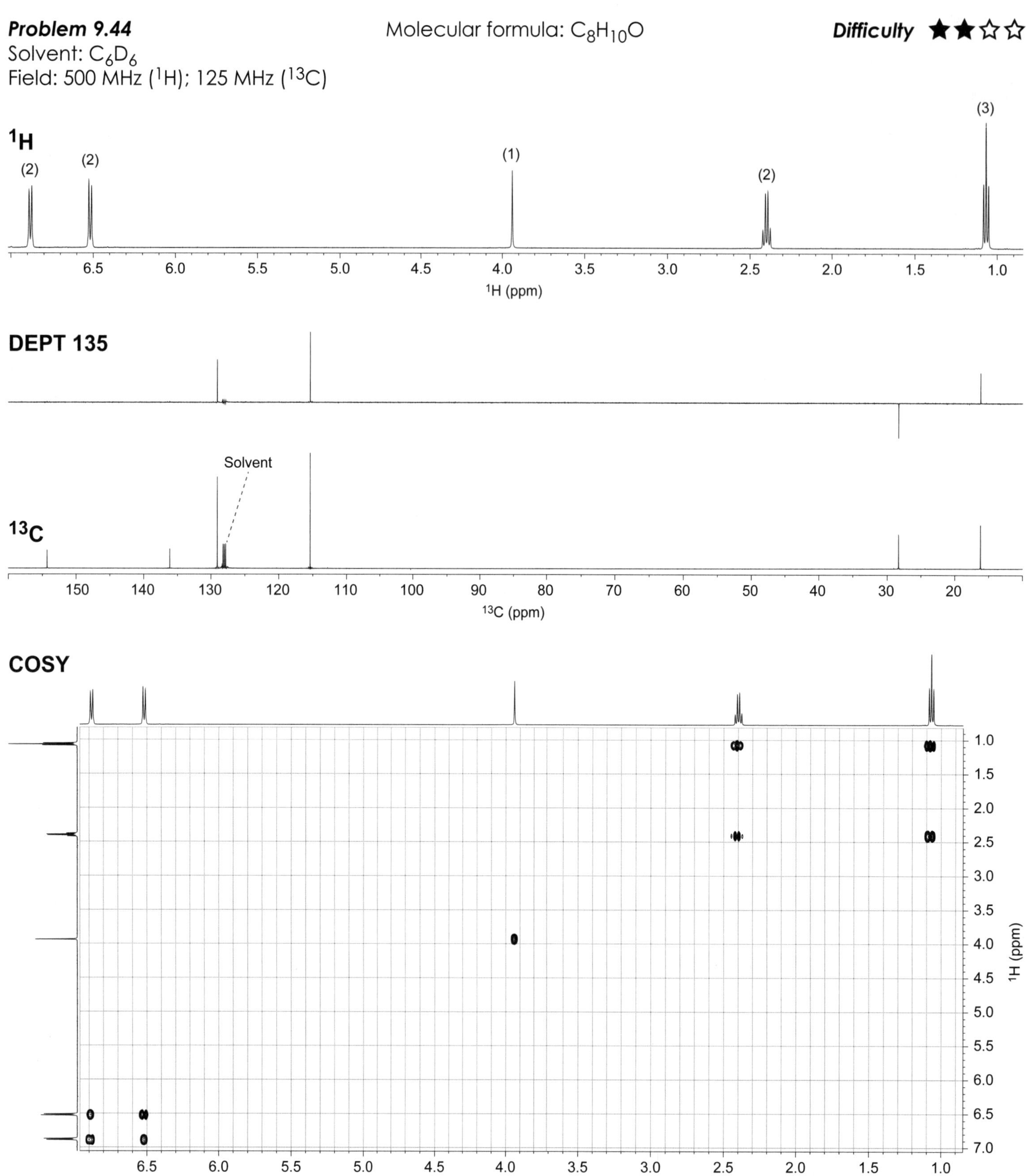

HSQC

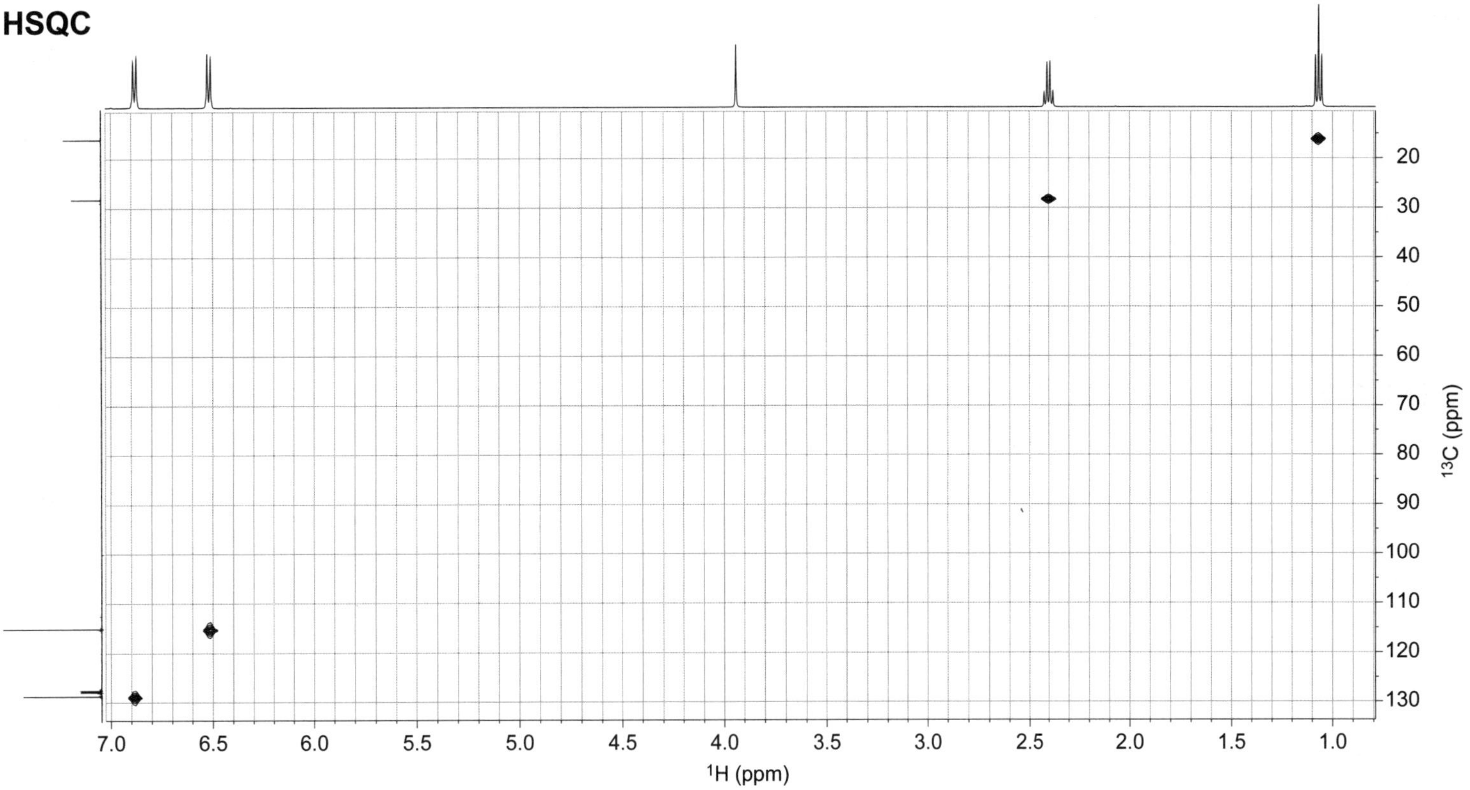

20 30 40 50 60 70 80 90 100 110 120 130

^{13}C (ppm)

7.0 6.5 6.0 5.5 5.0 4.5 4.0 3.5 3.0 2.5 2.0 1.5 1.0

^{1}H (ppm)

HMBC

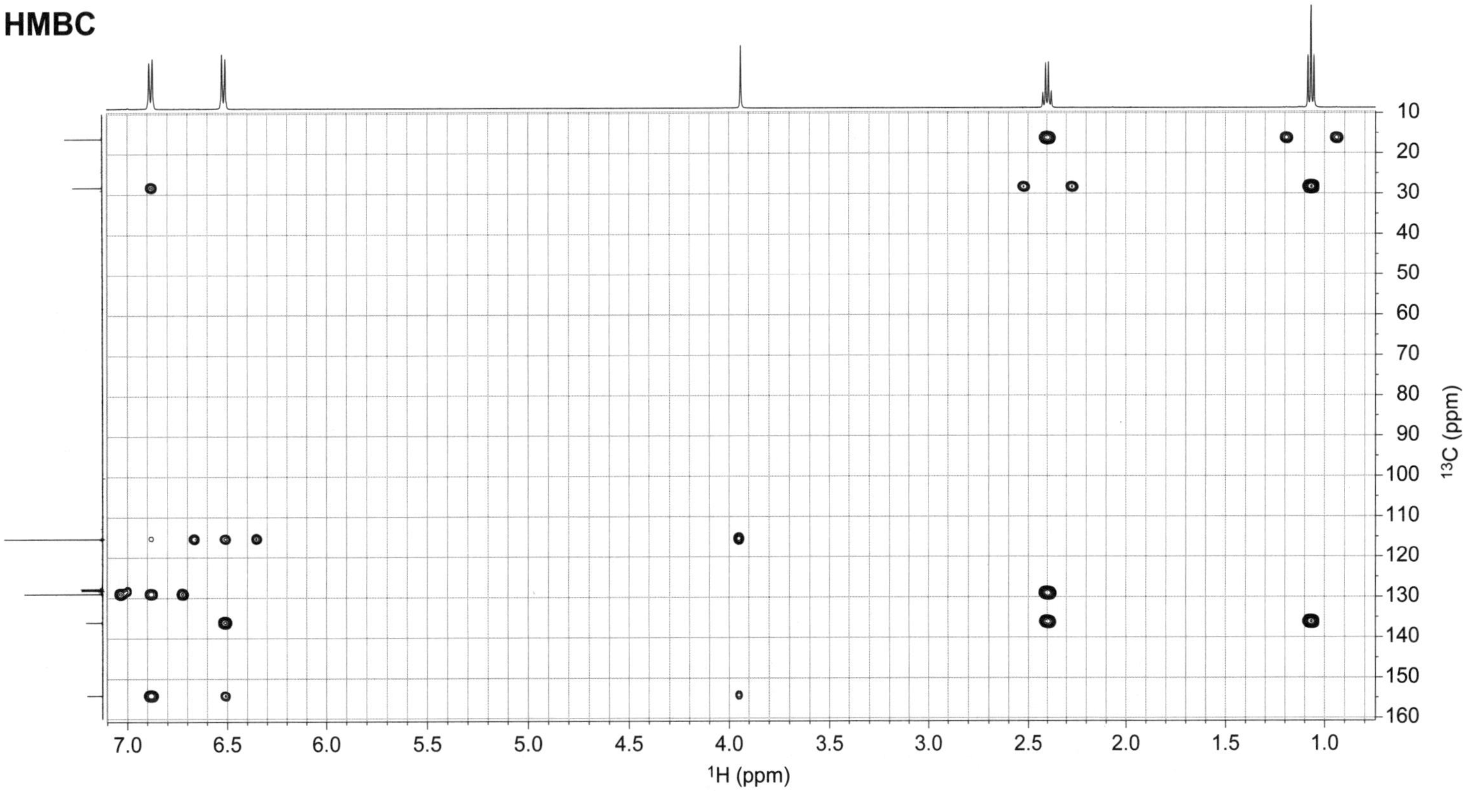

10 20 30 40 50 60 70 80 90 100 110 120 130 140 150 160

^{13}C (ppm)

7.0 6.5 6.0 5.5 5.0 4.5 4.0 3.5 3.0 2.5 2.0 1.5 1.0

^{1}H (ppm)

Problem 9.45
Solvent: CD_3Cl
Field: 500 MHz (^1H); 125 MHz (^{13}C)

Molecular formula: $C_{10}H_{12}O_2$

Difficulty ★★☆☆

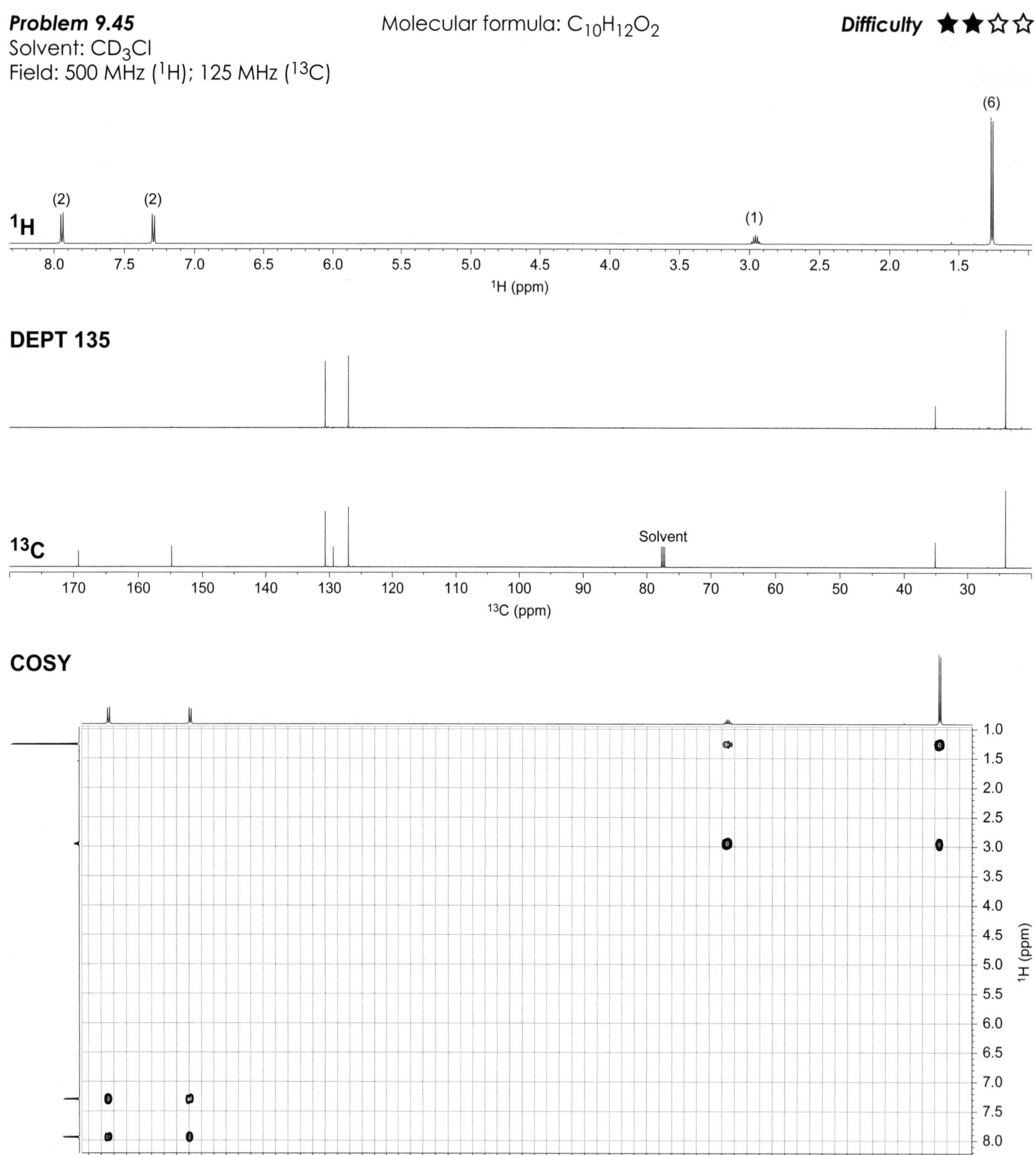

Difficulty ★★☆☆

HSQC

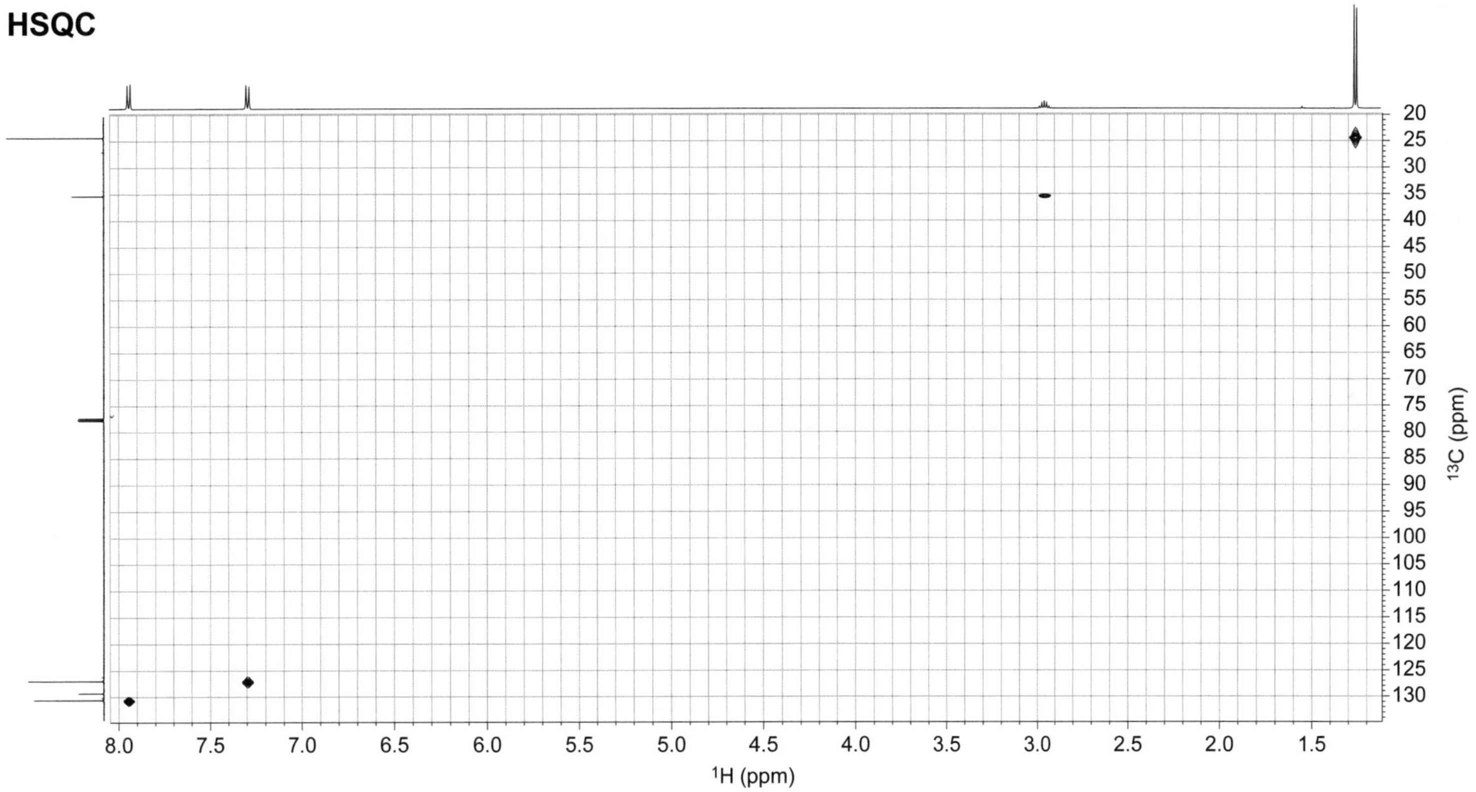

20 25 30 35 40 45 50 55 60 65 70 75 80 85 90 95 100 105 110 115 120 125 130

^{13}C (ppm)

8.0 7.5 7.0 6.5 6.0 5.5 5.0 4.5 4.0 3.5 3.0 2.5 2.0 1.5

^{1}H (ppm)

HMBC

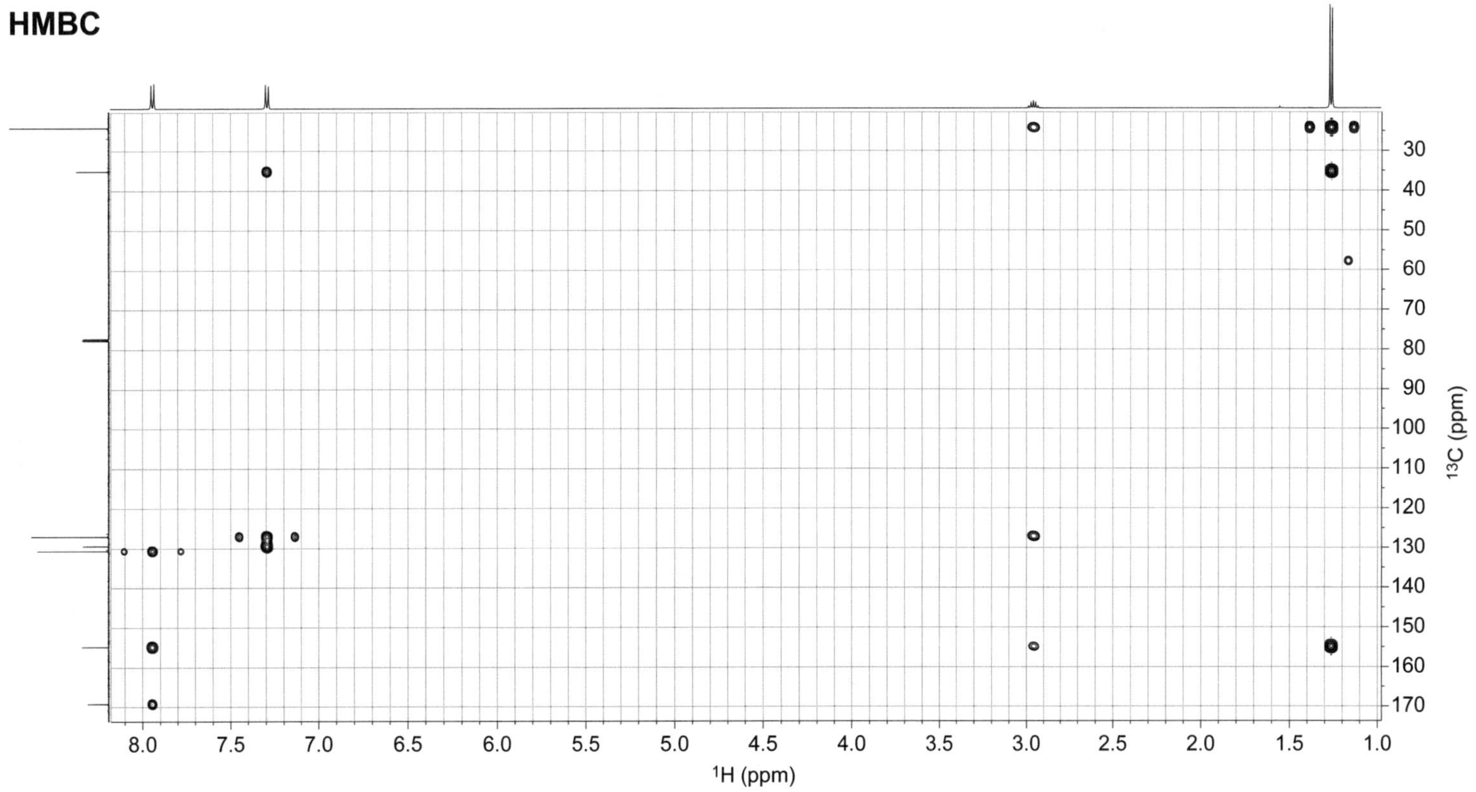

30 40 50 60 70 80 90 100 110 120 130 140 150 160 170

^{13}C (ppm)

8.0 7.5 7.0 6.5 6.0 5.5 5.0 4.5 4.0 3.5 3.0 2.5 2.0 1.5 1.0

^{1}H (ppm)

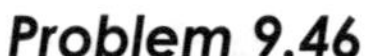

Problem 9.46
Solvent: CD_3-SO-CD_3
Field: 500 MHz (^1H); 125 MHz (^{13}C)

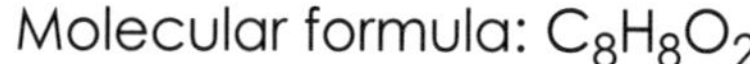

Molecular formula: $C_8H_8O_2$

Difficulty ★★☆☆

^{1}H

(3)
(2)
(2)
10.0
9.5
9.0
8.5
8.0
7.5
7.0
6.5
6.0
5.5
5.0
4.5
4.0
3.5
3.0
2.5
1H (ppm)

DEPT 135

^{13}C

Solvent
190
180
170
160
150
140
130
120
110
100
90
80
70
60
50
40
30
20
13C (ppm)

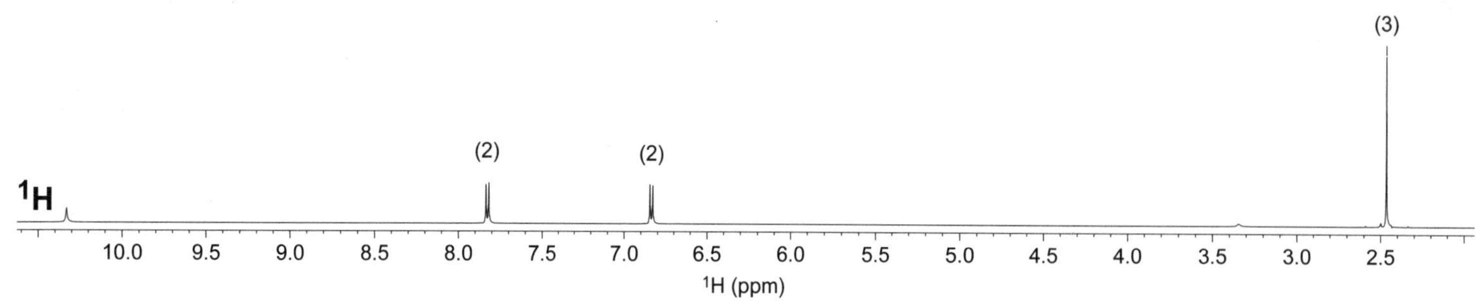

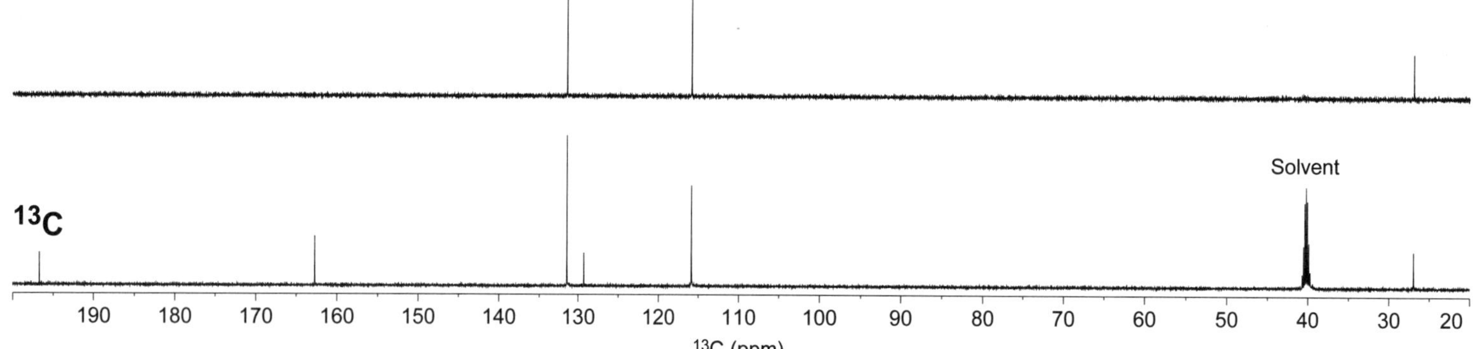

COSY

7.5
7.0
6.5
6.0
5.5
5.0
4.5
4.0
3.5
3.0
2.5
1H (ppm)
2.5
3.0
3.5
4.0
4.5
5.0
5.5
6.0
6.5
7.0
7.5
8.0
1H (ppm)

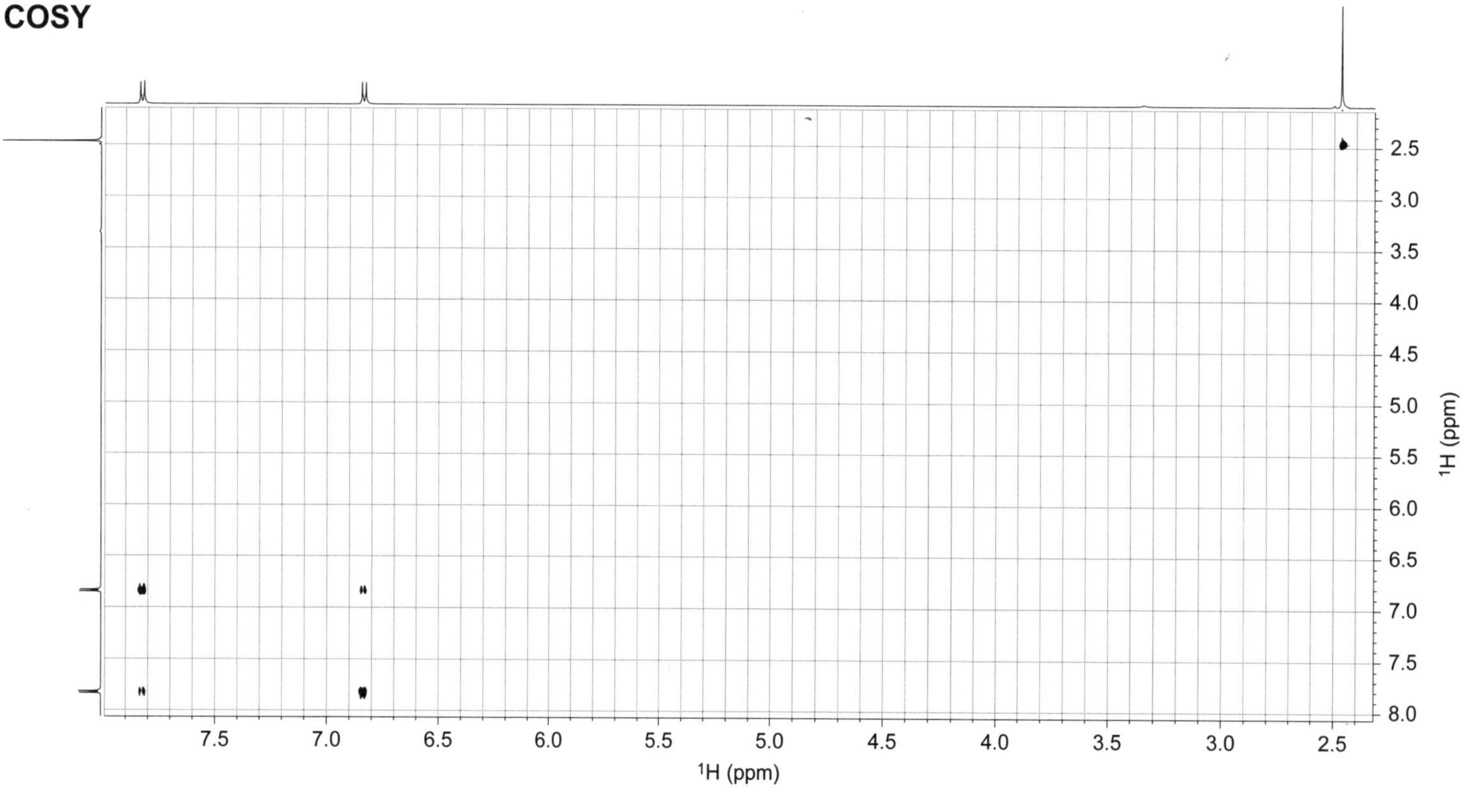

HSQC

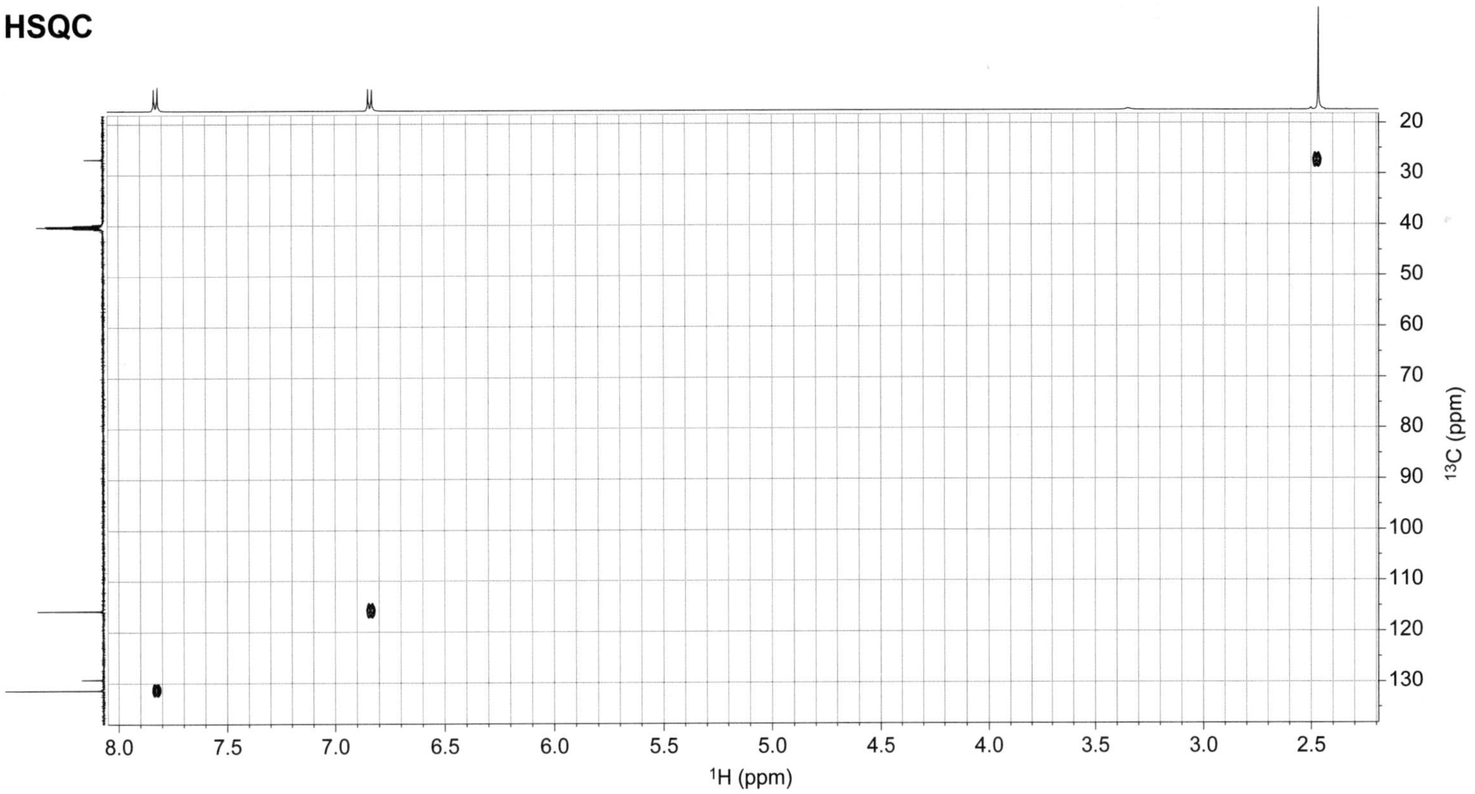

HMBC

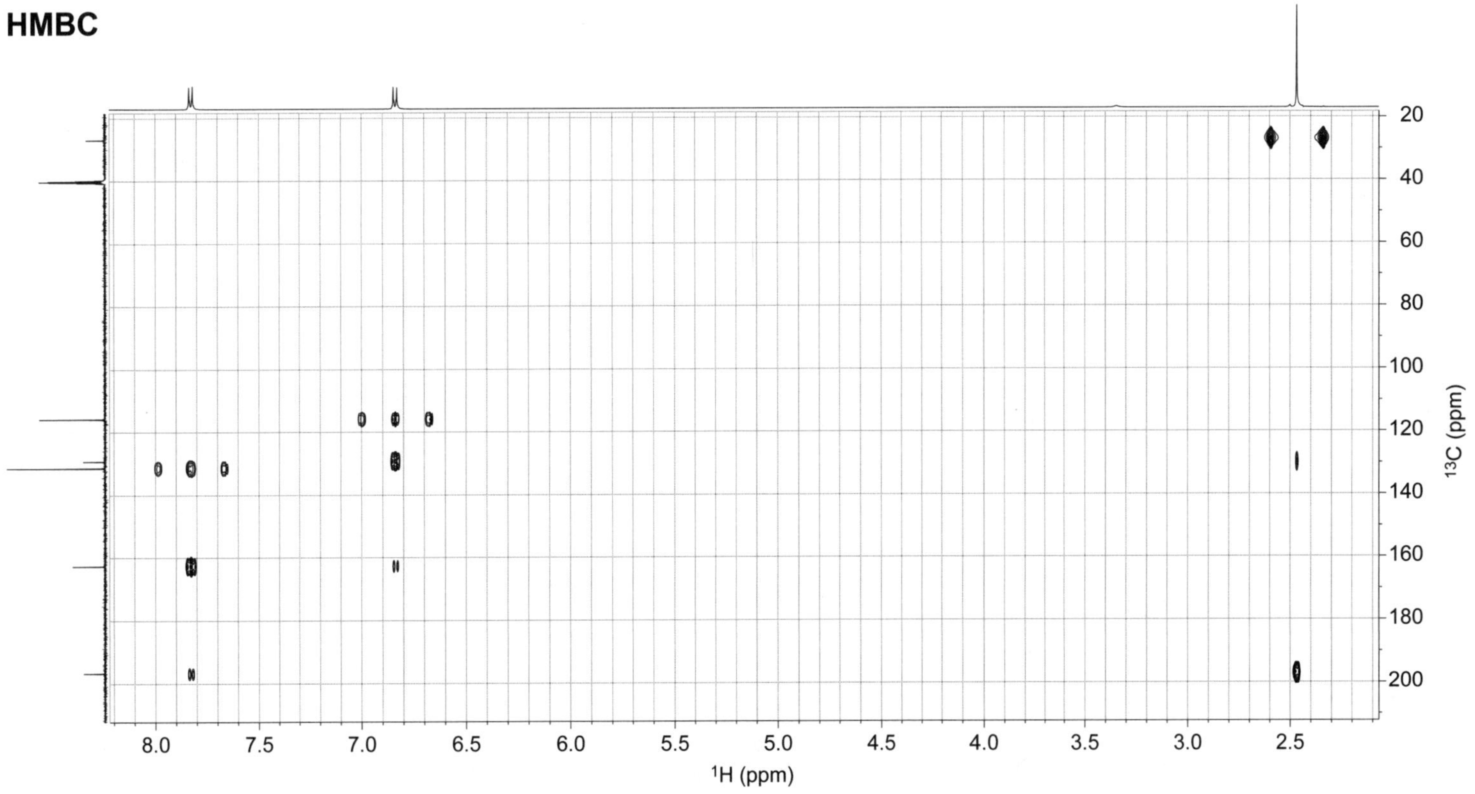

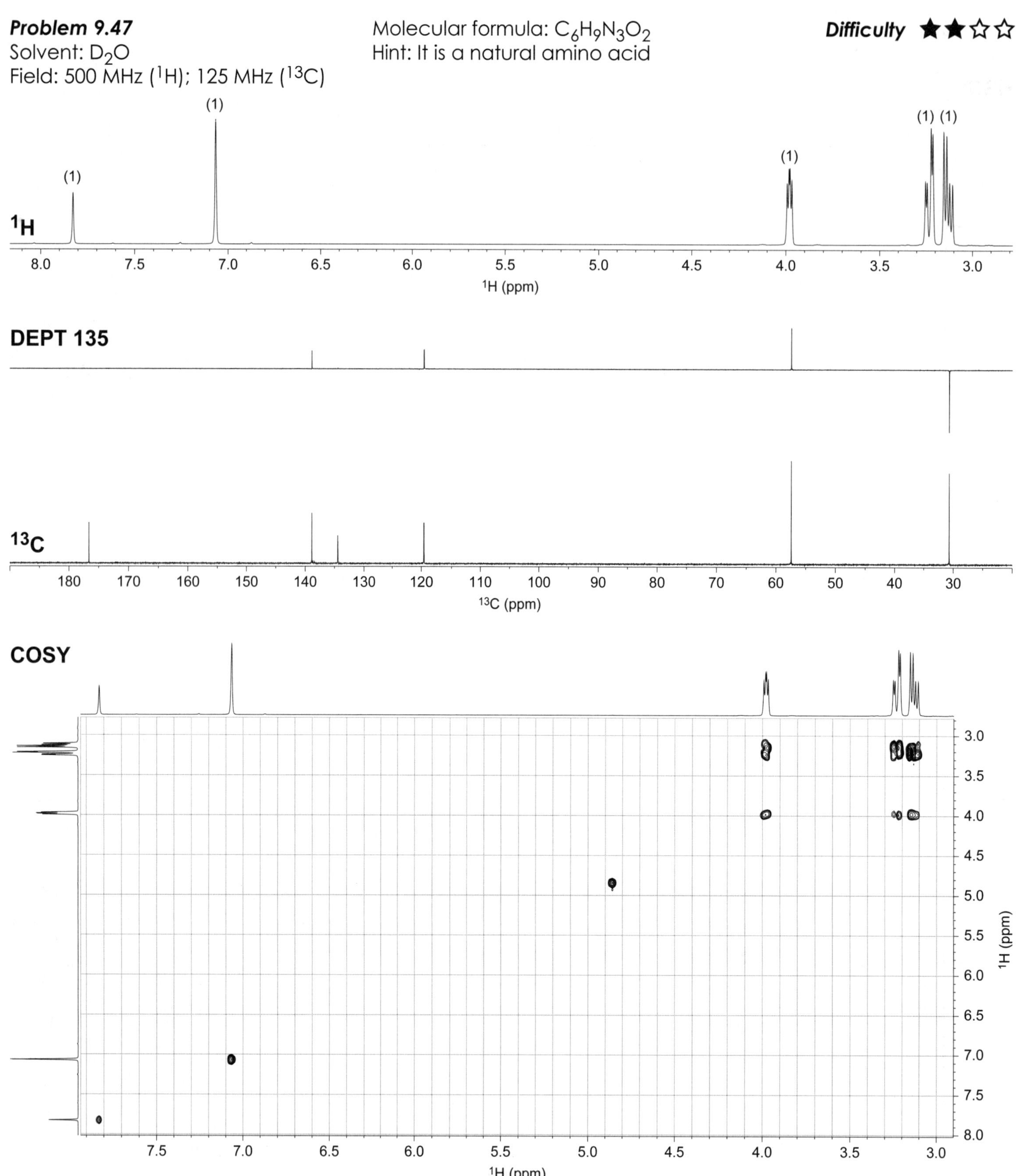
Problem 9.47
Solvent: D_2O
Field: 500 MHz (^{1}H); 125 MHz (^{13}C)
Molecular formula: $C_6H_9N_3O_2$
Hint: It is a natural amino acid
Difficulty ★★☆☆
(1)
(1)
(1)
(1) (1)
^{1}H
8.0 7.5 7.0 6.5 6.0 5.5 5.0 4.5 4.0 3.5 3.0
^{1}H (ppm)
DEPT 135
^{13}C
180 170 160 150 140 130 120 110 100 90 80 70 60 50 40 30
^{13}C (ppm)
COSY
7.5 7.0 6.5 6.0 5.5 5.0 4.5 4.0 3.5 3.0
^{1}H (ppm)
3.0 3.5 4.0 4.5 5.0 5.5 6.0 6.5 7.0 7.5 8.0
^{1}H (ppm)

HSQC

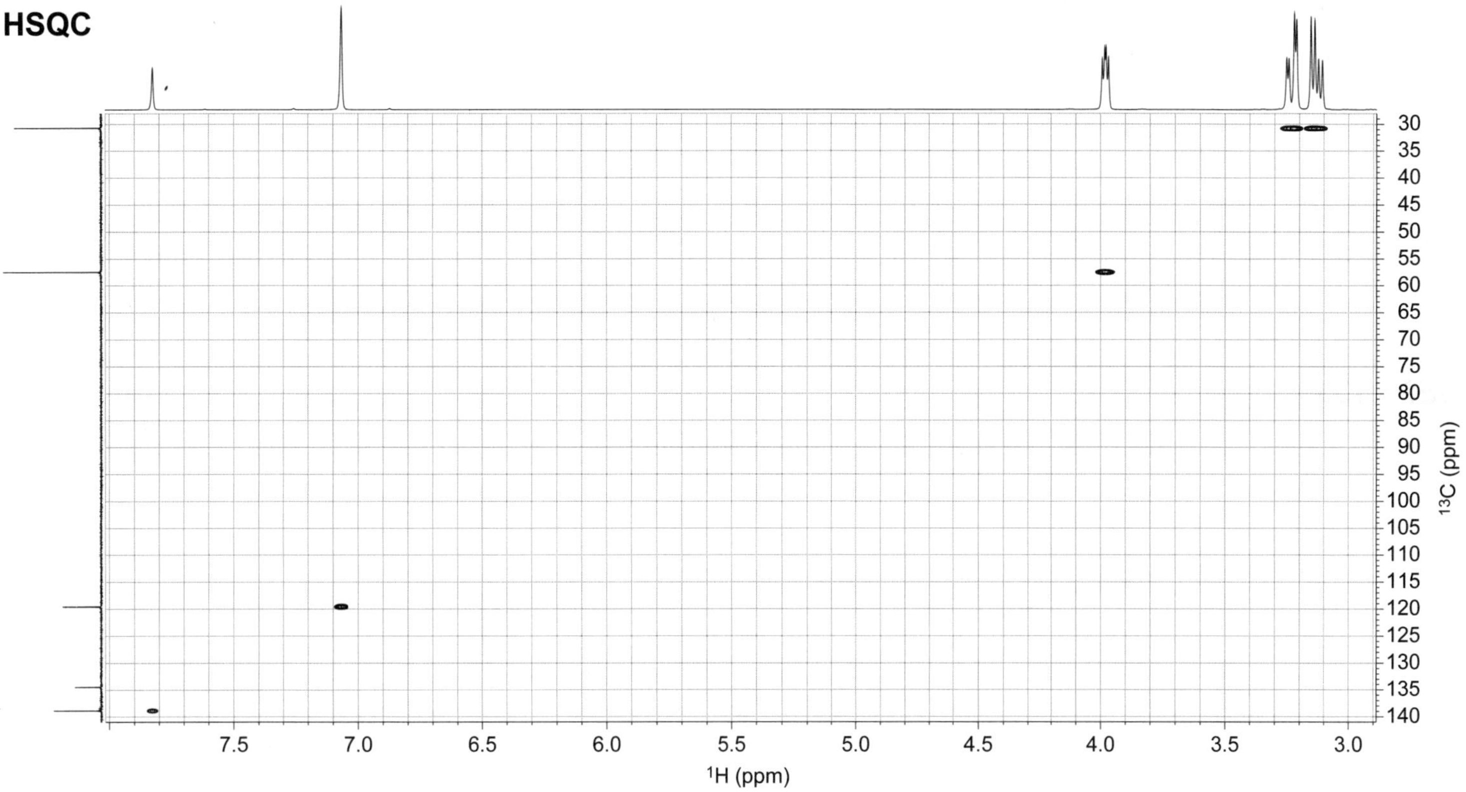

HMBC

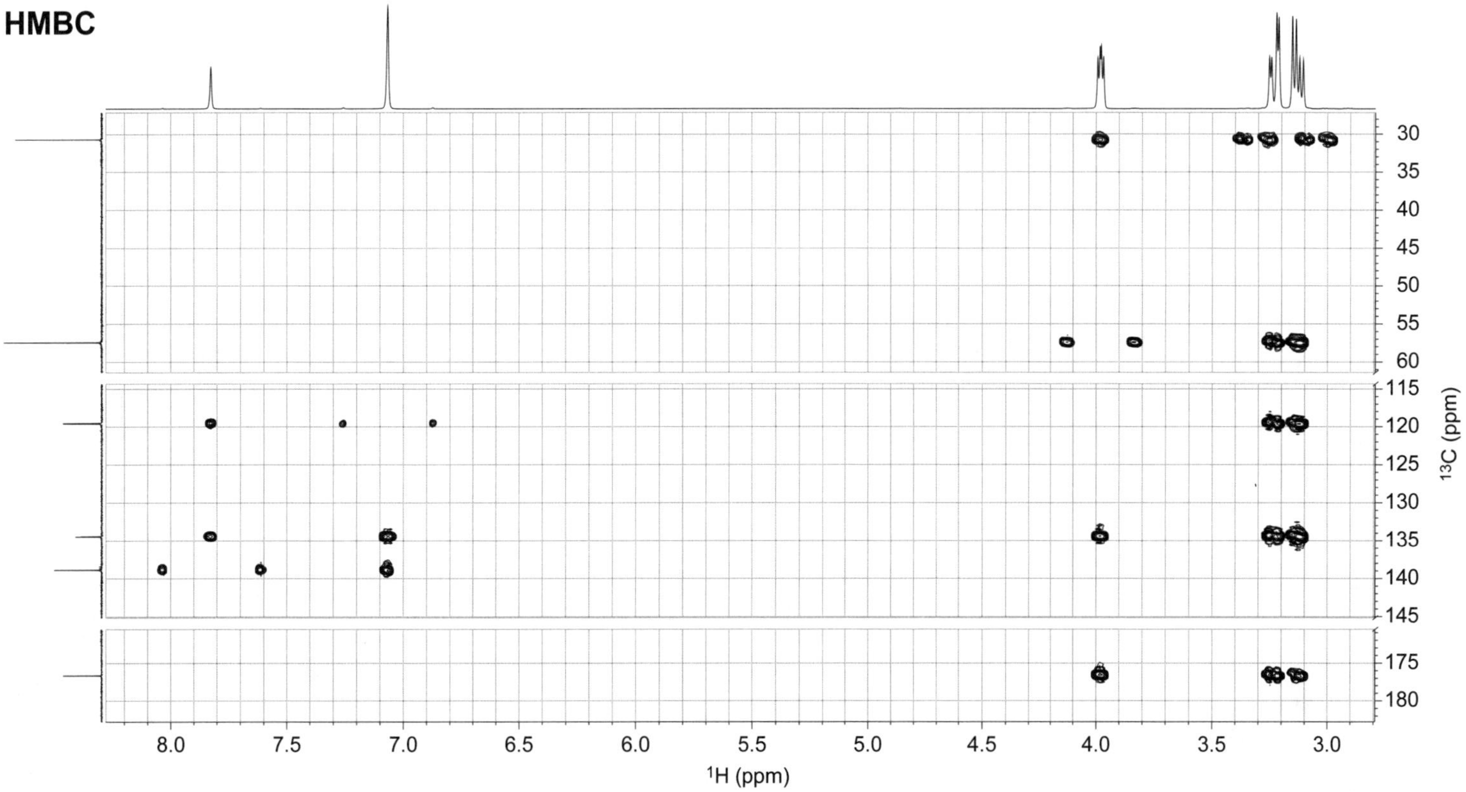

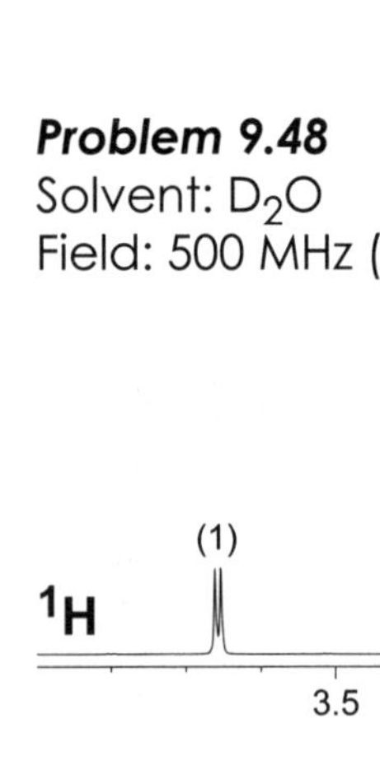

Problem 9.48
Solvent: D_2O
Field: 500 MHz (^{1}H); 125 MHz (^{13}C)

Molecular formula: $C_6H_{13}NO_2$
Hint: It is a natural amino acid

Difficulty 

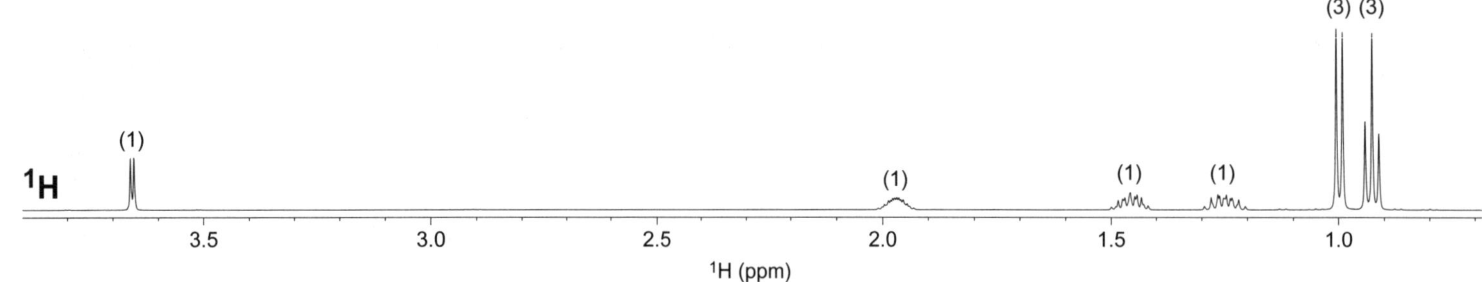

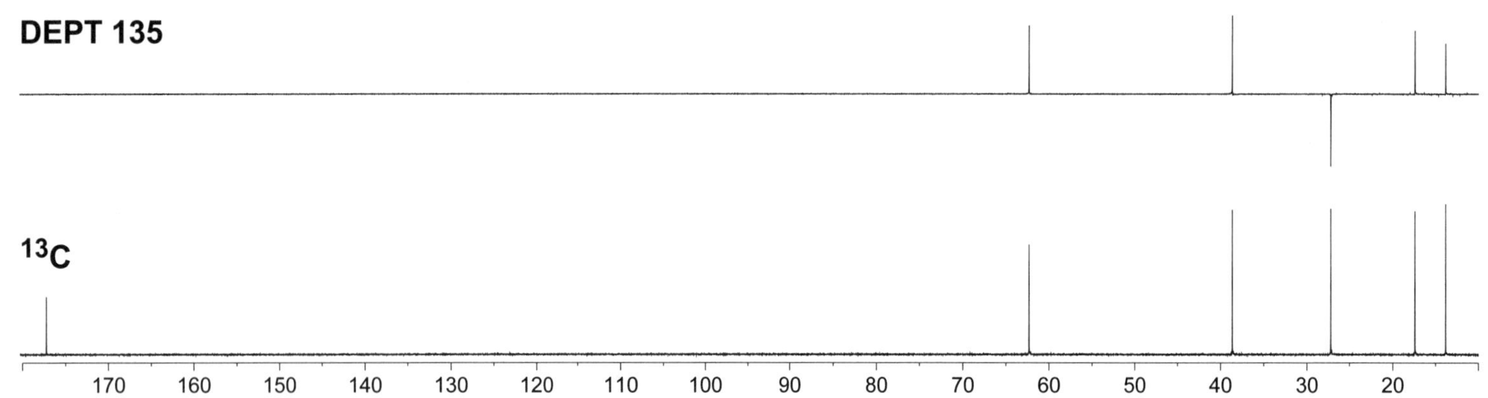

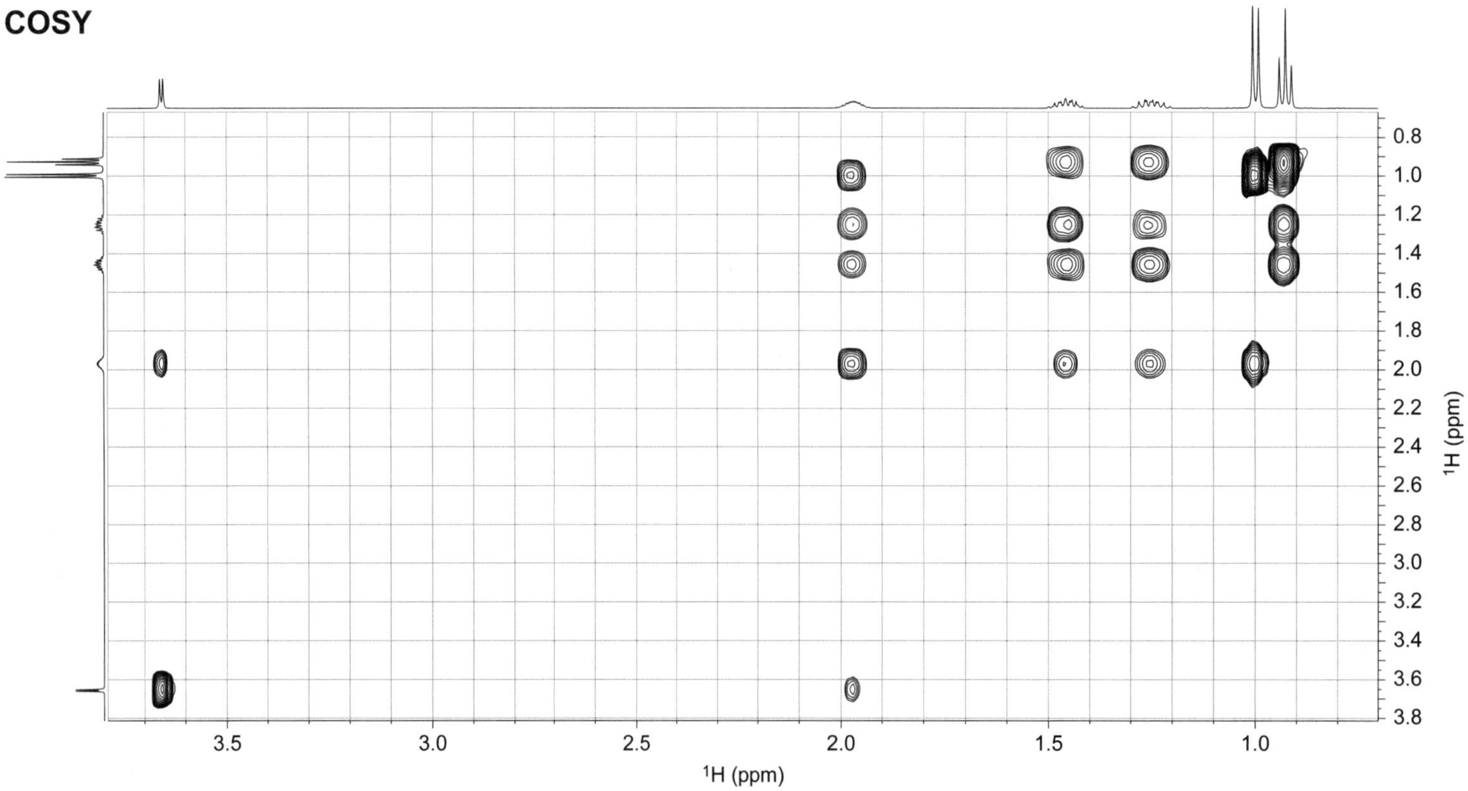

COSY

HSQC

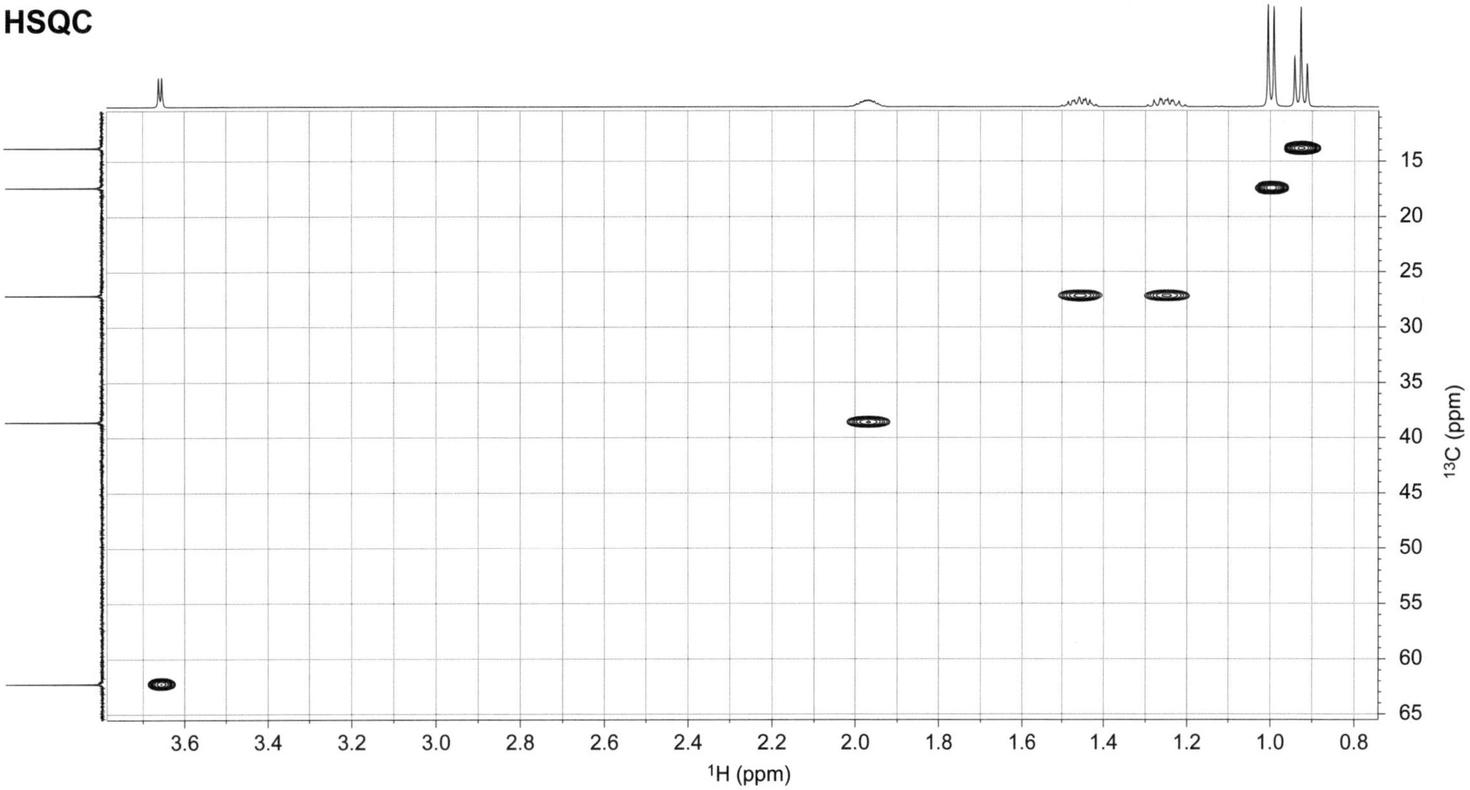
15
20
25
30
35
40
45
50
55
60
65
^{13}C (ppm)
3.6 3.4 3.2 3.0 2.8 2.6 2.4 2.2 2.0 1.8 1.6 1.4 1.2 1.0 0.8
^{1}H (ppm)

HMBC

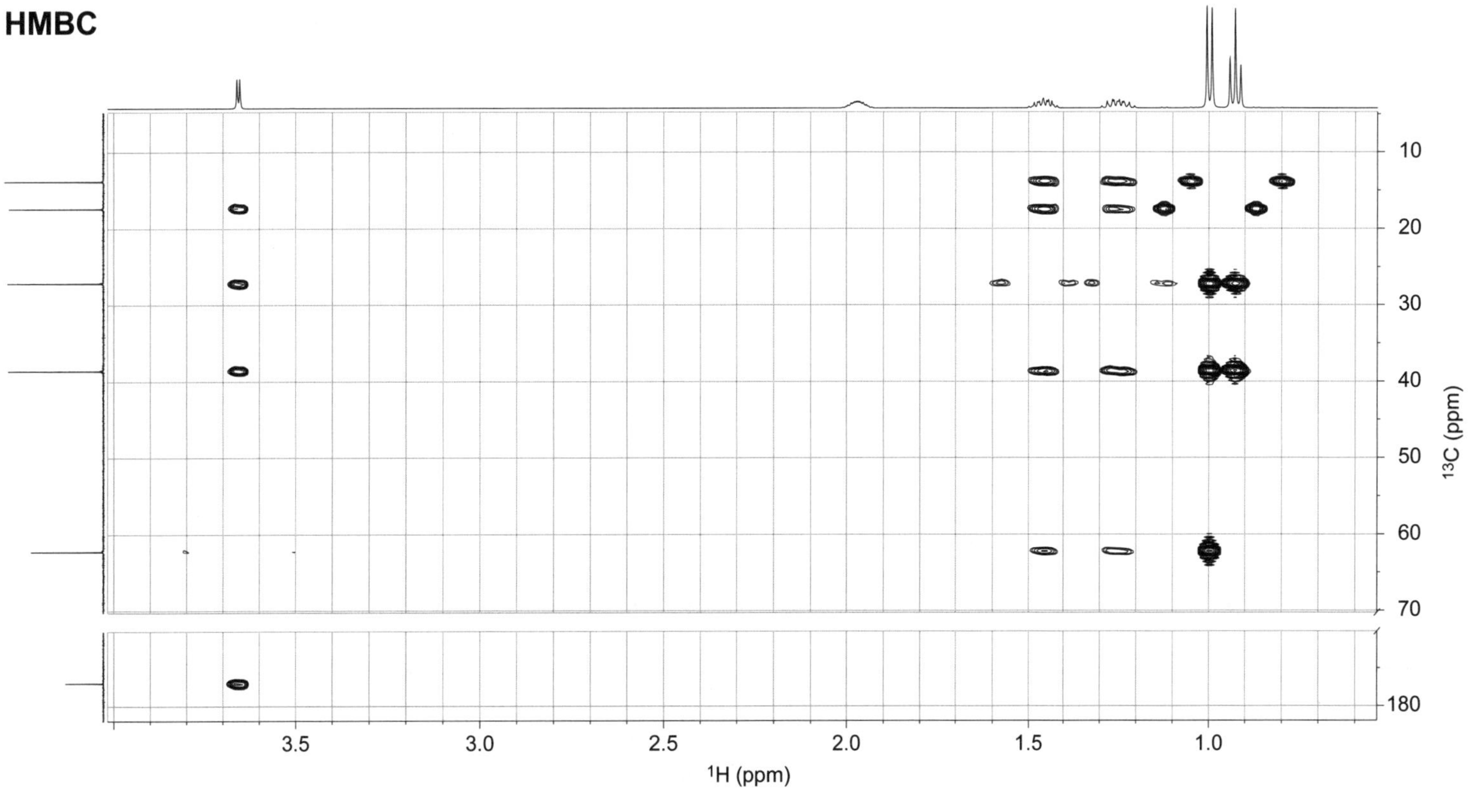
10
20
30
40
50
60
70
180
^{13}C (ppm)
3.5 3.0 2.5 2.0 1.5 1.0
^{1}H (ppm)

Problem 9.49 Molecular formula: $C_6H_{14}N_2O_2$ **Difficulty** ★★☆☆

Solvent: D_2O

Field: 500 MHz (^1H); 125 MHz (^{13}C)

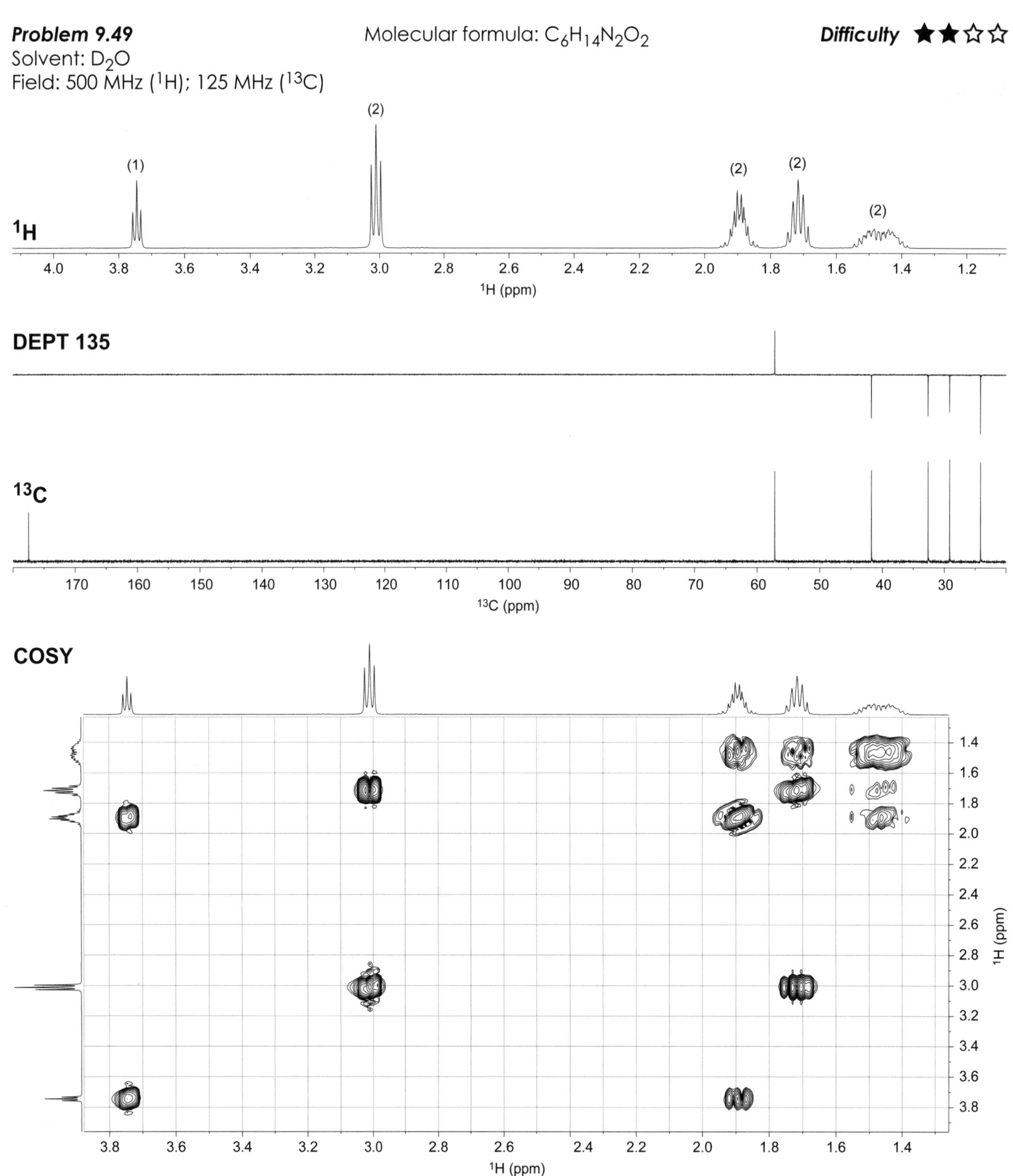

HSQC

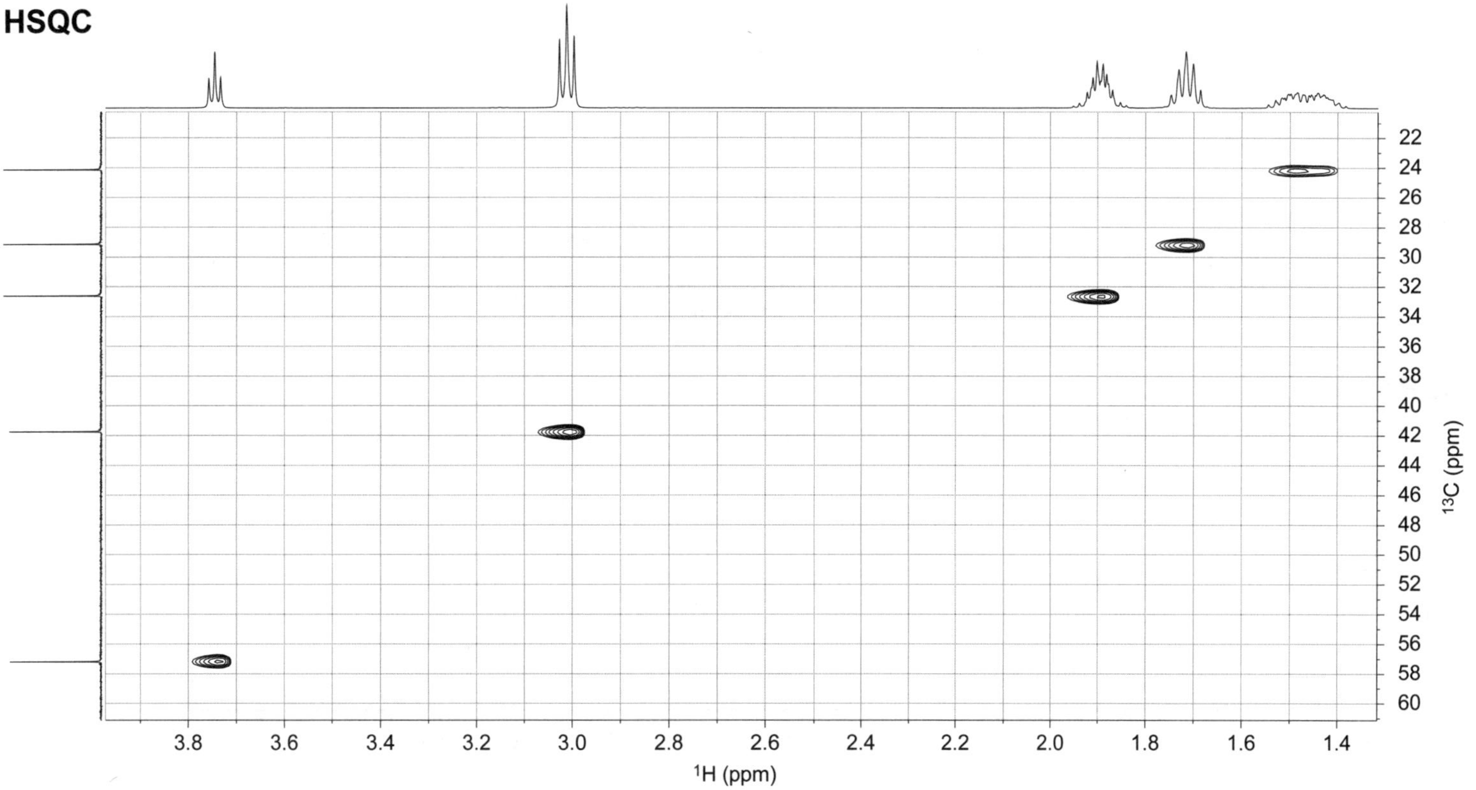

3.8 3.6 3.4 3.2 3.0 2.8 2.6 2.4 2.2 2.0 1.8 1.6 1.4
¹H (ppm)
22 24 26 28 30 32 34 36 38 40 42 44 46 48 50 52 54 56 58 60
¹³C (ppm)

HMBC

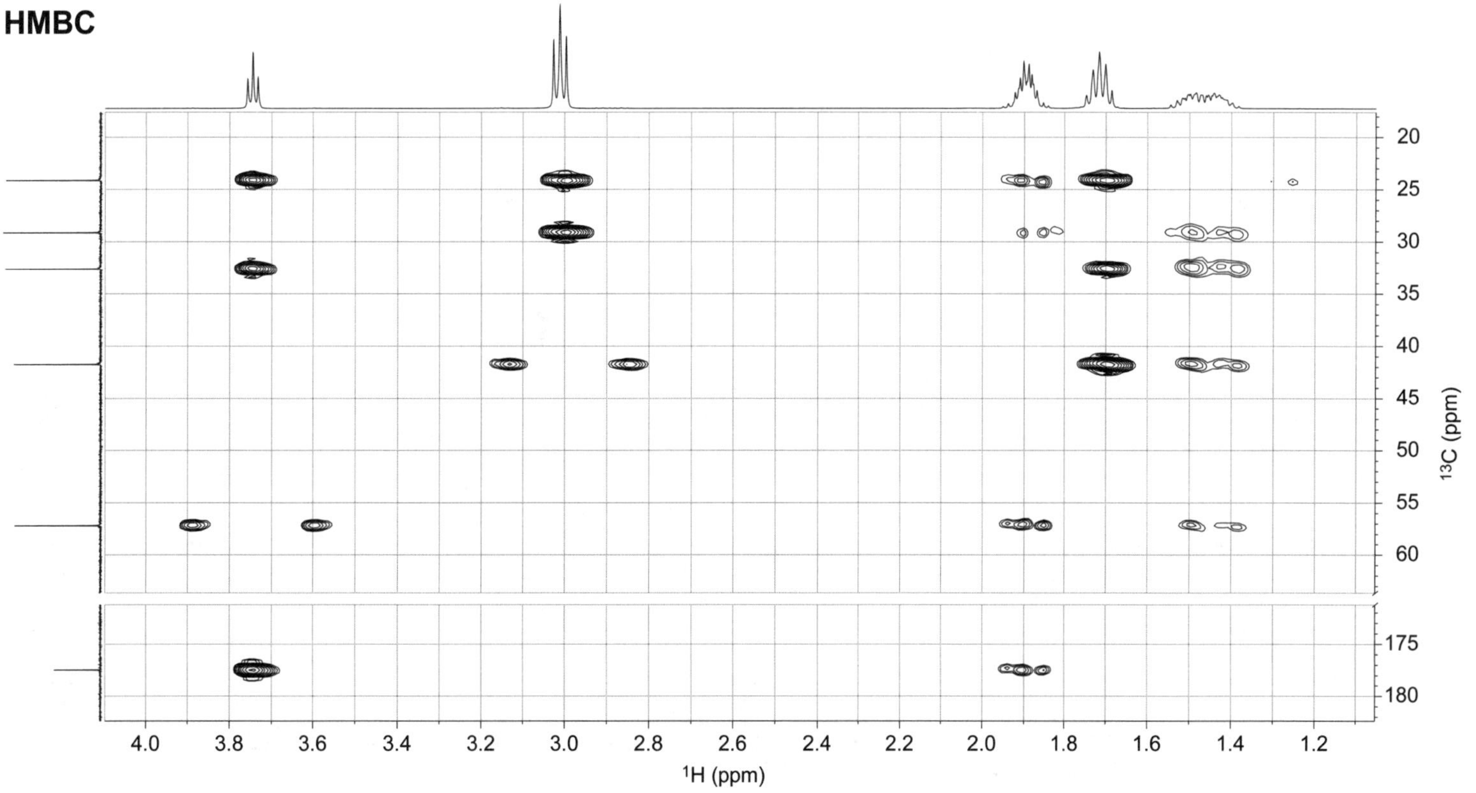

4.0 3.8 3.6 3.4 3.2 3.0 2.8 2.6 2.4 2.2 2.0 1.8 1.6 1.4 1.2
¹H (ppm)
20 25 30 35 40 45 50 55 60 175 180
¹³C (ppm)

Problem 9.50
Solvent: D_2O
Field: 500 MHz (^{1}H); 125 MHz (^{13}C)

Molecular formula: $C_5H_{11}NO_2S$
Hint: It is a natural amino acid

Difficulty ★★☆☆

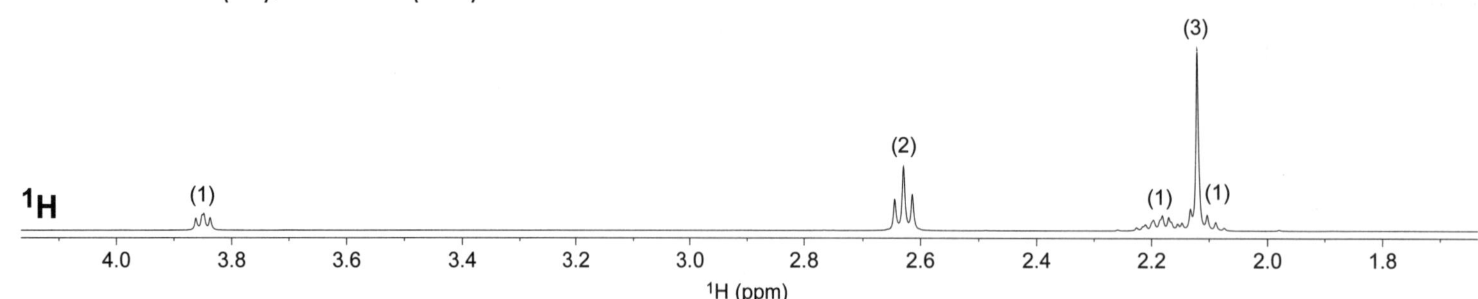

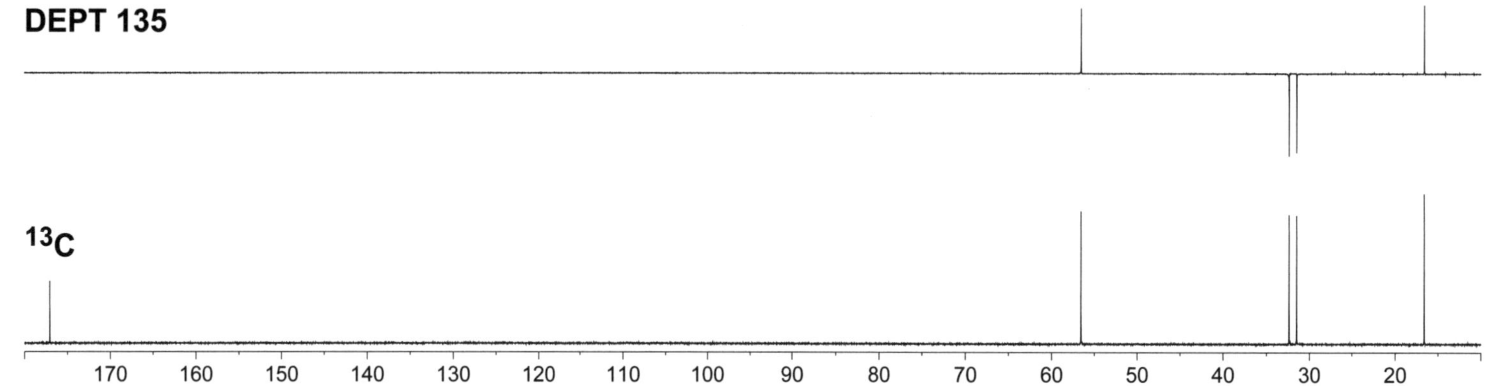

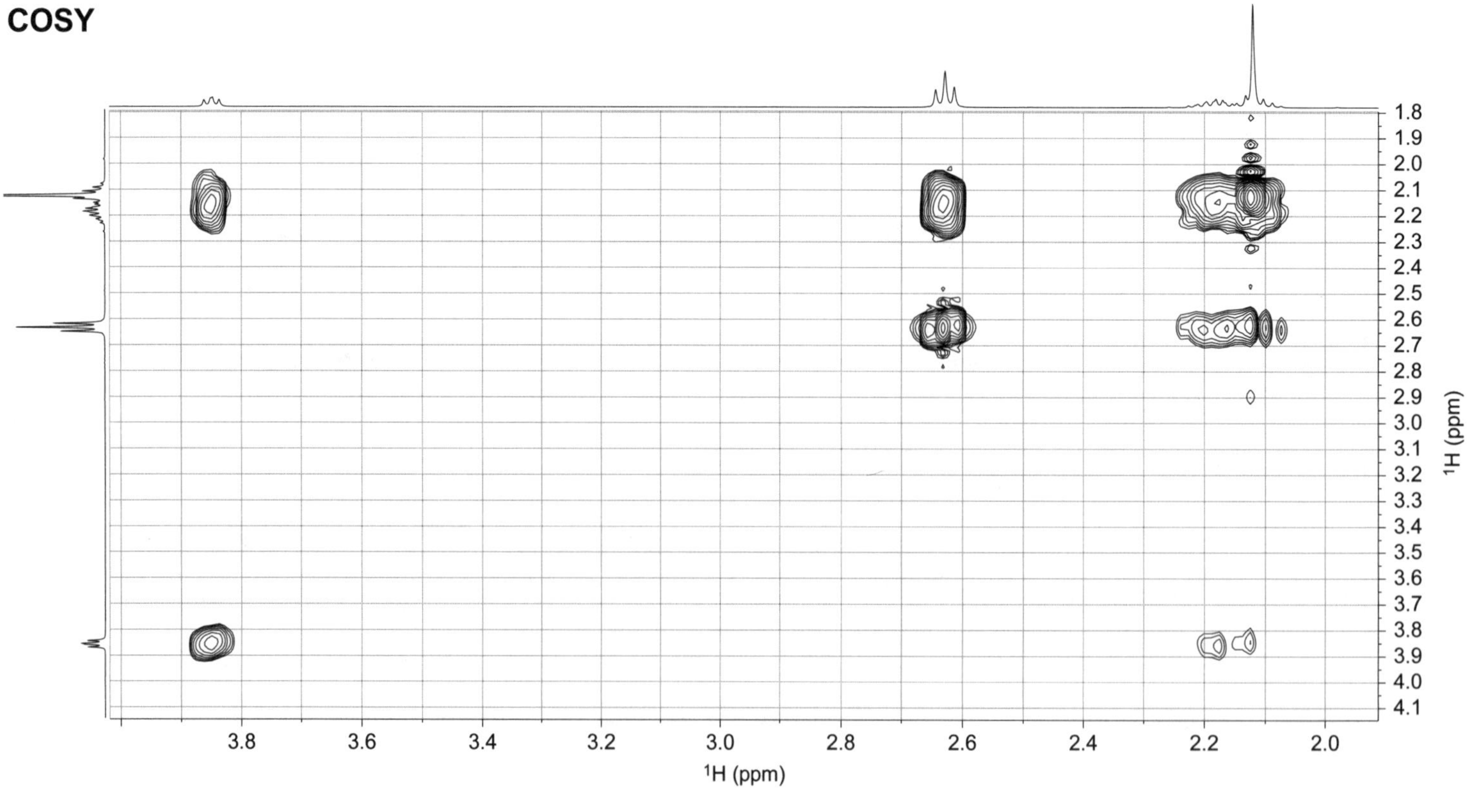

COSY

1H (ppm)

HSQC

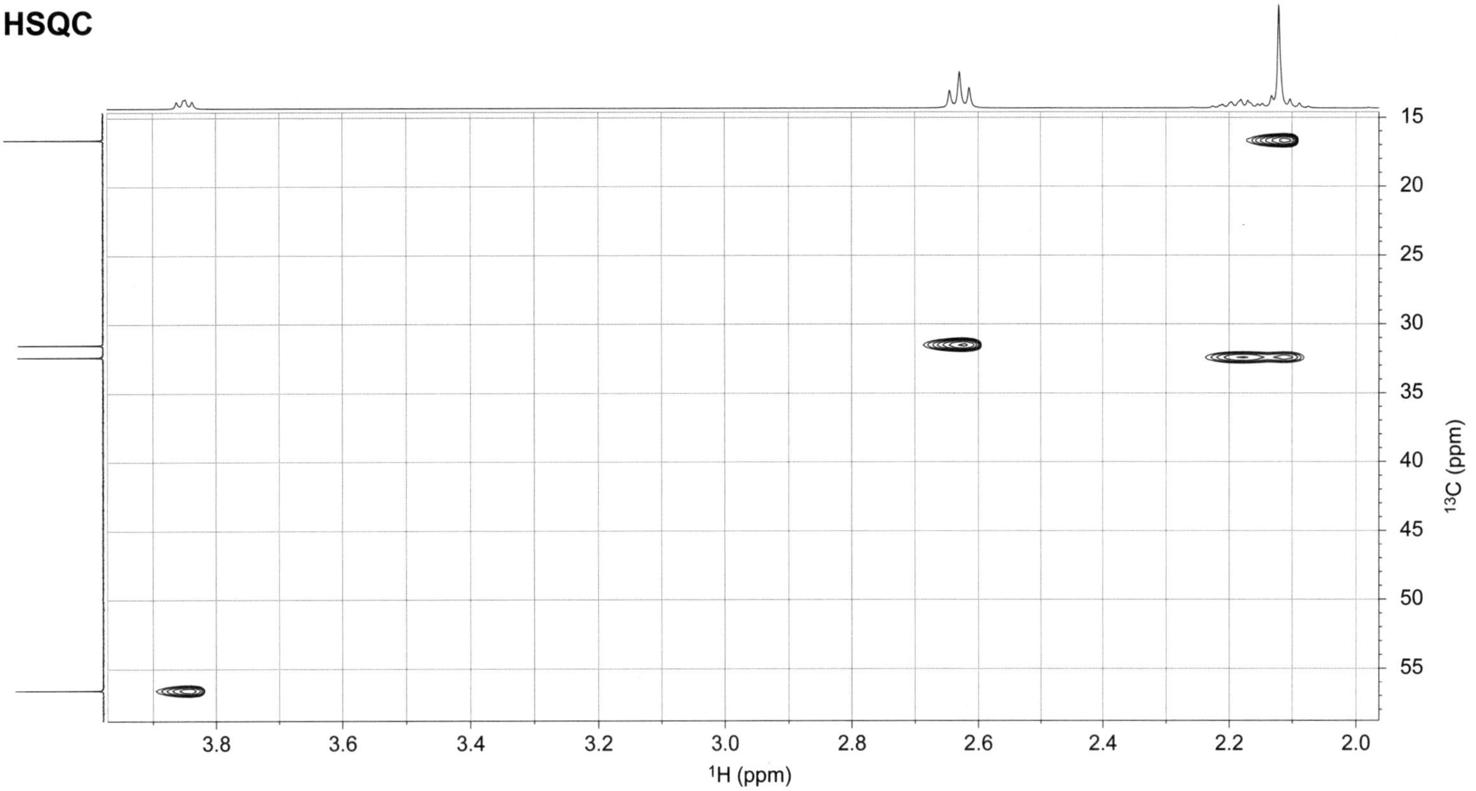

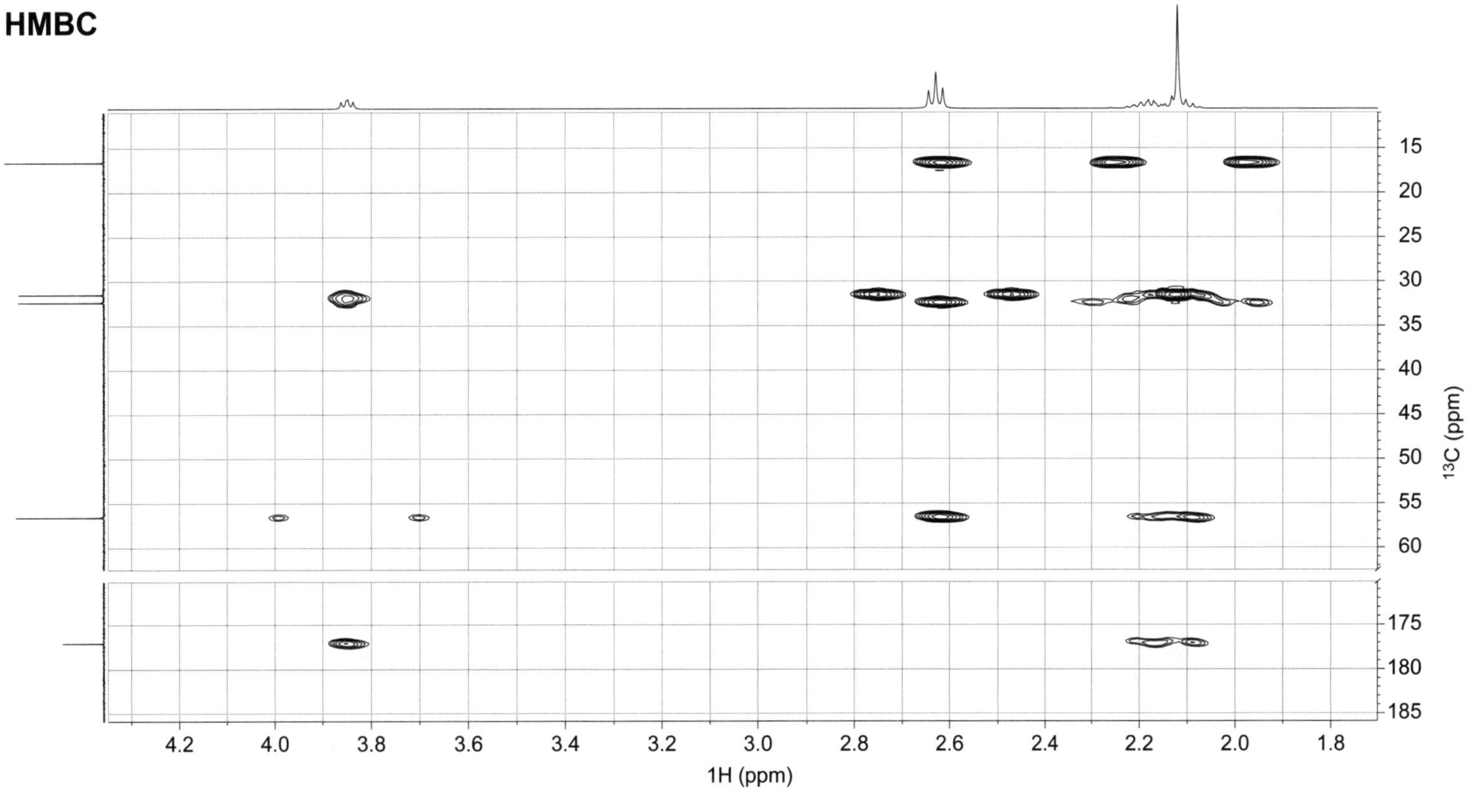

HMBC

Problem 9.51
Solvent: D_2O
Field: 400 MHz (^1H); 100 MHz (^{13}C)

Molecular formula: $C_5H_{12}N_2O_2$

Difficulty ★★☆☆

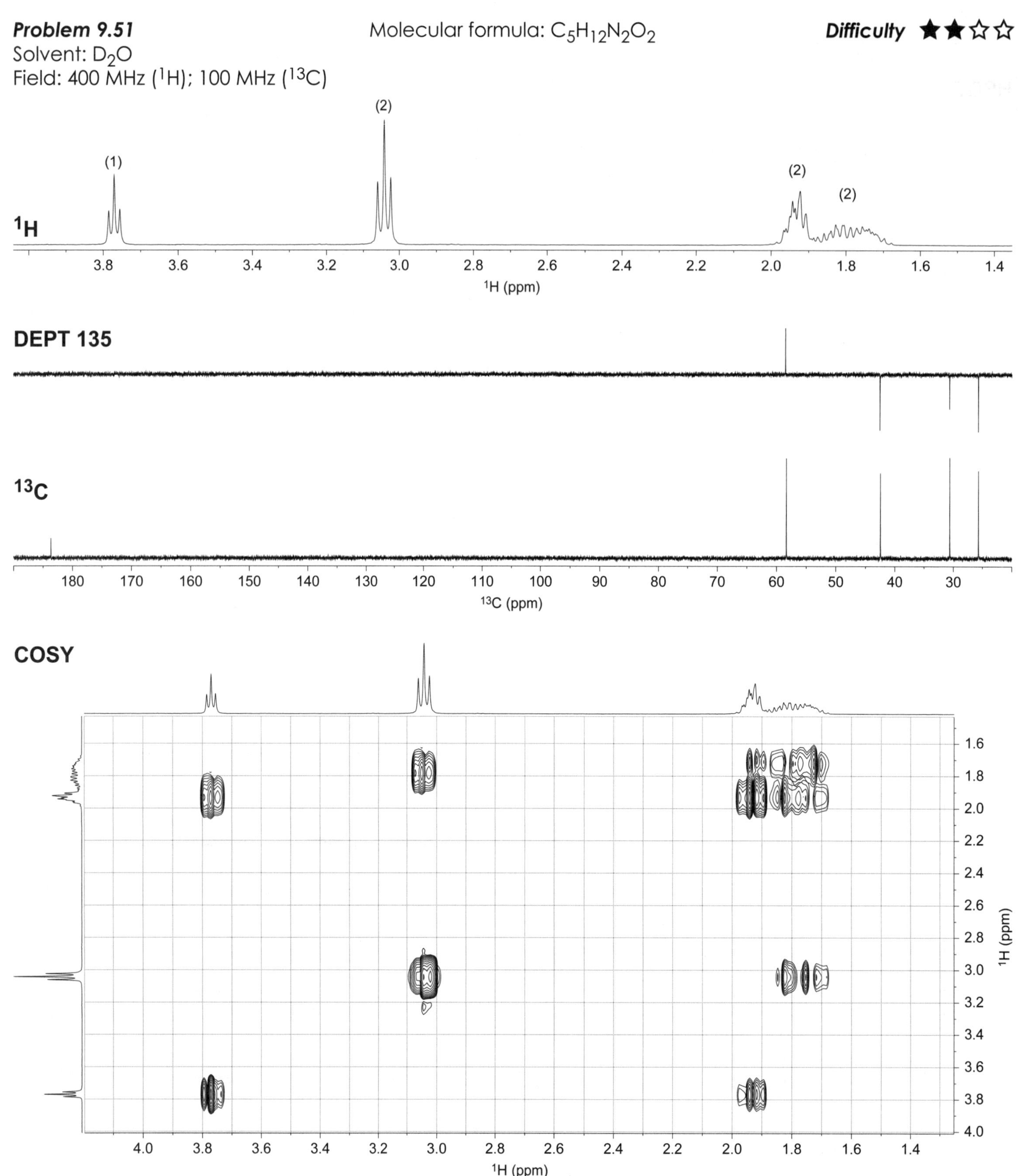

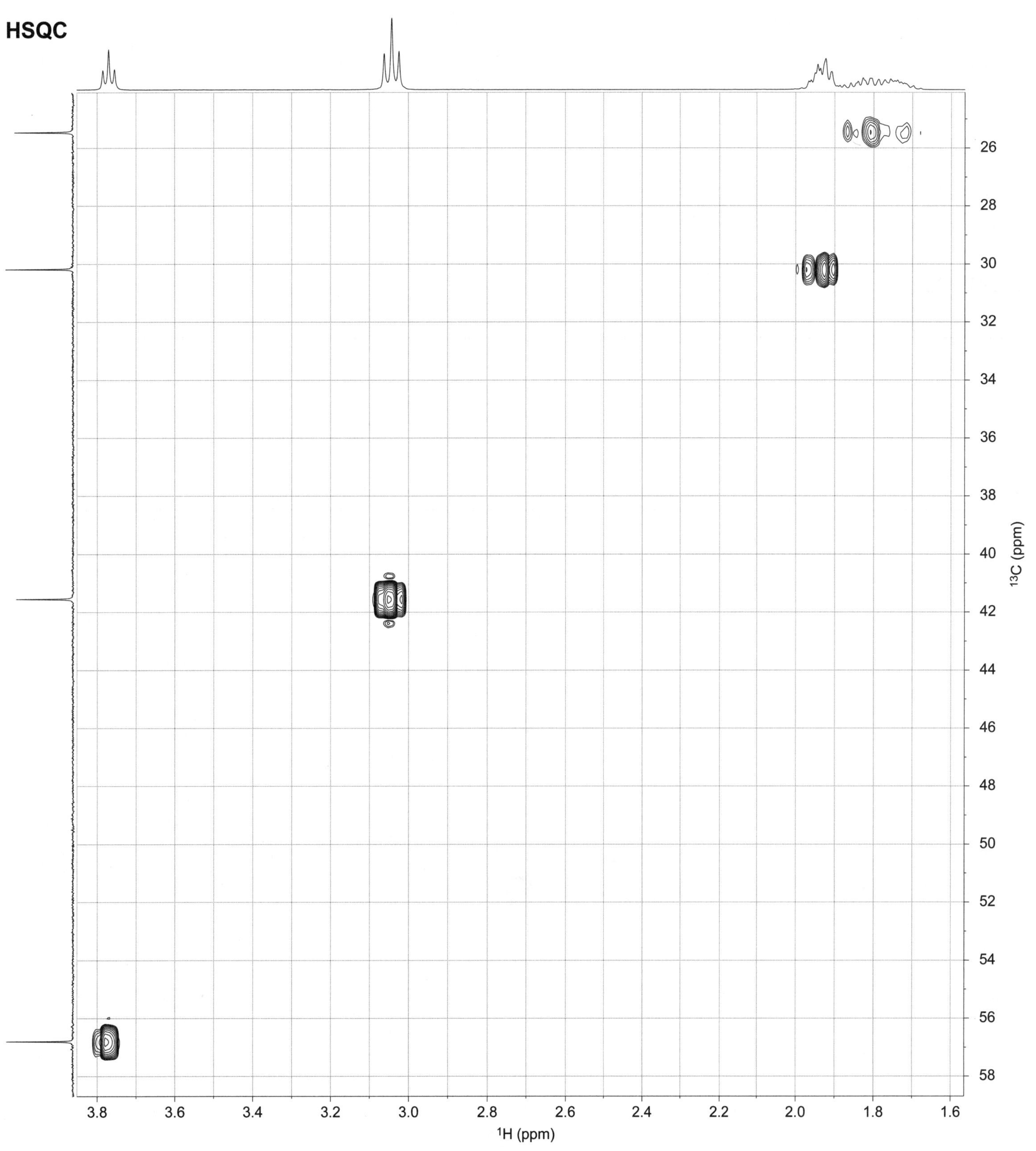
HSQC
26
28
30
32
34
36
38
40
42
44
46
48
50
52
54
56
58
^{13}C (ppm)
3.8
3.6
3.4
3.2
3.0
2.8
2.6
2.4
2.2
2.0
1.8
1.6
^{1}H (ppm)

Problem 9.52 Molecular formula: $C_8H_{10}O_2$ **Difficulty** ★★☆☆

Solvent: D_2O

Field: 400 MHz (^{1}H); 100 MHz (^{13}C)

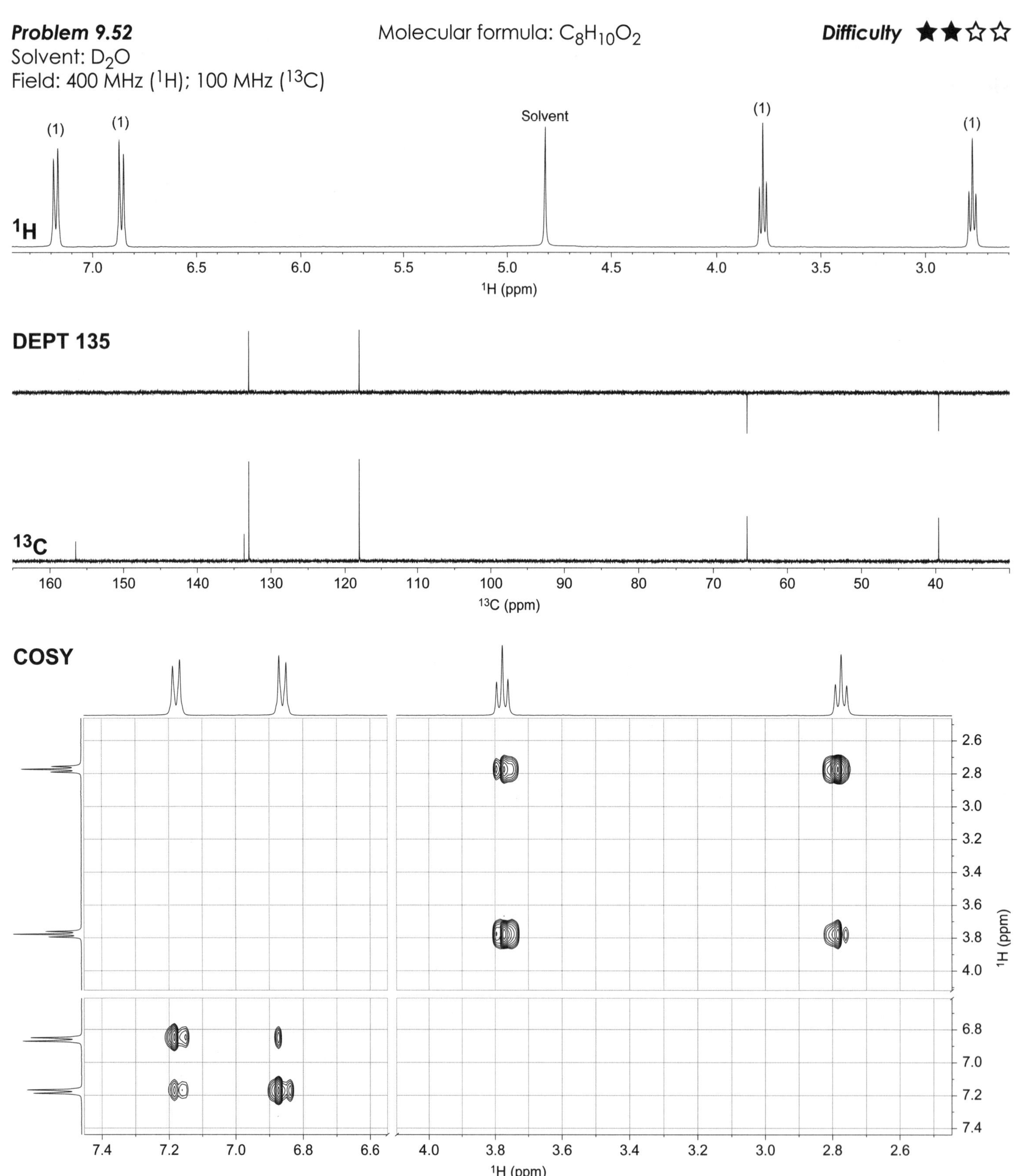

HSQC

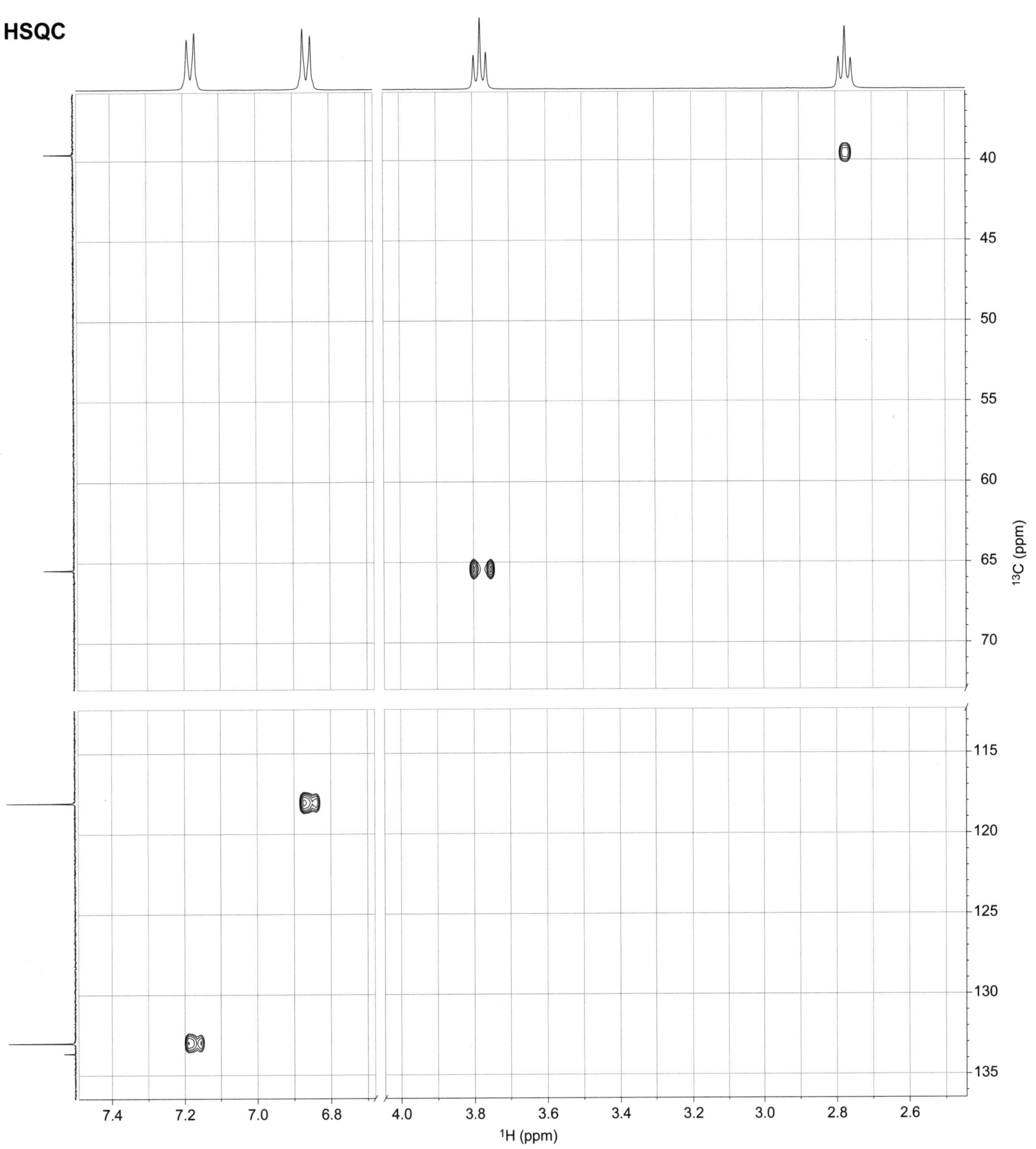

Problem 9.53 Molecular formula: C_7H_8O **Difficulty** ★★☆☆

Solvent: D_2O

Field: 500 MHz (^{1}H); 125 MHz (^{13}C)

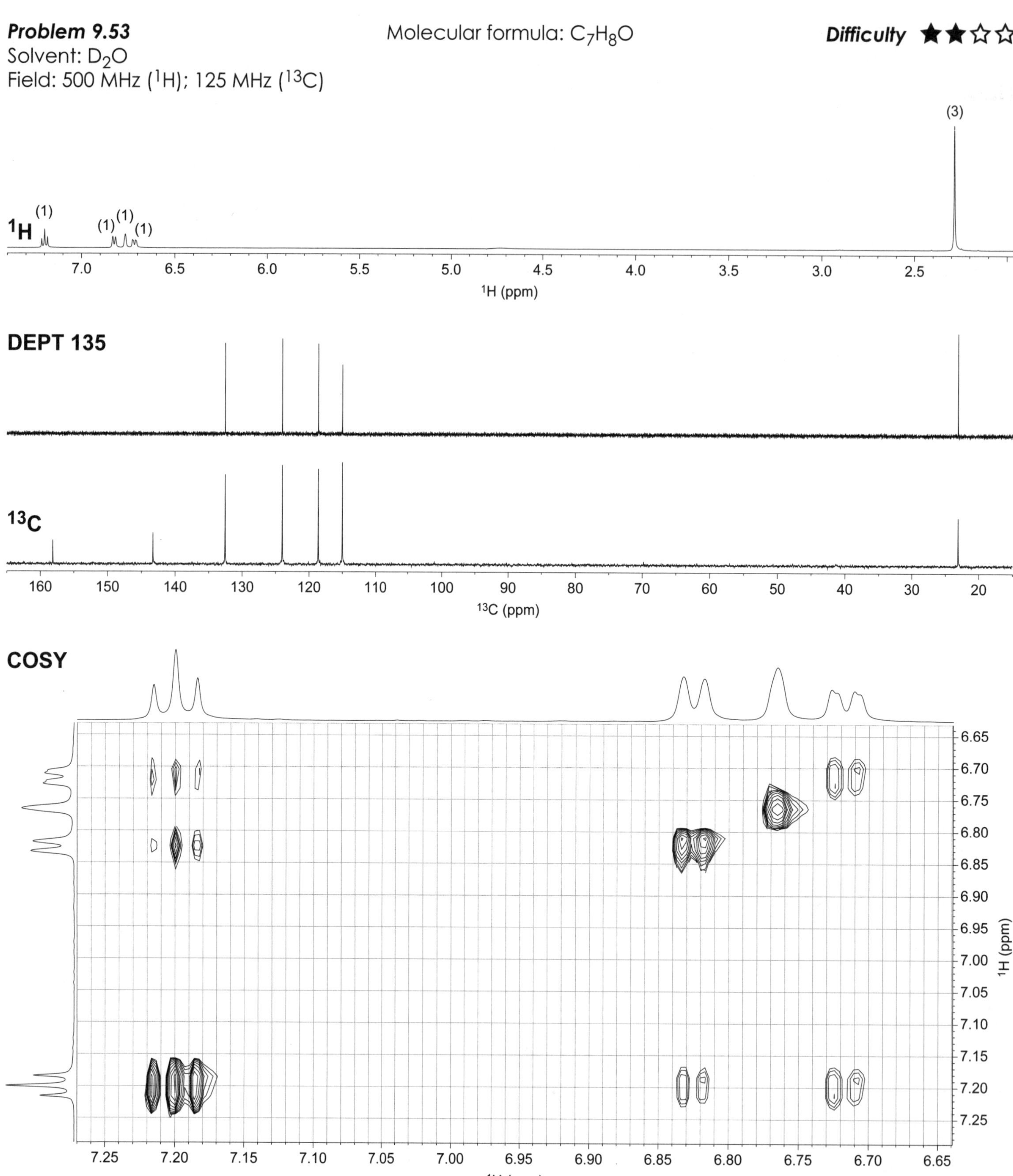

HSQC

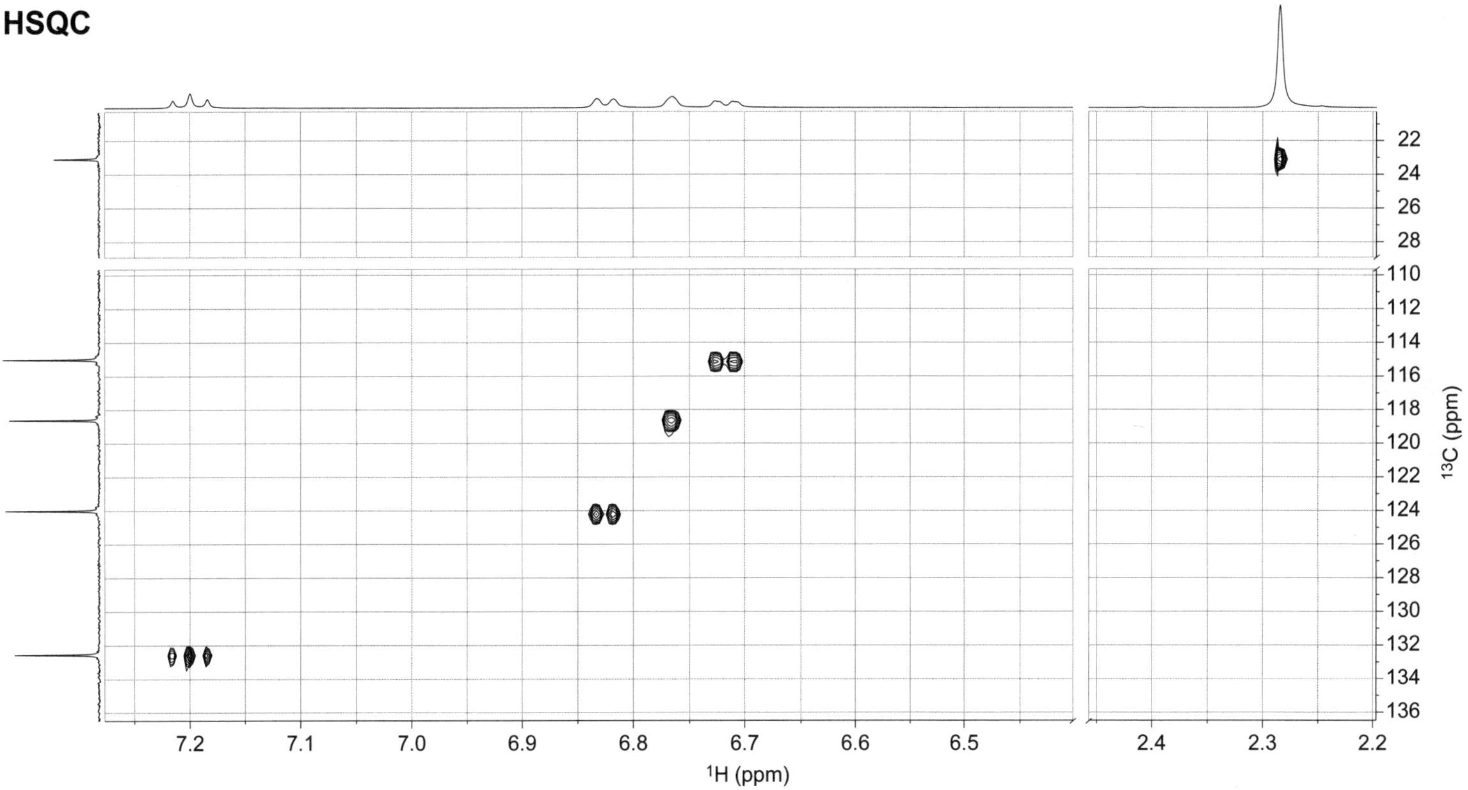

22
24
26
28
110
112
114
116
118
120
122
124
126
128
130
132
134
136
^{13}C (ppm)
7.2
7.1
7.0
6.9
6.8
6.7
6.6
6.5
2.4
2.3
2.2
^{1}H (ppm)

20
25
30
115
120
125
130
135
140
145
150
155
160
165
^{13}C (ppm)
7.6
7.4
7.2
7.0
6.8
6.6
6.4
2.6
2.4
2.2
2.0
1.8
^{1}H (ppm)

HMBC

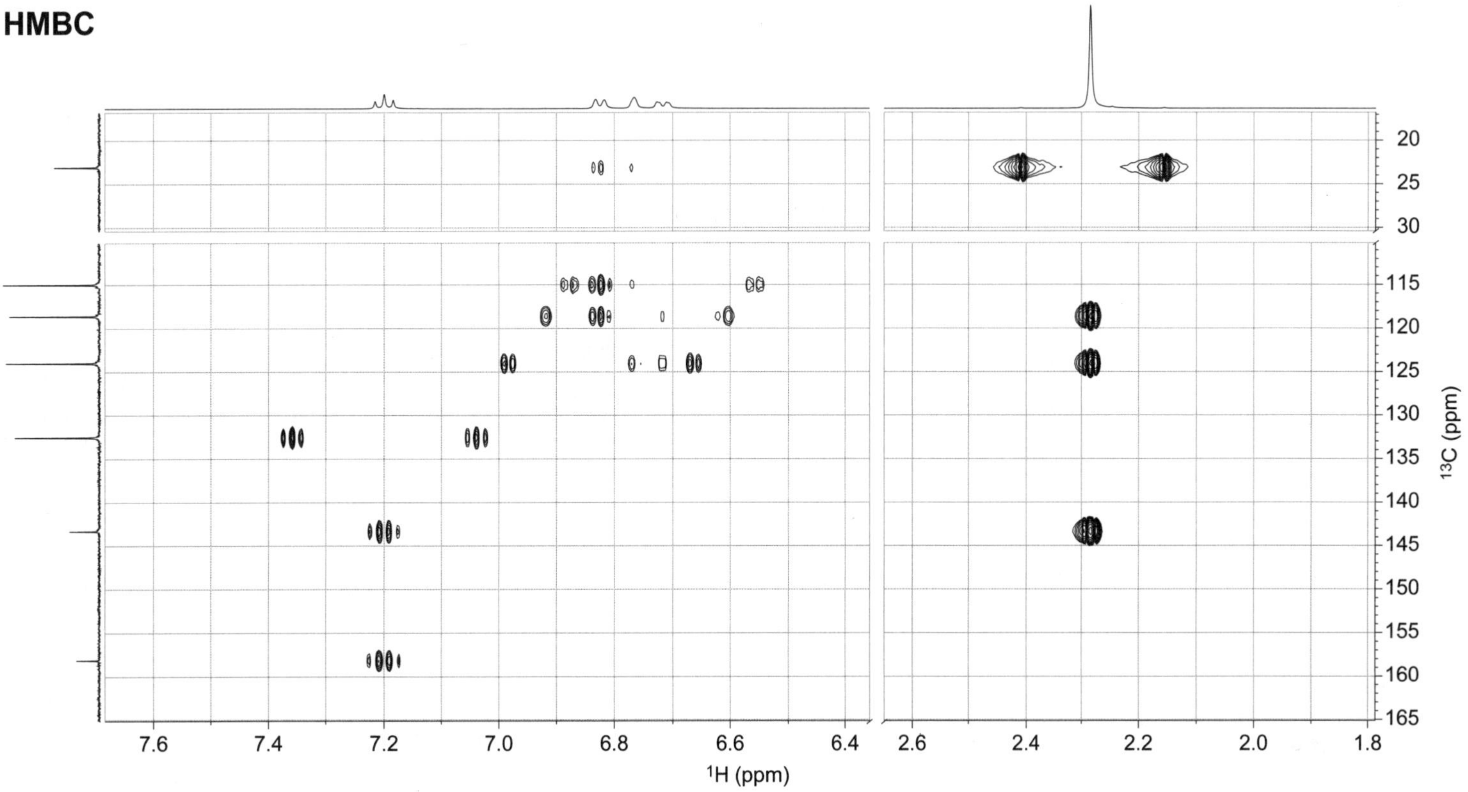

Problem 9.54
Solvent: D_2O
Field: 500 MHz (^1H); 125 MHz (^{13}C)

Molecular formula: $C_6H_{13}NO_2$

Difficulty ★★☆☆

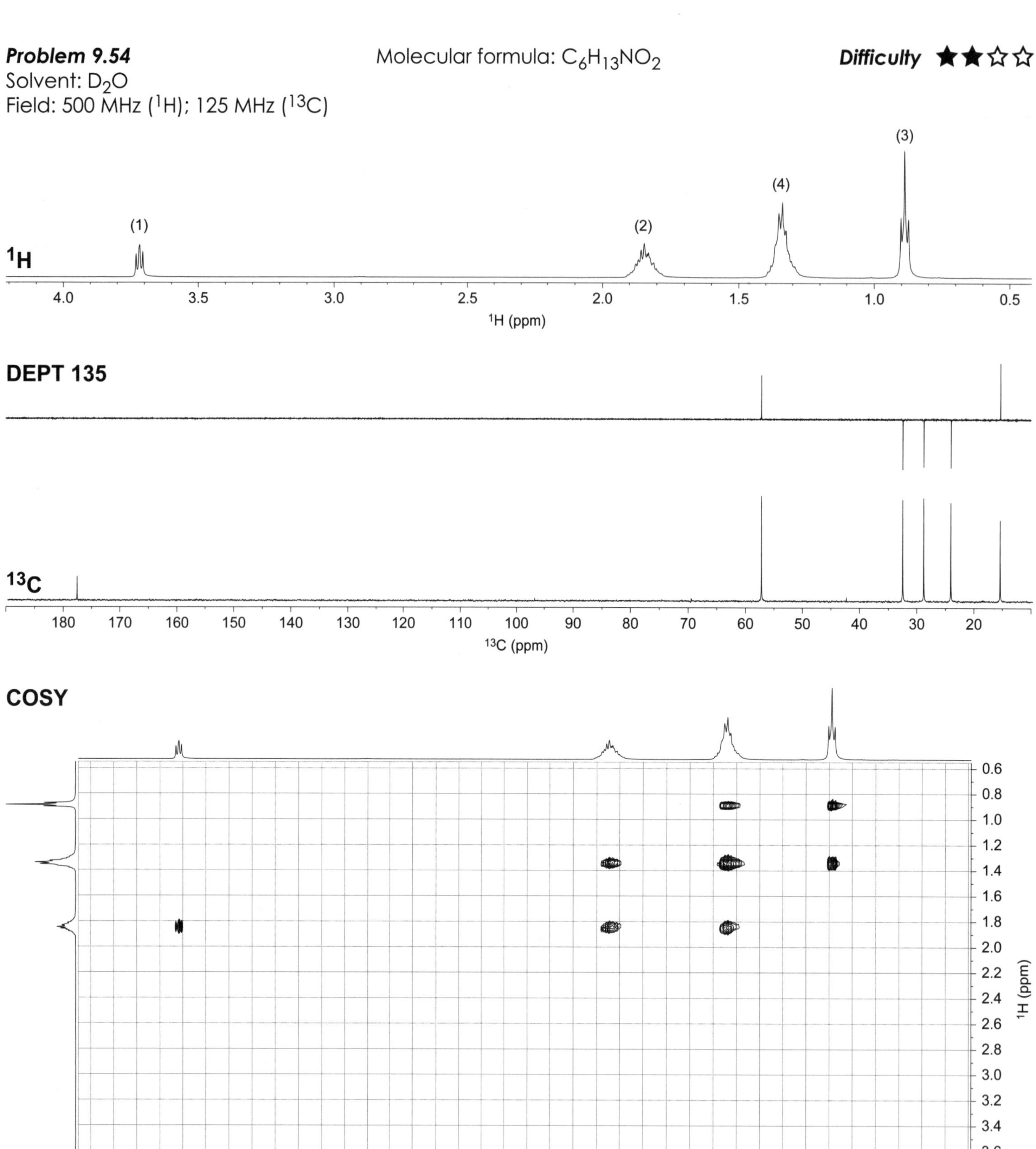

HSQC

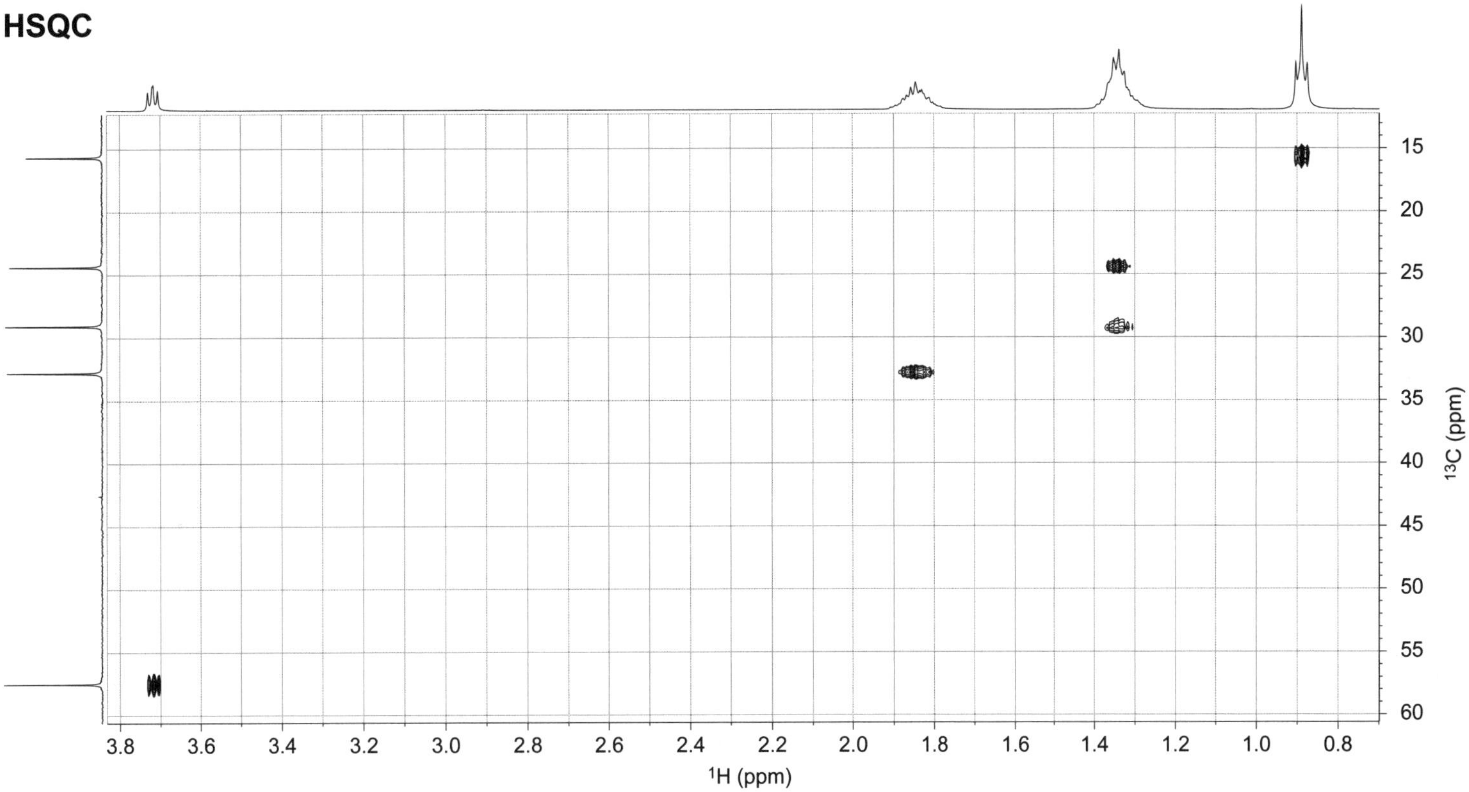
15
20
25
30
35
40
45
50
55
60
^{13}C (ppm)
3.8 3.6 3.4 3.2 3.0 2.8 2.6 2.4 2.2 2.0 1.8 1.6 1.4 1.2 1.0 0.8
^{1}H (ppm)

HMBC

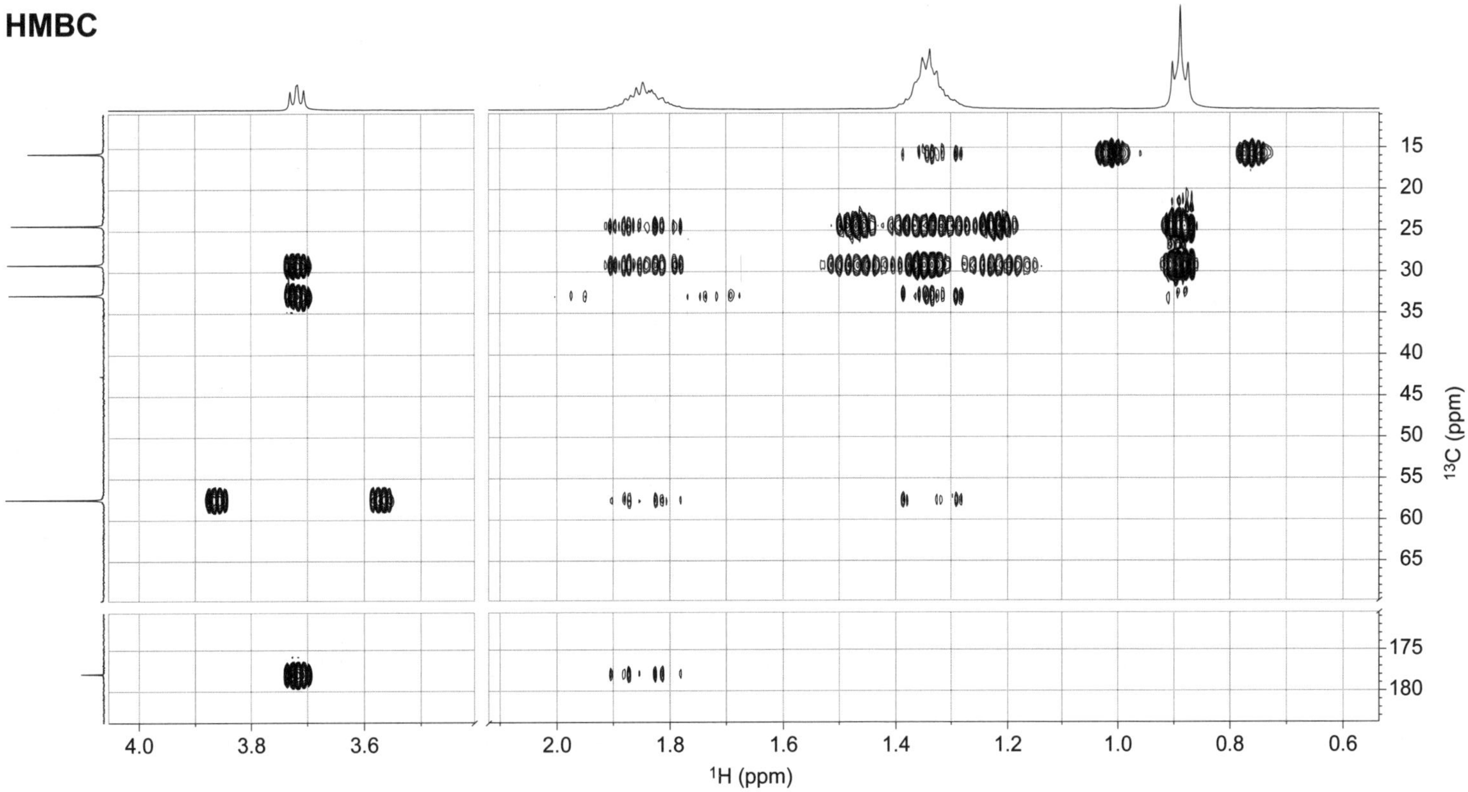
15
20
25
30
35
40
45
50
55
60
65
175
180
^{13}C (ppm)
4.0 3.8 3.6 2.0 1.8 1.6 1.4 1.2 1.0 0.8 0.6
^{1}H (ppm)

Problem 9.55
Solvent: D_2O
Field: 500 MHz (^1H); 125 MHz (^{13}C)

Molecular formula: $C_6H_6O_2$

Difficulty ★★☆☆

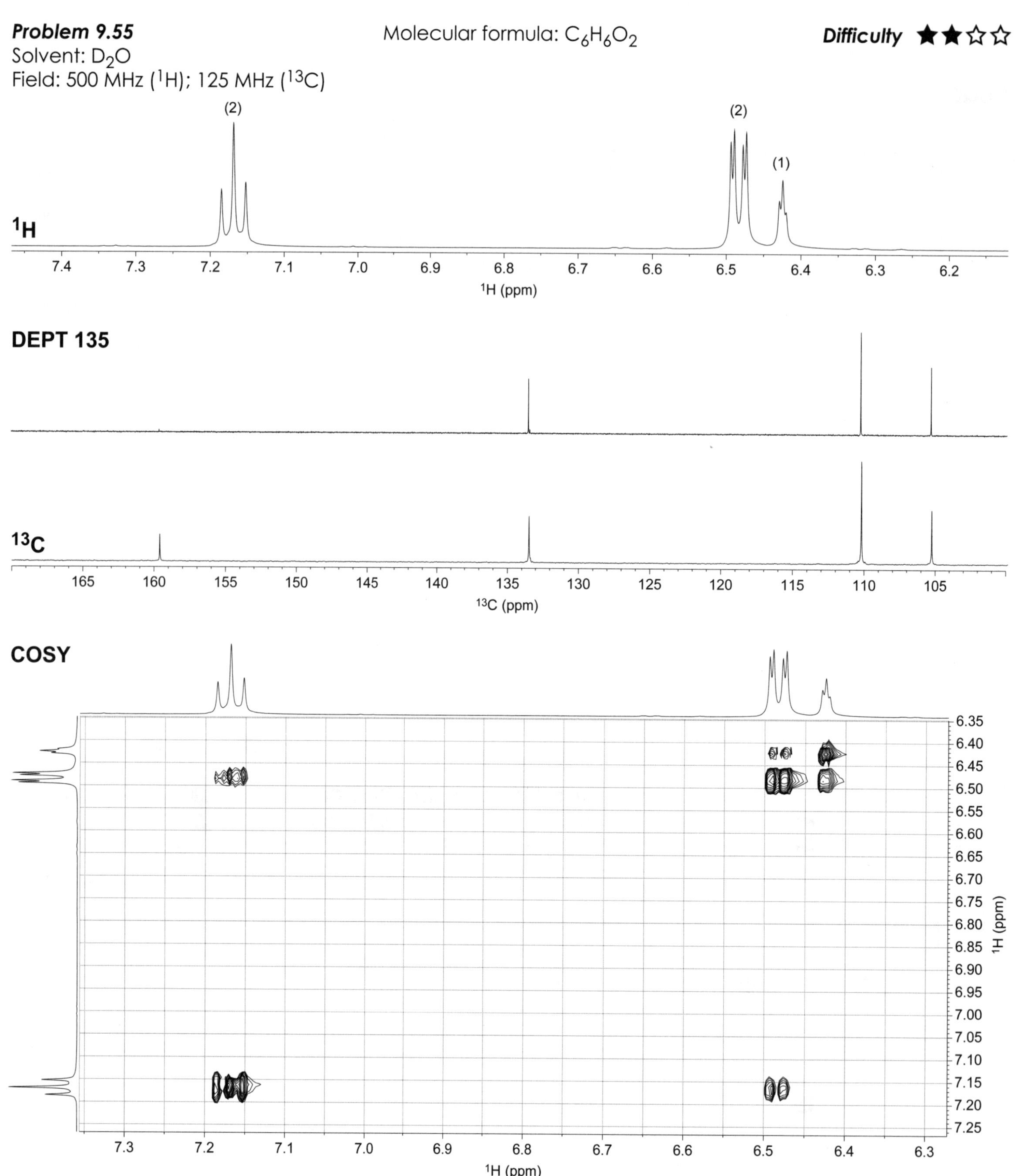

HSQC

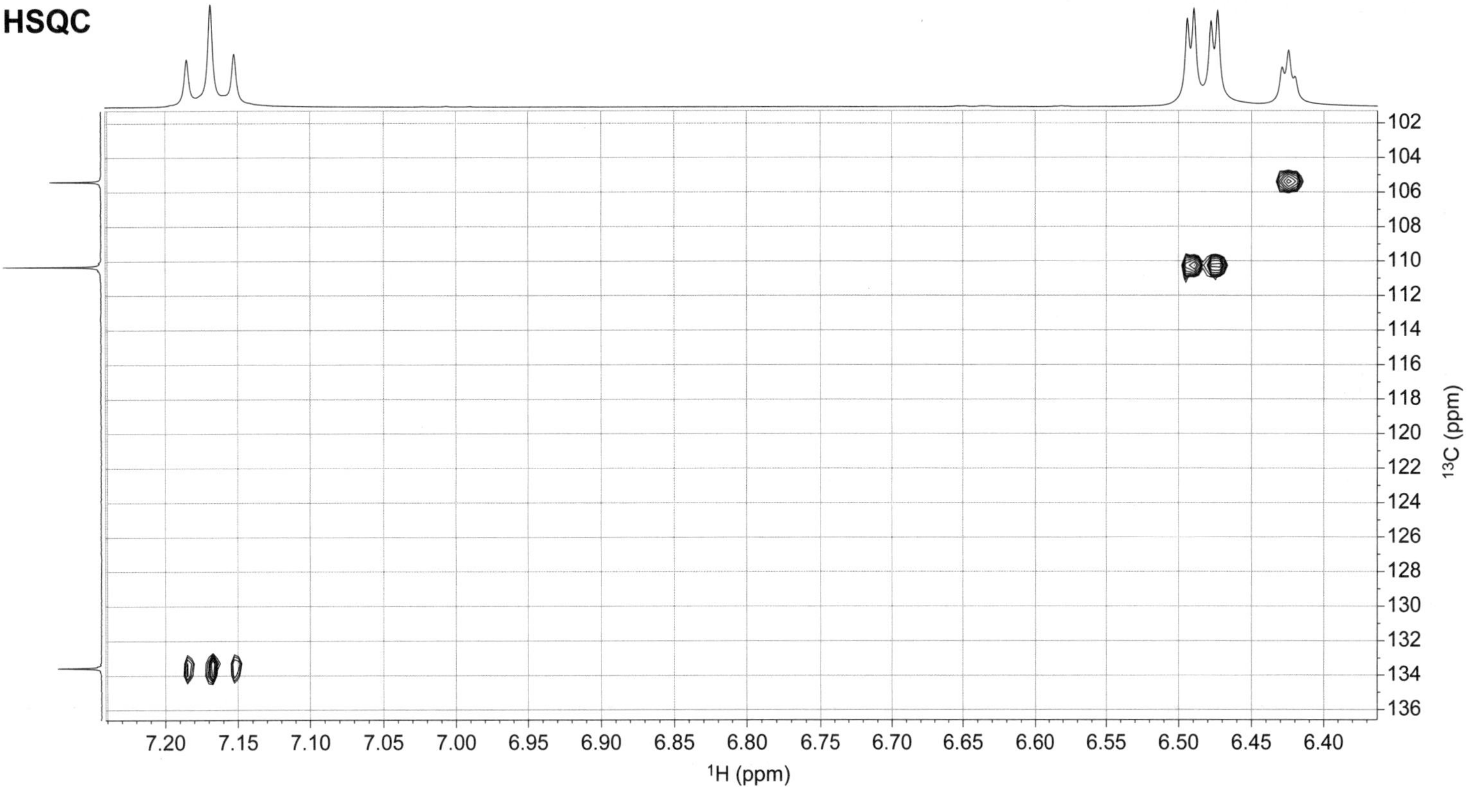
7.20
7.15
7.10
7.05
7.00
6.95
6.90
6.85
6.80
6.75
6.70
6.65
6.60
6.55
6.50
6.45
6.40
^{1}H (ppm)
102
104
106
108
110
112
114
116
118
120
122
124
126
128
130
132
134
136
^{13}C (ppm)

HMBC

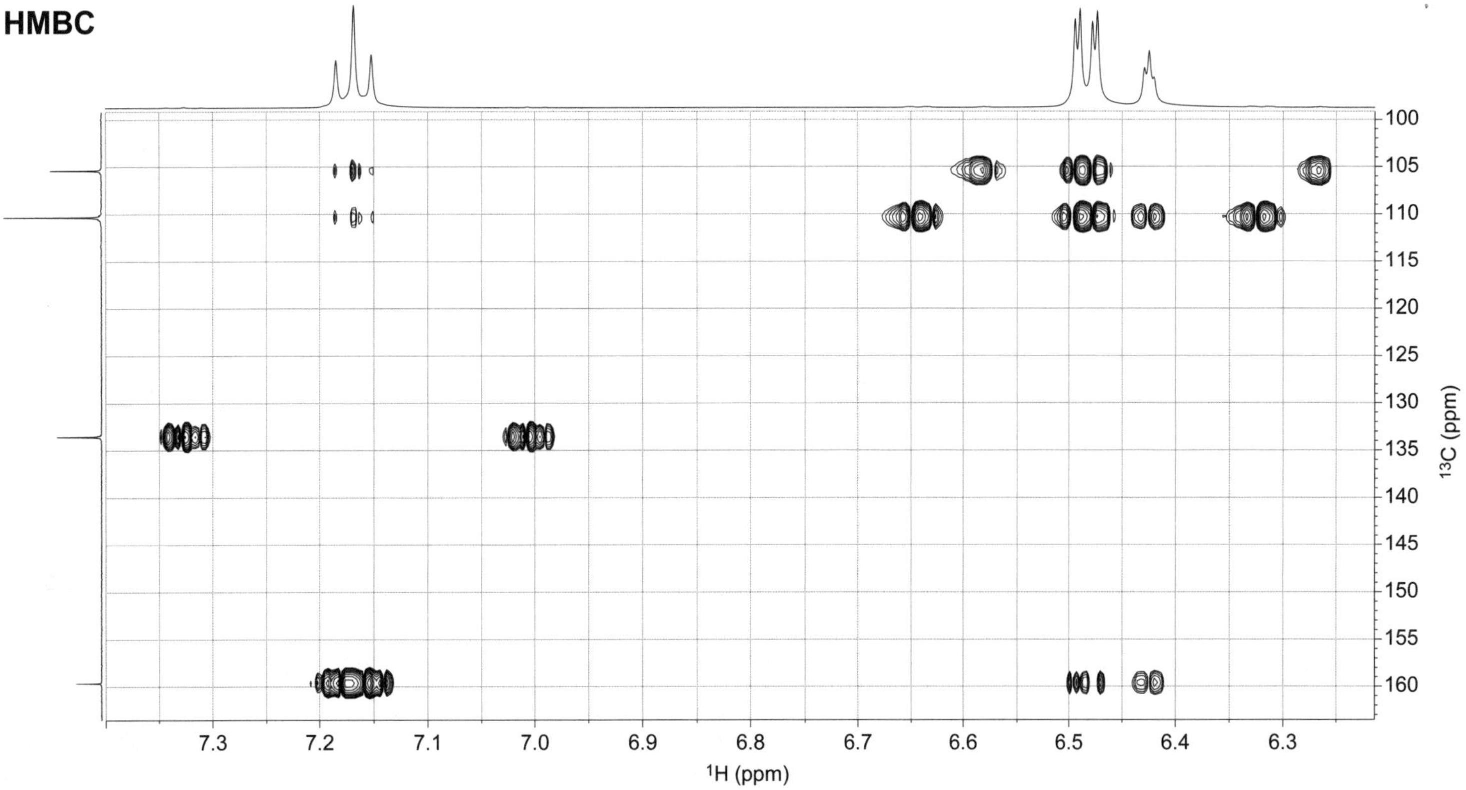
7.3
7.2
7.1
7.0
6.9
6.8
6.7
6.6
6.5
6.4
6.3
^{1}H (ppm)
100
105
110
115
120
125
130
135
140
145
150
155
160
^{13}C (ppm)

Problem 9.56 Molecular formula: $C_8H_{17}NO_2$ **Difficulty** ★★☆☆

Solvent: D_2O

Field: 500 MHz (^{1}H); 125 MHz (^{13}C)

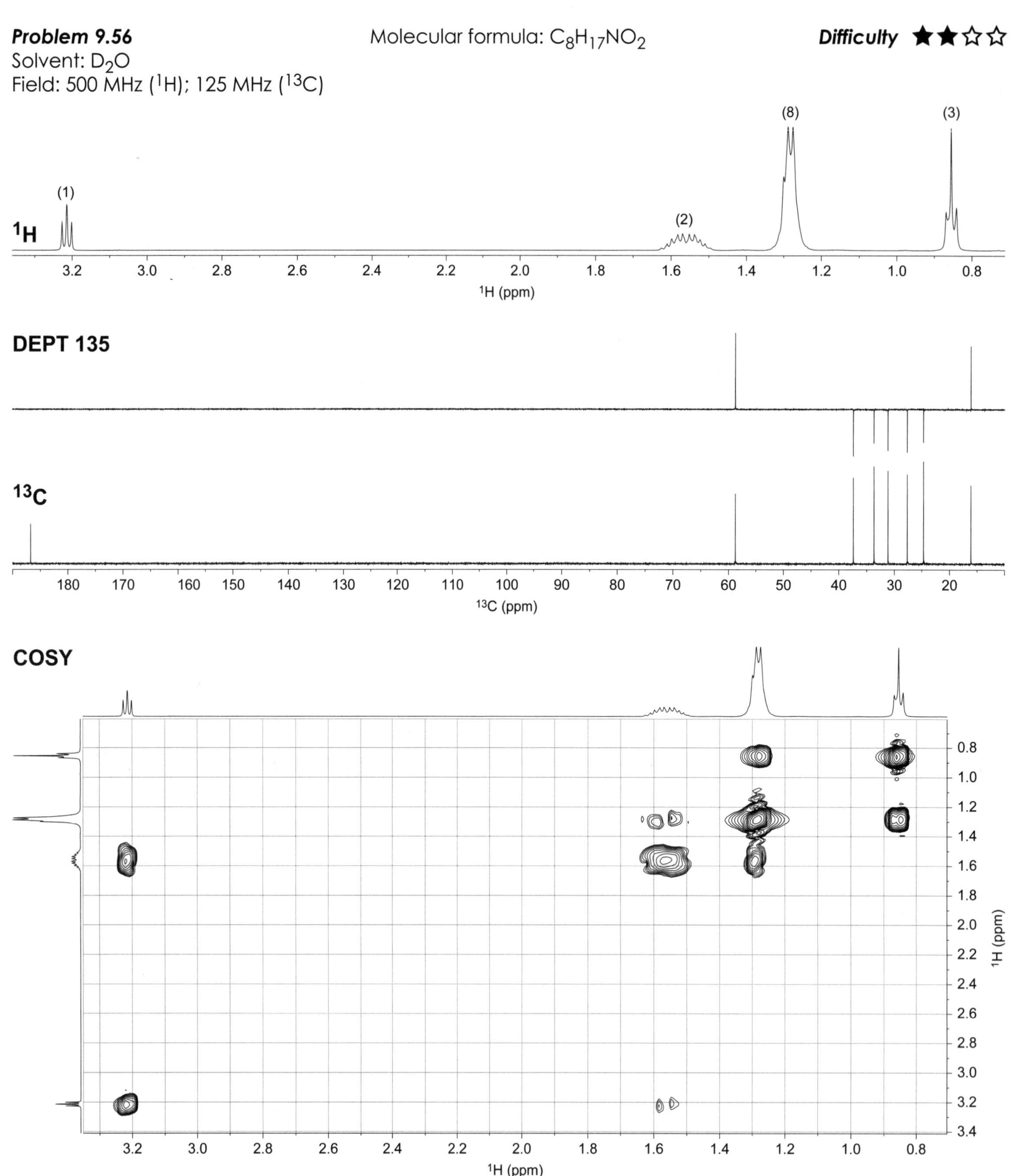

HSQC

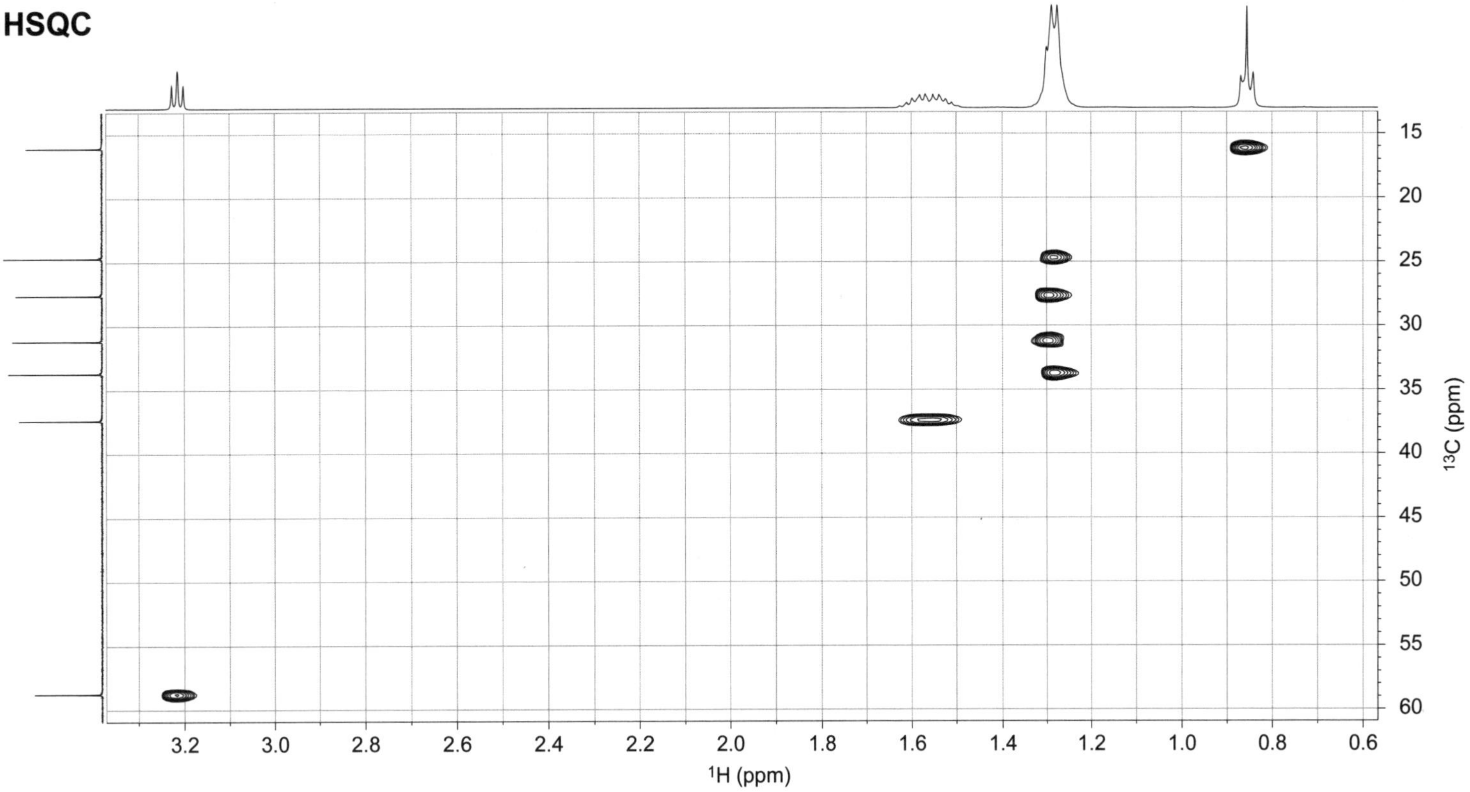

15
20
25
30
35
40
45
50
55
60
^{13}C (ppm)
3.2
3.0
2.8
2.6
2.4
2.2
2.0
1.8
1.6
1.4
1.2
1.0
0.8
0.6
^{1}H (ppm)

HMBC

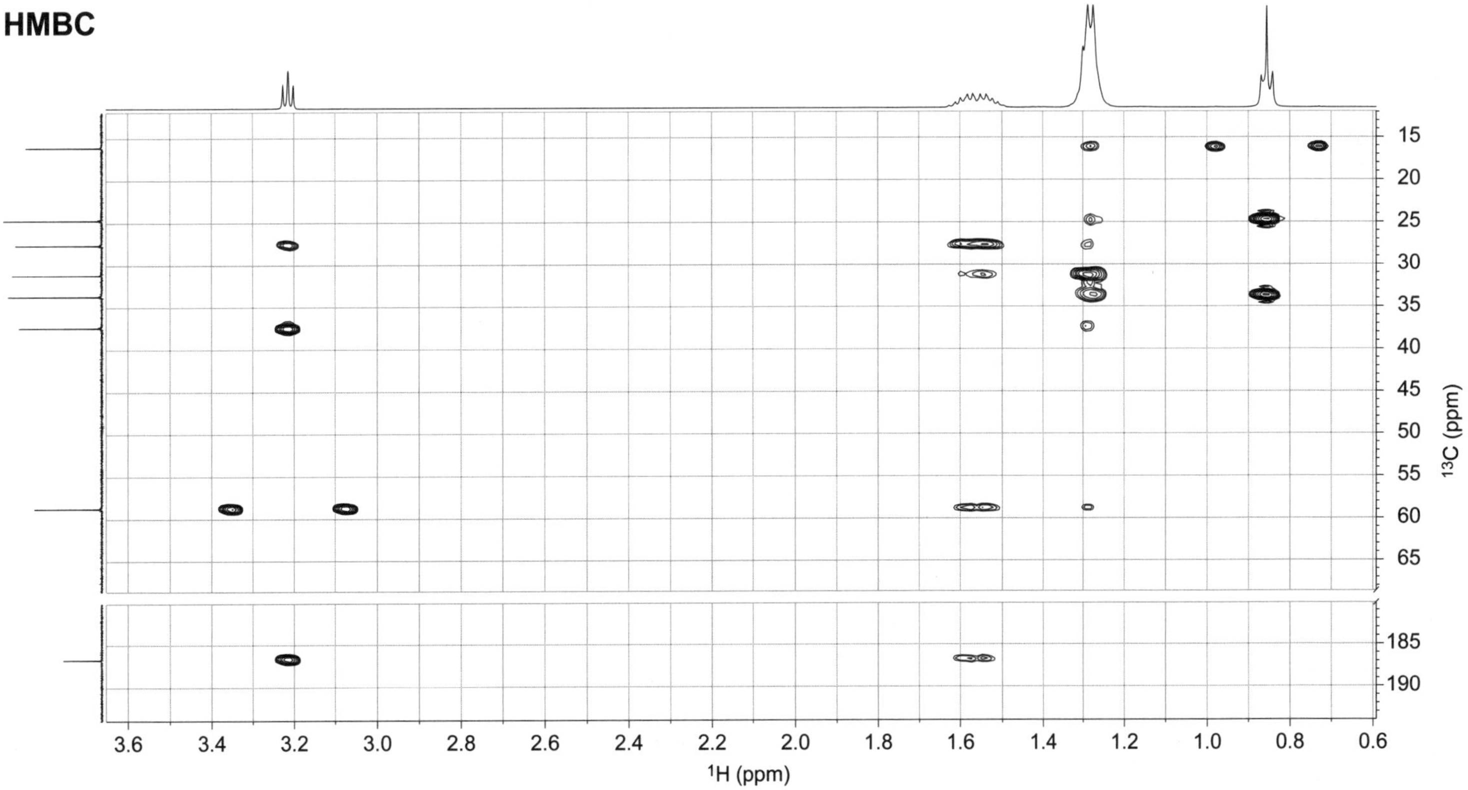

15
20
25
30
35
40
45
50
55
60
65
185
190
^{13}C (ppm)
3.6
3.4
3.2
3.0
2.8
2.6
2.4
2.2
2.0
1.8
1.6
1.4
1.2
1.0
0.8
0.6
^{1}H (ppm)

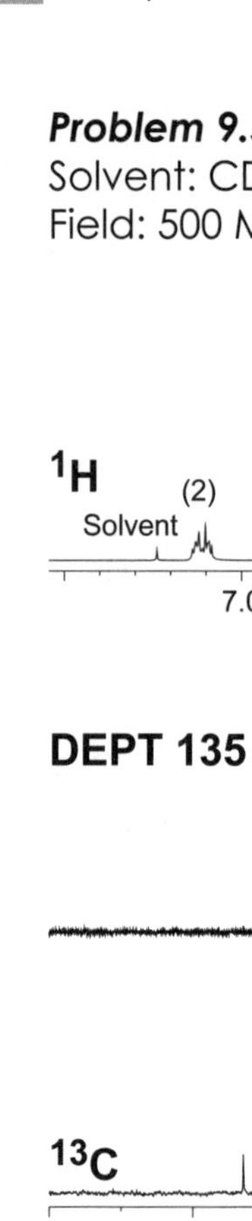

Problem 9.57 Molecular formula: C_8H_{10} ***Difficulty*** ★★☆☆

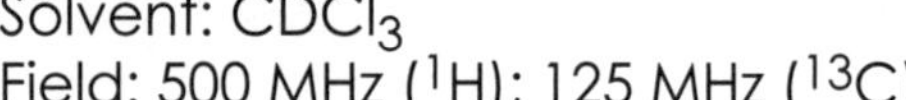

Solvent: $CDCl_3$

Field: 500 MHz (^1H); 125 MHz (^{13}C)

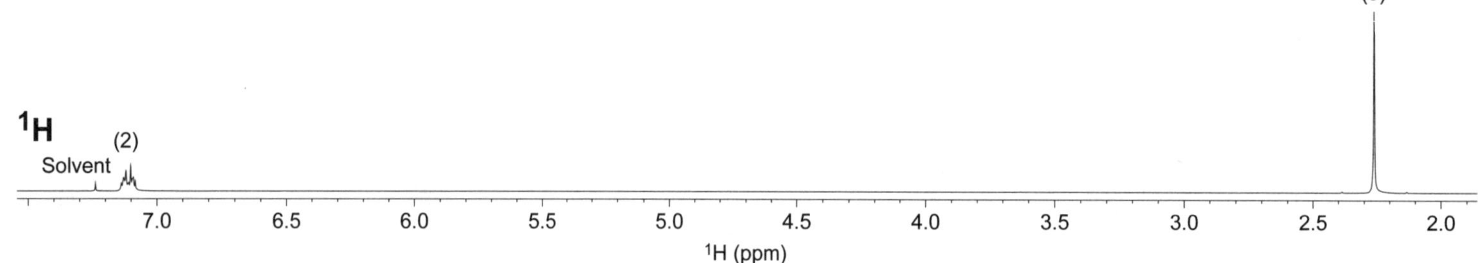

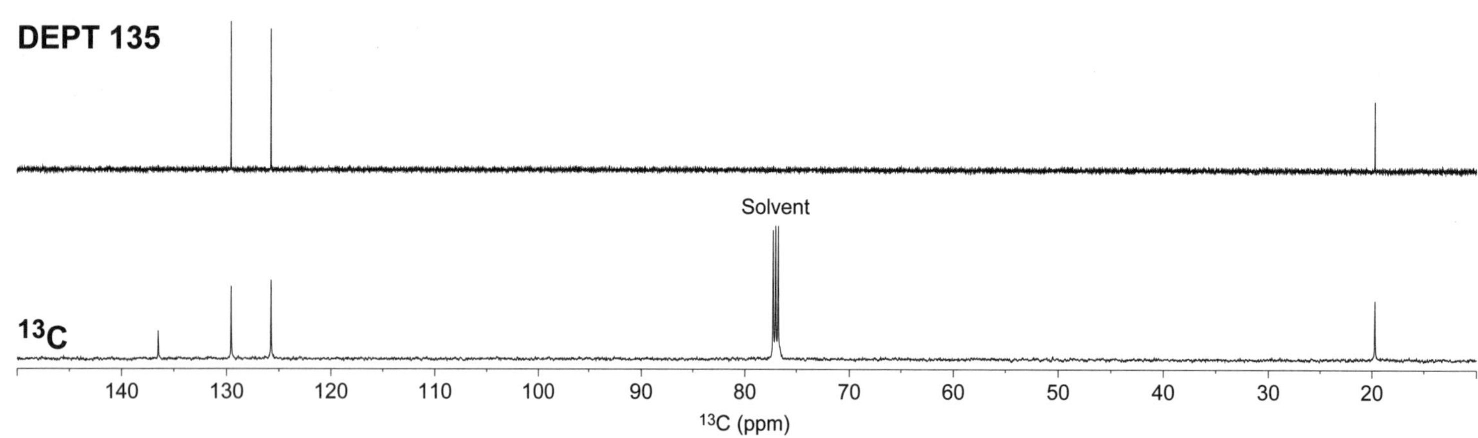

^1H

(3)
(2)
Solvent
7.0
6.5
6.0
5.5
5.0
4.5
4.0
3.5
3.0
2.5
2.0
1H (ppm)

DEPT 135

^{13}C

Solvent
140
130
120
110
100
90
80
70
60
50
40
30
20
13C (ppm)

COSY

2.5
3.0
3.5
4.0
4.5
5.0
5.5
6.0
6.5
7.0
1H (ppm)
7.0
6.5
6.0
5.5
5.0
4.5
4.0
3.5
3.0
2.5
2.0
1H (ppm)

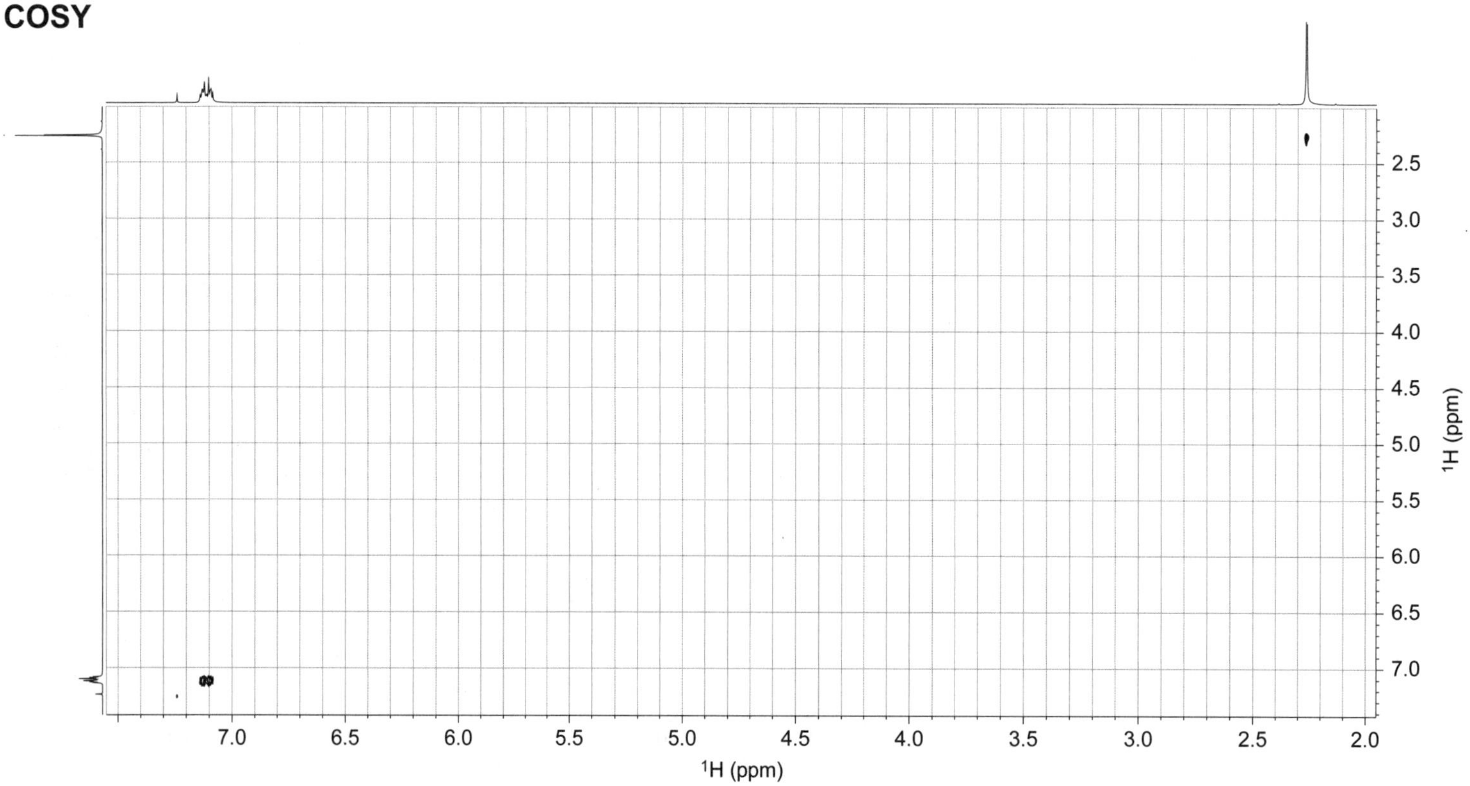

HSQC

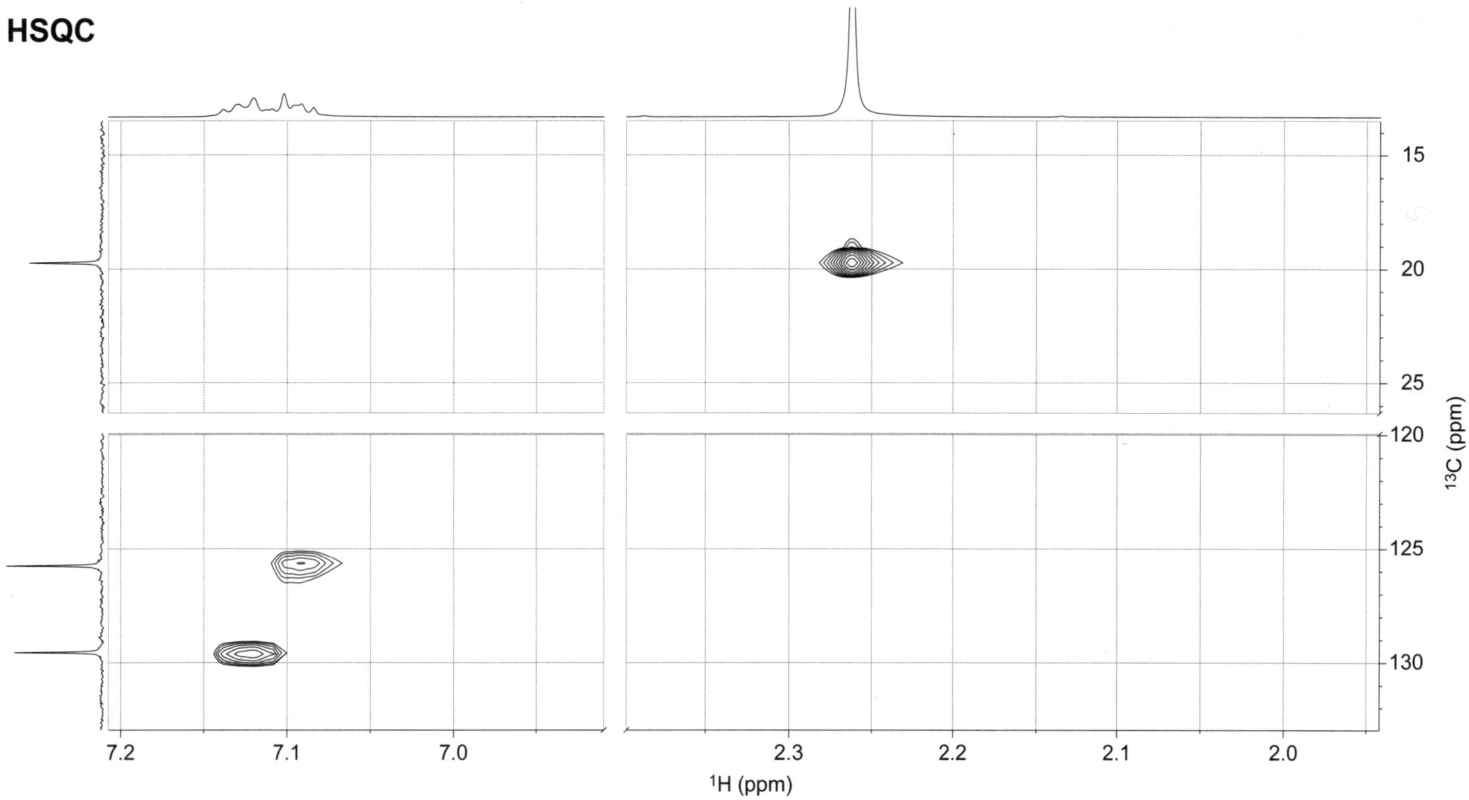
15
20
25
120
125
130
13C (ppm)
7.2
7.1
7.0
2.3
2.2
2.1
2.0
1H (ppm)

HMBC

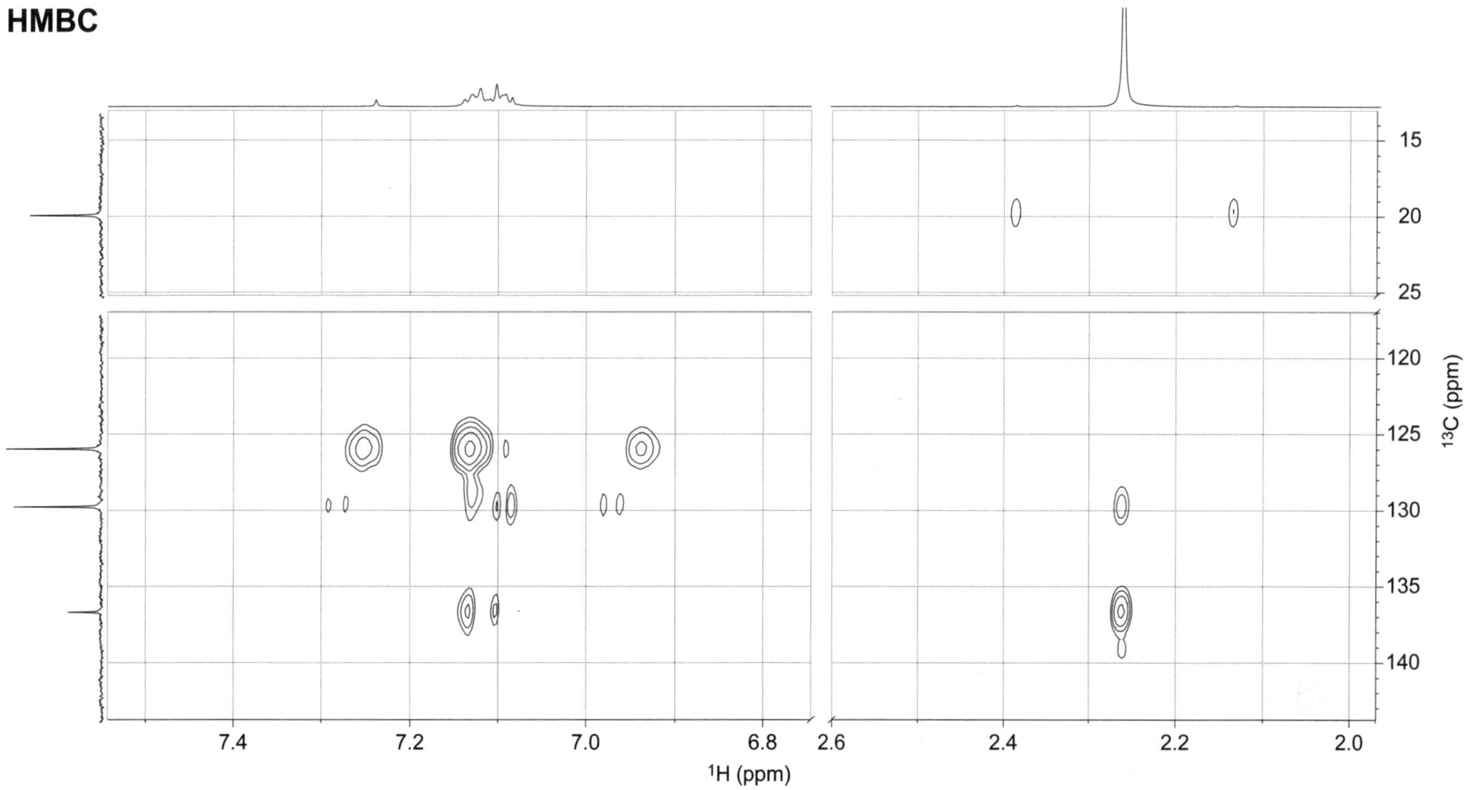
15
20
25
120
125
130
135
140
13C (ppm)
7.4
7.2
7.0
6.8
2.6
2.4
2.2
2.0
1H (ppm)

Problem 9.58
Solvent: $CDCl_3$
Field: 500 MHz (^1H); 125 MHz (^{13}C)

Molecular formula: $C_8H_{10}O$

Difficulty ★★☆☆

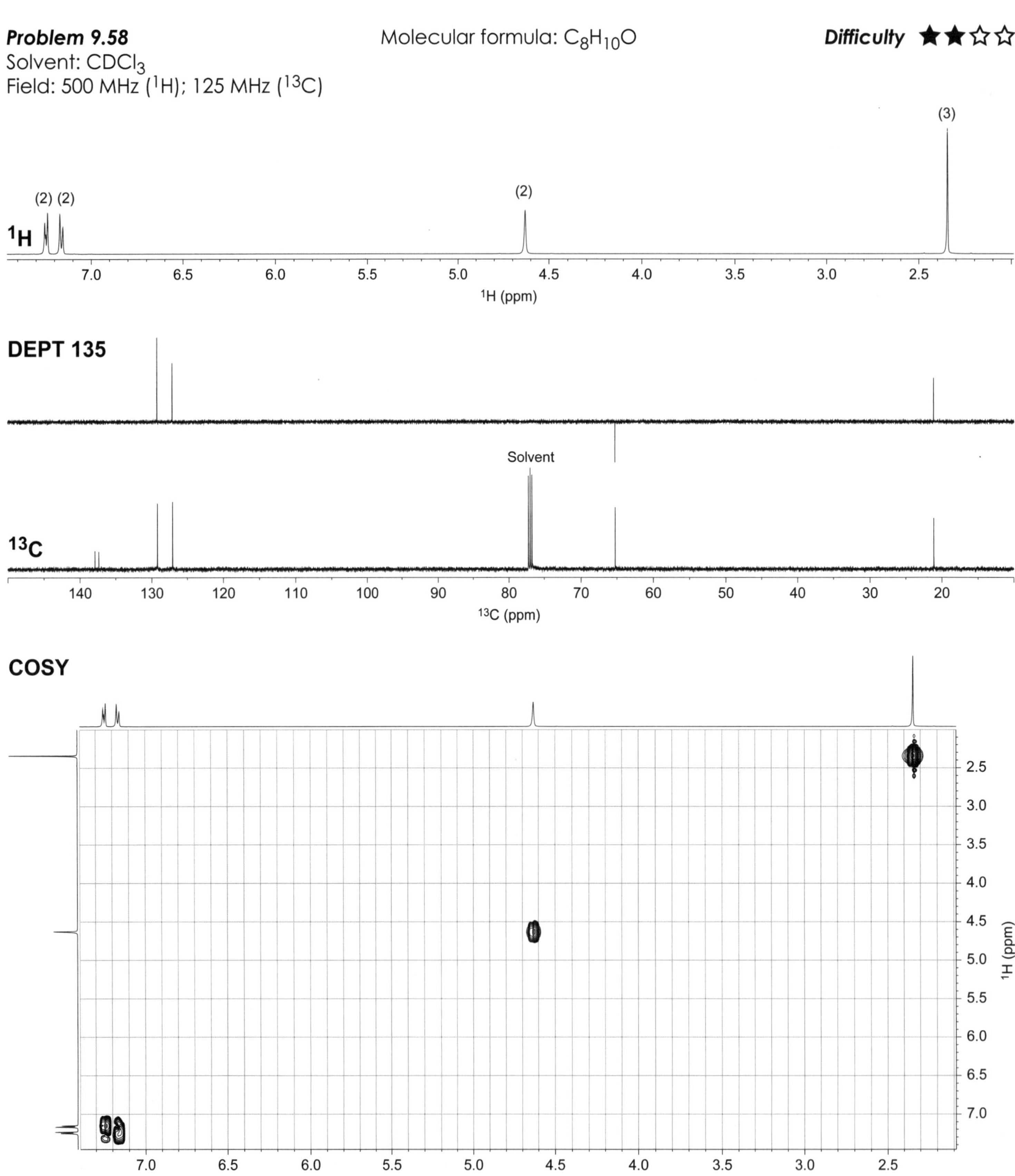

HSQC

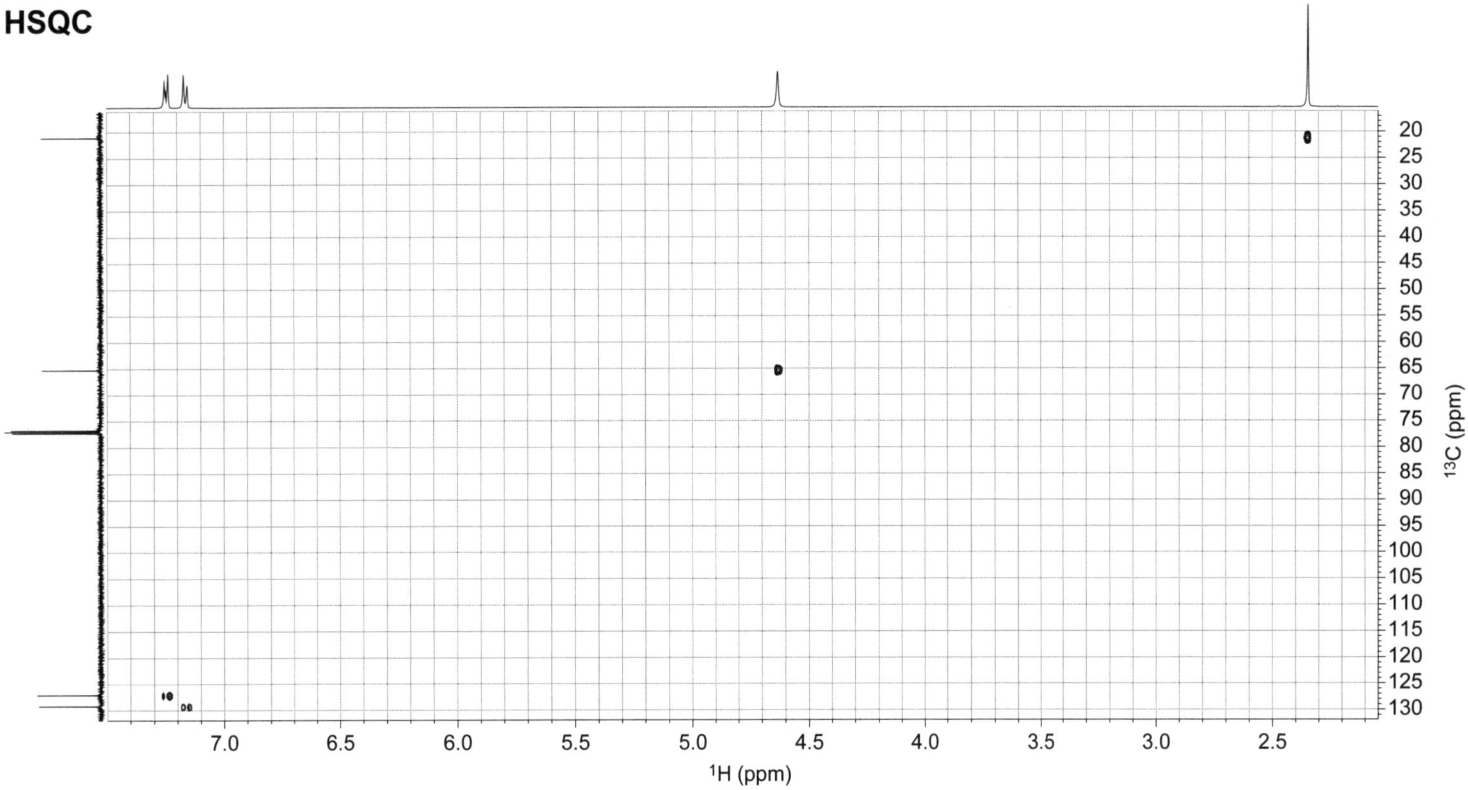

20 25 30 35 40 45 50 55 60 65 70 75 80 85 90 95 100 105 110 115 120 125 130

^{13}C (ppm)

7.0 6.5 6.0 5.5 5.0 4.5 4.0 3.5 3.0 2.5

^{1}H (ppm)

HMBC

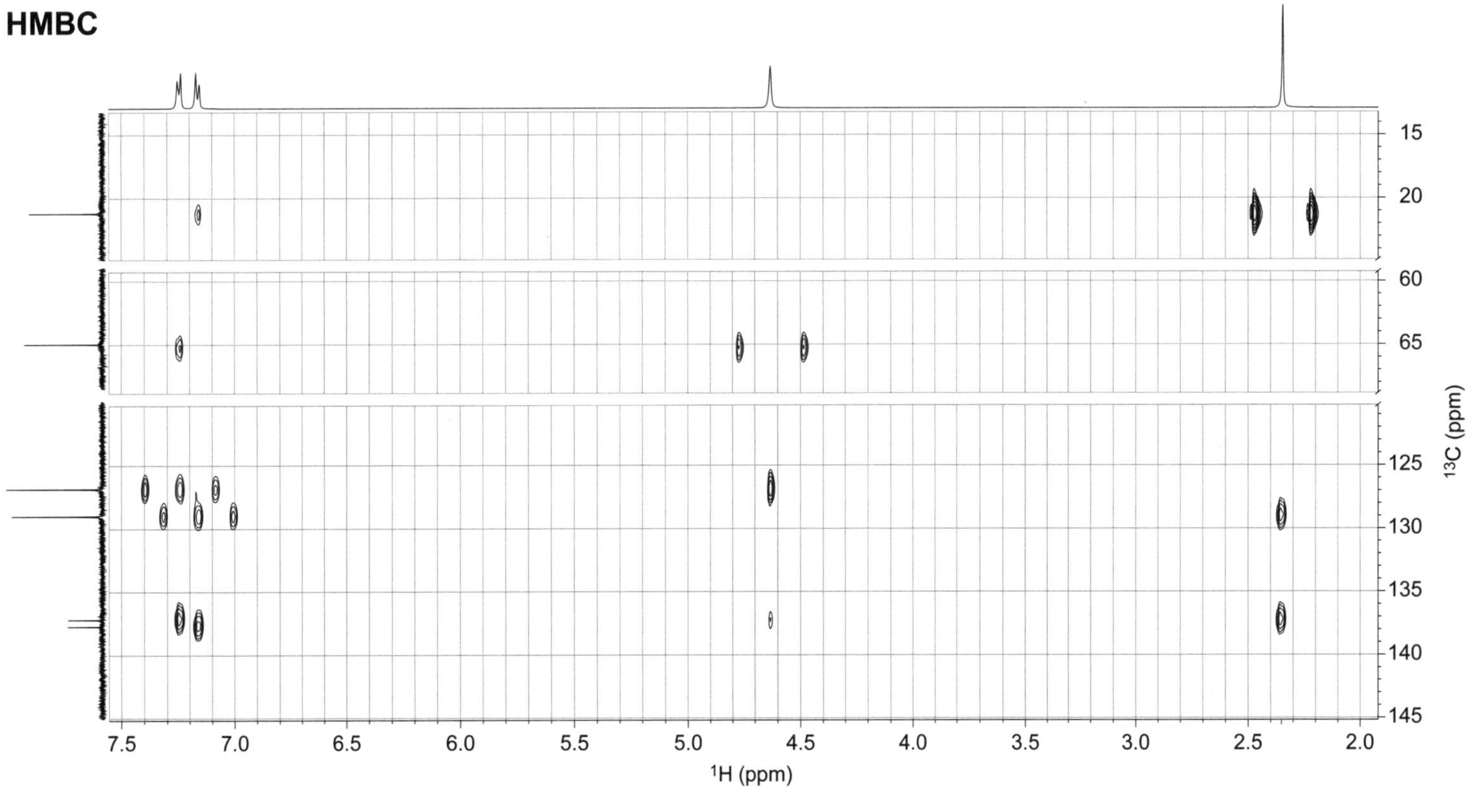

15 20 60 65 125 130 135 140 145

^{13}C (ppm)

7.5 7.0 6.5 6.0 5.5 5.0 4.5 4.0 3.5 3.0 2.5 2.0

^{1}H (ppm)

Problem 9.59
Solvent: D_2O
Field: 500 MHz (^1H); 125 MHz (^{13}C)

Molecular formula: $C_6H_{11}N_3O$
Hint: There is an imidazole moiety

Difficulty ★★☆☆

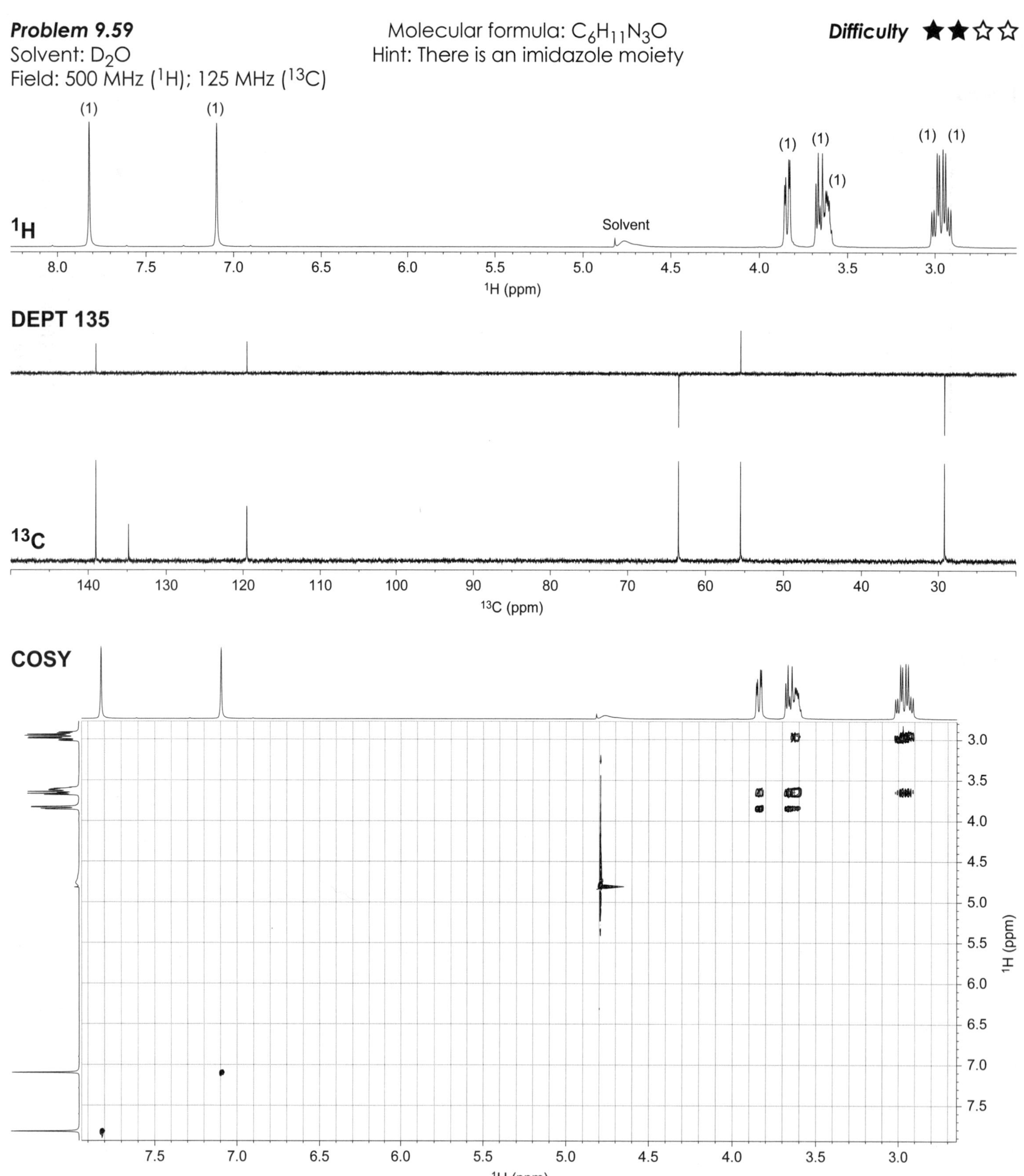

HSQC

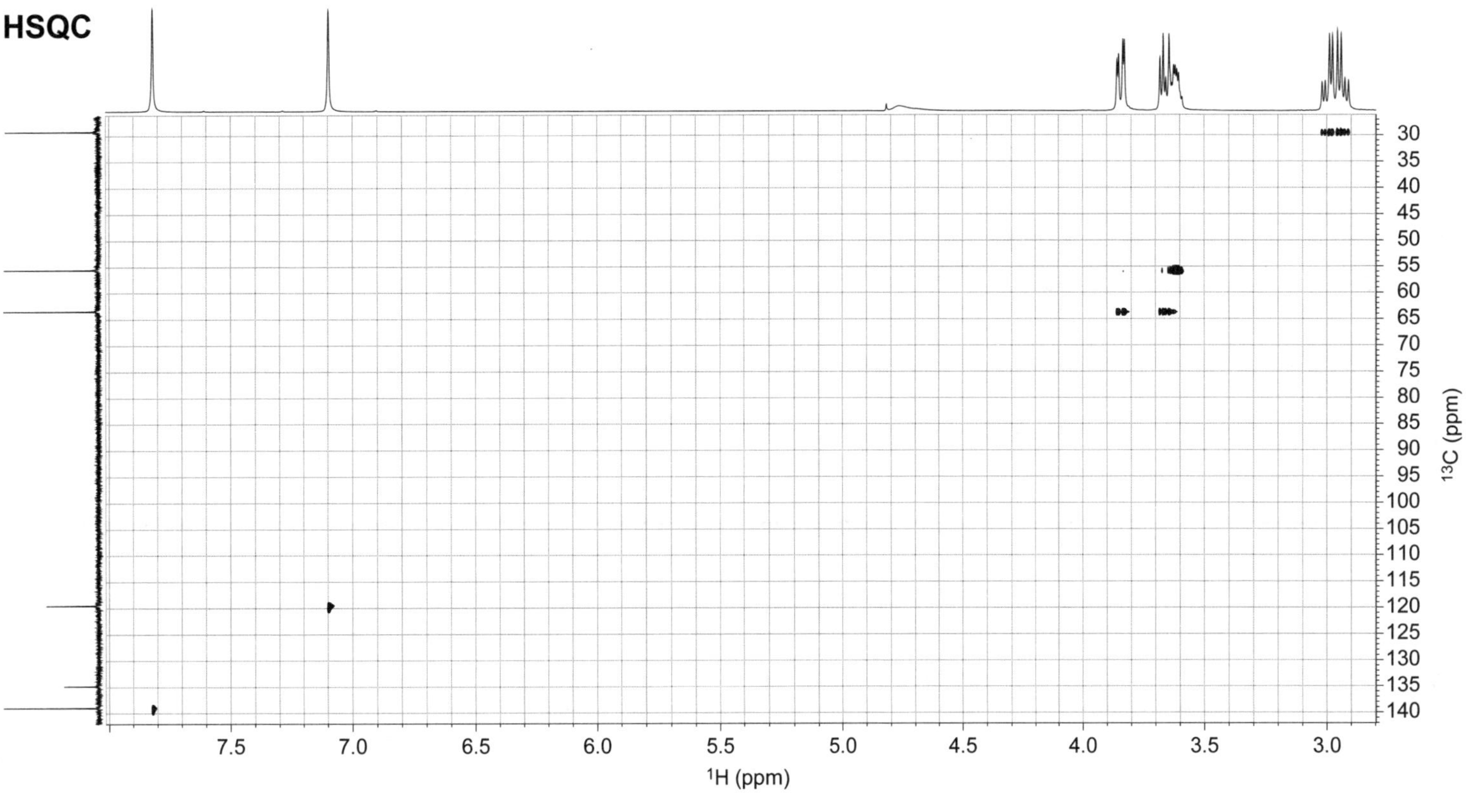

HMBC

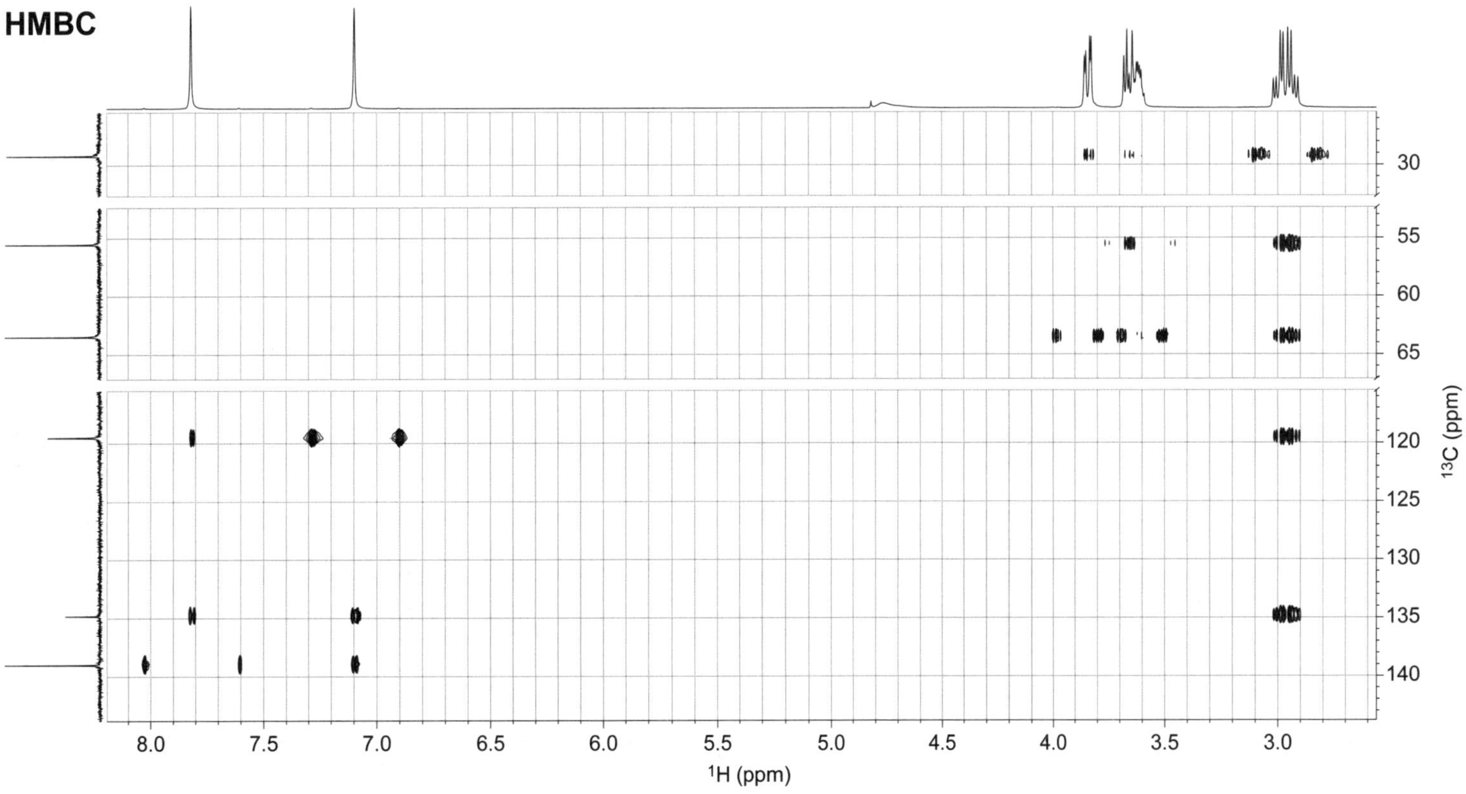

Problem 9.60 Molecular formula: C_7H_8O ***Difficulty*** ★★☆☆

Solvent: D_2O

Field: 500 MHz (^1H); 125 MHz (^{13}C)

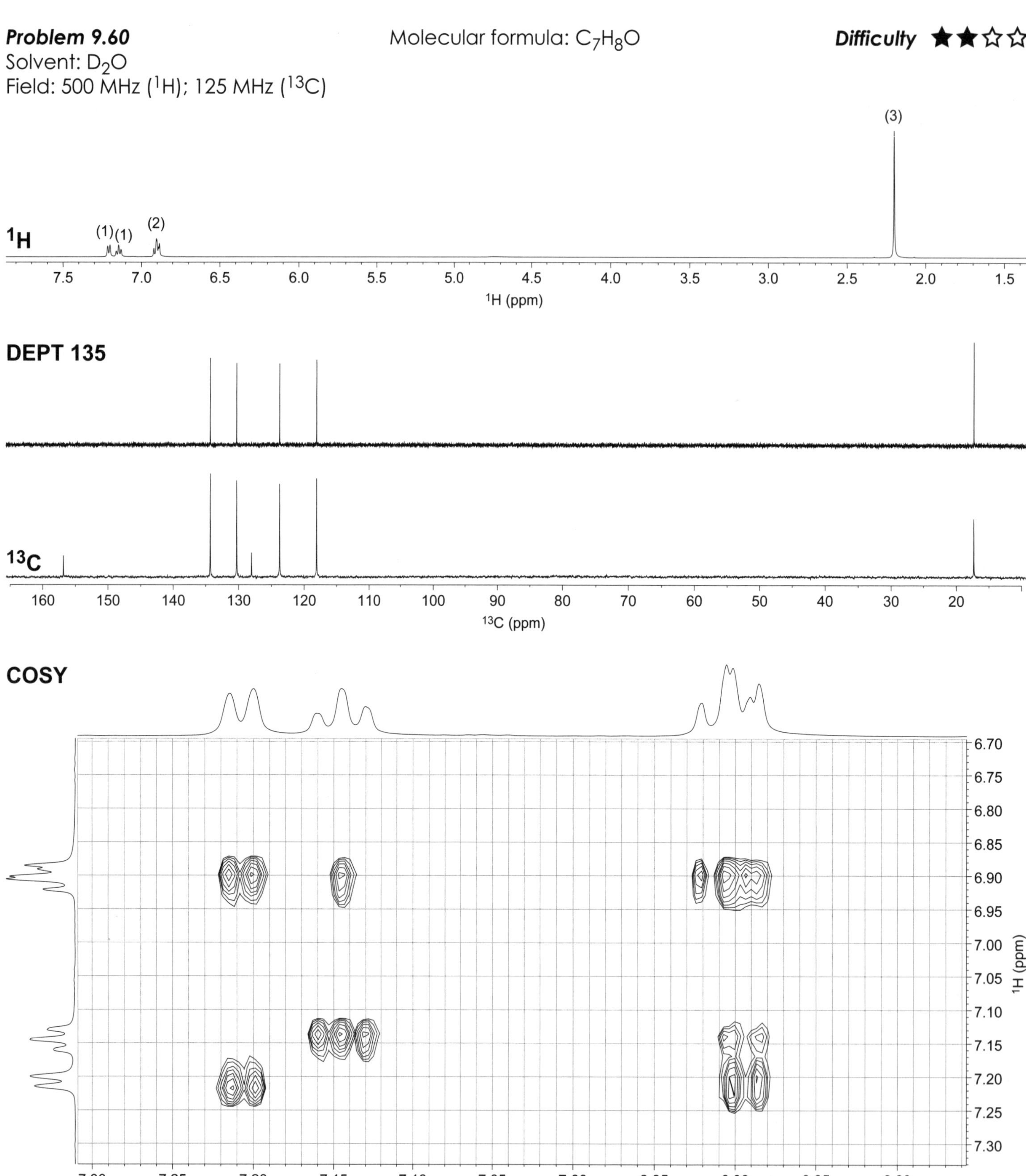

HSQC

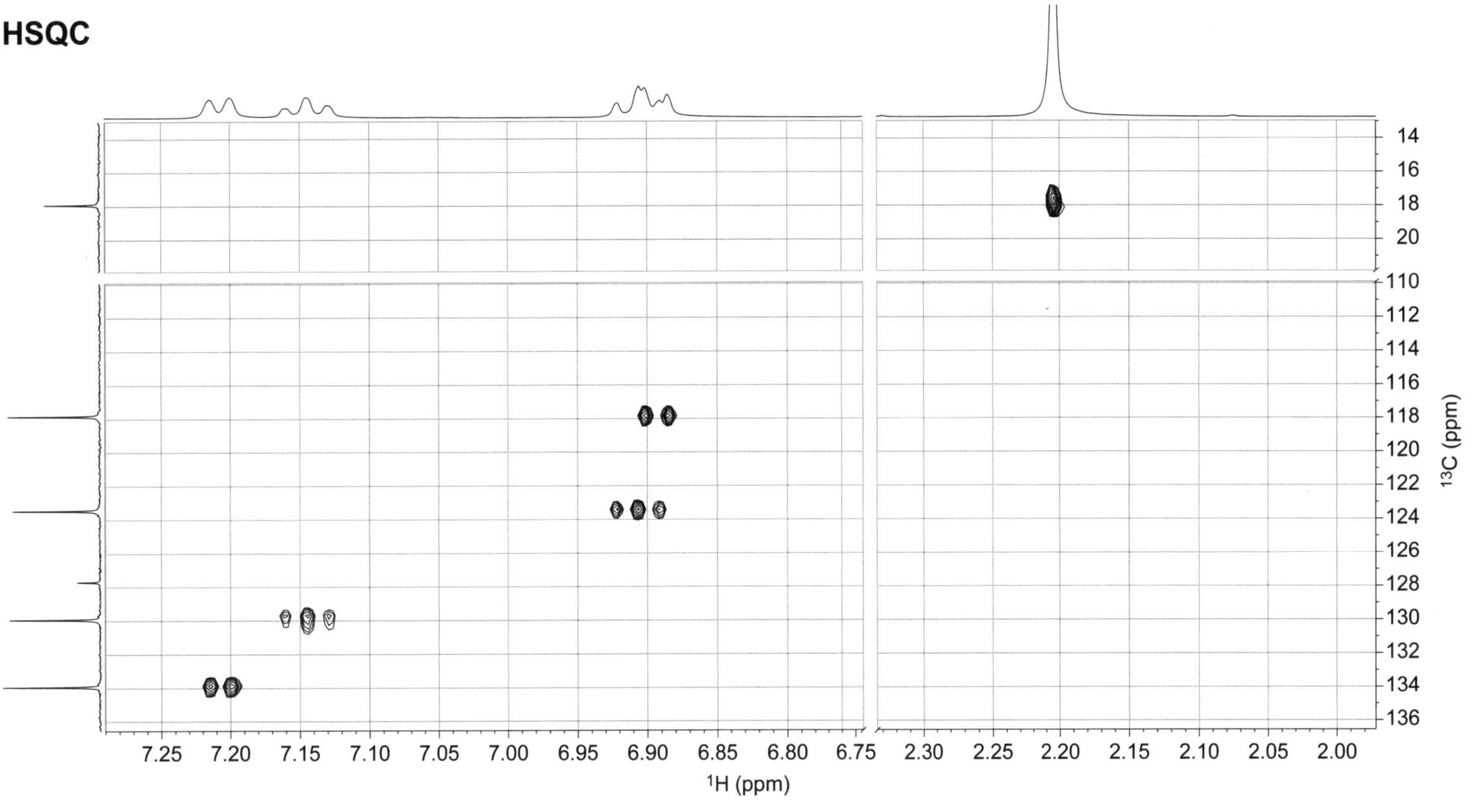

7.25 7.20 7.15 7.10 7.05 7.00 6.95 6.90 6.85 6.80 6.75 2.30 2.25 2.20 2.15 2.10 2.05 2.00
^{1}H (ppm)
14 16 18 20 110 112 114 116 118 120 122 124 126 128 130 132 134 136
^{13}C (ppm)

7.4 7.3 7.2 7.1 7.0 6.9 6.8 6.7 6.6 2.4 2.3 2.2 2.1 2.0
^{1}H (ppm)
15 20 115 120 125 130 135 140 145 150 155 160
^{13}C (ppm)

HMBC

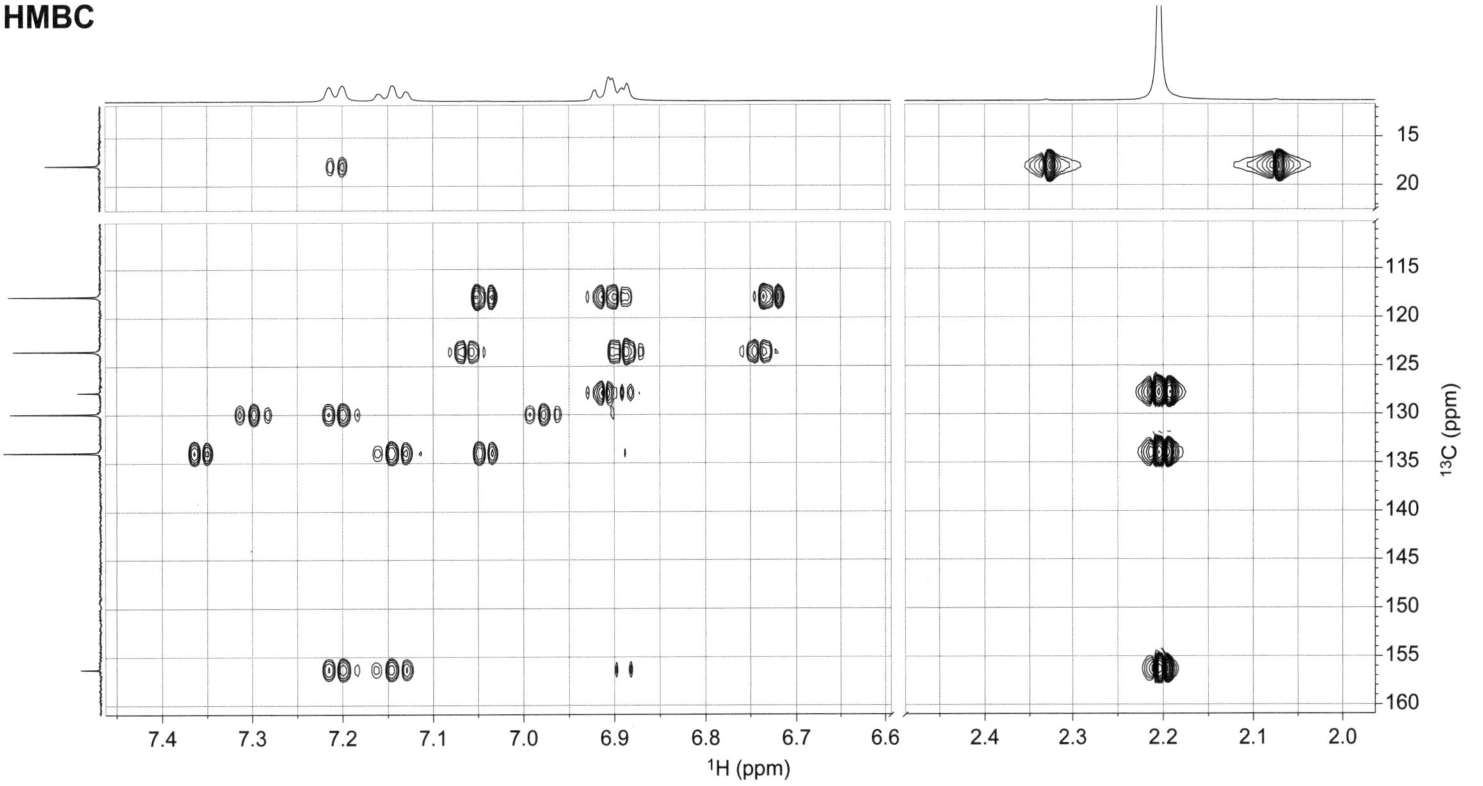

Problem 9.61 Molecular formula: $C_6H_{12}O_2$ ***Difficulty*** ★★☆☆

Solvent: $CDCl_3$

Field: 500 MHz (^1H); 125 MHz (^{13}C)

^1H

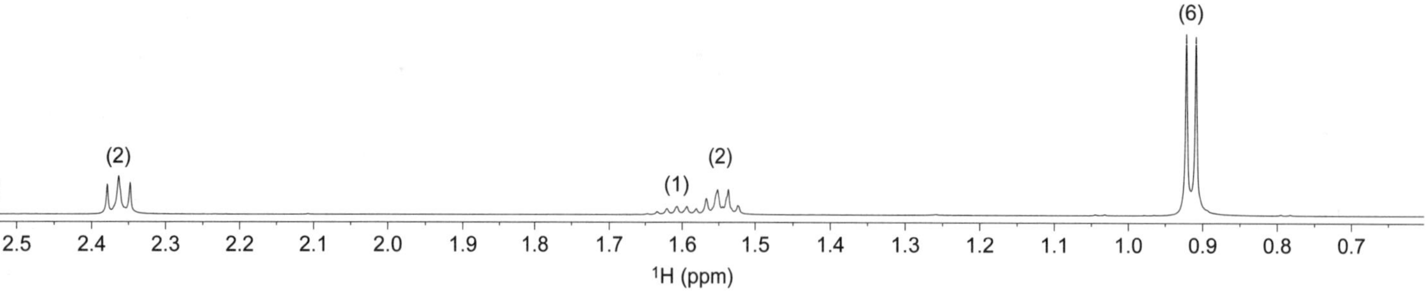

(6)
(2)
(2)
(1)
2.5 2.4 2.3 2.2 2.1 2.0 1.9 1.8 1.7 1.6 1.5 1.4 1.3 1.2 1.1 1.0 0.9 0.8 0.7
1H (ppm)

DEPT 135

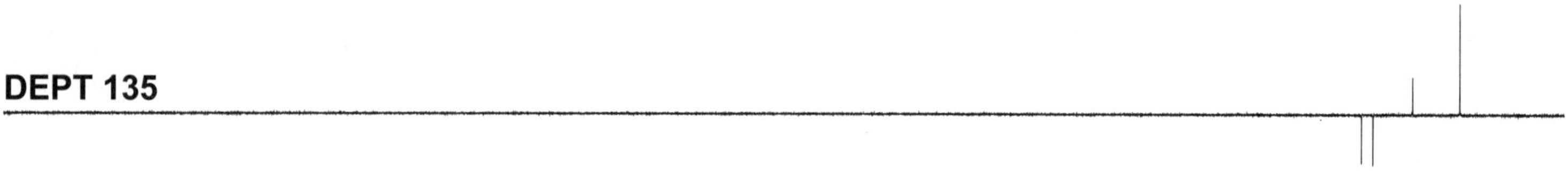

^{13}C

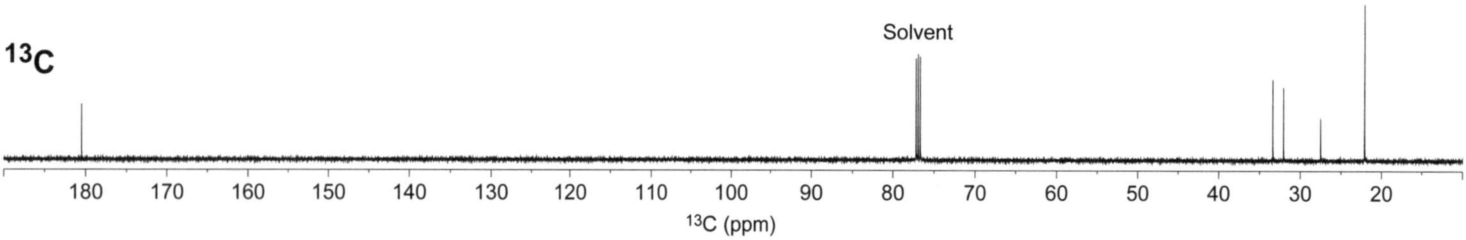

Solvent
180 170 160 150 140 130 120 110 100 90 80 70 60 50 40 30 20
13C (ppm)

COSY

2.5 2.4 2.3 2.2 2.1 2.0 1.9 1.8 1.7 1.6 1.5 1.4 1.3 1.2 1.1 1.0 0.9 0.8 0.7
1H (ppm)
0.6 0.7 0.8 0.9 1.0 1.1 1.2 1.3 1.4 1.5 1.6 1.7 1.8 1.9 2.0 2.1 2.2 2.3 2.4 2.5
1H (ppm)

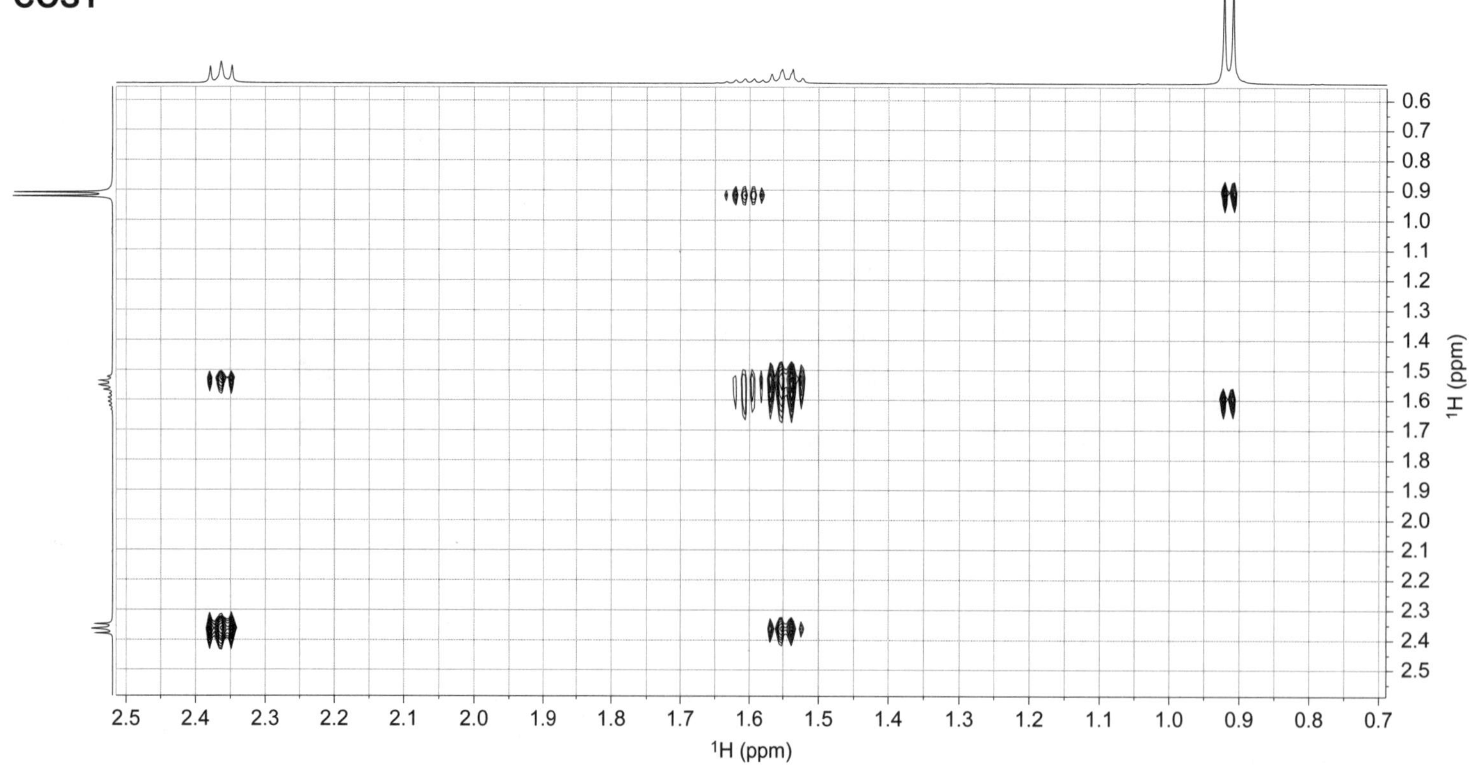

HSQC

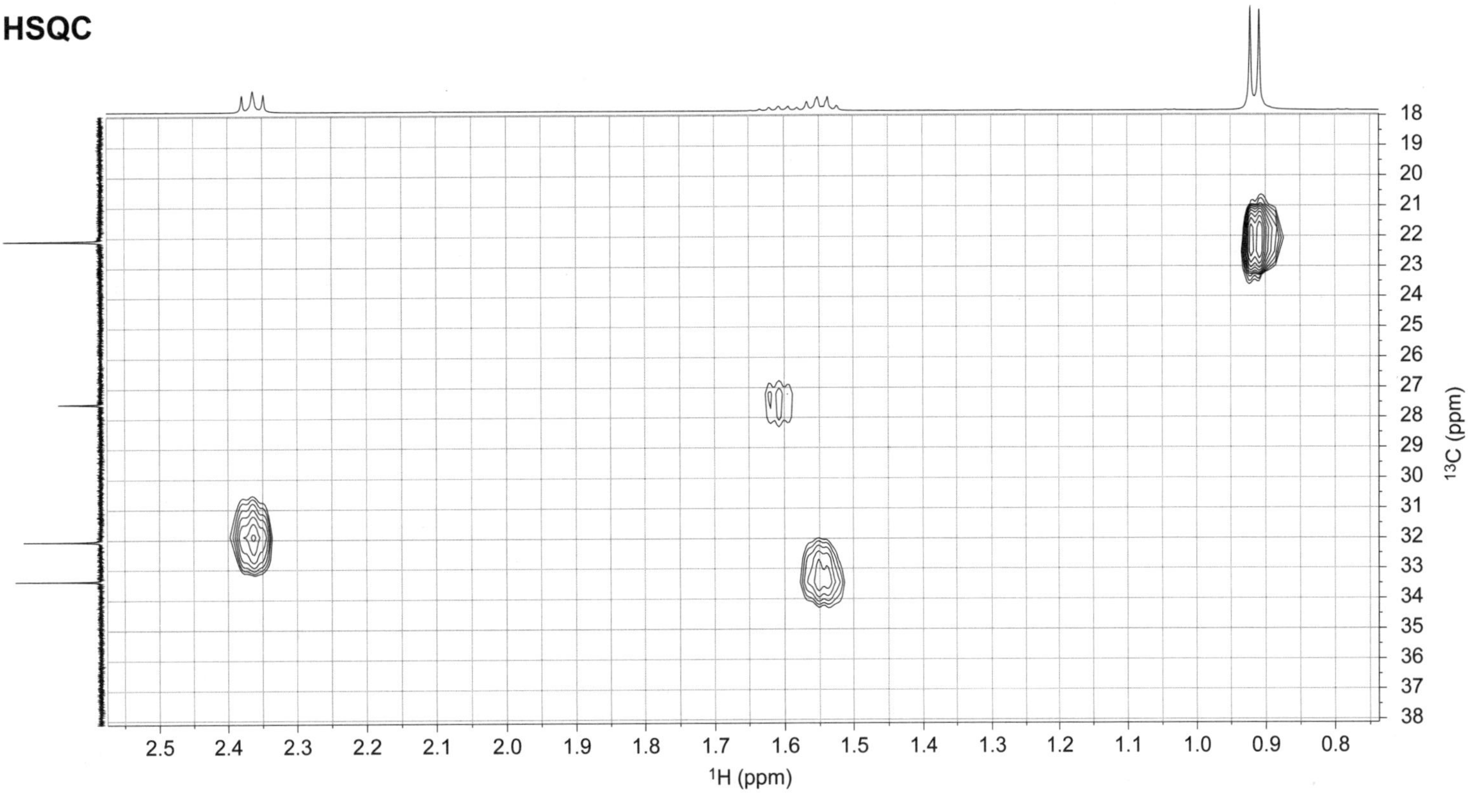

18
19
20
21
22
23
24
25
26
27
28
29
30
31
32
33
34
35
36
37
38
¹³C (ppm)
2.5
2.4
2.3
2.2
2.1
2.0
1.9
1.8
1.7
1.6
1.5
1.4
1.3
1.2
1.1
1.0
0.9
0.8
¹H (ppm)

HMBC

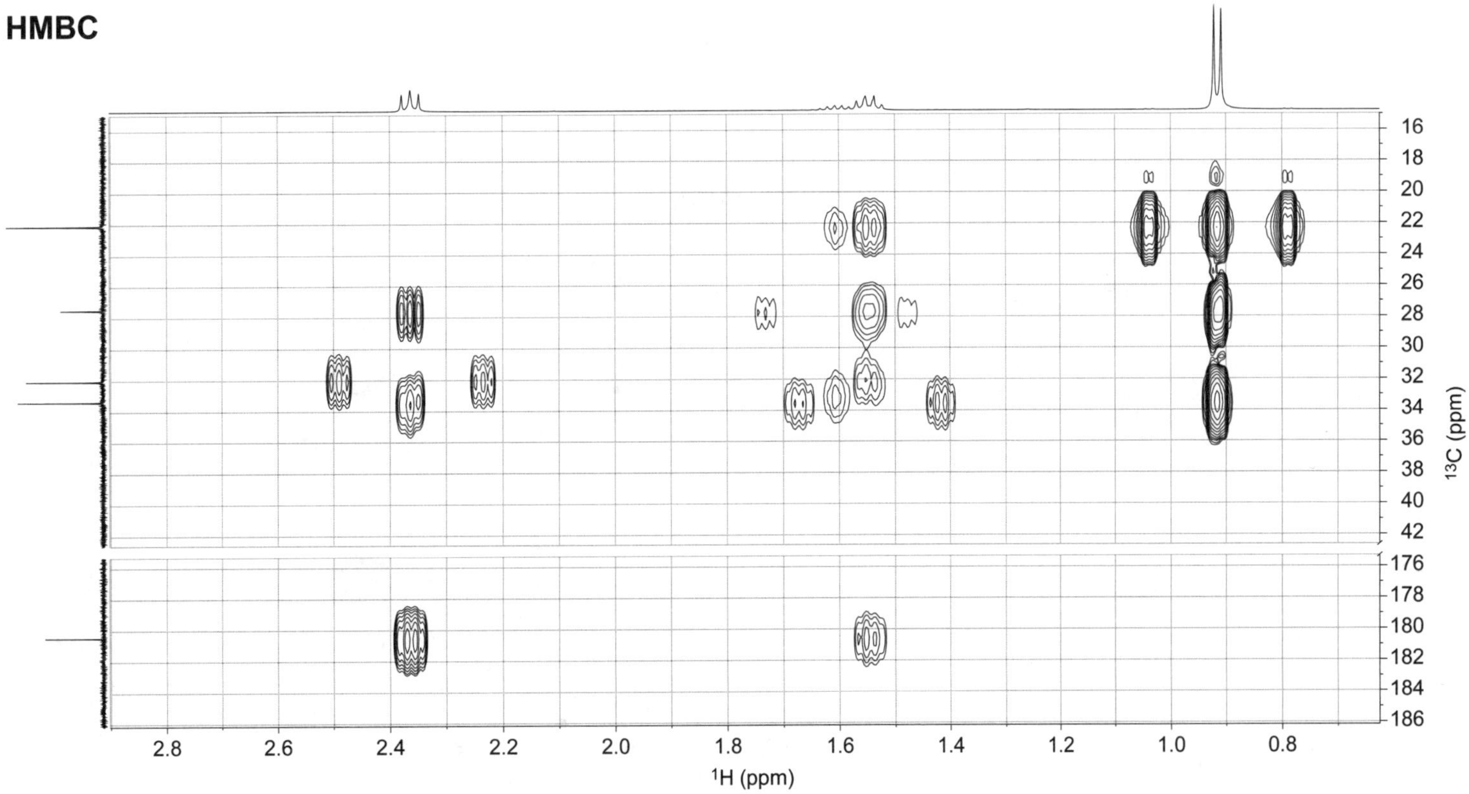

16
18
20
22
24
26
28
30
32
34
36
38
40
42
176
178
180
182
184
186
¹³C (ppm)
2.8
2.6
2.4
2.2
2.0
1.8
1.6
1.4
1.2
1.0
0.8
¹H (ppm)

Problem 9.62
Solvent: D_2O
Field: 500 MHz (^1H); 125 MHz (^{13}C)

Molecular formula: $C_6H_6O_3$

Difficulty ★★☆☆

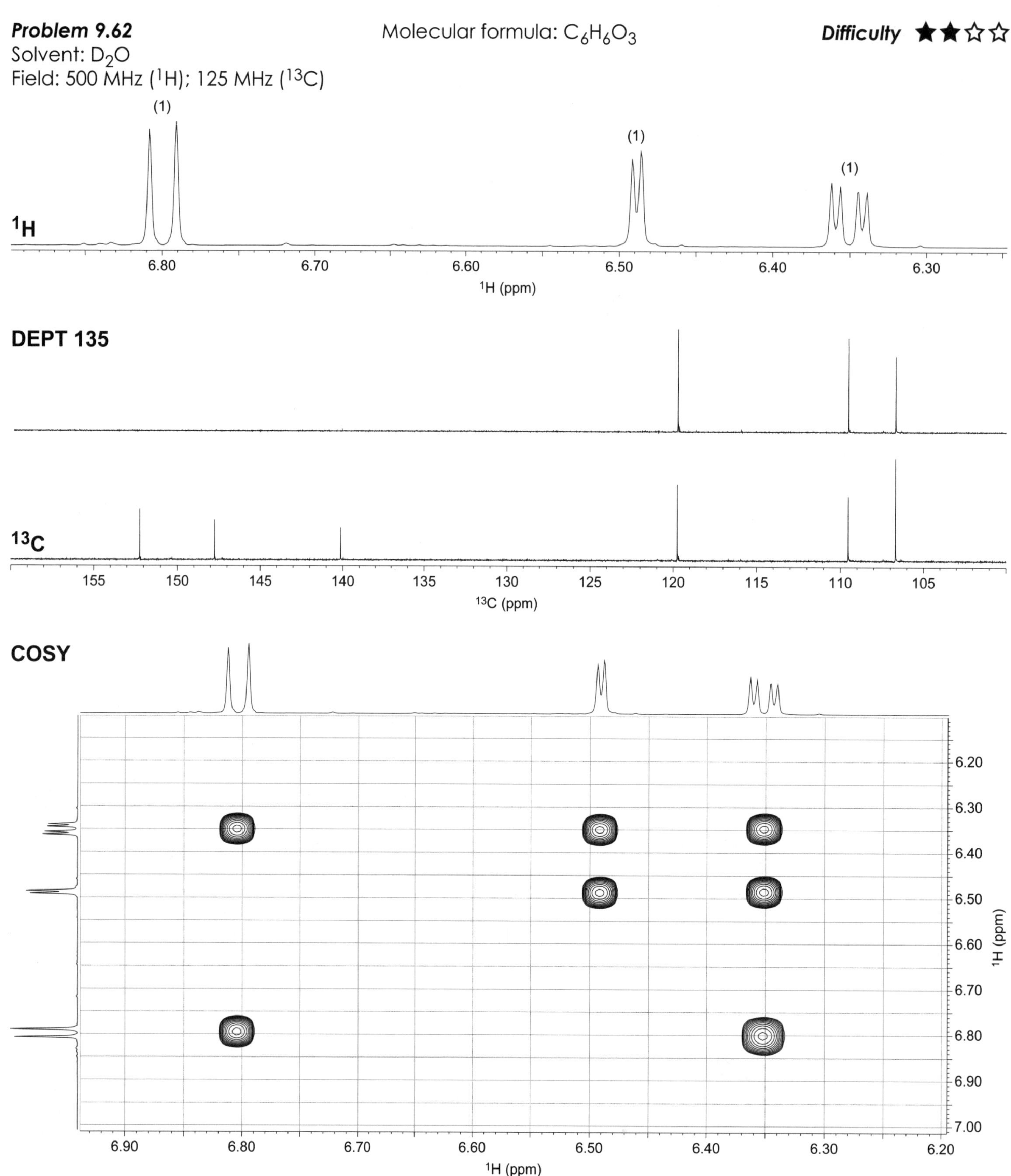

HSQC

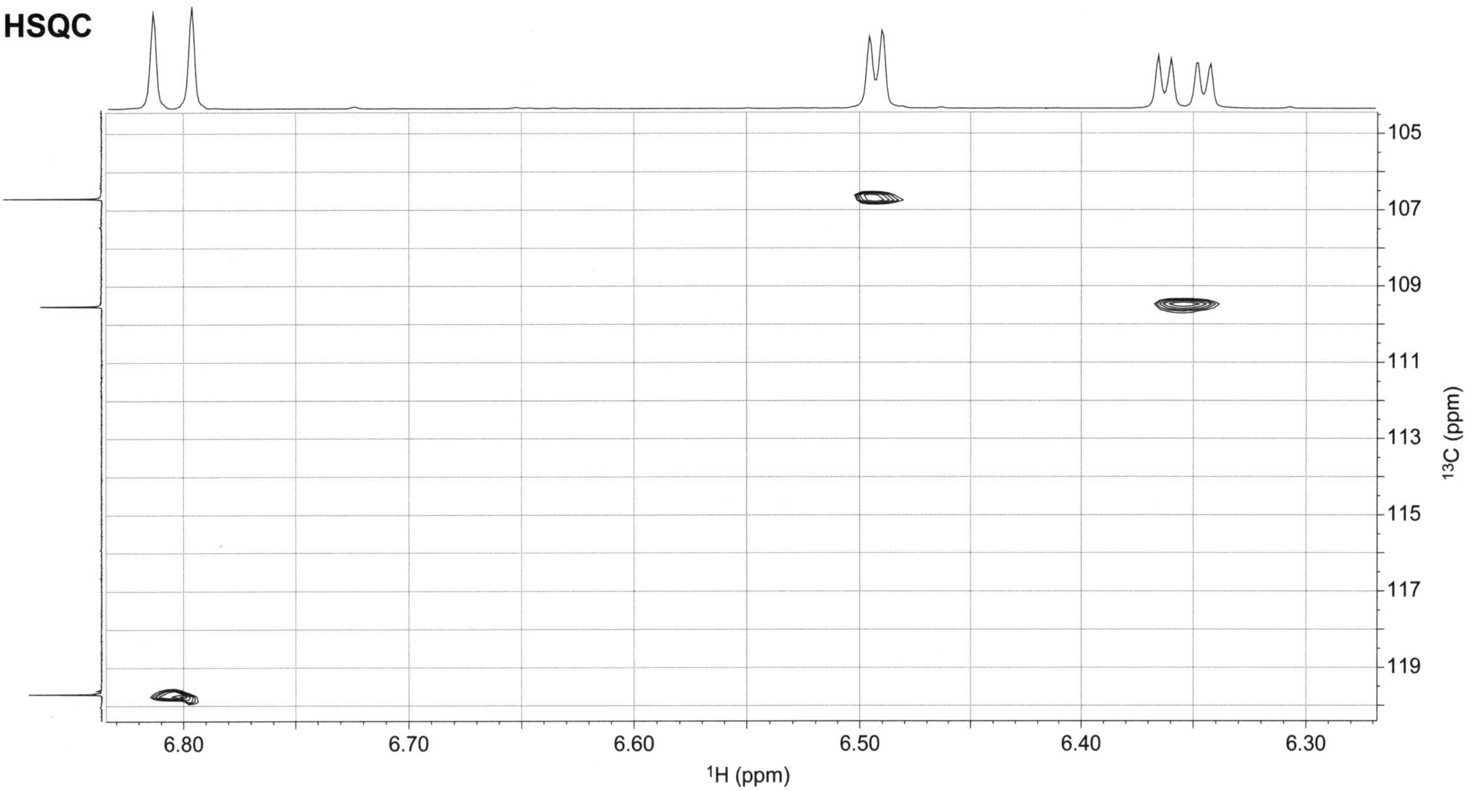
105
107
109
111
113
115
117
119
^{13}C (ppm)
6.80
6.70
6.60
6.50
6.40
6.30
^1H (ppm)

HMBC

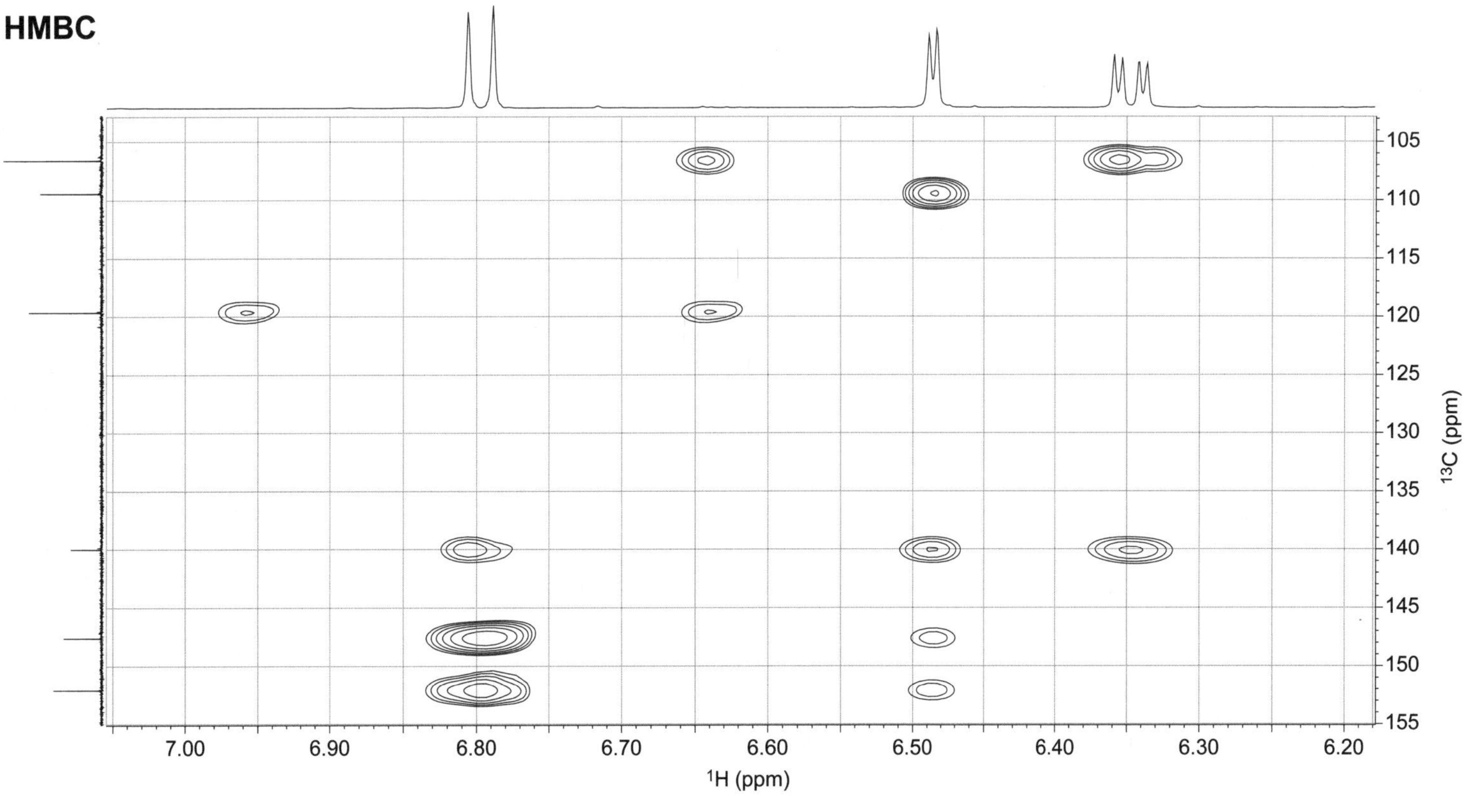
105
110
115
120
125
130
135
140
145
150
155
^{13}C (ppm)
7.00
6.90
6.80
6.70
6.60
6.50
6.40
6.30
6.20
^1H (ppm)

Problem 9.63 Molecular formula: $C_8H_8O_2$ **Difficulty** ★★☆☆

Solvent: CD_3OD

Field: 700 MHz (^1H); 175 MHz (^{13}C)

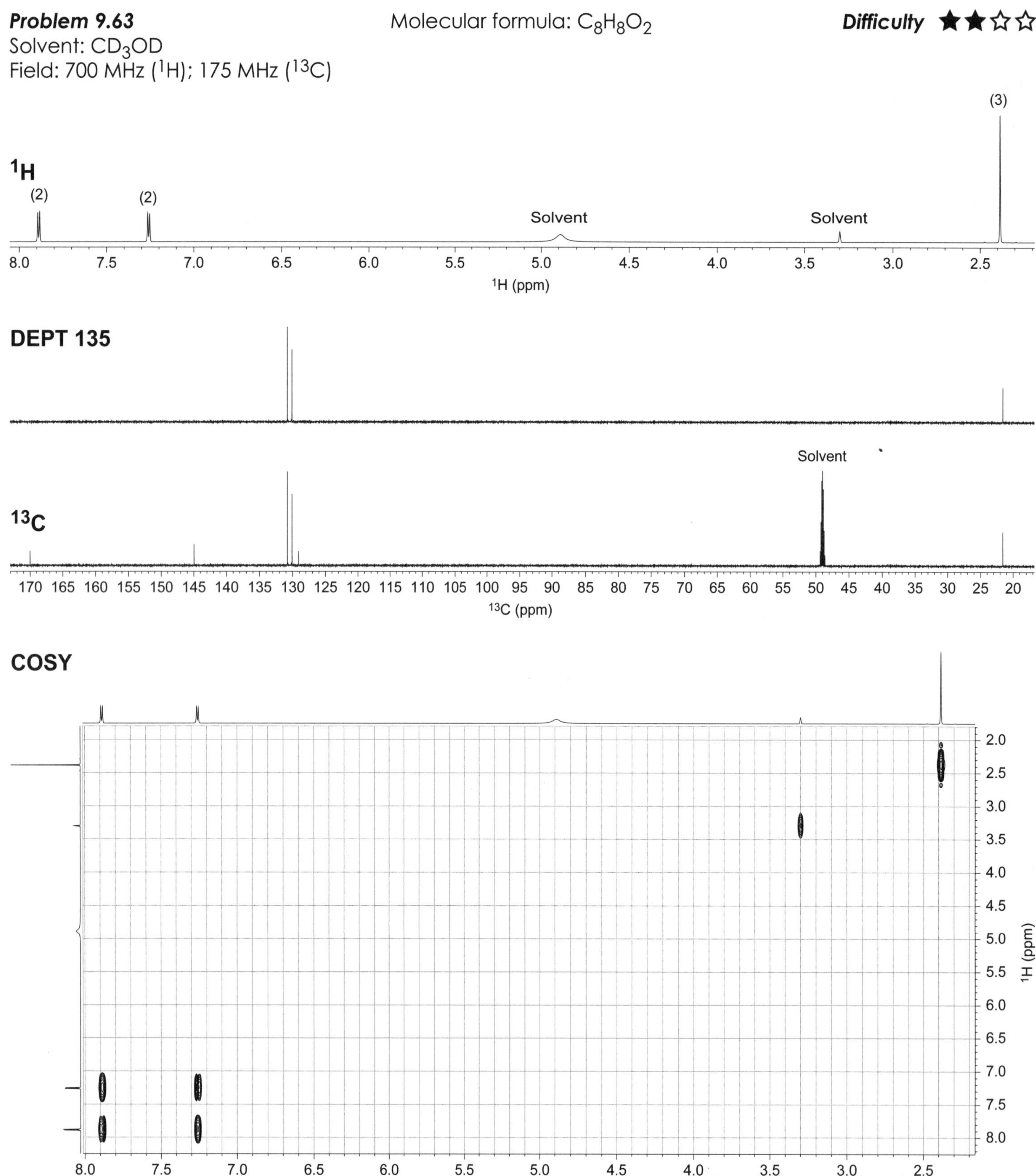

HSQC

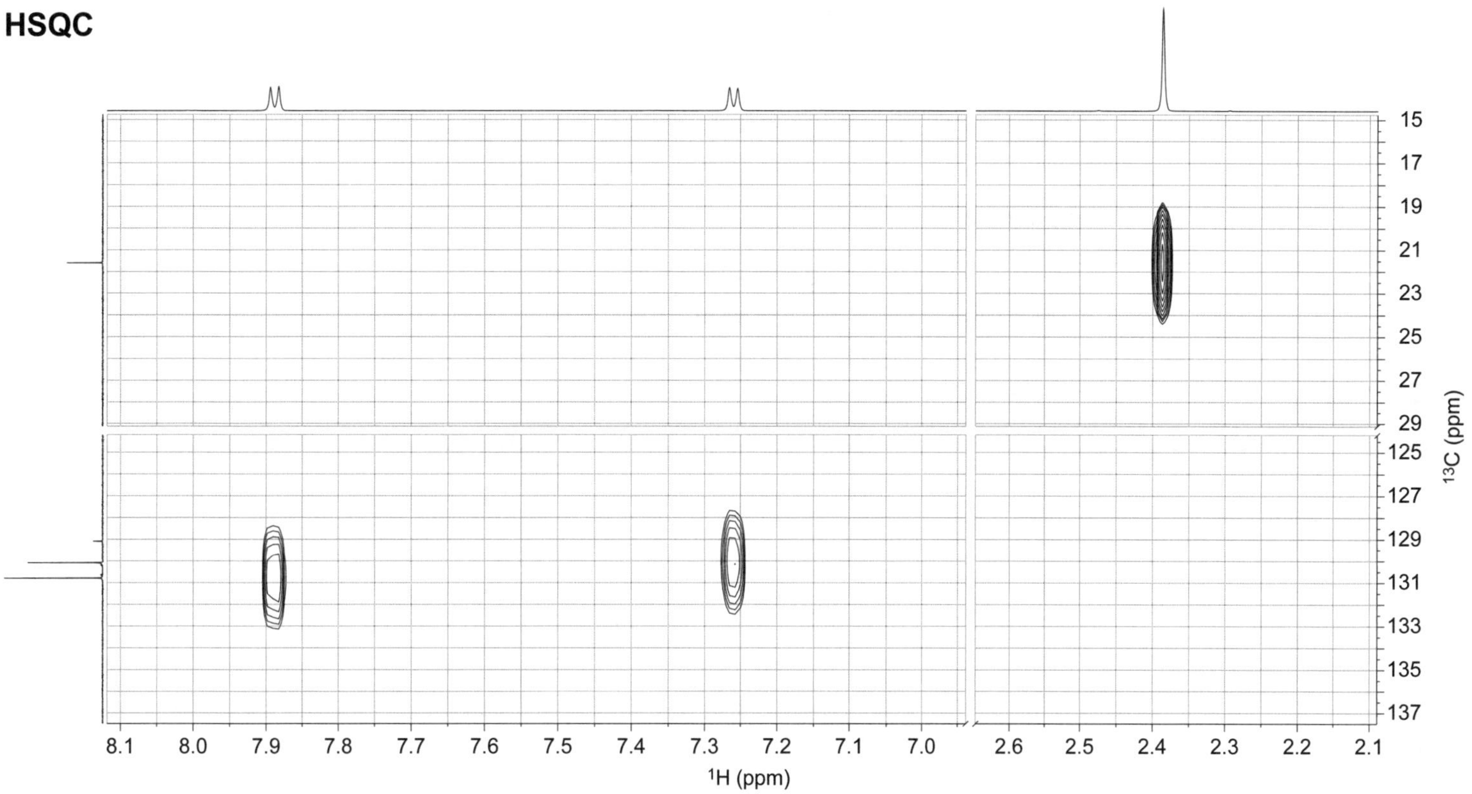
15
17
19
21
23
25
27
29
125
127
129
131
133
135
137
^{13}C (ppm)
8.1 8.0 7.9 7.8 7.7 7.6 7.5 7.4 7.3 7.2 7.1 7.0 2.6 2.5 2.4 2.3 2.2 2.1
^{1}H (ppm)

HMBC

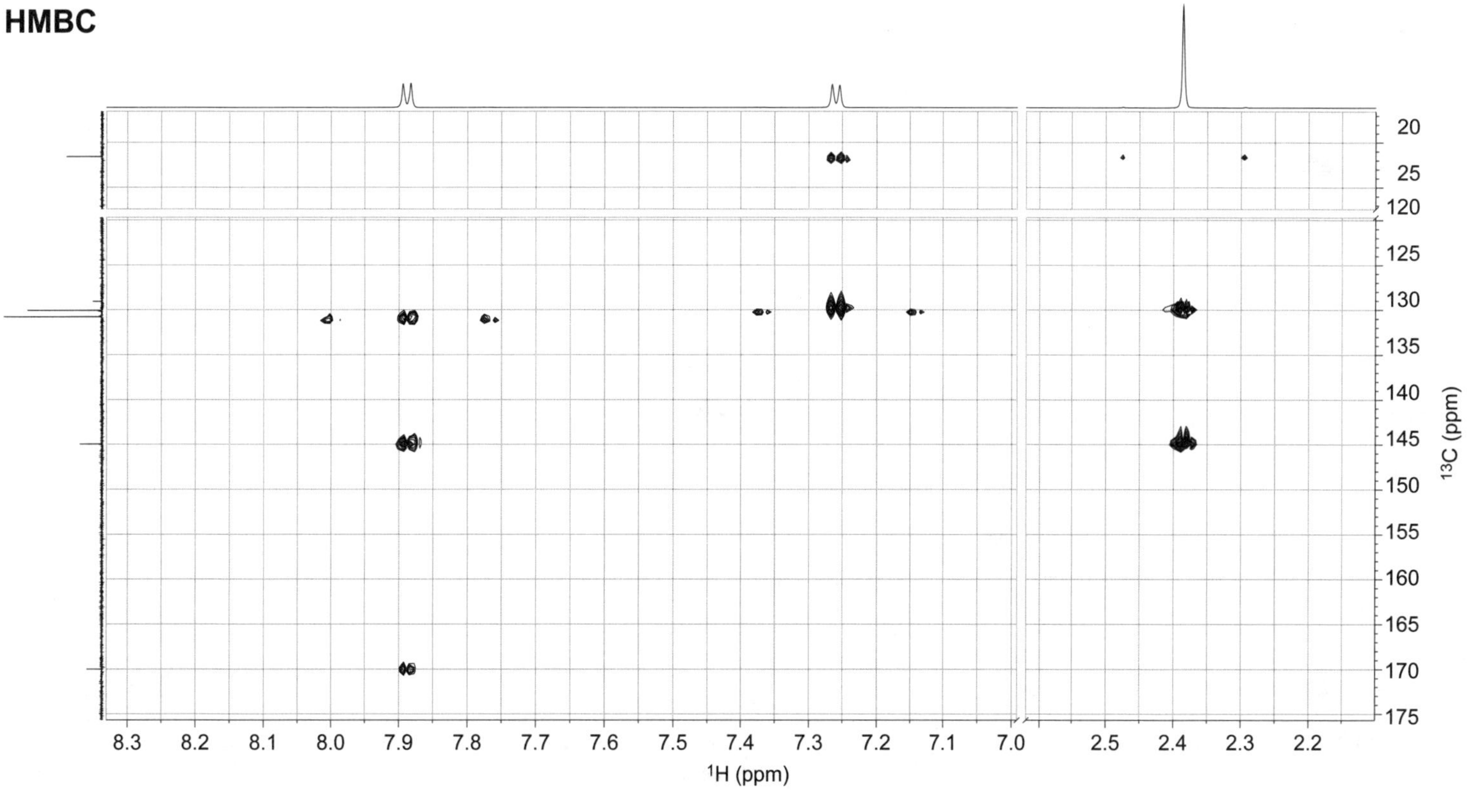
20
25
120
125
130
135
140
145
150
155
160
165
170
175
^{13}C (ppm)
8.3 8.2 8.1 8.0 7.9 7.8 7.7 7.6 7.5 7.4 7.3 7.2 7.1 7.0 2.5 2.4 2.3 2.2
^{1}H (ppm)

Problem 9.64
Solvent: CD_3OD
Field: 700 MHz (^1H); 175 MHz (^{13}C)

Molecular formula: $C_8H_{14}O_4$

Difficulty ★★☆☆

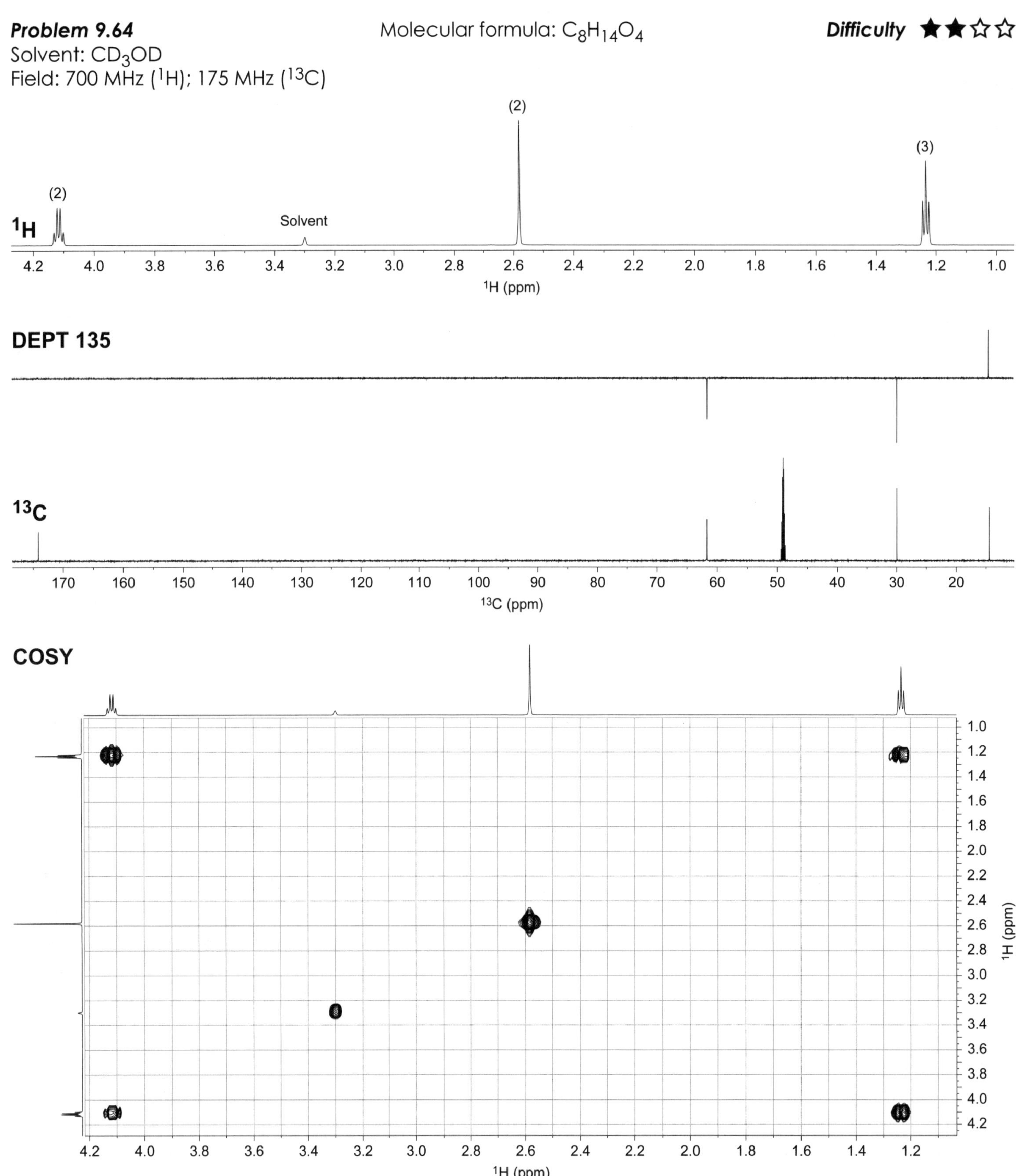

HSQC

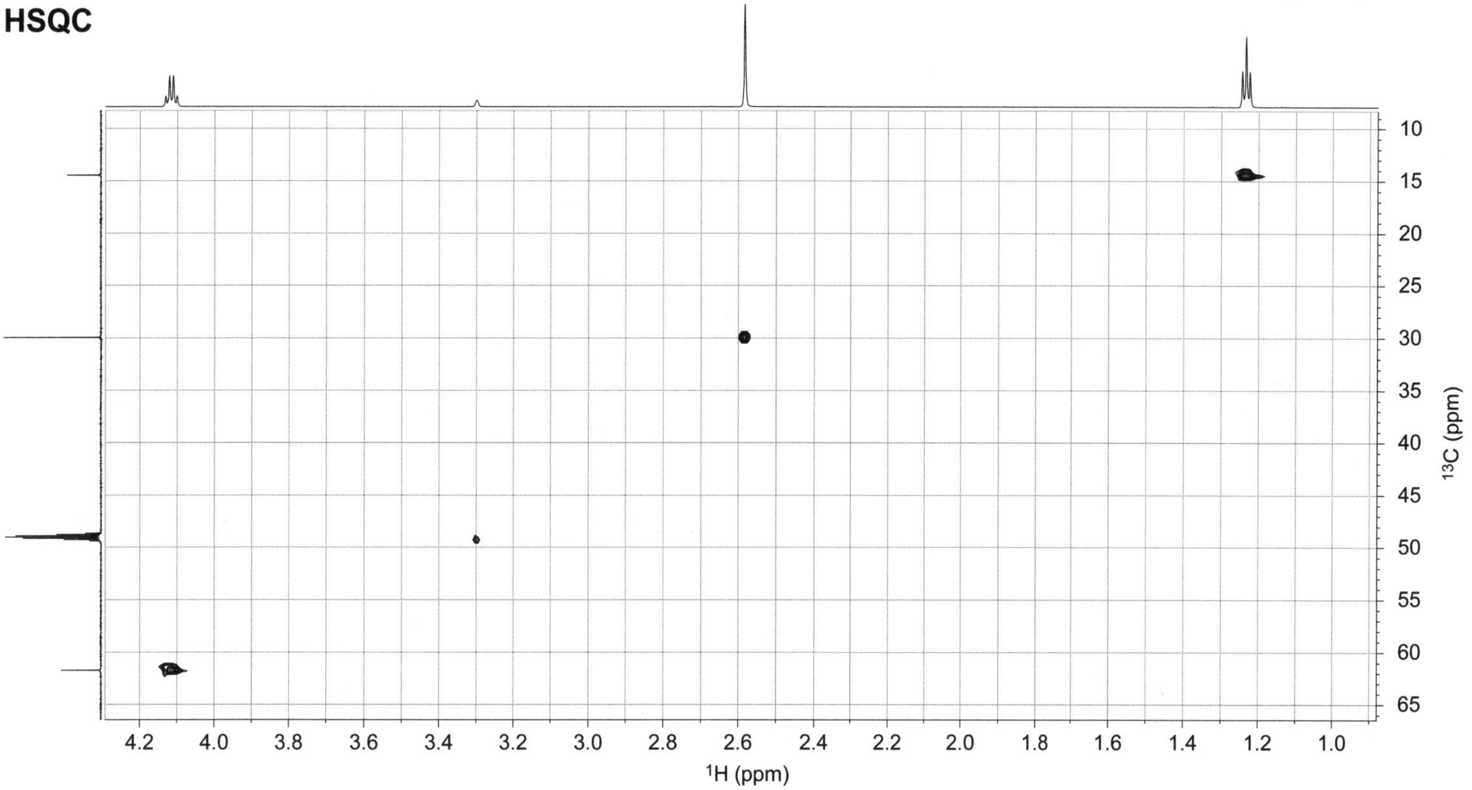

HMBC

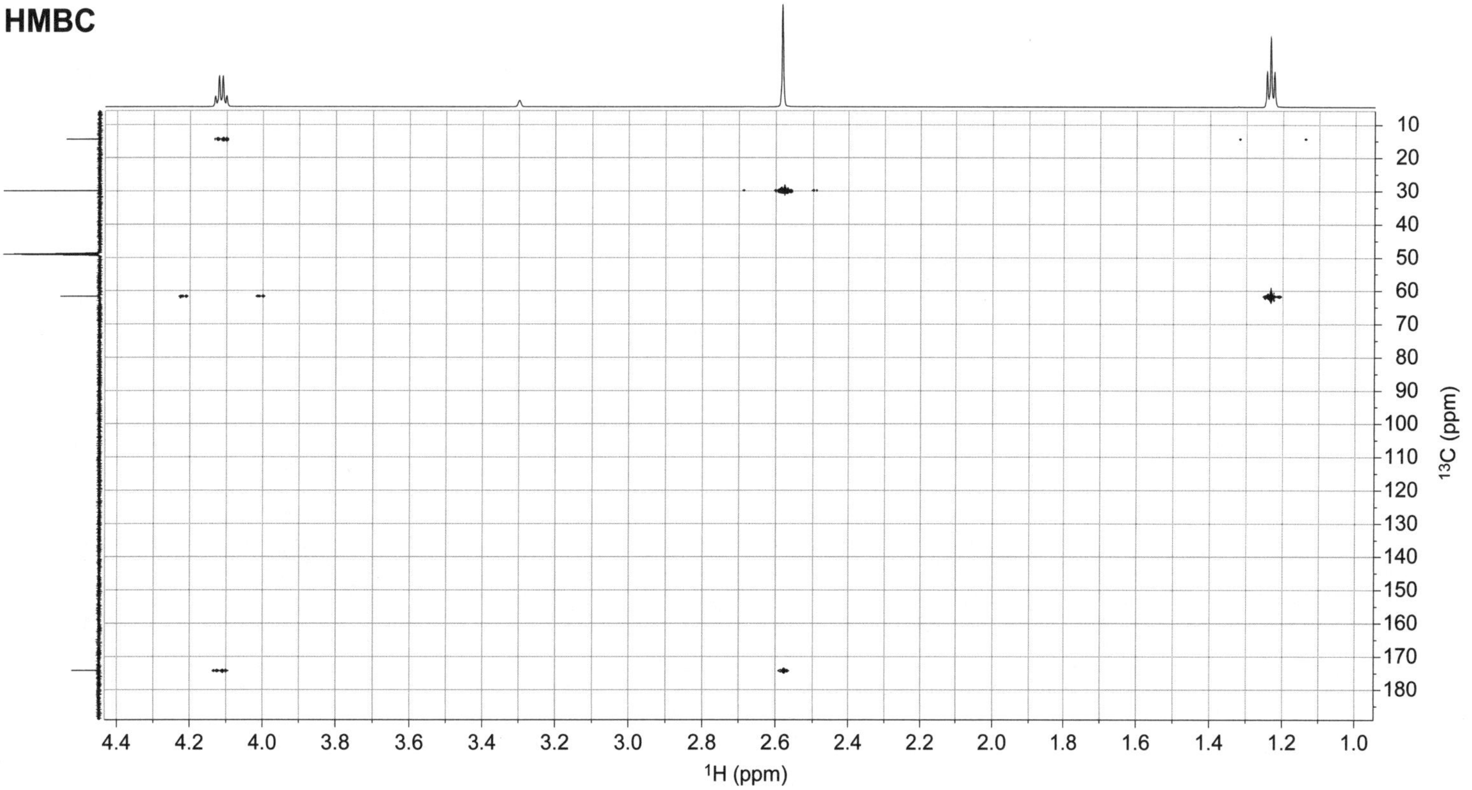

Problem 9.65
Solvent: CD_3OD
Field: 500 MHz (^1H); 125 MHz (^{13}C)

Molecular formula: $C_8H_9NO_2$
Hint: There is an amide group

Difficulty ★★☆☆

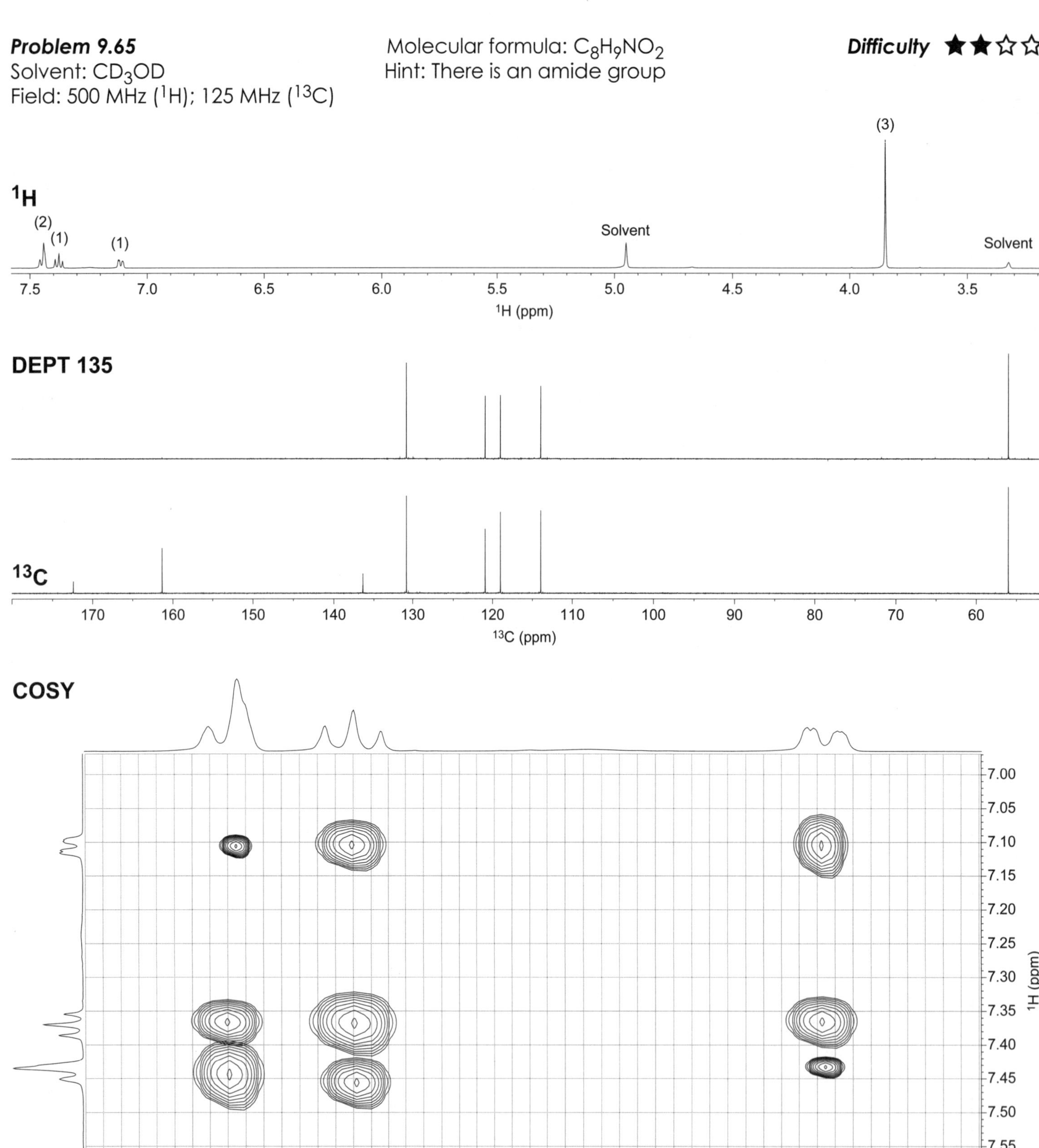

HSQC

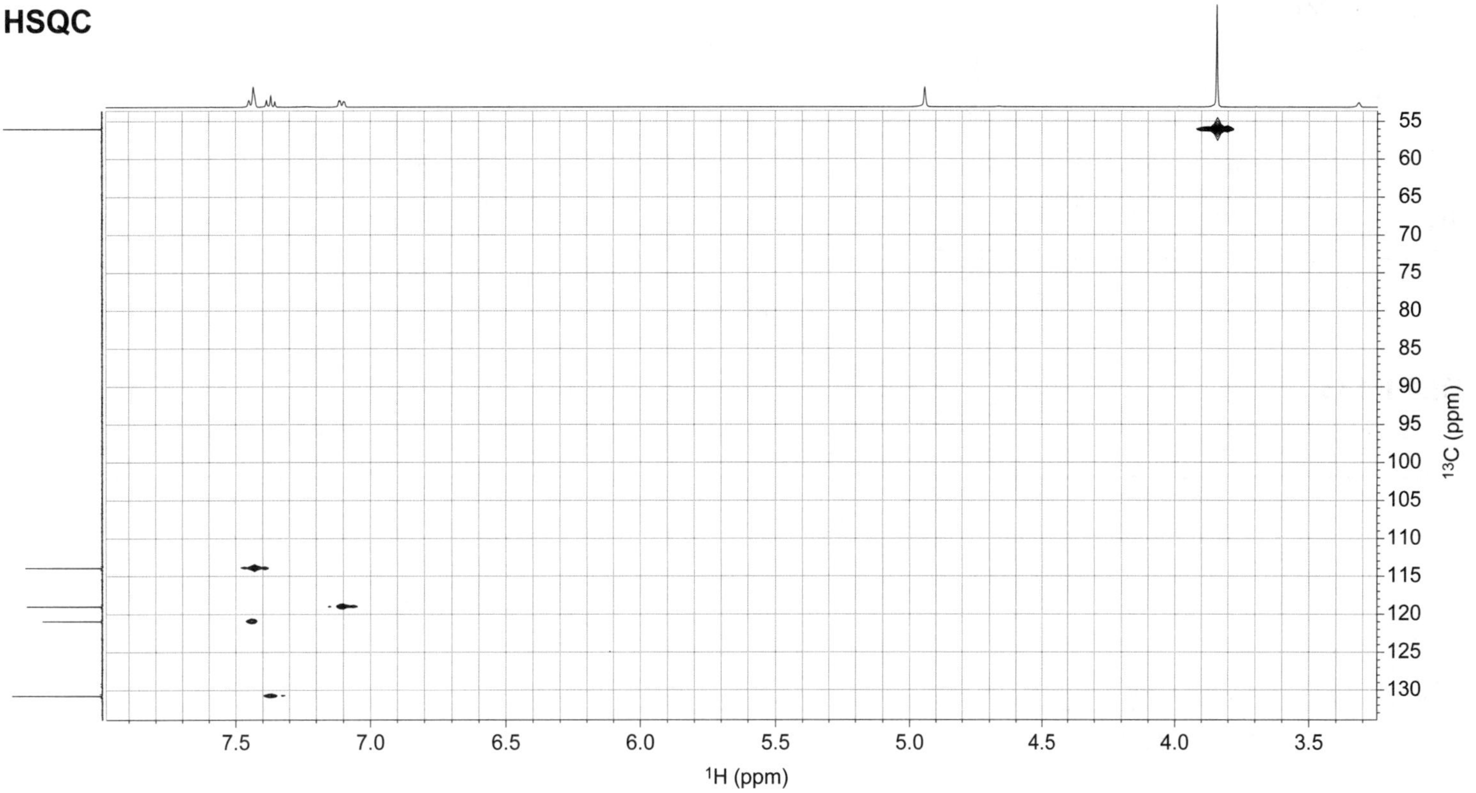
55
60
65
70
75
80
85
90
95
100
105
110
115
120
125
130
7.5
7.0
6.5
6.0
5.5
5.0
4.5
4.0
3.5
^{1}H (ppm)
^{13}C (ppm)

HMBC

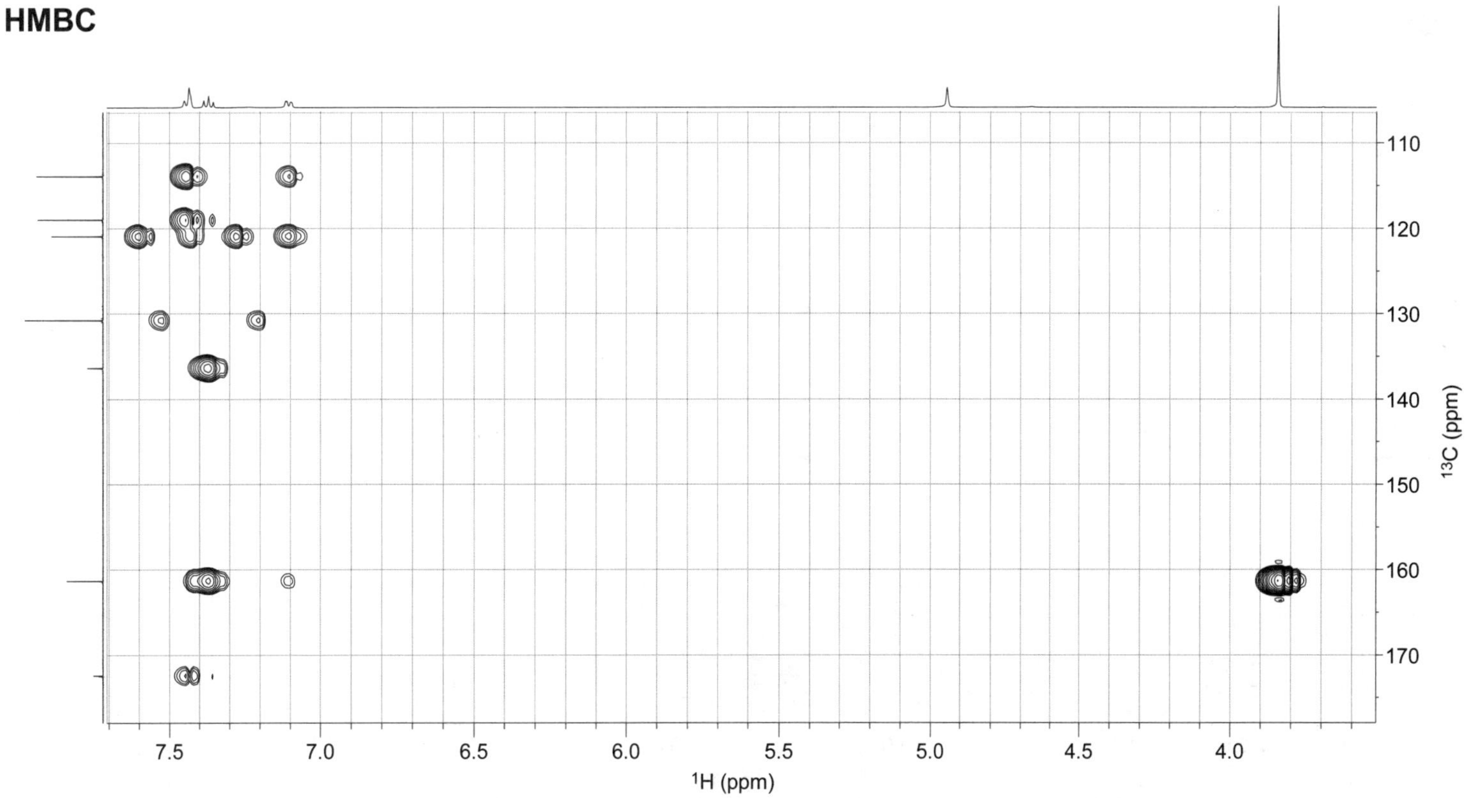
110
120
130
140
150
160
170
7.5
7.0
6.5
6.0
5.5
5.0
4.5
4.0
^{1}H (ppm)
^{13}C (ppm)

Problem 9.66 Molecular formula: $C_6H_{12}O$ **Difficulty** ★★★☆

Solvent: D_2O

Field: 500 MHz (^1H); 125 MHz (^{13}C)

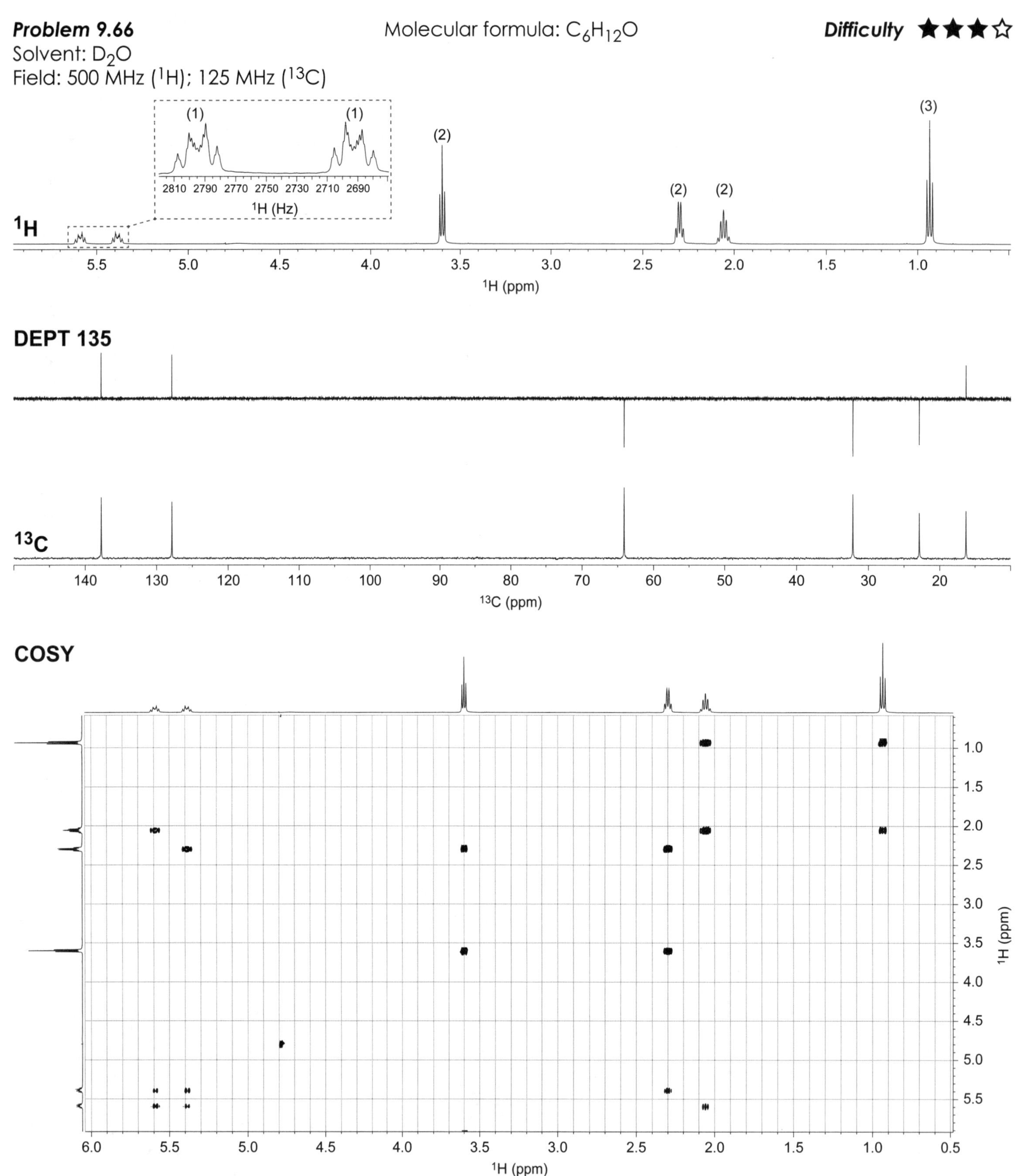

HSQC

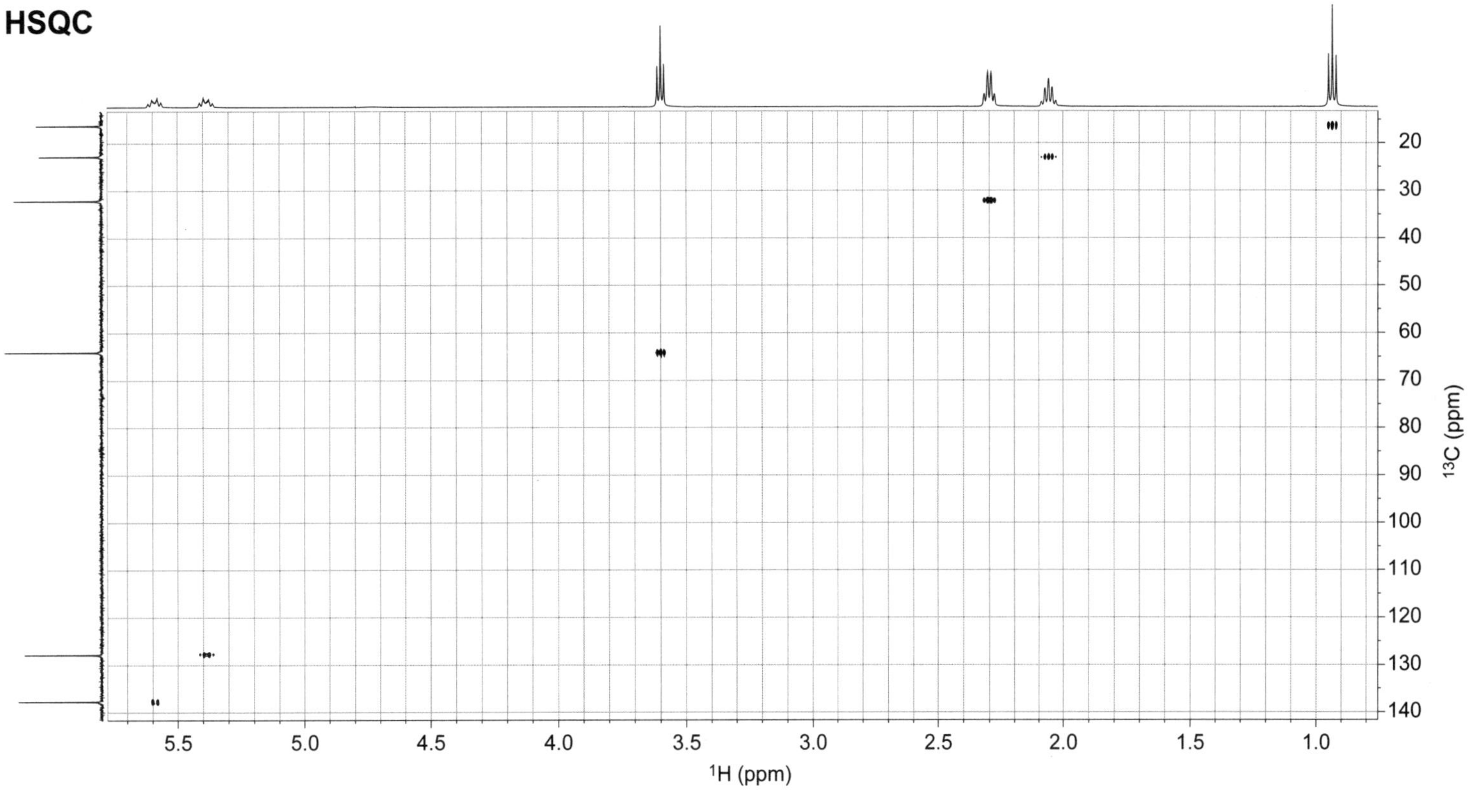

5.5
5.0
4.5
4.0
3.5
3.0
2.5
2.0
1.5
1.0
1H (ppm)
20
30
40
50
60
70
80
90
100
110
120
130
140
13C (ppm)

HMBC

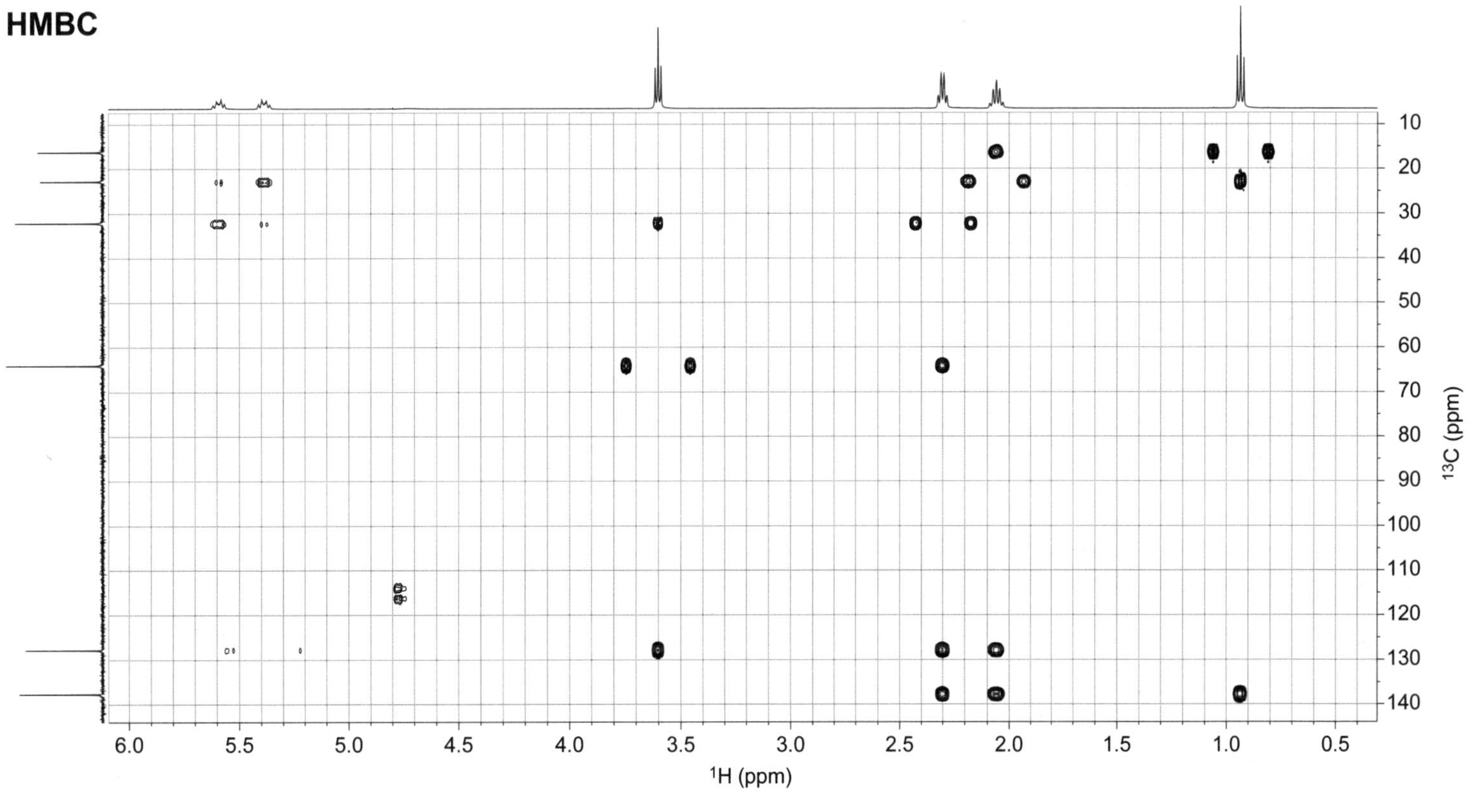

6.0
5.5
5.0
4.5
4.0
3.5
3.0
2.5
2.0
1.5
1.0
0.5
1H (ppm)
10
20
30
40
50
60
70
80
90
100
110
120
130
140
13C (ppm)

Problem 9.67
Solvent: D_2O
Field: 500 MHz (^1H); 125 MHz (^{13}C)

Molecular formula: $C_6H_{13}NO_2$

Difficulty ★★★☆

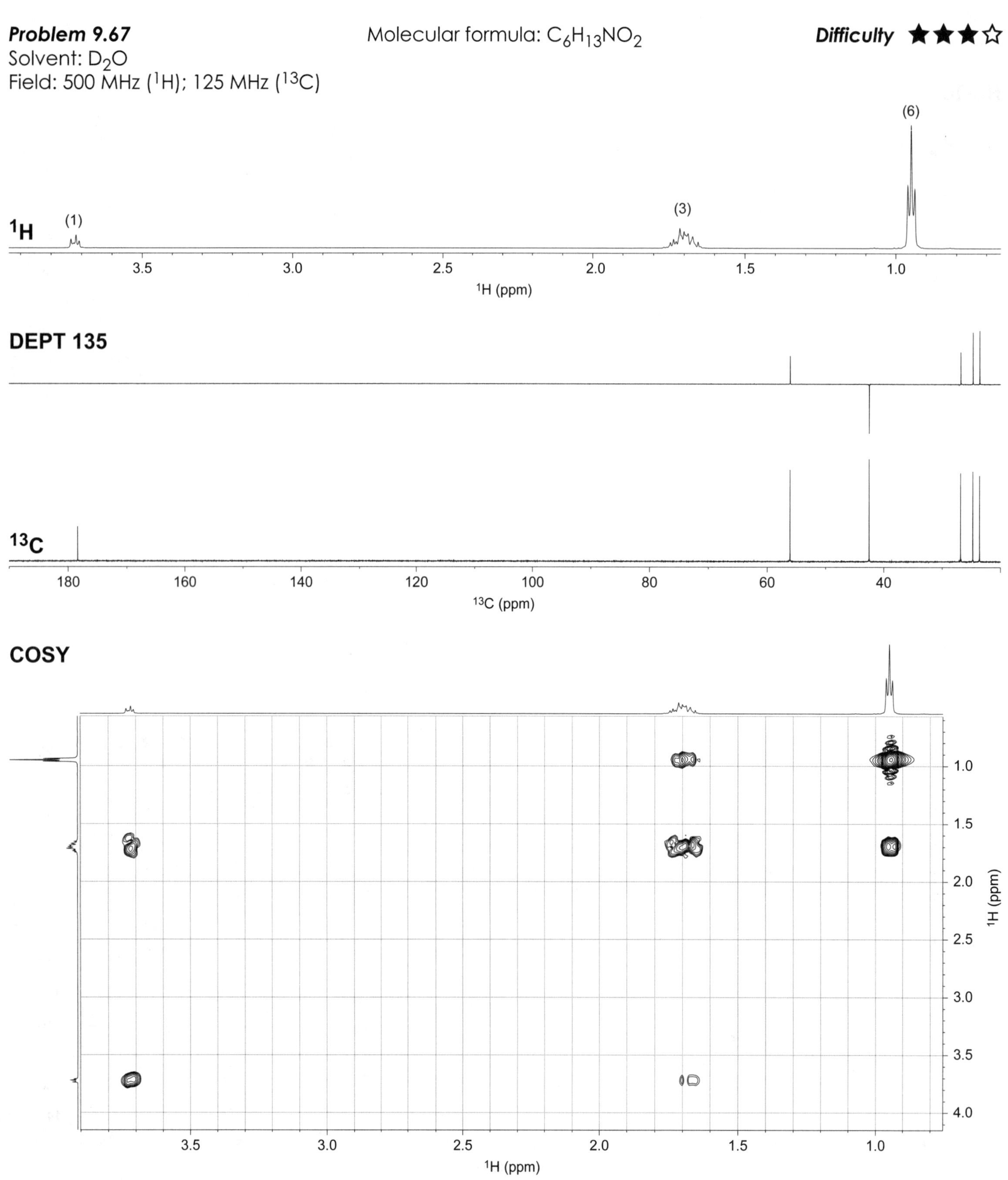

HSQC

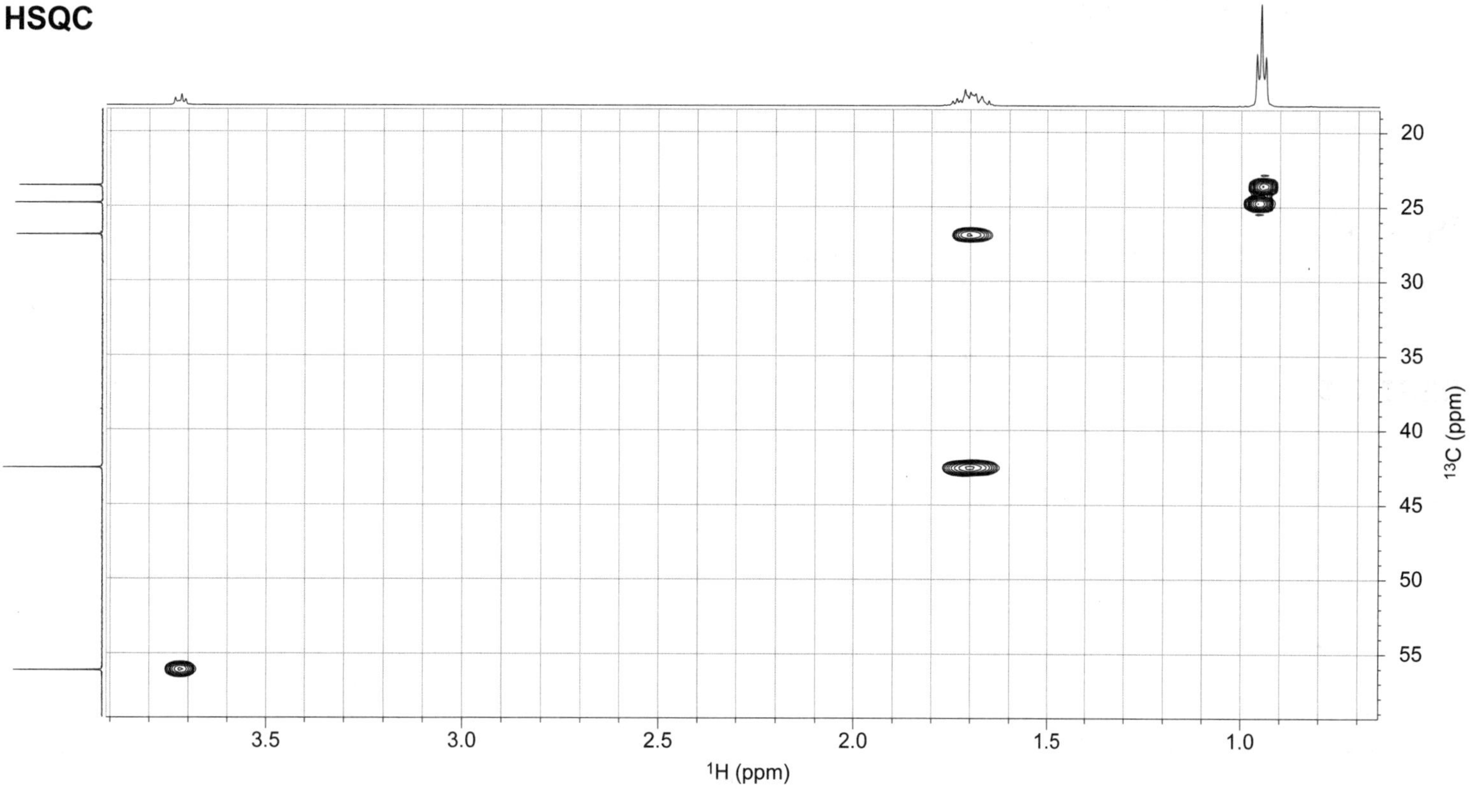

20
25
30
35
40
45
50
55
^{13}C (ppm)

3.5 3.0 2.5 2.0 1.5 1.0
^{1}H (ppm)

HMBC

20
25
30
35
40
45
50
55
60
175
180
^{13}C (ppm)

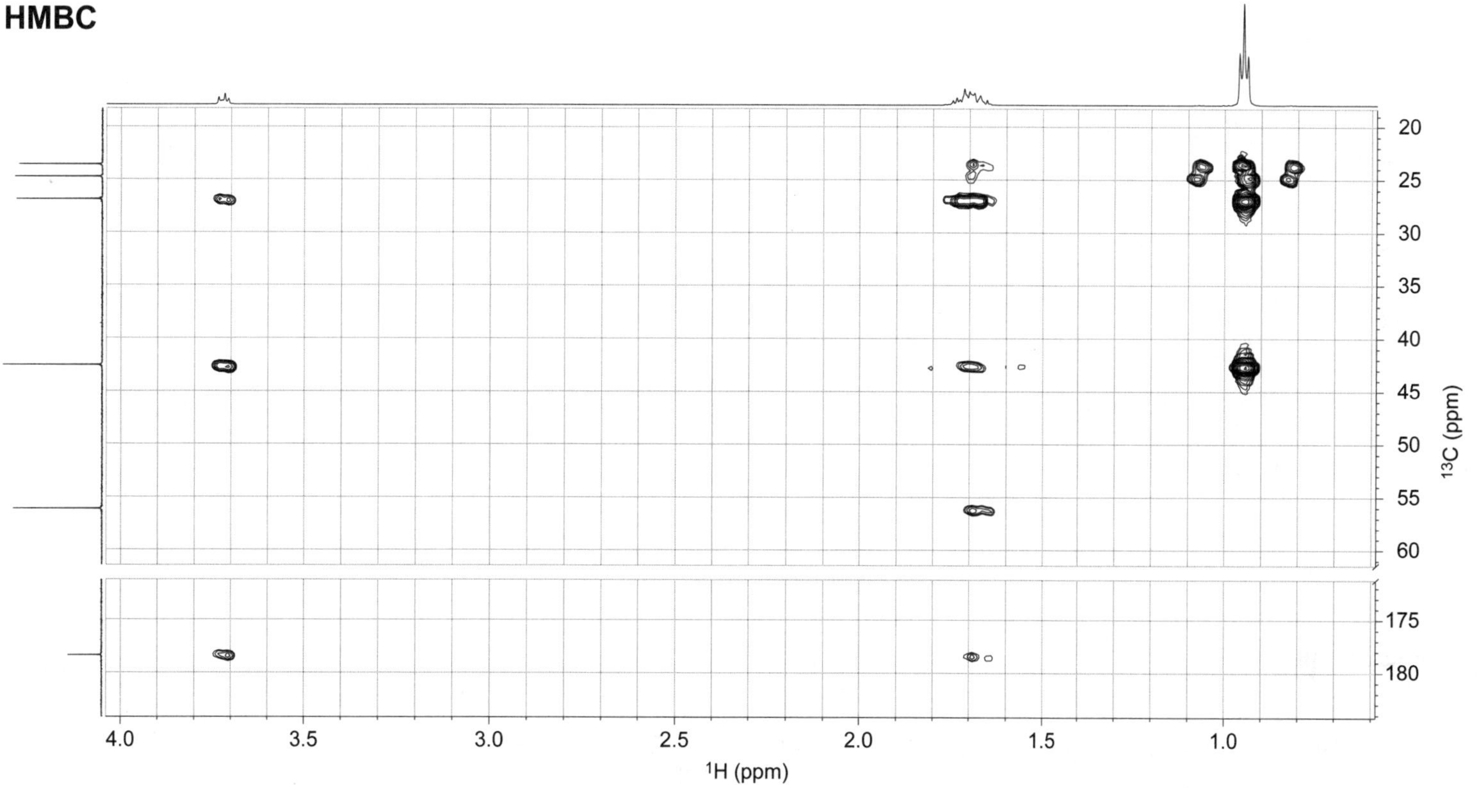

4.0 3.5 3.0 2.5 2.0 1.5 1.0
^{1}H (ppm)

Problem 9.68
Solvent: D_2O
Field: 500 MHz (^1H); 125 MHz (^{13}C)

Molecular formula: $C_9H_{11}NO_2$
Hint: It is a natural amino acid

Difficulty ★★★☆

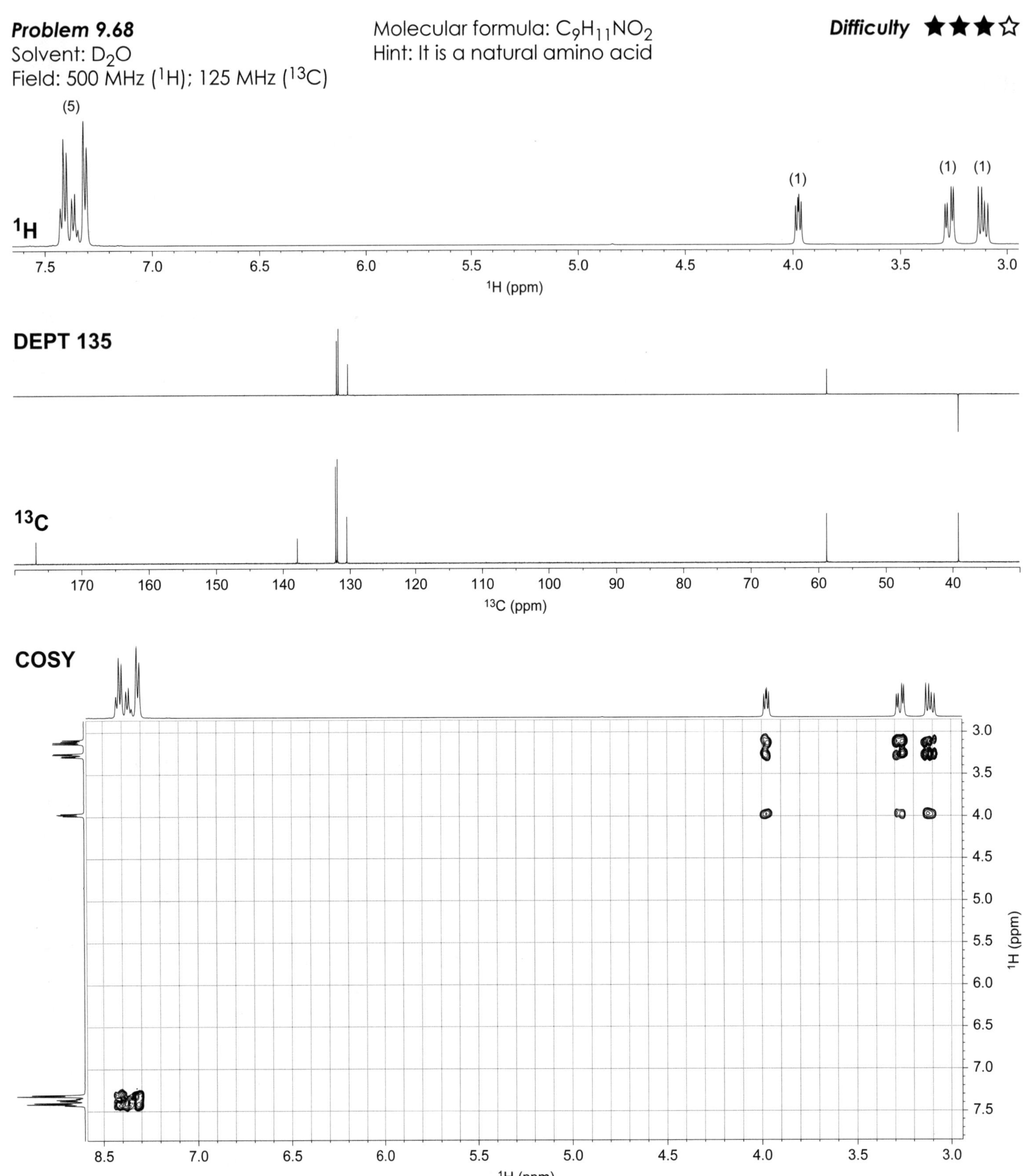

HSQC

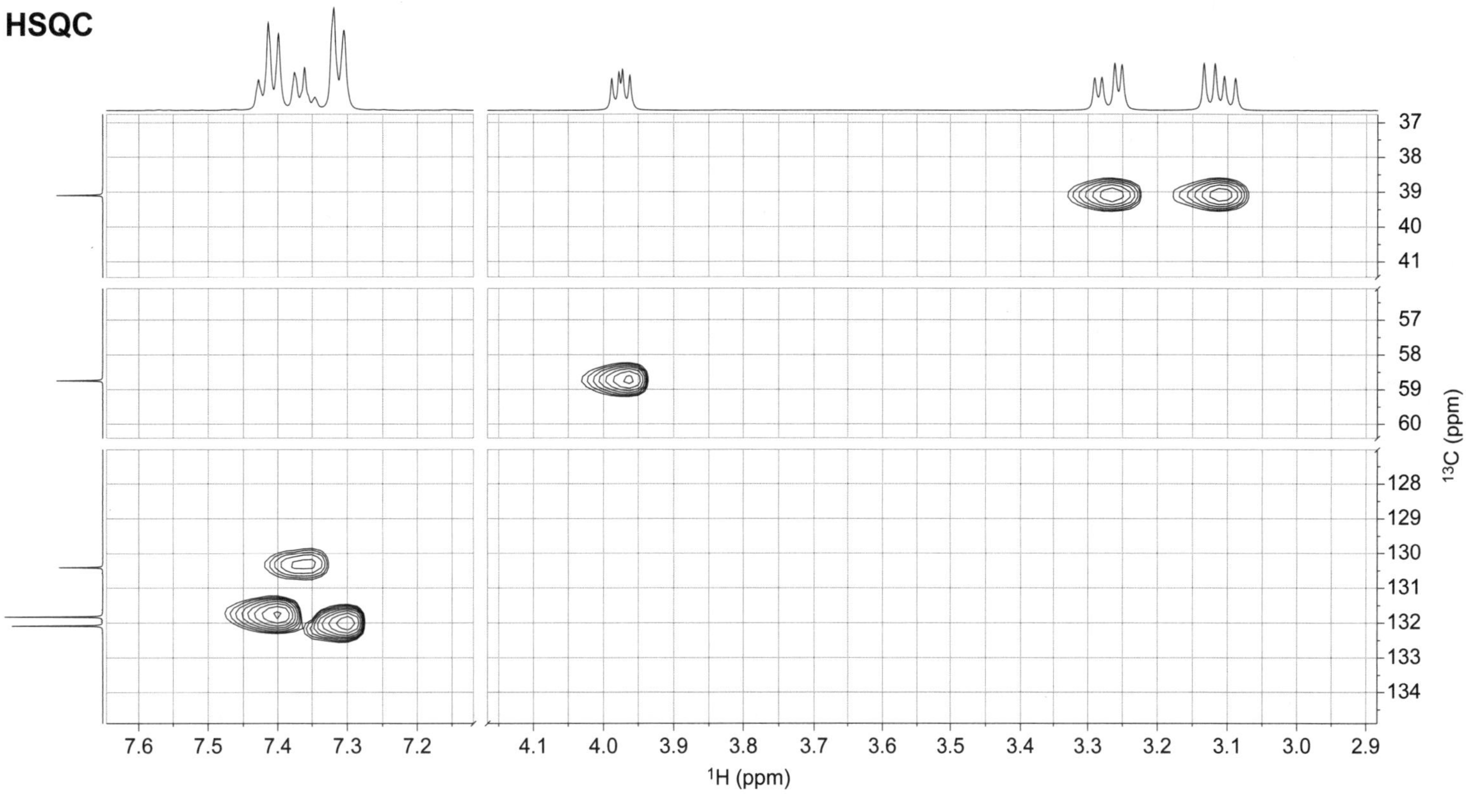

HMBC

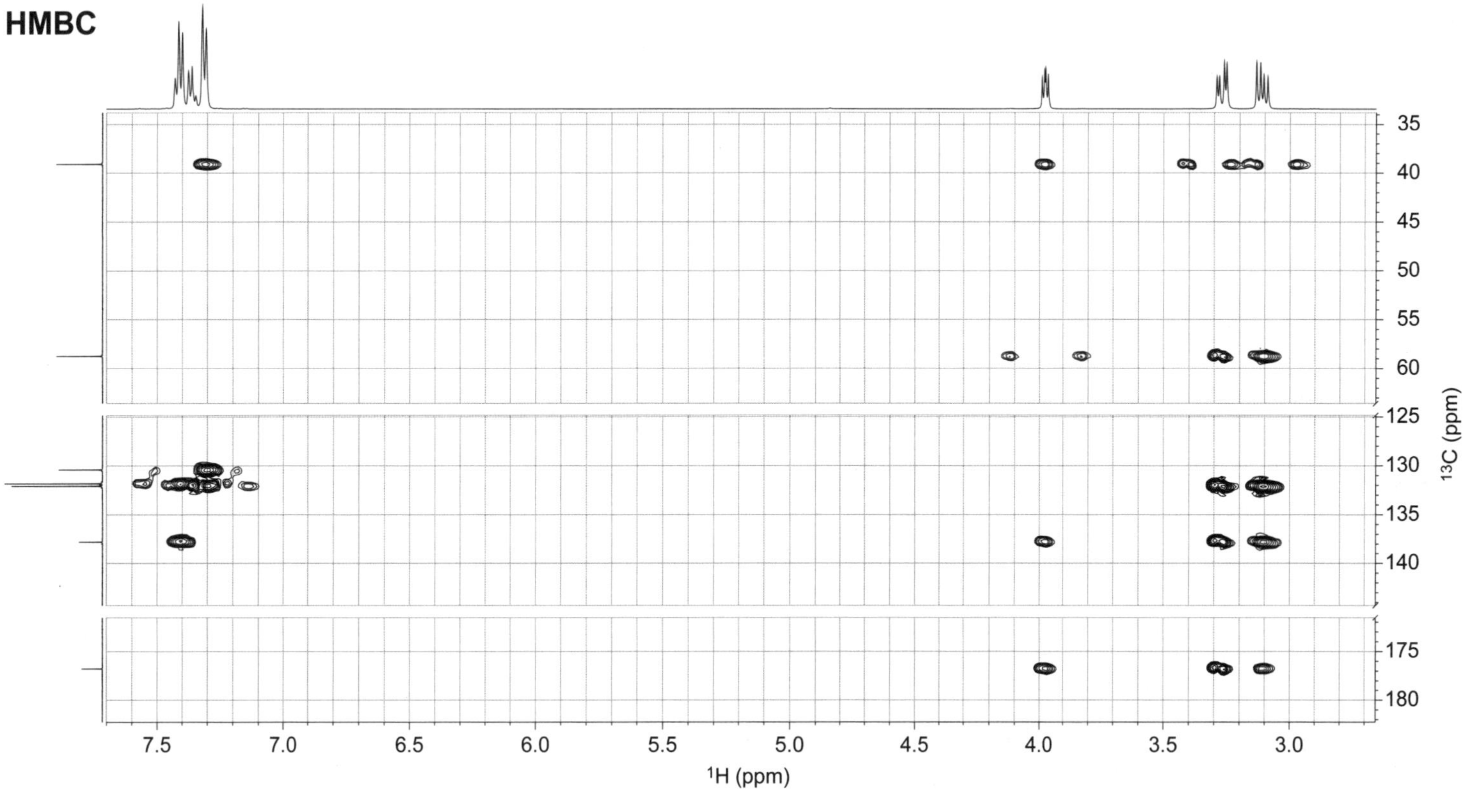

Problem 9.69
Solvent: D_2O
Field: 500 MHz (^1H); 125 MHz (^{13}C)

Molecular formula: $C_5H_9NO_2$
Hint: It is a natural amino acid

Difficulty ★★★☆

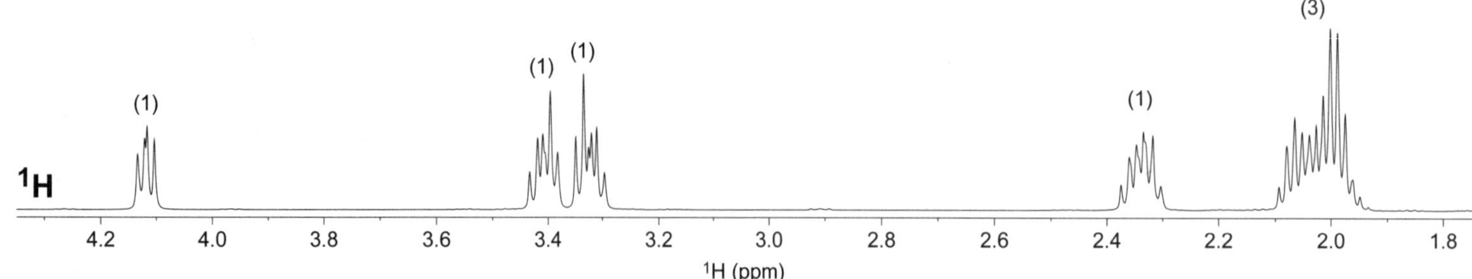

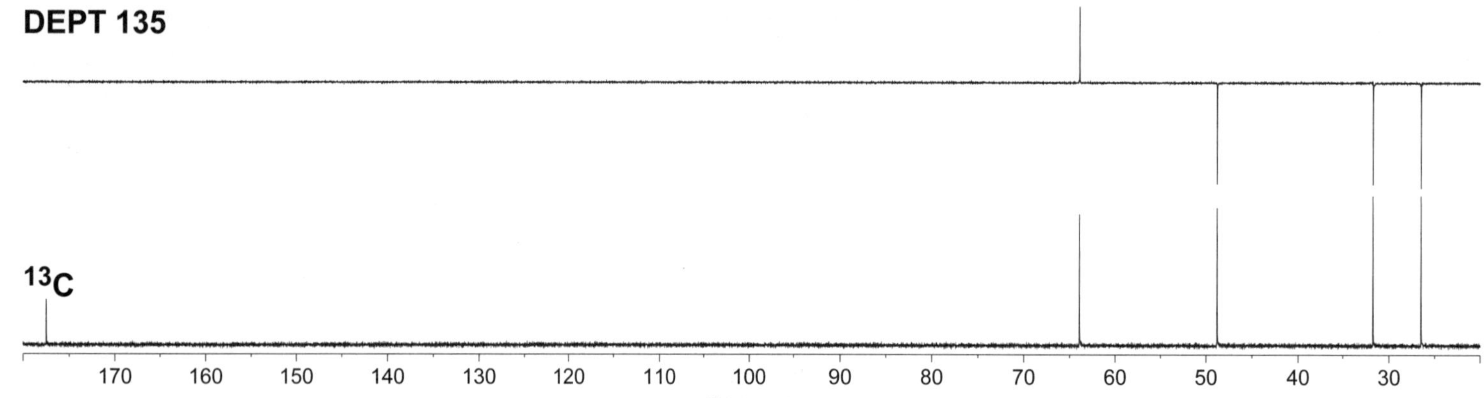

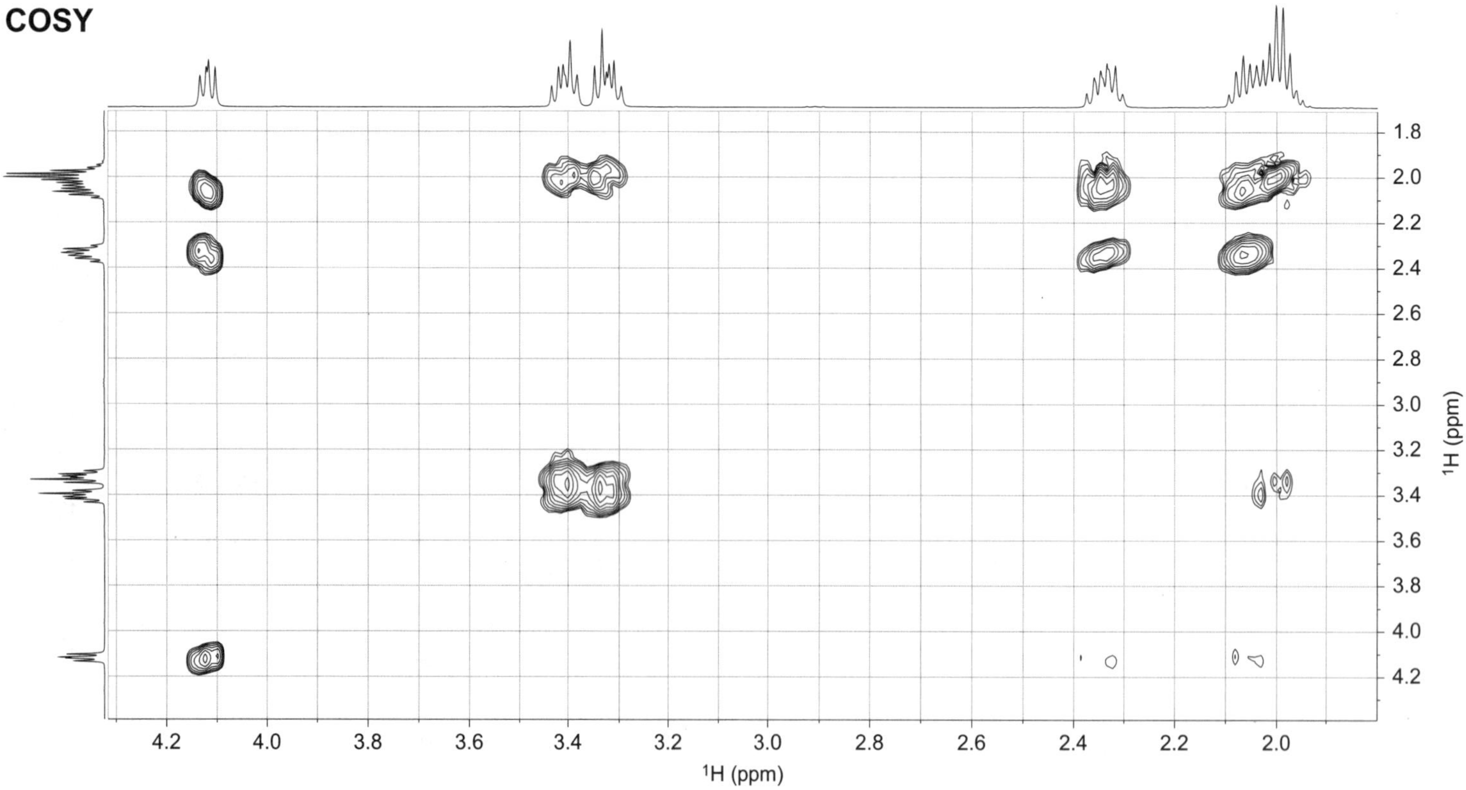

COSY

HSQC

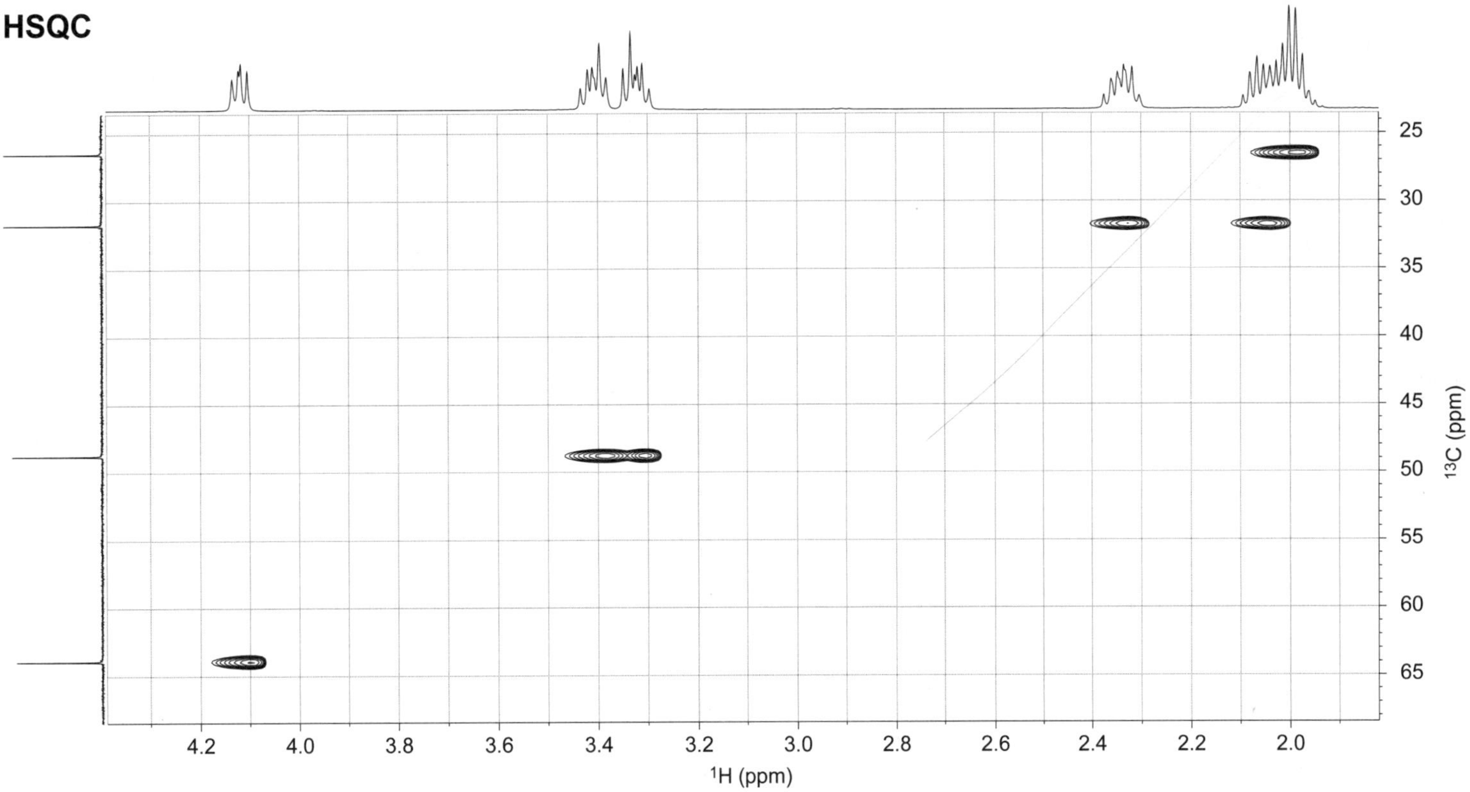
25
30
35
40
45
50
55
60
65
^{13}C (ppm)
4.2
4.0
3.8
3.6
3.4
3.2
3.0
2.8
2.6
2.4
2.2
2.0
^{1}H (ppm)

HMBC

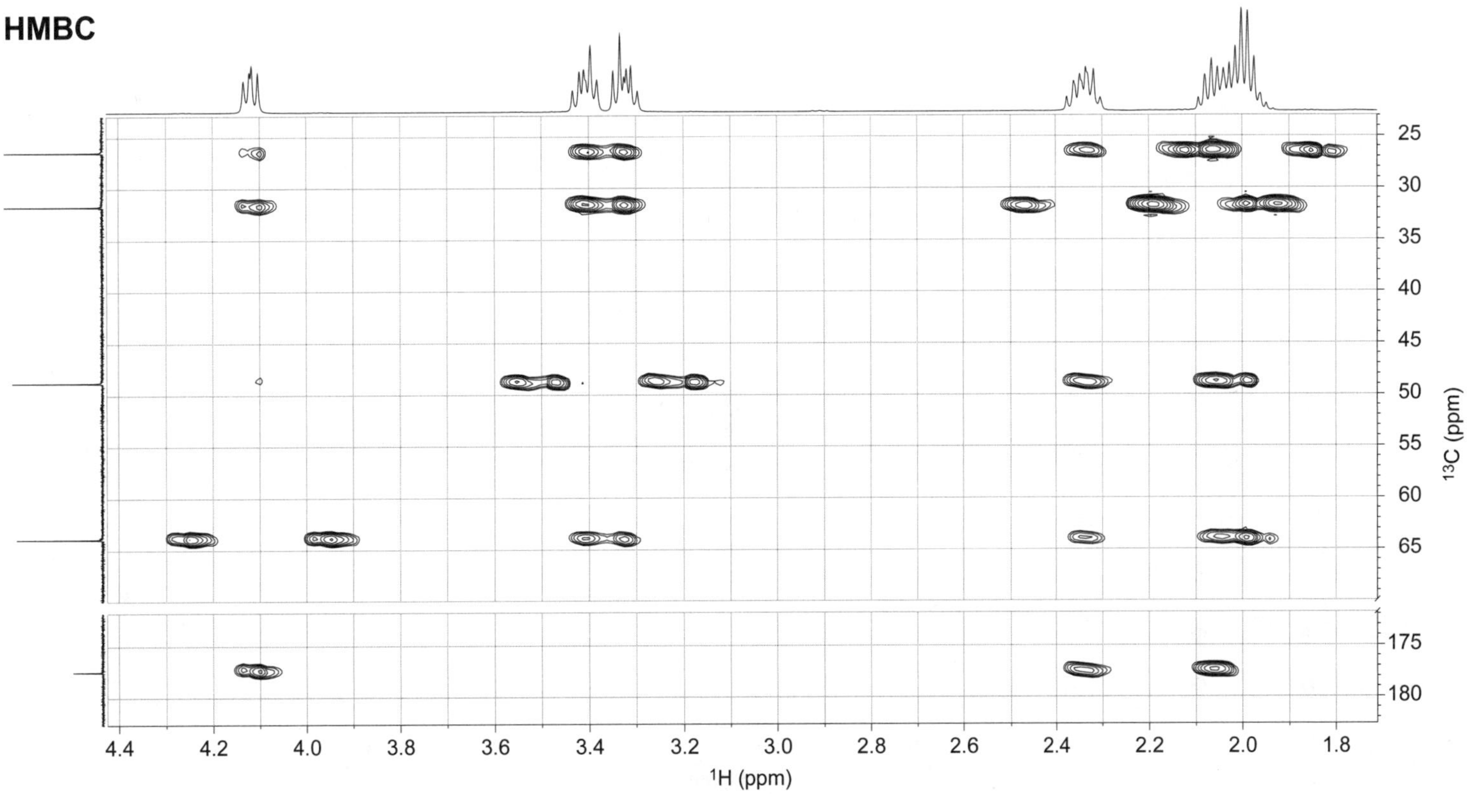
25
30
35
40
45
50
55
60
65
175
180
^{13}C (ppm)
4.4
4.2
4.0
3.8
3.6
3.4
3.2
3.0
2.8
2.6
2.4
2.2
2.0
1.8
^{1}H (ppm)

Problem 9.70
Solvent: D_2O
Field: 500 MHz (^1H); 125 MHz (^{13}C)

Molecular formula: $C_9H_{11}NO_3$
Hint: It is a natural amino acid

Difficulty ★★★☆

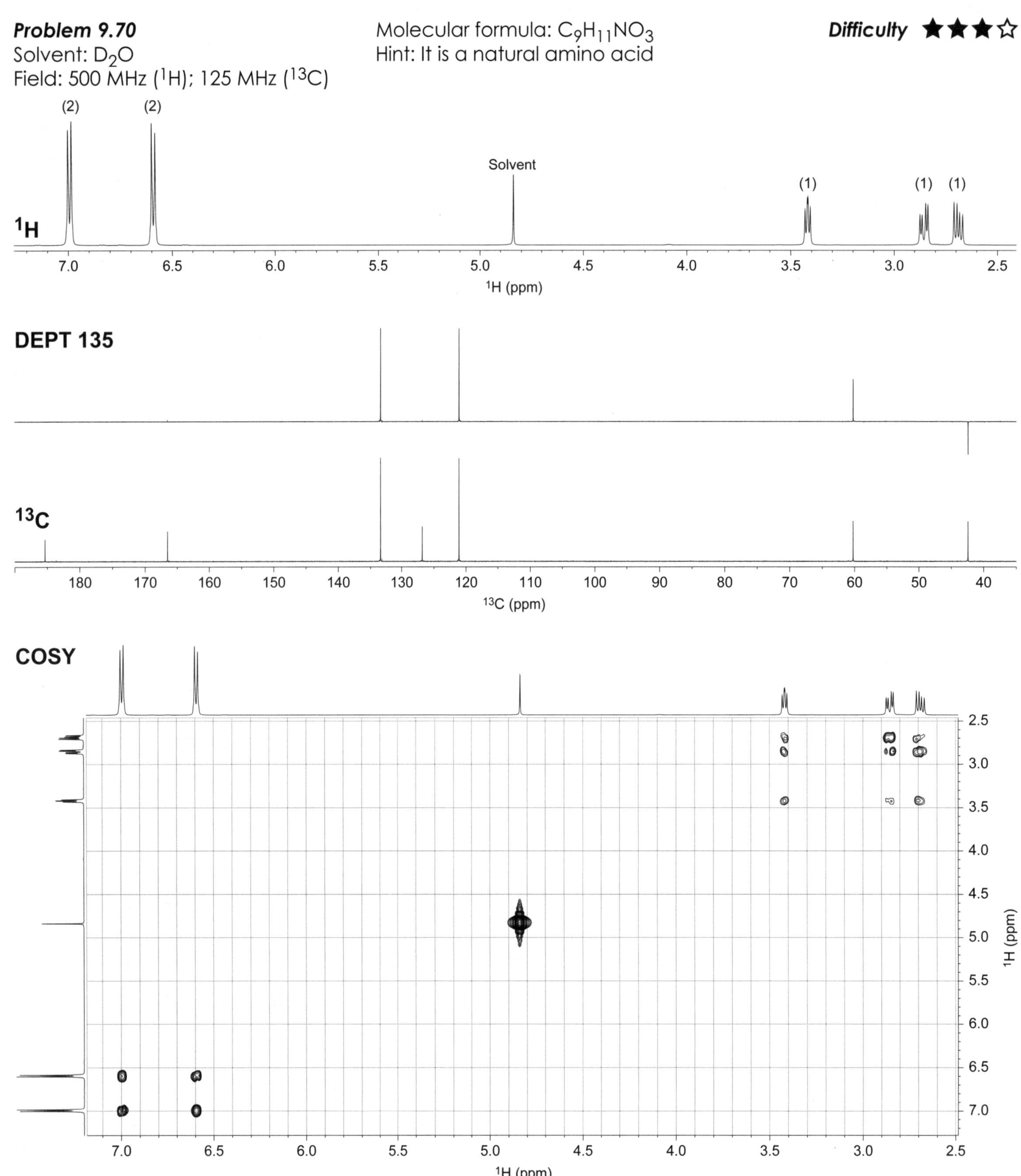

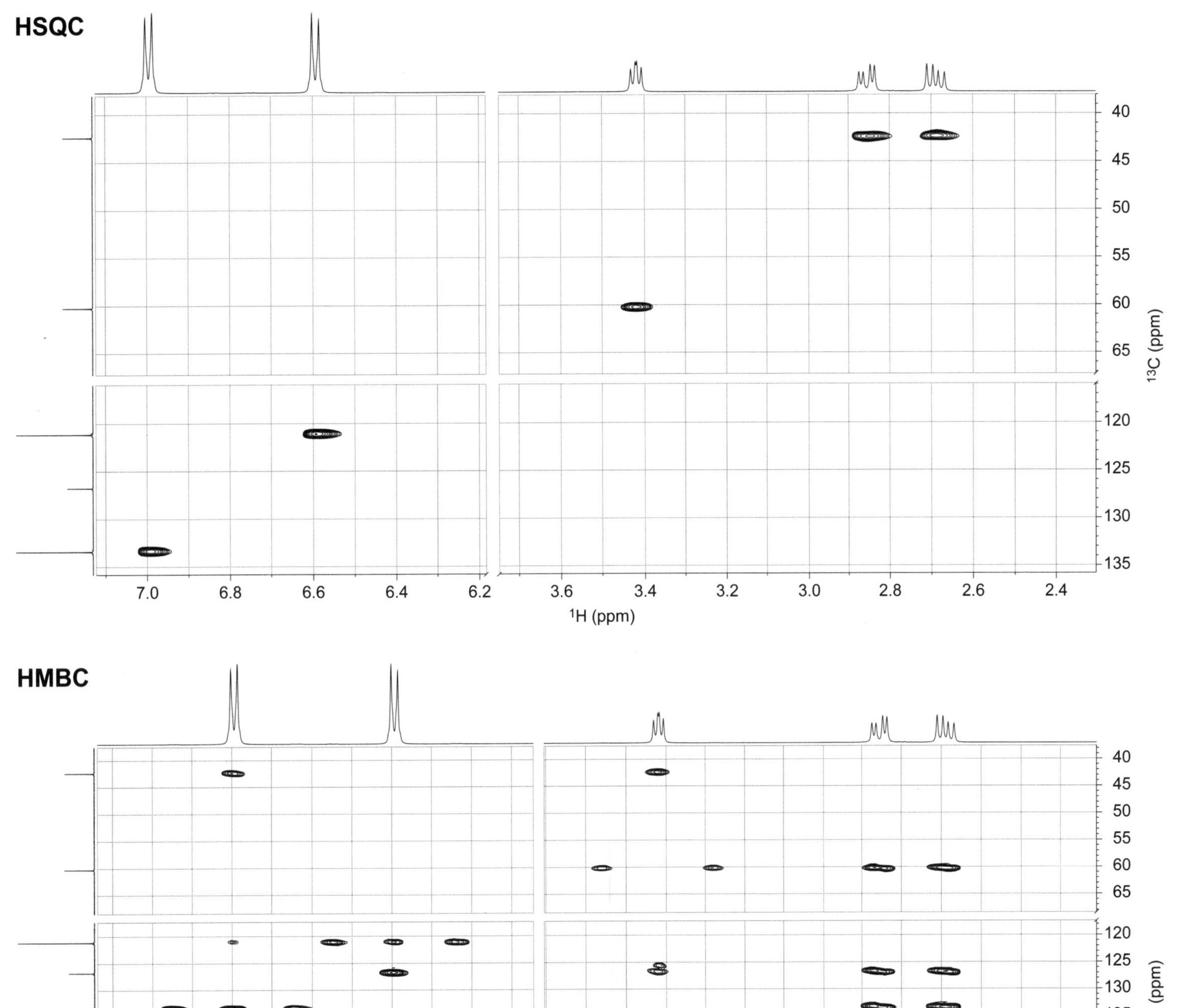
HSQC
HMBC
^{1}H (ppm)
^{13}C (ppm)

Problem 9.71
Solvent: D_2O
Field: 500 MHz (^1H); 125 MHz (^{13}C)

Molecular formula: $C_5H_9NO_3$
Hint: It is a modified amino acid

Difficulty ★★★☆

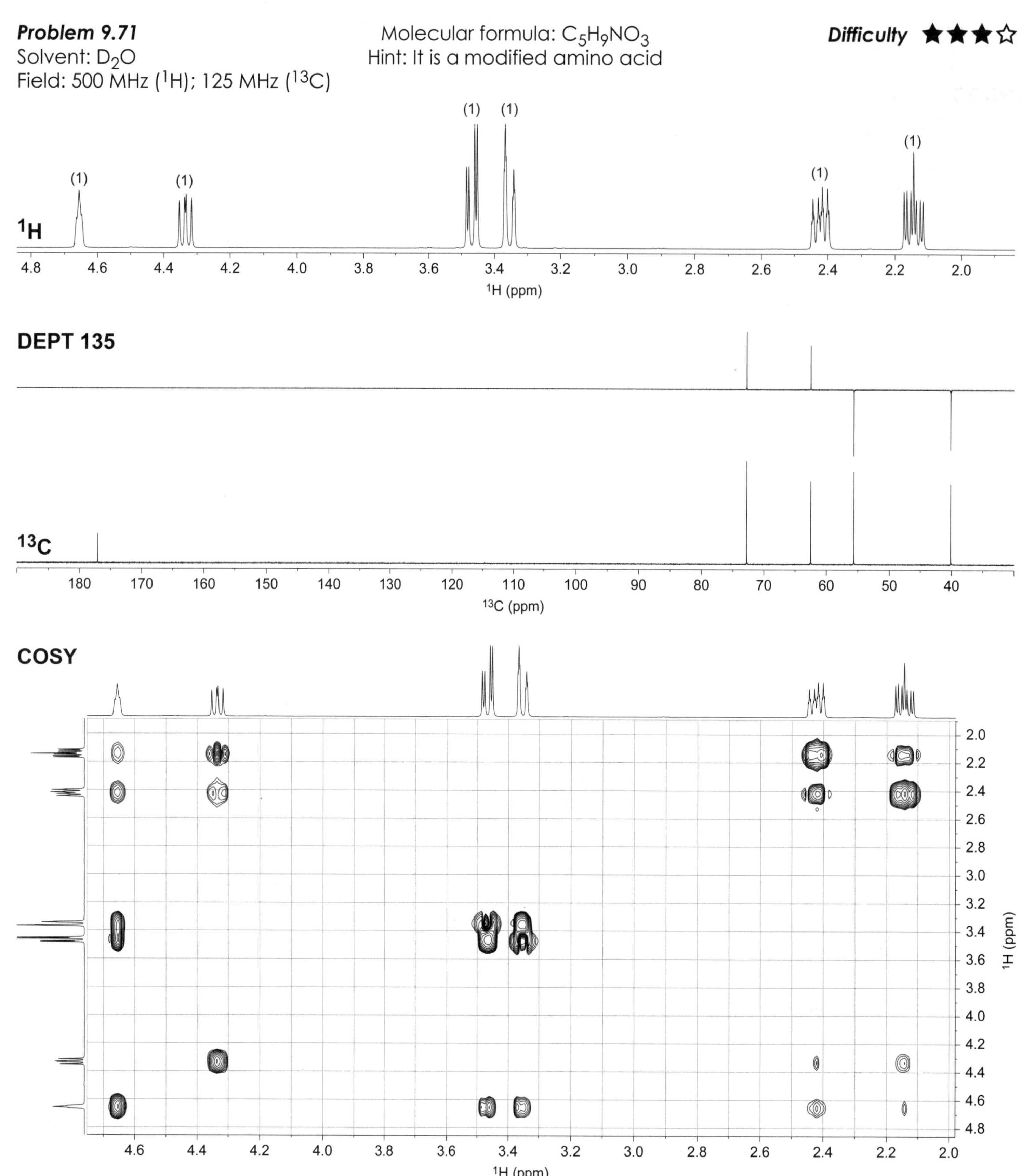

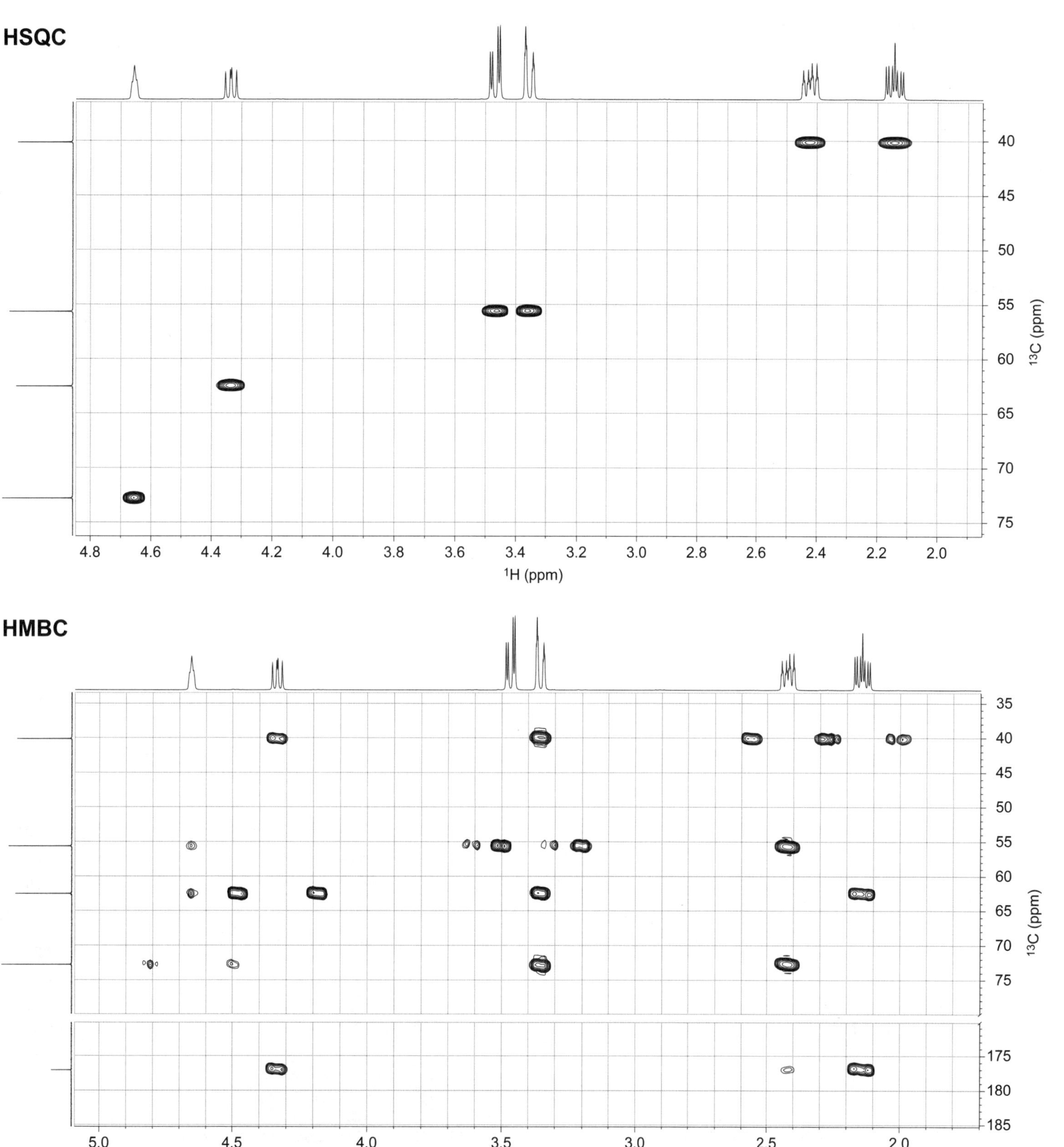
HSQC
40
45
50
55
60
65
70
75
^{13}C (ppm)
4.8
4.6
4.4
4.2
4.0
3.8
3.6
3.4
3.2
3.0
2.8
2.6
2.4
2.2
2.0
^{1}H (ppm)
HMBC
35
40
45
50
55
60
65
70
75
175
180
185
^{13}C (ppm)
5.0
4.5
4.0
3.5
3.0
2.5
2.0
^{1}H (ppm)

Problem 9.72 Molecular formula: $C_8H_{11}N$ ***Difficulty*** ★★★☆

Solvent: D_2O

Field: 500 MHz (^1H); 125 MHz (^{13}C)

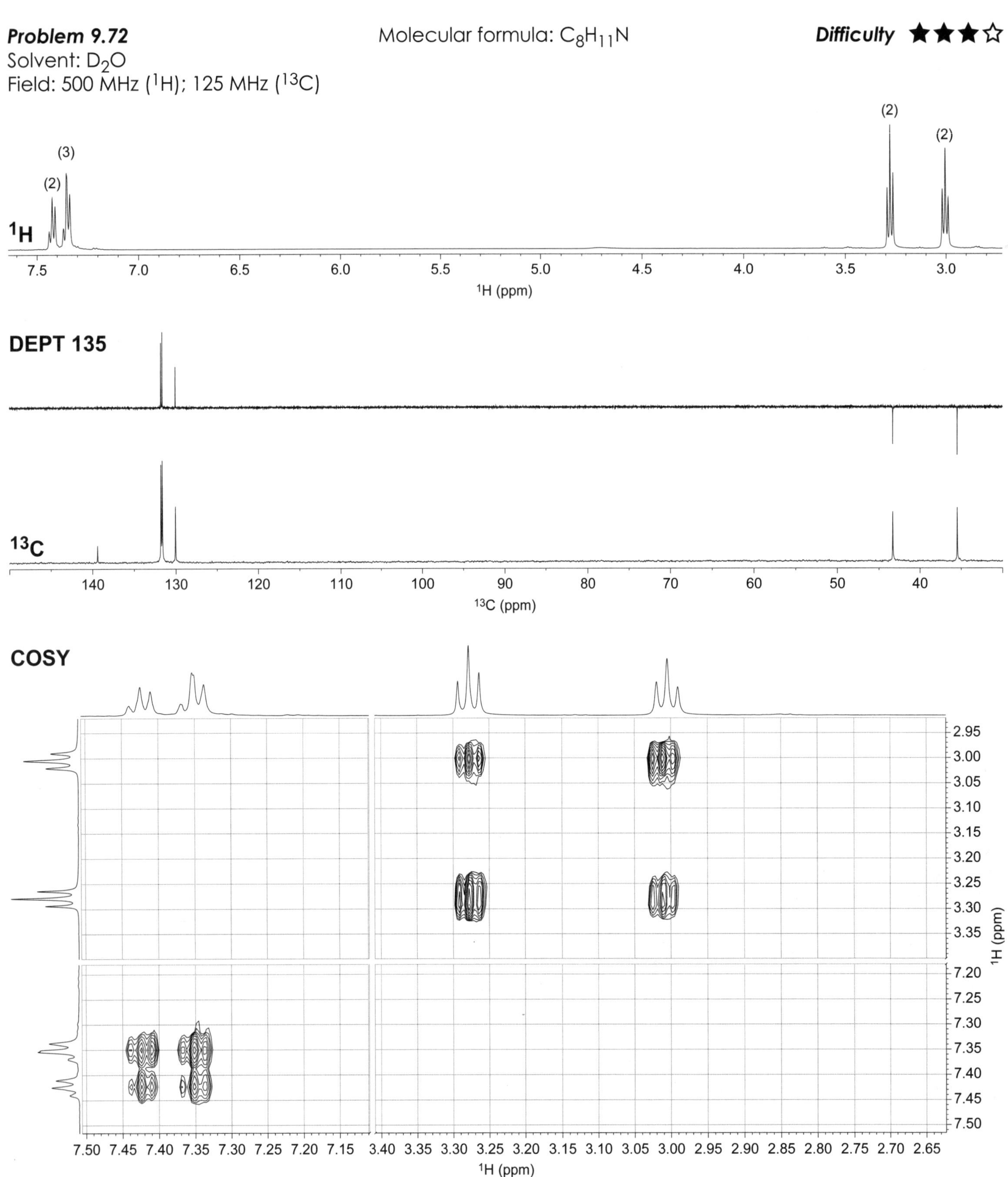

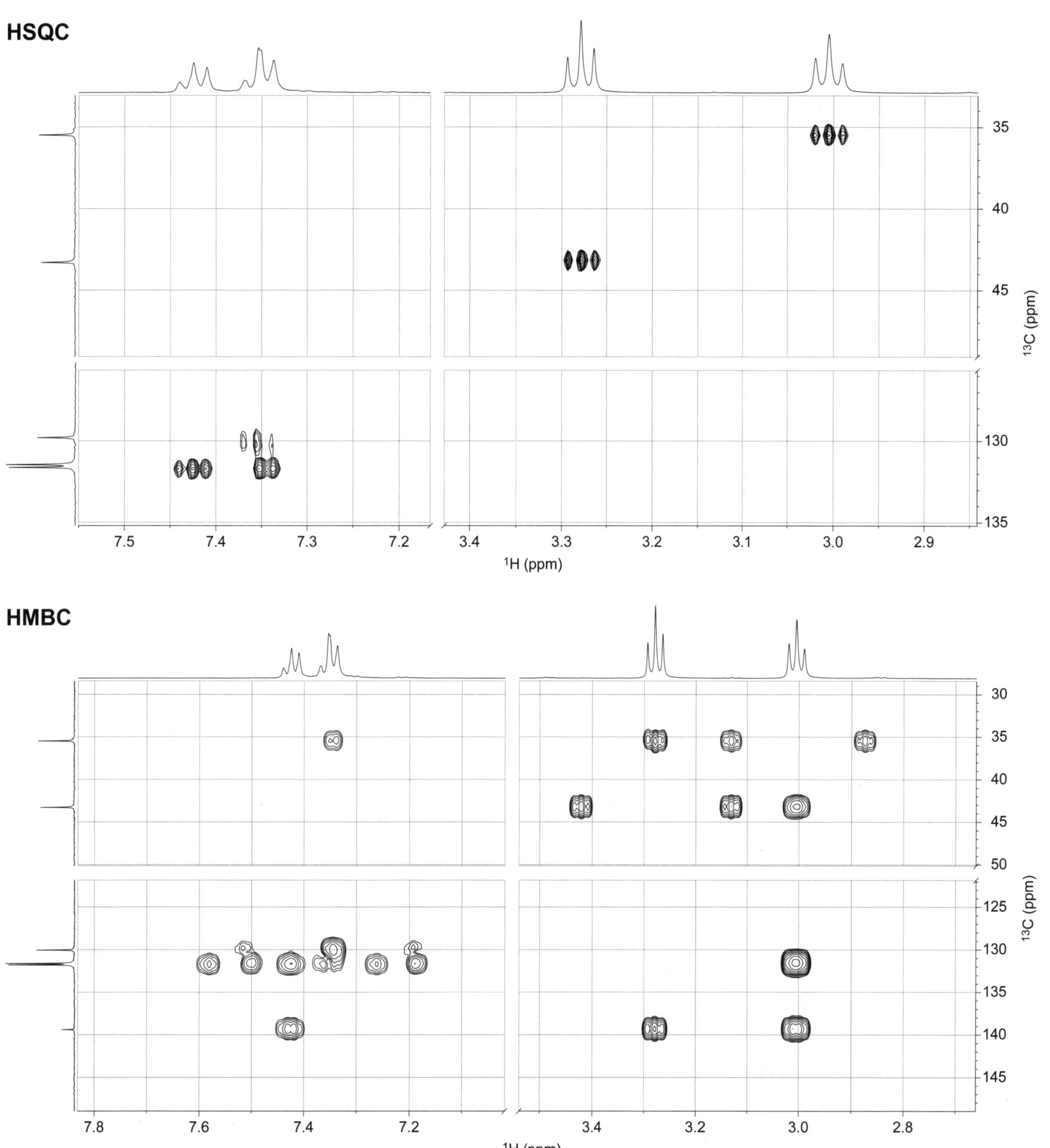
HSQC
35
40
45
130
135
^{13}C (ppm)
7.5
7.4
7.3
7.2
3.4
3.3
3.2
3.1
3.0
2.9
^{1}H (ppm)
HMBC
30
35
40
45
50
125
130
135
140
145
^{13}C (ppm)
7.8
7.6
7.4
7.2
3.4
3.2
3.0
2.8
^{1}H (ppm)

Problem 9.73
Solvent: D_2O
Field: 500 MHz (^1H); 125 MHz (^{13}C)

Molecular formula: $C_9H_8O_2$

Difficulty ★★★☆

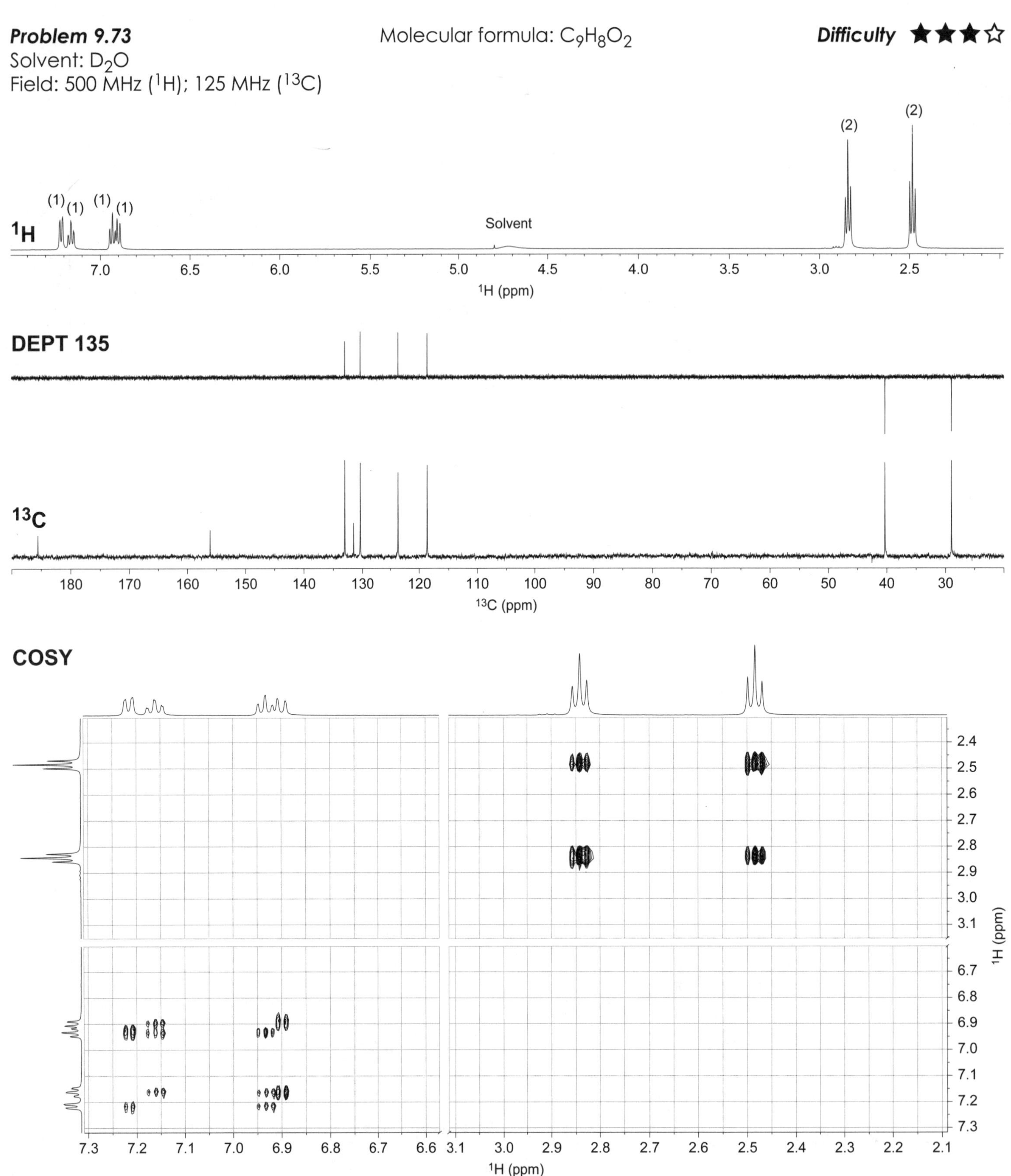

HSQC

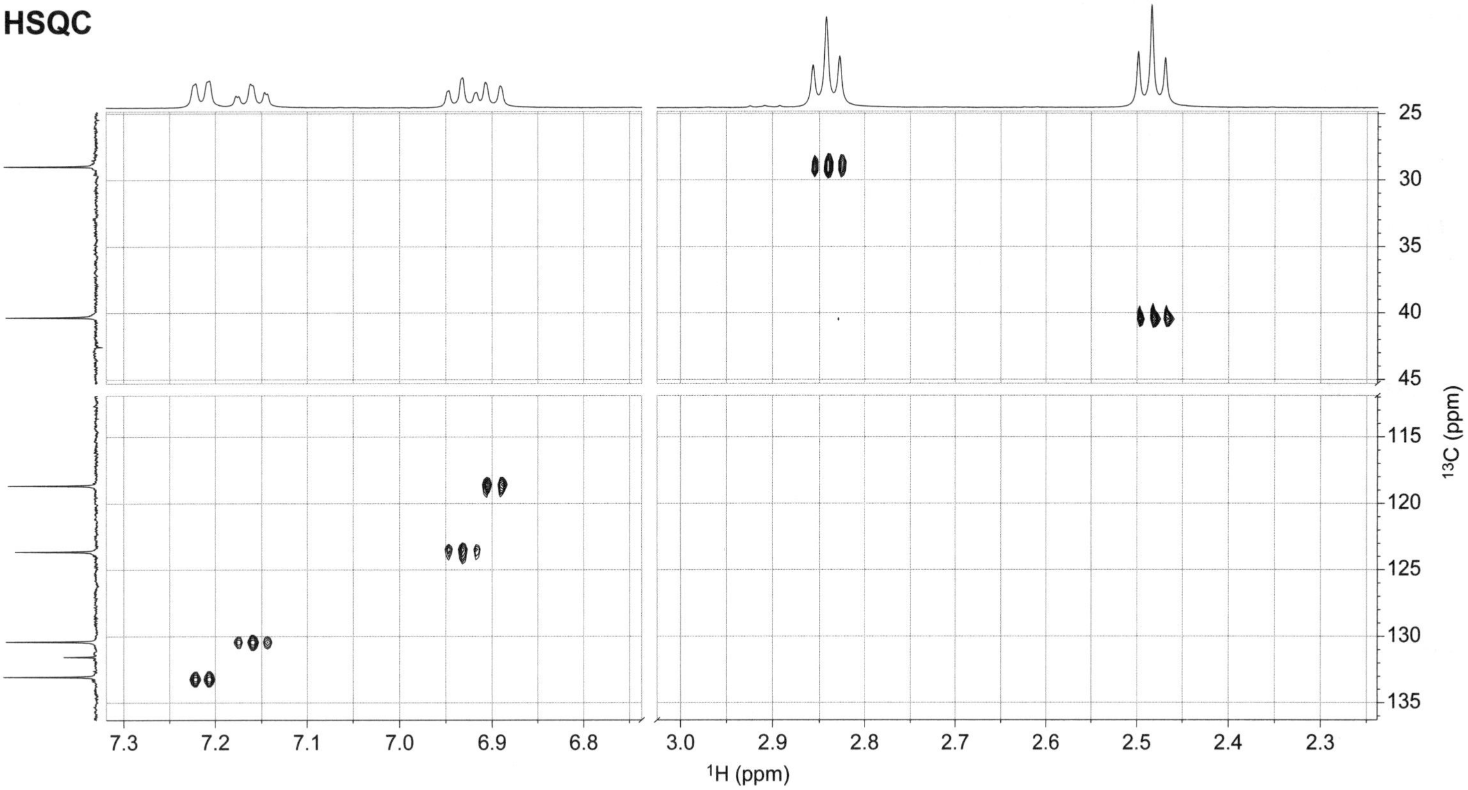

7.3 7.2 7.1 7.0 6.9 6.8
3.0 2.9 2.8 2.7 2.6 2.5 2.4 2.3
^{1}H (ppm)
25 30 35 40 45
115 120 125 130 135
^{13}C (ppm)

HMBC

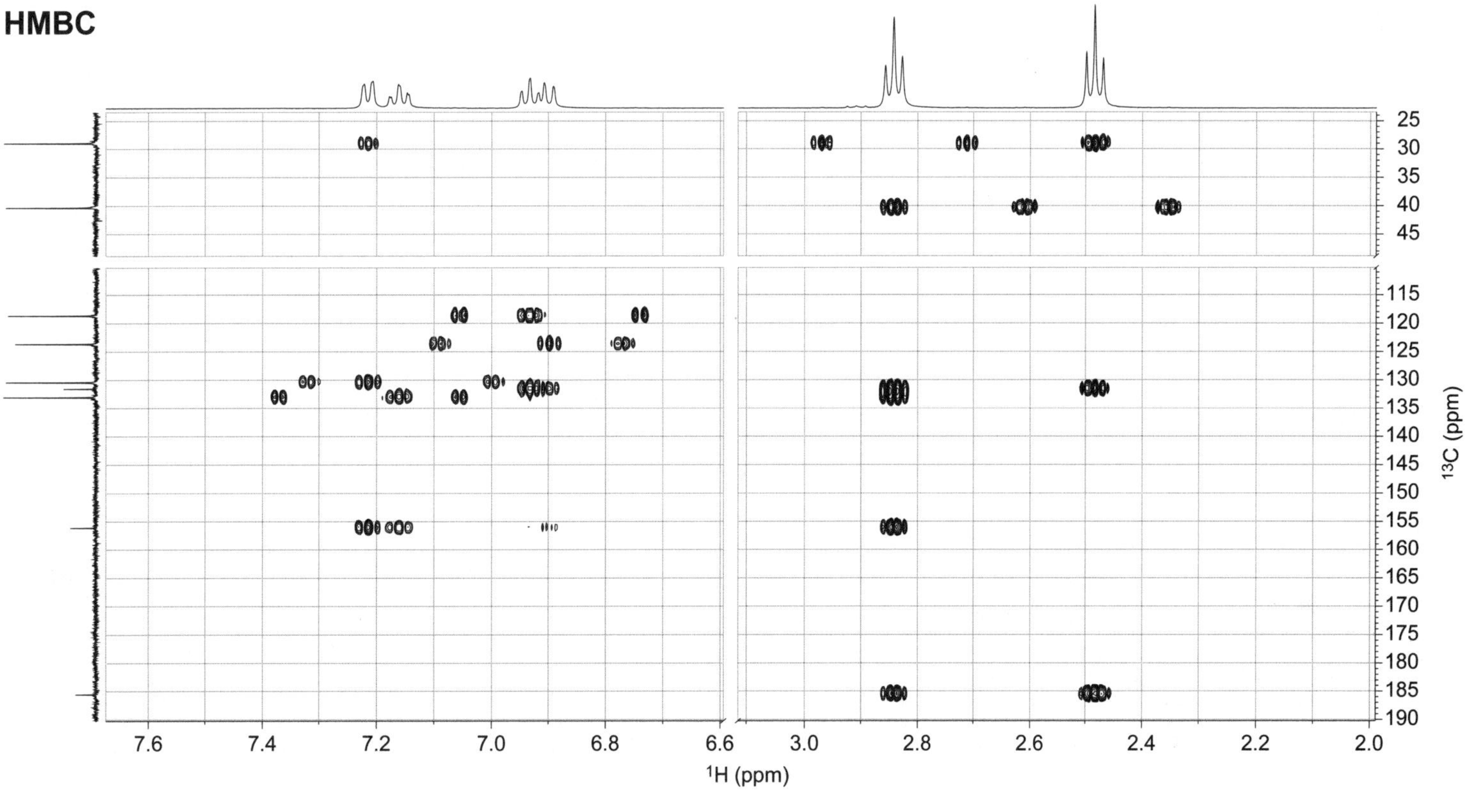

7.6 7.4 7.2 7.0 6.8 6.6
3.0 2.8 2.6 2.4 2.2 2.0
^{1}H (ppm)
25 30 35 40 45
115 120 125 130 135 140 145 150 155 160 165 170 175 180 185 190
^{13}C (ppm)

Problem 9.74
Solvent: D_2O
Field: 500 MHz (^1H); 125 MHz (^{13}C)

Molecular formula: $C_8H_9NO_3$

Difficulty ★★★☆

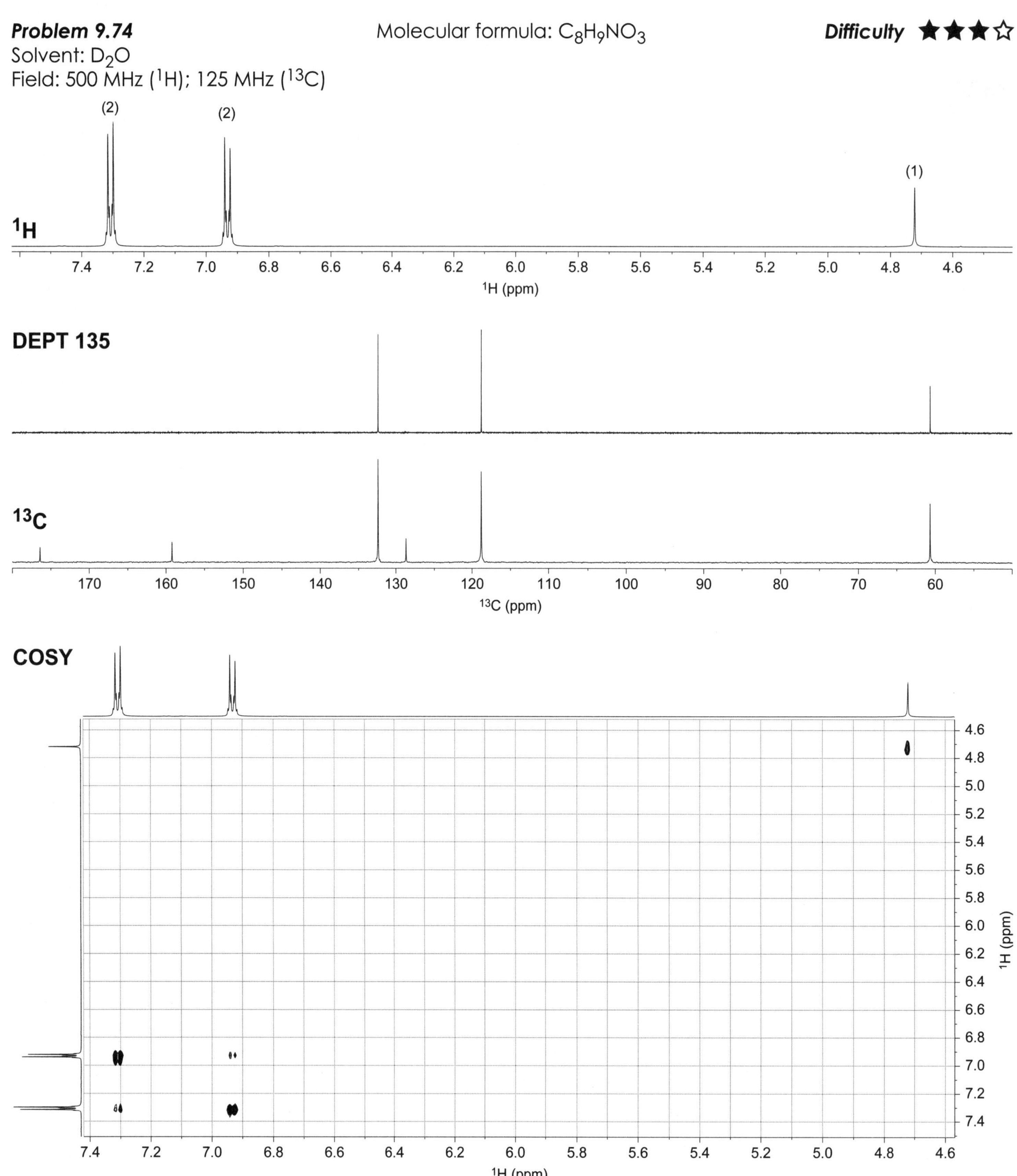

HSQC

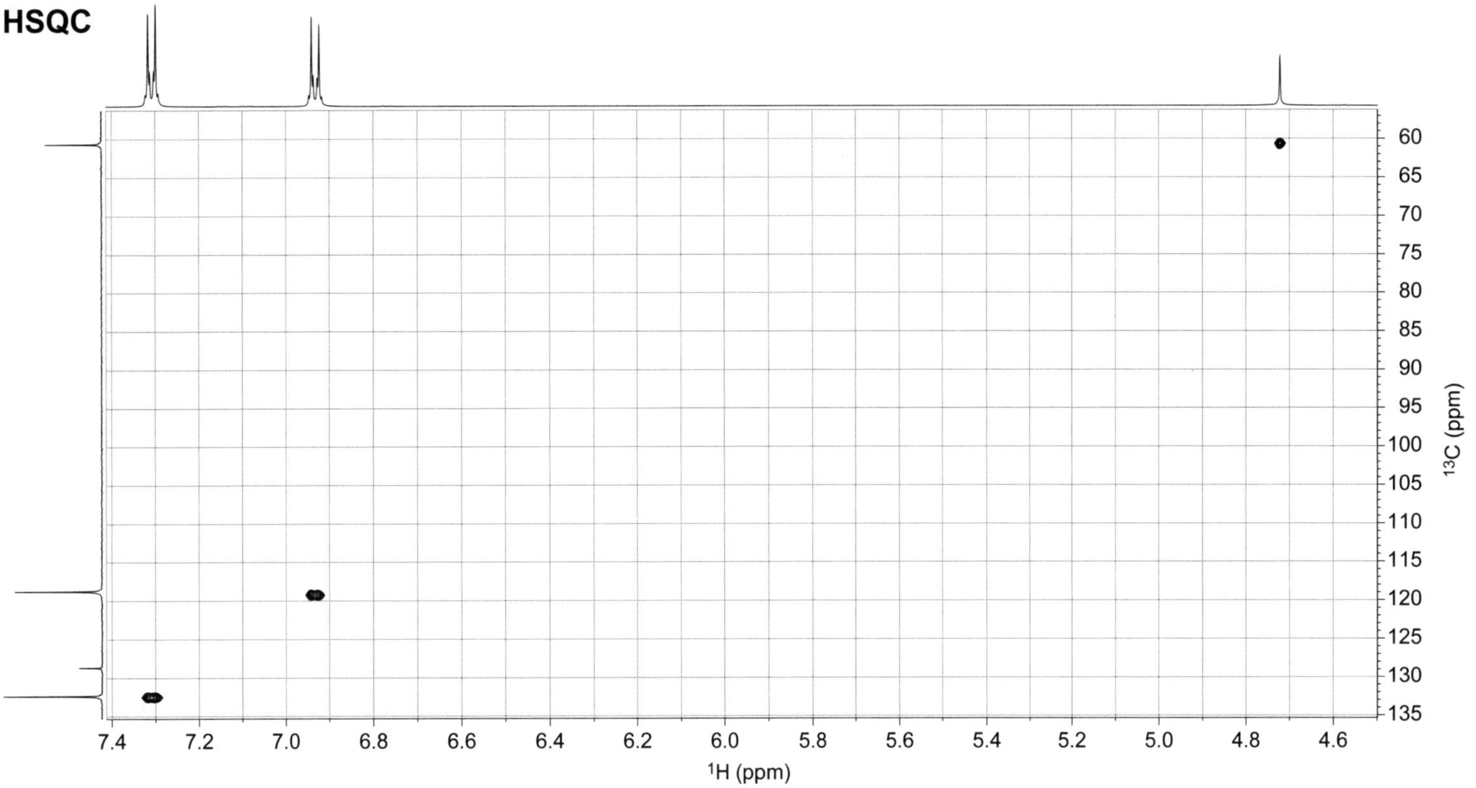

7.4 7.2 7.0 6.8 6.6 6.4 6.2 6.0 5.8 5.6 5.4 5.2 5.0 4.8 4.6
^{1}H (ppm)
60 65 70 75 80 85 90 95 100 105 110 115 120 125 130 135
^{13}C (ppm)

HMBC

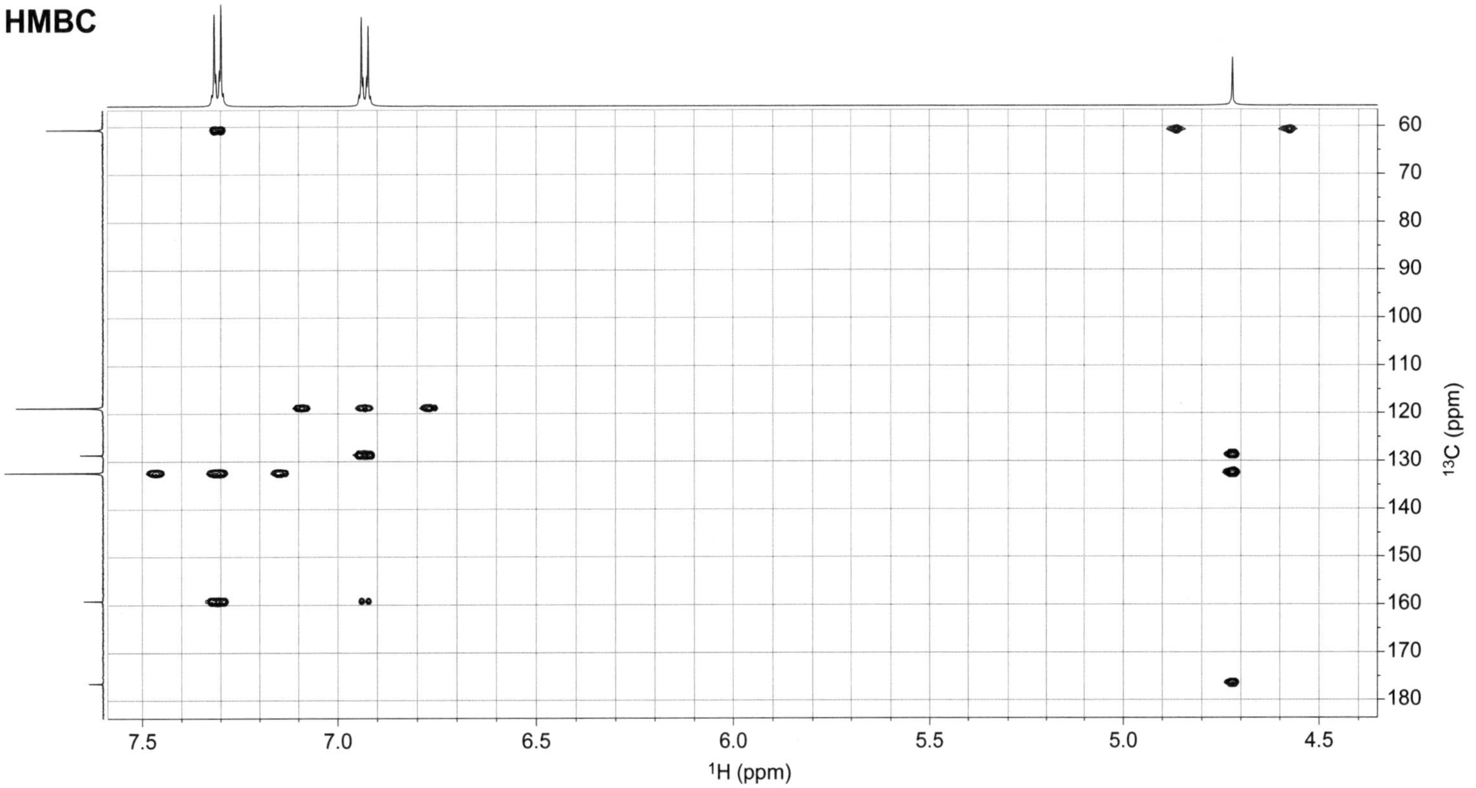

7.5 7.0 6.5 6.0 5.5 5.0 4.5
^{1}H (ppm)
60 70 80 90 100 110 120 130 140 150 160 170 180
^{13}C (ppm)

Problem 9.75 Molecular formula: C_9H_8O **Difficulty** ★★★☆

Solvent: $CDCl_3$

Field: 400 MHz (^1H); 100 MHz (^{13}C)

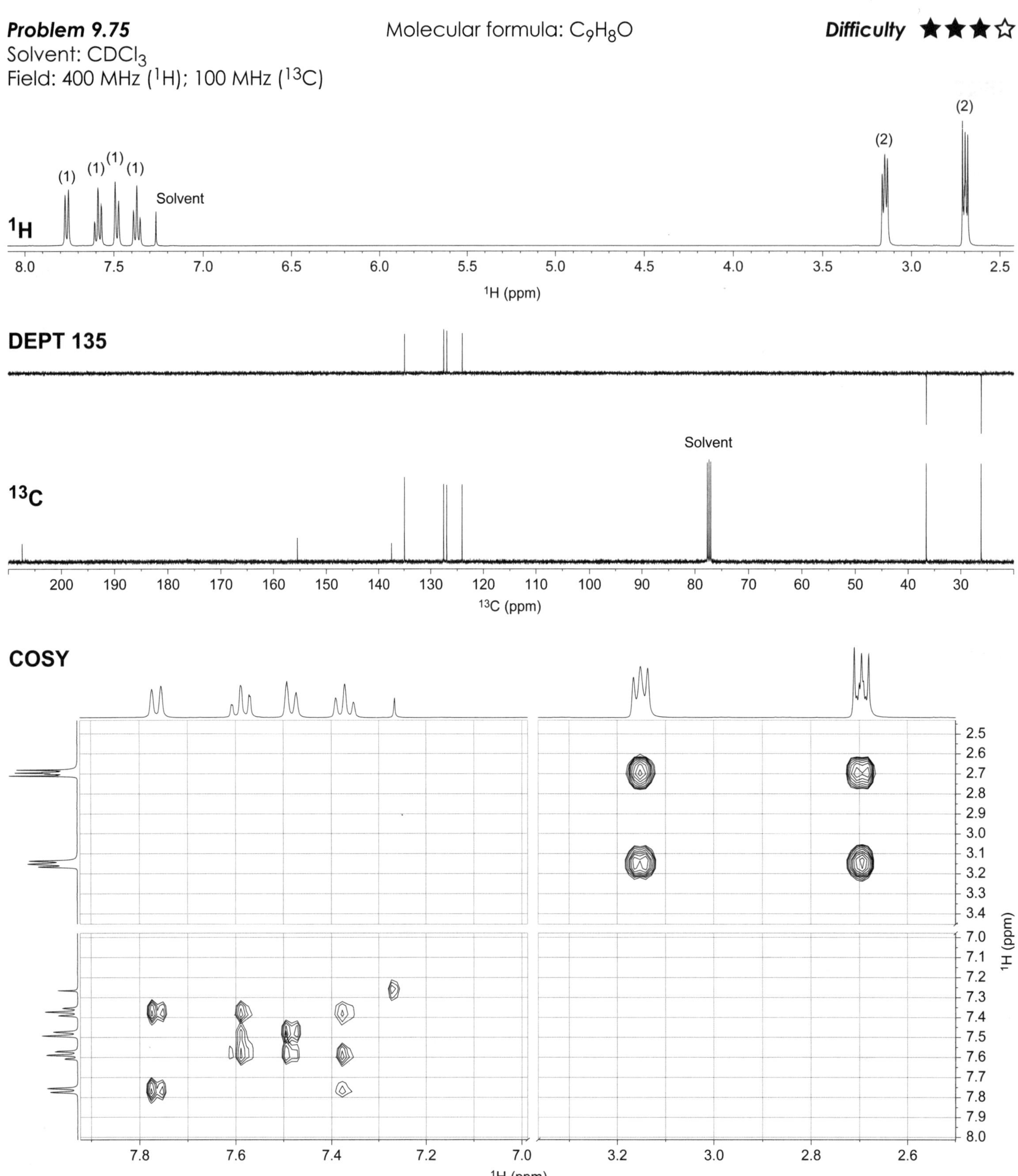

HSQC

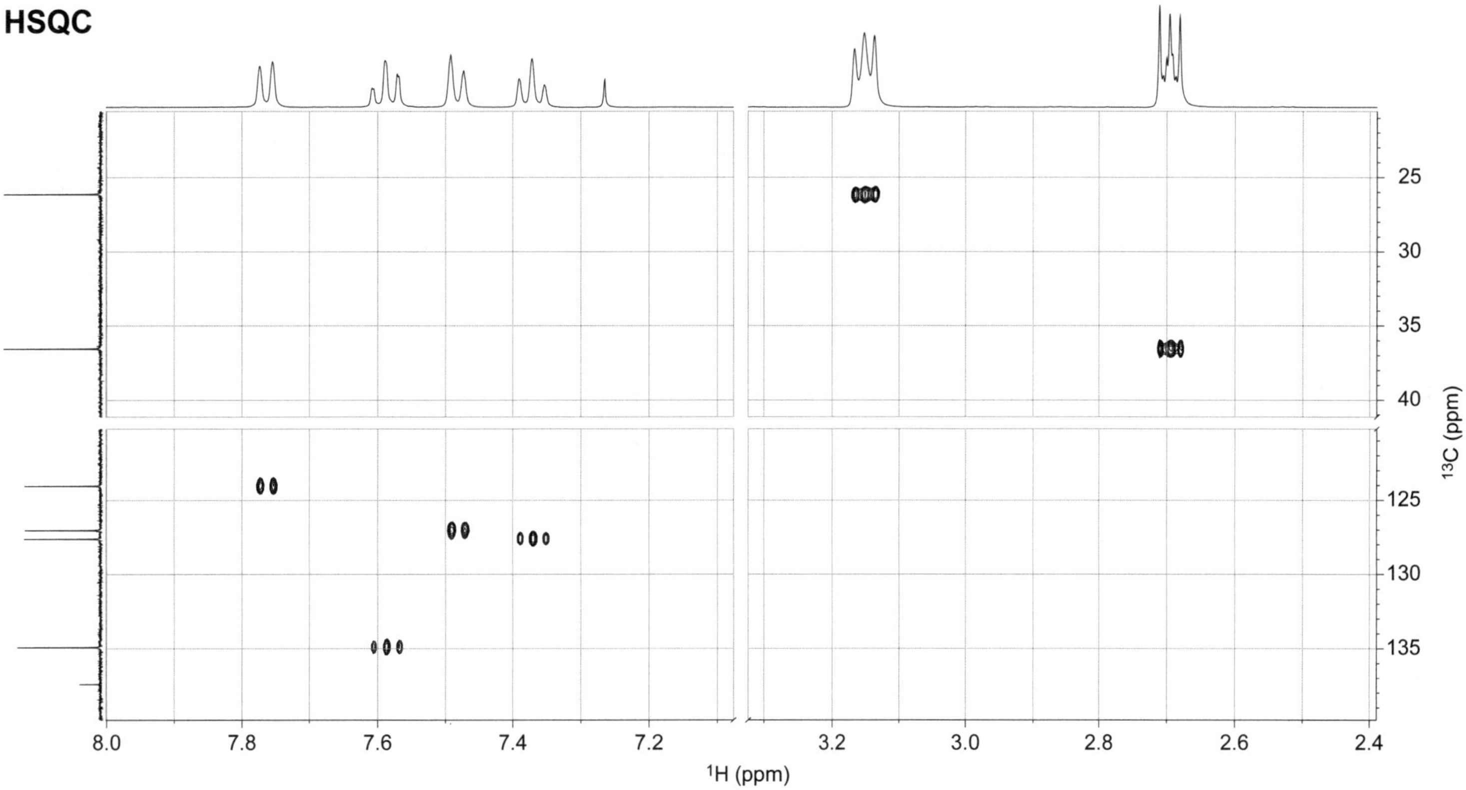
25
30
35
40
125
130
135
¹³C (ppm)
8.0
7.8
7.6
7.4
7.2
3.2
3.0
2.8
2.6
2.4
¹H (ppm)

HMBC

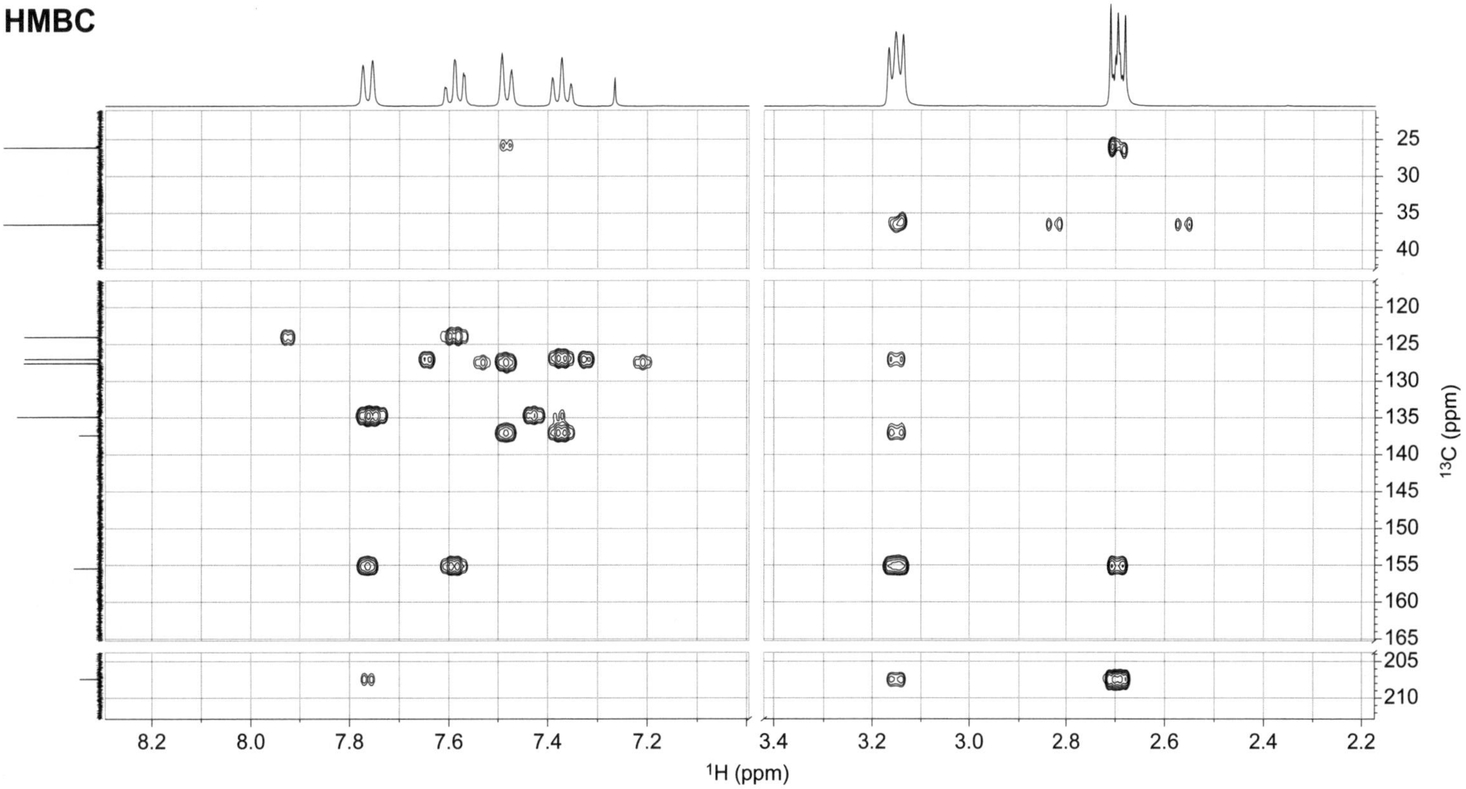
25
30
35
40
120
125
130
135
140
145
150
155
160
165
205
210
¹³C (ppm)
8.2
8.0
7.8
7.6
7.4
7.2
3.4
3.2
3.0
2.8
2.6
2.4
2.2
¹H (ppm)

Problem 9.76 Molecular formula: $C_8H_{10}O$ ***Difficulty*** ★★★☆

Solvent: D_2O

Field: 500 MHz (^{1}H); 125 MHz (^{13}C)

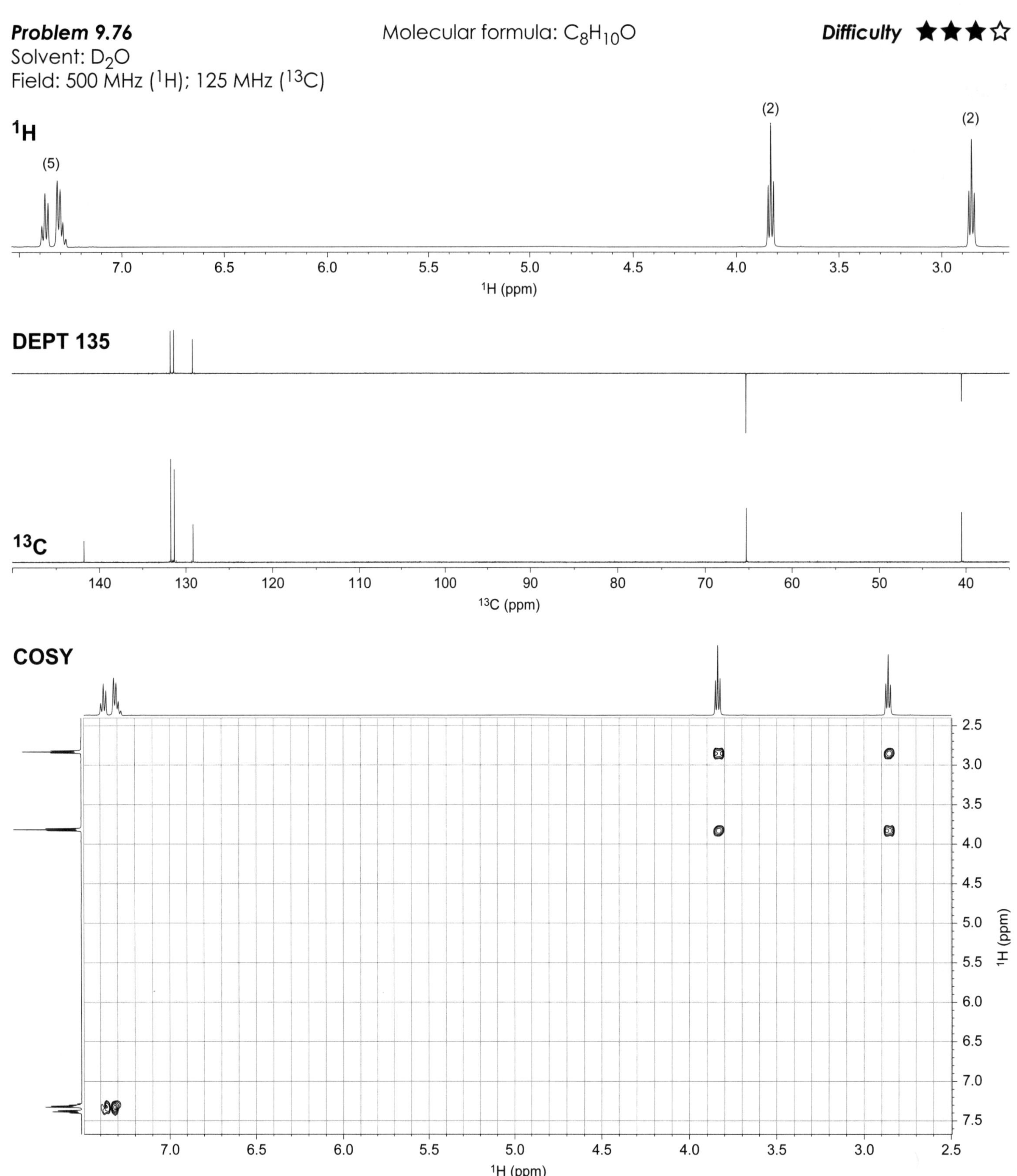

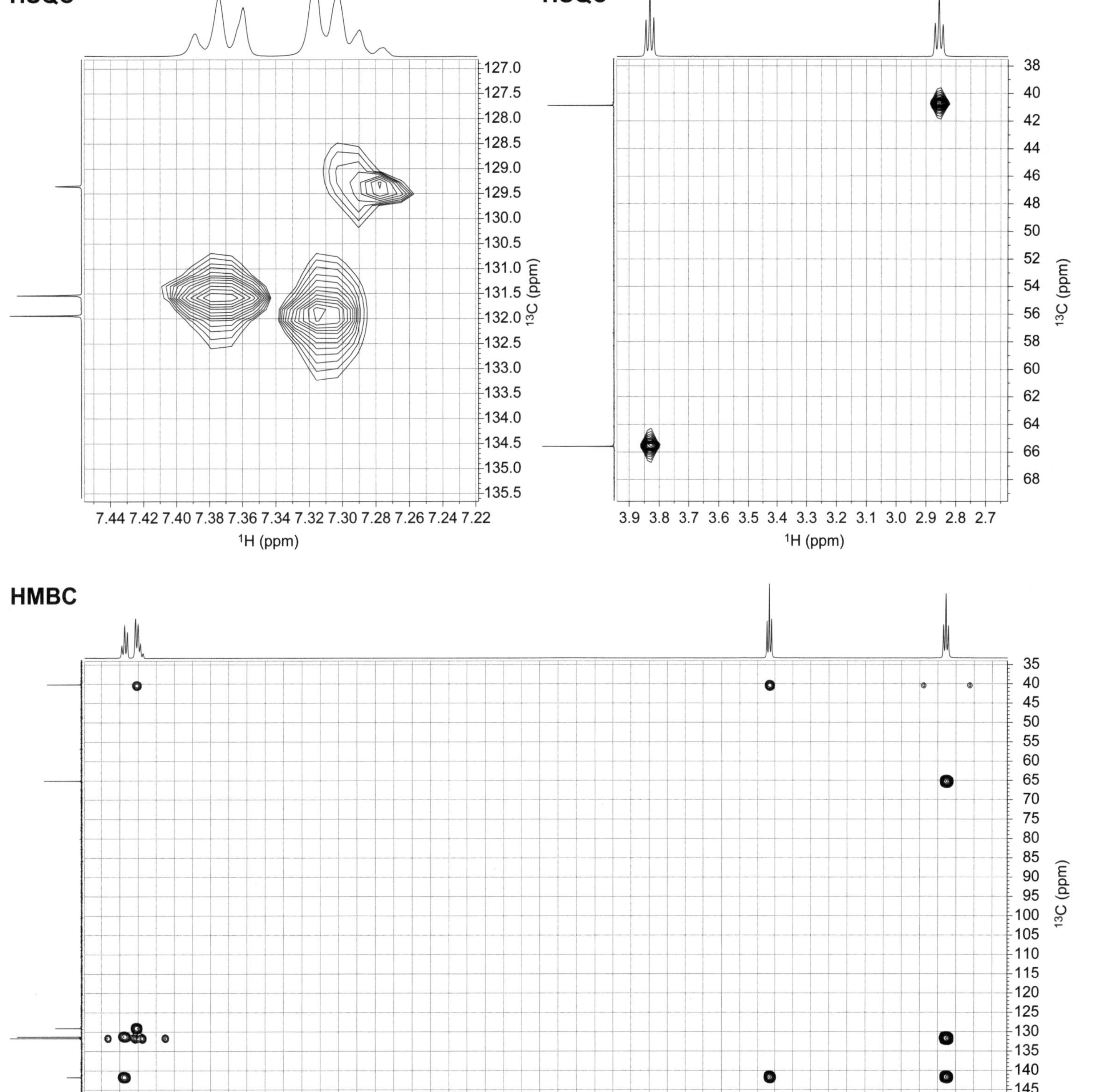
HSQC
HSQC
HMBC
1H (ppm)
13C (ppm)

Problem 9.77 Molecular formula: $C_7H_6O_2$ ***Difficulty*** ★★★☆

Solvent: C_6D_6

Field: 500 MHz (^{1}H); 125 MHz (^{13}C)

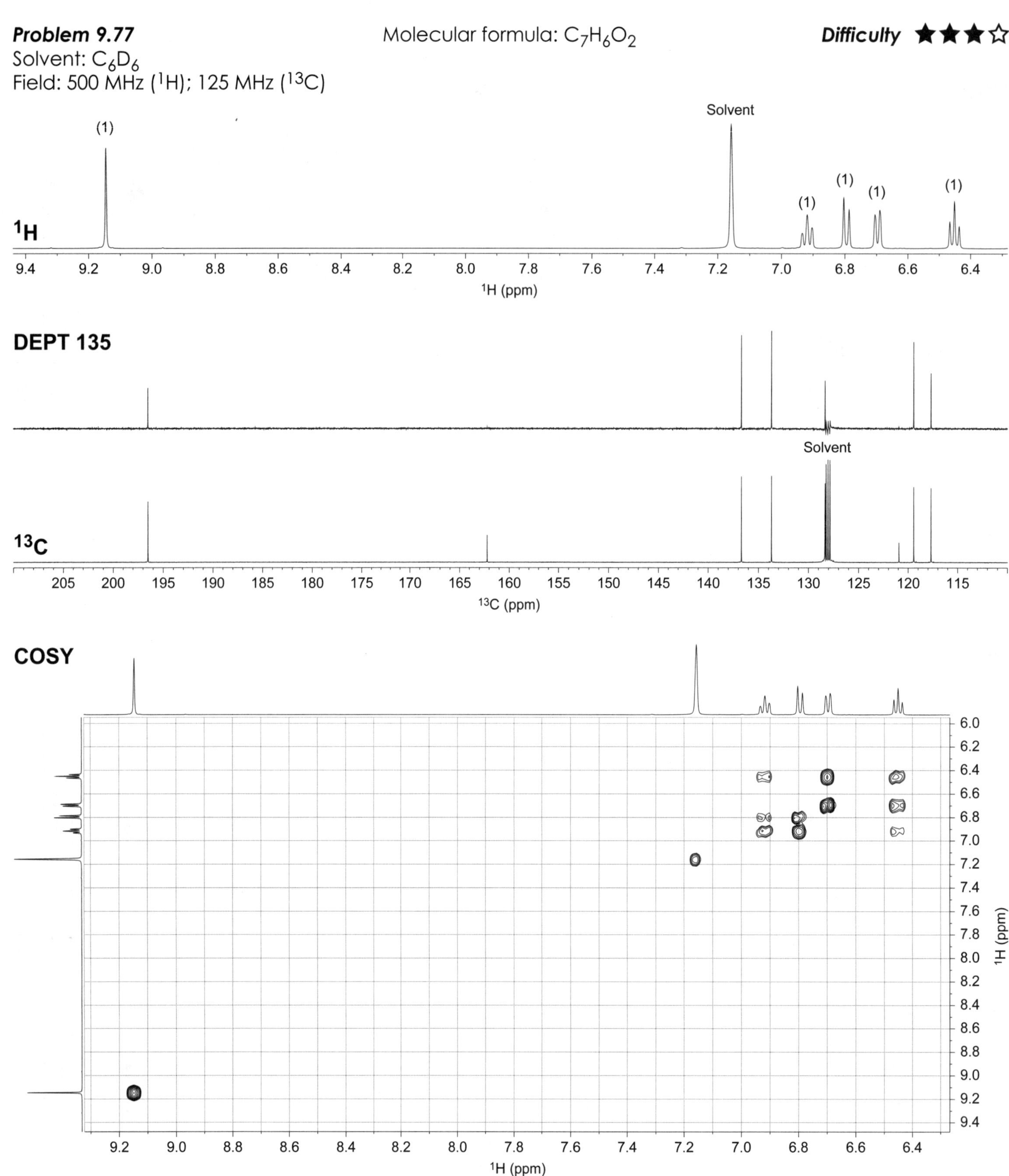

HSQC

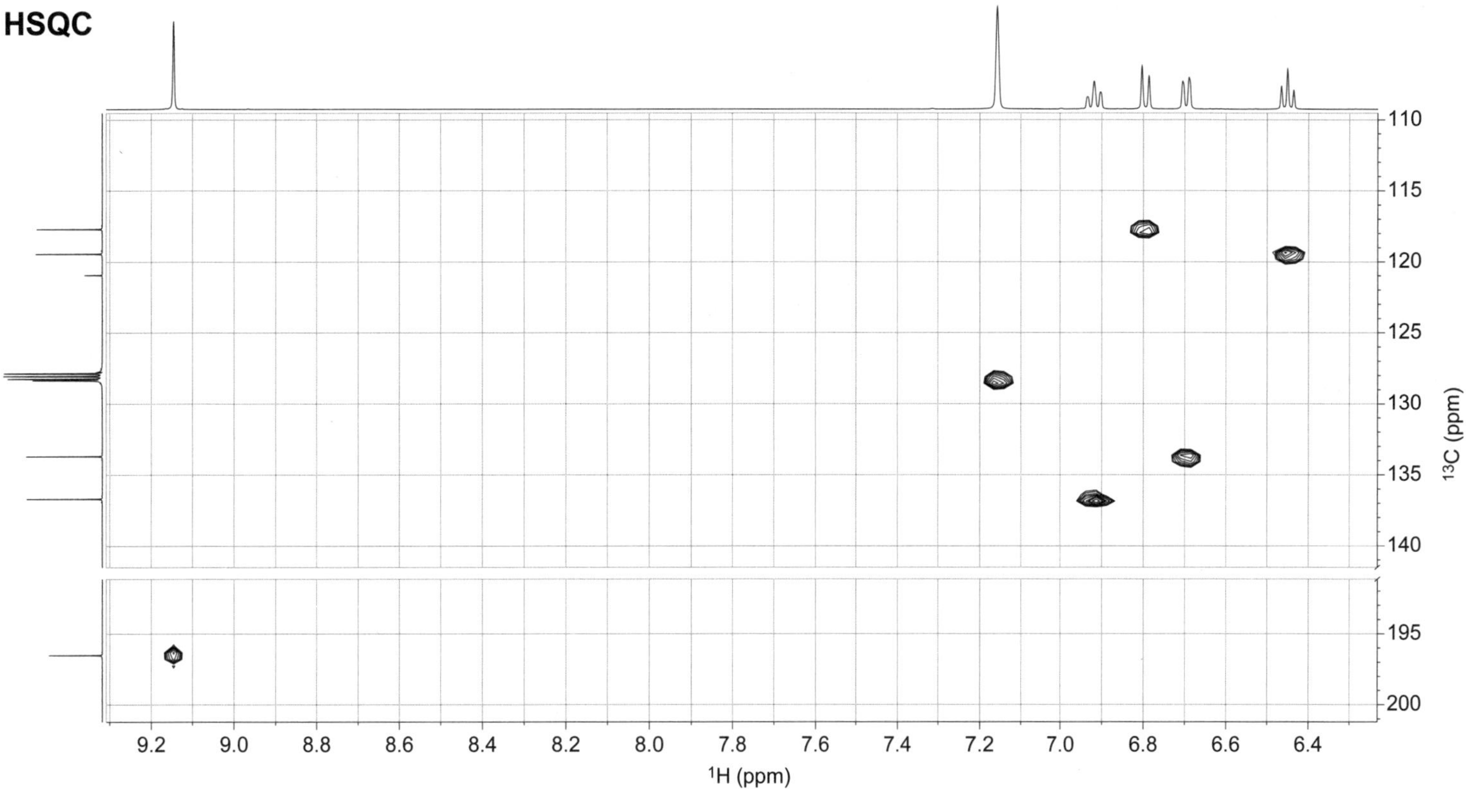
110
115
120
125
130
135
140
195
200
^{13}C (ppm)
9.2
9.0
8.8
8.6
8.4
8.2
8.0
7.8
7.6
7.4
7.2
7.0
6.8
6.6
6.4
^{1}H (ppm)

HMBC

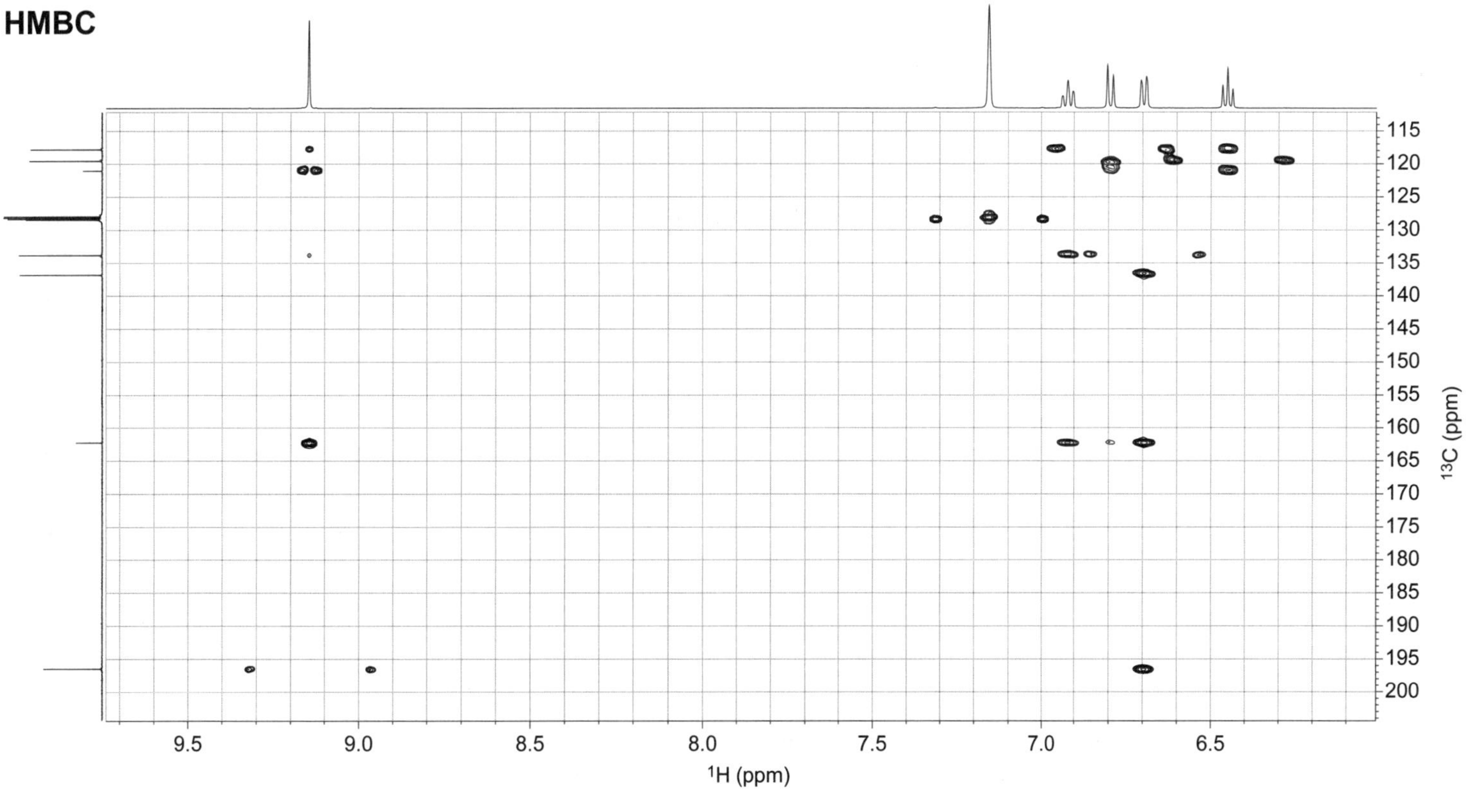
115
120
125
130
135
140
145
150
155
160
165
170
175
180
185
190
195
200
^{13}C (ppm)
9.5
9.0
8.5
8.0
7.5
7.0
6.5
^{1}H (ppm)

Problem 9.78 Molecular formula: $C_8H_9NO_3$ **Difficulty** ★★★☆

Solvent: CD_3OD
Field: 500 MHz (^1H); 125 MHz (^{13}C)

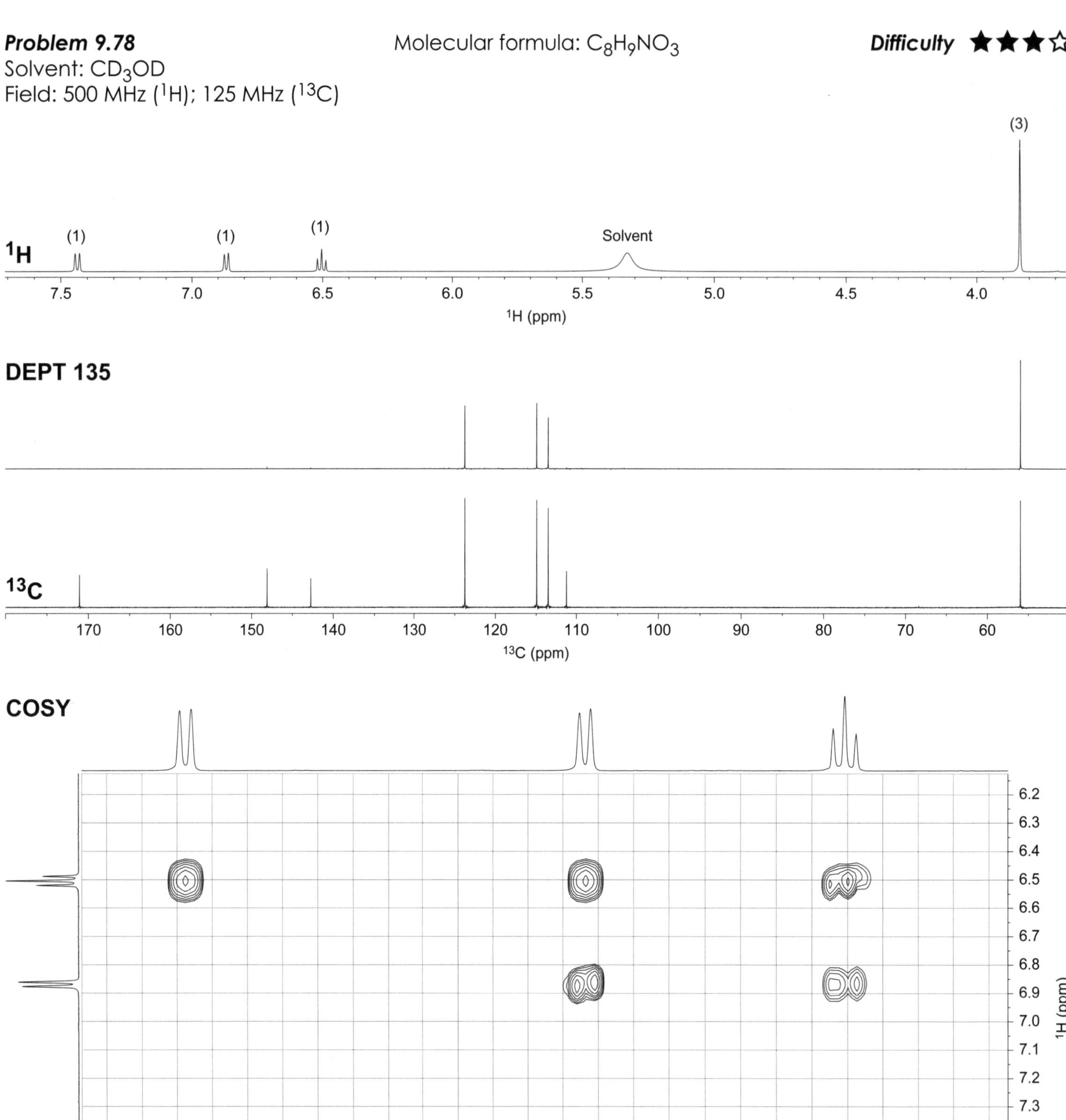

HSQC

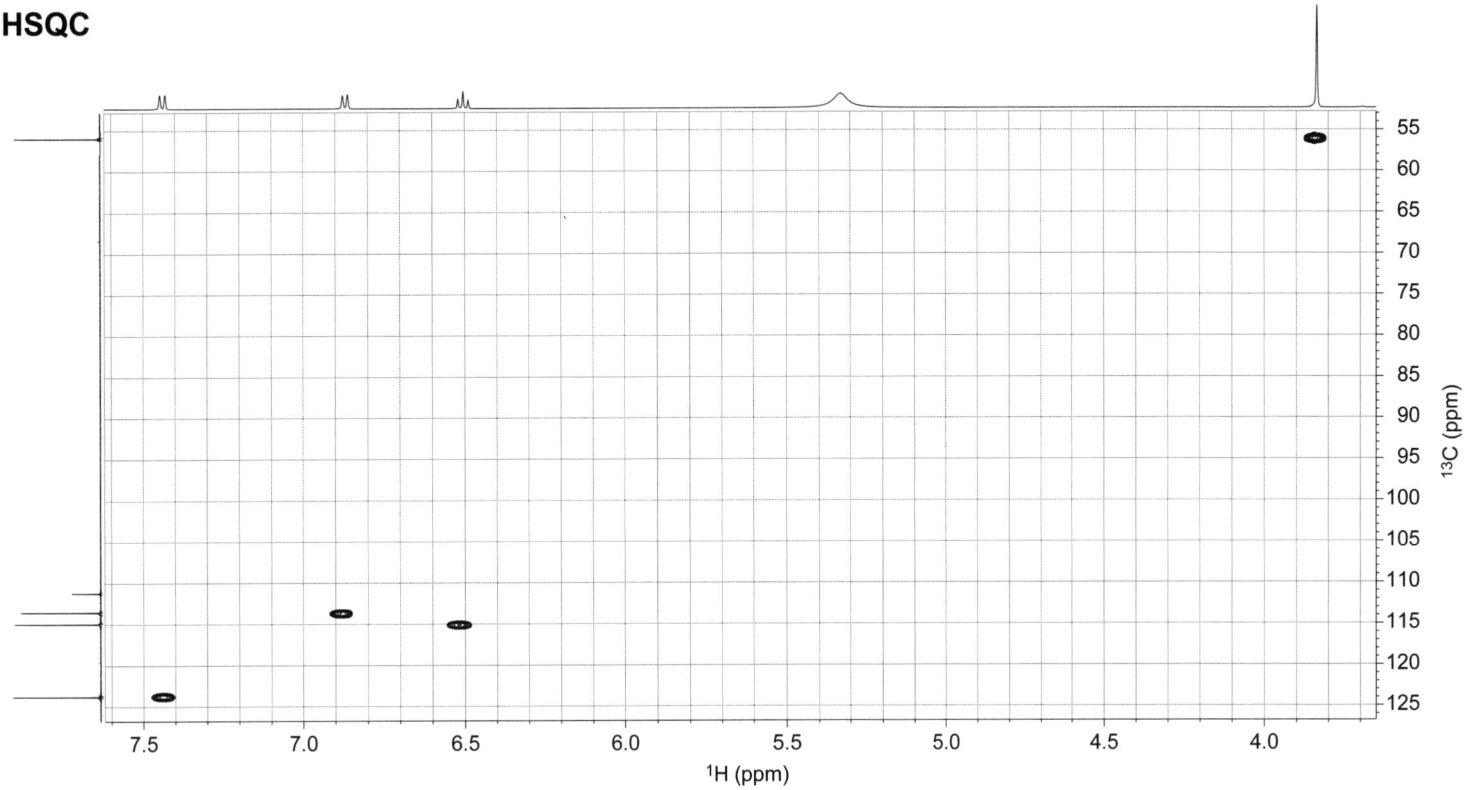
55
60
65
70
75
80
85
90
95
100
105
110
115
120
125
7.5
7.0
6.5
6.0
5.5
5.0
4.5
4.0
1H (ppm)
13C (ppm)

HMBC

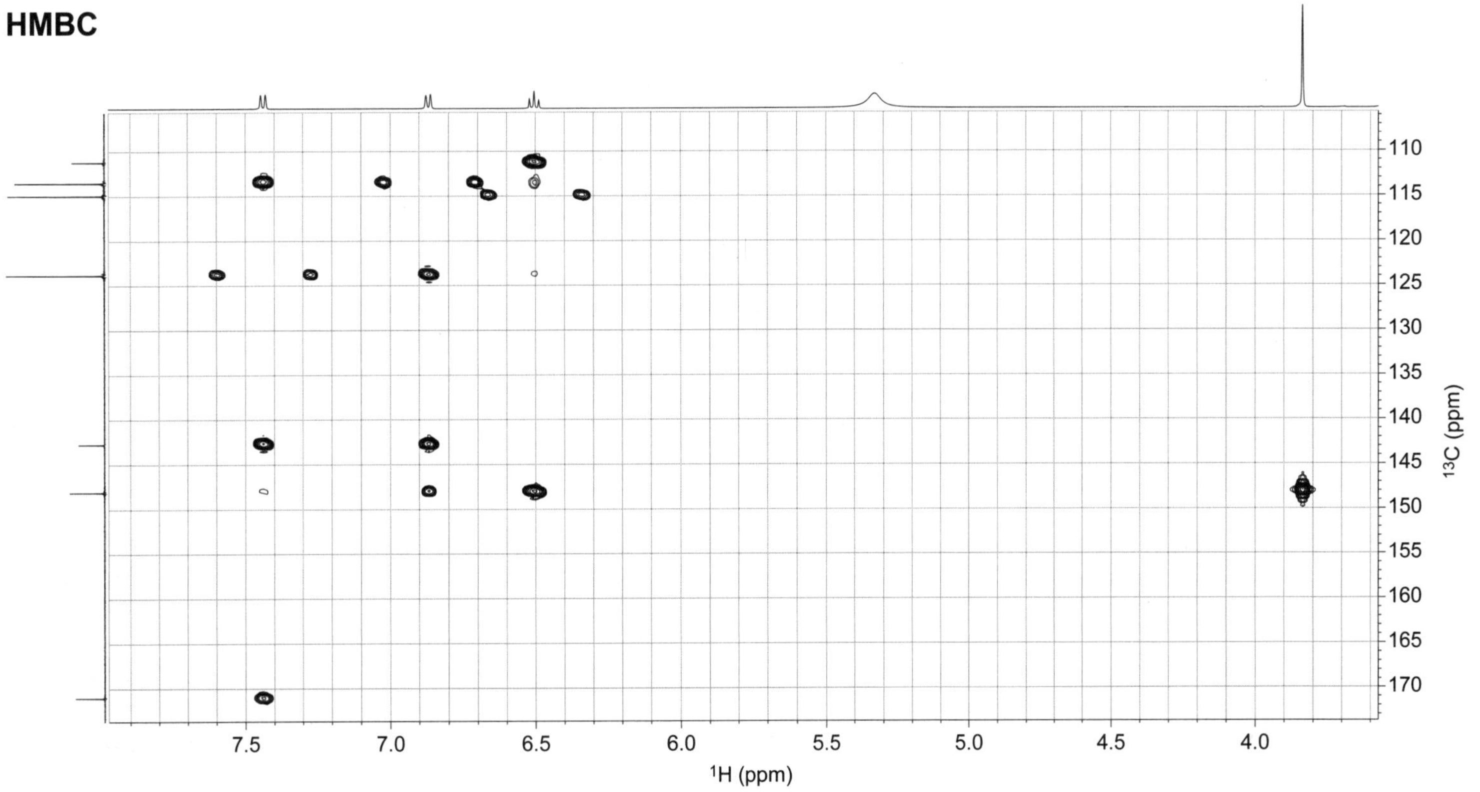
110
115
120
125
130
135
140
145
150
155
160
165
170
7.5
7.0
6.5
6.0
5.5
5.0
4.5
4.0
1H (ppm)
13C (ppm)

Problem 9.79 Molecular formula: $C_8H_{11}NO_2$ ***Difficulty*** ★★★☆

Solvent: D_2O

Field: 500 MHz (^1H); 125 MHz (^{13}C)

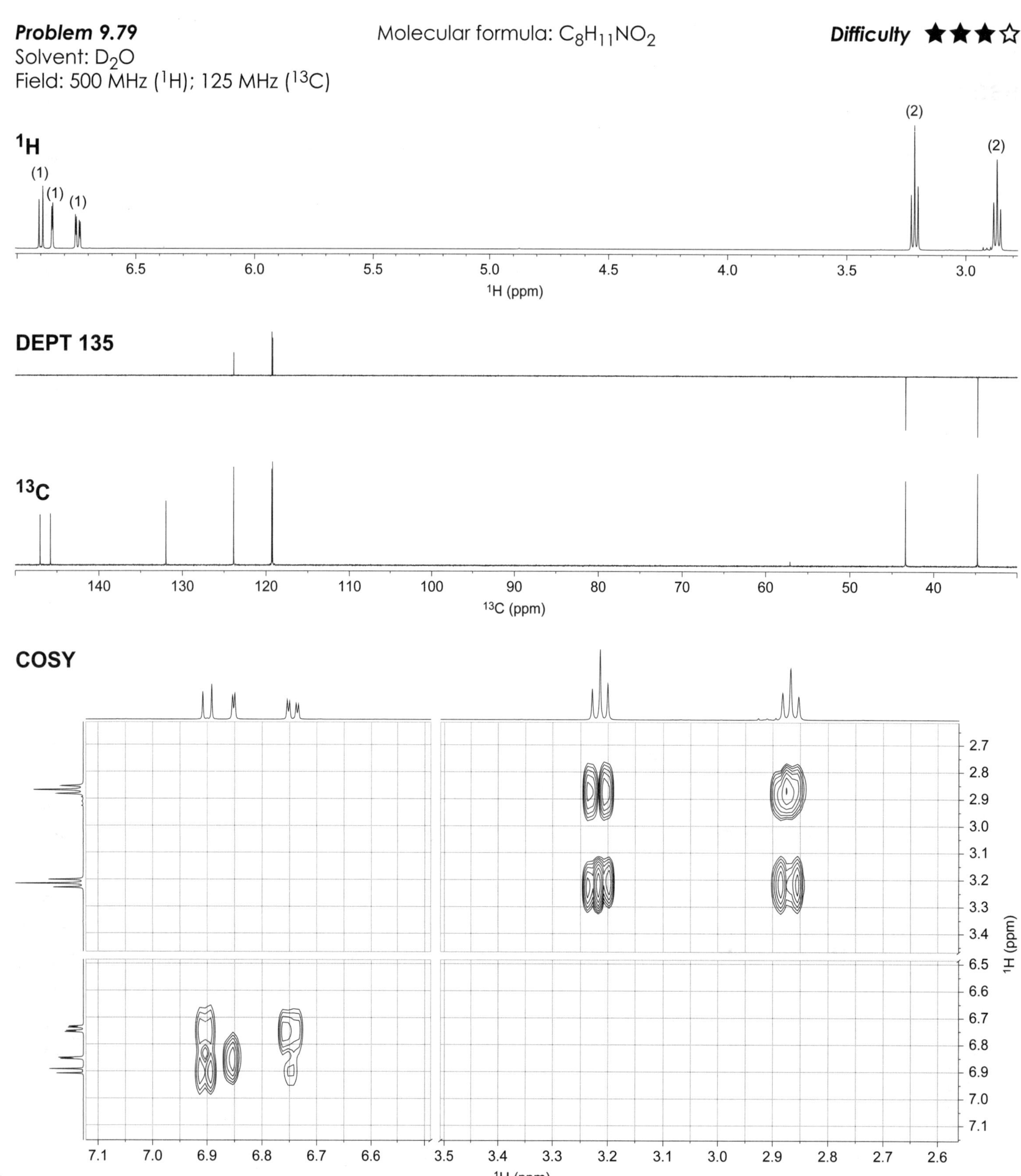

HSQC

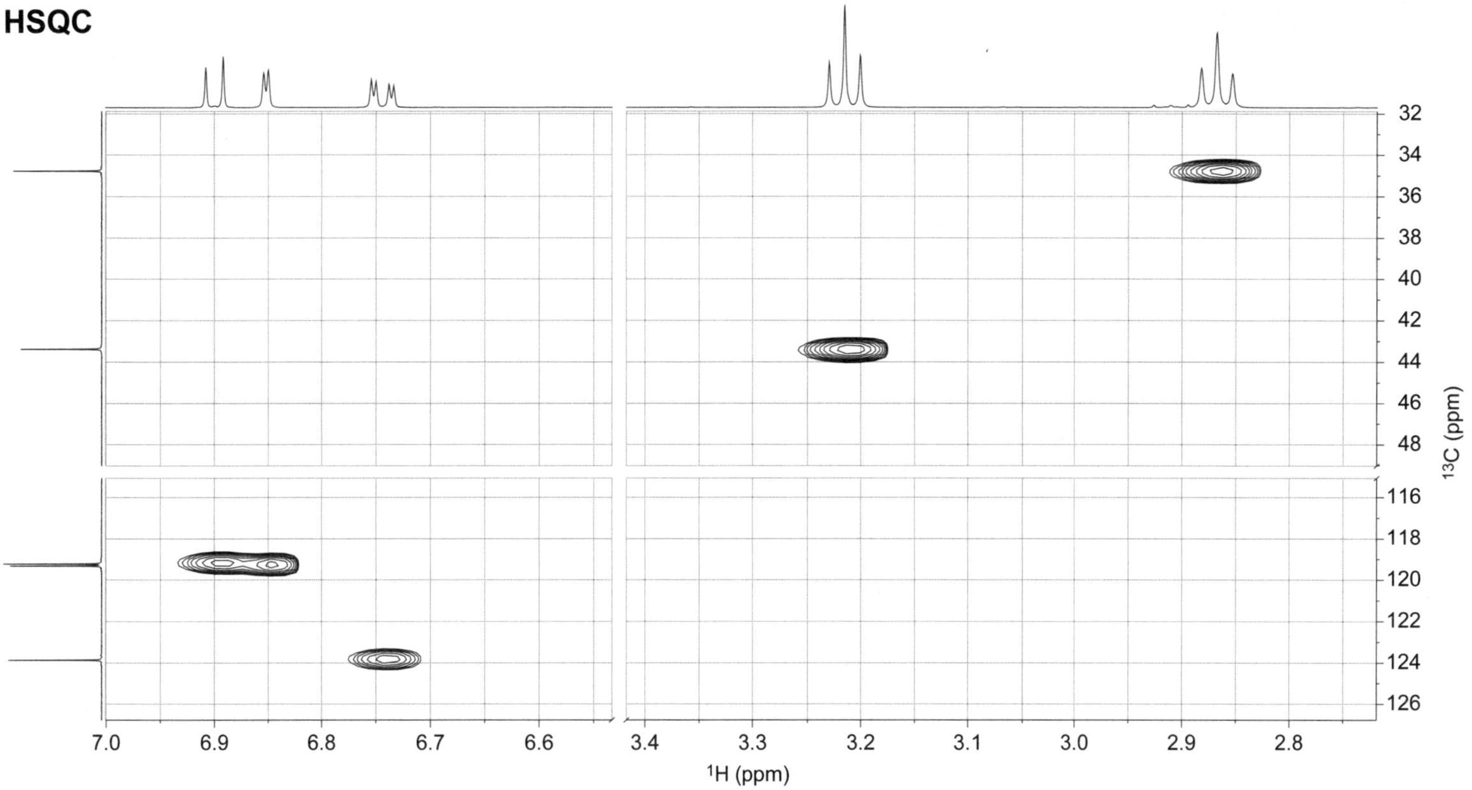
32
34
36
38
40
42
44
46
48
116
118
120
122
124
126
^{13}C (ppm)
7.0
6.9
6.8
6.7
6.6
3.4
3.3
3.2
3.1
3.0
2.9
2.8
^{1}H (ppm)

HMBC

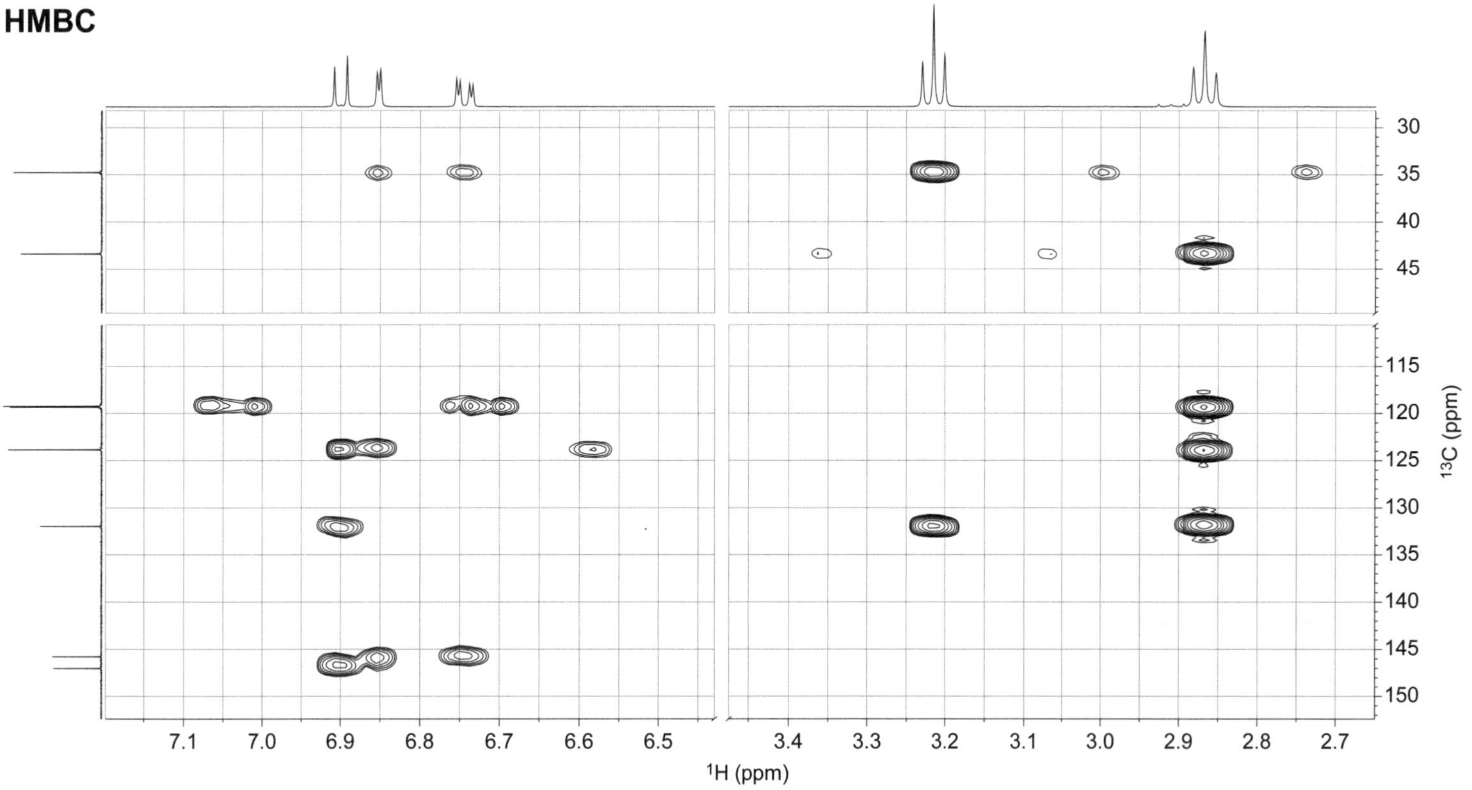
30
35
40
45
115
120
125
130
135
140
145
150
^{13}C (ppm)
7.1
7.0
6.9
6.8
6.7
6.6
6.5
3.4
3.3
3.2
3.1
3.0
2.9
2.8
2.7
^{1}H (ppm)

Problem 9.80 Molecular formula: $C_8H_8O_3$ ***Difficulty*** ★★★☆

Solvent: D_2O

Field: 500 MHz (^1H); 125 MHz (^{13}C)

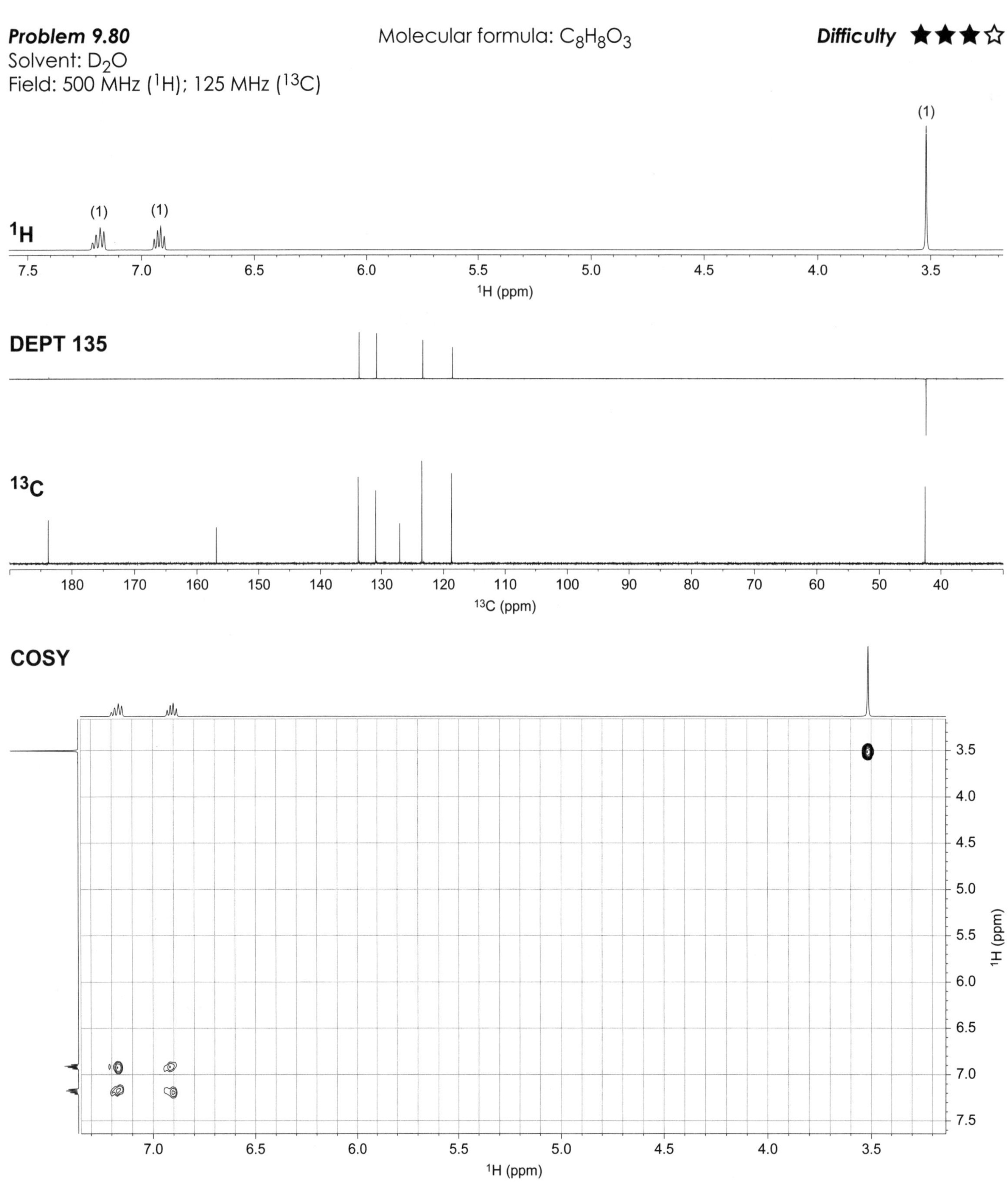

HSQC

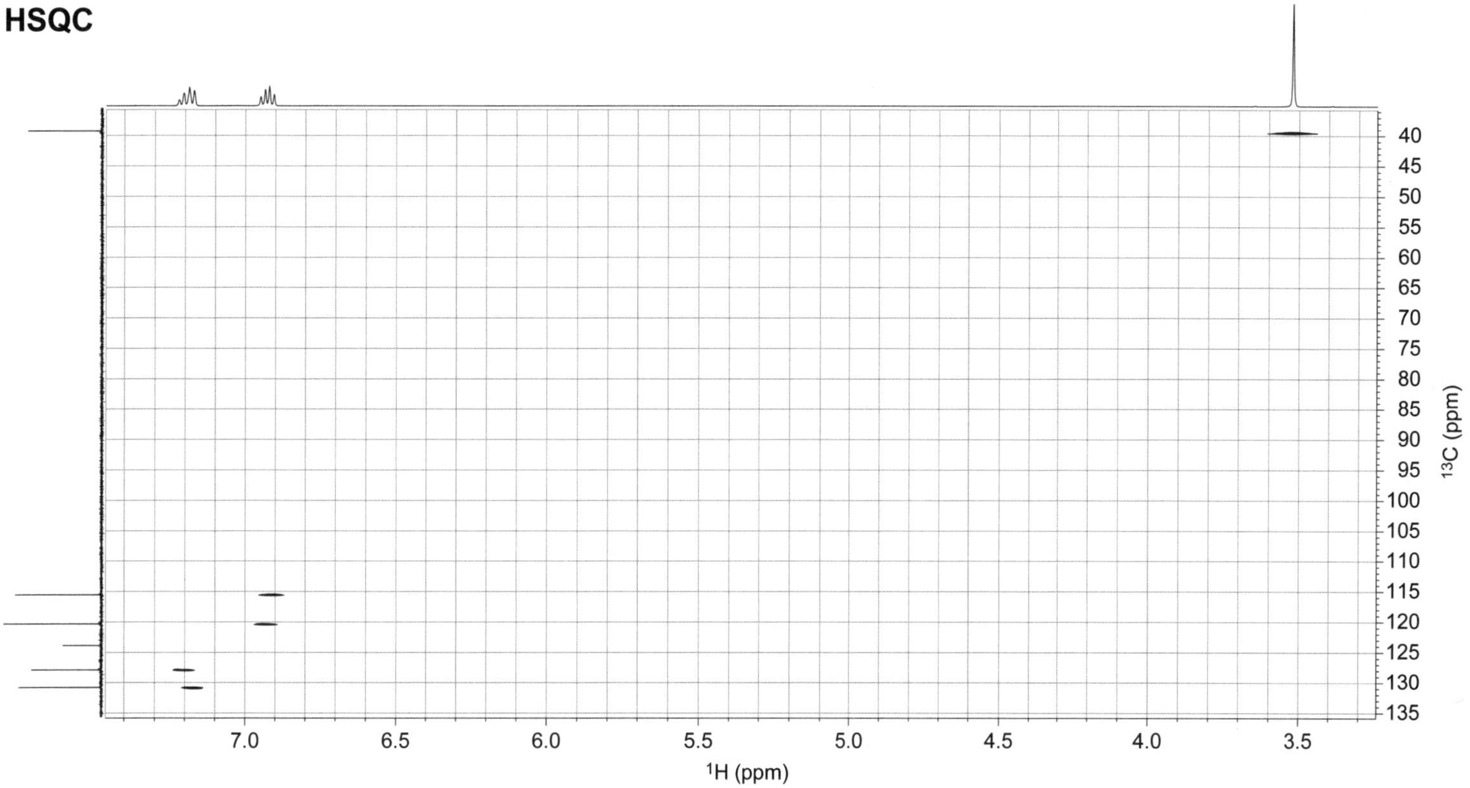

40
45
50
55
60
65
70
75
80
85
90
95
100
105
110
115
120
125
130
135
13C (ppm)
7.0
6.5
6.0
5.5
5.0
4.5
4.0
3.5
1H (ppm)

HMBC

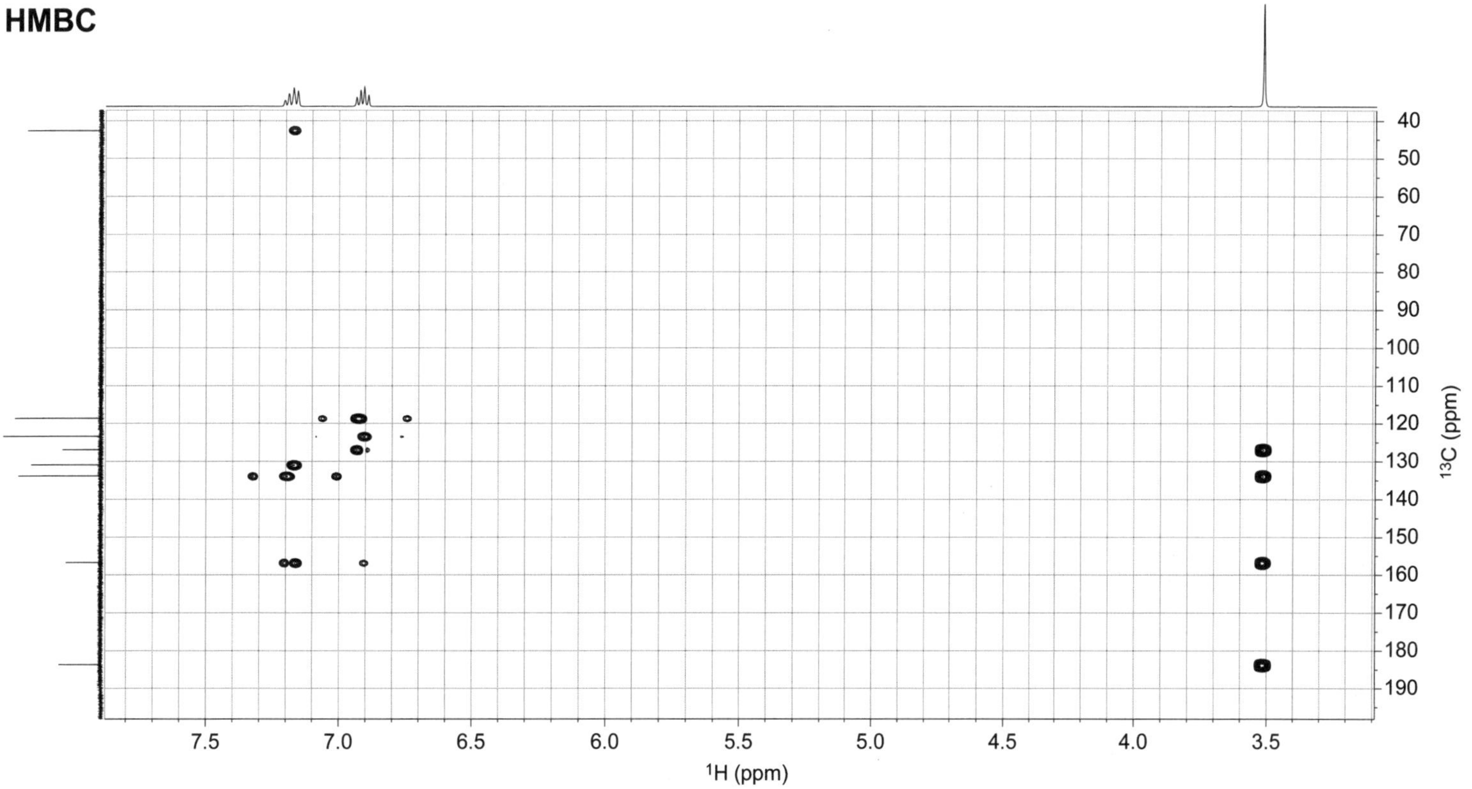

40
50
60
70
80
90
100
110
120
130
140
150
160
170
180
190
13C (ppm)
7.5
7.0
6.5
6.0
5.5
5.0
4.5
4.0
3.5
1H (ppm)

Problem 9.81 Molecular formula: $C_7H_8O_2$ **Difficulty** ★★★☆

Solvent: D_2O

Field: 500 MHz (^1H); 125 MHz (^{13}C)

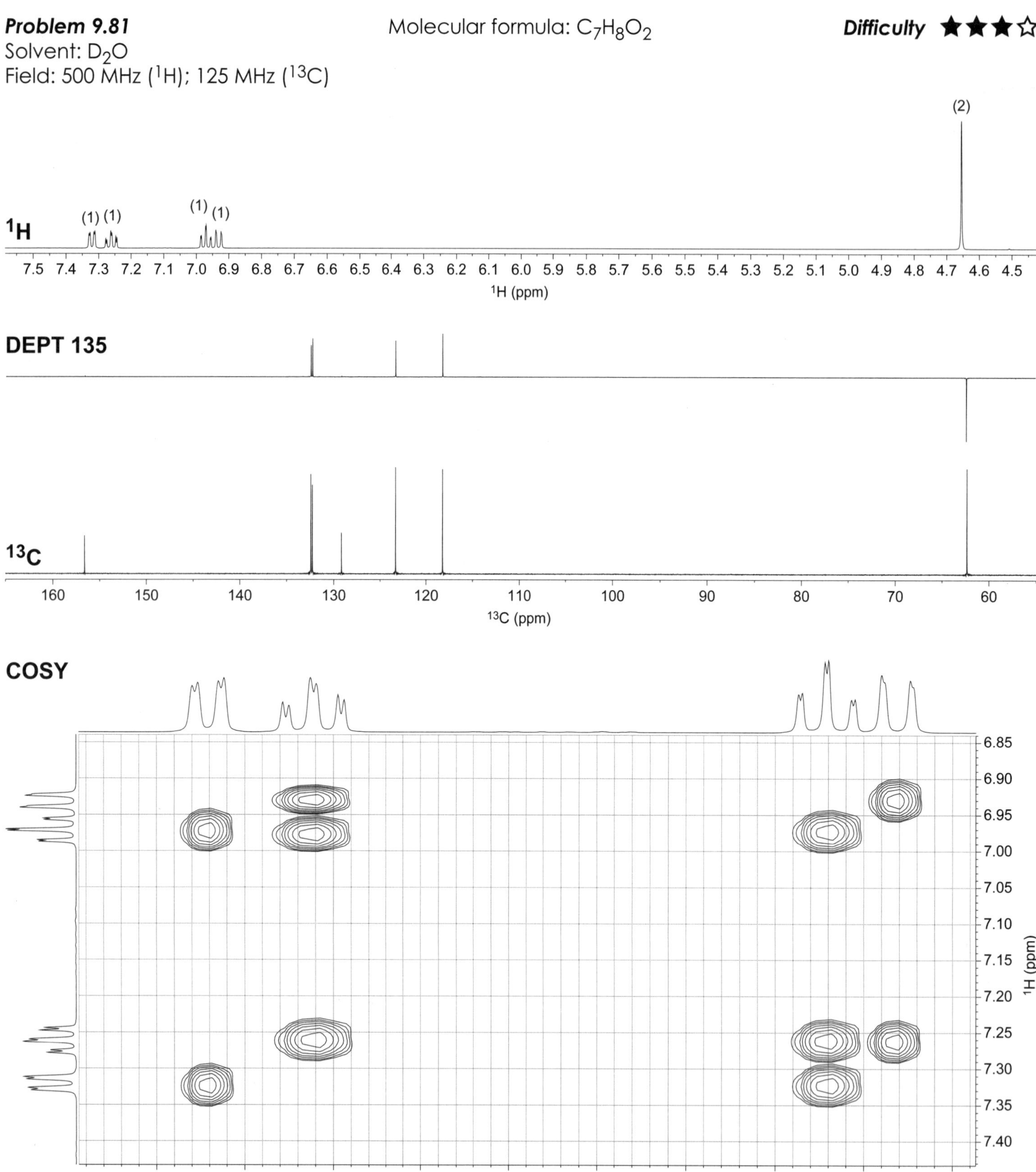

HSQC

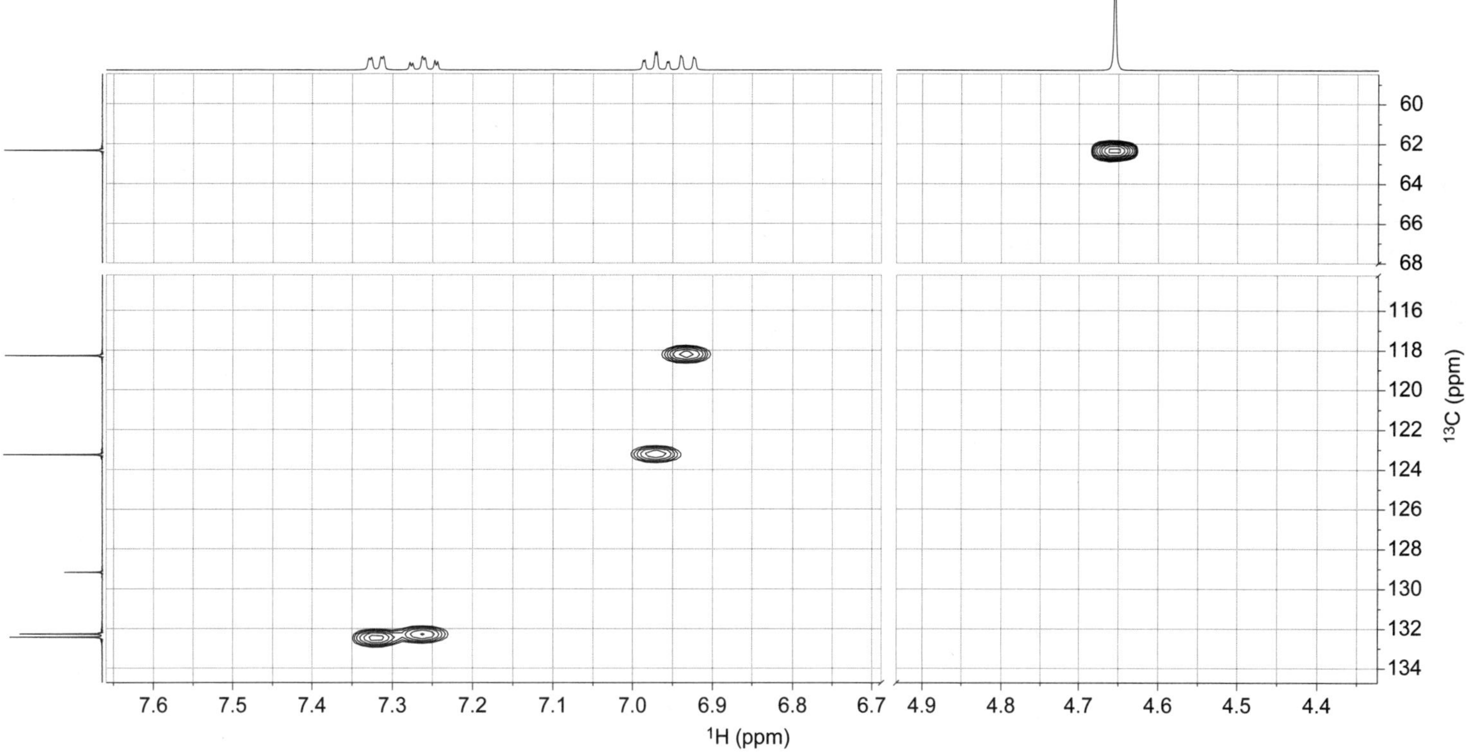

60
62
64
66
68
116
118
120
122
124
126
128
130
132
134
^{13}C (ppm)
7.6 7.5 7.4 7.3 7.2 7.1 7.0 6.9 6.8 6.7 4.9 4.8 4.7 4.6 4.5 4.4
^{1}H (ppm)

60
65
115
120
125
130
135
155
^{13}C (ppm)
7.7 7.6 7.5 7.4 7.3 7.2 7.1 7.0 6.9 6.8 6.7 4.8 4.7 4.6 4.5 4.4 4.3
^{1}H (ppm)

HMBC

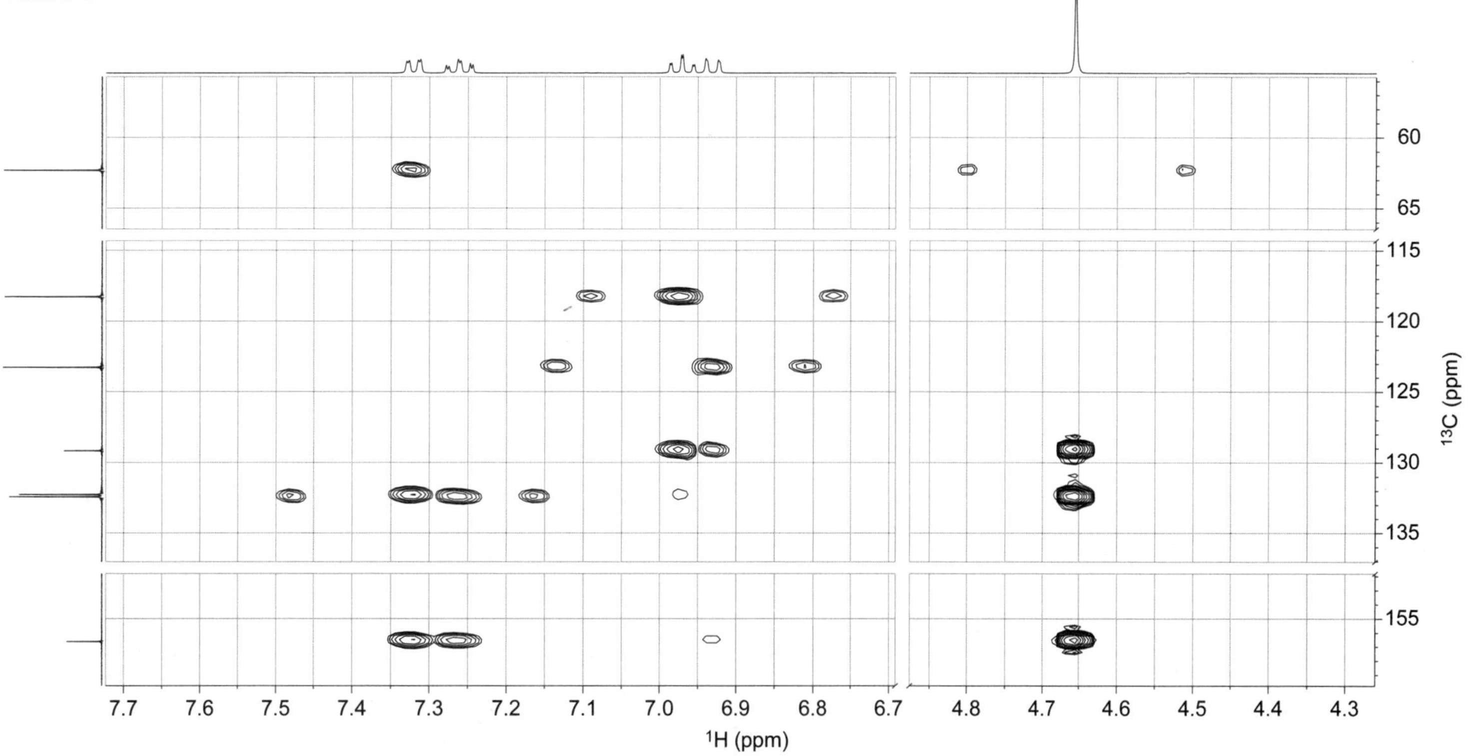

Problem 9.82
Solvent: D_2O
Field: 500 MHz (^1H); 125 MHz (^{13}C)

Molecular formula: $C_5H_9N_3$
Hint: There is a 5-membered heteronuclear aromatic ring

Difficulty ★★★☆

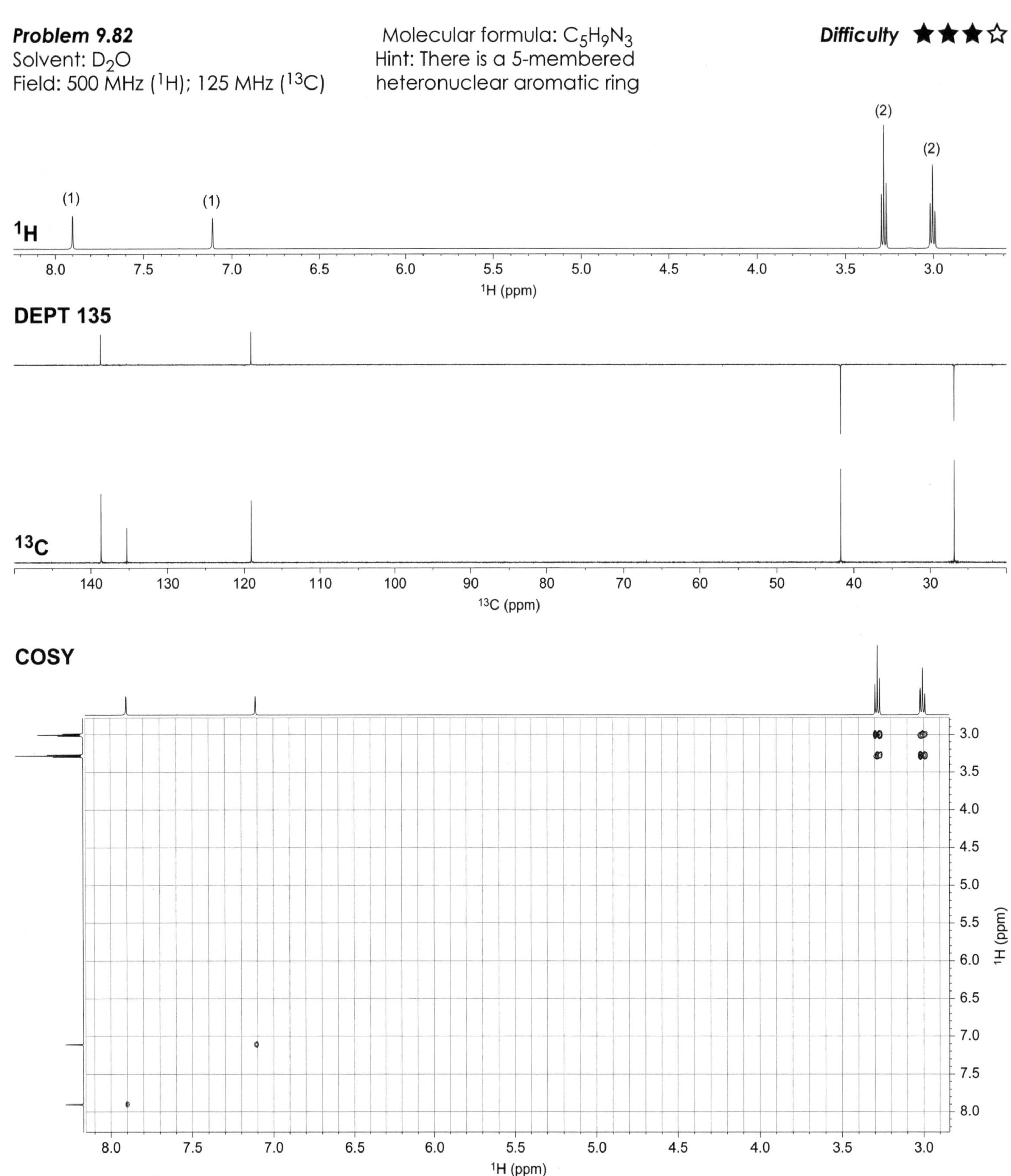

HSQC

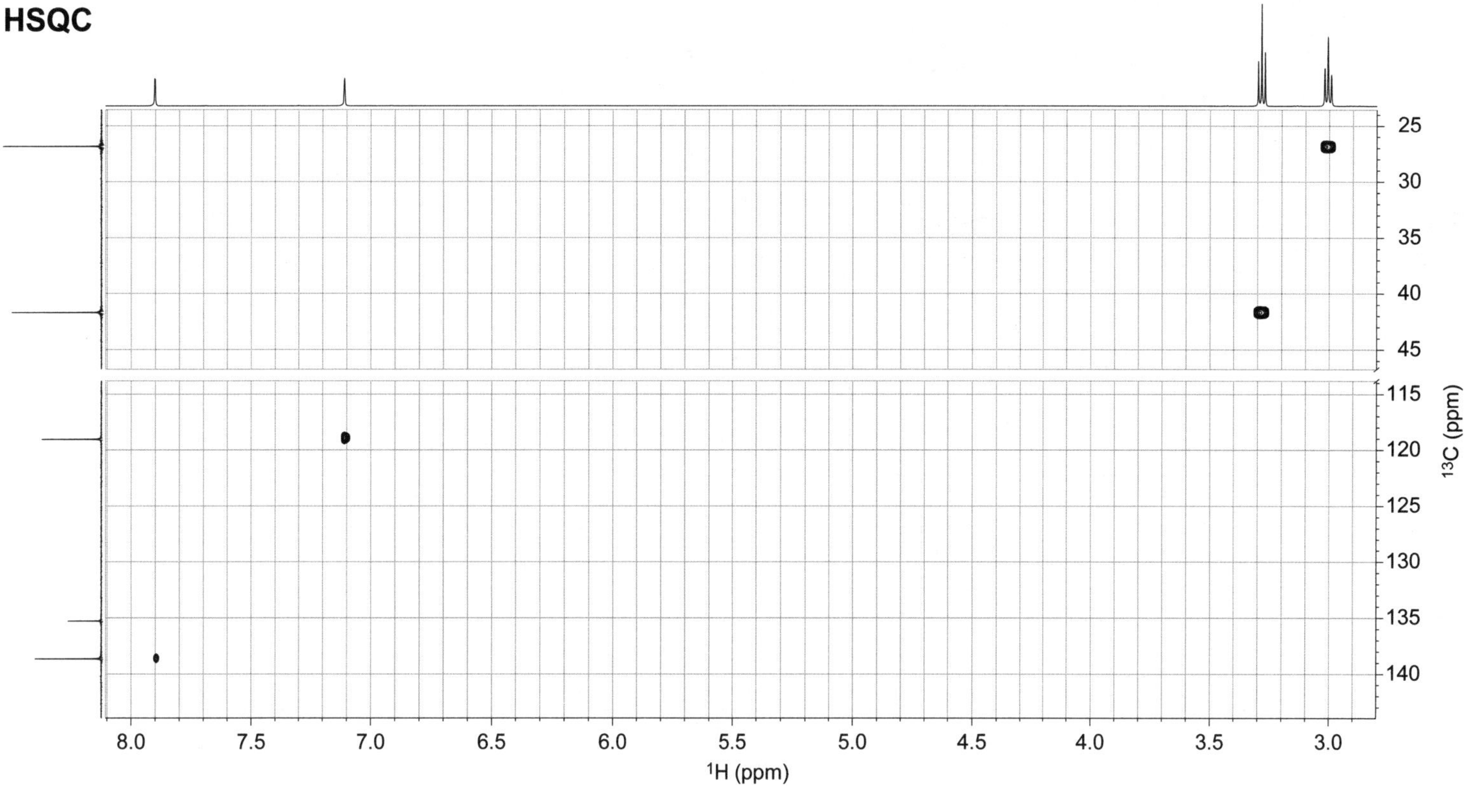

8.0 7.5 7.0 6.5 6.0 5.5 5.0 4.5 4.0 3.5 3.0

^{1}H (ppm)

25 30 35 40 45 115 120 125 130 135 140

^{13}C (ppm)

HMBC

8.0 7.5 7.0 6.5 6.0 5.5 5.0 4.5 4.0 3.5 3.0

^{1}H (ppm)

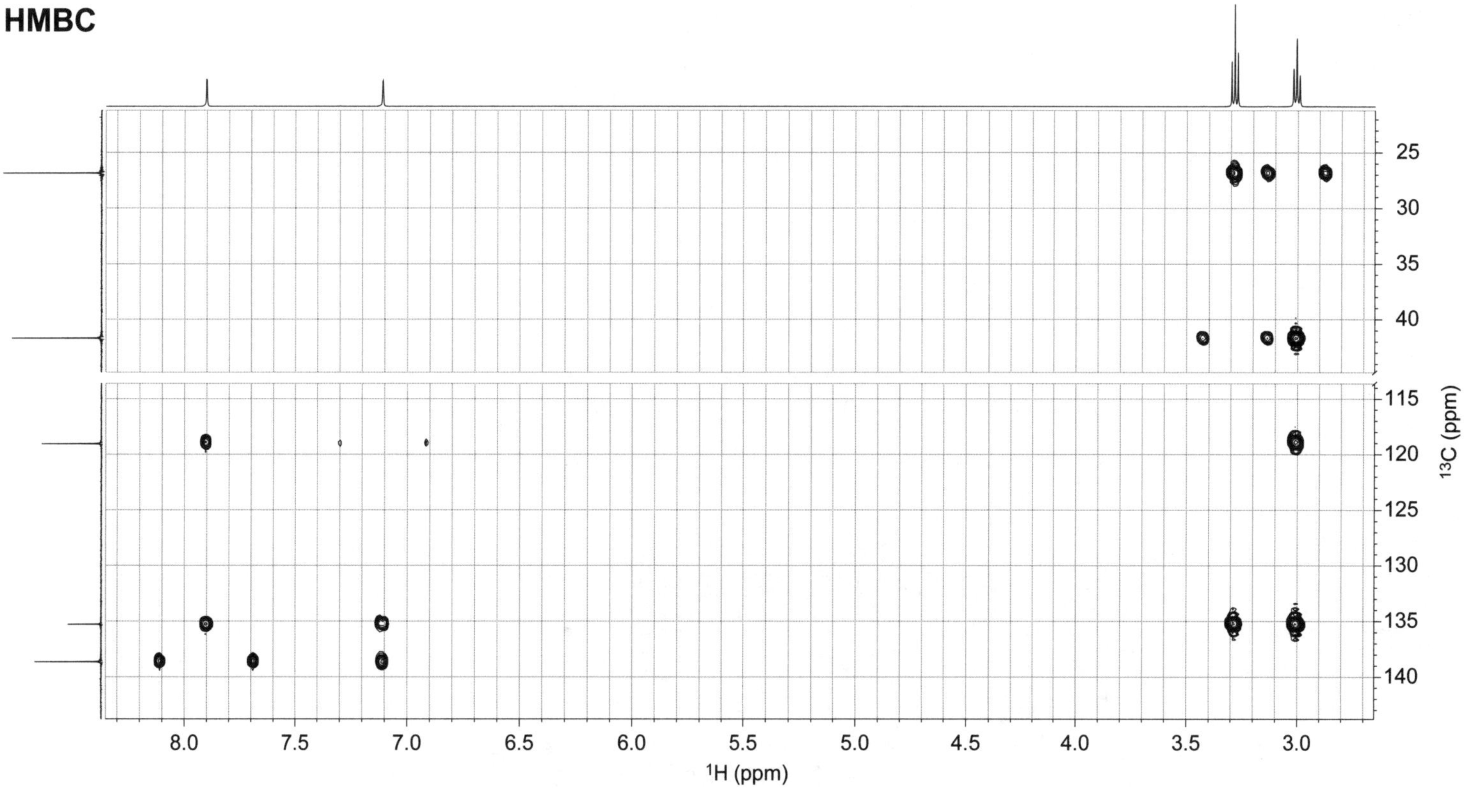

25 30 35 40 115 120 125 130 135 140

^{13}C (ppm)

Problem 9.83

Molecular formula: C_8H_9BrO

Difficulty ★★★☆

Solvent: CD_3OD

Field: 700 MHz (^{1}H); 175 MHz (^{13}C)

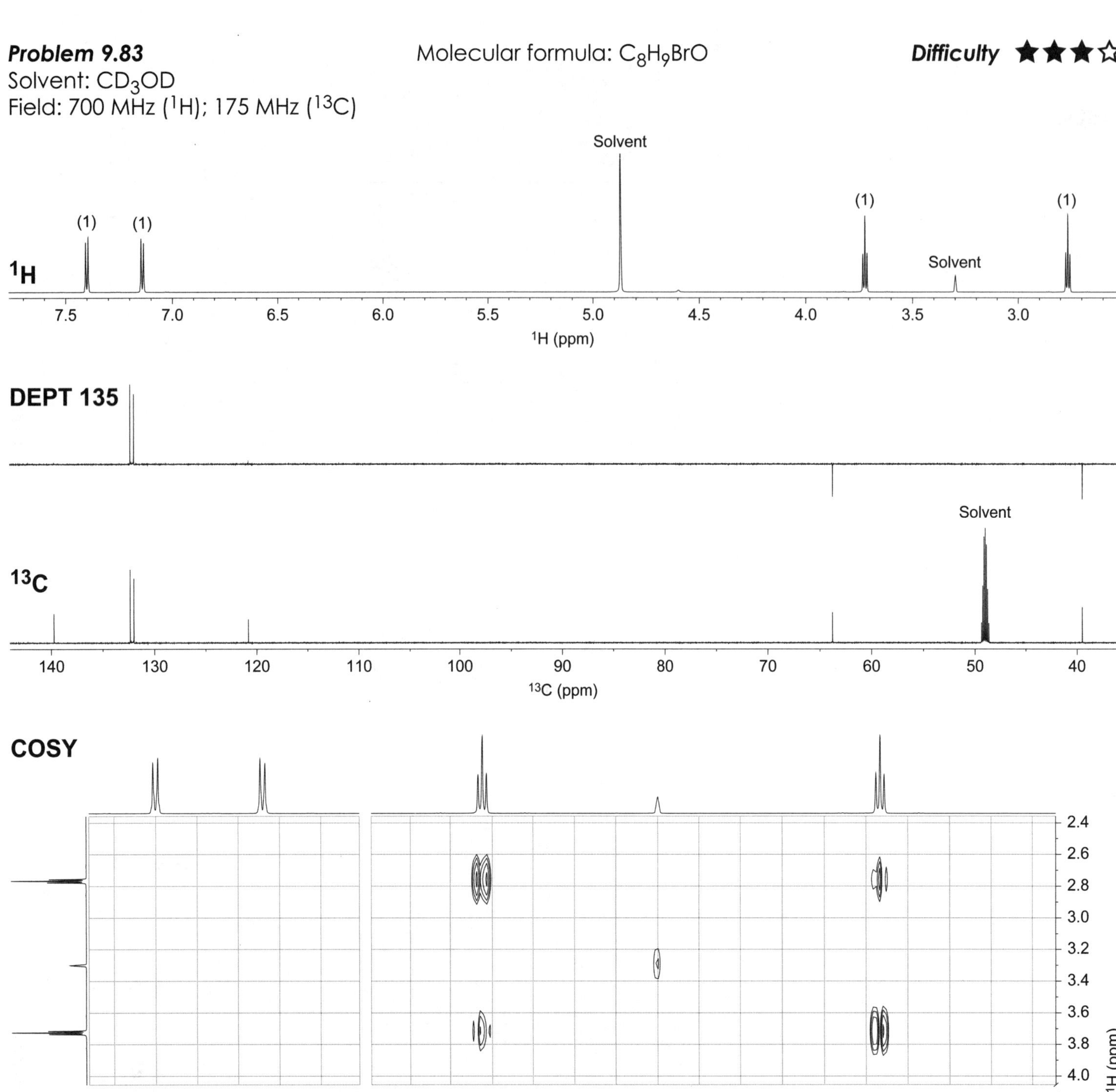

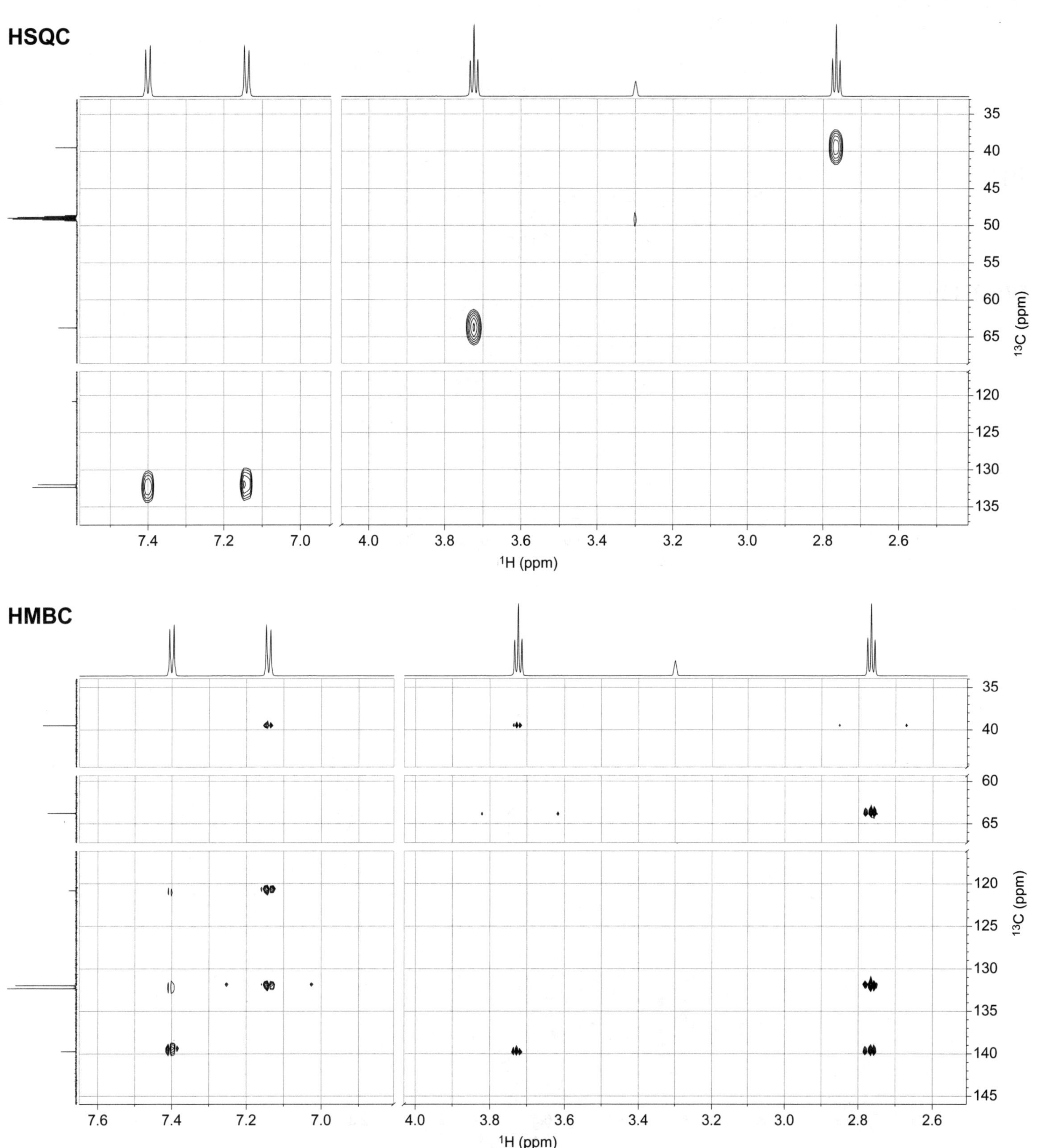
HSQC
35
40
45
50
55
60
65
120
125
130
135
^{13}C (ppm)
7.4
7.2
7.0
4.0
3.8
3.6
3.4
3.2
3.0
2.8
2.6
^{1}H (ppm)
HMBC
35
40
60
65
120
125
130
135
140
145
^{13}C (ppm)
7.6
7.4
7.2
7.0
4.0
3.8
3.6
3.4
3.2
3.0
2.8
2.6
^{1}H (ppm)

Problem 9.84 Molecular formula: $C_8H_{12}N_2$ ***Difficulty*** ★★★☆

Solvent: CD_3OD

Field: 700 MHz (^{1}H); 175 MHz (^{13}C)

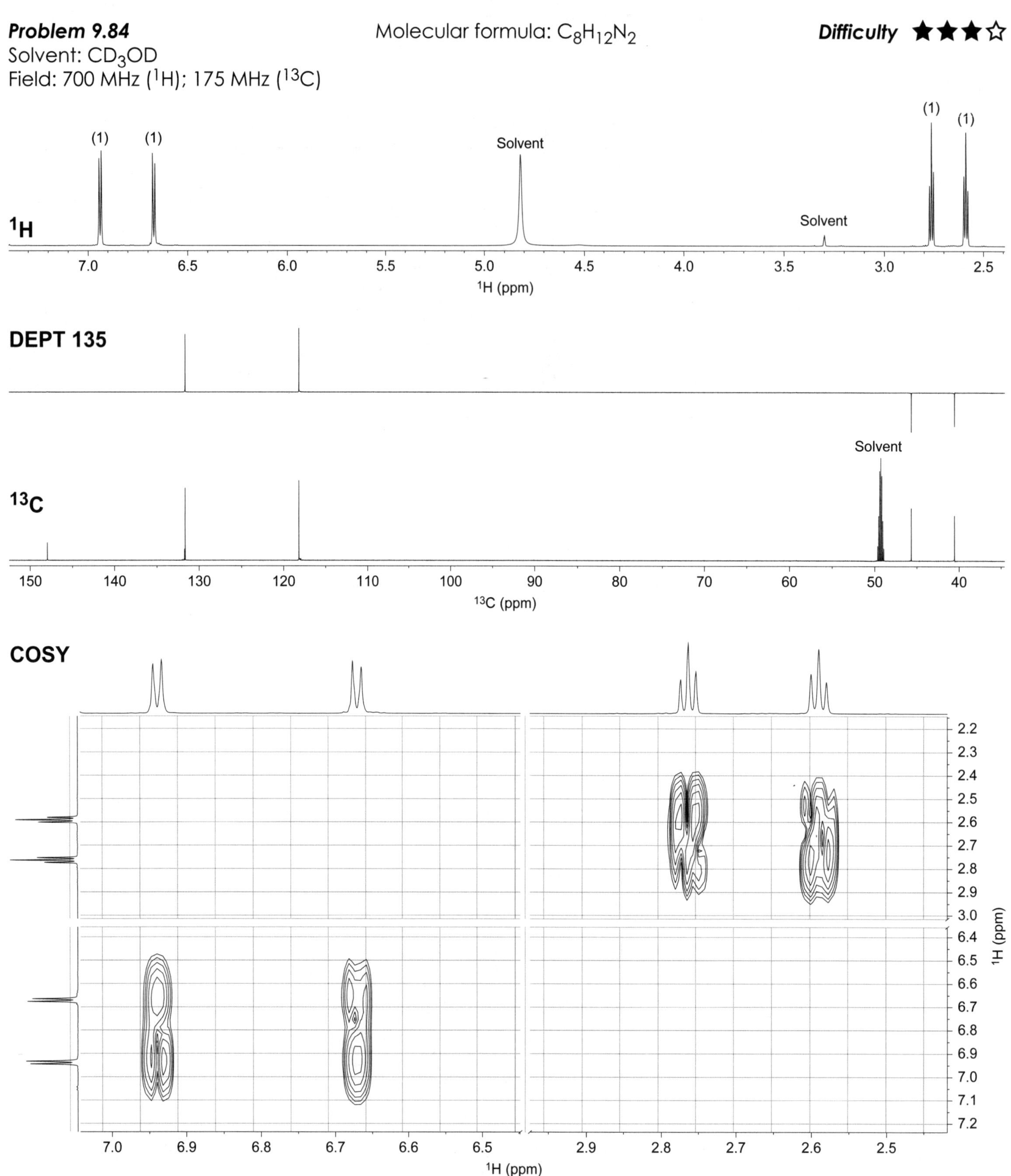

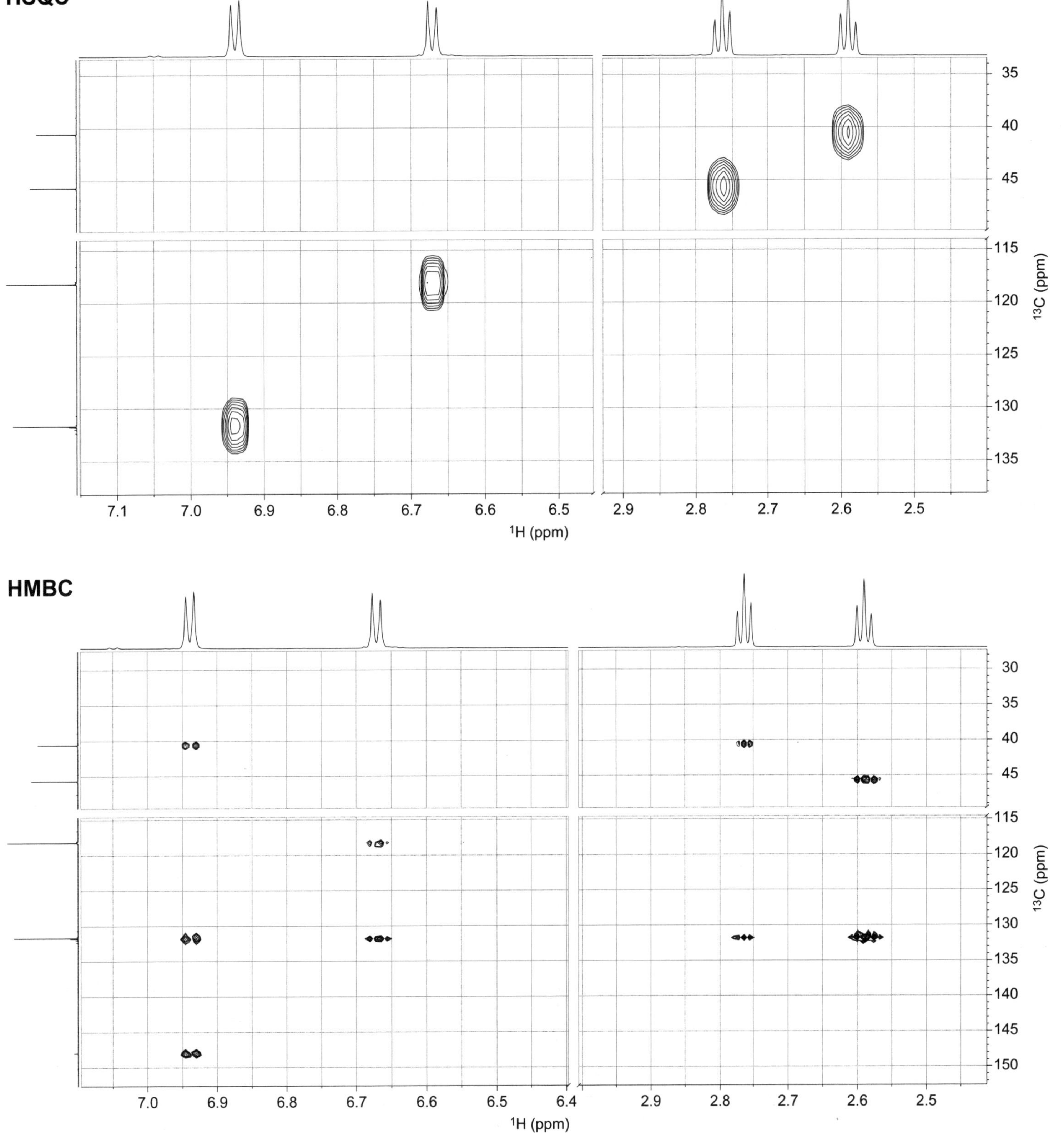
HSQC
13C (ppm)
1H (ppm)
HMBC
13C (ppm)
1H (ppm)

Problem 9.85 Molecular formula: $C_8H_{10}N_2O_2$ **Difficulty** ★★★☆

Solvent: CD_3OD

Field: 700 MHz (^1H); 175 MHz (^{13}C)

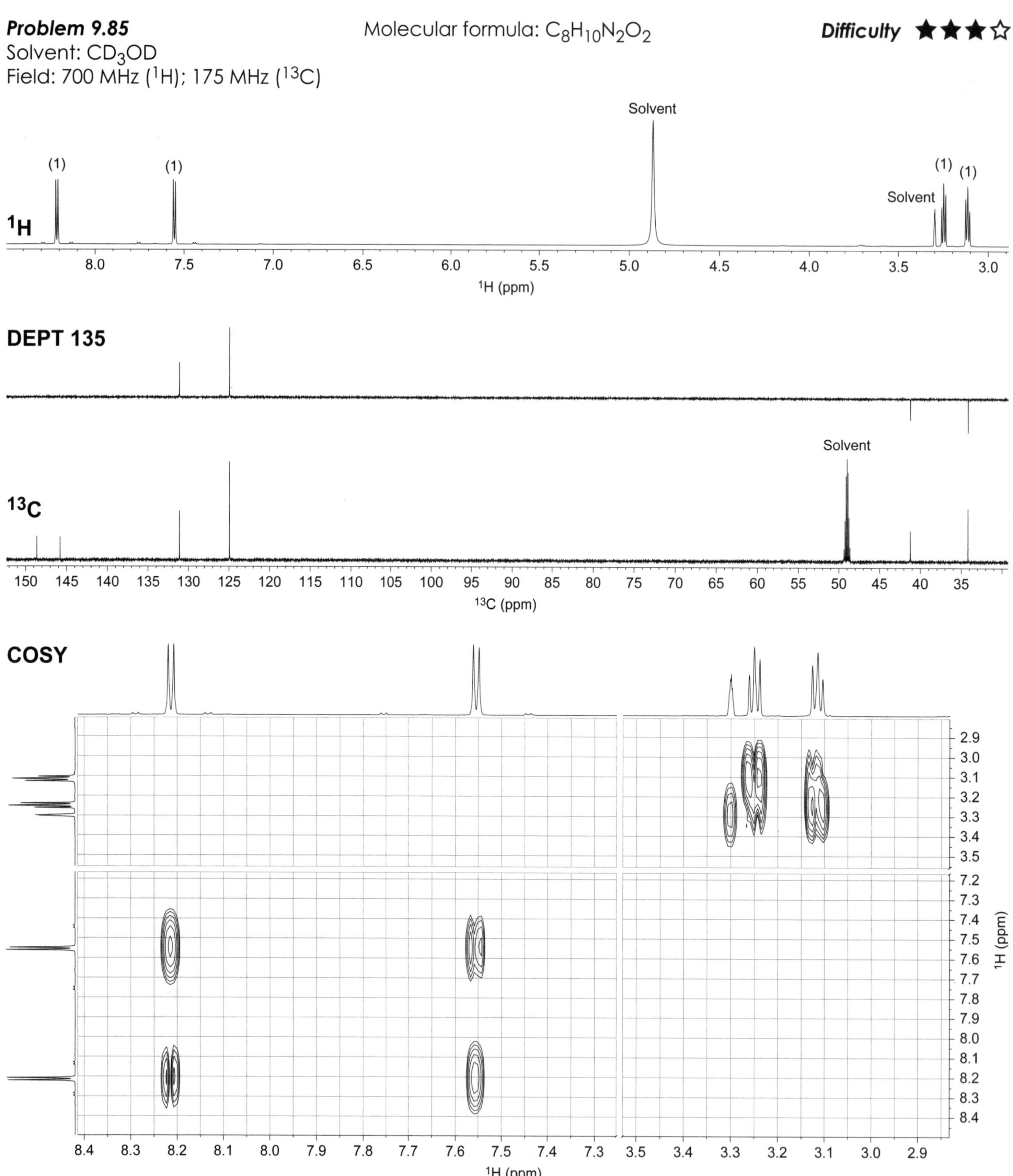

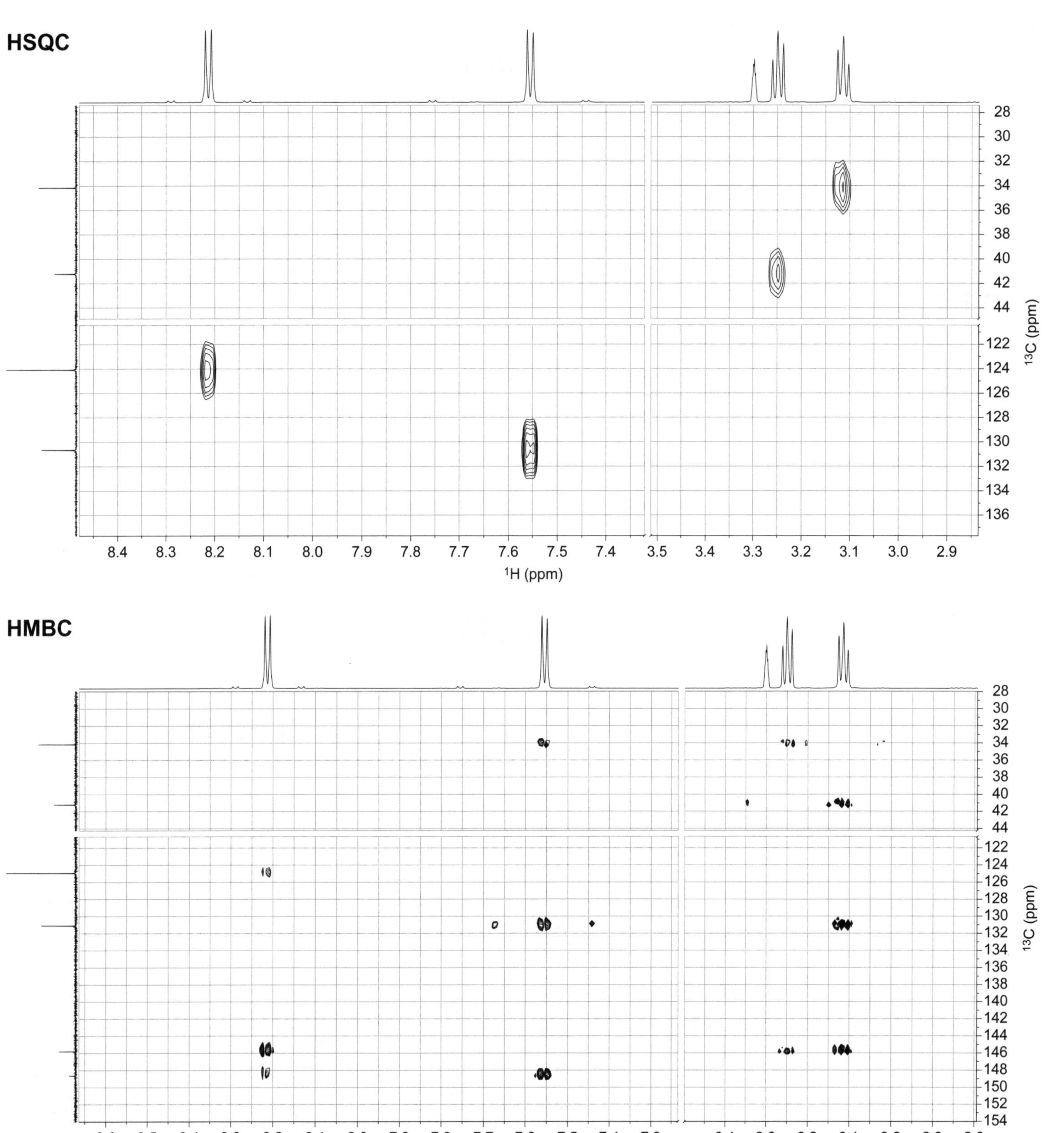
HSQC
28
30
32
34
36
38
40
42
44
122
124
126
128
130
132
134
136
^{13}C (ppm)
8.4 8.3 8.2 8.1 8.0 7.9 7.8 7.7 7.6 7.5 7.4 3.5 3.4 3.3 3.2 3.1 3.0 2.9
^{1}H (ppm)
HMBC
28
30
32
34
36
38
40
42
44
122
124
126
128
130
132
134
136
138
140
142
144
146
148
150
152
154
^{13}C (ppm)
8.6 8.5 8.4 8.3 8.2 8.1 8.0 7.9 7.8 7.7 7.6 7.5 7.4 7.3 3.4 3.3 3.2 3.1 3.0 2.9 2.8
^{1}H (ppm)

Problem 9.86
Solvent: D_2O
Field: 500 MHz (^1H); 125 MHz (^{13}C)

Molecular formula: $C_8H_6O_4$

Difficulty ★★★☆

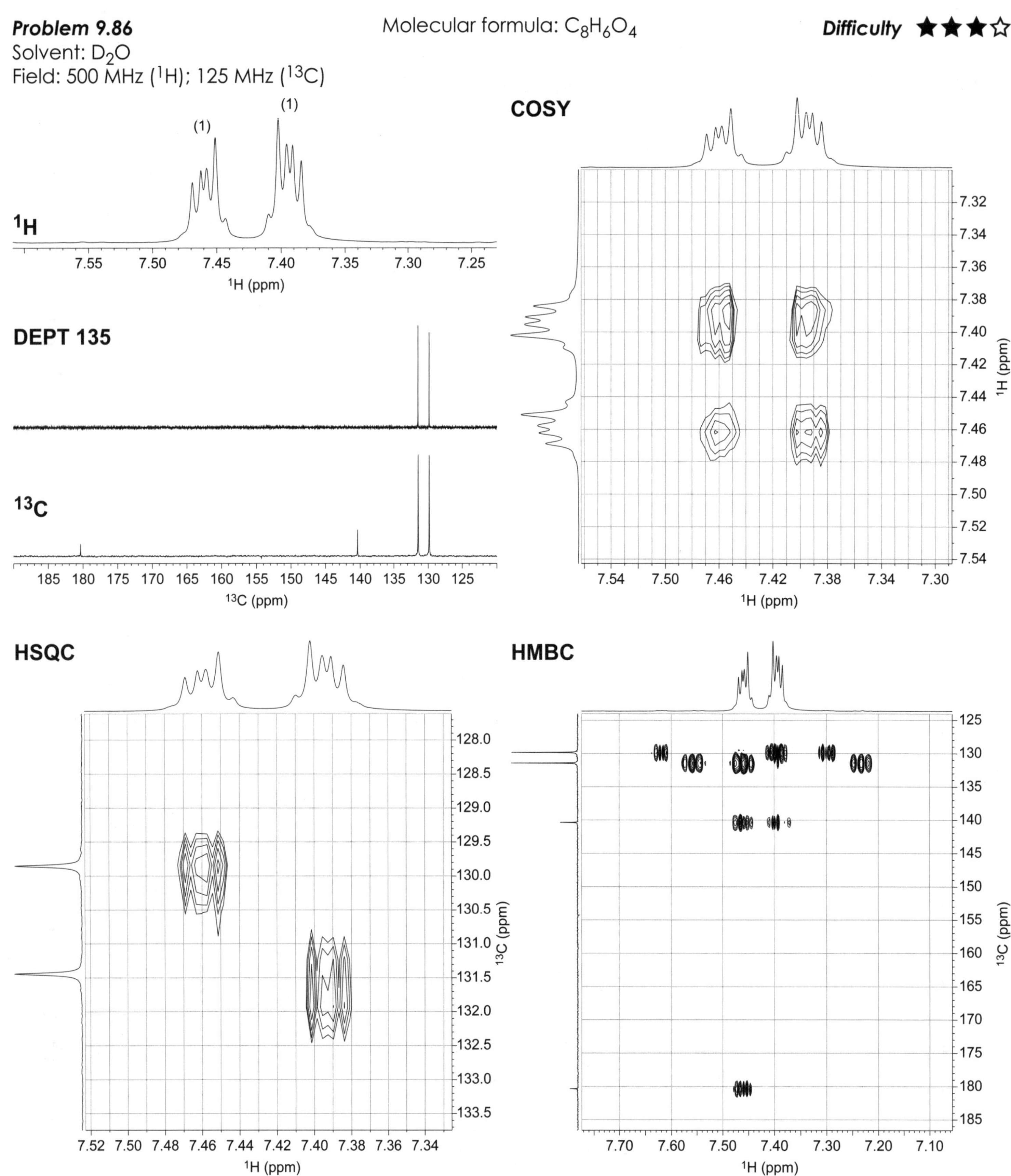

Problem 9.87
Solvent: D_2O
Field: 400 MHz (^1H); 100 MHz (^{13}C)

Molecular formula: $C_8H_8O_4$

Difficulty ★★★☆

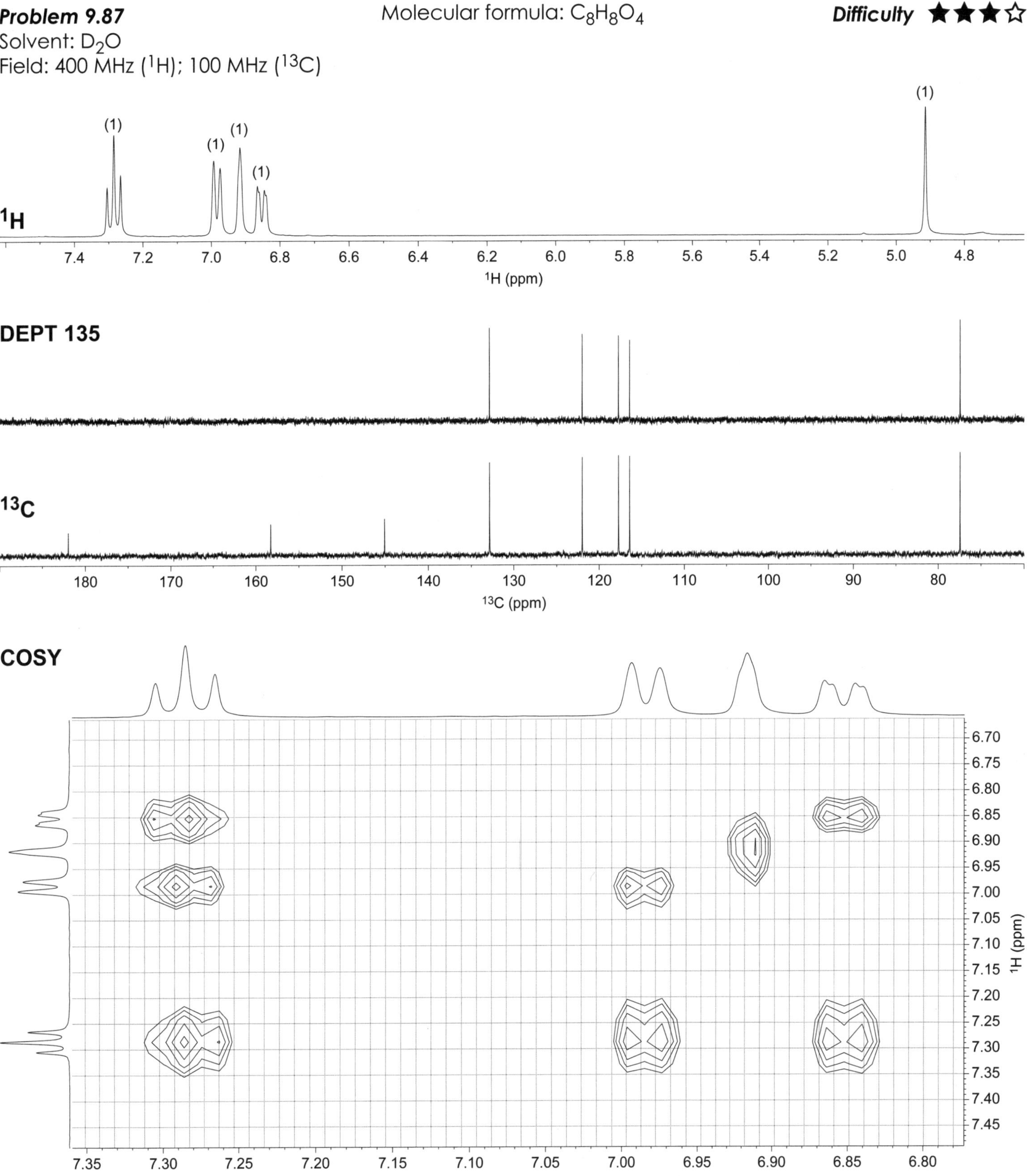

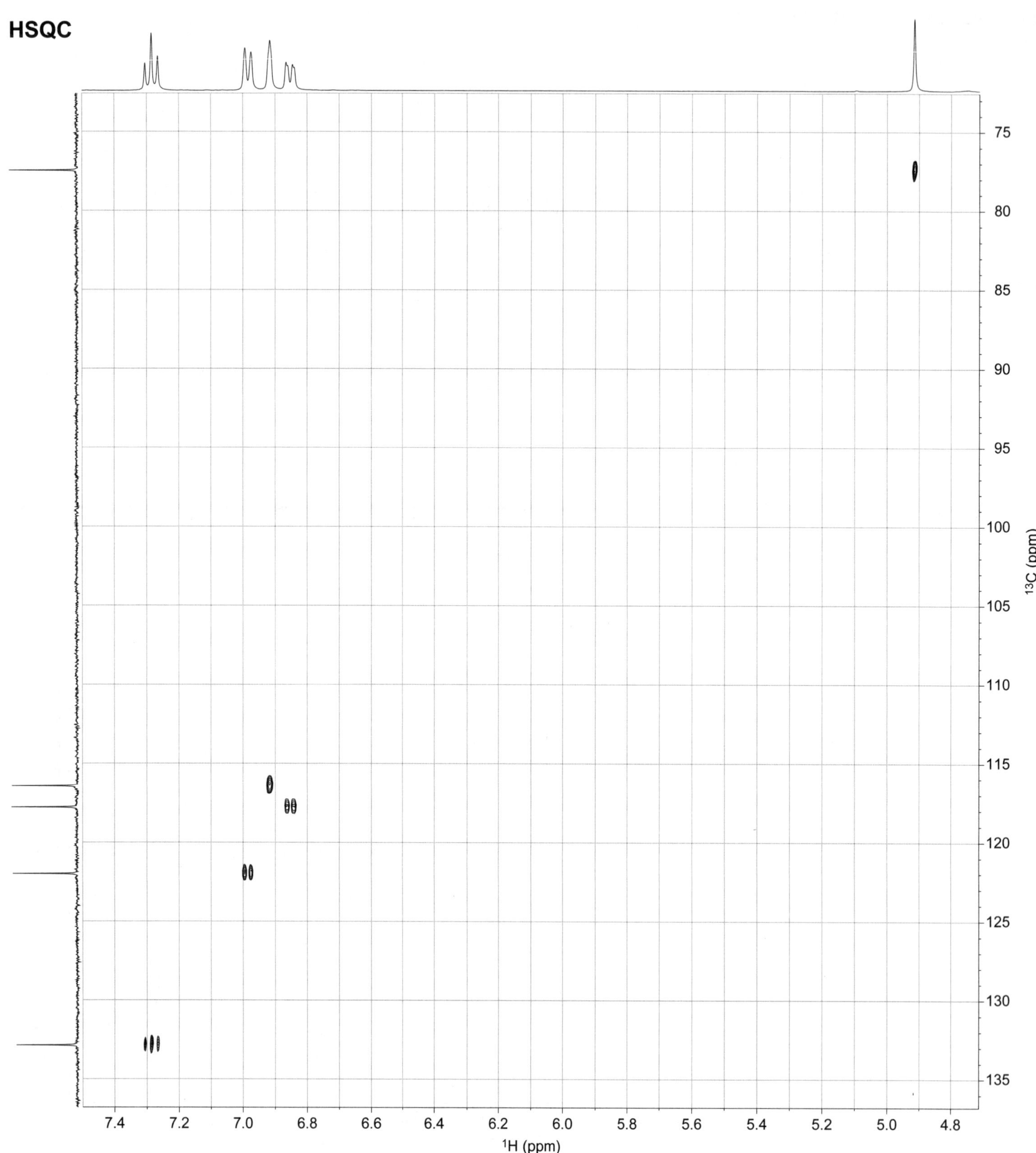
HSQC
75
80
85
90
95
100
105
110
115
120
125
130
135
^{13}C (ppm)
7.4
7.2
7.0
6.8
6.6
6.4
6.2
6.0
5.8
5.6
5.4
5.2
5.0
4.8
^{1}H (ppm)

HMBC

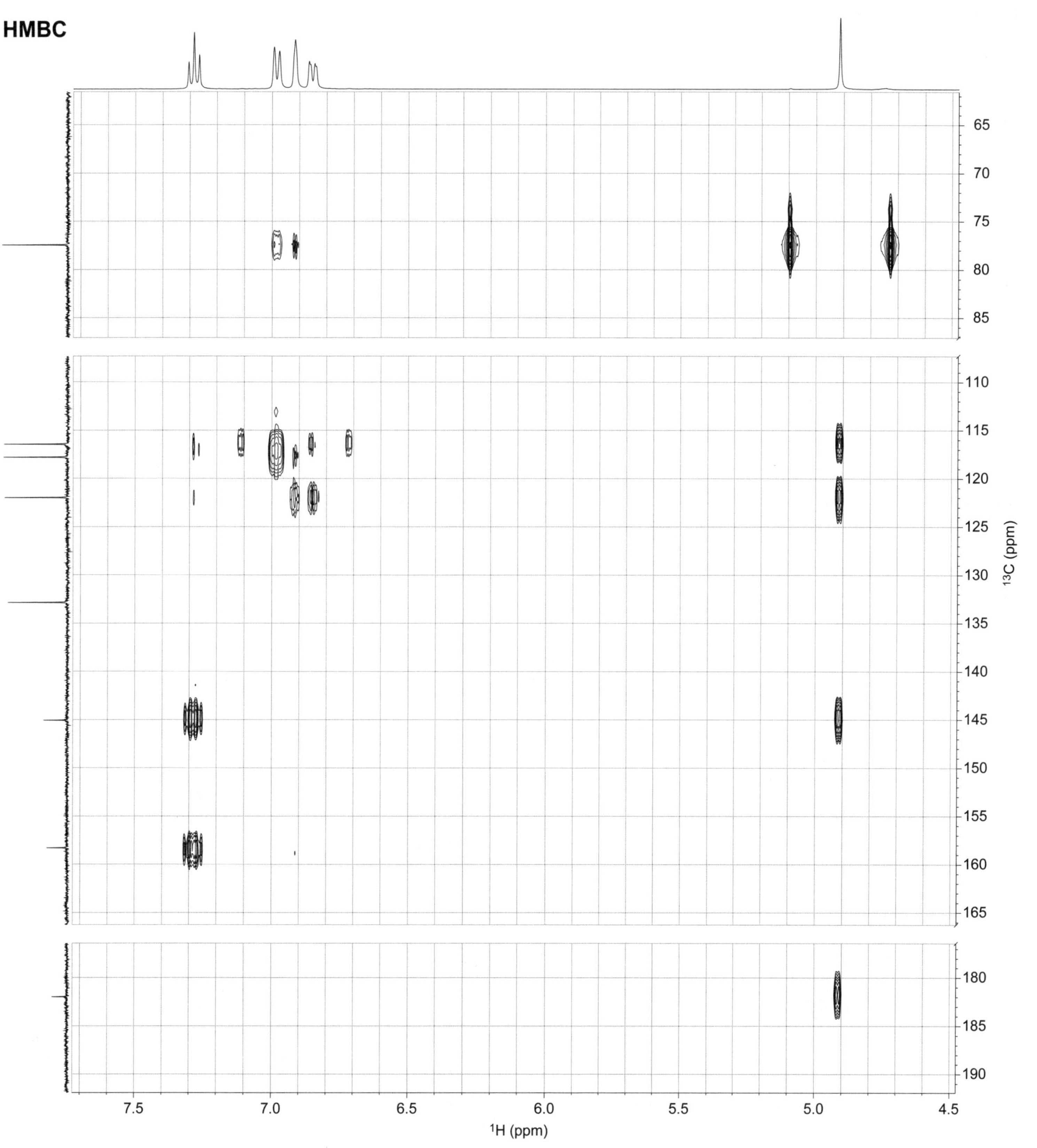

Problem 9.88 Molecular formula: $C_8H_{11}NO_3$ ***Difficulty*** ★★★★
Solvent: CD_3OD
Field: 700 MHz (^{1}H); 175 MHz (^{13}C)

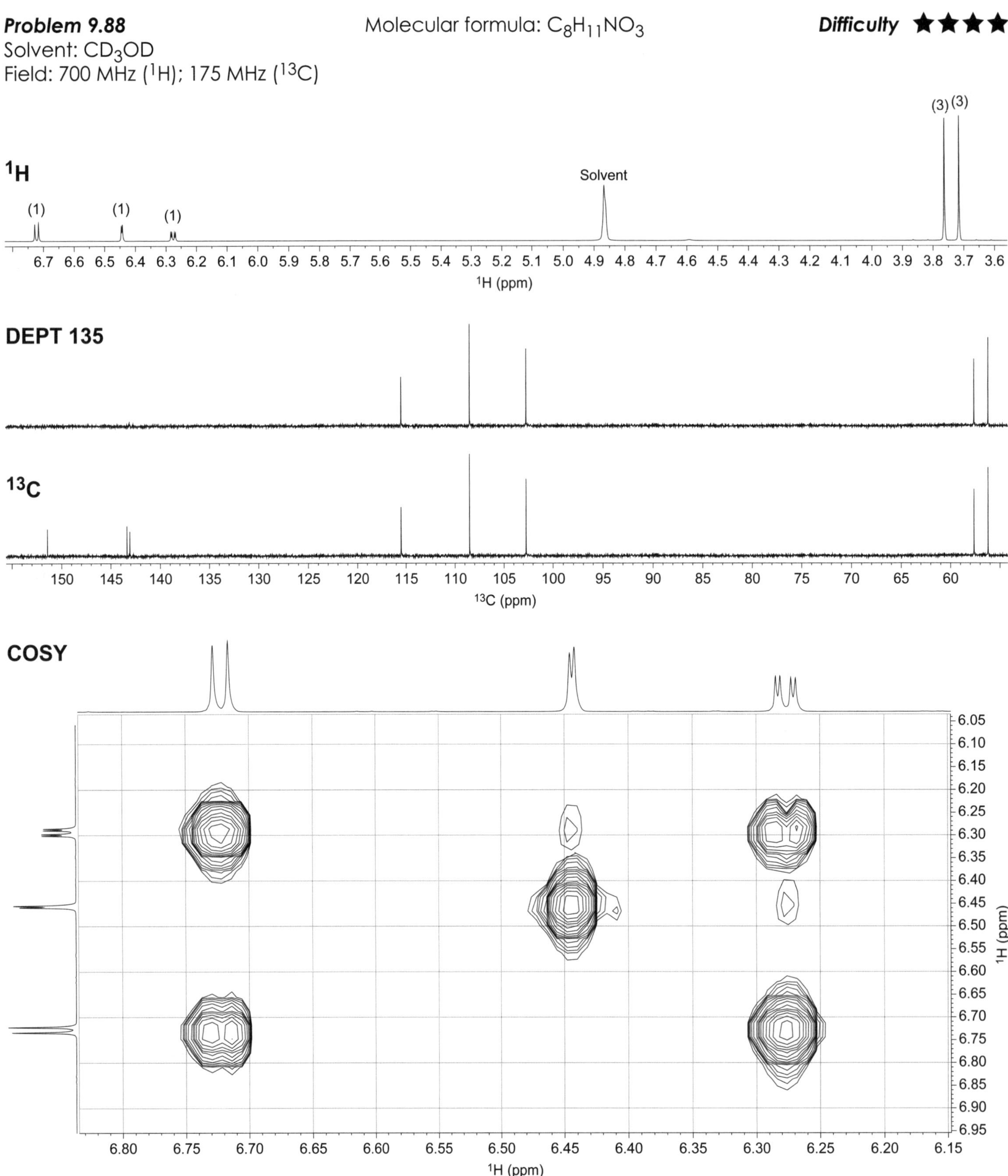

HSQC

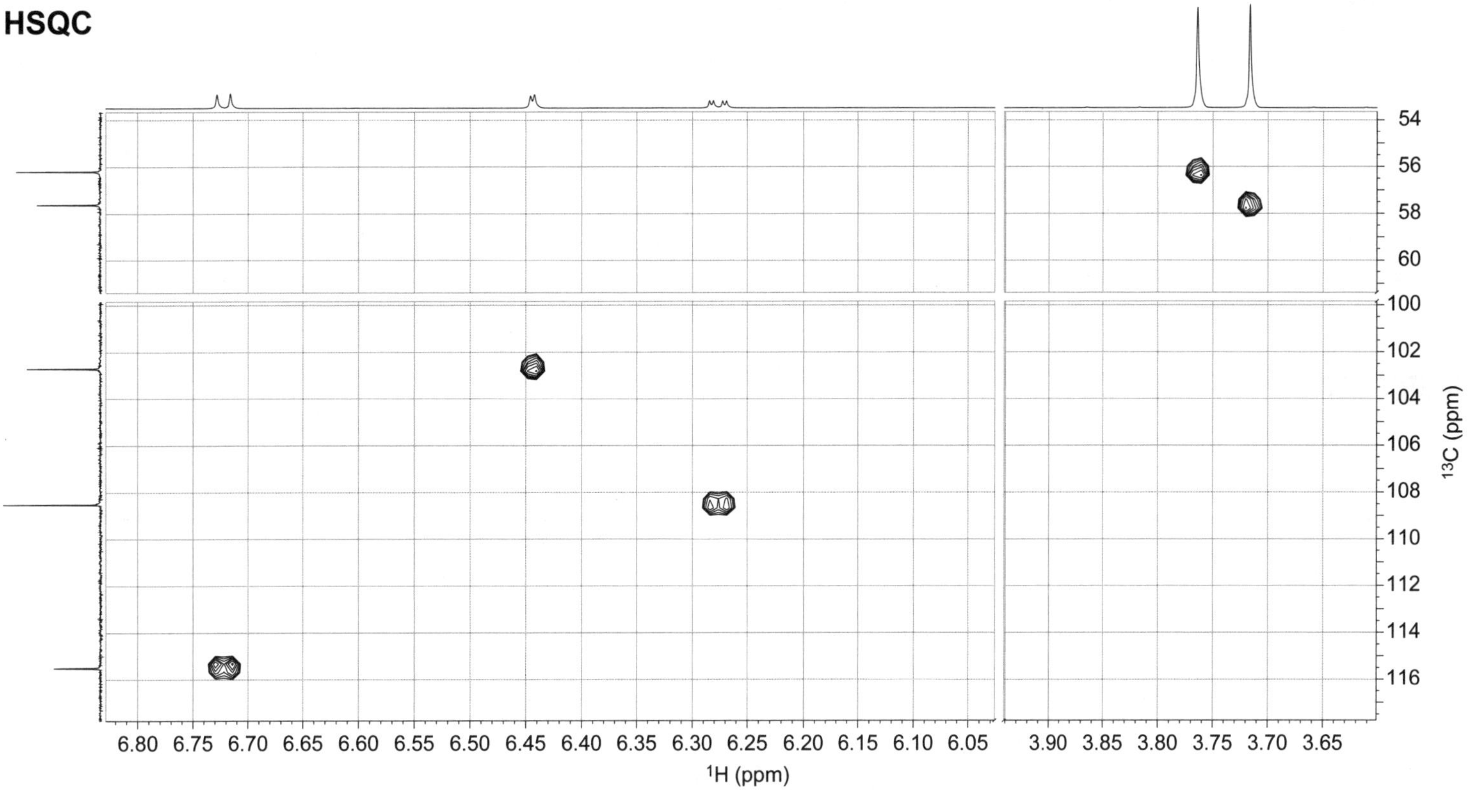

54
56
58
60
100
102
104
106
108
110
112
114
116
^{13}C (ppm)
6.80 6.75 6.70 6.65 6.60 6.55 6.50 6.45 6.40 6.35 6.30 6.25 6.20 6.15 6.10 6.05
3.90 3.85 3.80 3.75 3.70 3.65
^{1}H (ppm)

54
58
100
104
108
112
116
120
138
142
146
150
154
^{13}C (ppm)
7.0 6.9 6.8 6.7 6.6 6.5 6.4 6.3 6.2 6.1
3.9 3.8 3.7 3.6 3.5
^{1}H (ppm)

HMBC

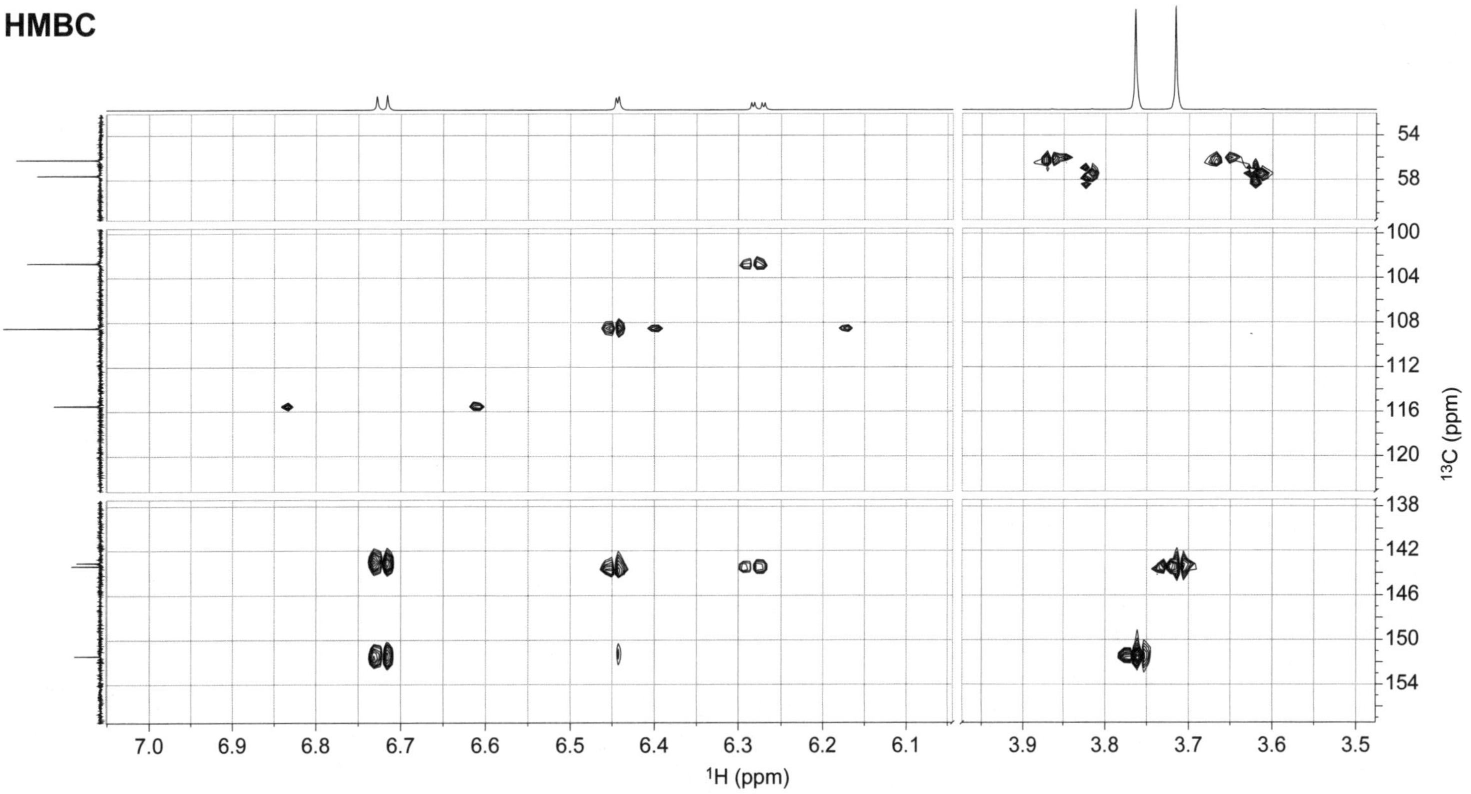

Problem 9.89 Molecular formula: $C_9H_8O_3$ ***Difficulty*** ★★★★

Solvent: D_2O

Field: 500 MHz (^1H); 125 MHz (^{13}C)

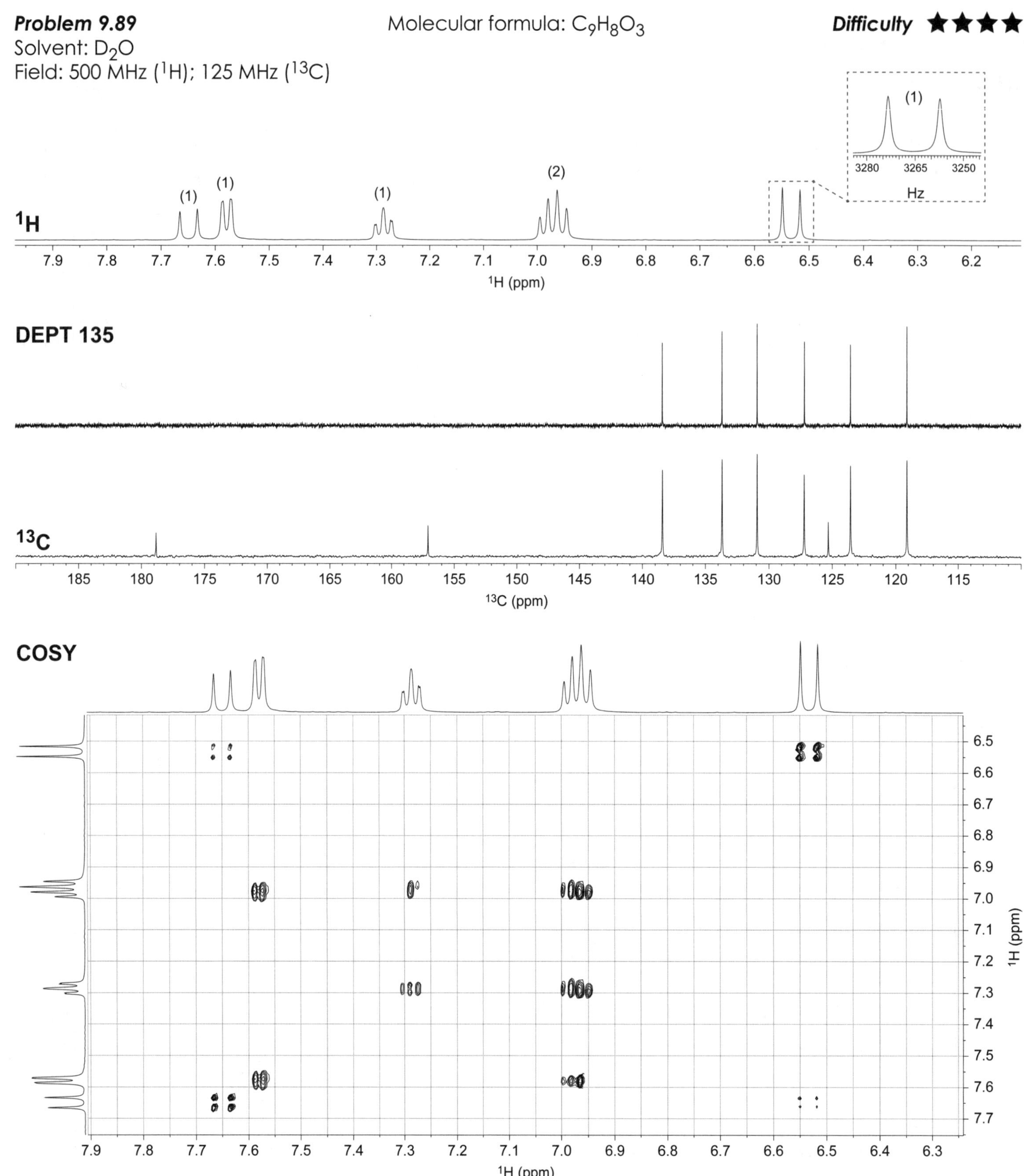

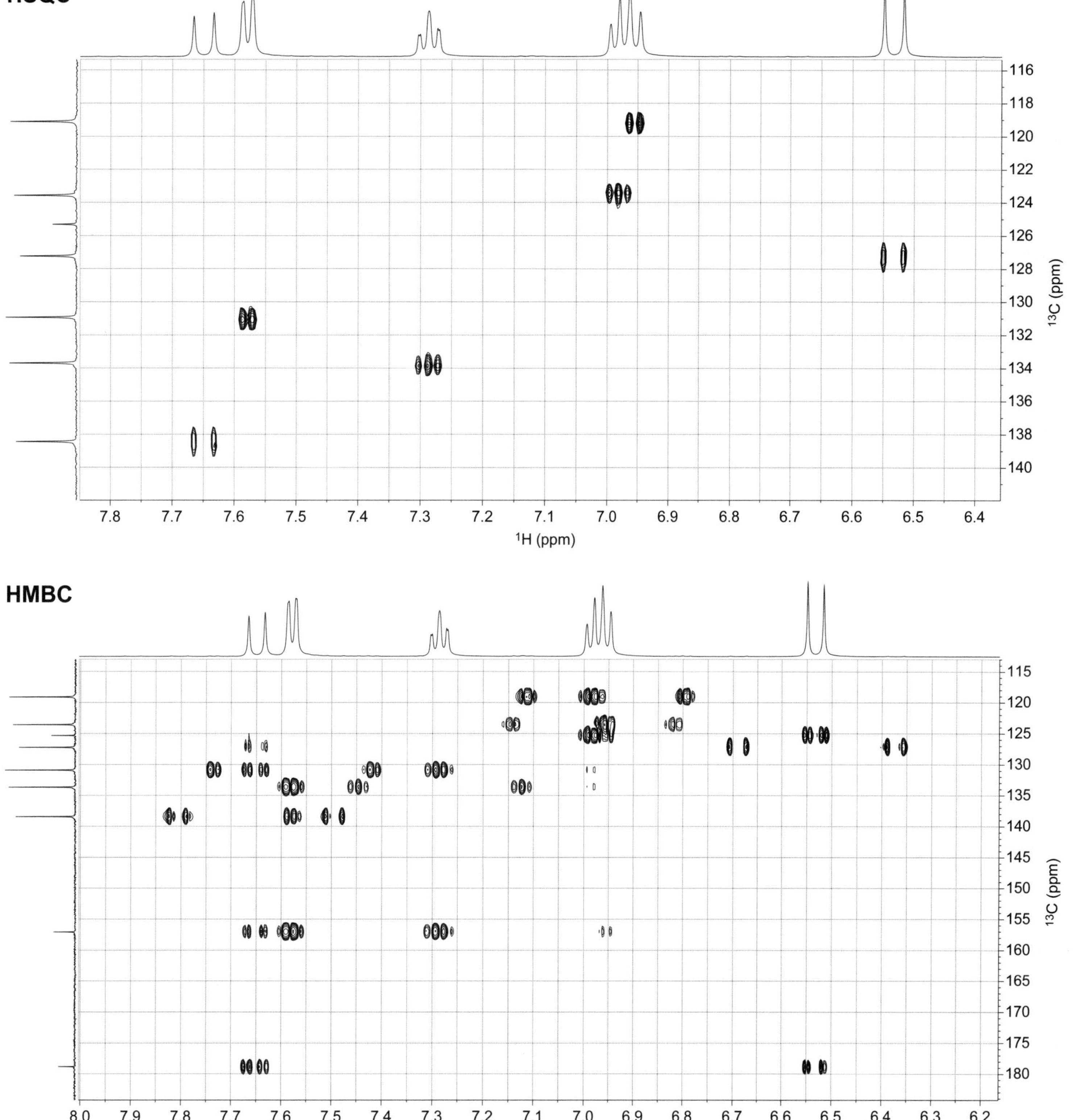
HSQC
116
118
120
122
124
126
128
130
132
134
136
138
140
^{13}C (ppm)
7.8
7.7
7.6
7.5
7.4
7.3
7.2
7.1
7.0
6.9
6.8
6.7
6.6
6.5
6.4
^{1}H (ppm)
HMBC
115
120
125
130
135
140
145
150
155
160
165
170
175
180
^{13}C (ppm)
8.0
7.9
7.8
7.7
7.6
7.5
7.4
7.3
7.2
7.1
7.0
6.9
6.8
6.7
6.6
6.5
6.4
6.3
6.2
^{1}H (ppm)

Problem 9.90 Molecular formula: $C_8H_8O_4$ **Difficulty** ★★★★

Solvent: D_2O

Field: 500 MHz (^1H); 125 MHz (^{13}C)

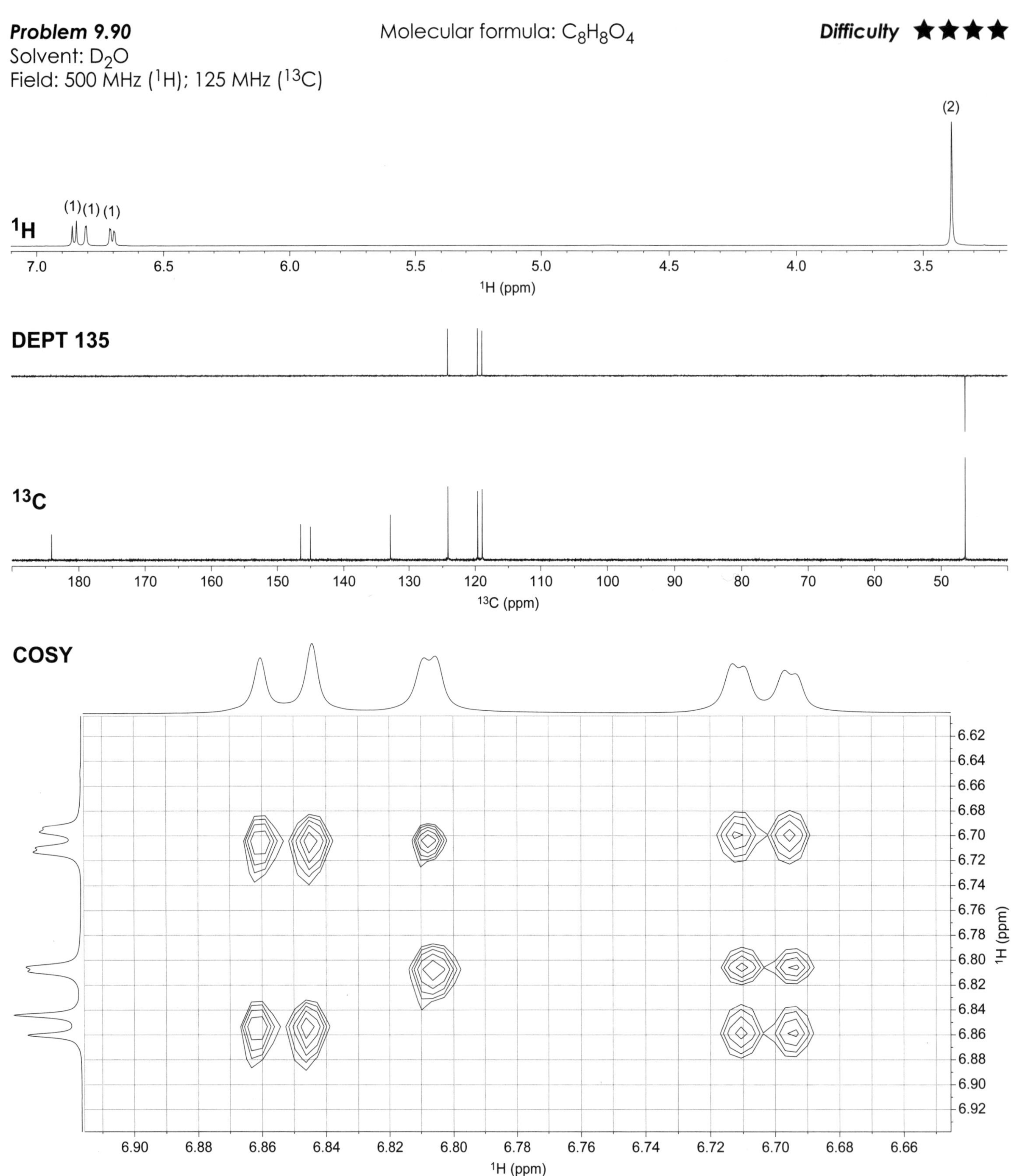

HSQC

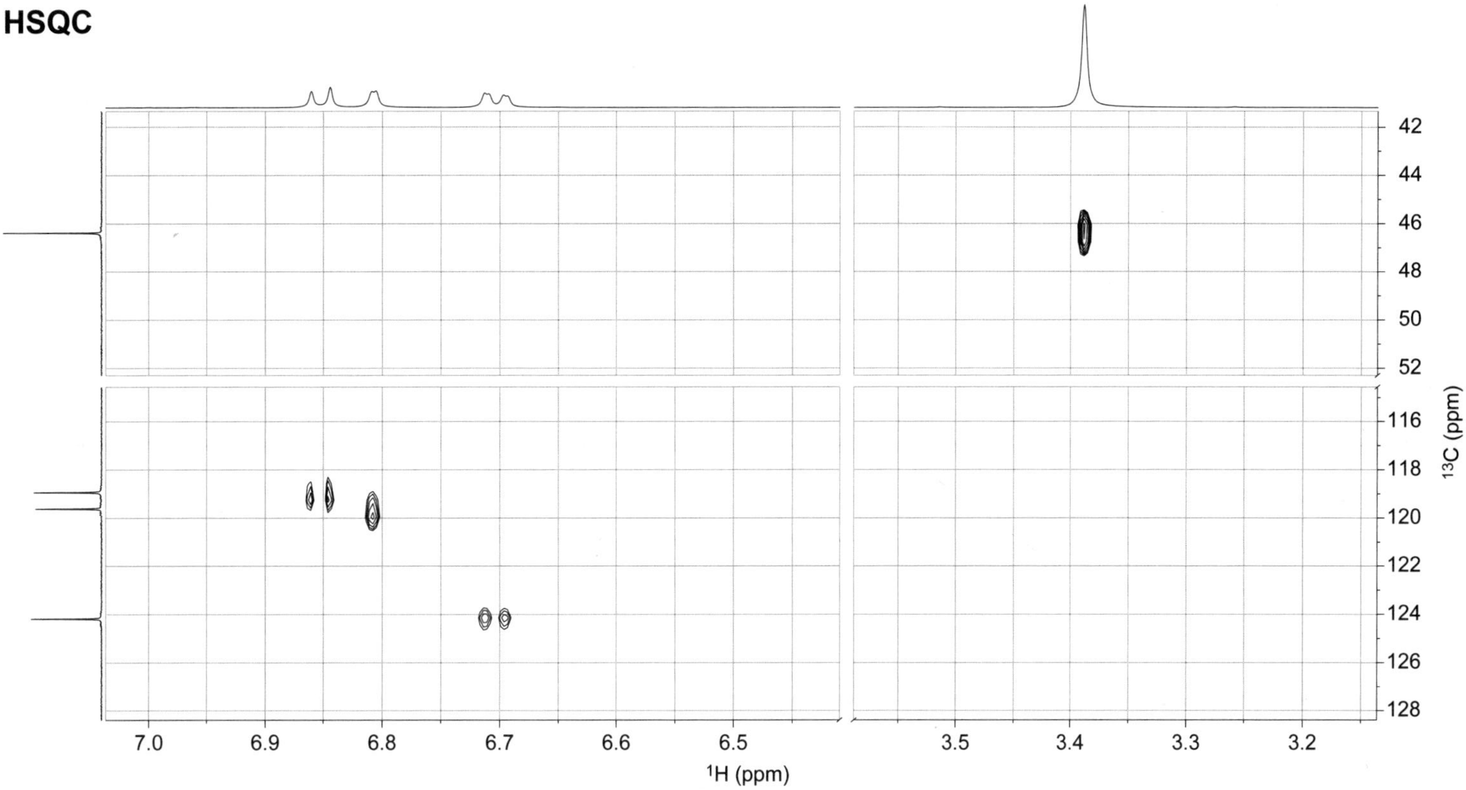
42
44
46
48
50
52
116
118
120
122
124
126
128
^{13}C (ppm)
7.0
6.9
6.8
6.7
6.6
6.5
3.5
3.4
3.3
3.2
^{1}H (ppm)

HMBC

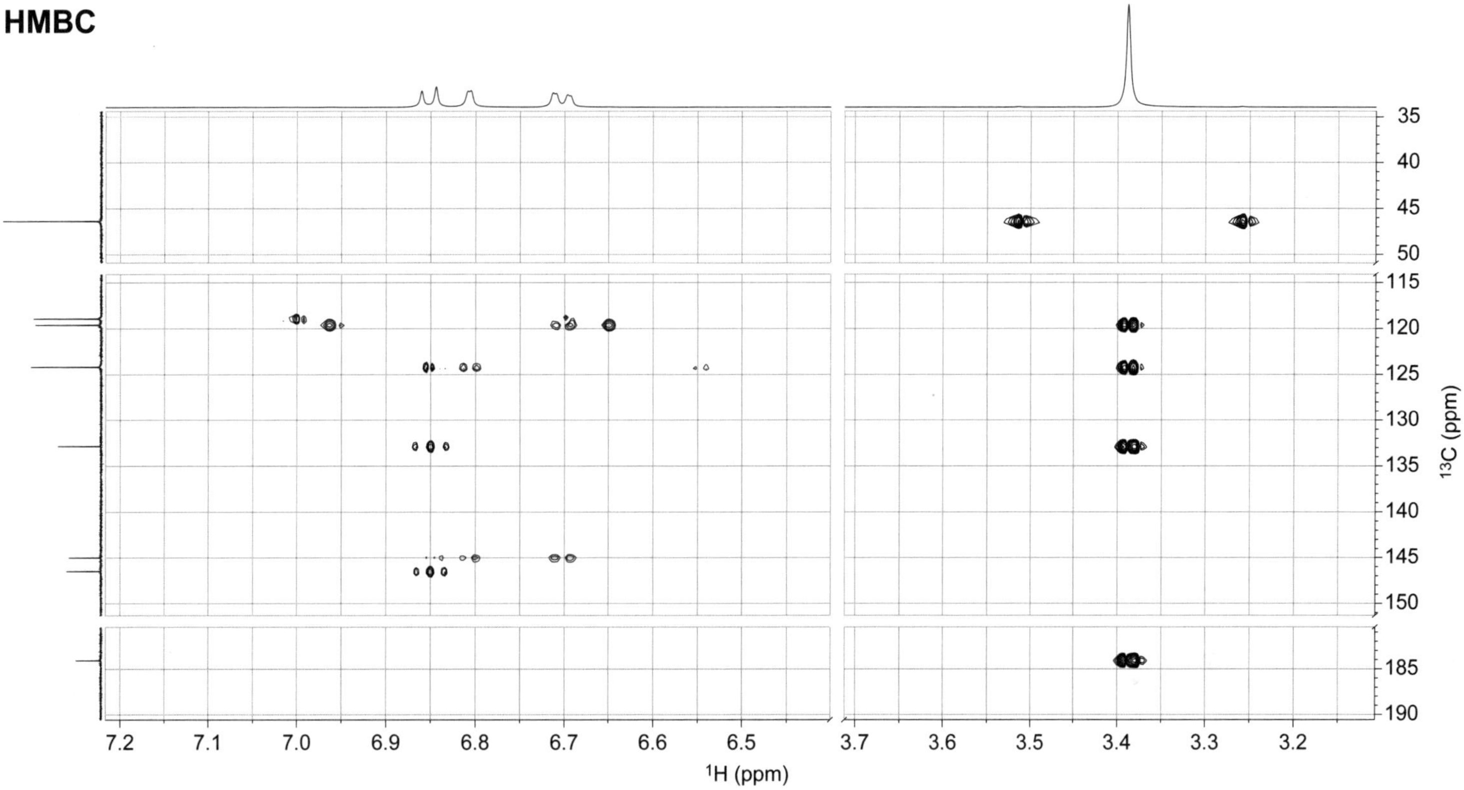
35
40
45
50
115
120
125
130
135
140
145
150
185
190
^{13}C (ppm)
7.2
7.1
7.0
6.9
6.8
6.7
6.6
6.5
3.7
3.6
3.5
3.4
3.3
3.2
^{1}H (ppm)

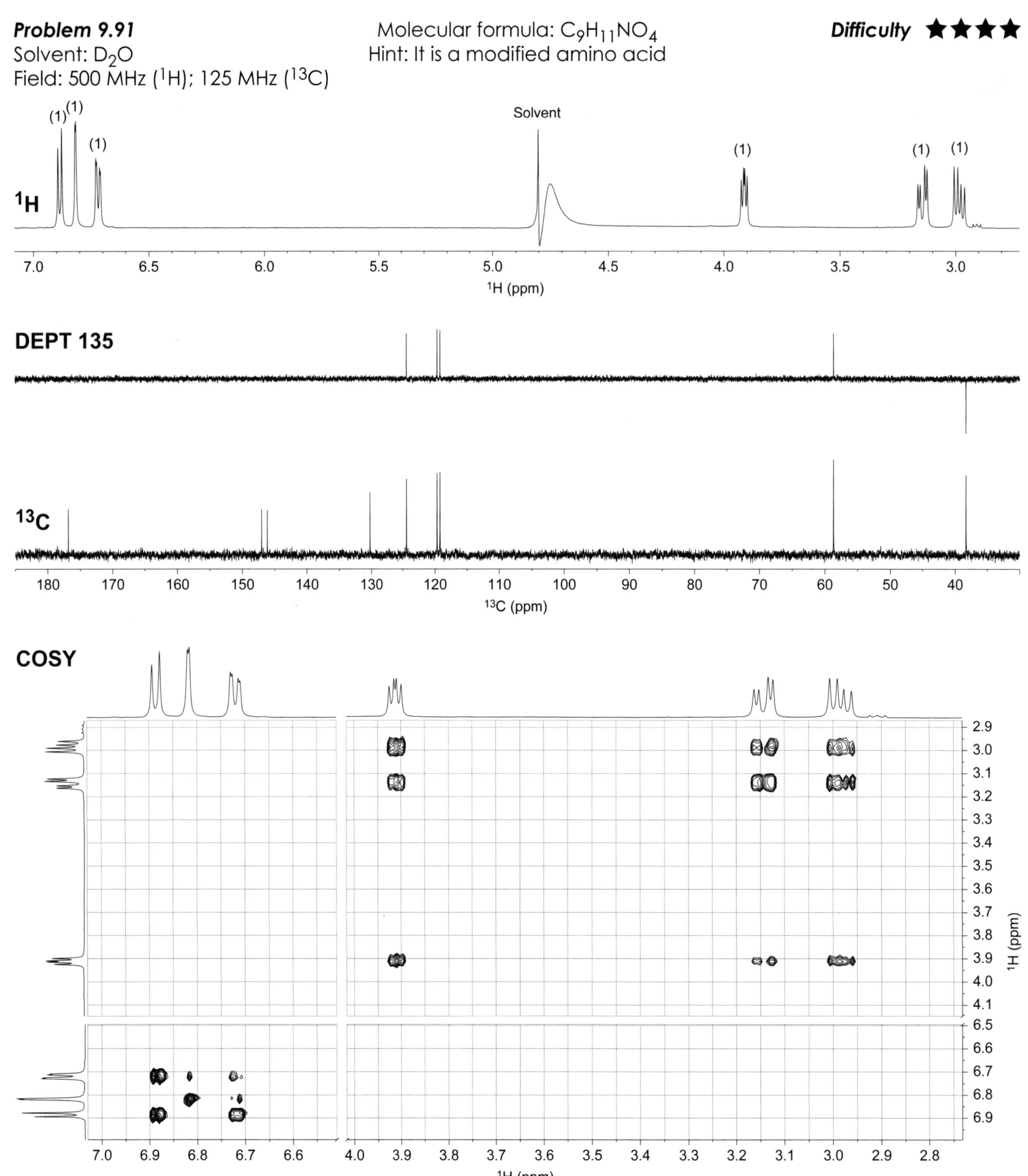
Problem 9.91
Solvent: D2O
Field: 500 MHz (1H); 125 MHz (13C)
Molecular formula: C9H11NO4
Hint: It is a modified amino acid
Difficulty ★★★★
1H
Solvent
(1) (1) (1) (1) (1) (1)
1H (ppm)
DEPT 135
13C
13C (ppm)
COSY
1H (ppm)

HSQC

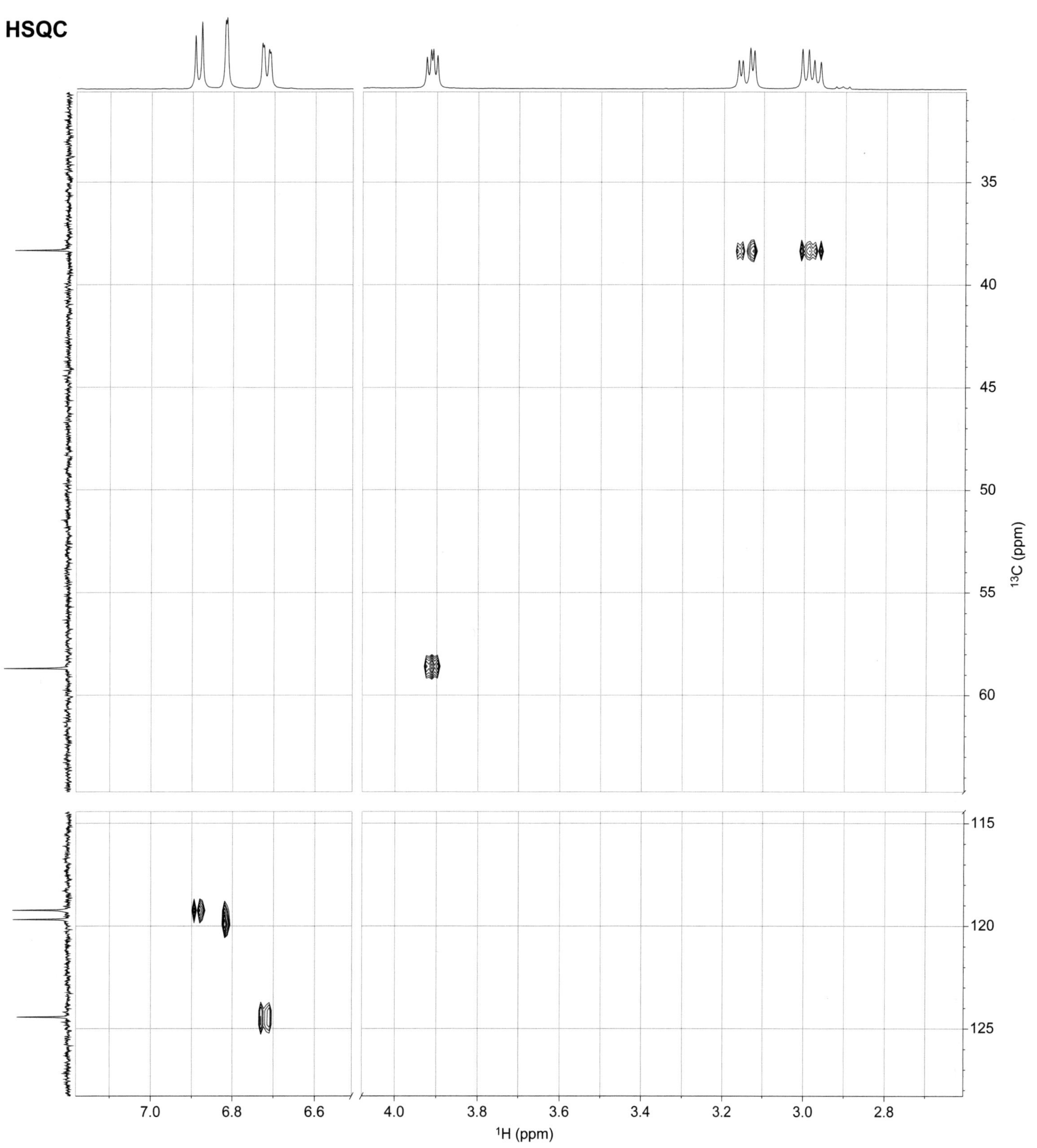

HMBC

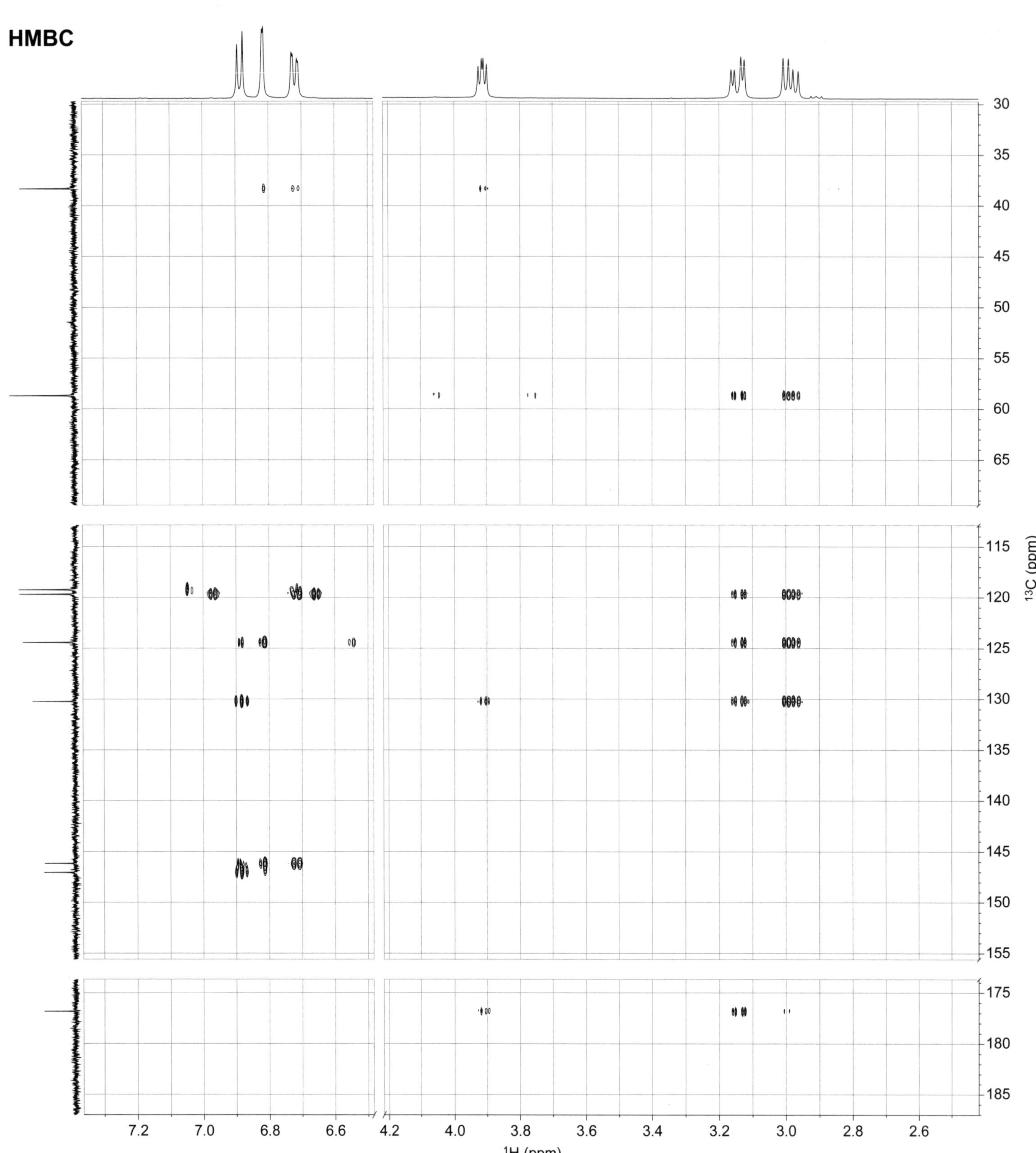

30
35
40
45
50
55
60
65
115
120
125
130
135
140
145
150
155
175
180
185
^{13}C (ppm)
7.2
7.0
6.8
6.6
4.2
4.0
3.8
3.6
3.4
3.2
3.0
2.8
2.6
^{1}H (ppm)

Problem 9.92 Molecular formula: $C_{10}H_{10}O_3$ ***Difficulty*** ★★★★

Solvent: D_2O

Field: 500 MHz (^1H); 125 MHz (^{13}C)

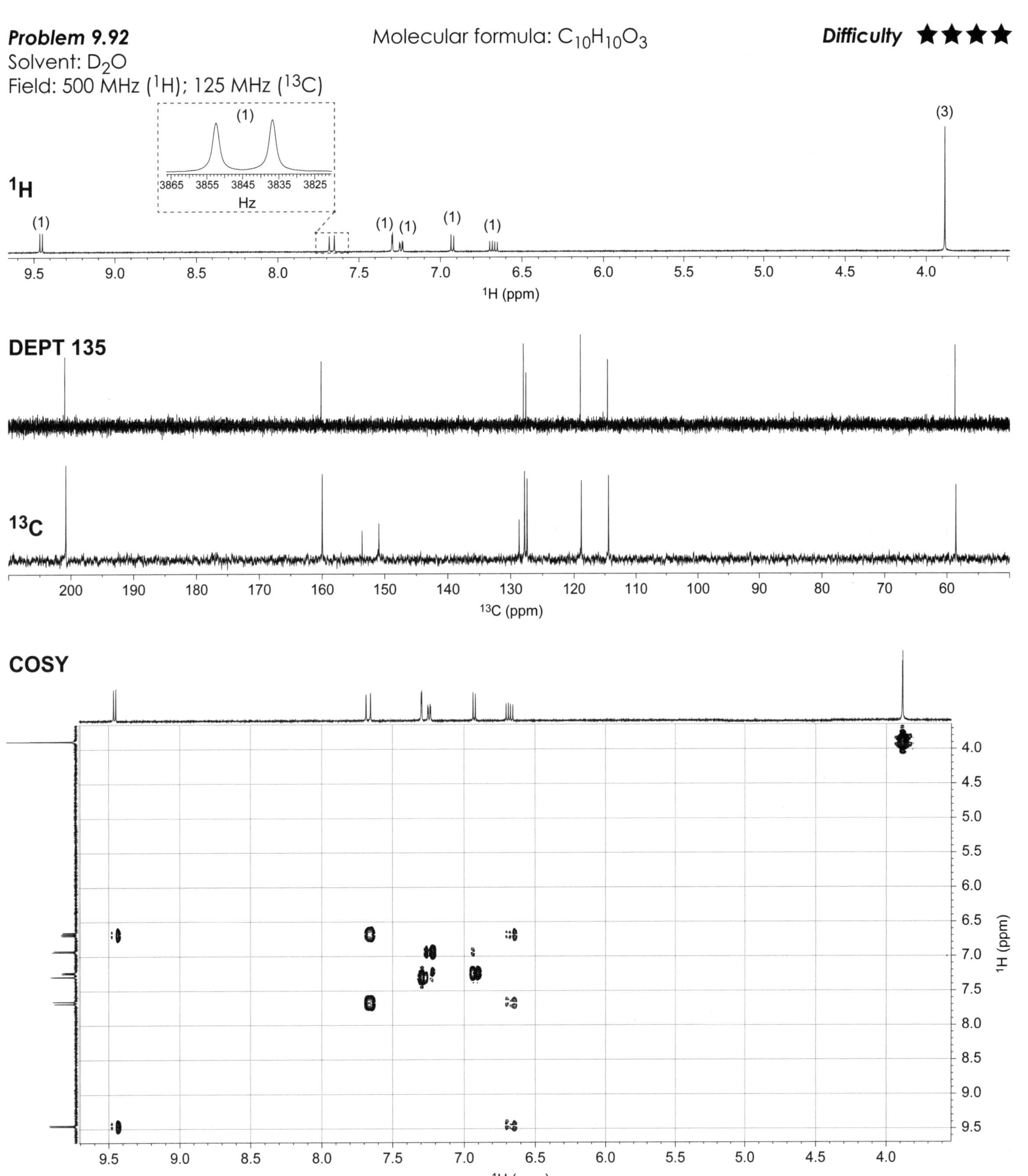

HSQC

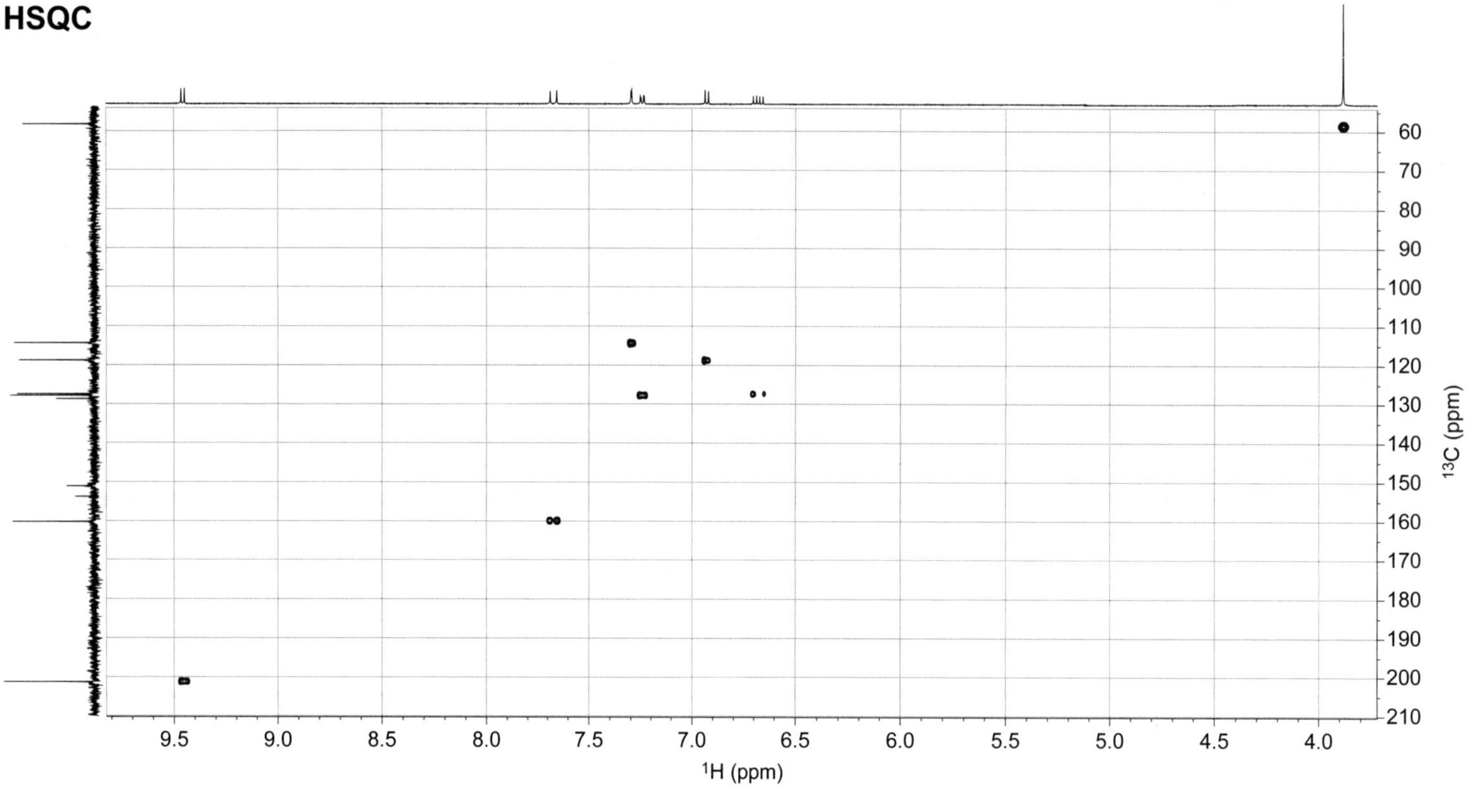

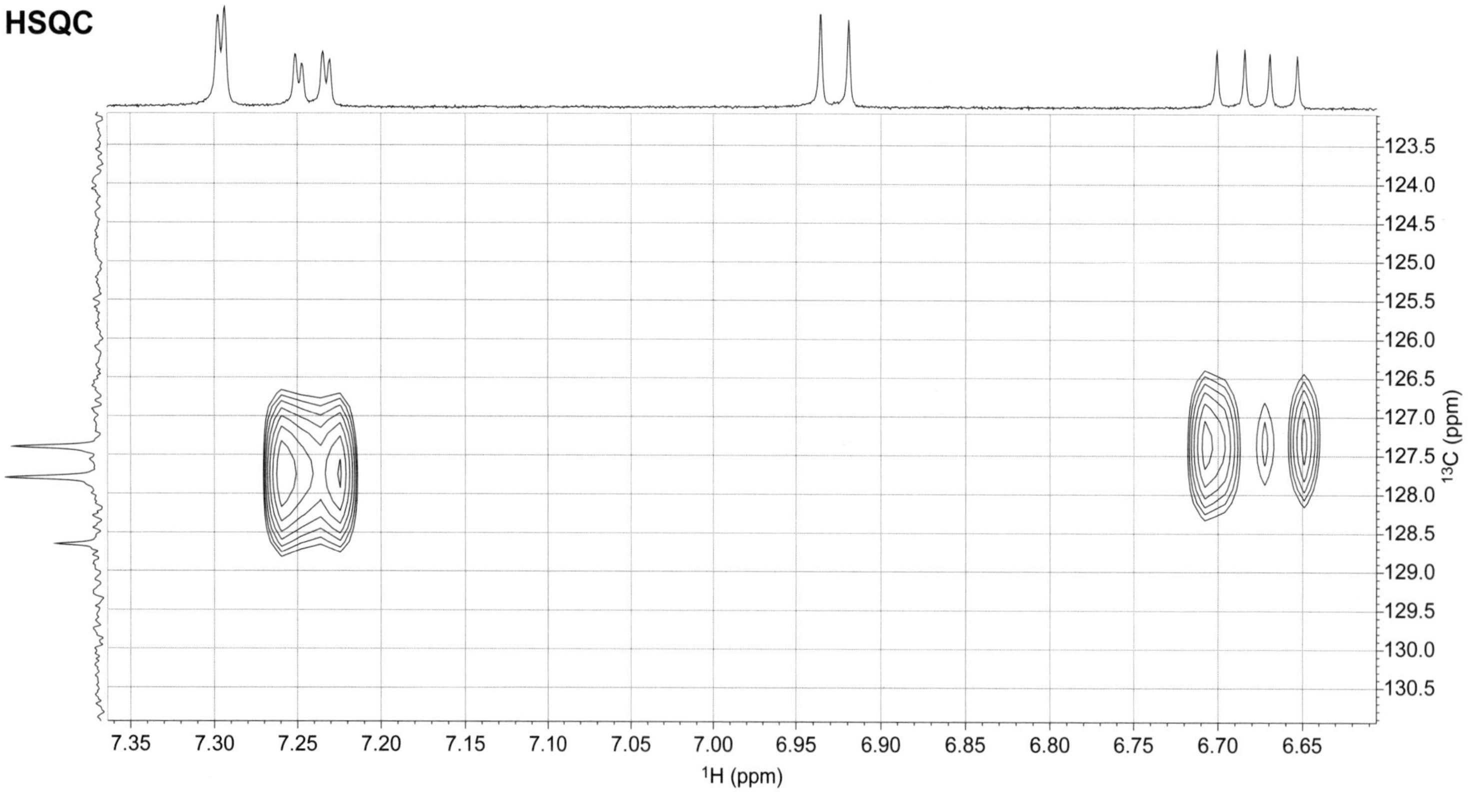

HSQC

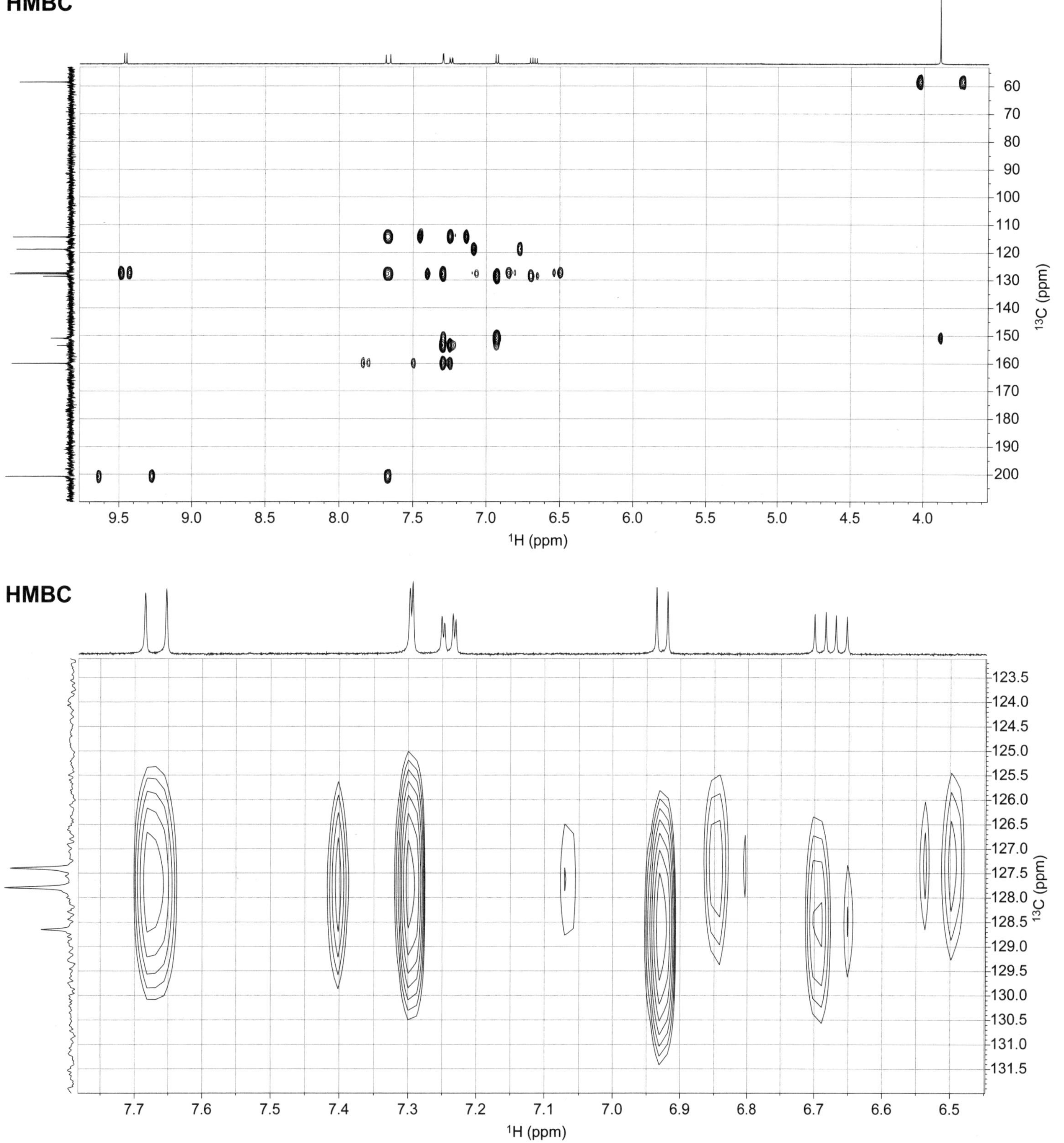

HMBC
60
70
80
90
100
110
120
130
140
150
160
170
180
190
200
^{13}C (ppm)
9.5
9.0
8.5
8.0
7.5
7.0
6.5
6.0
5.5
5.0
4.5
4.0
^{1}H (ppm)
HMBC
123.5
124.0
124.5
125.0
125.5
126.0
126.5
127.0
127.5
128.0
128.5
129.0
129.5
130.0
130.5
131.0
131.5
^{13}C (ppm)
7.7
7.6
7.5
7.4
7.3
7.2
7.1
7.0
6.9
6.8
6.7
6.6
6.5
^{1}H (ppm)

Problem 9.93 Molecular formula: $C_8H_{11}NO_3$ ***Difficulty*** ★★★★

Solvent: D_2O

Field: 500 MHz (^1H); 125 MHz (^{13}C)

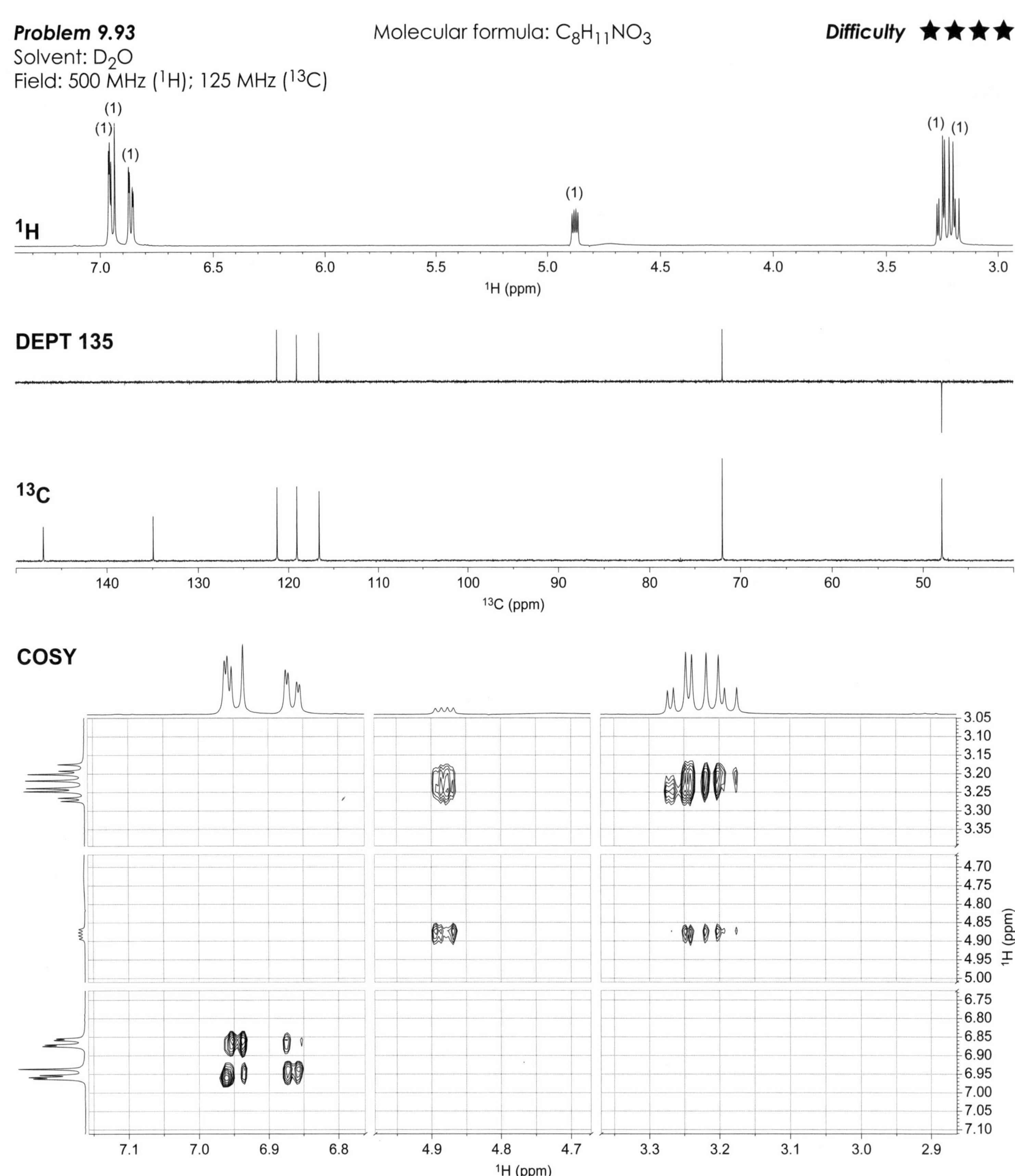

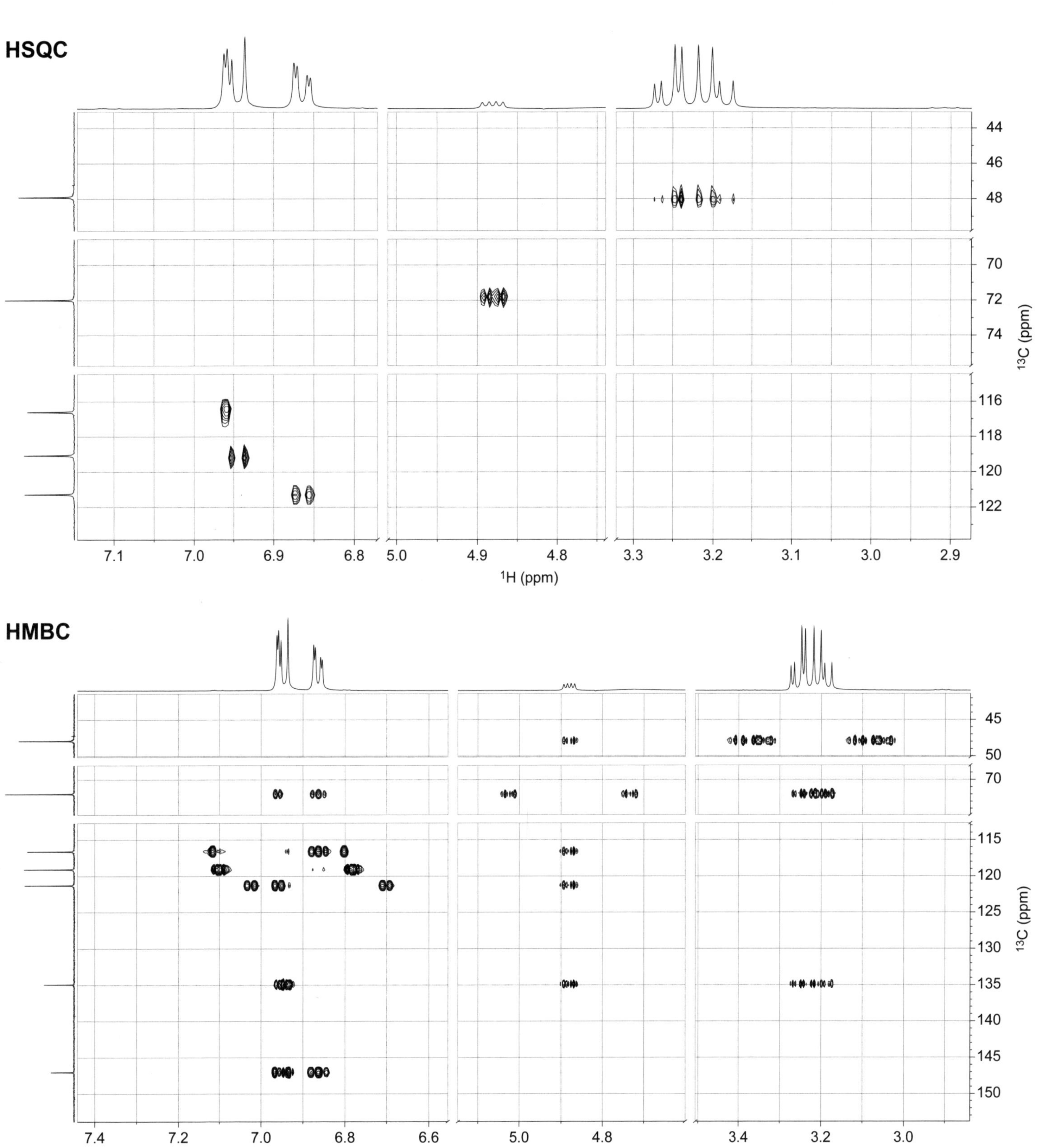
HSQC
44
46
48
70
72
74
116
118
120
122
13C (ppm)
7.1
7.0
6.9
6.8
5.0
4.9
4.8
3.3
3.2
3.1
3.0
2.9
1H (ppm)
HMBC
45
50
70
115
120
125
130
135
140
145
150
13C (ppm)
7.4
7.2
7.0
6.8
6.6
5.0
4.8
3.4
3.2
3.0
1H (ppm)

Problem 9.94
Solvent: CD_3-SO-CD_3
Field: 500 MHz (^{1}H); 125 MHz (^{13}C)

Molecular formula: $C_9H_{10}O_3$

Difficulty ★★★★

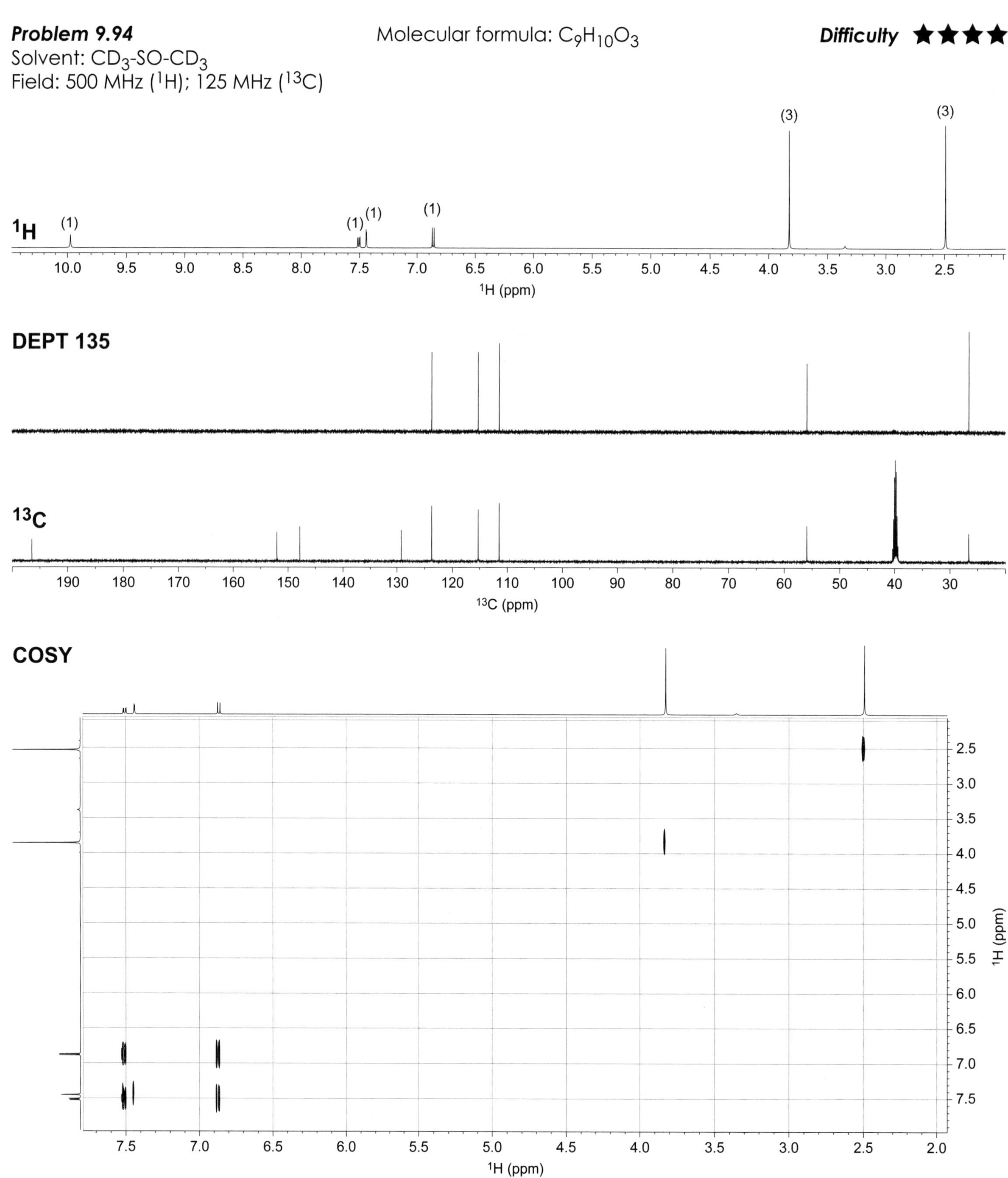

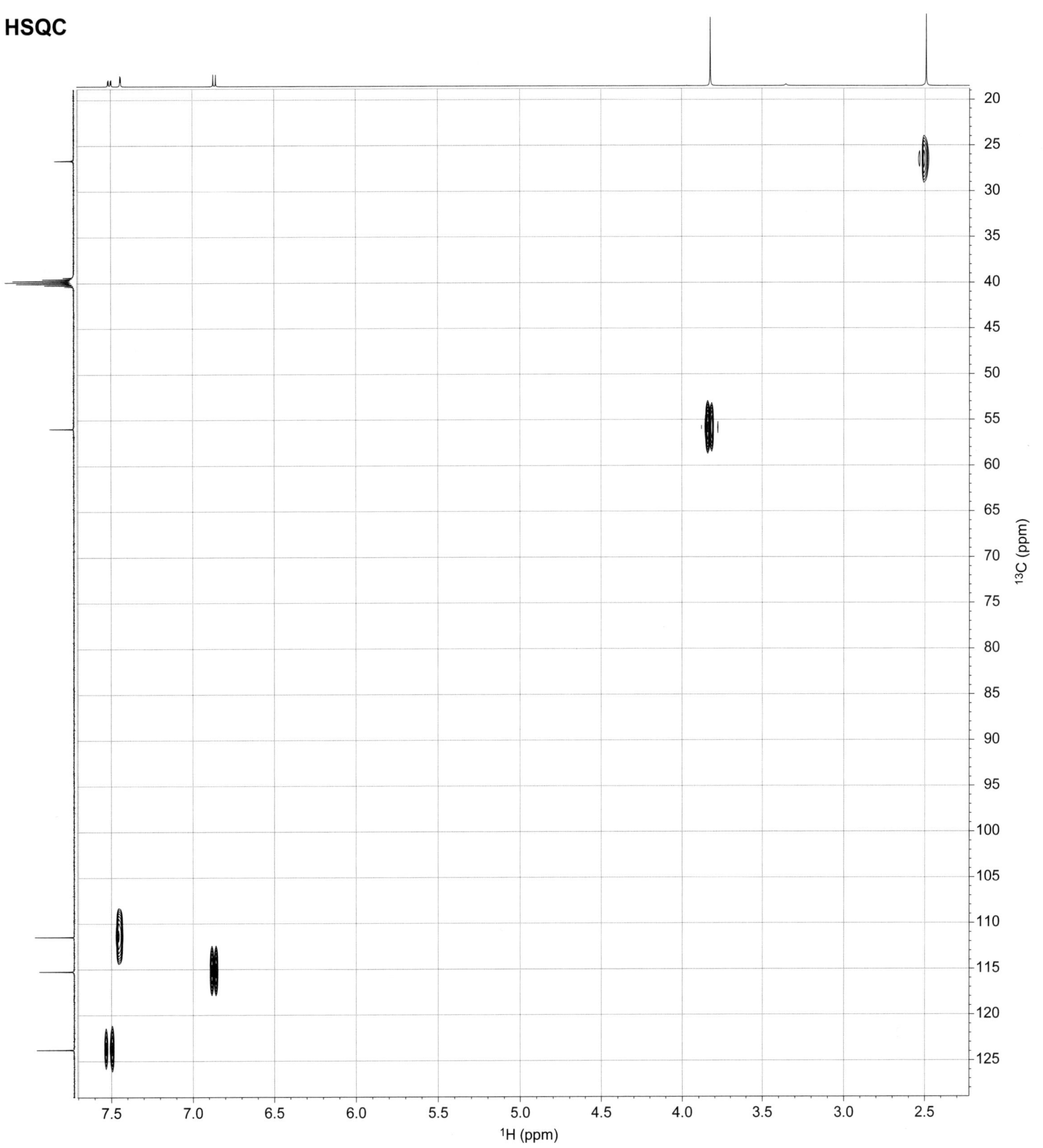
HSQC
20
25
30
35
40
45
50
55
60
65
70
75
80
85
90
95
100
105
110
115
120
125
13C (ppm)
7.5
7.0
6.5
6.0
5.5
5.0
4.5
4.0
3.5
3.0
2.5
1H (ppm)

HMBC

20
30
40
50
60
70
80
90
100
110
120
130
140
150
160
170
180
190
200
¹³C (ppm)
7.5
7.0
6.5
6.0
5.5
5.0
4.5
4.0
3.5
3.0
2.5
2.0
¹H (ppm)

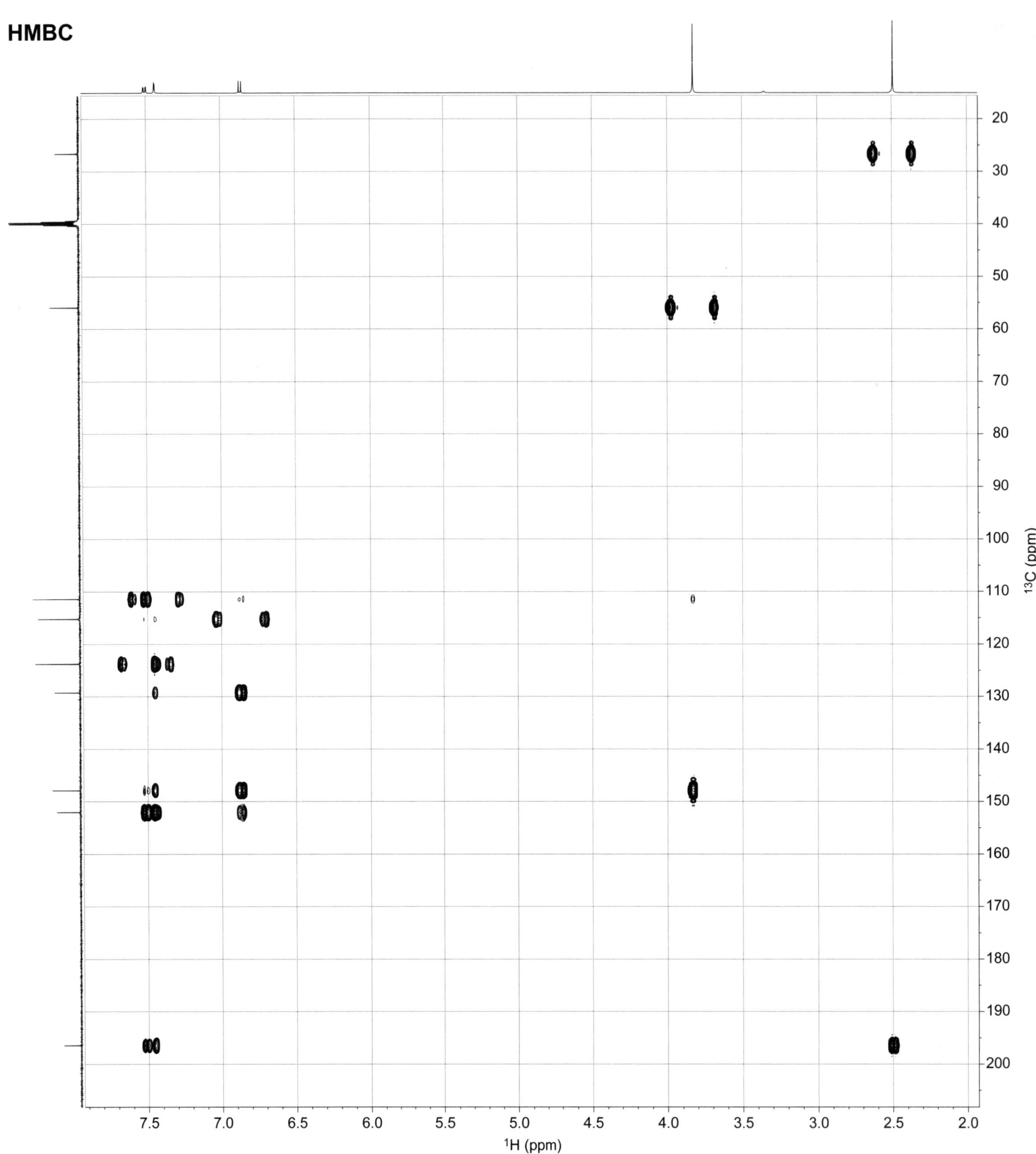

Problem 9.95 Molecular formula: $C_{10}H_{14}O_3$ ***Difficulty*** ★★★★

Solvent: CD_3-SO-CD_3

Field: 500 MHz (^1H); 125 MHz (^{13}C)

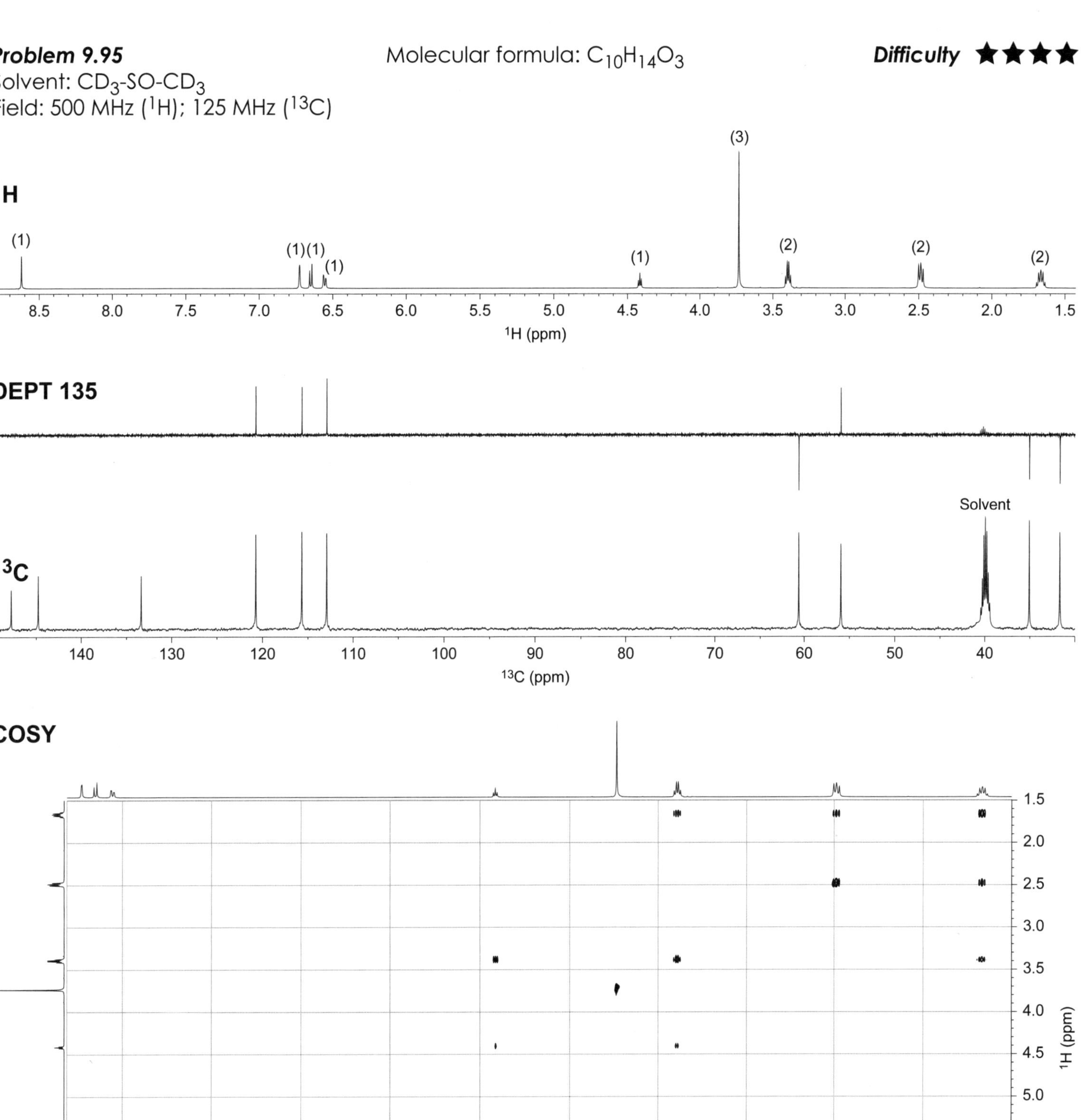

HSQC

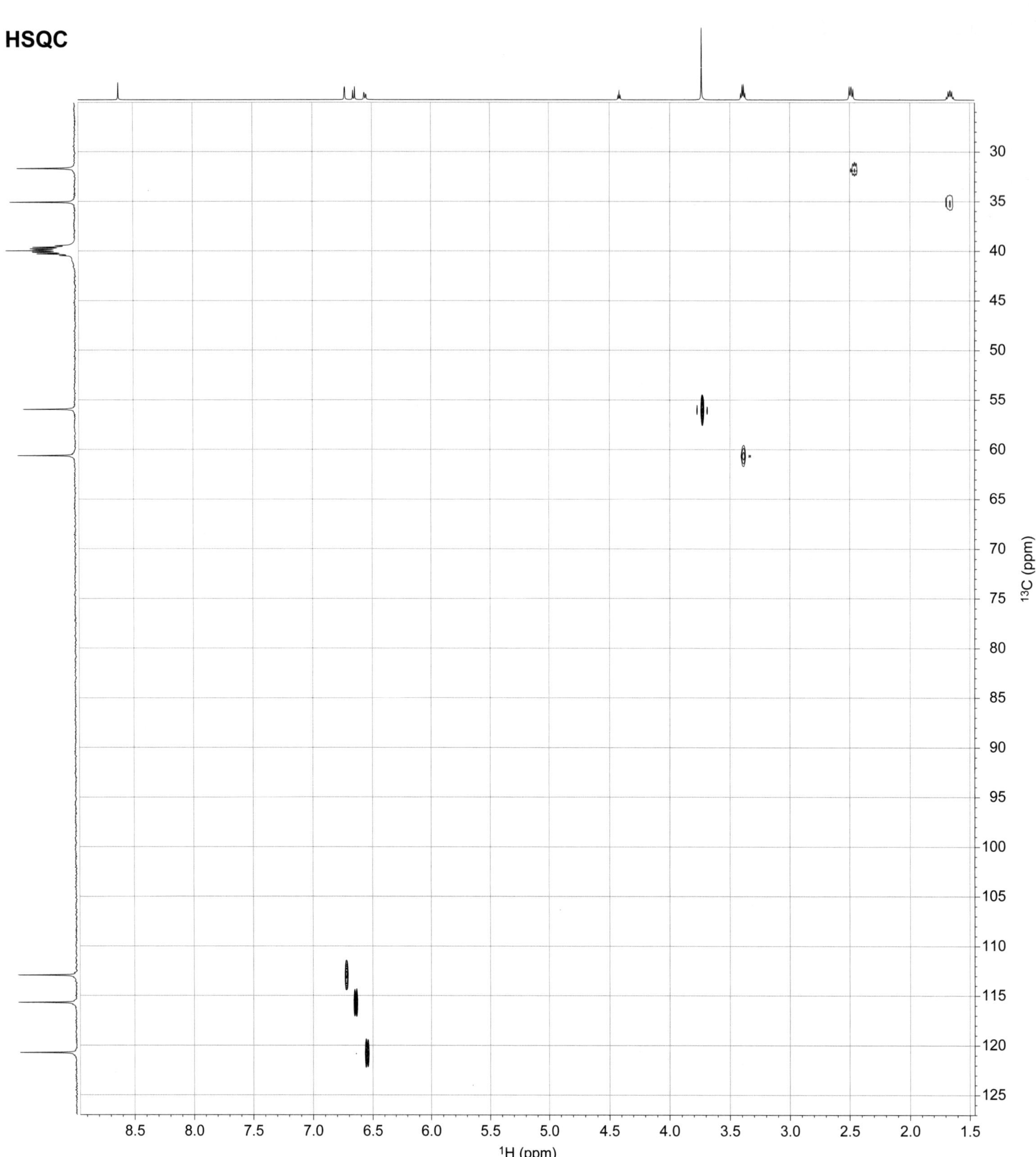

HMBC

8.5
7.0
6.5
4.5
4.0
3.5
3.0
2.5
2.0
1.5
^{1}H (ppm)
30
35
40
45
50
55
60
110
115
120
125
130
135
140
145
150
^{13}C (ppm)

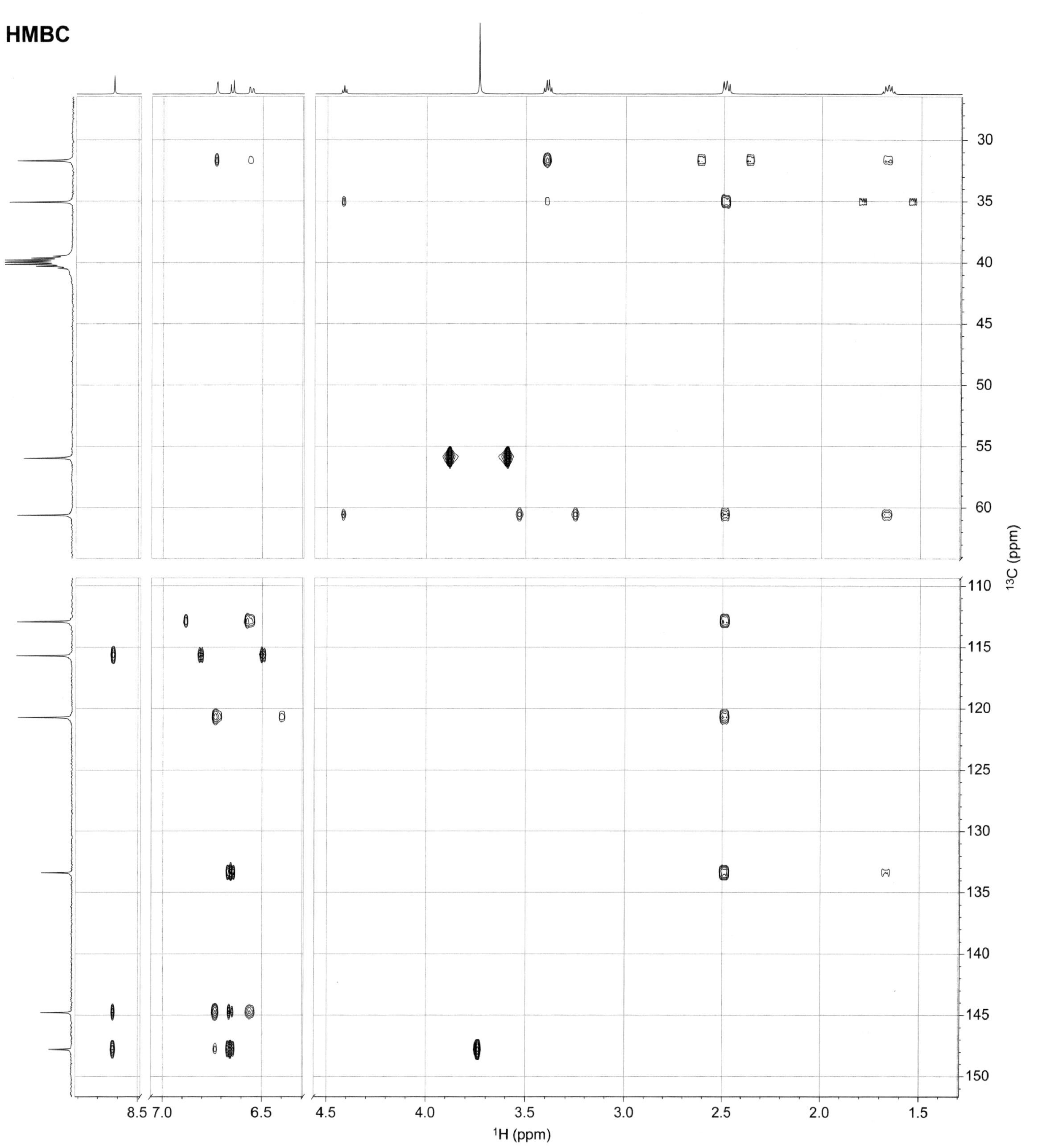

Problem 9.96
Solvent: CD_3-SO-CD_3
Field: 500 MHz (^{1}H); 125 MHz (^{13}C)

Molecular formula: $C_{10}H_{10}O_4$

Difficulty ★★★★

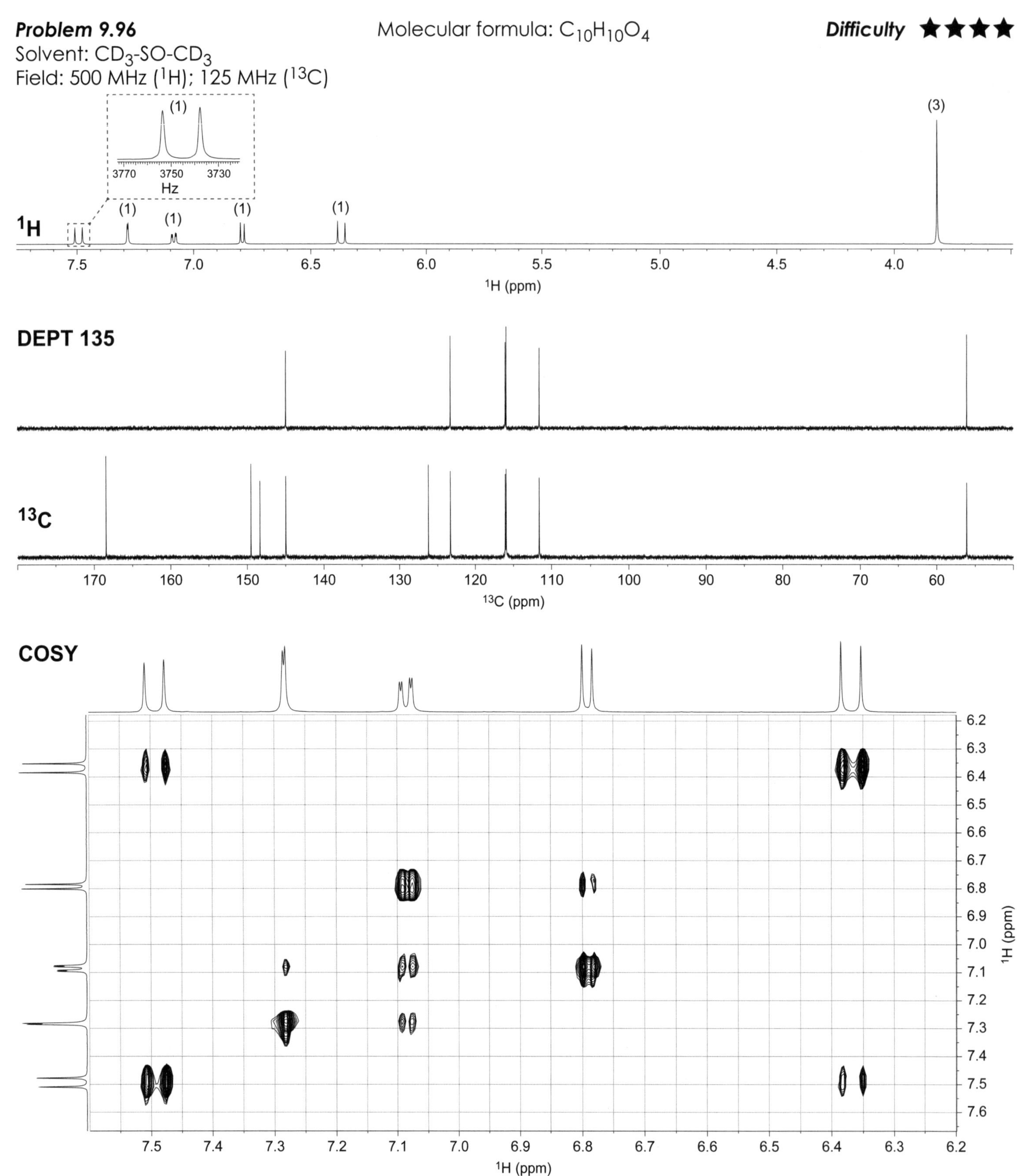

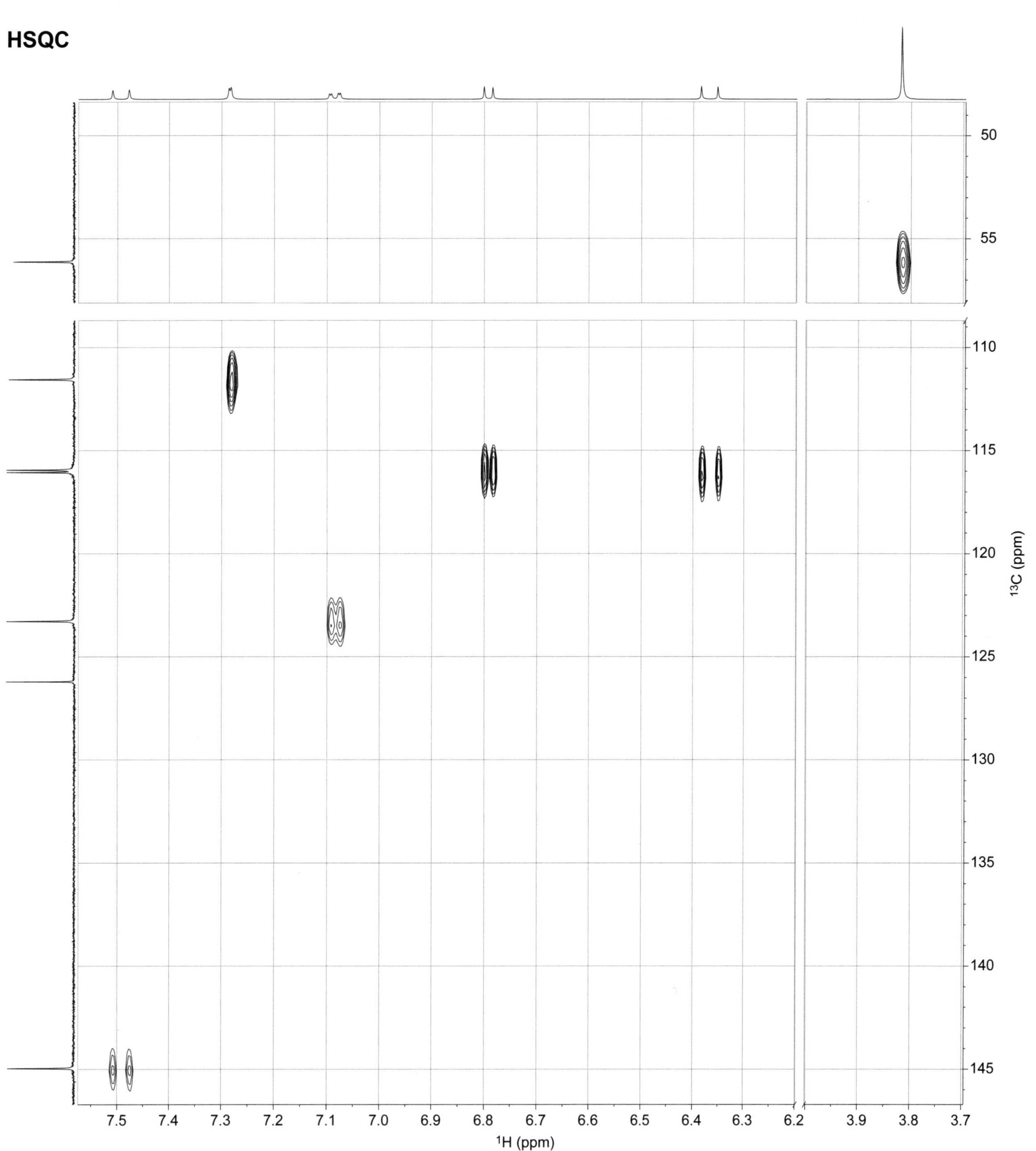
HSQC
7.5
7.4
7.3
7.2
7.1
7.0
6.9
6.8
6.7
6.6
6.5
6.4
6.3
6.2
3.9
3.8
3.7
¹H (ppm)
50
55
110
115
120
125
130
135
140
145
¹³C (ppm)

HMBC

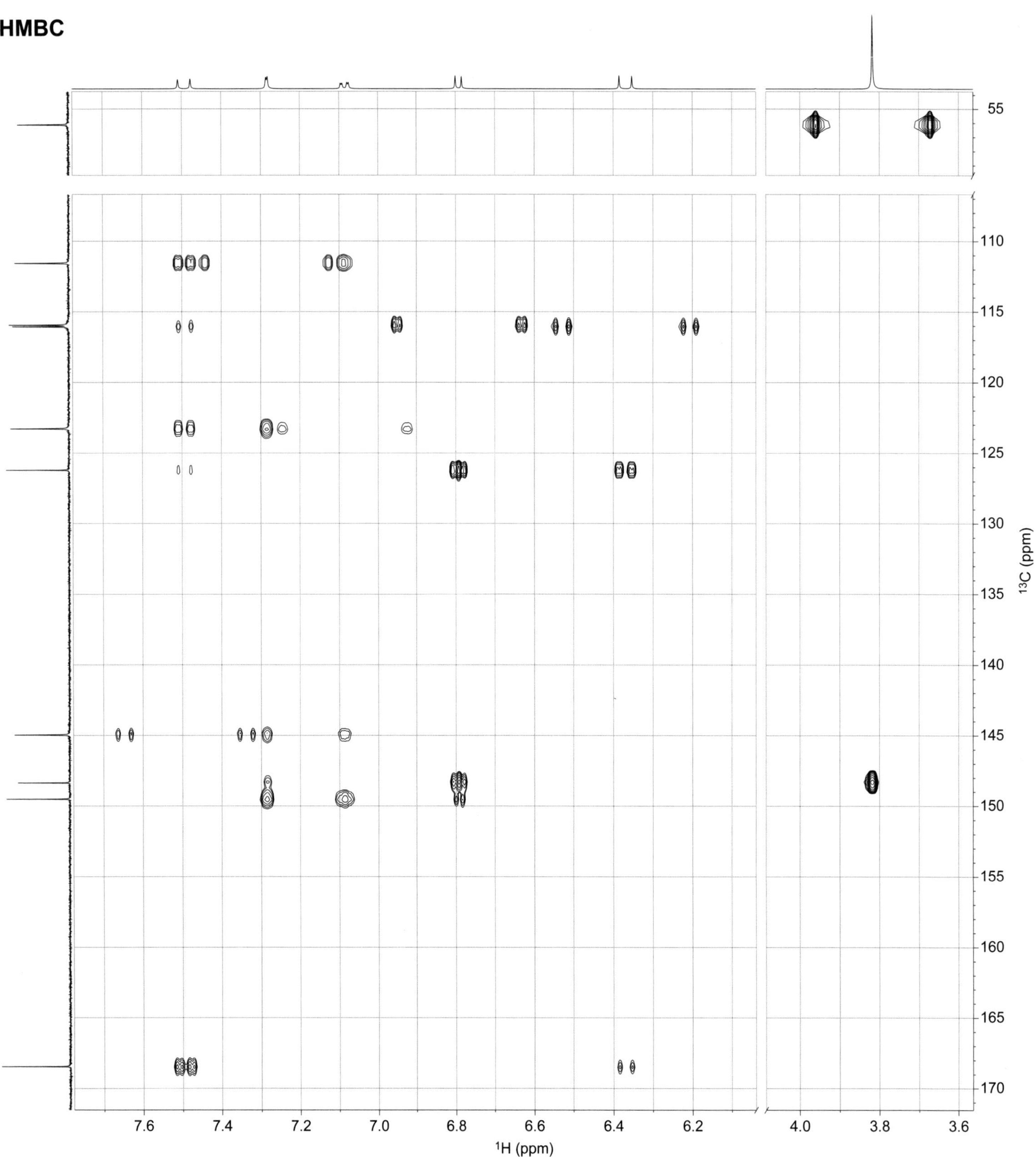

Problem 9.97

Solvent: CD_3-SO-CD_3

Field: 500 MHz (^1H); 125 MHz (^{13}C)

Molecular formula: $C_{11}H_{12}O_4$

Difficulty ★★★★

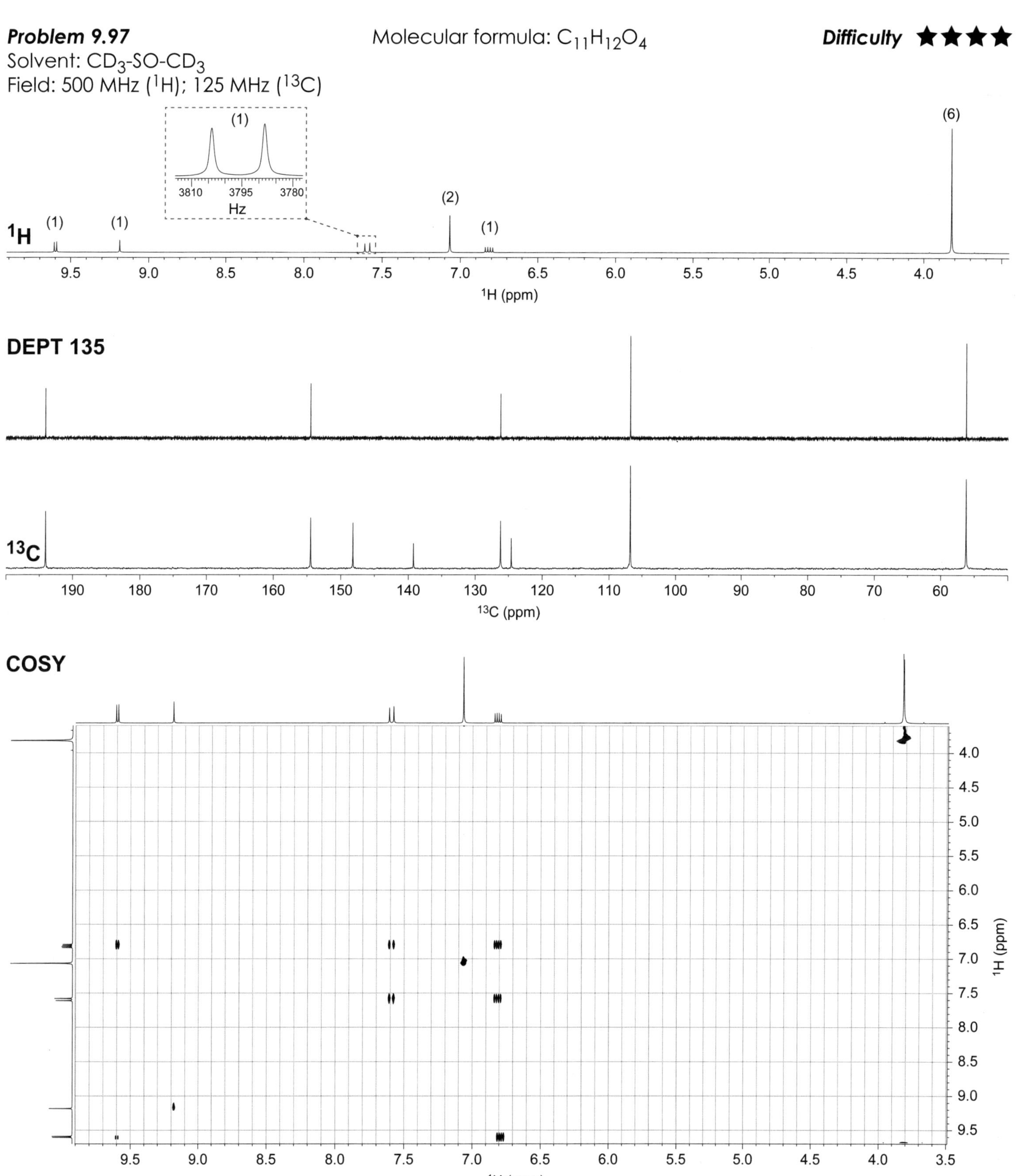

HSQC

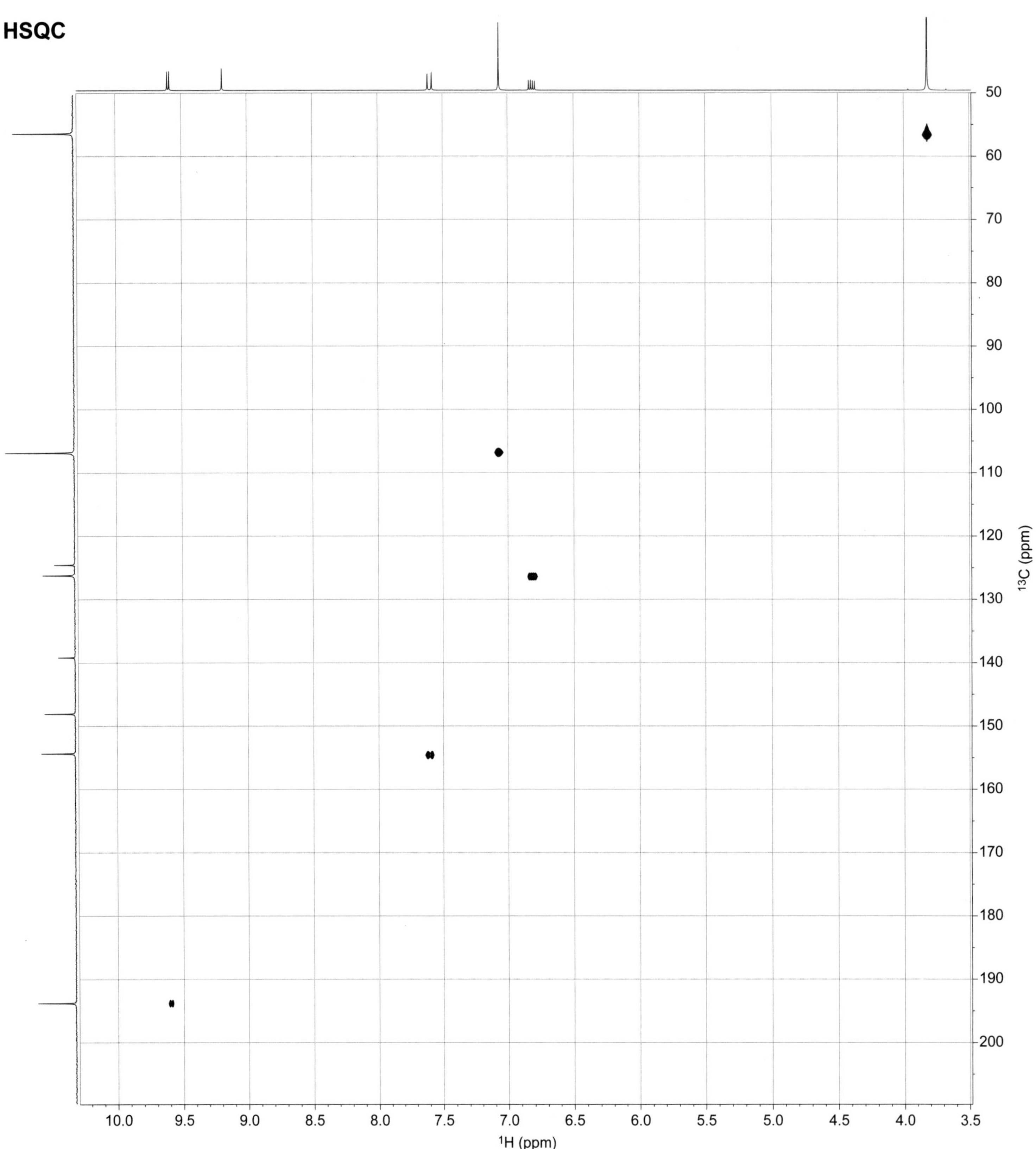

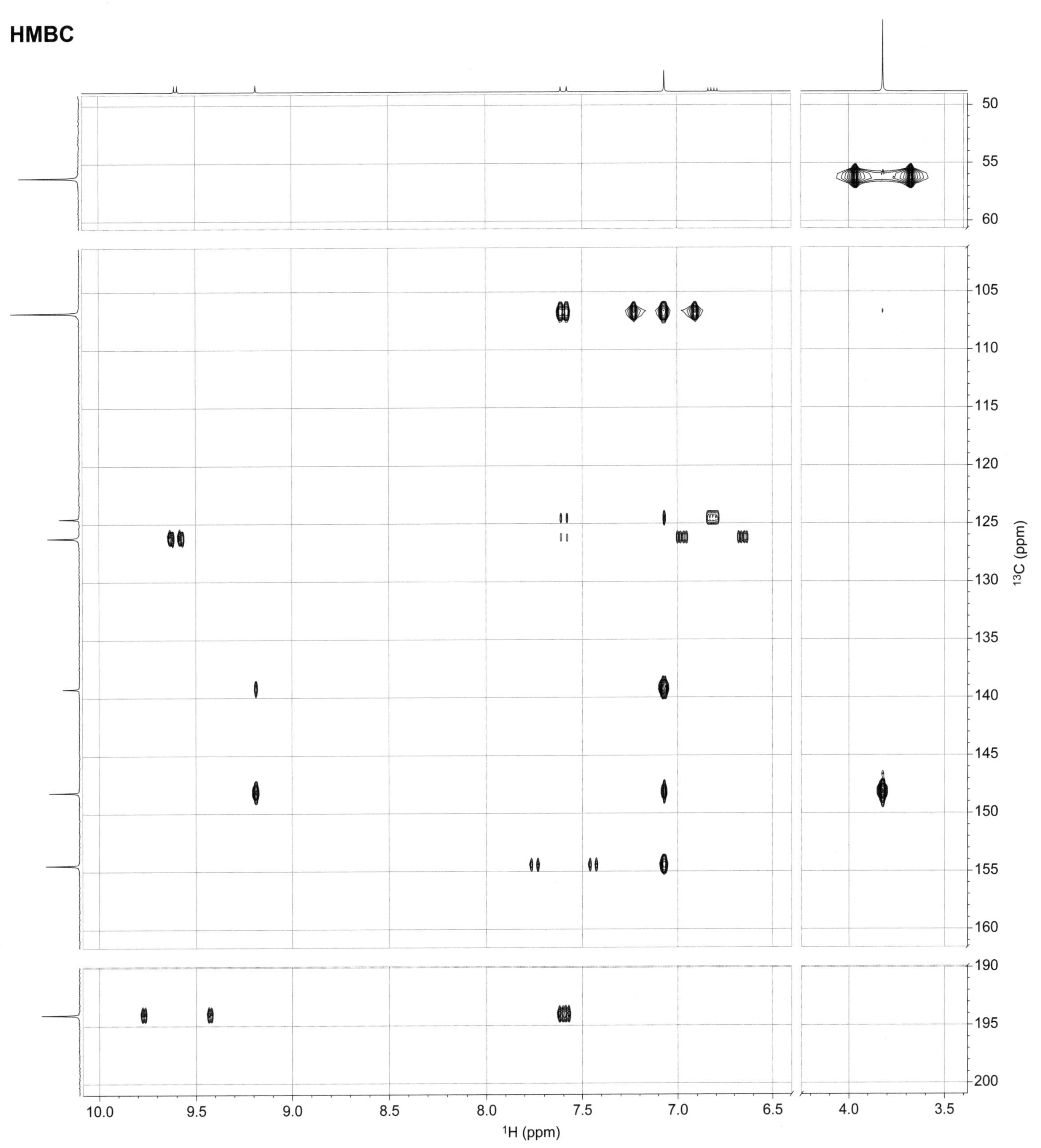
HMBC
50
55
60
105
110
115
120
125
130
135
140
145
150
155
160
190
195
200
^{13}C (ppm)
10.0
9.5
9.0
8.5
8.0
7.5
7.0
6.5
4.0
3.5
^{1}H (ppm)

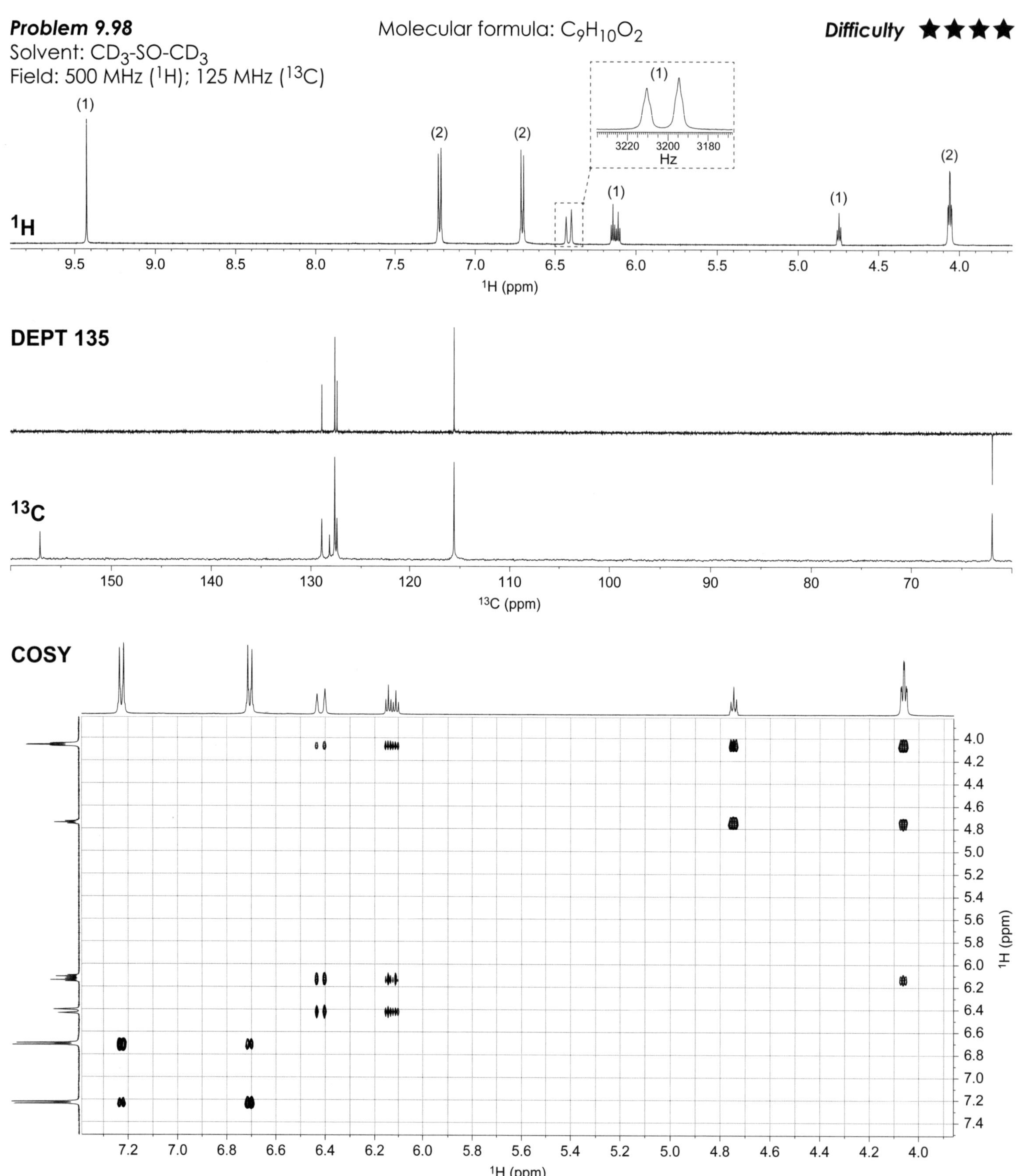

Problem 9.98
Solvent: CD3-SO-CD3
Field: 500 MHz (1H); 125 MHz (13C)
Molecular formula: C9H10O2
Difficulty ★★★★
1H
(1)
(2)
(2)
(1)
(1)
(1)
(2)
3220
3200
3180
Hz
9.5
9.0
8.5
8.0
7.5
7.0
6.5
6.0
5.5
5.0
4.5
4.0
1H (ppm)
DEPT 135
13C
150
140
130
120
110
100
90
80
70
13C (ppm)
COSY
1H (ppm)
1H (ppm)

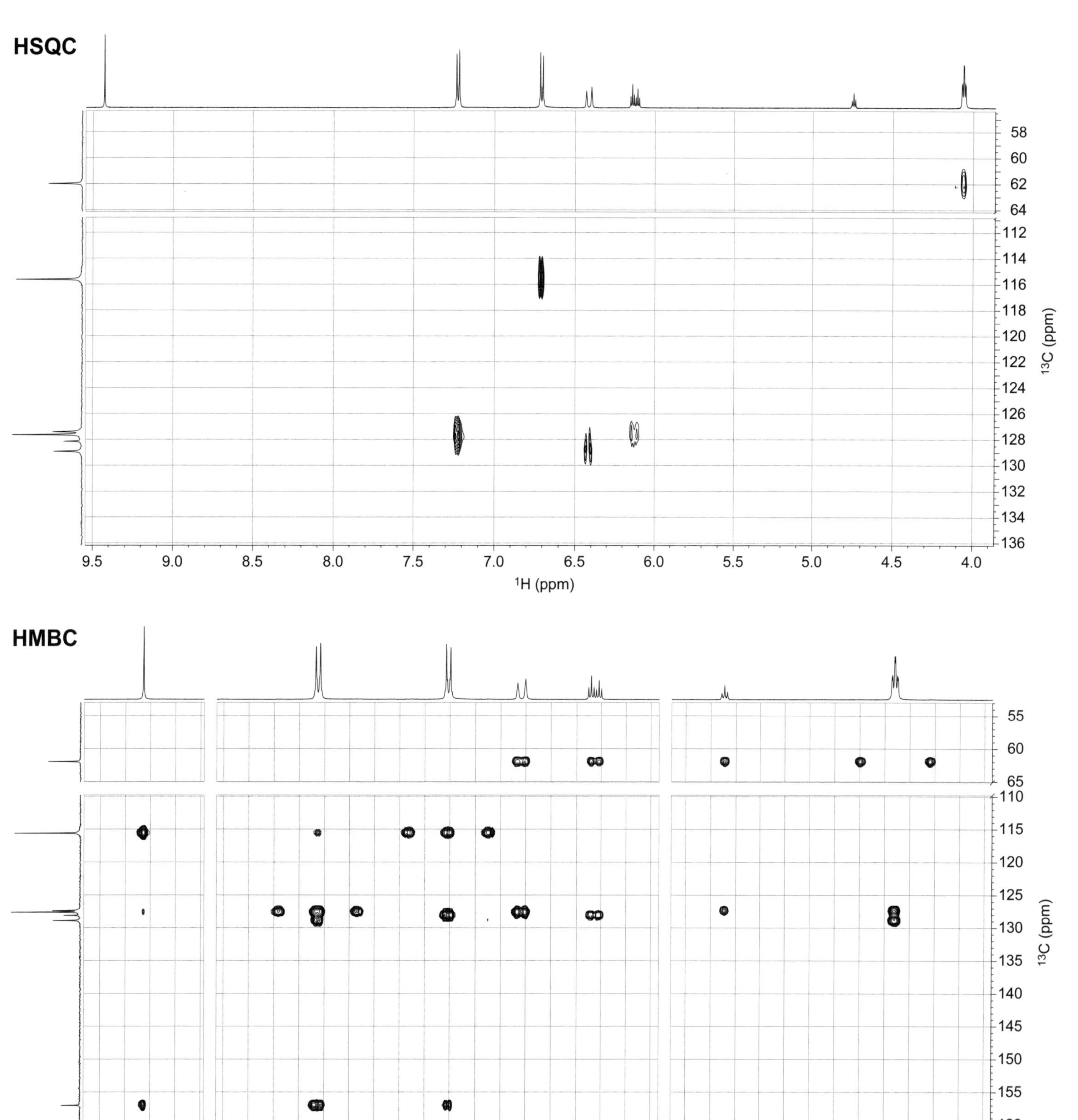
HSQC
¹H (ppm)
¹³C (ppm)
HMBC
¹H (ppm)
¹³C (ppm)

Problem 9.99
Solvent: CD_3OD
Field: 700 MHz (^1H); 175 MHz (^{13}C)

Molecular formula: $C_{12}H_{18}$

Difficulty ★★★★

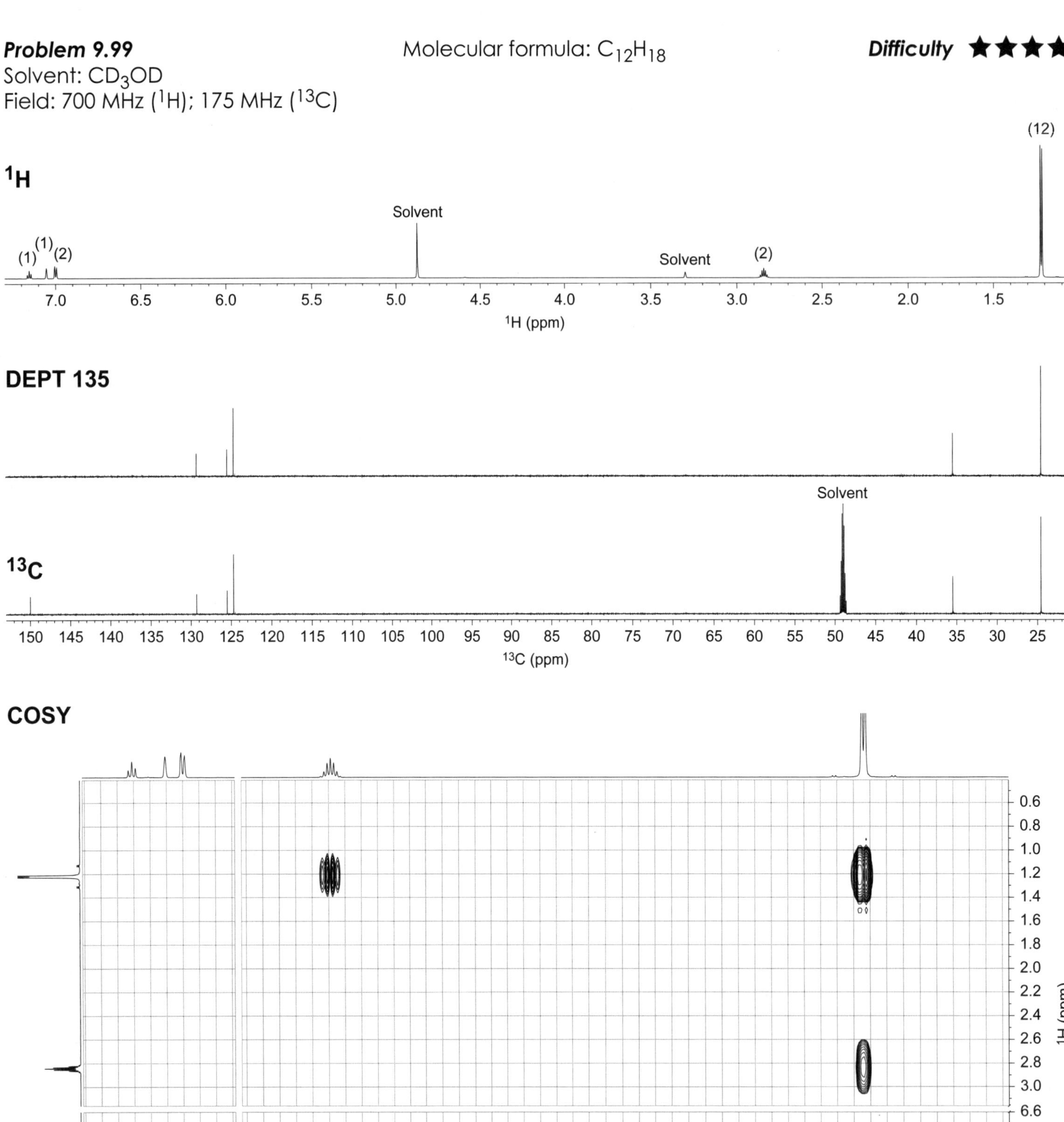

HSQC

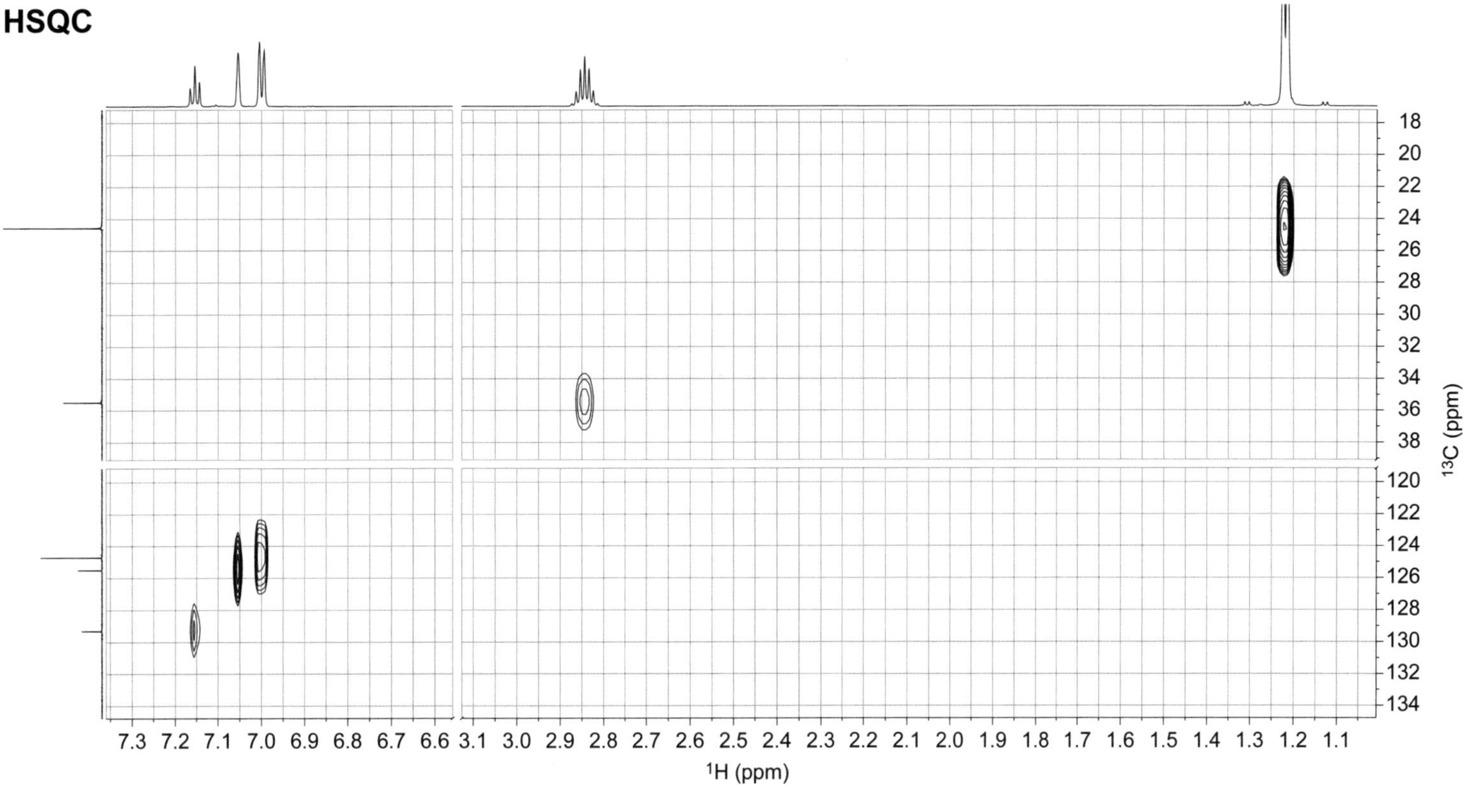

7.3 7.2 7.1 7.0 6.9 6.8 6.7 6.6 3.1 3.0 2.9 2.8 2.7 2.6 2.5 2.4 2.3 2.2 2.1 2.0 1.9 1.8 1.7 1.6 1.5 1.4 1.3 1.2 1.1

^{1}H (ppm)

^{13}C (ppm): 18 20 22 24 26 28 30 32 34 36 38 120 122 124 126 128 130 132 134

HMBC

7.4 7.2 7.0 3.0 2.8 2.6 2.4 2.2 2.0 1.8 1.6 1.4 1.2 1.0 0.8

^{1}H (ppm)

^{13}C (ppm): 15 20 25 30 35 40 125 130 135 140 145 150 155

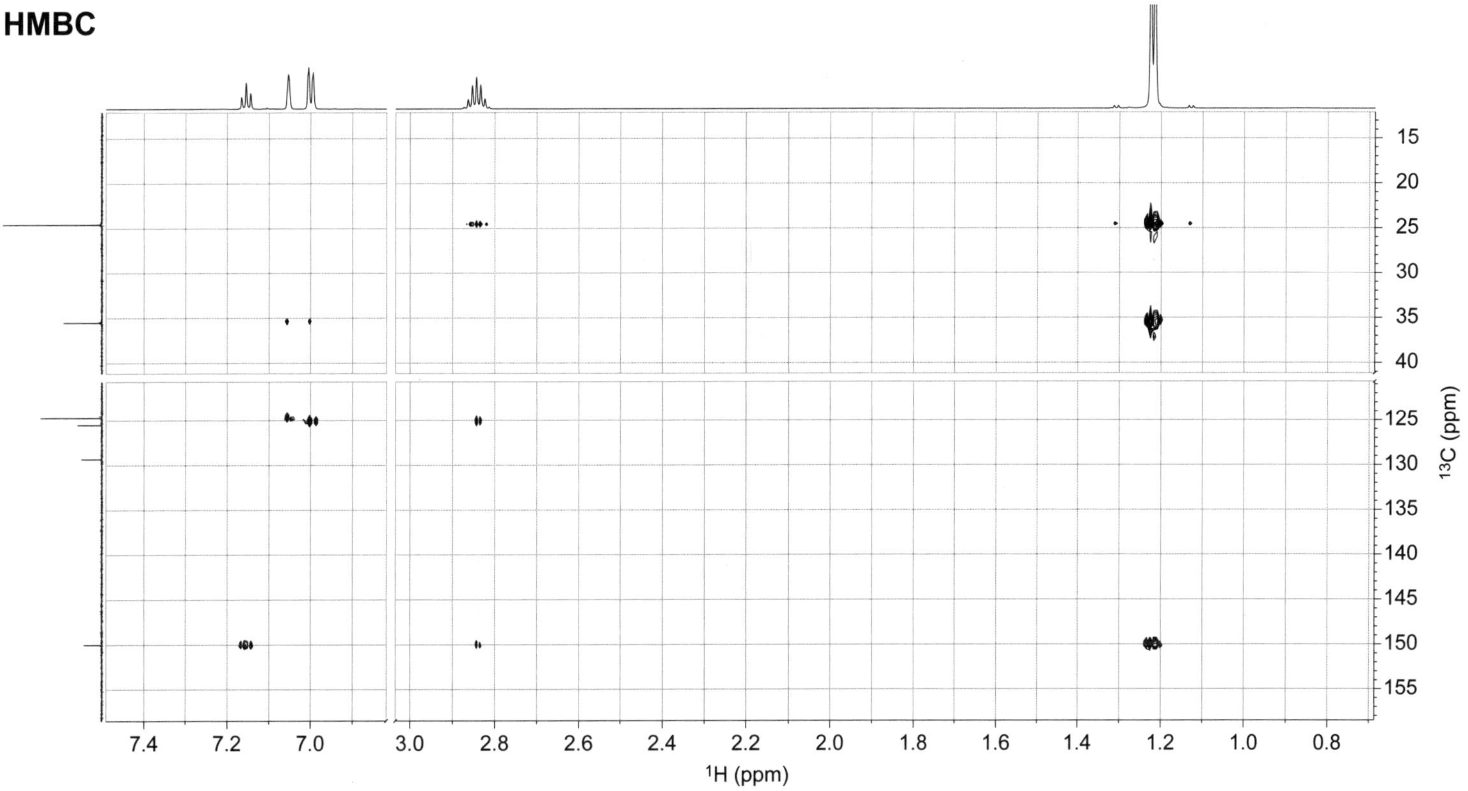

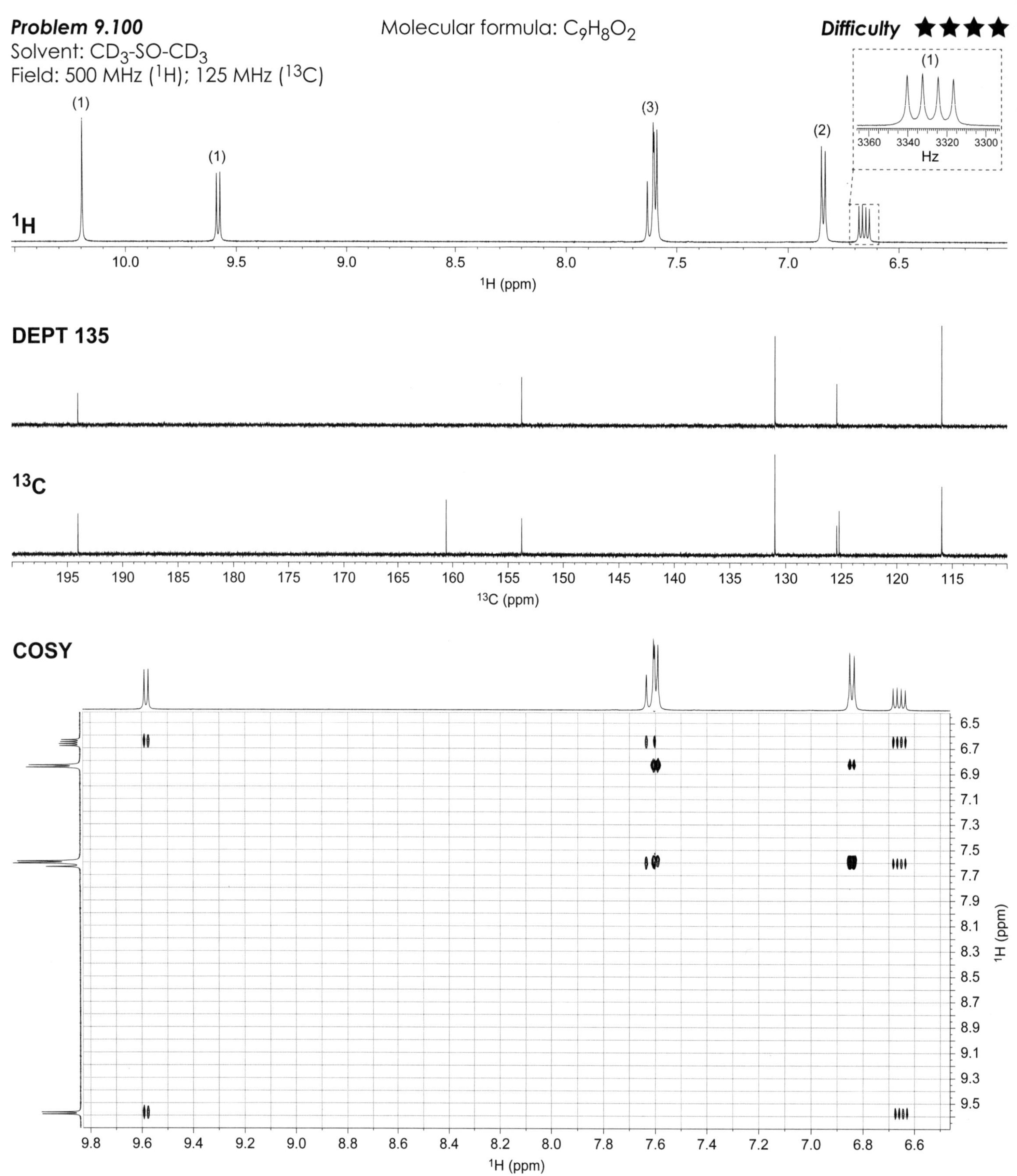

Problem 9.100
Molecular formula: $C_9H_8O_2$
Difficulty ★★★★
Solvent: CD_3-SO-CD_3
Field: 500 MHz (^1H); 125 MHz (^{13}C)
(1)
(1)
(3)
(2)
(1)
3360 3340 3320 3300
Hz
^1H
10.0 9.5 9.0 8.5 8.0 7.5 7.0 6.5
^1H (ppm)
DEPT 135
^{13}C
195 190 185 180 175 170 165 160 155 150 145 140 135 130 125 120 115
^{13}C (ppm)
COSY
6.5 6.7 6.9 7.1 7.3 7.5 7.7 7.9 8.1 8.3 8.5 8.7 8.9 9.1 9.3 9.5
^1H (ppm)
9.8 9.6 9.4 9.2 9.0 8.8 8.6 8.4 8.2 8.0 7.8 7.6 7.4 7.2 7.0 6.8 6.6
^1H (ppm)

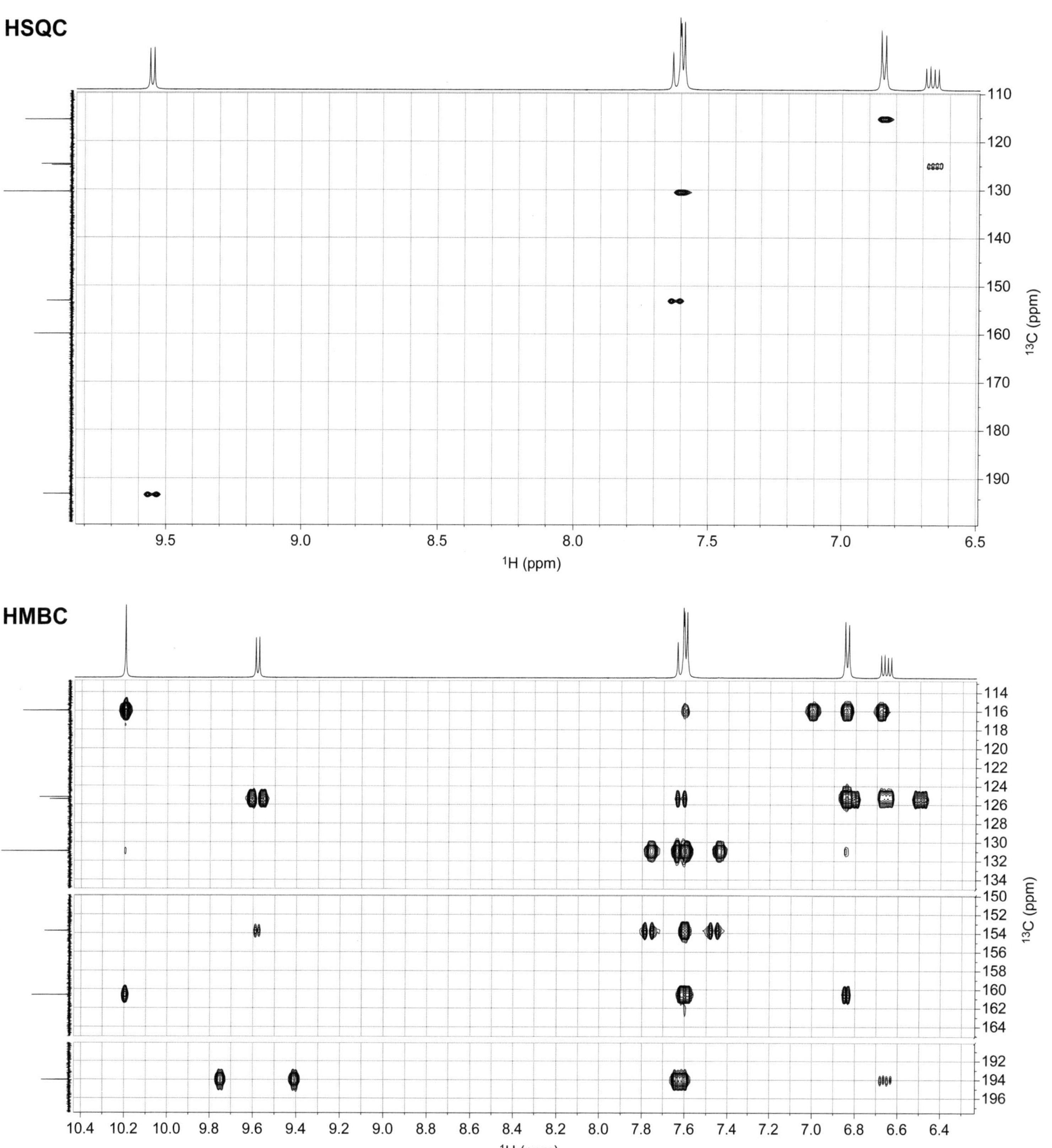
HSQC
9.5
9.0
8.5
8.0
7.5
7.0
6.5
1H (ppm)
110
120
130
140
150
160
170
180
190
13C (ppm)
HMBC
10.4 10.2 10.0 9.8 9.6 9.4 9.2 9.0 8.8 8.6 8.4 8.2 8.0 7.8 7.6 7.4 7.2 7.0 6.8 6.6 6.4
1H (ppm)
114
116
118
120
122
124
126
128
130
132
134
150
152
154
156
158
160
162
164
192
194
196
13C (ppm)

Problem 9.101 Molecular formula: $C_{11}H_{12}O_4$ **Difficulty** ★★★★

Solvent: D_2O

Field: 500 MHz (^1H); 125 MHz (^{13}C)

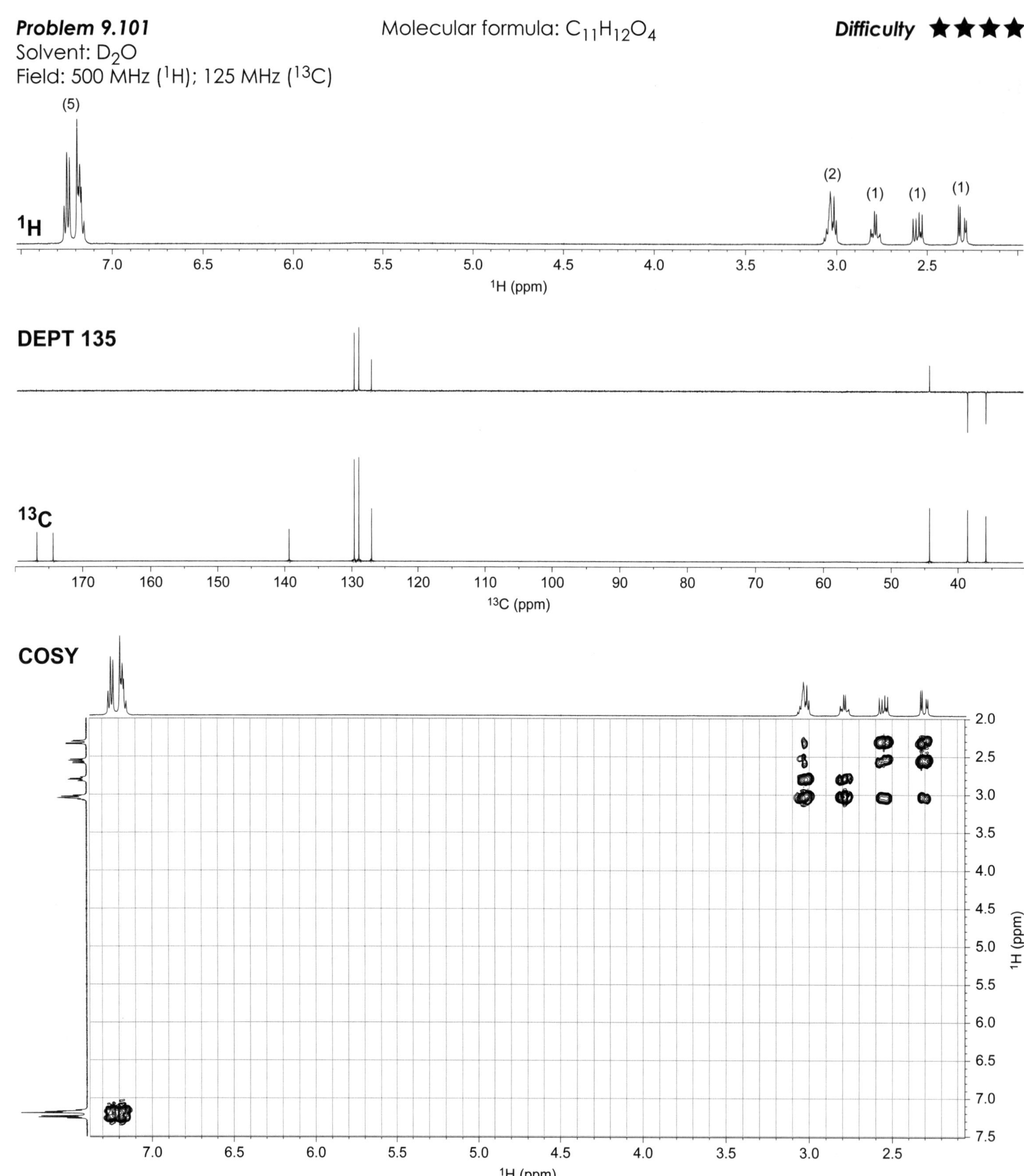

HSQC

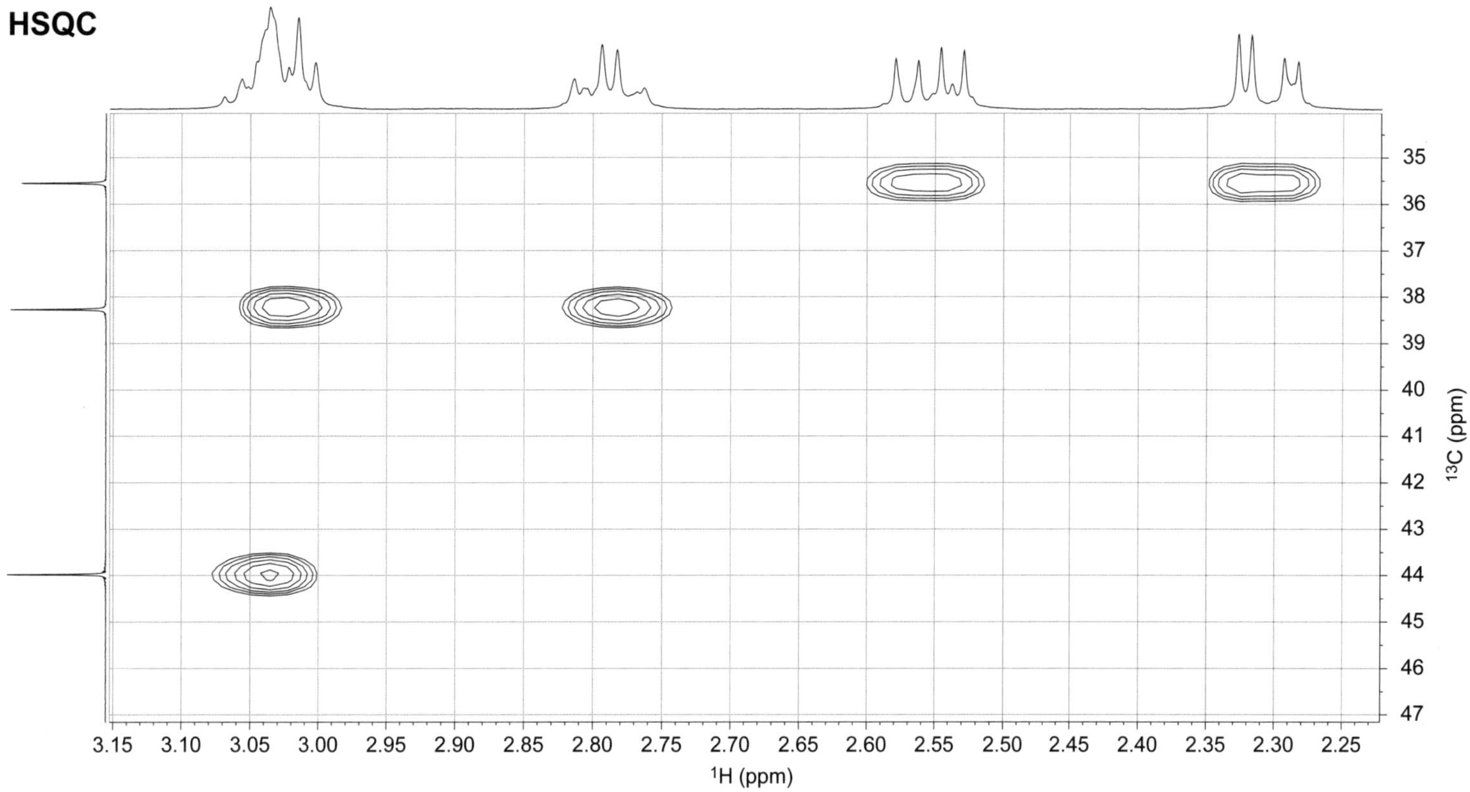

HSQC

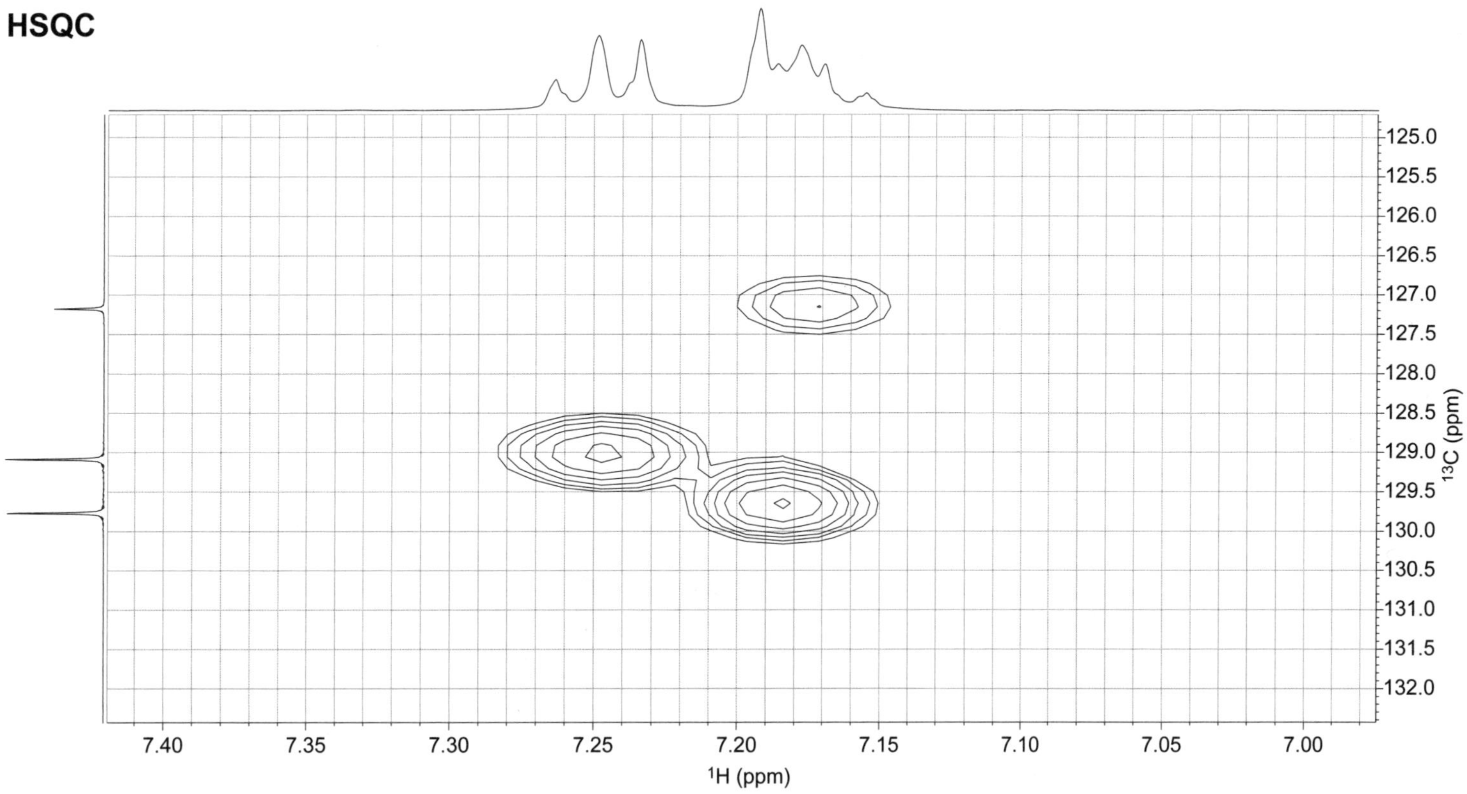

HMBC

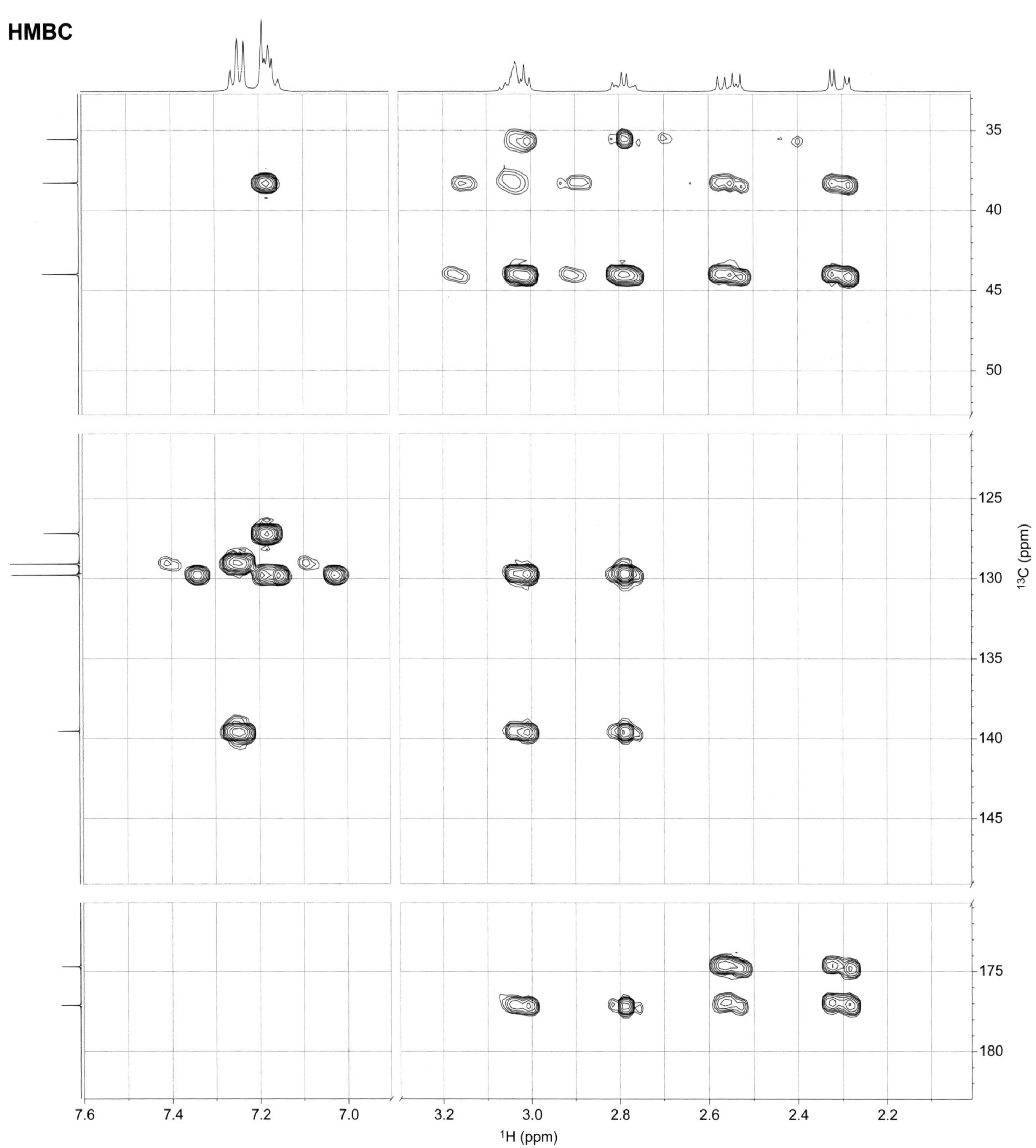

Problem 9.102
Solvent: D_2O
Field: 500 MHz (^{1}H); 125 MHz (^{13}C)

Molecular formula: $C_{10}H_{12}N_2O$
Hint: It is an indole derivative

Difficulty ★★★★

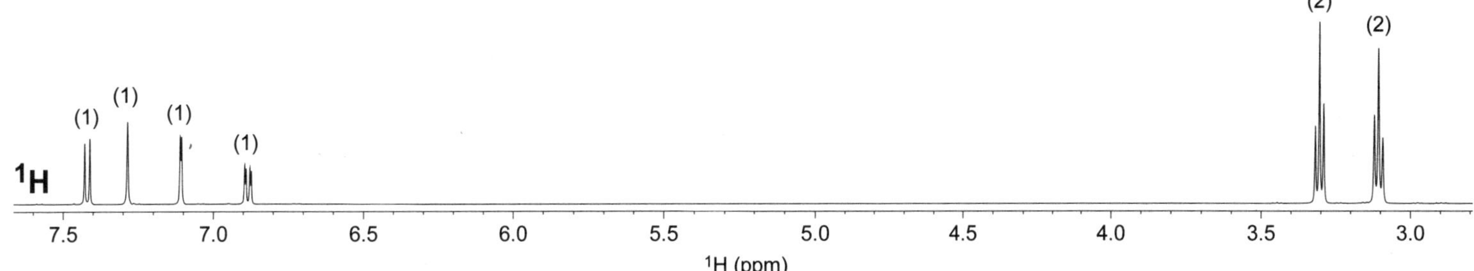

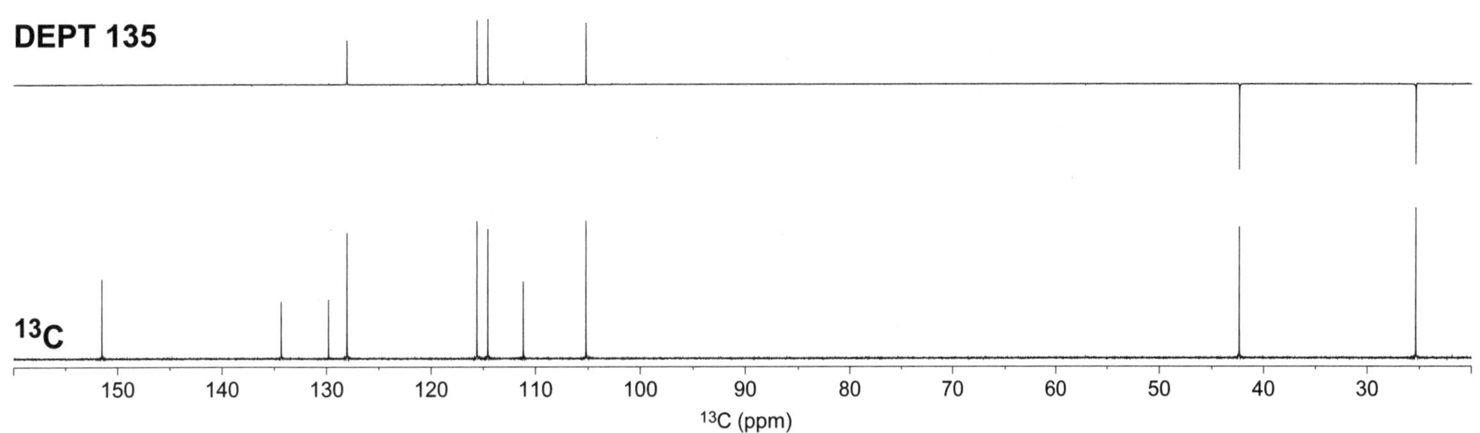

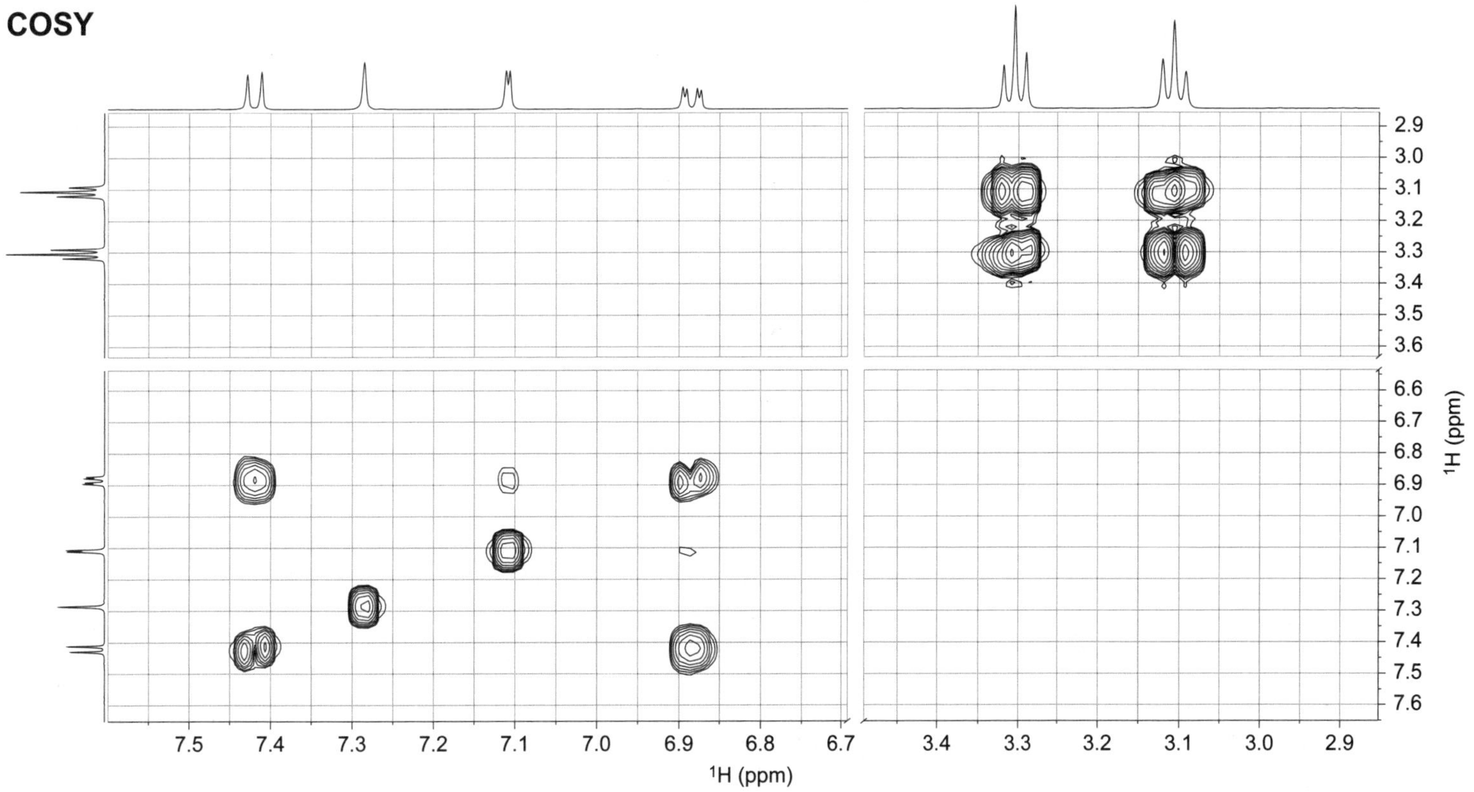

COSY

HSQC

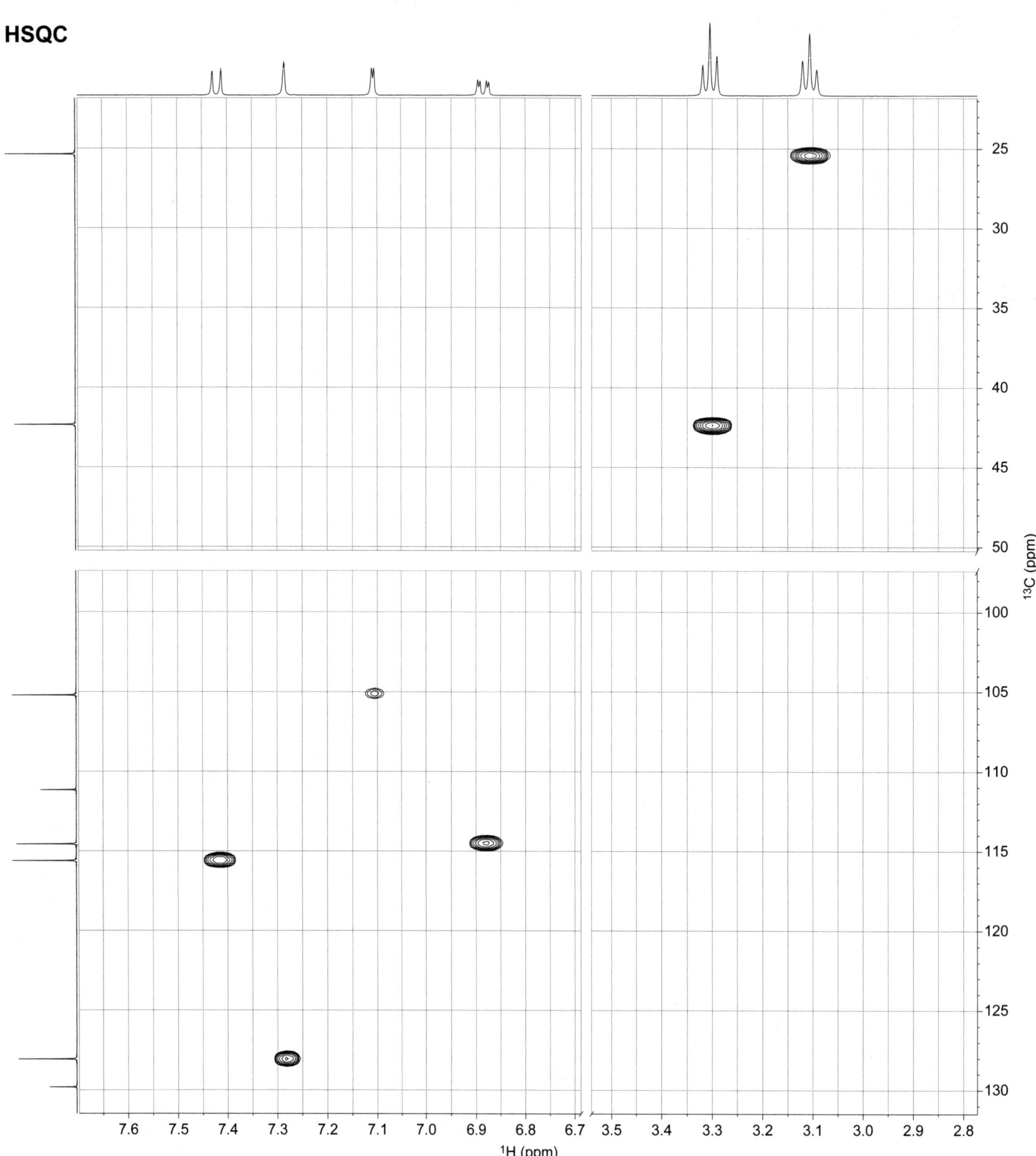

7.6
7.5
7.4
7.3
7.2
7.1
7.0
6.9
6.8
6.7
3.5
3.4
3.3
3.2
3.1
3.0
2.9
2.8
¹H (ppm)
25
30
35
40
45
50
100
105
110
115
120
125
130
¹³C (ppm)

HMBC

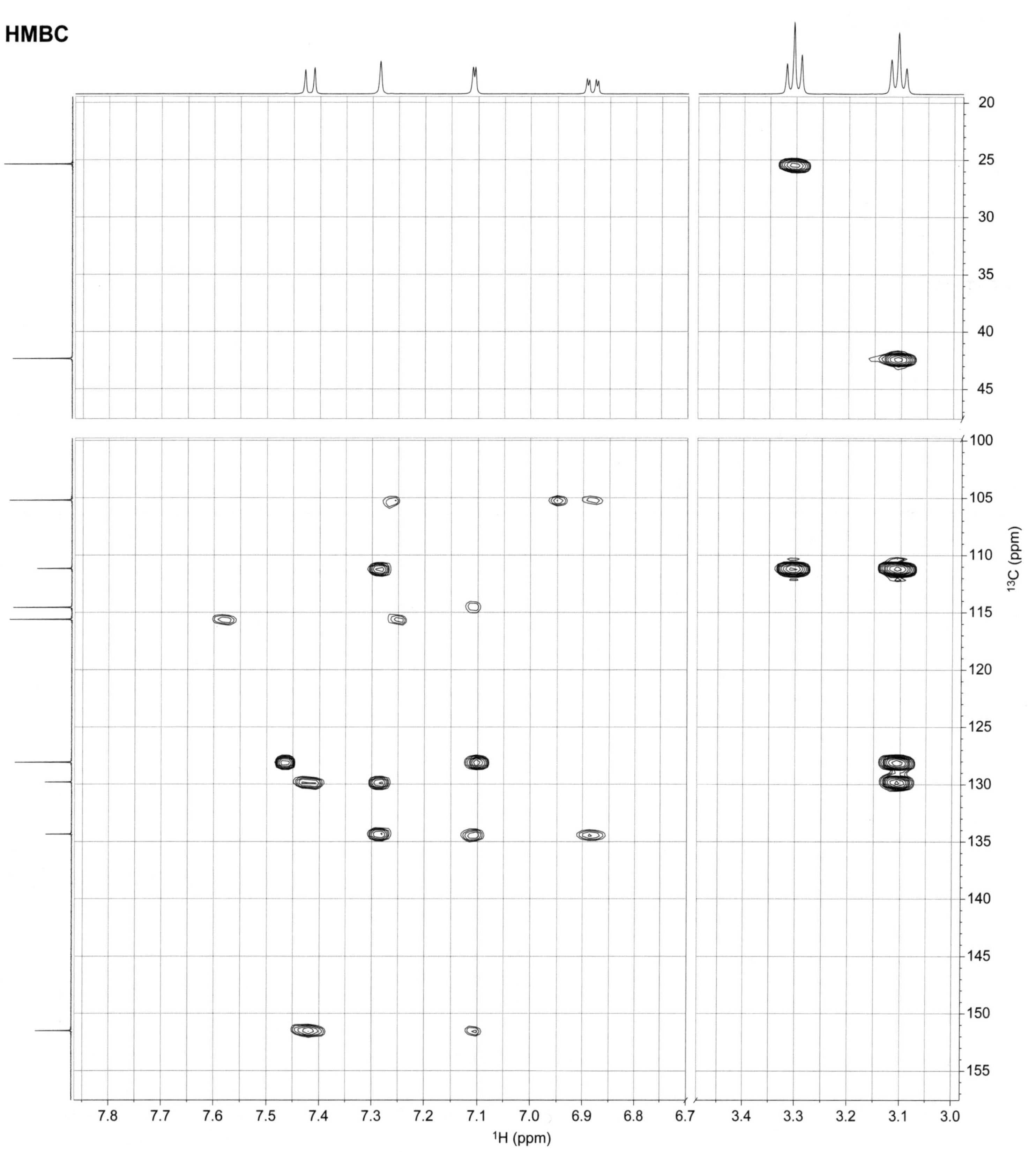

Problem 9.103 Molecular formula: $C_9H_{10}O_4$ **Difficulty** ★★★★

Solvent: CD_3-SO-CD_3

Field: 500 MHz (^{1}H); 125 MHz (^{13}C)

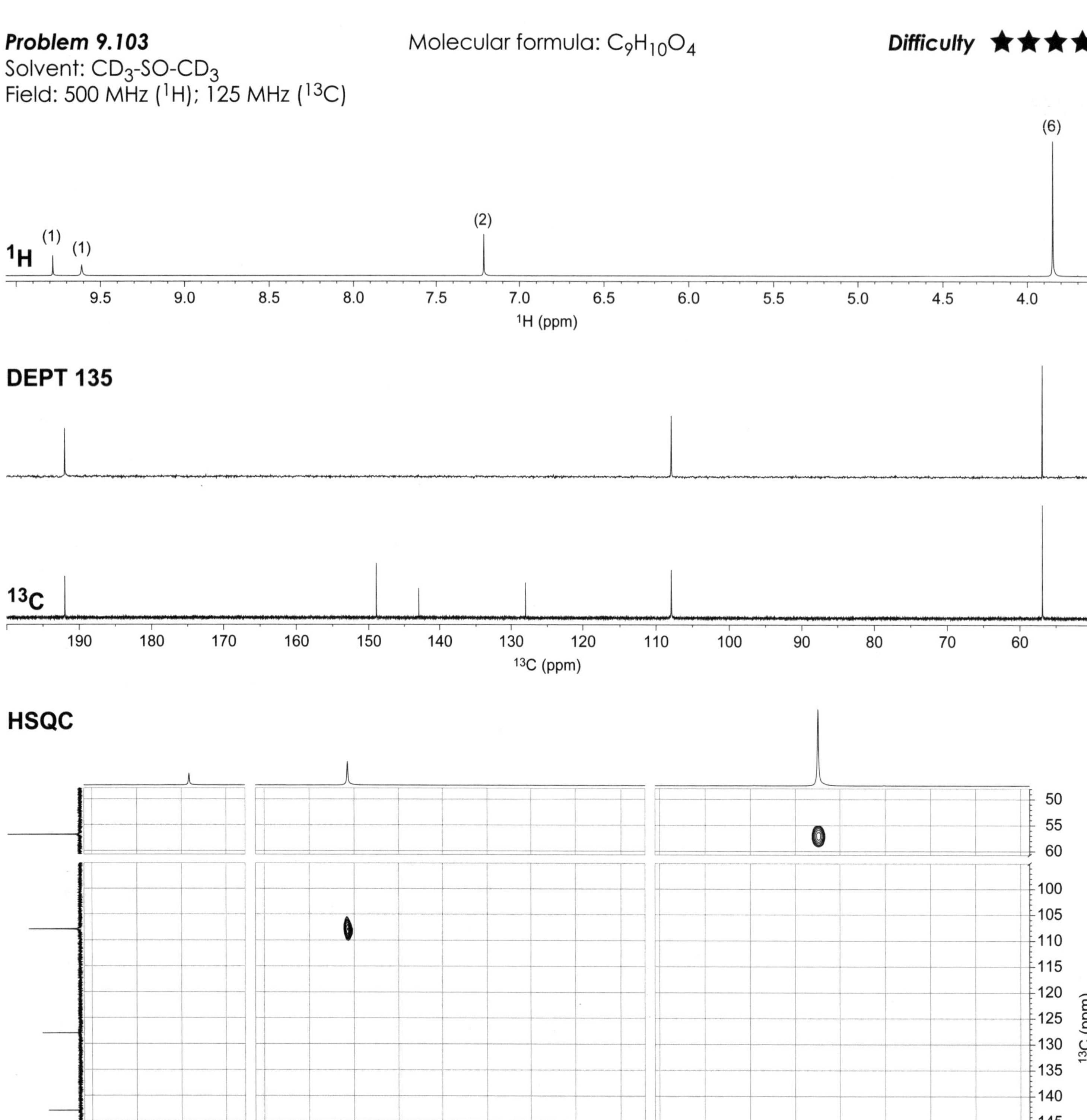

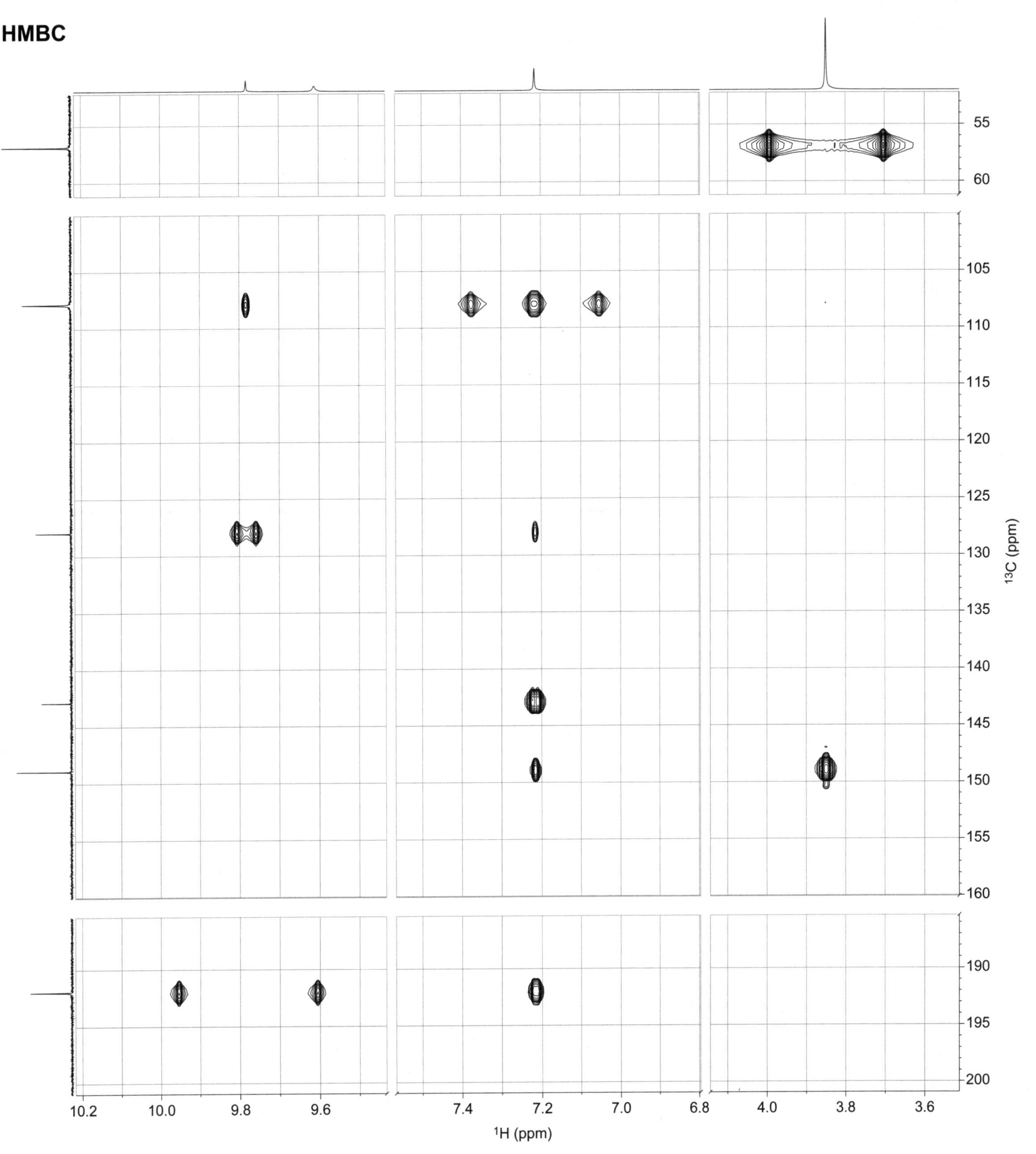
HMBC
55
60
105
110
115
120
125
130
135
140
145
150
155
160
190
195
200
^{13}C (ppm)
10.2
10.0
9.8
9.6
7.4
7.2
7.0
6.8
4.0
3.8
3.6
^{1}H (ppm)

Problem 9.104
Solvent: CD_3OD
Field: 700 MHz (^1H); 175 MHz (^{13}C)

Molecular formula: $C_{11}H_{10}ClNO_2$
Hint: It is an indole derivative

Difficulty ★★★★

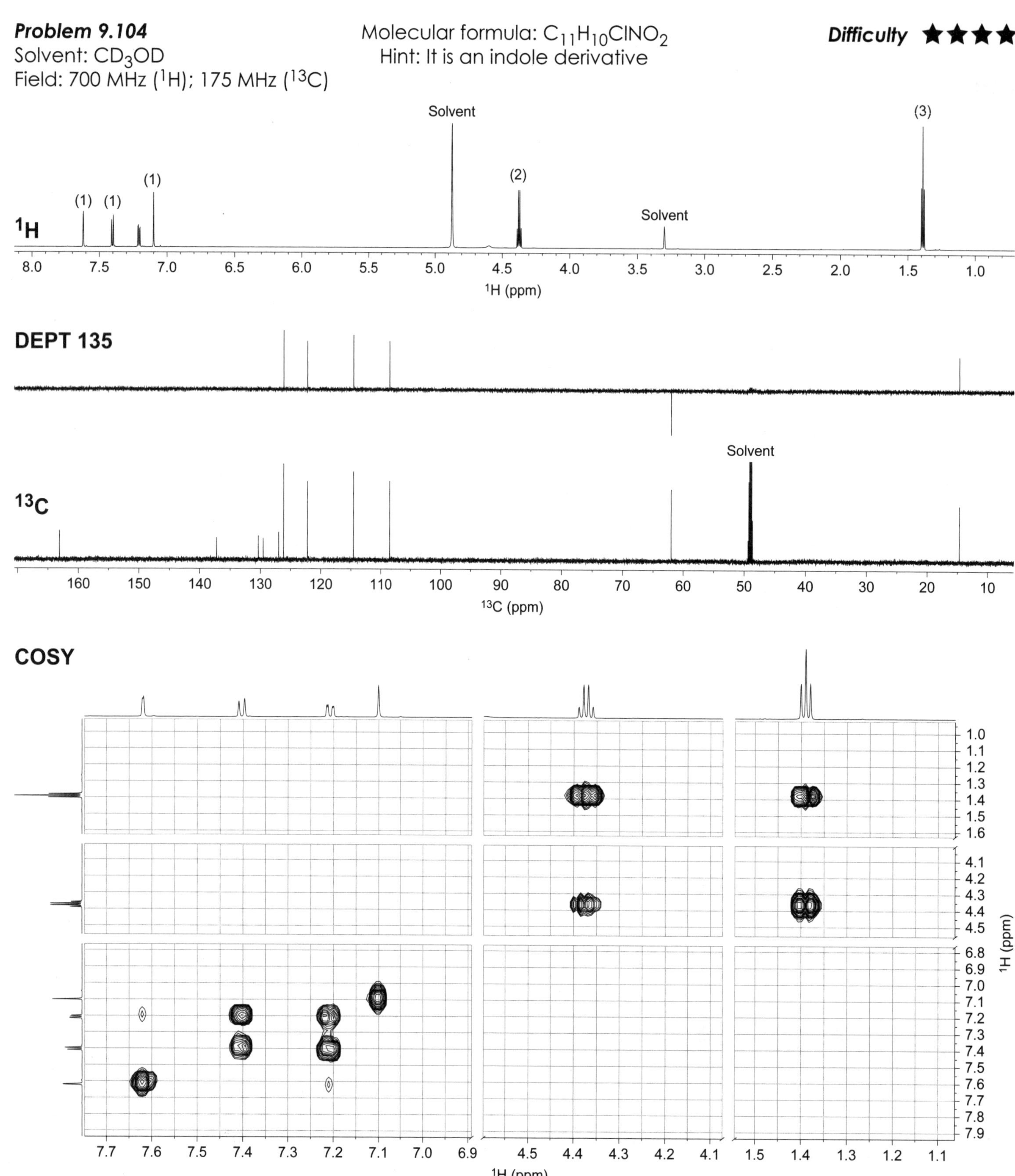

HSQC

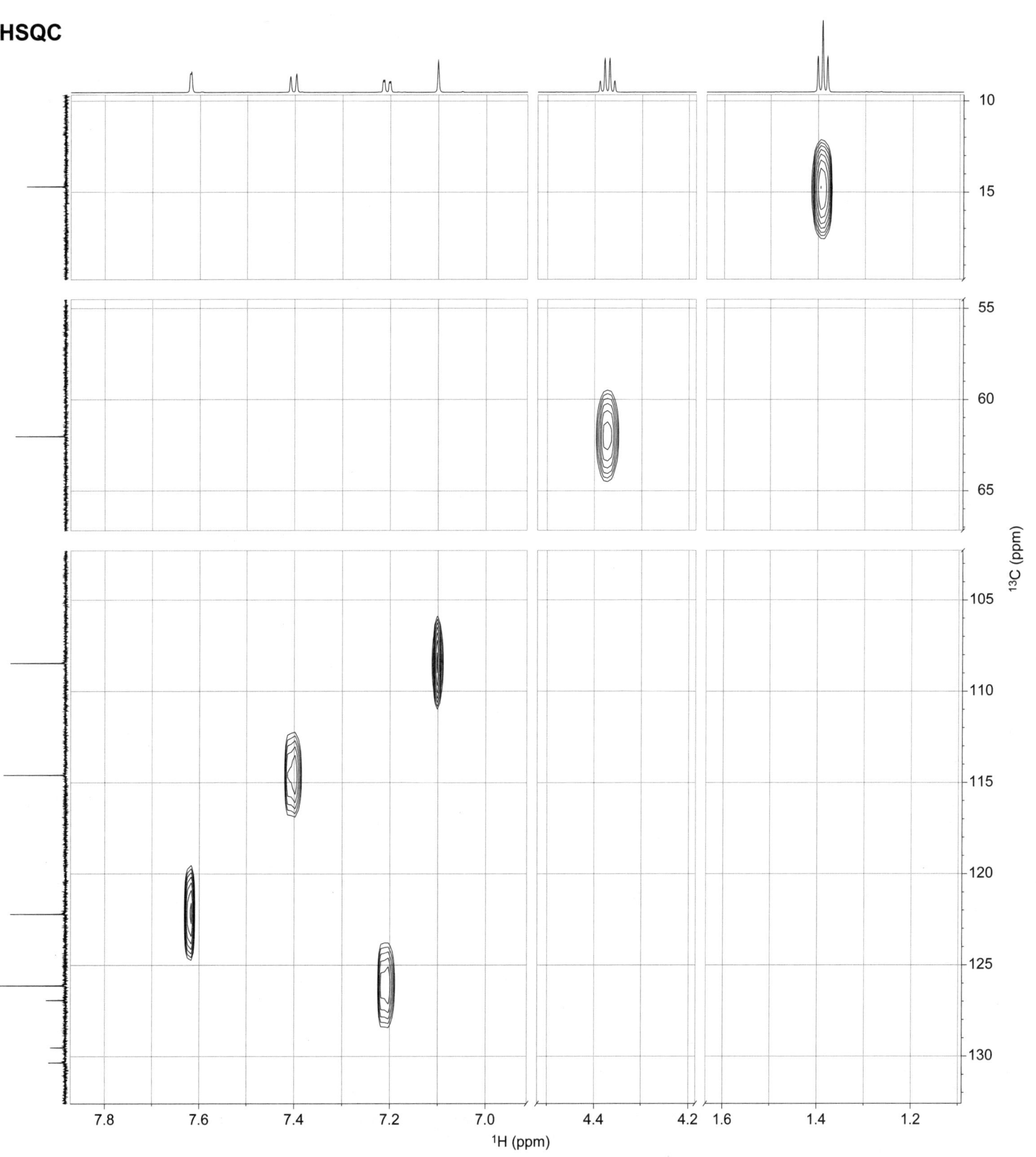

10
15
55
60
65
105
110
115
120
125
130
^{13}C (ppm)
7.8
7.6
7.4
7.2
7.0
4.4
4.2
1.6
1.4
1.2
^{1}H (ppm)

HMBC

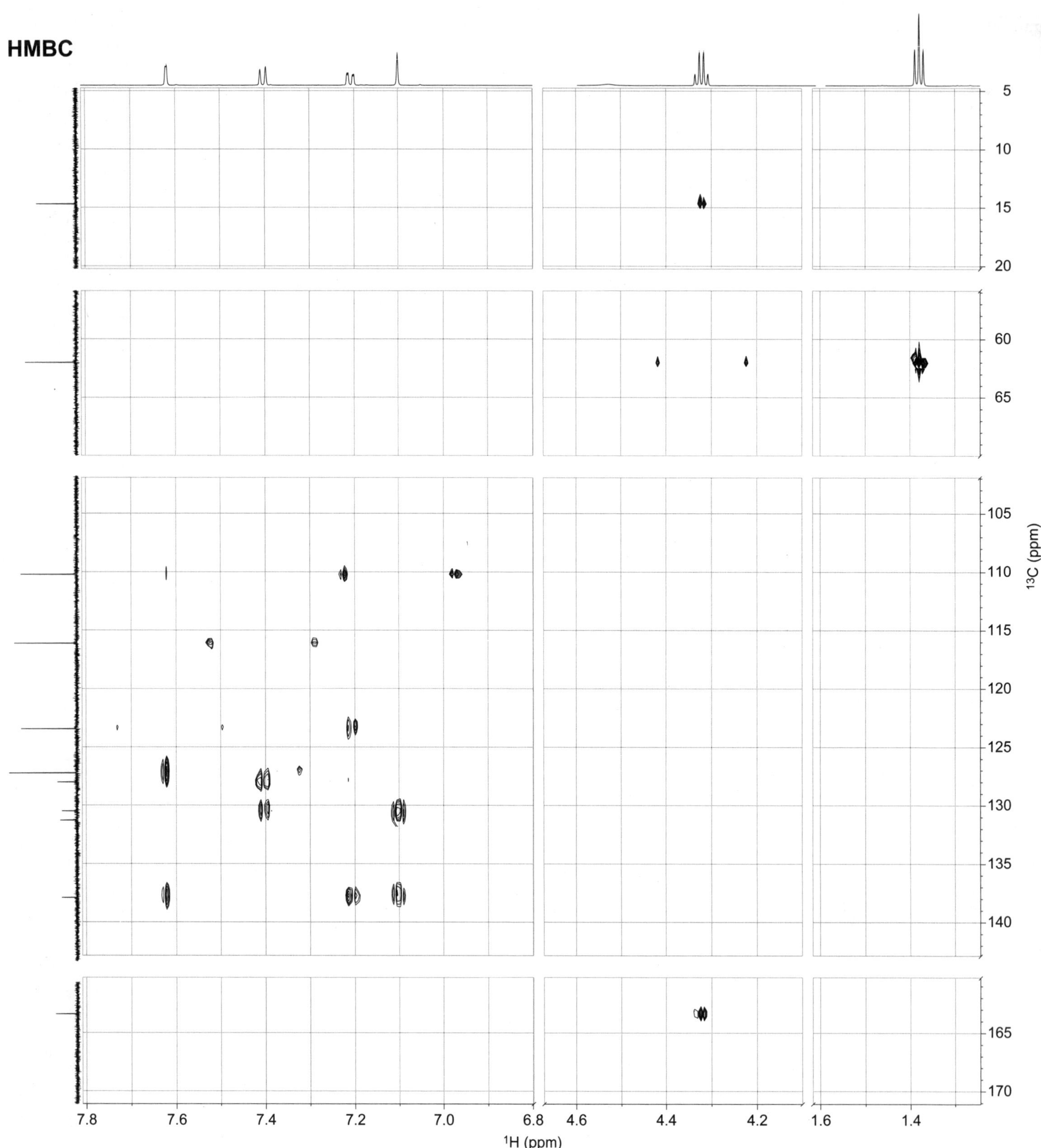

7.8
7.6
7.4
7.2
7.0
6.8
4.6
4.4
4.2
1.6
1.4
¹H (ppm)
5
10
15
20
60
65
105
110
115
120
125
130
135
140
165
170
¹³C (ppm)